SOME COMMON CONVERSIONS

MASS	$1\ u = 1.66 \times 10^{-27}\ kg$
LENGTH	$1\ m = 39.4\ in = 3.28\ ft$
	$1\ mi = 1.61\ km = 5280\ ft$
	$1\ in = 2.54\ cm$ (exact)
VOLUME	$1\ L = 10^3\ cm^3 = 0.0353\ ft^3$
	$1\ ft^3 = 0.0283\ m^3 = 7.480\ gal$
TIME	$1\ y = 3.15 \times 10^7\ s$
SPEED	$1\ mi/h = 1.47\ ft/s = 0.447\ m/s = 1.61\ km/h$
FORCE	$1\ lb = 4.45\ N$
ENERGY	$1\ J = 0.239\ cal = 0.738\ ft \cdot lb$
	$1\ kcal = 4186\ J$
	$1\ kWh = 3.6 \times 10^6\ J$
	$1\ MeV = 10^6\ eV = 1.602 \times 10^{-13}\ J$
POWER	$1\ W = 0.738\ ft \cdot lb/s$
	$1\ hp = 550\ ft \cdot lb/s = 746\ W$
PRESSURE	$1\ torr = 133\ Pa = 0.0193\ lb/in^2$
	$1\ atm = 760\ torr = 1.01 \times 10^5\ Pa$

SOME COMMON CONVERSIONS

PHYSICS IN SCIENCE AND INDUSTRY

PHYSICS
IN SCIENCE AND INDUSTRY

Alan Cromer

Professor of Physics
Northeastern University

McGRAW-HILL BOOK COMPANY

New York   St. Louis
San Francisco   Auckland
Bogotá   Hamburg
Johannesburg   London   Madrid
Mexico   Montreal   New Delhi
Panama   Paris   São Paulo
Singapore   Sydney
Tokyo   Toronto

**Physics
in Science
and
Industry**

1 2 3 4 5 6 7 8 9 0 D O D O 8 9 8 7 6 5 4 3 2 1 0

This book was set in Optima by York Graphic Services, Inc.
The editors were C. Robert Zappa and J. W. Maisel;
the designer was Nicholas Krenitsky;
the production supervisor was Dominick Petrellese.
The drawings were done by Felix Cooper.
Cover photo by Werner H. Müller/Peter Arnold, Inc.
R. R. Donnelley & Sons Company was printer and binder.

Library of Congress Cataloging in Publication Data

Cromer, Alan H   date
    Physics in science and industry.

    Includes index.
    1. Physics. I. Title.
QZ21.2.C76   530   79-16726
ISBN 0-07-014437-0

To my
Little Brother
Randy Stone

CONTENTS

PART III WAVES

PART IV ELECTRICITY AND MAGNETISM

PART V MODERN PHYSICS

This book explains the basic principles of physics in the context of modern technology. It is written for a broad class of technically oriented students (architects, engineering technologists, technicians, mechanics, etc.) who need a general knowledge of physics and its relation to their work. Realistic applications of physics to science and industry are used throughout the book, both to clarify the physical principles and to explain important aspects of modern technology. In this way, physics is shown to provide the concepts needed to understand the complex technological world in which we live.

PREFACE

To make the book useful to students who are still developing their mathematical skills, the mathematics is limited to elementary algebra and trigonometry. Use of the electronic calculator is emphasized from the beginning, so that any deficiency in computational skills will not be a block to learning physics. Once students develop confidence with the calculator, they are better able to handle sophisticated problems involving logarithms and exponentials. The calculator allows students to work with more advanced concepts than they could otherwise handle.

The divisions of the book—Part, Chapter, section, subsection, and sub-subsection—are used to display the logical relationships among the topics. For example,

PART IV   ELECTRICITY AND MAGNETISM
Chapter 16   Direct Current
SECTION 16.5   ELECTROCHEMISTRY
Contemporary batteries
Carbon battery

shows the relation of the topic *carbon battery* to the broader categories of electricity and magnetism. The placing of a topic in a sub-subsection does not imply that the topic is less important than other topics, but only that it is logically subordinate to them.

Each section is generally a fairly self-contained discussion of a major topic, which can serve as the basis for several lectures. The division of sections into subsections and sub-subsections makes it easy to adjust the time spent on each section. For example, Sec. 16.5 (Electrochemistry) is divided into five subsections: Electrolysis, Faraday's law, Voltaic cell, Daniell cell, Contemporary batteries. The instructor can adjust the time spent on electrochemistry by decreasing or increasing the number of subsections covered.

The Guide to Major Topics at the end of each chapter groups the subject matter of the chapter into specific study units; each topic listed in the guide is correlated with a set of equations, examples, and problems that is the basis for studying the topic. The instructor can use the Guide to make problem assignments on the basis of the specific topics within each chapter that are to be emphasized. The student can use the Guide to find the equations relevant to an assigned problem and to locate other related problems for further study.

In order to gain first-hand information about modern industrial practice, I visited a number of factories and research laboratories in different parts of the country. From talking with engineers and production-line supervisors, I learned what aspects of physics are used on a day-to-day basis. In the course of these visits, I also became aware of some serious gaps that exist between academic physics and industrial physics. For instance, although the usual textbook discussion of the dc motor (Sec. 17.2) gives a clear picture of how electrical energy can be converted into mechanical energy, it does not provide a useful description of any motor of commercial significance. Therefore, in this book, a more detailed discussion of motors is given in Sec. 18.5, in order to bridge the gap between the textbook world and the real world of technology.

It is a special privilege to acknowledge the people who have assisted me with this book. I want to thank George Cromer of General Motors, John Driscoll of Inland Steel, and George Sasdi of New England Electric for being so generous with their time and making my visits to their companies so interesting and informative. I am especially indebted to my friend George Sasdi for patiently and efficiently fulfilling my many requests for photographs, diagrams, and equipment specifications. I hope that some of George's dedication and enthusiasm for engineering has found its way onto the pages of this book.

Parts of the book have been critically reviewed by Robert Boughton (Northeastern University), Russ Brengelman (Morehead State University), Pearley Cunningham (Allegheny Community College), Paul Everett (University of Kentucky), William Hall (Community College of Allegheny County), Philip Hetland (North Dakota State University), Thomas Hughes (College of Alameda), Tom Sandin (North Carolina Agricultural and Technical State University), and George Sasdi (New England Electric Company). The comments of these reviewers have been of great value in helping me correct errors and clarify ambiguities. The errors and ambiguities that remain are entirely my responsibility.

Invaluable assistance in the day-to-day preparation of the book was provided by my photographic researcher, Carole Frohlich, and my typist, Geraldine Crehan. Only another author can fully appreciate my gratitude for the skill and intelligence they brought to their work.

Alan Cromer

PART I

MECHANICS

Mechanics is the study of the conditions under which objects remain at rest (*statics*) and the laws governing objects in motion (*dynamics*). The basic concepts of mechanics—force, mass, energy, etc.—are fundamental to all branches of physics, so the study of mechanics is a necessary preparation for the study of subjects such as thermodynamics, electricity and magnetism, and nuclear physics. Furthermore, mechanics has direct application to all fields of engineering. For example, the static conditions are used to calculate the forces that the structural members of a bridge or building must be able to support when fully loaded. Similarly, the laws of dynamics determine the forces on objects in circular, oscillatory, or

Early vacuum pump
(Hauxbee, c. 1708).
(*London, Science Museum.*)

other types of accelerated motion and the power
required to operate various machines and pumps.

T he human mind ascribes many different attributes to people and things, such as length, weight, color, beauty, and patriotism. Some attributes are clearly measurable, and others are not. Thus there exist well-defined procedures for measuring length and weight, but not beauty or patriotism. [Color is an intermediate case, for while a numerical value can be assigned to every color (Sec. 13.5), color cannot be ranked in order.] *Physics* is the study of the measurable attributes of things. The basic concepts of physics are defined in terms of measurements, and the purpose of physical theories is to correlate the results of measurements. A physical theory, no matter how abstractly it may be stated, is ultimately a statement about concrete operations that can be performed in a laboratory or factory.

1 MEASURE-MENT

1.1 Science and Industry

Throughout history the study of natural phenomena has engaged the attention of scholars and academics. Every great civilization of the past—Babylonia, Egypt, China, India, Greece—has produced treatises on mathematics, physics, and natural history. But these works, as important as they were, did little to affect the civilizations that produced them, and even less to affect the life of the average person. They were written by scholars for scholars and, with few exceptions, were never intended to produce practical results.

Science as we know it today started to develop around A.D. 1500 in a few countries in Western Europe. The European scientists of that time felt themselves to be backward and ignorant in comparison with the great scholars of antiquity, but they developed a new method for scientific discovery that was more powerful than anything known before. Indeed, it was the potency of their methodology that propelled Europe from a backward territory on the fringes of civilization to a position of world dominance.

Historians like to speculate about the special circumstances that led to the development of Western science. Why, for instance, was William Harvey (1578–1657) able to discover the circulation of the blood in the 1620s, when countless generations of anatomists before him had failed to do so? It was not simply better observation and experimentation. Dissection had been practiced since the time of the great Greek anatomist Galen (c. A.D. 130–200), and for 1400 years thereafter, each generation of physicians was taught anatomy according to Galen. The anatomy professor would read Galen to the students while an assistant performed the dissection. Failure to confirm a statement of Galen was interpreted as a failure of the assistant, not of Galen.*

But if this was the pattern of learning for over a thousand

---

*There is nothing shocking about this. If you fail to get an expected result in your physics laboratory, no one suspects that the laws of physics are wrong. Most of the time we accept as valid only those observations which confirm our expectations. This is why a new discovery is so rare and so wonderful.

years, what happened to suddenly produce a change? The answer, I believe, was the diffusion of technical knowledge from the workshop to the academy.

In previous periods of history, the practical work of the artisan and technician was unknown to the academicians, but with the invention of printing, knowledge of the mechanical arts began to spread out from the shop. Georgius Agricola [Georg Bauer (1494–1555)], a practicing physician from Bohemia, wrote extensively about the mining and metallurgical practices of his day. His classic work *De re metallica* (1556) became the principal mining and metallurgical textbook for the next 200 years. Figure 1.1 is an engraving from *De re metallica* which shows a simple suction pump being used to drain water from a mine. Agricola's description of this device clearly states the function and operation of the valves it used.

It was knowledge of this sort that gave Harvey the edge over his predecessors. For when Harvey dissected a heart, he could see that the heart had one-way valves similar to the one-way valves in a pump. He saw what no one else had seen, because he was no longer isolated from the practical knowledge of his day.

The suction pump fascinated other scientists of the sixteenth and seventeenth centuries. Moreover, because the principles on which it operated were not understood, it inspired Galileo, Torricelli, and Pascal to investigate the nature of the vacuum. This in turn led to the invention of the air pump, and then to a study of steam and the development of the steam engine. Efforts to improve the efficiency of the steam engine resulted in fundamental discoveries in thermodynamics.

It is this back-and-forth exchange of ideas between science and technology that is the basis of our modern technological civilization. Neither science nor technology can flourish for long if it is isolated from the other. It is the purpose of this book to help strengthen the bond between science and technology by showing technology students the broad generalizing principles upon which their discipline is based. At the same time this book serves as an introduction to modern technology by applying these principles to practical problems in structural mechanics, machinery, hydraulics, heat engineering, acoustics, photometry, optics, electric circuitry, electric machinery, electronics, and nuclear engineering.

1.2 Units

Physics studies the things that can be measured. What can be measured at any time depends on the available technology. For instance, x-rays could not be discovered before photographic plates were available to detect them. The scope of physics continually increases as new inventions expand the range of possible measurements.

Every physical quantity is measured by comparing its magnitude with that of a standard sample. The most familar example of this is the measurement of length, in which the length of an object is compared to a standard length, usually 1 foot or 1 meter. A meterstick is a good instrument for this purpose, but for

FIGURE 1.1

Suction pump from Agricola's *De Re Metallica* (1556). The man on the left is making a pipe by boring a tree trunk with an auger. (*Dover Publications.*)

a   b   FIGURE 1.2

(a) Vernier caliper. (*Sargent-Welch.*) (b) Micrometer. (*Central Scientific Co.*)

precision measurements of lengths under 5 centimeters, a vernier caliper or micrometer is used (Fig. 1.2). These instruments have jaws which clamp firmly to the object being measured and a marker which indicates the length of the opening of the jaw against a ruled scale.

Measurements are always made relative to some standard, called the *unit*. In the International System of Units (called SI units), the standard of length is the meter (abbreviated m), whereas in the English engineering system, the standard of length is the foot (ft). Both systems of units are in common practice today, so you must be familar with both. In addition, there are many other customary units of length, such as the inch, mil, yard, and mile. Appendix II contains a number of tables which give the relation between SI, English, and customary units.

Most industries are currently adopting SI units, so this system is emphasized in this book. The conventions to be observed when using this system are described in Appendix I. For now it is only necessary to note that the SI system uses prefixes to designate multiples and submultiples of a unit. The most commonly used prefixes are shown in Table 1.1, and a complete list of prefixes is given in Appendix I.

TABLE **1.1**

Prefix	Symbol	Value
mega-	M	1,000,000, or 10^6
kilo-	k	1,000, or 10^3
centi-	c	0.01, or 10^{-2}
milli-	m	0.001, or 10^{-3}
micro-	μ	0.000001, or 10^{-6}

Common SI prefixes

According to Table 1.1, a kilometer (kilo + meter), abbreviated km, is 1000 m:

1 km = 1000 m

A centimeter (centi + meter), abbreviated cm, is 0.01 m:

1 cm = 0.01 m

Other multiples and submultiples are formed similarly.

Every measurement must specify both the magnitude and the unit. For instance, if the diameter d of a ball bearing measured with a micrometer calibrated in centimeters is found to be 2.617, the result should be written

$d = 2.617$ cm

This means the length is 2.617 times 1 cm, or

$d = (2.617)(1$ cm$)$

Although this second way of writing the result is never used in practice, it is useful to think of the unit this way when you have to convert from one unit to another.

Example 1.1 The diameter of a ball bearing is 2.617 cm. Express the diameter in meters.

Since 1 cm = 0.01 m, we can write

$$d = 2.617 \text{ cm} = (2.617)(1 \text{ cm})$$
$$= (2.617)(0.01 \text{ m})$$
$$= 0.02617 \text{ m} \qquad \square$$

Conversions between SI units are easy because they only involve moving the decimal point so many places to the left or right. Conversions between SI and English or customary units involve a multiplication or division, which may be done on a calculator. Appendix II gives the conversion factors between SI and other commonly used units.

Example 1.2 Convert 2.617 cm to feet.

From Appendix II we find 1 m = 3.28 ft, so

$$1 \text{ cm} = 0.01 \text{ m} = (0.01)(1 \text{ m})$$
$$= (0.01)(3.28 \text{ ft})$$
$$= 0.0328 \text{ ft}$$

Therefore, we have

$$2.617 \text{ cm} = (2.617)(1 \text{ cm})$$
$$= (2.617)(0.0328 \text{ ft})$$
$$= 0.08584 \text{ ft} \qquad \square$$

There are many cases of interest in which the direct measurement of a length using a meterstick or micrometer is not possible, and indirect methods must be employed. However, even in an indirect measurement, a direct measurement must be made at some stage.

For example, to measure the distance d between two points A and B on opposite sides of a river (Fig. 1.3), a surveyor's transit (Fig. 1.4) is used. A third point C is picked on the surveyor's side

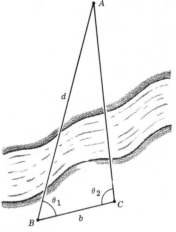

FIGURE 1.3

The distance d between two points A and B on opposite sides of a river is found by measuring the distance b and the angles θ_1 and θ_2.

FIGURE 1.4

Surveyor's transit. (*Berger Instruments.*)

of the river, and sightings are taken with the transit at B and C to measure the angles* θ_1 and θ_2. The length b of the baseline BC is measured directly by stretching a chain between B and C and measuring its length with a meterstick. From these measurements the length d can be found using the law of sines (Sec. 2.5). However, a problem like this can also be solved *graphically* by making a scale drawing.

Example 1.3 What is the distance d in Fig. 1.3, given that $b = 0.50$ km, $\theta_1 = 80°$, and $\theta_2 = 85°$?

A scale drawing is begun by choosing a suitable scale, say 10 cm = 1 km. With this scale the baseline BC would be drawn 5 cm long. A protractor is next used to draw lines at 80 and 85° to BC from points B and C, respectively. These lines are extended until they intersect at point A. Finally, the distance AB is measured with a ruler and converted back into the original units. In this case, AB will be found to be about 19.2 cm. According to our scale, 1 cm = 0.1 km, so

$$19.2 \text{ cm} = (19.2)(0.1 \text{ km}) = 1.92 \text{ km} \qquad \square$$

The graphical method is not as accurate as the trigonometric method, but it is conceptually much simpler and is accurate enough for now. Some surveying problems are given at the end of this chapter to prepare you for vectors in the next chapter.

Large distances on earth are measured by a sequence of such small-distance *triangulations,* and the size of the earth itself is ultimately determined in this way. Once the size of the earth is known, the distance to the sun can be found by measuring the angle of the sun from two different points on the earth at the same time and using the known distance between these points as the baseline (Fig. 1.5). The distance to the sun is then used to measure the distance to a (nearby) star by taking the diameter of the earth-sun orbit as the baseline (Fig. 1.6). Thus even astronomical measurements are related, often through a large number of intermediate steps, to a direct measurement of distance with chain and meterstick.

Other concepts require other methods of measurement. Time is particularly subtle. To measure time, a device is needed that continually repeats some event, so that the interval between two events can be taken as the unit of time. A clock is a mechanical device which produces a periodic event to mark time. The daily rotation of the earth on its axis is a naturally periodic event, and until 1967 the unit of time (the second) was defined as 1/86,400 of a mean solar day. In the 1950s atomic clocks were developed that were even more accurate than the earth, which does not rotate with absolute uniformity because of tidal friction between the oceans and the ocean floor. Since 1967 the standard *second* (s) has been defined as the duration of 9,192,631,770 oscillations of a certain electron transition in the cesium atom.† This standard is maintained by the Bureau International de

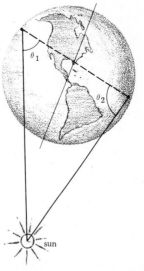

FIGURE 1.5

The distance to the sun is found by measuring the angles θ_1 and θ_2 from two points on the earth that are a known distance apart.

FIGURE 1.6

The distance to a star is found by using the diameter of the earth-sun orbit as a base line.

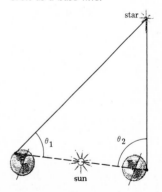

*θ is the Greek (lowercase) letter theta.
†Likewise the meter is now defined as 1,650,763.73 times the wavelength of the orange light emitted by the gas krypton when electrically excited.

FIGURE 1.7

The atomic clock (NBS-6) at the National Bureau of Standards. This clock has an accuracy of one part in 10^{13}, which is equivalent to an error of less than 1 s in 370,000 years. (*National Bureau of Standards.*)

l'Heure (BIH) in Paris, from data supplied to it from cesium clocks run by several national standard laboratories, such as the National Bureau of Standards in Washington (Fig. 1.7).

Because the rate of rotation of the earth continues to decrease, a mean solar day is no longer exactly 86,400 (atomic) seconds. To correct for this, the BIH adds an extra second (leap second) to the year from time to time. The first leap second was added on June 30, 1972.

Length and time are two fundamental quantities in physics. Only three other fundamental quantities (mass, temperature, and charge) will be introduced;* all the others will be defined in terms of these. For instance, the (average) speed v of a car in an auto race is the distance d the car travels (obtained by measuring the racetrack and the number of circuits made) divided by the total elapsed time t (measured with a stopwatch). In general, average speed is defined by the equation

$$v = \frac{d}{t}$$

Definition The *dimensions* of a physical quantity are the factors of the fundamental (or derived) quantities which define it. We denote the dimensions of a quantity by putting the factors in brackets. Thus the dimensions of length and time are simply $[l]$ and $[t]$, and the dimensions of speed are obtained from its definition as length divided by time, or $[l/t]$.

The dimensions of area and volume are also related to $[l]$. The area A of a rectangle (Fig. 1.8) with sides a and b is

$$A = ab$$

*A sixth fundamental quantity, *luminous intensity,* is used only in the specialized field of photometry (Sec. 13.6).

so its dimension is

$$[A] = [a][b] = [l][l] = [l^2]$$

Example 1.4 What is the area of a rectangle which is 5 m long and 3 m wide (Fig. 1.8)?

$$A = ab = (3\,\text{m})(5\,\text{m}) = 15\,\text{m}^2$$

The unit in this case is the square meter (m²), which is the SI unit of area. Note that the unit of a quantity is obtained from the units in the factors for that quantity by the ordinary rules of algebra. □

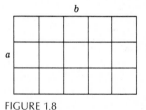

FIGURE 1.8

Rectangle with sides a and b.

The volume V of a rectangular solid (Fig. 1.9) with sides a, b, and c is

$$V = abc$$

so its dimension is

$$[V] = [a][b][c] = [l][l][l] = [l^3]$$

Example 1.5 What is the volume of a rectangular solid with sides $a = 3$ m, $b = 5$ m, and $c = 2$ m (Fig. 1.9)?

$$V = abc = (3\,\text{m})(5\,\text{m})(2\,\text{m}) = 30\,\text{m}^3$$

The unit in this case is the cubic meter (m³), which is the SI unit of volume. □

Various other units are used in practice to measure area and volume. In the English engineering system, the units of area and volume are the square foot (ft²) and the cubic foot (ft³). Other customary units are acre, square mile, gallon, and barrel. All problems in this book are to be done using either SI or engineering units. If the data are given in other units, they should be converted into engineering or SI units before proceeding with the problem. Once the answer is found, it can be converted to a customary unit if desired.

Example 1.6 Water flows into a bathtub at the rate of 8.5 gal/min. How long will it take to fill a tub 5 ft long and 2 ft wide to a depth of 1.5 ft?

From Appendix II we find that the conversion from gallons to cubic feet is

$$1\,\text{gal} = 0.1337\,\text{ft}^3$$

FIGURE 1.9

Rectangular solid with sides a, b, and c.

so the rate of flow R is

$$\begin{aligned} R &= 8.5\,\text{gal/min} \\ &= (8.5)(0.1337\,\text{ft}^3/\text{min}) \\ &= 1.14\,\text{ft}^3/\text{min} \end{aligned}$$

The volume to be filled is

$$V = abc = (5\,\text{ft})(2\,\text{ft})(1.5\,\text{ft}) = 15\,\text{ft}^3$$

The time t required to fill this volume at rate R is

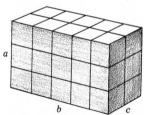

$$t = \frac{V}{R} = \frac{15 \text{ ft}^3}{1.14 \text{ ft}^3/\text{min}} = 13.1 \text{ min}$$

Notice that cubic feet (ft³) cancel in the last equation, and that minutes (min), which is in the denominator of the denominator, can be brought up into the numerator. The units will always come out properly in a correct calculation, and this is often a good way to check your result. □

1.3 Scientific Notation

Exponential notation is commonly used in science to express very large and very small numbers. Thus, instead of writing 1 million or 1,000,000, we write 10^6. The superscript, or *exponent*, denotes the number of zeros after the 1 in the conventional expression of the number. Similarly, instead of writing 1 millionth, or

$$\frac{1}{1,000,000} = 0.000,001$$

we write

$$\frac{1}{10^6} = 10^{-6}$$

A negative exponent denotes the number of places the decimal point is from the right side of 1 in the decimal expression of the number.

Exponential notation comes from algebra, where the expression a^n, called a *power*, means a multiplied by itself n times:

$$a^n = \underbrace{a \cdot a \cdots a}_{n \text{ factors}}$$

For instance:

$$a^3 = a \cdot a \cdot a$$
$$a^6 = a \cdot a \cdot a \cdot a \cdot a \cdot a$$

In particular, if $a = 10$, we have

$$10^n = \underbrace{10 \cdot 10 \cdots 10}_{n \text{ factors}}$$

so

$$10^3 = 10 \cdot 10 \cdot 10 = 1000$$
$$10^6 = 10 \cdot 10 \cdot 10 \cdot 10 \cdot 10 \cdot 10 = 1,000,000$$

A number like 375,000 can be written as the product of a simple decimal number and a power of 10. Thus we can write

$$375,000 = 37,500.0 \times 10$$

or

$$375,000 = 3750.00 \times 10^2$$

or

$$375,000 = 3.75 \times 10^5$$

This last expression, in which the number is expressed as a decimal factor between 1 and 10 times a power of 10, is called *scientific notation*.

> **Remark** Scientific calculators have a key labeled $\boxed{\text{EE}}$ or $\boxed{\text{EEX}}$ for entering the exponent of a power of 10. To enter 3.75×10^5 on such a calculator, press the keys in the sequence
>
> $\boxed{3}$ $\boxed{\cdot}$ $\boxed{7}$ $\boxed{5}$ $\boxed{\text{EE}}$ $\boxed{5}$
>
> The number will be displayed as
>
> 3.75   05
>
> In the display of a scientific calculator, the two digits on the far right always denote the exponent in the scientific notation of the number.

Multiplication and division of powers are particularly easy because of the following theorem.

Law of Exponents *The exponent of the product of two powers is the sum of the exponents of the factors:*

$$a^n a^m = a^{n+m}$$

and the exponent of the quotient of two powers is the difference between the exponents of the dividend and the divisor:

$$\frac{a^n}{a^m} = a^{n-m}$$

This theorem is easily proved from the meaning of the exponential notation. Thus we have

$$a^n a^m = \underbrace{(a \cdot a \cdots a)}_{n \text{ factors}}\underbrace{(a \cdot a \cdots a)}_{m \text{ factors}}$$

$$= \underbrace{(a \cdot a \cdots a)}_{n + m \text{ factors}} = a^{n+m}$$

and

$$\frac{a^n}{a^m} = \frac{\overbrace{\cancel{a} \cdot \cancel{a} \cdots a}^{n \text{ factors}}}{\underbrace{\cancel{a} \cdot \cancel{a} \cdots a}_{m \text{ factors}}} = \underbrace{a \cdot a \cdots a}_{n - m \text{ factors}} = a^{n-m}$$

The meaning of a^0 follows at once from the law of exponents, because on the one hand we know that

$$\frac{a^n}{a^n} = 1$$

(because anything divided by itself is 1), while on the other hand we know from the law of exponents that

$$\frac{a^n}{a^n} = a^{n-n} = a^0$$

so we must have

$$a^0 = 1$$

From this we can understand the meaning of negative exponents, because we can write

$$\frac{1}{a^n} = \frac{a^0}{a^n} = a^{0-n} = a^{-n}$$

That is, a^{-n} is the reciprocal of a^n. For instance, since 10^6 is 1 million, 10^{-6} is 1 millionth.

Example 1.7 Find the reciprocal of 2.5×10^7.

$$\frac{1}{2.5 \times 10^7} = \frac{1}{2.5} \times \frac{1}{10^7} = 0.4 \times 10^{-7} = 4 \times 10^{-8}$$

Note that in the last expression the numerical factor was increased by a factor of 10 by moving the decimal point one place to the right, while the power of 10 was decreased by a factor of 10 by increasing the negative exponent from -7 to -8.

To perform this calculation on a calculator with a reciprocal $\boxed{1/x}$ key, press the keys in the sequence

$\boxed{2}$ $\boxed{\cdot}$ $\boxed{5}$ $\boxed{\text{EE}}$ $\boxed{7}$
$\boxed{1/x}$

The display will be

4. -08 □

Scientific notation is very useful in calculations requiring the multiplication and division of large numbers. Thus consider the product

$$P = (375,000)(9,230,000)$$

In scientific notation this is

$$P = (3.75 \times 10^5)(9.23 \times 10^6)$$
$$= (3.75 \times 10^5 \times 9.23 \times 10^6)$$

Rearranging terms in the product we get

$$P = (3.75 \times 9.23 \times 10^5 \times 10^6)$$
$$= (3.75 \times 9.23)(10^5 \times 10^6)$$
$$= (34.6)(10^{11})$$
$$= 3.46 \times 10^{12}$$

The law of exponents was used to find the product of 10^5 and 10^6, and in the last expression the decimal factor was decreased by a factor of 10, while the power of 10 was increased by a factor of 10.

Remark On a scientific calculator, such as the Texas Instruments SR-40 (Fig. 1.10a), which uses *algebraic logic*, the last calculation would be performed by pressing the keys in sequence

$\boxed{3}$ $\boxed{\cdot}$ $\boxed{7}$ $\boxed{5}$ $\boxed{\text{EE}}$ $\boxed{5}$
$\boxed{\times}$
$\boxed{9}$ $\boxed{\cdot}$ $\boxed{2}$ $\boxed{3}$ $\boxed{\text{EE}}$ $\boxed{6}$
$\boxed{=}$

a   b   FIGURE 1.10

Scientific calculators: (a) Texas Instruments SR-40 (*Texas Instruments*) and (b) Hewlett-Packard HP-21 (*Hewlett-Packard.*)

On a scientific calculator, such as the Hewlett-Packard HP-21 (Fig. 1.10*b*), which uses *reverse Polish logic,* the calculation would be performed by pressing the keys in the sequence

| 3 | | · | | 7 | | 5 | | EE | | 5 |

| ENTER |

| 9 | | · | | 2 | | 3 | | EE | | 6 |

| × |

In either case, the display would be

3.46125   12

which should be written as 3.46×10^{12}. The last three digits (125) are dropped because they are not significant.

You can tell at once which logic your calculator uses by observing whether it has an | ENTER | or an | = | key.

Example 1.8   An earth-filled dam requires 430,000 m³ of fill to construct. The fill is brought by trucks with a capacity of 20 m³. How many truckloads of fill are required to complete the project?

The volume V of fill required is $V = 430,000 \text{ m}^3 = 4.3 \times 10^5 \text{ m}^3$ and the capacity c of a truck is

$$c = 20 \text{ m}^3 = 2 \times 10^1 \text{ m}^3$$

so the number N of truckloads required is

$$N = \frac{V}{c} = \frac{4.3 \times 10^5 \text{ m}^3}{2 \times 10^1 \text{ m}^3}$$

$$= \frac{4.3}{2} \times \frac{10^5}{10^1} = 2.15 \times 10^4 = 21,500 \qquad \square$$

1.4 Precision, Accuracy, and Experimental Error

PRECISION AND ACCURACY

The precision and accuracy of a measurement depend on the quality of the measuring instrument and the skill of the operator. Although the terms *precision* and *accuracy* are often used interchangeably, they refer to quite different concepts.

Definition The *precision* of an instrument or measurement is the degree to which it can detect differences between quantities. For instance, with a steel ruler graduated in millimeters, a length can be measured to the nearest millimeter, or 0.1 cm. The ruler cannot detect a difference in the length of two cylinders which differ by less than 0.1 cm. With a vernier caliper (Fig. 1.2a), the length of each cylinder can be measured to the nearest 0.01 cm, and with a micrometer (Fig. 1.2b), each length can be measured to the nearest 0.001 cm. The micrometer is a more precise instrument than the vernier caliper, and the caliper is more precise than the ruler.

For instance, the results of measurements of the length of a cylinder made with each instrument might be

Instrument	Length, cm
Steel ruler	3.4
Vernier caliper	3.43
Micrometer	3.437

The more precise the measurement, the more digits there are in the result.

Definition The *accuracy* of an instrument or measurement is the degree to which it gives the true value. A steel ruler is generally more accurate than a plastic measuring tape because the tape can be stretched and thus give a false reading. Both instruments have the same precision—that is, both can be read to 0.1 cm—but the graduations on the plastic tape are not truly 0.1 cm apart. It is important not to be misled into thinking that an instrument with good precision is necessarily accurate. A micrometer that has been dropped on the floor a couple of times may still be precise—that is, it can still be read to the nearest 0.001 cm—but it may no longer be accurate.

THE VERNIER CALIPER

The vernier caliper achieves its increased precision over the ruler by use of a vernier scale. The main scale on the stationary body of the caliper is graduated in 1-mm intervals, whereas the vernier scale on the movable jaw of the caliper is graduated in 0.9-mm intervals. The opening of the jaw is indicated by the position of the first line of the vernier scale against the main scale.

Figure 1.2a shows the situation when the opening of the caliper is 2.8+ cm. Note that the third line on the vernier scale is aligned exactly with the 3.0-cm line on the main scale. The distance between the first and third lines of the vernier scale is

$$2(0.9\,\text{mm}) = 1.8\,\text{mm} = 0.18\,\text{cm}$$

because it contains exactly two 0.9-mm intervals. Consequently, the position of the first line of the vernier scale is

$$3.00\,\text{cm} - 0.18\,\text{cm} = 2.82\,\text{cm}$$

The vernier scale allows one to determine the fractional distance between two divisions of the main scale. The general procedure is to find the line on the vernier scale that most nearly is aligned with a line on the main scale. The number of spaces between the first vernier line and the aligned line is the next digit in the measurement. In Fig. 1.2a, the first line of the vernier scale is at 2.8+ cm, and there are two spaces between it and the line aligned with the 3.0-cm mark. Thus the measurement is 2.82 cm.

THE MICROMETER

A micrometer (Fig. 1.2b) has an accurately machined screw with a thread lead of exactly 0.5 mm; that is, the thread advances 1 mm when it is rotated *twice*. The barrel of the micrometer is graduated into 50 intervals, so in rotating the barrel twice, 100 intervals pass the main scale and the thread advances 1 mm. Thus each interval on the barrel corresponds to 0.01 mm, or 0.001 cm.

To use a micrometer, the shaft is closed on the object being measured. To avoid damaging the instrument by closing it too tightly, always turn the shaft by the small rachet on the end of the barrel. The main scale on the barrel shows the length to the nearest millimeter. For instance, the micrometer in Fig. 1.2b reads 13+ mm. The fractional part of a millimeter is read on the circular scale. In Fig. 1.2b, this is 5, which represents 0.05 mm, so the opening is 13.05 mm, or 1.305 cm.

If the barrel is more than halfway between two divisions on the main scale, the barrel has been turned more than one full revolution, so 0.50 mm must be added to the reading. That is, the reading in Fig. 1.2b could be either 13.05 or 13.55 mm, depending on whether the barrel is less than or more than halfway between the 13- and 14-mm marks. In the case shown, it is less than halfway, so 13.05 mm is the correct reading.

EXPERIMENTAL ERROR

There are two different kinds of errors that can affect a measurement: random errors and systematic errors.

Definition A *random error* is an error caused by unpredictable variations in the measuring process. For instance, when a micrometer is used to measure the diameter of a cylinder, the shaft may be closed with different pressures on different occasions, giving rise to slightly different readings each time. This is a random error because it is equally likely to make the reading larger or smaller than the true value.

The magnitude of the random error in a measuring process is found by repeating the measurement many times. Table 1.2 gives the results of 10 micrometer measurements of the diameter d of a steel cylinder. The average, or *mean*, $\bar{d}$ of these N measurements is found by adding them and dividing by N. The formula for this is

$$\bar{d} = \frac{\Sigma d}{N} \qquad\qquad 1.1$$

where the symbol* Σ is used to indicate a sum of all the diameters. For the data in Table 1.2 we have

$$\bar{d} = \frac{\Sigma d}{N} = \frac{13.750 \text{ cm}}{10} = 1.375 \text{ cm}$$

This is 0.003 cm larger than the smallest reading (1.372 cm) and 0.003 cm smaller than the largest reading (1.378 cm).

In practice, the extreme variation is not used to indicate the error. Instead one uses the *standard deviation*† σ, which is the variation in which about two-thirds of the data occur. For the data in Table 1.2, the standard deviation is $\sigma = 0.002$ cm, because 6 out of the 10 measurements fall between

$$\bar{d} - \sigma = 1.373 \text{ cm} \qquad \text{and} \qquad \bar{d} + \sigma = 1.377 \text{ cm}$$

The overall result of these 10 measurements is written

$$d = 1.375 \pm 0.002 \text{ cm}$$

to show both the mean (average) and the standard deviation.

> **REMARK** The technical definition of the standard deviation is
>
> $$\sigma = \sqrt{\frac{\Sigma(d - \bar{d})^2}{N}} \qquad\qquad 1.2$$
>
> To use this formula, you first must calculate the mean $\bar{d}$. Then you find the difference $d - \bar{d}$ between each measurement and the mean, square these differences, add the squares together, divide by the number of measurements, and finally take the square root. Table 1.2 shows the calculation for the 10 measurements of the diameter of the cylinder.
>
> For a large number of measurements with a normally distributed random error, 68 percent of the measurements fall between $\bar{d} - \sigma$ and $\bar{d} + \sigma$, and 32 percent fall outside this range.

Definition A *systematic* error is a consistent bias in a measurement caused by a flaw in the instrument or the experimental procedure. For instance, if the shaft of the micrometer is always closed too tightly by turning the barrel instead of the rachet, then every measurement will be smaller than the true value. Known sources of systematic error must be avoided whenever possible. When this is not possible, the measurement can be corrected by estimating the magnitude of the error. Unknown sources of systematic error are found only after the same measurement is repeated using different equipment or a different procedure.

---

*Σ is the Greek (capital) letter sigma.

†σ is the Greek (lowercase) letter sigma.

TABLE **1.2**

d, cm	$d - \bar{d}$, cm	$(d - \bar{d})^2$, cm²
1.376	0.001	1×10^{-6}
1.375	0	0
1.373	−0.002	4×10^{-6}
1.375	0	0
1.372	−0.003	9×10^{-6}
1.378	0.003	9×10^{-6}
1.376	0.001	1×10^{-6}
1.378	0.003	9×10^{-6}
1.375	0	0
1.375	−0.003	9×10^{-6}
$\Sigma d = 13.750$		$\Sigma(d - \bar{d})^2 = 42 \times 10^{-6}$

The results of 10 measurements made with a micrometer of the diameter of a steel cylinder

This table also shows the procedure for calculating the average and the standard deviation of these measurements.

$$\bar{d} = \frac{1}{N}\Sigma d = 1.375 \text{ cm}$$

$$\sigma = \sqrt{\frac{1}{N}\Sigma(d - \bar{d})^2}$$
$$= \sqrt{4.2 \times 10^{-6} \text{ cm}^2}$$
$$= 0.002 \text{ cm}$$

SIGNIFICANT FIGURES

It is often unnecessary to quote the error associated with a measurement since, by convention, the number of digits used to express the number roughly indicates the error. If the length of this book is quoted as 24.2 cm, this indicates that something is known about the value of the last digit. That is, it tells us that the error is less than 1.0 cm and greater than 0.05 cm. For if the error were greater than 1.0 cm, the length would have been given as 24 cm, and if the error were less than 0.05 cm, the length would have been given as 24.20 cm.

If the standard deviation of the measurement is 0.2 cm, the true length of the book lies between 24.0 and 24.4 cm with reasonable certainty.* The last digit in the number 24.2 is thus significant, because it is probably either 0, 1, 2, 3, or 4, and not 5, 6, 7, 8, or 9. It would be incorrect to quote the measurement as 24.20 cm, because nothing is known about the last digit. It could as well be 3 (if the true length were 24.23 cm) as 9 (if the length were 22.19 cm). A quoted result should give only the reliably known digits, or *significant figures,* in the measurement.

In the course of a calculation, the number of digits can increase beyond the significance of the result. In such a case, the final number should be rounded off to leave only the significant figures.

Example 1.9 Suppose again that the length a of this book is found to be 24.2 ± 0.2 cm, and the width b is found to be 19.5 ± 0.2 cm. Calculate the area of the book and express it with the correct number of significant figures.

From these measurements, the area A of the book is calculated to be

$$A = ab = (24.2 \text{ cm})(19.5 \text{ cm}) = 471.90 \text{ cm}^2$$

*The probability that the true value lies outside the range given by the standard error is 0.37. That is, about one-third of the time the true value will be outside the range. However, it is seldom very far from the range.

But not all the digits in the final result can be significant. At one extreme, the length may really be 24.4 cm and the width may be 19.7 cm, so that the area is about 481 cm². At the other extreme, the length may be only 24.0 cm and the width only 19.3 cm, so the area is only about 463 cm². Thus the value of A is known to only two significant figures, so it should be written

$$A = 4.7 \times 10^2 \, cm^2$$

As a general rule, the result of a calculation cannot have more significant figures than the number of digits in the least significant figure used in the calculation. In this example the two numbers used in the calculation each had three significant figures, but the final result has only two significant figures. (In the problems and examples in this book, answers are often given to three significant figures, even when the data are given to only two. This is to aid you in checking your work.) ☐

THE SPEED OF LIGHT

Some quantities that are fundamental to physics are remeasured every few years, as better techniques for improving the precision become available. For example, the speed of light in empty space (vacuum) is a universal constant of nature, and its exact determination is important in many areas of physics and astronomy. The speed of light is usually expressed in kilometers per second (km/s) and is found by measuring the time it takes a pulse of light to travel a known distance. In 1862 Foucault found the speed to be $298,000 \pm 500$ km/s, or, as we should write it, $(2.980 \pm 0.005) \times 10^5$ km/s. Note that the first 0 after the 8 is significant and the others are not.

Since 1862 the speed of light has been measured with increasing precision by a large number of investigators. Table 1.3 lists some of these measurements to show how the number of significant figures has increased over the years. The newer measurements generally do not correct the older measurements but only increase the precision of the result.

TABLE **1.3**

Date	Investigator	Measurement, km/s	Speed of light measured by various investigators
1676	Römer	$220,000 \pm ?$	
1849	Fizeau	$313,300 \pm ?$	
1862	Foucault	$298,000 \pm 500$	
1875	Cornu	$299,990 \pm 200$	
1880	Michelson	$299,910 \pm 50$	
1883	Newcomb	$299,860 \pm 30$	
1926	Michelson	$299,796 \pm 4$	
1928	Helstaedt	$299,778 \pm 10$	
1935	Michelson, Pease, and Pearson	$299,774 \pm 2$	
1941	Anderson	$299,776 \pm 6$	
1949	Aslakson	$299,792 \pm 3.5$	
1950	Essen	$299,792.5 \pm 1.0$	
1952	Froome	$299,792.6 \pm 0.7$	
1953	MacKenzie	$299,792.4 \pm 0.5$	
1957	Bergstrand	$299,792.85 \pm 0.16$	
1958	Froome	$299,792.50 \pm 0.10$	
1967	Grosse	$299,792.50 \pm 0.05$	

An interesting exception to this is the 1935 measurement of Michelson, Pease, and Pearson. Their result differs from the 1967 measurement of Grosse by 18 km/s, or 9 times their estimated error. They had repeated their measurement over 2000 times to reduce the random error to only 2 km/s. Unfortunately, there was a large unsuspected systematic error in their experiment which became apparent only when still more precise measurements were made in the 1950s.

Every experiment is haunted by the possibility of serious systematic error. A good researcher is aware of this and does everything possible to check the equipment and procedure. But the final check comes only after the experiment is repeated independently by others.

1.5   Area, Volume, and Density

AREA

The *area A* of a surface is the number of unit squares that fill the surface. Table 1.4 gives the formulas for the areas of several common surfaces. Notice that each formula contains two factors of length, so that in each case the unit will be a length squared.

Table **1.4**

Surface		Area A	Areas of common surfaces
Rectangle		ab	
Triangle		$\frac{1}{2}hb$	
Circle		πr^2	
Sphere		$4\pi r^2$	
Curved surface of a circular cylinder		$2\pi rL$	

Example 1.10 What is the surface area of a sphere with a radius of 0.25 m?

From the formula for the area of a sphere given in Table 1.4 we have

$A = 4\pi r^2 = 4(3.14)(0.25 \text{ m})^2 = 0.79 \text{ m}^2$ □

VOLUME

The *volume* V of a solid is the number of unit cubes that fill the solid. Table 1.5 gives the formulas for the volumes of several common solids. Here each formula contains three factors of length, so that in each case the unit will be a length cubed. (In the formula for the volume of an arbitrary right cylinder, the area A of the base is a length squared, so Ah is a length cubed.)

Example 1.11 What is the volume of a right cylinder with a base area A of 0.15 m² and a height h of 0.35 m?

From the formula in Table 1.5 we have

$V = Ah = (0.15 \text{ m}^2)(0.35 \text{ m}) = 0.053 \text{ m}^3$ □

TABLE **1.5**

Solid		Volume V	**Volume of common solids**
Solid rectangle		abc	
Cone		$\frac{1}{3}\pi r^2 h$	
Circular cylinder		$\pi r^2 h$	
Arbitrary right cylinder		Ah	
Sphere		$\frac{4\pi}{3} r^3$	

The volume of a liquid or solid substance is unaffected by changes in the shape or the degree of subdivision of the substance. For instance, when a liquid is poured from one container to another, its volume remains the same regardless of the shape of the containers. Likewise, when a solid or a liquid is separated into smaller portions, the combined volume of the portions is equal to the original volume.

Example 1.12 A quantity of water fills a cylindrical glass vessel to a height h of 5 cm. The area A of the base of the cylinder is 30 cm². What will be the height h' of the water if it is poured into another cylinder with a base area $A' = 12$ cm² (Fig. 1.11)?

The unit of length in this problem is the centimeter (cm), so the unit of volume is the cubic centimeter (cm³). From the formula for the volume of a cylinder, the volume V of the water in the original cylinder is

$$V = Ah = (30 \text{ cm}^2)(5 \text{ cm}) = 150 \text{ cm}^3$$

Since this will also be the volume of the water in the second cylinder, we have

$$V' = A'h' = 150 \text{ cm}^3$$

or

$$h' = \frac{V'}{A'} = \frac{150 \text{ cm}^3}{12 \text{ cm}^2} = 12.5 \text{ cm} \qquad \square$$

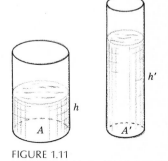

FIGURE 1.11

Two cylindrical vessels containing the same volume of water.

Example 1.13 In the carburetor of an automobile, 27 cm³ of gasoline is atomized into spherical droplets which each have a radius of 2×10^{-3} cm. (a) Find the number N of droplets produced. (b) Find the total surface area of these droplets, and compare it with the surface area the gasoline had before it was divided.

(a) The volume V' of each droplet is

$$V' = \frac{4\pi}{3}r'^3 = (4.19)(2 \times 10^{-3} \text{ cm})^3$$

$$= 33.5 \times 10^{-9} \text{ cm}^3$$

If there are N droplets, their total volume is NV', and this is equal to the orginal volume V of the gasoline. Thus we have

$$NV' = V$$

or

$$N = \frac{V}{V'} = \frac{27 \text{ cm}^3}{33.5 \times 10^{-9} \text{ cm}^3} = 8.06 \times 10^8$$

(b) The surface area A' of each droplet is

$$A' = 4\pi r'^2 = (12.57)(2 \times 10^{-3} \text{ cm})^2$$

$$= 5.0 \times 10^{-5} \text{ cm}^2$$

so the total surface area of all N droplets is

$$NA' = (8.06 \times 10^8)(5.0 \times 10^{-5} \text{ cm}^2) = 4.0 \times 10^4 \text{ cm}^2$$

The surface area of the gasoline before it was atomized depends somewhat on the shape it had. If we assume the shape was a cube of length a on each side, then

$$V = a^3 = 27 \text{ cm}^3$$

so

$$a = \sqrt[3]{27 \text{ cm}^3} = 3 \text{ cm}$$

Each of the six sides of the cube has an area a^2, so the total surface area of the cube is

$$A = 6a^2 = 6(3 \text{ cm})^2 = 54 \text{ cm}^2$$

The total area NA' of the droplets is greater than the area of the unatomized gasoline by the factor

$$\frac{NA'}{A} = \frac{4.0 \times 10^4 \text{ cm}^2}{54 \text{ cm}^2} = 741$$

Atomizing the gasoline increases the surface area by a factor of 741. This increased area promotes the rapid vaporization of the fuel. □

DENSITY

Definition The *mass density* ρ of a substance is the ratio of the mass m of the substance to its volume V:

$$\rho = \frac{m}{V} \qquad 1.3$$

The *weight density* d of a substance is the ratio of the weight W of the substance to its volume V:

$$d = \frac{W}{V} \qquad 1.4$$

Density is a characteristic property of a substance, independent of its volume. Although mass and weight are different physical concepts (Sec. 4.3), this difference is not important now.

TABLE **1.6**

Substance	Mass density, kg/m³	Weight density, lb/ft³	Densities of some common substances
Air (20°C)	1.20	7.5×10^{-2}	
Aluminum	2.7×10^3	1.7×10^2	
Bone	1.6×10^3	1.0×10^2	
Copper	8.9×10^3	5.5×10^2	
Glass	2.6×10^3	1.6×10^2	
Granite	2.7×10^3	1.7×10^2	
Ice	0.917×10^3	0.57×10^2	
Iron	7.7×10^3	4.8×10^2	
Lead	11.3×10^3	7.1×10^2	
Mercury	13.6×10^3	8.5×10^2	
Silicon	2.33×10^3	1.46×10^2	
Steam (100°C)	0.596	3.7×10^{-2}	
Steel	7.7×10^3	4.8×10^2	
Water			
Pure (0°C)	1.000×10^3	62.4	
Pure (20°C)	0.998×10^3	62.4	
Sea (15°C)	1.025×10^3	64.0	
Wood, maple	0.7×10^3	44	

UNITS Mass density is used exclusively in the International System, and weight density is used exclusively in the English engineering system. The SI unit of mass is the kilogram (kg), so the SI unit of mass density is the kilogram per cubic meter (kg/m^3). The English engineering unit of weight is the pound (lb), so the English engineering unit of weight density is the pound per cubic foot (lb/ft^3). Table 1.6 gives the densities of some common substances in both units.

Example 1.14 A truck can carry a load of 20 metric tons (1 metric ton = 1000 kg = 1 Mg). What is the volume of rock the truck can carry?

The mass of the rock is

$$m = 20 \text{ ton} = (20)(1000 \text{ kg}) = 2 \times 10^4 \text{ kg}$$

From Table 1.6 the density of granite is

$$\rho = 2.7 \times 10^3 \text{ kg/m}^3 = \frac{m}{V}$$

so the volume V of the rock is

$$V = \frac{m}{\rho} = \frac{2 \times 10^4 \text{ kg}}{2.7 \times 10^3 \text{ kg/m}^3} = 7.4 \text{ m}^3 \qquad \square$$

Example 1.15 The steel ball of a wrecking crane is 2 ft in diameter. How much does the ball weigh?

The radius of the ball is 1 ft, so its volume is

$$V = \frac{4\pi}{3} r^3 = (4.19)(1 \text{ ft})^3 = 4.19 \text{ ft}^3$$

From Table 1.6 the weight density of the ball is

$$d = 480 \text{ lb/ft}^3 = \frac{W}{V}$$

so the weight of the ball is

$$W = dV = (480 \text{ lb/ft}^3)(4.19 \text{ ft}^3) = 2.0 \times 10^3 \text{ lb} \qquad \square$$

GUIDE TO MAJOR TOPICS

Topic	Reference	Problems
Units and unit conversion	Sec. 1.2; Examples 1.1 and 1.2	1 to 4
Area and volume	Secs. 1.2 and 1.5; Examples 1.4 to 1.6, 1.8, 1.11 to 1.13; Tables 1.4 and 1.5	5 to 8, 25 to 30
*Graphical solution of trigonometric problems	Sec. 1.2; Example 1.3	9 to 12
Scientific notation	Sec. 1.3; Example 1.7	13 to 20
Precision, accuracy, random and systematic error	Sec. 1.4	
Significant figures	Sec. 1.4	21 and 22
*Average and standard deviation	Sec. 1.4; Eqs. 1.1 and 1.2; Table 1.2	23 and 24
Density	Sec. 1.5; Eqs. 1.3 and 1.4; Examples 1.14 and 1.15; Table 1.6	31 to 35

*Supplemental material, which may be omitted.

PROBLEMS

1 (a) Convert 40 ft to inches. (b) Convert 15 m to feet. (c) Convert 83 ft to meters. (d) Convert 35 mi to kilometers.
Ans. (a) 480 in; (b) 49.2 ft; (c) 25.3 m; (d) 56.3 km

2 (a) Convert 45 m to centimeters. (b) Convert 2.5 cm to meters. (c) Convert 7.8 km to meters. (d) Convert 63 mm to meters.

3 (a) Convert 39 min to seconds. (b) Convert 2492 s to hours. (c) Convert 53 m/s to kilometers per hour.
Ans. (a) 2340 s; (b) 0.692 h; (c) 191 km/h

4 (a) Find the conversion from square feet to square inches. (b) Find the conversion from cubic feet to cubic inches.

5 A room is 16.5 ft long and 12.3 ft wide. Find the area of the room in (a) square feet and (b) square meters.
Ans. (a) 203 ft^2; (b) 18.9 m^2

6 A rectangular plot of land is 3530 ft long and 1750 ft wide. Find the area of the plot in acres (1 acre = 43,560 ft^2).

7 A shipping container is 6.1 m long, 2.0 m wide, and 2.6 m high. Find the volume of the container in (a) cubic meters and (b) cubic feet.
Ans. (a) 31.7 m^3; (b) 1120 ft^3

8 A *cord* is a unit of wood cut for fuel equal to a stack 8 ft long, 4 ft high, and 4 ft deep. How many cords are in a stack 20 ft long, 6 ft high, and 8 ft deep?

9 A surveyor wishing to find the distance from point B to point A on the opposite side of a river (Fig. 1.3) measures the baseline BC and the angles θ_1 and θ_2. Make a scale drawing and find the distance BA, given that $\theta_1 = 86°$, $\theta_2 = 83°$, and $BC = 0.15$ km.
Ans. 0.78 km

10 A hiker walks north for 2 km and then, turning 45°, walks northeast for 3 km. What is the straight-line distance between the beginning and end of the walk?

11 An obelisk (Fig. 1.12) casts a shadow of length $d = 12$ m. At the same time, the angle θ is found to be 75°. What is the height of the obelisk?
Ans. 45 m

12 The top of a 20-ft ladder leaning against a wall is 19 ft above the ground. How far is the base of the ladder from the wall?

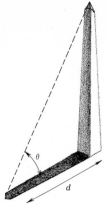

FIGURE 1.12

Problem 11.

13 Express the following numbers in scientific notation:
(a) 10,000   (b) 100,000,000
(c) 700,000   (d) 34,400
Ans. (a) 10^4; (b) 10^8; (c) 7 × 10^5; (d) 3.64 × 10^4

14 Express the following numbers in conventional notation:
(a) 10^7 (b) 2 × 10^3
(c) 4.76 × 10^4 (d) 1.4862 × 10^3

15 Express the following numbers in scientific notation:
(a) 0.00479   (b) 173.28
(c) 378,300   (d) 0.00000305
Ans. (a) 4.79 × 10^{-3}; (b) 1.7328 × 10^2; (c) 3.783 × 10^5; (d) 3.05 × 10^{-6}

16 Express the following numbers in conventional notation:
(a) 2 × 10^{-3} (b) 1.46 × 10^{-5}
(c) 76.254 × 10^{-4} (d) 578.935 × 10^2

17 Calculate the following expressions:
(a) $(1.3 \times 10^5)(3.5 \times 10^3)$
(b) $(2.4 \times 10^{14})(5.0 \times 10^9)$
(c) $\dfrac{8.4 \times 10^8}{2.1 \times 10^2}$ (d) $\dfrac{7.5 \times 10^{19}}{2.5 \times 10^{25}}$
Ans. (a) 4.55 × 10^8; (b) 1.2 × 10^{24}; (c) 4.0 × 10^6; (d) 3.0 × 10^{-6}

18 Calculate the following expressions:
(a) $(9.2 \times 10^{-3})(4.1 \times 10^7)$
(b) $(3.9 \times 10^{-17})(6.1 \times 10^{-5})$
(c) $\dfrac{4.5 \times 10^{25}}{1.5 \times 10^{-10}}$ (d) $\dfrac{6.0 \times 10^{-17}}{1.2 \times 10^{13}}$

19 The following exercises are to be done with a calculator:

(a) $\dfrac{(3.15)(4.69)}{6.05}$ (b) $\dfrac{(7.86)(5.20)}{9.15}$

(c) $\dfrac{6.17}{(8.15)(1.05)}$ (d) $\dfrac{9.05}{(3.17)(5.71)}$

Ans. (a) 2.44; (b) 4.47; (c) 0.721; (d) 0.500

20 The following exercises are to be done with a calculator:

(a) $\dfrac{(6.07 \times 10^{13})(29.8)}{149}$

(b) $\dfrac{(183.25)(1.08 \times 10^{-17})}{6.15 \times 10^{16}}$

(c) $\dfrac{(7.96 \times 10^{7})(6.76 \times 10^{29})}{9.15 \times 10^{15}}$

(d) $\dfrac{(3.14)(6.47 \times 10^{9})^{2}}{4.81 \times 10^{-27}}$

21 A man's height is measured to be 5 ft 11 in, with an uncertainty of 0.5 in. Express his height in feet, keeping only significant figures.
Ans. 5.92 ft

22 The length of a rectangular piece of land is found by surveying to be 1235 ± 25 ft, and its width is found to be 736 ± 15 ft. Calculate the area of the land in square feet, and express the result so that the number of significant figures is displayed properly.

23 Twelve students measured the length of the same cylinder with a steel ruler. Their results were:

5.2 cm	5.3 cm	5.1 cm
5.1 cm	5.3 cm	5.2 cm
5.2 cm	5.2 cm	5.2 cm
5.0 cm	5.4 cm	5.4 cm

Find the average and the standard deviation of these measurements.
Ans. (5.22 ± 0.11) cm

24 In a quality-control study the diameters of 100 similar ball bearings were measured to determine their variation in size. The results were:

Number of bearings	Diameter, cm
4	0.251
11	0.252
22	0.253
28	0.254
19	0.255
13	0.256
3	0.257
100	

Find the average and standard deviation of the diameters of these bearings.

25 In hydrology (the science of water distribution on land), a common unit of volume is the *acre-foot,* which is the volume of water which would cover 1 acre to a depth of 1 ft. (a) Given that 1 acre $= 43,560$ ft², find the conversion from acre-feet to cubic feet. (b) A thunderstorm drops 1.5 in of rain over a 50-mi² area. How many acre-feet of water is this?
Ans. (b) 4000 acre-feet.

26 What is the volume of a cylindrical tree trunk that is 0.7 m in diameter and 20 m long?

27 (a) Find the cross-sectional area of the I beam shown in Fig. 1.13. (b) If the beam is 18 ft long, what is its volume?
Ans. (a) 0.236 ft²; (b) 4.25 ft³

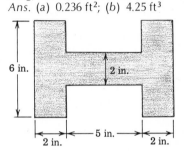

FIGURE 1.13

Problems 27 and 31.

28 Coal is piled into a cone-shaped pile 25 ft high and 50 ft in diameter at the base. What is the volume of the pile?

29 A gallon of paint covers 250 ft². What is the thickness of the applied paint?
Ans. 6.4×10^{-3} in

30 How many ball bearings 2.0 cm in diameter can be made from an ingot of steel 0.8 m wide, 0.2 h high, and 1.5 m long?

31 What is the weight (in pounds) of the I beam described in Prob. 27 (use Table 1.6).
Ans. 2.04×10^{3} lb

32 The mass of a metal cylinder 5 cm in diameter and 32 cm long is 5.3 kg. (a) What is the density of the cylinder? (Watch out for units here.) (b) What metal is the cylinder made of?

33 What is the weight (in pounds) of an acre-foot of water? (See Prob. 25.)
Ans. 2.71×10^{6} lb

34 An aluminum cable used for electric power transmission is 0.823 cm in diameter and 50 m long. What is the mass of the cable in kilograms?

35 If 1500 metric tons (1 metric ton $=$ 1000 kg) of iron ingots were stacked in a pile 8 m long and 12 m wide, how high would the pile be?
Ans. 2.0 m

BIBLIOGRAPHY

FISHER, ARTHUR: "The Riddle of the Leap Second," *Popular Science* **110:**(March 1973). A popular account of the leap second that is used to bring astronomical time into agreement with atomic-clock time.

NATIONAL BUREAU OF STANDARDS: *A Walk Through Time*. This pamphlet, available from the National Bureau of Standards, describes ancient and modern methods of measuring time. It explains the significance of the leap second, introduced in 1972 in connection with the establishment of Coordinated Universal Time (UTC).

PLEDGE, H. T.: *Science since 1500,* The Science Museum, London, 1946. A short history of mathematics, physics, chemistry, and biology. Chapter 2 discusses the relation of pumps and William Harvey's discovery of the circulation of blood.

SINGER, CHARLES, E. J. HOLMYARD, A. R. HALL, and TREVOR J. WILLIAMS: *A History of Technology*, Oxford University Press, London, 1957, 5 vols. The most comprehensive history of technology in the English language. Agricola is discussed in several chapters of Vol. 3, especially Chaps. 2 and 13.

T he behavior of an object—whether the object moves, rotates, bends, or breaks—depends on the forces acting on it. In designing a physical structure, such as a bridge truss, it is important to calculate the forces that will be exerted on each member, so that the member can be made strong enough to resist bending or breaking. This chapter and the next develop the laws of *statics,* which are the conditions under which an object will remain at rest. From these conditions the forces exerted on each part of a complex structure can be calculated.

2

FORCE

The laws of statics are a special case of Newton's laws of motion, which are the fundamental laws that describe the effect of forces on the motion of objects. *Dynamics,* the study of the motion of objects, is treated in Chap. 4.

2.1 Properties of Force

Force is an influence which acting on an object causes the object to change its state of motion. We exert a force whenever we push or pull on something. Forces are also exerted by the pull of gravity, by the physical contact of one object with another, and by the tension in a stretched cord. Despite their different origins, all forces have the following four properties:

Property 1 *A force is always applied by one material object on another.* Thus the hand in Fig. 2.1 exerts a force **F** (by means of the rope) on the block.

Property 2 *A force is characterized by both its magnitude and the direction in which it acts.* That is, both the magnitude and the direction are required to specify a force completely.

The direction of a force is the direction that the force would tend to move the object to which it is applied, in the absence of other forces. The direction is indicated by an arrow like that in Fig. 2.1, which shows that the force exerted by the hand is along the rope. The direction of a force is not always evident from inspection, but flexible ropes always transmit forces along their length. The magnitude of a force is the size of the force expressed in suitable units.

> **UNITS** The unit of force in the English engineering system is the pound (lb); the SI unit of force is the newton (N). A force of 1 lb is the force required to support a 1-lb object. The conversion between pounds and newtons is
>
> 1 lb = 4.45 N

Definition A quantity characterized by both a magnitude and a direction is called a *vector.* An arrow can be used to represent the magnitude of a vector at the same time that it gives its direction. In a given problem one simply adopts a scale, say 1 cm = 5 N, and then a force of 10 N is represented by an arrow 2 cm long. Everything one needs to know about the force is conveniently represented by such arrows. The vector **F** in Fig. 2.1

FIGURE 2.1

A force **F** applied by the hand on the block.

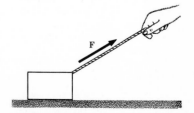

indicates a force of 6 N, with the scale 1 cm = 5 N, since its length is 1.2 cm.

The symbol for a force is written in boldface type (**F**) to emphasize that it is a vector quantity. The same symbol written in italic (*F*) refers to the magnitude of the force alone.

Property 3 (Newton's Third Law of Motion) *Whenever an object A exerts a force* **F** *on an object B, object B simultaneously exerts a force* **R** *on object A.* The force **R** is equal in magnitude to **F**, but it has the opposite direction. Thus it can be said that forces always come in pairs.

For example, while the hand in Fig. 2.2 is applying the force **F** on the block, the block simultaneously applies the force **R** on the hand. These two forces have the same magnitude (6 N), as indicated by their having the same length, but they point in opposite directions.

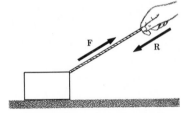

FIGURE 2.2

The force **R** applied by the block on the hand is the reaction to the force **F** applied by the hand on the block.

The two forces in any pair are sometimes termed the *action* and the *reaction* forces. This terminology is somewhat misleading, since it implies that one of the forces comes into existence in response to the other. This is not the case. Both forces come into existence simultaneously, and it is arbitrary which force is called the action and which the reaction. Usually the force acting on the object of primary interest is called the action, but this is only an arbitrary convention. In Fig. 2.2, if the block were the object of main interest, **F** would be the action and **R** the reaction, but if the hand were the focus of attention, **R** might be called the action and **F** the reaction.

> **REMARK** It is of *extreme importance* to realize that *the action and reaction forces always act on two different objects.* Property 1 states that a force is always applied by one object to another. Newton's third law goes further and says that each object exerts a force on the other.

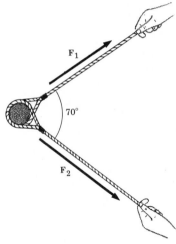

FIGURE 2.3

An object subjected to the two forces **F**₁ and **F**₂.

Property 4 *If two (or more) forces act simultaneously on the same object, their effect is the same as that of a single force equal to* the vector sum *of the individual forces.*

Figure 2.3 shows an object subjected to two forces **F**₁ and **F**₂. The vector sum, or *resultant,* **S** of these forces is found in the following way (Fig. 2.4). Draw an arrow to represent **F**₁, and from the head of **F**₁ draw a second arrow to represent **F**₂. Connect the *tail* of **F**₁ to the *head of* **F**₂ by a third arrow. This third arrow represents the resultant **S** of **F**₁ and **F**₂. In other words, the meaning of the equation

$$S = F_1 + F_2$$

is that the three vectors **S**, **F**₁, and **F**₂ form a closed triangle when drawn end to end as in Fig. 2.4.

> **REMARK** In going from Fig. 2.3 to Fig. 2.4, the vector **F**₂ was moved so that its tail touched the head of **F**₁. For the purpose of adding vectors it is permissible to move a vector from one position to another as long as its length and direction are not changed.

Property 4 allows us to find the combined effect of several forces acting on an object. Of course, we have not yet said what

FIGURE 2.4

The force **S** is the vector sum of **F**₁ and **F**₂.

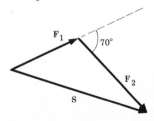

the effect of a force is, i.e., what a force does. This information is given by Newton's first and second laws. The first law, which is the basis of statics, is sufficient to enable us to solve a number of interesting problems. The second law, which is the basis of dynamics, is treated in Chap. 4.

Newton's First Law of Motion (Special Case) *For an object to remain at rest, i.e., to be in equilibrium, it is necessary that the vector sum of all the forces acting on the object be zero.*

Stated this way, Newton's first law is only a necessary condition for an object to be at rest. That is, if an object is to remain at rest, the resultant force on it must be zero, and if the resultant force is not zero, the object cannot remain at rest. However, it is possible for an object to move with constant speed, or even to rotate, even though the resultant force on it is zero.

Complete understanding of the properties of force and Newton's laws of motion is possible only after studying how they are applied to specific problems. The remainder of this chapter is devoted to utilizing the properties of force and Newton's first law in a variety of situations. While studying these applications you should constantly refer to Properties 1 through 4 and Newton's law to make sure you understand their relevance to the problem at hand.

REMARK The three laws of motion were first stated by Isaac Newton (1642–1727) in his great book *Principia Mathematica,* published in 1687. We have chosen to list the third law of motion as a property of force, since it does not depend on the motion of the objects involved.

2.2 Some Specific Forces

The last section discussed the general properties common to all forces. In this section we discuss the special properties of the forces that occur most frequently in everyday engineering practice.

FORCE OF GRAVITY

The earth exerts a force on all objects near its surface. This force is called *gravity,* or *weight.* For a given object, say a block of lead, this force has approximately the same magnitude everywhere on the earth's surface. It does vary by as much as 0.5 percent between either of the poles and the equator, but we shall neglect this small variation. The force of gravity on a given object decreases steadily as the object moves away from the earth's surface, and this effect must be taken into account in dealing with earth-orbiting satellites and interplanetary space vehicles.

The direction of the force is toward the center of the earth, so the direction of the force of gravity on the lead block at the pole is different from the direction at the equator (Fig. 2.5). The lead block is said to be attracted toward the center of the earth by the force of gravity. Therefore, by Newton's third law, the block must attract the earth with a force of equal magnitude and opposite direction to the force exerted by the earth on the

FIGURE 2.5

The force of gravity on a lead block at the pole and at the equator.

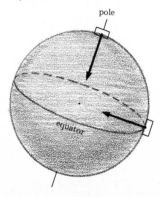

block. This reaction force acts on the earth and is not a second force acting on the block.

SPRING FORCE

Figure 2.6a shows a spring suspended from a bar and hanging vertically alongside a ruler. When a 1-N weight is attached to the lower end of the spring, the spring stretches a certain distance, say 0.75 cm (Fig. 2.6b). When a 2-N weight is attached to the spring, the spring stretches 1.5 cm, or twice as much as with 1 N. In general, it is found that the amount that the spring is stretched (up to a limit) is proportional to the force of gravity on the weight. The magnitude F_g of this force is thus related to the distance x that the spring is stretched by

$$F_g = kx$$

where k is a characteristic constant of the spring, called the *spring constant*. In the present case,

$$k = \frac{1\,\text{N}}{0.75\,\text{cm}} = \frac{2\,\text{N}}{1.5\,\text{cm}} = 1.33\,\text{N/cm}$$

Since a block hanging from the spring is at rest, Newton's first law says that the total force on it must be zero. But the earth is exerting a force of gravity $\mathbf{F}_g$ on it, directed downward. Therefore, there must be another force $\mathbf{F}_k$ on the block to cancel $\mathbf{F}_g$. This force is exerted by the spring and arises from the molecular rearrangements produced in the metal when the spring is stretched.

To study the forces acting on an object, it is useful to focus attention on the object by isolating it from its surroundings. First draw a dotted line around the object, as in Fig. 2.7a. Then redraw the object alone (Fig. 2.7b). Wherever the dotted line cut something in contact with the object, draw an arrow to represent the force exerted on the object at the point of contact. In addition, draw an arrow to represent the force of gravity. The result, shown in Fig. 2.7b, is called a *free-body diagram*.

The unknown force $\mathbf{F}_k$ is found by applying Newton's first law to the isolated object. Since the object is at rest, the vector sum

FIGURE 2.6

(a) Unstretched spring. (b) The spring stretched a distance x by a weight suspended from it.

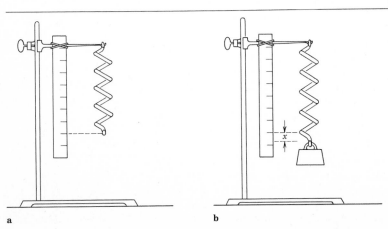

a   b

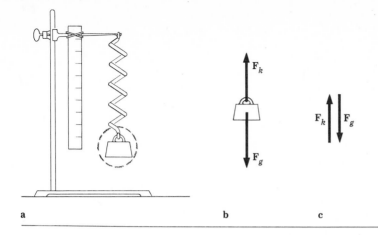

a   b   c

FIGURE 2.7

(a) The weight on the spring is isolated by drawing a dotted line around it. (b) Free-body diagram of the weight, showing the forces that act on it. (c) The forces $\mathbf{F}_k$ and $\mathbf{F}_g$ drawn head to tail to show that their vector sum is zero.

of all the forces on it must be zero. In this case, we have just two forces, so

$$\mathbf{F}_g + \mathbf{F}_k = 0 \qquad 2.1$$

To add these forces, Property 4 says that $\mathbf{F}_k$ must first be drawn with its tail on the head of $\mathbf{F}_g$. The sum of these vectors is then the vector drawn from the tail of $\mathbf{F}_g$ to the head of $\mathbf{F}_k$. Since this sum is to be zero, the tail of $\mathbf{F}_g$ must coincide with the head of $\mathbf{F}_k$. Figure 2.7c shows this: $\mathbf{F}_k$ must have the same magnitude as $\mathbf{F}_g$ and be oppositely directed.

We can also use the normal rules of algebra to solve Eq. 2.1 for $\mathbf{F}_k$. When $\mathbf{F}_g$ is transposed to the right-hand side of the equation, we get

$$\mathbf{F}_k = -\mathbf{F}_g$$

What does the minus sign in front of the vector $\mathbf{F}_g$ mean? We know that $\mathbf{F}_k$ is equal in magnitude and opposite in direction to $\mathbf{F}_g$. This suggests the general rule: *For any force $\mathbf{F}$, the force $-\mathbf{F}$ has the same magnitude as $\mathbf{F}$ and is oppositely directed.* Thus $\mathbf{F}_k$ is equal in magnitude but antiparallel (oppositely directed) to $\mathbf{F}_g$, so

$$F_k = F_g = kx \qquad 2.2$$

A spring scale uses the distance a spring is stretched to measure forces. Such scales are suitable for simple lecture demonstrations, but they are not accurate enough for most commercial purposes. Figure 2.8 shows a typical spring scale.

FIGURE 2.8

A spring scale. (*Sargent-Welch.*)

NORMAL FORCE

A block sitting on a table is at rest, and so again, by Newton's first law, there must be another force acting on it in addition to the force of gravity. The surface of the table is slightly deformed by the block and, as a consequence, exerts an upward force $\mathbf{F}_n$

on the block. This force, which is directed perpendicular to the surface that produces it, is called the *normal force*.* Since the sum of $\mathbf{F}_g$, the force of gravity on the block, and $\mathbf{F}_n$, the normal force on the block, is zero by Newton's first law, it follows that $\mathbf{F}_n$ must be directed opposite to $\mathbf{F}_g$, or

$$\mathbf{F}_n = -\mathbf{F}_g$$

Normal forces are exerted whenever solid objects come in contact. They are real forces and are accompanied by small distortions in the surfaces of the bodies that produce them. In fact, a normal force differs from a spring force only in the degree to which distortion occurs. A spring is designed to produce a noticeable change of shape when it applies a force, whereas only minute changes occur in the surface of a solid body exerting a normal force. Thus, in a sense, a solid body acts like a very stiff spring. Consequently, a solid body can exert normal forces of different magnitudes in different circumstances without any noticeable change in its appearance. However, this should not mislead you into thinking that these forces are any less real.

Figure 2.9 shows the two forces $\mathbf{F}_g$ and $\mathbf{F}_n$ that act on the block on the table. Because the magnitude of $\mathbf{F}_n$ adjusts itself to equal the magnitude of $\mathbf{F}_g$, $\mathbf{F}_n$ is sometimes called a reactive force. We shall not use this terminology, however, because "reactive" is easily confused with "reaction." $\mathbf{F}_n$ is *not* the reaction of $\mathbf{F}_g$, even though it is equal in magnitude and opposite in direction to $\mathbf{F}_g$. The relation $\mathbf{F}_n = -\mathbf{F}_g$ is a consequence of Newton's first law, not the third. Furthermore, $\mathbf{F}_n$ and $\mathbf{F}_g$ act on the same object, while action-reaction forces always act on different objects.

The reaction to $\mathbf{F}_g$ is the gravitational force $\mathbf{R}_g$ that the block exerts on the earth: it acts on the earth (at its center). The normal force $\mathbf{F}_n$ also has a reaction force $\mathbf{R}_n$, which is a normal force and acts on whatever is producing $\mathbf{F}_n$. Since the table produces $\mathbf{F}_n$, the reaction to $\mathbf{F}_n$ is the force $\mathbf{R}_n$ the block exerts on the table. Figure 2.10 shows all these forces. Remember, in applying Newton's first law, only the forces acting on the object in question must sum to zero.

FRICTION

Friction, like a normal force, is a force which a surface applies to an object in contact with it. However, whereas the normal force is always perpendicular to the surface, friction is always parallel to the surface. *Static friction* is the parallel force the surface exerts on a stationary object, and *kinetic friction* is the parallel force the surface exerts on a moving object. Friction generally acts to oppose any externally applied force. Thus, again consider a block on a table and imagine that a small force $\mathbf{F}_a$ is applied to it, parallel to the surface of the table (Fig. 2.11). If $\mathbf{F}_a$ is small

Normal means "perpendicular."

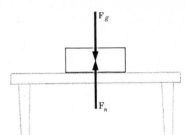

FIGURE 2.9

The force of gravity $\mathbf{F}_g$ and the normal force $\mathbf{F}_n$ acting on a block at rest on a table.

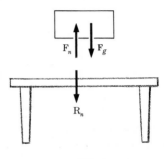

FIGURE 2.10

The reaction to $\mathbf{F}_g$ is a force $\mathbf{R}_g$ that acts on the earth, and the reaction to $\mathbf{F}_n$ is a force $\mathbf{R}_n$ that acts on the table.

FIGURE 2.11

The force of friction $\mathbf{F}_f$ on a block opposes a force $\mathbf{F}_a$ applied parallel to the surface.

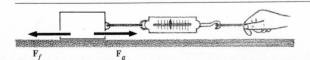

enough, the block will not move, so there must be some other force acting on it; this is the force of static friction $\mathbf{F}_f$. As long as the block remains at rest, Newton's first law requires that $\mathbf{F}_f = -\mathbf{F}_a$.

The sum of the normal and friction forces, $\mathbf{F}_n + \mathbf{F}_f$, is the total *surface force* $\mathbf{F}_s$ that the surface exerts on an object (Fig. 2.12). The magnitude of $\mathbf{F}_f$ can change to equal F_a, just as the magnitude of $\mathbf{F}_n$ can change to equal F_g. However, there is a maximum value to the magnitude of $\mathbf{F}_f$, and if F_a exceeds this value, the force of friction no longer balances it and the block will move. ($\mathbf{F}_n$ also has a maximum value, namely, the value at which the surface breaks. This value is not often achieved in everyday life, because material objects, both natural and manufactured, are designed not to break under the forces to which they are normally subjected.)

The maximum force of static friction depends on the nature of the two surfaces involved. It is greater for a wood block on a wood table than for a waxed ski on ice. It usually also depends on the magnitude of the normal force $\mathbf{F}_n$. For simple problems, we assume that the maximum frictional force is just proportional to F_n:

Maximum force of static friction $= \mu_s F_n$ 2.3

As long as the body remains at rest, the actual force of friction on it will be less than (or equal to) this, so

$$F_f \le \mu_s F_n \qquad\qquad 2.4$$

The symbol $\le$ means "less than or equal to," and Eq. 2.4 says that the force of friction is less than or equal to a certain fraction of the contact force. The constant* μ_s is called the *coefficient of static friction,* and Table 2.1 gives its value for some common materials.

Example 2.1 A 50-lb block of ice is sitting on a frozen lake. If the coefficient of static friction μ_s between the block and the surface is 0.05, what is the maximum force of static friction between the block and the surface? Will the block move if a horizontal force of 2 lb is applied to the block?

Since the block is at rest, the normal force $\mathbf{F}_n$ that the surface exerts on it is equal to $-\mathbf{F}_g$, so $F_n = 50$ lb. Then by Eq. 2.3 the maximum force of static friction is

$$F_{f,\text{max}} = \mu_s F_n = (0.05)(50\,\text{lb}) = 2.5\,\text{lb}$$

If a 2-lb horizontal force is applied to the block, the actual force of friction will be 2 lb directed opposite to the applied force, so the block will not move. If the applied force exceeds 2.5 lb, it will exceed the maximum force of static friction. In this case, the total horizontal force will not be zero and the block will not remain at rest. $\quad\square$

The term *static* refers to the fact that Eq. 2.4 holds only as long as the body is at rest. Once F_a exceeds $\mu_s F_n$, the body is no longer in equilibrium, and so it will start to slide. The frictional force

*μ is the Greek (lowercase) letter mu.

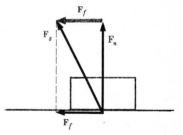

FIGURE 2.12

The surface force $\mathbf{F}_s$ is the vector sum of the normal force $\mathbf{F}_n$ and the force of friction $\mathbf{F}_f$.

TABLE **2.1**

Coefficients of static friction of some common materials

Material	Coefficient of static friction μ_s
Ice on ice	0.05–0.15
Leather, on wood	0.3–0.4
On metal	0.6
Ski wax on dry snow	0.04
Steel on steel, dry	0.6
Lubricated	0.10
Wood, on wood, dry	0.25–0.50
On metal, dry	0.2–0.6

now changes in character and depends on the speed of the object. For simple problems this dependence is neglected, and the moving friction is taken to be a constant, also proportional to F_n. For moving bodies we have

$$F_f = \mu_k F_n \qquad \qquad 2.5$$

where μ_k, the *coefficient of kinetic friction*, is usually less than μ_s. Table 2.2 gives the coefficients of kinetic friction of some common substances.

COMPRESSION AND TENSION

A solid block that has two opposing forces $\mathbf{F}_1$ and $\mathbf{F}_2 = -\mathbf{F}_1$ pressing it on either side (Fig. 2.13a) will remain at rest. Nevertheless, it clearly differs in some sense from a block that has no forces acting on it. When the two forces press in, the block is said to be *compressed* or in a state of *compression;* the magnitude C of the compression is equal to the magnitude of either force acting on it. That is, $C = F_1 = F_2$. (Remember we use italic symbols to indicate only the magnitude of the force.)

Likewise, a block at rest could have two opposing forces pulling on it, as in Fig. 2.13b. In this case, the block is said to be in a state of *tension,* and the magnitude T of the tension is again equal to the magnitude of either force acting on it ($T = F_1 = F_2$).

FLEXIBLE CORDS

A flexible cord, such as a string, rope, or cable, has a number of special properties:

1 It can be in a state of tension but not compression.

2 It can transmit a force only along its length. (This is in contrast to a solid rod, e.g., a golf club or baseball bat, which can exert a force both along it length and perpendicular to it.)

3 In the absence of friction, the tension is the same at all points along a cord.

> **REMARK** Cords usually weigh much less than the forces they transmit, so their own weight can be neglected. When this is not the case, the tension will vary in a vertically hanging cord.

Example 2.2 An 80-lb weight hangs from a cord that passes over a frictionless pulley. The other end of the cord attaches to a heavy hatch cover (Fig. 2.14). Find the magnitude and direction of the force that the cord exerts on the cover.

Figure 2.15a shows the free-body diagram of the 80-lb weight. The forces on it are the force of gravity $\mathbf{F}_g$ and the force $\mathbf{F}_c$ exerted by the cord. Since the weight is at rest, we know from Newton's first law that

$$\mathbf{F}_c = -\mathbf{F}_g \qquad \text{or} \qquad F_c = F_g = 80 \text{ lb}$$

Since the cord exerts an 80-lb force upward on the weight, by Newton's third law the weight must exert an 80-lb force downward on the cord (Fig. 2.15b). That is, the cord is pulled at one end with a force of 80 lb, so the tension in this part of the cord is 80 lb. But since the cord passes over a frictionless pulley, the tension is the same everywhere in the cord (Property 3). Thus

TABLE **2.2**

Coefficients of kinetic friction of some common materials

Material	Coefficient of kinetic friction μ_k
Brass on ice	0.02
Ice on ice	0.02
Rubber on concrete	1.02
Waxed hickory:	
On dry snow	0.04
On wet snow	0.14

FIGURE 2.13

(a) A block compressed by two opposing forces pressing on it. (b) A block in tension from two opposing forces pulling on it.

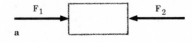

a

b

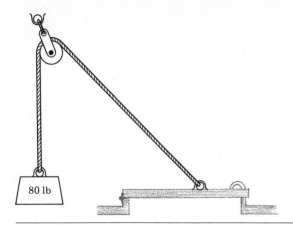

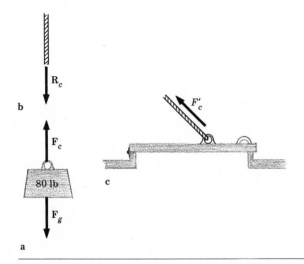

FIGURE 2.14

A cord attached at one end to a hatch cover and at the other end to an 80-lb weight.

FIGURE 2.15

(a) Free-body diagram of the 80-lb weight in Fig. 2.14. (b) The force $\mathbf{R}_c$ applied to the end of the cord is the reaction to $\mathbf{F}_c$. (c) The force $\mathbf{F}'_c$ applied to the hatch cover is directed along the cord.

the cord exerts an 80-lb force on the hatch cover. By Property 2, the direction of this force is along the part of the cord attached to the cover (Fig. 2.15c) □

Cords and pulleys are often used in mechanical systems to change the direction of an applied force. In the last example, the cord-and-pulley arrangement was used to convert the downward force of gravity on the 80-lb weight into an upwardly inclined force of 80 lb acting on the hatch cover. The purpose of such an arrangement is to aid someone lifting the cover.

FIGURE 2.16

(a) Two forces $\mathbf{F}_1$ and $\mathbf{F}_2$ with the same direction. (b) The sum $\mathbf{S}$ of $\mathbf{F}_1$ and $\mathbf{F}_2$.

2.3   Examples Involving Forces in a Line

Two forces are said to lie in a line if they are parallel (have the same direction) or antiparallel (have opposite directions). The forces $\mathbf{F}_1$ and $\mathbf{F}_2$ in Fig. 2.16a have the same direction; Fig. 2.16b shows that the magnitude of their sum $\mathbf{S} = \mathbf{F}_1 + \mathbf{F}_2$ is equal to the sum of the magnitudes of $\mathbf{F}_1$ and $\mathbf{F}_2$. For instance, if $F_1 = 8$ N and $F_2 = 5$ N, then $S = 13$ N.

Figure 2.17a shows two forces $\mathbf{F}_1$ and $\mathbf{F}_2$ that have opposite

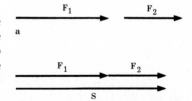

directions. In this case, the magnitude of their sum $S = F_1 + F_2$ is equal to the *difference* of the magnitudes of F_1 and F_2 (Fig. 2.17b). For instance, if $F_1 = 8\,N$ and $F_2 = 5\,N$, then $S = 3\,N$.

The general rule for the magnitude S of the sum of two forces F_1 and F_2 lying in a line is

$S = F_1 + F_2$ parallel   2.6a

$S = F_1 - F_2$ antiparallel   2.6b

If two forces do not lie in a line, their sum is not equal to either the sum or the difference of the magnitudes of the individual forces, as can be seen from Fig. 2.4. The general case of adding forces that are not in a line is treated in Sec. 2.4.

In this section we shall give several examples of problems involving forces in a line. Examples of problems involving forces not in a line are given in Sec. 2.5.

Example 2.3 What does the scale in Fig. 2.18 read when block B, which weighs 10 N, is placed on top of block A, which weighs 5 N?

REMARK Everyone knows that the scale will read 15 N. The same principle applies when you stand on a scale. The scale reads the sum of the gravitational forces on all the parts of your body, even though only your feet are in contact with the scale. The more one thinks about these common everyday experiences, the more mysterious they seem. There must be some fundamental law of nature which accounts for them. This is Newton's third law of motion. In fact, the best evidence of the validity of Newton's third law is its ability to explain some of the most commonplace expreriences of everyday life.

To analyze a problem involving two objects, we must make a separate free-body diagram of each. Figure 2.19a shows the free-body diagram for block B. Since there is the 10-N force of gravity F_g acting downward on it, by Newton's first law there is an upward force of 10 N as well. The only possible source for this second force is the normal force F_n that A exerts on B. Therefore F_n is a 10-N force directed upward.

Block A is acted upon by the 5-N force of gravity F_g' and by the upward normal force F_n' exerted by the scale (Fig. 2.19b). However, in this case, the magnitude of F_n' does not equal 5 N, because F_g' and F_n' are not the only forces acting on A. Newton's third law says there is a reaction force R_n to $F_n:R_n$ is the 10-N normal force that B exerts against A. Newton's first law requires that the sum $F_g' + F_n' + R_n$ be zero. The sum $F_g' + R_n$ is a 15-N force directed downward, so F_n' is a 15-N force directed upward.

Finally, by Newton's third law again, the reaction R_n' to F_n' is a 15-N force exerted downward by A on the scale. It is this force on the scale that the scale registers. It is Newton's third law that guarantees that the force on the scale (which is a normal force and not a gravitational force) is equal in magnitude to the total gravitational force on all the objects sitting on the scale. ☐

It is particularly important to note that in this last example the upward normal force F_n' on the lower block A is greater than the upward normal force F_n on the upper block B. In any vertical

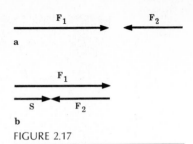

a

b

FIGURE 2.17

(a) Two forces F_1 and F_2 with opposite directions. (b) The sum S of F_1 and F_2.

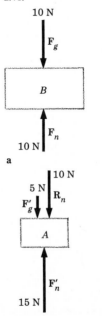

FIGURE 2.18

Two blocks sitting on a scale.

FIGURE 2.19

(a) The free-body diagram of block B in Fig. 2.18. (b) The free-body diagram of block A in Fig. 2.18.

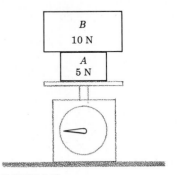

a

b

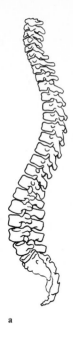

a

b

FIGURE 2.20

The size of a vertical structure increases from top to bottom.
(a) The human spinal column.
(b) The Eiffel Tower (*French Em-
bassy Press and Information
Division*).

structure the normal force on a part near the bottom of the
structure is greater than the normal force on a part near the top
because each part supports the entire weight of all the parts
above it. In many structures, both natural and fabricated, the
lower parts are larger than the higher parts in order to be strong
enough to bear the larger force. In the human spinal column, for
instance, the vertebrae increase steadily in size from top to
bottom in order to support an upright posture (Fig. 2.20*a*). Likewise, in steel structures such as the Eiffel Tower (Fig. 2.20*b*), the
base is made larger and stronger than the top.

Example 2.4 With what force **F** must the hand in Fig. 2.21 pull
on the cord in order to hold the weight W in place?

The free-body diagram for the weight (Fig. 2.22*a*) shows that
the force **F'** exerted by cord 1 is equal to W. Since the tension T_1
is equal to the force **F'** that the cord exerts, we have

$$T_1 = F' = W$$

Figure 2.22*b* shows the free-body diagram for the lower pulley. In order to isolate this pulley, cord 2 had to be cut in two
places, and cord 1 in one place. At each cut the cord pulls on the
pulley with a force equal to the cord's tension. Thus, cord 2
exerts a force T_2 *twice* on the pulley. For the pulley to remain at
rest, the magnitude of the total upward force $2T_2$ must equal the

magnitude of the total downward force T_1:

$$2T_2 = T_1 \quad \text{or} \quad T_2 = \tfrac{1}{2}T_1 = \tfrac{1}{2}W$$

The tension in cord 2 has to be only one-half the weight being supported, because cord 2 applies its force twice. The hand pulls on cord 2 with a force $\mathbf{F}$ equal to the cord's tension:

$$F = T_2 = \tfrac{1}{2}W$$

Pulleys are used in this way to change both the magnitude and direction of an applied force. $\square$

Example 2.5 A 40-N block hangs from a cord that passes over a frictionless pulley and attaches to a 100-N block sitting on a horizontal surface (Fig. 2.23). What is the smallest value of the coefficient of static friction between the 100-N block and the horizontal surface that will prevent the block from sliding?

This problem involves two objects and a number of forces. To analyze such a complex mechanical system, we must proceed systematically. First we isolate the blocks and draw a free-body diagram for each.

Figure 2.24a shows the free-body diagram for the 40-N block. The forces on it are its weight $\mathbf{F}_g$ and the force exerted by the cord $\mathbf{F}_c$. We assume that the blocks are not moving, so Newton's first law can be applied separately to each block. Thus for the 40-N block to remain at rest, we must have

$$\mathbf{F}_c = -\mathbf{F}_g \quad \text{or} \quad F_c = F_g = 40 \text{ N}$$

Since the cord exerts a 40-N force on the 40-N block, the block exerts a 40-N force on the cord (action-reaction). That is, the cord is being pulled at one end with a force of 40 N, so the cord has a tension of 40 N. The tension is the same everywhere in the cord because the cord passes over a frictionless pulley.

Figure 2.24b shows the free-body diagram for the 100-N block. Although there are four forces acting on this block, the vertical and horizontal forces can be treated independently, since they must separately sum to zero if the block is to remain at rest. Property 2 of cords tells us that the cord applies a force $\mathbf{F}_c'$ that is parallel to the cord. The force of friction $\mathbf{F}_f$ is in the opposite direction to $\mathbf{F}_c'$. Thus, for the block to remain at rest, we must have

$$\mathbf{F}_c' + \mathbf{F}_f = 0 \quad \text{or} \quad F_c' = F_f$$

But F_c' is equal to the tension in the cord, so

$$F_f = F_c' = 40 \text{ N}$$

The vertical forces also must sum to zero, so

$$F_n = F_g' = 100 \text{ N}$$

Then from Eq. 2.4, we have

$$F_f \le \mu_s F_n$$

so

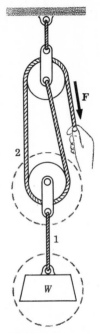

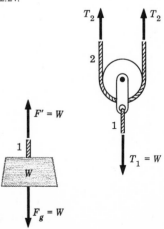

FIGURE 2.21

Pulley and cord arrangement for lifting a weight.

FIGURE 2.22

(a) Free-body diagram of the weight in Fig. 2.21. (b) Free-body diagram of the lower pulley in Fig. 2.21.

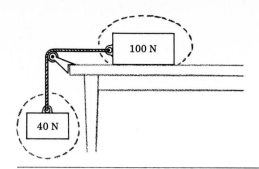

FIGURE 2.23

A vertically-hanging block connected by a cord to a second block
sitting on a horizontal surface.

$$\mu_s \geq \frac{F_f}{F_n}$$

or

$$\mu_s \geq \frac{40 \text{ N}}{100 \text{ N}} = 0.4$$

The coefficient of friction must be greater than or equal to 0.4 for
the 100-N block not to slide under the pull of the 40-N block. ☐

Example 2.6   Figure 2.25 shows two blocks connected by a cord.
Block A weighs 30 N and block B weighs 20 N. The coefficient of
static friction μ_s between the blocks and the table is 0.4. What is
the minimum force $\mathbf{F}_a$ that must be applied to block B to move
the entire assembly? What is the tension T in the connecting
cord when the assembly just starts to move?

To analyze this problem, we must consider the forces acting
on each block separately.

Figure 2.26a shows the free-body diagram for block A. The
forces on it are its weight $\mathbf{F}_g$, the normal force $\mathbf{F}_n$, the force $\mathbf{F}_c$
exerted by the cord, and the frictional force $\mathbf{F}_f$. In order for the
block not to move, the horizontal and vertical forces must separately sum to zero. Thus for the vertical forces we have

FIGURE 2.24

(a) Free-body diagram of the vertically-hanging block in Fig. 2.23.
(b) Free-body diagram of the block
sitting on the horizontal surface in
Fig. 2.23.

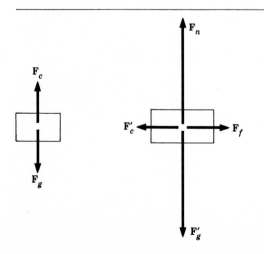

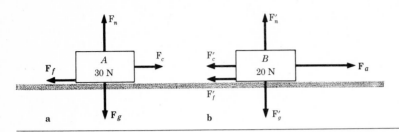

FIGURE 2.25

A force $\mathbf{F}_a$ applied to two blocks connected by a cord.

FIGURE 2.26

(a) Free-body diagram of block A in Fig. 2.25. (b) Free-body diagram of block B in Fig. 2.25.

$$\mathbf{F}_g + \mathbf{F}_n = 0 \quad \text{or} \quad F_n = F_g = 30 \text{ N}$$

and for the horizontal forces we have

$$\mathbf{F}_f + \mathbf{F}_c = 0 \quad \text{or} \quad F_c = F_f \qquad\qquad 2.7$$

The maximum value of F_f is given by Eq. 2.3:

$$F_{f,\text{max}} = \mu_s F_n = (0.4)(30 \text{ N}) = 12 \text{ N}$$

So block A will start to slide when F_c just exceeds 12 N, for then Eq. 2.7 can no longer be satisfied. But F_c is equal to the tension in the cord, so $T = 12$ N when the block starts to slide.

Figure 2.26b shows the free-body diagram for block B. In order for this block not to move, the vertical forces must satisfy the relation

$$\mathbf{F}'_g + \mathbf{F}'_n = 0 \quad \text{or} \quad F'_n = F'_g = 20 \text{ N}$$

and the horizontal forces must satisfy the relation

$$\mathbf{F}_a + \mathbf{F}'_c + \mathbf{F}'_f = 0 \quad \text{or} \quad F_a = F'_c + F'_f \qquad\qquad 2.8$$

When block A just starts to move, the tension in the cord is 12 N, so the force F'_c which the cord exerts on block B is also 12 N. If block B is also just starting to move, the frictional force on it will be

$$F'_f = F'_{f,\text{max}} = \mu_s F'_n = (0.4)(20 \text{ N}) = 8 \text{ N}$$

Then from Eq. 2.8 we see that the applied force F_a must be

$$F_a = F'_c + F'_f = 12 \text{ N} + 8 \text{ N} = 20 \text{ N}$$

Notice that a 20-N force is applied to block B, but only 12 N is applied to block A. The 8-N frictional force on block B reduces the force transferred to block A. Friction is often used in this way to reduce an applied force. $\square$

2.4 Vector Addition

GRAPHICAL METHOD

The rule for the addition of two or more forces is based on the geometric construction outlined under Property 4 in Sec. 2.1. Each force is represented by an arrow whose length represents the magnitude of the force according to an appropriately chosen scale. The direction of the arrow is the same as the direction of the force it represents. When these arrows are drawn head to tail, the sum of the forces is represented by the arrow drawn from the tail of the first arrow to the head of the last.

Definition A *vector* is a physical quantity, such as force, which has both magnitude and direction, and which obeys the preceding rule of addition. Other vector quantities which occur frequently in physics are displacement, velocity, acceleration, and momentum. The rule of addition is most easily visualized for displacement.

Definition When an object moves from point A to point B along any path, the *displacement* $\mathbf{d}_{AB}$ is the vector whose magnitude is the straight-line distance from A to B and whose direction points from A to B.

Path 1 in Fig. 2.27 is a possible path between A and B. The displacement for all such paths between A and B is $\mathbf{d}_{AB}$. Likewise, path 2 is a possible path between B and C. The displacement for all such paths between B and C is $\mathbf{d}_{BC}$. Paths 1 and 2, taken together, are a possible path between A and C. The displacement for all such paths between A and C is $\mathbf{d}_{AC}$. Thus $\mathbf{d}_{AC}$ is the displacement for the same paths as $\mathbf{d}_{AB}$ and $\mathbf{d}_{BC}$ together. In this sense we can consider $\mathbf{d}_{AC}$ to be the sum of $\mathbf{d}_{AB}$ and $\mathbf{d}_{BC}$. That is, we can write

$$\mathbf{d}_{AC} = \mathbf{d}_{AB} + \mathbf{d}_{BC}$$

with the understanding that this equation means that $\mathbf{d}_{AC}$ is related to $\mathbf{d}_{AB}$ and $\mathbf{d}_{BC}$ by a triangular construction like that in Fig. 2.27.

Example 2.7 A truck travels north 5 mi, then east 3 mi, and finally southeast 2 mi. Draw an arrow to represent the displacement of each segment of the journey. Find the truck's overall displacement.

Choose a convenient scale, such as 1 mi = 1 cm, and draw lines to represent the north-south and east-west directions (Fig. 2.28a). The first displacement $\mathbf{d}_1$ is represented by an arrow 5 cm long drawn parallel to the north line. The second displacement $\mathbf{d}_2$ is represented by an arrow 3 cm long drawn parallel to the east line. The third displacement $\mathbf{d}_3$ is represented by an arrow 2 cm long drawn 45° south of the east line.

The overall, or *resultant,* displacement is the displacement that goes directly from the initial to the final position of the truck. It is fround by adding the three individual displacements according to the rule for vector addition. This requires only that the tail of each arrow representing a displacement start at the head of the preceding one (Fig. 2.28b). When this is done, the tail of $\mathbf{d}_1$ will be at the initial position and the head of $\mathbf{d}_3$ will be at the final position of the truck. The resultant $\mathbf{d}$ is represented by the arrow drawn from the tail of $\mathbf{d}_1$ to the head of $\mathbf{d}_3$. This is in accordance with the rule of vector addition and with the commonsense notion that the resultant displacement goes directly from the initial to the final position of the truck.

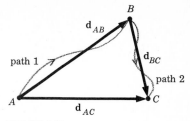

FIGURE 2.27

The displacement $\mathbf{d}_{AC}$ from A to C is the vector sum of the displacement $\mathbf{d}_{AB}$ from A to B and the displacement $\mathbf{d}_{BC}$ from B to C.

FIGURE 2.28

(a) Arrows representing the displacements of the truck in Example 2.7. (b) Vector addition of the displacements in (a).

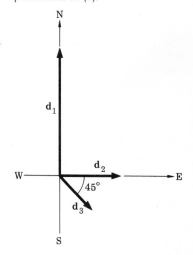

a

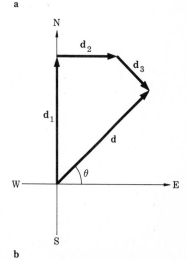

b

The length of **d** is measured with a ruler and found to be 5.7 cm, so the magnitude is

$$d = (5.7\text{ cm})(1\text{ mi/cm}) = 5.7\text{ mi}$$

The angle θ is measured with a protractor and found to be 39°, so **d** is directed 39° north of east.

> **REMARK** Study Fig. 2.28 carefully, paying particular attention to how the arrows, which are drawn tail-to-tail in Fig. 2.28a, are drawn tail-to-head in Fig. 2.28b. Note also that the resultant **d** goes from the tail of **d**$_1$ to the head of **d**$_3$ and is *not* head-to-tail like the other displacements. ☐

Example 2.8 Find the sum of the three forces shown in Fig. 2.29a.

The arrows in Fig. 2.29a are not drawn to scale, nor are the angles drawn accurately. However, the figure specifies the magnitude and direction of each force relative to a coordinate system. To find their sum, it is necessary to redraw the arrows to scale in a head-to-tail fashion.

Choose the scale 5 N = 1 cm. Redraw the coordinate axes (Fig. 2.29b). With a protractor draw a line at 60° to the x axis. Measure a distance of 4 cm along this line to represent the 20-N force **F**$_1$. From the head of **F**$_1$ draw a line parallel to the y axis. Measure a distance of 2 cm along this line to represent the 10-N force **F**$_3$. From the head of **F**$_3$ draw a line parallel to the x axis. Draw another line at 15° to the last one. Measure a distance of 5 cm along this line to represent the 25-N force **F**$_2$.

The sum, or resultant, **F** of these forces is represented by the arrow drawn from the tail of **F**$_1$ to the head of **F**$_2$. The length of this arrow is found to be 3.9 cm, so the magnitude of **F** is

$$F = (3.9\text{ cm})(5\text{ N/cm}) = 19.5\text{ N}$$

The angle θ is found to be 44°.

> **REMARK** Figure 2.29b appears at first sight to be more complicated than Fig. 2.28b, but this is so only because the arrows cross at one point. This has no consequence and is simply the result of the order in which we chose to arrange the arrows. The sum **F** does not depend on the order in which the forces are added. Figure 2.30 shows the same forces added in the order **F**$_3$, **F**$_2$, and **F**$_1$. The magnitude and direction of the resultant are the same as in Fig. 2.29b. ☐

TRIGONOMETRIC METHOD

The graphical method of vector addition shows directly the relation between vectors and their sum. It is easy to learn and with reasonable care gives results that are precise enough for many purposes. However, it is rather tedious and the precision is limited to two significant figures. With the aid of a scientific calculator it is quicker and more precise to use trigonometry to solve vector problems.

Trigonometry is based on the fact that the ratios of corresponding sides of similar triangles are equal. For example, consider the two right triangles in Fig. 2.31. They both have a right (90°) angle and a 30° angle. The third angle ϕ must therefore also be the same, because the sum of the angles of a triangle is 180°. Thus for each triangle we have

$$90° + 30° + \phi = 180° \qquad \text{or} \qquad \phi = 180° - 120° = 60°$$

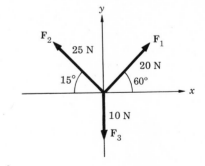

a

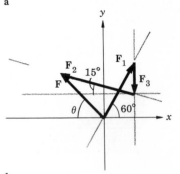

b

FIGURE 2.29

(a) Arrows representing the three forces in Example 2.8. (b) Vector addition of the forces in (b).

FIGURE 2.30

Vector addition of the forces in Fig. 2.29a. The three forces are added in a different order than in Fig. 2.29b, but the resultant **F** is the same.

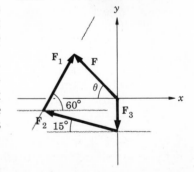

Triangles with the same angles are similar; i.e., they have the same shape. Consequently, the ratio a/c of two sides of the small triangle is equal to the ratio a'/c' of the corresponding sides of the large triangle. In all 30° right triangles the ratio a/c is the same, namely, 0.866. Likewise, in all 30° right triangles the ratio b/c is 0.500 and the ratio b/a is 0.577. These ratios, which depend on the angle of the triangle, are called *trigonometric functions*.

The sides of a right triangle are labeled in Fig. 2.32. The side opposite the right angle is called the *hypotenuse,* the side opposite the angle θ is called the *opposite side,* and the side adjacent to θ is called the *adjacent side.* In terms of this notation, the three trigonometric functions of importance to us are defined as follows:

Sine $= \dfrac{\text{opposite}}{\text{hypotenuse}}$ *written* $\sin \theta$ 2.9a

Cosine $= \dfrac{\text{adjacent}}{\text{hypotenuse}}$ *written* $\cos \theta$ 2.9b

Tangent $= \dfrac{\text{opposite}}{\text{adjacent}}$ *written* $\tan \theta$ 2.9c

Tables are available which give the values of these functions for different angles, but they can be found more easily with a scientific calculator.

Up to now we have specified a vector by giving its magnitude and direction. Another way to specify a vector is to give the lengths of the sides of the right triangle of which the vector is the hypotenuse. These lengths are called the *components* of the vector.

Figure 2.33 shows a vector **F** which makes an angle θ with the positive x axis. This vector is the hypotenuse of the right triangle with sides parallel to the x and y axes. The side parallel to the x axis is called the x component F_x, and the side parallel to the y axis is called the y component F_y. From Eq. 2.9b we have

$$\cos \theta = \frac{\text{adjacent}}{\text{hypotenuse}} = \frac{F_x}{F}$$

or

$$F_x = F \cos \theta \qquad\qquad 2.10a$$

and from Eq. 2.9a we have

$$\sin \theta = \frac{\text{opposite}}{\text{hypotenuse}} = \frac{F_y}{F}$$

or

$$F_y = F \sin \theta \qquad\qquad 2.10b$$

Equations 2.10a and b determine the components of a vector in terms of its magnitude F and the angle θ it makes with the x axis. If a component lies along a negative axis, it is given a minus sign.

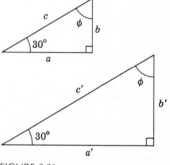

FIGURE 2.31

Two 30° right triangles.

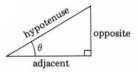

FIGURE 2.32

The sides of a right triangle.

FIGURE 2.33

A vector **F** and its x and y components F_x and F_y.

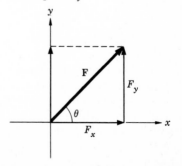

Example 2.9 Find the components of the vectors F_1 and F_2 in Fig. 2.34.

From Eq. 2.10a the x component of F_1 is

$$F_{1x} = F_1 \cos \theta = (30\text{ N}) \cos 42°$$
$$= (30\text{ N})(0.743) = 22.3\text{ N}$$

On a calculator with algebraic logic, press the keys in the sequence*:

| 4 | | 2 | | cos | | × | | 3 | | 0 | | = |

Display   22.29434476

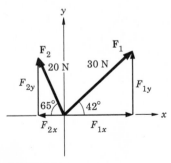

The displayed result should be rounded off to three significant figures, yielding $F_{1x} = 22.3\text{ N}$.

From Eq. 2.10b the y component of F_1 is

$$F_{1y} = F_1 \sin \theta = (30\text{ N}) \sin 42°$$
$$= (30\text{ N})(0.669) = 20.1\text{ N}$$

FIGURE 2.34

The two forces in Examples 2.9 and 2.10.

On a calculator with reverse Polish logic, press the keys in the sequence

| 4 | | 2 | | ENTER | | SIN | | 3 | | 0 | | × |

Display   20.07391819

When this is rounded off, the result is $F_{1y} = 20.1\text{ N}$.

The components of F_2 are found in the same way, except that since the x component lies along the negative x axis, F_{2x} is given a minus sign. You should perform the required calculation and obtain $F_{2x} = -8.45\text{ N}$ and $F_{2y} = 18.1\text{ N}$. ☐

The sum F of two or more vectors is easily found once the components of the vectors are known, because F_x is equal to the sum of the x components and F_y is equal to the sum of the y components. That is, if F is the sum of F_1 and F_2

$$F = F_1 + F_2$$

then the x component of F is just the sum of the x components of F_1 and F_2

$$F_x = F_{1x} + F_{2x} \qquad\qquad 2.11a$$

and the y component of F is just the sum of the y components of F_1 and F_2

$$F_y = F_{1y} + F_{2y} \qquad\qquad 2.11b$$

The components of a vector are ordinary numbers, not vectors. Equations 2.11a and b involve ordinary addition, even though some of the components may be negative.

In order to demonstrate Eqs. 2.11a and b, the vectors in Fig. 2.34 have been redrawn in Fig. 2.35 so that the tail of F_2 coincides with the head of F_1. The sum F of these vectors is then obtained by the usual graphical method. In moving F_2 we have also

*When working in degrees, the R-D switch must be set on D.

moved its components. A positive component is represented by an arrow pointing in the positive direction, and a negative component is represented by an arrow pointing along the negative direction.

Although Fig. 2.35 is rather complex, it is important that you study it carefully. Note that F_y, the y component of **F**, is equal to the sum of the y components of F_1 and F_2:

$$F_y = F_{1y} + F_{2y} = 20.1 \text{ N} + 18.1 \text{ N} = 38.2 \text{ N}$$

On the other hand, F_x, the x component of **F**, appears to be equal to the difference between the x components of F_1 and F_2. Actually F_x is also equal to the sum when we take into account that F_{2x} is negative. Thus, using the values for F_{1x} and F_{2x} found in Example 2.9, we get

$$F_x = F_{1x} + F_{2x} = 22.3 \text{ N} + (-8.45 \text{ N})$$
$$= 22.3 \text{ N} - 8.45 \text{ N} = 13.85 \text{ N}$$

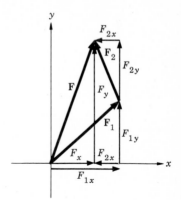

FIGURE 2.35

Vector addition of the two forces in Fig. 2.34.

The magnitude and direction of a vector can be found from knowledge of the components. Figure 2.33 shows that the magnitude F of a vector is the hypotenuse of a right triangle in which the components F_x and F_y are the sides. The hypotenuse of a triangle is related to its sides by the pythagorean theorem:

$$F^2 = F_x^2 + F_y^2$$

so the magnitude of a vector is given in terms of its components by

$$F = \sqrt{F_x^2 + F_y^2} \qquad\qquad 2.12a$$

The ratio F_y/F_x is just the tangent of θ, so

$$\tan \theta = \frac{F_y}{F_x}$$

or

$$\theta = \arctan \frac{F_y}{F_x} \qquad\qquad 2.12b$$

The arctangent (arctan) of a number is the angle whose tangent is the number. These calculations are easily performed on a calculator.

Example 2.10 Find the magnitude F and the angle θ of the sum of the two vectors in Fig. 2.34.

We have already found that the components of **F** are

$$F_x = 13.85 \text{ N} \qquad \text{and} \qquad F_y = 38.2 \text{ N}$$

From Eq. 2.12a the magnitude of **F** is

$$F = \sqrt{F_x^2 + F_y^2} = \sqrt{(13.85 \text{ N})^2 + (38.2 \text{ N})^2}$$
$$= \sqrt{192 \text{ N}^2 + 1459 \text{ N}^2} = \sqrt{1651 \text{ N}^2} = 40.6 \text{ N}$$

On a calculator with algebraic logic, press the keys in the sequence:

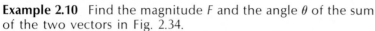

TABLE **2.3**

Components of a vector:

$$F_x = F \cos \theta \qquad \text{2.10a}$$
$$F_y = F \sin \theta \qquad \text{2.10b}$$

Components of a sum of vectors:

If $\qquad \mathbf{F} = \mathbf{F}_1 + \mathbf{F}_2 + \cdots + \mathbf{F}_n$

then $\qquad F_x = F_{1x} + F_{2x} + \cdots + F_{nx} \qquad$ 2.11a

and $\qquad F_y = F_{1y} + F_{2y} + \cdots + F_{ny} \qquad$ 2.11b

Magnitude and direction of a vector:

$$F = \sqrt{F_x^2 + F_y^2} \qquad \text{2.12a}$$

$$\theta = \arctan \frac{F_y}{F_x} \qquad \text{2.12b}$$

Vector equations

The quantities used in these equations are shown graphically in Fig. 2.33.

Display   40.63326839

With a scientific calculator it is not necessary to write down any intermediate steps.
From Eq. 2.12*b* the angle θ between **F** and the *x* axis is

$$\theta = \arctan \frac{F_y}{F_x} = \arctan \frac{38.2\ \text{N}}{13.85\ \text{N}}$$

$$= \arctan 2.76 = 70.1°$$

On a calculator, press the keys in the sequence:

| 3 | 8 | · | 2 | ÷ | 1 | 3 | · | 8 | 5 |
| = | arc | tan |

Display   70.0711052   ☐

For your convenience Table 2.3 lists the equations needed for finding the components of a vector, the sum of components, and the magnitude and direction of a vector. Each equation summarizes the sequence of steps you must take with your calculator to get the indicated quantity.

Example 2.11   Find the magnitude and direction of the sum of the three vectors in Fig. 2.36.
Since $\mathbf{F}_1$ is parallel to the *x* axis, its *y* component is zero and its *x* component is equal to the magnitude of $\mathbf{F}_1$:

$$F_{1x} = F_1 = 20\ \text{N} \qquad F_{1y} = 0$$

This also follows from Eqs. 2.10a and *b* and the fact that $\cos 0° = 1$ and $\sin 0° = 0$.
The components of $\mathbf{F}_2$ and $\mathbf{F}_3$ are calculated from Eqs. 2.10a and *b*:

$$F_{2x} = (10\ \text{N}) \cos 45° = 7.07\ \text{N} \qquad F_{2y} = (10\ \text{N}) \sin 45° = 7.07\ \text{N}$$

FIGURE 2.36

The three forces in Example 2.11.

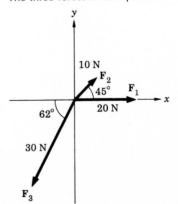

$$F_{3x} = -(30 \text{ N}) \cos 62° = -14.1 \text{ N}$$
$$F_{3y} = -(30 \text{ N}) \sin 62° = -26.5 \text{ N}$$

Both the x and y components of $\mathbf{F}_3$ are negative because each component lies along a negative axis.

From Eq. 2.11 the components of the sum of these vectors are

$$F_x = F_{1x} + F_{2x} + F_{3x} = 20 \text{ N} + 7.07 \text{ N} + (-14.1 \text{ N})$$
$$= 13.0 \text{ N}$$

$$F_y = F_{1y} + F_{2y} + F_{3y} = 0 + 7.07 \text{ N} + (-26.5 \text{ N})$$
$$= -19.4 \text{ N}$$

Since F_x is positive and F_y is negative, $\mathbf{F}$ must lie in the lower right-hand quadrant, as shown in Fig. 2.37. Its magnitude is given by Eq. 2.12a:

$$F = \sqrt{(13.0 \text{ N})^2 + (-19.4 \text{ N})^2} = 23.4 \text{ N}$$

The negative component causes no trouble in this last formula because it becomes a positive number when it is squared.

To get a positive value for the angle θ between $\mathbf{F}$ and the x axis, we must use the positive value of the components in Eq. 2.12b. That is, we use 19.4 N instead of -19.4 N for F_y. Then the angle is

$$\theta = \arctan \frac{19.4 \text{ N}}{13.0 \text{ N}} = 56.2°$$

Equation 2.12b does not tell us in which quadrant the vector lies. To determine this you must draw the components along the appropriate axes and sketch in the vectors, as is done in Fig. 2.37. ☐

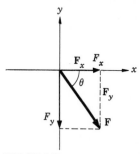

FIGURE 2.37

The sum **F** of the three vectors in Fig. 2.36.

2.5 Problems Involving Forces in a Plane

The problem in statics is to determine an unknown force acting on an object at rest given all the other forces. This is done by using Newton's first law of motion and the rule of vector addition. If only two forces act on an object, they must necessarily be equal and opposite in order for their sum to be zero. For example, we know that when a 50-N block sits on a table there is a gravitational force of 50 N acting downward on the block. Consequently, since the block is at rest, the table must exert a 50-N force upward on the block.

The situation is more interesting when three or more forces act on an object, because now the forces can sum to zero without any two of them being along the same line. Consider, for instance, the situation illustrated in Fig. 2.38a. Three cords are attached to a ring. Known weights hang from two of these cords, so the forces these cords exert on the ring are known. The problem is to determine the force exerted by the third cord. The magnitude of this force is equal to the tension in that cord. It is important to determine such an unknown force when designing a mechanical system in order to know how strong the cord should be.

The free-body diagram of the ring is shown in Fig. 2.38b. Each

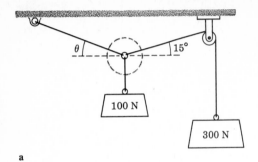

a

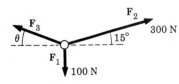

b

FIGURE 2.38

(a) Three cords attached to a ring.
(b) Free-body diagram of the ring.

force is given a symbol and all the information known about each force is indicated on the diagram. From Newton's first law of motion we know that the sum of the three forces is zero:

$$\mathbf{F}_1 + \mathbf{F}_2 + \mathbf{F}_3 = 0 \quad \text{or} \quad \mathbf{F}_3 = -(\mathbf{F}_1 + \mathbf{F}_2) = -\mathbf{F}$$

where $\mathbf{F} = \mathbf{F}_1 + \mathbf{F}_2$. The unknown force $\mathbf{F}_3$ is the negative of the sum $\mathbf{F}$ of $\mathbf{F}_1$ and $\mathbf{F}_2$.

The graphical solution of this problem is shown in Fig. 2.39. The known forces $\mathbf{F}_1$ and $\mathbf{F}_2$ are drawn to scale, with the tail of $\mathbf{F}_2$ coincident with the head of $\mathbf{F}_1$. The sum $\mathbf{F}$ of these forces is represented by the arrow drawn from the tail of $\mathbf{F}_1$ to the head of $\mathbf{F}_2$. Since $\mathbf{F}_3$ equals $-\mathbf{F}$, it points from the head of $\mathbf{F}_2$ to the tail of $\mathbf{F}_1$. Study Fig. 2.39 carefully. It is important that you understand why $\mathbf{F}$ and $\mathbf{F}_3$ are drawn the way they are. When all the forces acting on an object at rest are drawn head to tail, they will form a closed figure. If they did not, their sum would not be zero. From measurements of the length of $\mathbf{F}_3$ and the angle θ, we find $F_3 = 290$ N and $\theta = 4°$.

The trigonometric solution of this problem requires us to find the components of $\mathbf{F}_1$ and $\mathbf{F}_2$. From Fig. 2.38b we see that these components are

$$F_{1x} = 0 \quad F_{1y} = -100 \text{ N}$$

$$F_{2x} = (300 \text{ N}) \cos 15° = 290 \text{ N} \quad F_{2y} = (300 \text{ N}) \sin 15° = 77.6 \text{ N}$$

The components of their sum $\mathbf{F}$ are

$$F_x = F_{1x} + F_{2x} = 290 \text{ N} \quad \text{and} \quad F_y = F_{1y} + F_{2y} = -22.4 \text{ N}$$

Since $\mathbf{F}_3$ equals $-\mathbf{F}$, its components are the negative of the components of $\mathbf{F}$:

$$F_{3x} = -F_x = -290 \text{ N} \quad F_{3y} = -F_y = 22.4 \text{ N}$$

The vector $\mathbf{F}_3$ is shown in Fig. 2.40. Its magnitude is

$$F_3 = \sqrt{(-290 \text{ N})^2 + (22.4 \text{ N})^2} = 291 \text{ N}$$

and the angle θ is

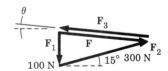

FIGURE 2.39

Graphical construction required to find the unknown force $\mathbf{F}_3$ in Fig. 2.38.

FIGURE 2.40

Determination of the unknown force $\mathbf{F}_3$ in Fig. 2.38 from its components.

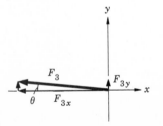

$$\theta = \arctan \frac{22.4\,N}{290\,N} = 4.4°$$

The results of the trigonometric method are more precise than the results of the graphical method. Note that we used the positive values of the components to find θ, but we used the signs of the components to find the quadrant in which $\mathbf{F}_3$ lies.

In statics problems of this type, two unknowns can be found. Here we found the magnitude and direction of a single force. In other problems the two unknowns could be the magnitudes of two forces of known directions or the directions of two forces of known magnitudes.

Example 2.12 Find the angle θ and the tension T of the cord supporting the pulley in Fig. 2.41a.

The tension in the long cord is $1000\,N = 1\,kN$ throughout, so this cord exerts two 1-kN forces on the pulley. The free-body diagram of the pulley is shown in Fig. 2.41b.

This is a problem in which the unknowns are the magnitude and direction of a single force. To solve it we have to find the sum of the other forces. The graphical solution is shown in Fig. 2.42.

To use the trigonometric method, we first find the components of the known forces:

$$F_{1x} = 0 \qquad F_{1y} = -1\,kN$$
$$F_{2x} = -(1\,kN)\cos 35° = -0.819\,kN$$
$$F_{2y} = -(1\,kN)\sin 35° = -0.574\,kN$$

Since the sum of $\mathbf{F}_1$, $\mathbf{F}_2$, and $\mathbf{F}_3$ is zero, the sum of their x and y components is zero. That is, from Newton's first law we can write

$$F_{1x} + F_{2x} + F_{3x} = 0 \qquad \text{so} \qquad F_{3x} = -F_{1x} - F_{2x} = 0.819\,kN$$

and

$$F_{1y} + F_{2y} + F_{3y} = 0 \qquad \text{so} \qquad F_{3y} = -F_{1y} - F_{2y} = 1.574\,kN$$

The magnitude of $\mathbf{F}_3$, then, is

$$F_3 = \sqrt{(0.819\,kN)^2 + (1.574\,kN)^2} = 1.77\,kN$$

which is equal to the tension T in the small cord. The angle of $\mathbf{F}_3$, which is the same as the angle of the cord, is

$$\theta = \arctan \frac{1.574\,kN}{0.819\,kN} = 62.5° \qquad \square$$

Example 2.13 Figure 2.43a shows a 60-N block sitting on a board inclined 25° to the horizontal. Find the magnitudes of the normal and frictional forces which the board exerts on the block. What is the minimum value of the coefficient of friction between the board and the block required to prevent the block from sliding?

This is an example of a problem in which the magnitudes of two forces are the unknowns. The normal force $\mathbf{F}_n$ is the force

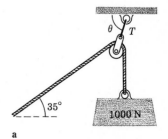

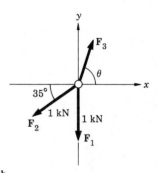

FIGURE 2.41

(a) Pulley supporting a cord and weight. (b) Free-body diagram of the pulley.

FIGURE 2.42

Graphical construction required to find the unknown force $\mathbf{F}_3$ in Fig. 2.41.

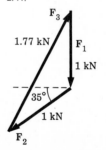

perpendicular to the board that the board exerts on the block. The frictional force $\mathbf{F}_f$ is the force parallel to the board that the board exerts on the block.

The free-body diagram of the block is shown in Fig. 2.43b. Since there is a 60-N force of gravity $\mathbf{F}_g$ acting downward on the block, the total surface force $\mathbf{F}_s$ must be a 60-N force upward on the block. In a sense this is a simple two-force problem. The unknowns are just the components of $\mathbf{F}_s$ parallel and perpendicular to the board.

We choose an x axis parallel to the inclined board and a y axis perpendicular to the board. The x axis therefore makes an angle of 25° to the horizontal and an angle

$$\theta = 90° - 25° = 65°$$

to the vertical ($\mathbf{F}_s$). The magnitude of $\mathbf{F}_f$ is just the x component of $\mathbf{F}_s$:

$$F_f = F_s \cos \theta = (60\text{ N}) \cos 65° = 25.4\text{ N}$$

The magnitude of $\mathbf{F}_n$ is the y component of $\mathbf{F}_s$:

$$F_n = F_s \sin \theta = (60\text{ N}) \sin 65° = 54.4\text{ N}$$

This method works because $\mathbf{F}_f$ and $\mathbf{F}_n$ are perpendicular to each other, and so their directions can form an x-y coordinate system.

From Eq. 2.3 the maximum force of friction is

$$F_{f,\text{max}} = \mu_s F_n$$

The value of 25.4 N found for F_f must be less than $F_{f,\text{max}}$ in order for the block not to slide. That is, we must have

$$F_f \leq \mu_s F_n$$

or

$$\mu_s \geq \frac{F_f}{F_n} = \frac{25.4\text{ N}}{54.4\text{ N}} = 0.47$$

If μ_s is less than 0.47, the block will not remain at rest on the board. □

Example 2.14 Find the tensions in the two cords in Fig. 2.44a.
The free-body diagram is shown in Fig. 2.44b. This is another example of a problem in which the two unknowns are the magnitudes of two forces. But in this case the forces are not perpendicular to each other, so the method of Example 2.13 cannot be used. Because of the importance of this type of problem, we describe three methods of solution.

1 Trigonometric Method The components of the three forces acting on the block are

$$F_{1x} = 0 \qquad F_{1y} = -200\text{ N}$$

$$F_{2x} = F_2 \cos 30° = 0.866\, F_2 \qquad F_{2y} = F_2 \sin 30° = 0.50\, F_2$$

$$F_{3x} = -F_3 \sin 40° = -0.766\, F_3 \qquad F_{3y} = F_3 \sin 40° = 0.643\, F_3$$

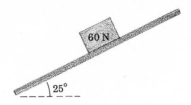

a

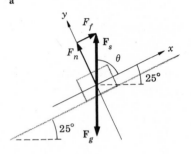

b

FIGURE 2.43

(a) Block resting on an inclined plane. (b) Free-body diagram of the block. The magnitudes of the normal force $\mathbf{F}_n$ and the frictional force $\mathbf{F}_f$ are the components of the surface force $\mathbf{F}_s$ perpendicular and parallel to the surface.

FIGURE 2.44

(a) A weight suspended from two cords. (b) The free-body diagram of the weight.

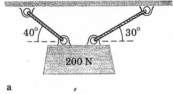

a

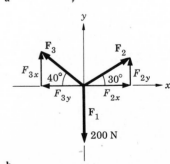

b

The sums of the x and y components are separately equal to zero, so we have the two equations

$$F_{1x} + F_{2x} + F_{3x} = 0 + 0.866\,F_2 - 0.766\,F_3 = 0$$

$$F_{1y} + F_{2y} + F_{3y} = -200\,\text{N} + 0.50\,F_2 + 0.643\,F_3 = 0$$

in the two unknowns F_2 and F_3. These unknowns are found by using the standard algebraic techniques for solving two equations in two unknowns.

Solve the first equation for F_2

$$F_2 = \frac{0.766}{0.866}\,F_3 = 0.885\,F_3 \qquad\qquad 2.13$$

and substitute into the second equation:

$$-200\,\text{N} + 0.50(0.885\,F_3) + 0.643\,F_3 = 0$$

Combining the terms in F_3 we get

$$-200\,\text{N} + 0.442\,F_3 + 0.643\,F_3 = 0$$

or

$$-200\,\text{N} + 1.085\,F_3 = 0$$

so

$$F_3 = \frac{200\,\text{N}}{1.085} = 184\,\text{N}$$

This value for F_3 is then substituted back into Eq. 2.13 to find F_2:

$$F_2 = 0.885\,F_3 = 0.885\,(184\,\text{N}) = 163\,\text{N}$$

The tensions in the cords are equal to the forces $\mathbf{F}_2$ and $\mathbf{F}_3$ which they exert.

2 Graphical Method This problem is easily solved graphically using the construction shown in Fig. 2.45. First the known force $\mathbf{F}_1$ is drawn to scale. Then lines in the directions of $\mathbf{F}_2$ and $\mathbf{F}_3$ are drawn from the head and tail of $\mathbf{F}_1$ (Fig. 2.45a). The intersection of these lines determines a force triangle. Arrowheads are put on the lines so the three arrows point in a head-to-tail fashion (Fig. 2.45b). The lengths of the arrows $\mathbf{F}_2$ and $\mathbf{F}_3$ determine the magnitude of these forces.

This intersecting-line method simultaneously determines the length of two vectors and so, in effect, solves two simultaneous equations. It is a simple and effective method of solving problems of this type.

3 Law-of-Sines Method This method combines the simplicity of the graphical method with the precision of the trigonometric method. First sketch the graphical solution of the problem and determine all the interior angles of the force triangle (Fig. 2.46a). The *law of sines* states that for any triangle (Fig. 2.46b) with sides F_1, F_2, and F_3 and opposite angles θ_1, θ_2, θ_3, we have the relations

$$\frac{F_1}{\sin\theta_1} = \frac{F_2}{\sin\theta_2} = \frac{F_3}{\sin\theta_3} \qquad \textit{law of sines} \qquad 2.14$$

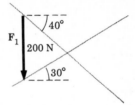

a

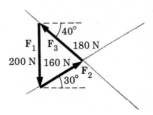

b

FIGURE 2.45

Graphical construction required to find the tensions in the cords in Fig. 2.44a. (a) From the head and tail of the known force $\mathbf{F}_1$, lines are drawn in the directions of the unknown forces $\mathbf{F}_2$ and $\mathbf{F}_3$. (b) The intersection of these lines determines the magnitudes of $\mathbf{F}_1$ and $\mathbf{F}_2$.

FIGURE 2.46

(a) Sketch of the graphical solution for the tensions in the cords in Fig. 2.44a. (b) Relations between the angles and sides of a triangle used in the statement of the law of sines (Eq. 2.14).

a

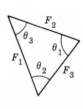

b

Comparing Fig. 2.46a and b we see that

$$\frac{F_3}{\sin 60°} = \frac{200 \text{ N}}{\sin 70°} \quad \text{or} \quad F_3 = \frac{\sin 60°}{\sin 70°}(200 \text{ N}) = 184 \text{ N}$$

and

$$\frac{F_2}{\sin 50°} = \frac{200 \text{ N}}{\sin 70°} \quad \text{or} \quad F_2 = \frac{\sin 50°}{\sin 70°}(200 \text{ N}) = 163 \text{ N} \quad \square$$

TRUSSES

Definition A *truss* is a rigid frame composed of long slender *members* rivoted together at *joints*. Figure 2.47 shows examples of trusses used in the construction of roofs, bridges, and towers. Because the members are slender, they cannot exert much force perpendicular to their length. We shall assume that they exert no perpendicular force; that is, we assume that, like a stretched cord, a member exerts a force parallel to its length.

Unlike a cord, however, a member can be either in tension or compression. When the member is in tension (Fig. 2.48a), it is pulled at either end with equal forces. At the same time, the reactions to these forces are forces pulling *away* from each joint. When the member is in compression (Fig. 2.48b), it is pushed at either end with equal forces. At the same time, the reactions to these forces are forces pushing *toward* each joint. When designing a truss, an engineer must determine the magnitude of the tension or compression each member will have under maximum loading conditions.

Figure 2.49 shows how a bridge is constructed from two trusses. The trusses span the space. Crossbeams connected to the lower joints of the trusses support the road. All the weight carried by the road is transferred to the joints by the crossbeams.

Consider the simple truss in Fig. 2.50. This truss is designed to carry a maximum load of 5000 lb at joint B. When this load is applied, the supports at either end will exert upward normal forces of 2500 lb each on joints A and C. This is necessary because the total force on the truss as a whole must be zero.

To determine the forces in the various members, we must isolate a joint at which only two unknown forces act. Joint A is an example of such a joint. Figure 2.51a shows the free-body diagram of this joint. To isolate this joint, members 1 and 2 were cut. Each of these members exerts a force of unknown magnitude parallel to its length. Since it is not yet known whether these members are in tension or compression, arrowheads have been omitted on F_1 and F_2.

When lines are drawn from the head and tail of the 2500-lb force in the direction of F_1 and F_2, we get the force triangle shown in Fig. 2.51b. Arrowheads are placed on F_1 and F_2 in such a way that the three forces are arranged in head-to-tail fashion. This determines that F_1 is directed toward the joint and F_2 is directed away from the joint (Fig. 2.51c). Thus member 1 is in compression and member 2 is in tension.

The magnitude of these forces is most easily found from the law of sines (Eq. 2.14):

a

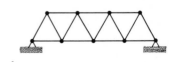

b

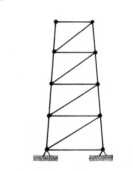

c

FIGURE 2.47

Typical trusses: (a) roof truss (*Howe*), (b) bridge truss (*Warren*), and (c) tower truss.

FIGURE 2.48

(a) A truss member under tension pulls on its joints. (b) A truss member in compression pushes on its joints.

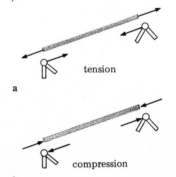

a

tension

compression

b

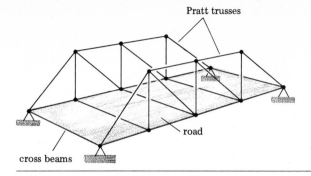

FIGURE 2.49

A bridge constructed from two
Pratt bridge trusses.

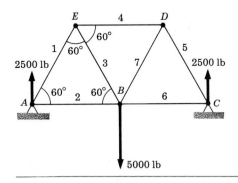

FIGURE 2.50

A truss with a 5000-lb load at
joint B.

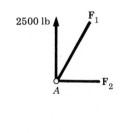

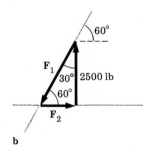

FIGURE 2.51

(a) Free-body diagram of joint A in
Fig. 2.50. (b) Graphical construction
for the forces in the members connected to A. (c) Direction of the
forces in the members connected
to A.

$$\frac{F_1}{\sin 90°} = \frac{2500 \text{ lb}}{\sin 60°} \quad \text{or} \quad F_1 = 2887 \text{ lb (compression)}$$

$$\frac{F_2}{\sin 30°} = \frac{2500 \text{ lb}}{\sin 60°} \quad \text{or} \quad F_2 = 1443 \text{ lb (tension)}$$

Example 2.15 Find the forces in all the other members of the
truss in Fig. 2.50.

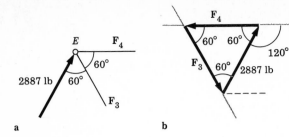

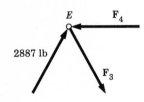

a　　　　　　　　　　b

c

FIGURE 2.52

(a) Free-body diagram of joint E in Fig. 2.50. (b) Graphical construction for the forces in the members of the truss connected to E. (c) Direction of the forces in the members connected to E.

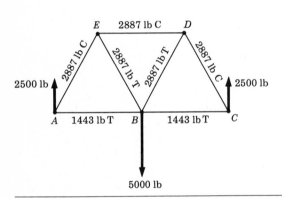

FIGURE 2.53

The forces in the members of the truss in Fig. 2.51.

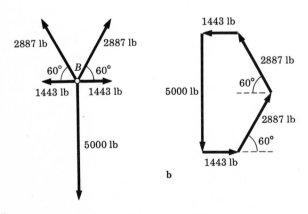

a　　　　　　　　　　b

FIGURE 2.54

(a) Free-body diagram of joint B in Fig. 2.54. (b) Vector sum of the forces on joint B, showing that they sum to zero.

We have determined that member 1 has a compression of 2887 lb, so it exerts a 2887-lb force toward joint E. Figure 2.52a shows the free-body diagram of joint E. When F_3 and F_4 are drawn from the head and tail of the 2887-lb force, we get the force triangle in Fig. 2.52b. Arrowheads are placed on F_3 and F_4 so that the three forces are arranged in head-to-tail fashion. From this we see that F_3 is directed away from E (tension) and F_4 is directed toward E (compression). Since the angles of the triangle are all 60°, the three sides are equal:

$F_3 = 2887$ lb (tension)

$F_4 = 2887$ lb (compression)

The forces in the first four members of the truss are now known. Members 5, 6, and 7 have the same relation to each other as members 1, 2, and 3, so they have the same forces. All the forces are shown in Fig. 2.53.

An engineer needs to know how much tension or compression each member has under a given load in order to know how strong to make the member. As we see in Fig. 2.52, some members must support a larger force than others.

As check on these calculations, we examine joint B. The free-body diagram of joint B is shown in Fig. 2.54a. When these forces are drawn head to tail, we get the closed polygon shown in Fig. 2.54b. This shows that the sum of these five forces is zero. If the figure had not closed, joint B would not be in equilibrium under the action of the forces we had calculated. This would indicate a mistake in our calculation. ☐

GUIDE TO MAJOR TOPICS

Topic	References	Problems
Newton's first and third laws of motion	Sec. 2.1	1 and 2
Forces in a line	Secs. 2.2–2.3	
Gravitational and normal forces	Example 2.3	1 and 2
Spring force	Eqs. 2.1 and 2.2	
Friction	Eqs. 2.3 and 2.4; Examples 2.1, 2.5 and 2.6	3 to 6
Tension in cords	Examples 2.2, 2.4 to 2.6	7 to 10
Vector addition	Sec. 2.4	
Graphical method	Examples 2.7 and 2.8	11, 13, 15, 17
Trigonometric method	Eqs. 2.10 to 2.12; Examples 2.9 to 2.11; Table 2.3	12, 14, 16, 18
Forces in a plane	Sec. 2.5	
Unknown magnitude and direction	Example 2.12	19 to 24
Two unknown magnitudes	Example 2.13 and 2.14	25 to 30
*Trusses	Example 2.15	31 to 34

*Supplemental material, which may be omitted.

PROBLEMS

1 A 25-N block sits on a table. (a) What is the magnitude of the force of gravity on the block? What is the direction of this force? What object exerts this force? (b) What is the normal force exerted on the block? What object exerts this force? (c) What is the reaction to the force

in a? On what object does this force act? (d) What is the reaction to the force in b? On what object does it act?
Ans. (a) 25 N down, earth; (b) 25 N up, table; (c) 25 N up, earth; (d) 25 N down, table.

2   A 12-lb block sits on top of a 4-lb block that sits on a scale. Give the magnitude and directions of the following forces: (a) force of gravity on the 4-lb block, (b) normal force exerted by the scale on the 4-lb block, (c) normal force exerted by the 12-lb block on the 4-lb block, (d) normal force exerted by the 12-lb block on the scale, and (e) normal force exerted by the 4-lb block on the 12-lb block. (f) Which of these forces are action-reaction pairs?

3   A 2-N block of wood on a wooden table just starts to slide when a horizontal force of 0.8 N is applied to it. (a) What is the coefficient of friction between the block and the table? (b) A 5-N weight is placed on top of the block. What is the magnitude of the horizontal force required to move the block now?
Ans. (a) 0.40; (b) 2.8 N

4   A 110-lb skier requires a 6-lb push to start moving on a horizontal snow-covered surface. What is the magnitude of the push required to start a 180-lb skier moving?

5   Two blocks are connected by a cord, as shown in Fig. 2.25. Block A weighs 20 N, and the coefficient of static friction between it and the surface is 0.4; block B weighs 10 N, and the coefficient of static friction between it and the surface is 0.5. (a) What is the magnitude of the minimum force $\mathbf{F}_a$ that must be applied to block B to move the entire assembly? (b) What is the tension T in the connecting cord when the blocks just start to move?
Ans. (a) 13 N; (b) 8 N

6   A 10-lb block rests on top of a 20-lb block that is sitting on a table. The coefficient of static friction is 0.30 between the two blocks and 0.50 between the 20-lb block and the table. (a) What is the minimum force that must be applied to the 20-lb block to slide the two blocks along the table? (b) What is the maximum force that can be applied to the 10-lb block without its sliding off the 20-lb block?

7   What are the tensions T_1 and T_2 in the cords in Fig. 2.55?
Ans. 3 and 11 N

8   What is the tension in the cord in Fig. 2.56?

9   (a) Find the tensions T_1, T_2, and T_3 in the three cords in Fig. 2.57. (b) What force must be applied to the cord by the hand to support the 50-lb weight?

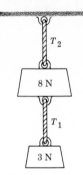

FIGURE 2.55

Problem 7.

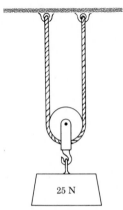

FIGURE 2.56

Problem 8.

FIGURE 2.57

Problem 9.

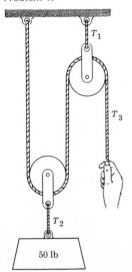

Ans. (a) 50, 50, and 25 lb; (b) 25 lb

10 Find the tensions T_1, T_2, and T_3 in the three cords in Fig. 2.58.

11 A woman walks 500 m northeast, 800 m west, and 300 m north. Use the graphical method to find her overall displacement.
Ans. 780 m, 41° north of west

12 Repeat Prob. 11 using the trigonometric method.

13 Figure 2.59 shows three forces. Use the graphical method to find (a) the sum of $\mathbf{F}_1$ and $\mathbf{F}_2$; (b) the sum of $\mathbf{F}_2$ and $\mathbf{F}_3$; and (c) the sum of $\mathbf{F}_1$, $\mathbf{F}_2$, and $\mathbf{F}_3$.
Ans. (a) 8.66 lb; (b) 8.26 lb; (c) 2.27 lb

14 Use the trigonometric method to find (a) the x and y components of each of the three forces in Fig. 2.59; (b) the components of the sum $\mathbf{F} = \mathbf{F}_1 + \mathbf{F}_2 + \mathbf{F}_3$; (c) the magnitude of $\mathbf{F}$; and (d) the angle $\mathbf{F}$ makes with the x axis.

15 Use the graphical method to find the resultant force $\mathbf{F}$ which the traction device in Fig. 2.60 exerts on the leg.
Ans. 9.2 lb, 15°

16 Repeat Prob. 15 using the trigonometric method.

17 Cords attached to a cleat exert the three forces shown in Fig. 2.61. Use the graphical method to find the resultant of these forces.
Ans. 245 lb, 81.7°

18 Repeat Prob. 17 using the trigonometric method.

19 Find the angle θ and the tension T of the cord being pulled by the pulley system in Fig. 2.62.
Ans. 27.5°, 27.7 lb

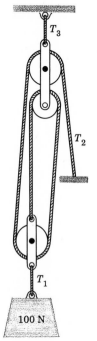

FIGURE 2.58

Problem 10.

FIGURE 2.59

Problems 13 and 14.

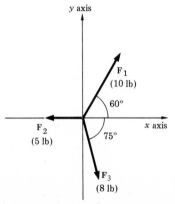

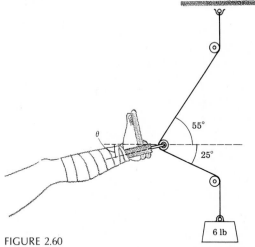

FIGURE 2.60

Problems 15 and 16.

FIGURE 2.61

Problems 17 and 18.

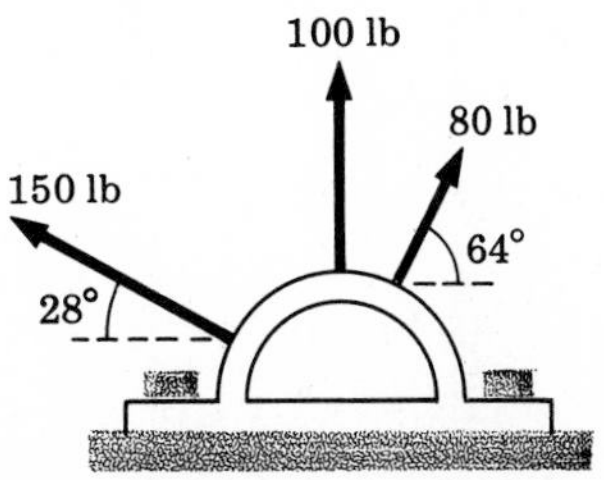

57

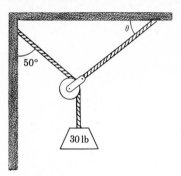

FIGURE 2.62

Problem 19.

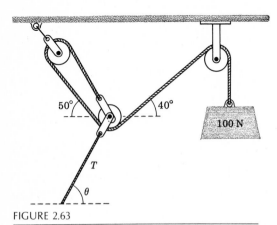

FIGURE 2.63

Problem 20.

20 Find the angle θ and the tension T of the cord being pulled by the pulley system in Fig. 2.63.

21 The block in Fig. 2.64 rests on a frictionless inclined plane. What is the tension in the cord? *Ans.* 40.8 lb

FIGURE 2.64

Problem 21.

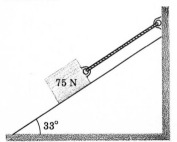

22 The inclined-plane apparatus shown in Fig. 2.65 was designed by NASA scientists to study the process of walking on the moon, where the force of gravity is one-sixth that of earth. The long cable is attached at one end to a harness on the astronaut and at the other end to a slider that runs along a track. The cable holds the astronaut perpendicular to the plane. Find the angle θ required to make the normal force on the astronaut one-sixth of his weight W. What is the tension in the cable?

FIGURE 2.65

Problem 22.

23 A 20-lb weight is suspended by two spring scales as shown in Fig. 2.66. One scale registers 15 lb and is inclined 54° to the horizontal. Find the reading on the other scale and the angle θ it makes with the horizontal.
Ans. 11.8 lb, 41.7°

FIGURE 2.66

Problem 23.

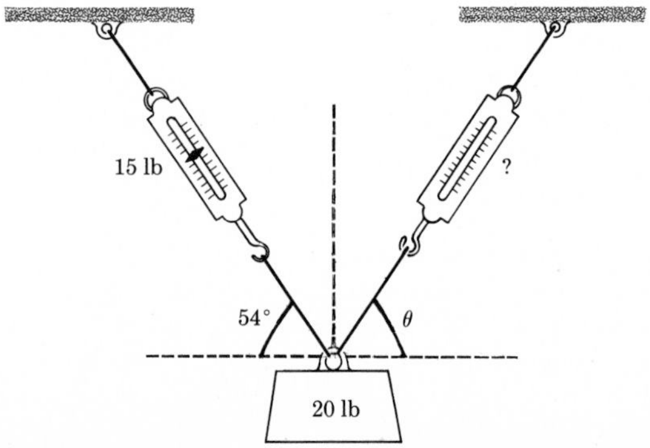

58

24 If the tension T_1 in Fig. 2.67 is 200 N, what are T_2 and θ?

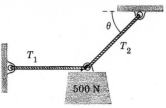

FIGURE 2.67

Problems 24 and 25.

25 If the angle θ in Fig. 2.67 is 40°, what are the tensions T_1 and T_2?
Ans. 596 N, 778 N

26 Find the tensions T_1 and T_2 in the cords supporting the weight in Fig. 2.68.

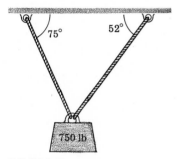

FIGURE 2.68

Problem 26.

27 A 30-N block rests on a plane inclined 28° to the horizontal. (a) Find the magnitudes of the normal force and the force of friction on the block. (b) Given that the block remains at rest, what is the minimum value of the coefficient of friction between the block and the plane?
Ans. (a) 26.5, 14.1 N; (b) 0.53

28 A method for determining the coefficent of friction μ_s between a block and a surface is to tilt the surface until the block just starts to slide. Show that the angle θ between the inclined surface and the horizontal when the block starts to slide is related to μ_s by $\mu_s = \tan \theta$.

29 A 400-N crate is to be pushed along the floor by a force **F** exerted at 30° to the horizontal (Fig. 2.69). The coefficient of friction between the crate and the floor is 0.4. (a) What minimum value of F is needed to move the crate? (b) Show that if the angle is increased to 70°, the crate cannot be moved no matter how large F is.
Ans. (a) 240 N

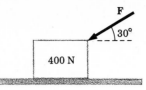

FIGURE 2.69

Problem 29.

30 The coefficient of friction between a mop and the floor is μ_s. Show that if the mop handle makes an angle less than arctan μ_s with the vertical, the mop cannot be moved no matter how hard one pushes on the handle. Neglect the weight of the mop.

31 Find the tension or compression in members 1, 2, 3, and 4 of the truss in Fig. 2.70.
Ans. 8000 lb (*T*), 11,314 lb (*C*), 8000 lb (*T*), 8000 lb (*T*)

32 Find the tension or compression in members 5, 6, 7, 8, and 9 of the truss in Fig. 2.70. Use the results of Prob. 31.

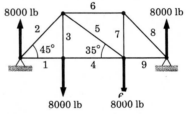

FIGURE 2.70

Problems 31 and 32.

33 Find the tension or compression in members 1, 2, 3, and 4 of the truss in Fig. 2.71.
Ans. 11,719 lb (*C*), 19,035 lb (*T*), 15,025 lb (*C*), 12,731 lb (*C*)

34 Find the tension or compression in members 10, 11, 12, and 13 of the truss in Fig. 2.71.

FIGURE 2.71

Problems 33 and 34.

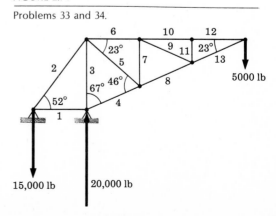

BIBLIOGRAPHY

BEER, FERDINAND P., and E. RUSSELL JOHN-
STON, JR.: *Mechanics for Engineers: Statics
and Dynamics,* 3d ed., McGraw-Hill, New
York, 1976. A junior-level mechanics textbook for engineering students. Chapter 6
describes different types of trusses and explains the special assumptions that are made
in analyzing them.

ANALYSIS OF STRUCTURES

Newton's first law of motion is sufficient to determine the forces in a system only when those forces all act at the same point. An additional equilibrium condition (the *torque condition*) is needed to analyze structures in which the forces act at different points. The torque condition and Newton's first law of motion together can determine three unknowns in a statics problem.

The purpose of such analyses is to calculate the force in each part of a structure so that the part can be made strong enough to support this force. The factors that determine the ability of a structural element to support applied forces are discussed in Secs. 3.4 and 3.5.

3.1 Rotational Equilibrium

Newton's first law is a necessary condition for an object to remain at rest. That is, if an object is known to be at rest, the sum of the forces acting on it must be zero. But the sum of the forces on an object can be zero without the object being at rest. For example, consider the two forces F_1 and F_2 acting on the block in Fig. 3.1. Even if $F_2 = -F_1$, so that the total force on the block is zero, the block will move. In fact it will rotate. The condition $F_1 + F_2 = 0$ only ensures that one point of the block (its center of gravity) remains at rest. A second condition is required to ensure that the block does not start to rotate about this point.

The tendency of a force to cause rotation about an axis depends on the magnitude of the force and its distance from the axis. This fact is in accord with one's experience on a seesaw. When two children sit on either end of a seesaw, the force each exerts on the board tends to rotate it in opposite senses (Fig. 3.2). From Newton's first and third laws and the discussion in Example 2.3, we know that if the 40-lb boy on the right is at rest, he exerts a downward force of 40 lb on the board, which tends to rotate the board clockwise about an axis perpendicular to the pivot point, or *fulcrum,* of the seesaw. (This force is the reaction to the normal force that the board exerts on the boy.) Likewise, the 60-lb force the 60-lb girl on the left exerts on the board tends to rotate it counterclockwise about the axis. In spite of the inequality of these forces, the seesaw can be balanced (in equilibrium) if the boy sits farther from the axis than the girl.

The rule is that the seesaw balances if the force the boy exerts times his distance from the axis is equal to the force the girl exerts times her distance from the axis. Thus if the boy sits 6 ft from the axis, he can balance the girl if she sits 4 ft from the axis, since

$$(40\ lb)(6\ ft) = 240\ lb \cdot ft = (60\ lb)(4\ ft)$$

To generalize this rule to more complex situations, the concept of *torque* is introduced by considering the force **F** and the pivot point O shown in Fig. 3.3. In this and in all subsequent figures, the axis of rotation is assumed to be perpendicular to the plane of the figure.

FIGURE 3.1

Two equal and opposite forces acting on a block.

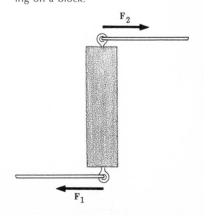

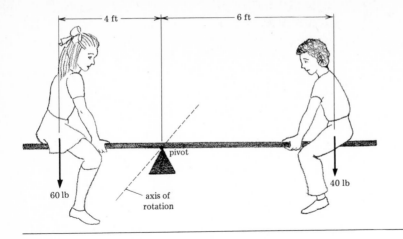

FIGURE 3.2

A 40-lb boy and a 60-lb girl balanced on a seesaw.

Definition The *torque* * τ exerted by a force **F** about a point O is equal to the magnitude of **F** times its perpendicular distance d from O:

$$\tau = Fd \qquad\qquad 3.1$$

The sign of τ is taken to be positive if **F** tends to produce counterclockwise rotation about O and negative if **F** tends to produce clockwise rotation. Torque is a quantitative measure of the tendency of a force to produce rotation about a point.

> **UNITS** In the English engineering system the unit of torque is the pound-foot (lb·ft), and in the International System it is the newton-meter (N·m).

There are two very important characteristics of torque:

1 The magnitude and sign of the torque produced by a given force depend on the point O about which it is calculated.

2 The distance d in Eq. 3.1 is the perpendicular distance from the point O to the *line of action* of the force. The line of action is the straight line in the direction of the force passing through the point where the force is applied.

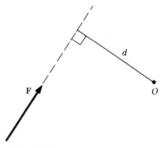

FIGURE 3.3

A force **F** exerting a torque about an axis through point O.

Example 3.1 The truss shown in Fig. 3.4 is supported at joints A and D and is designed to carry a 12,000-lb load at joint B and an 8000-lb load at joint C. Calculate the torque exerted by each load about joint D.

The perpendicular distance d_B from D to $\mathbf{F}_B$ is 50 ft, so the torque τ_B about D due to $\mathbf{F}_B$ is

$$\tau_B = F_B d_B = (12{,}000 \text{ lb})(50 \text{ ft}) = 6 \times 10^5 \text{ lb} \cdot \text{ft}$$

The torque is positive because if $\mathbf{F}_B$ were the only force on the truss, the truss would rotate counterclockwise about D.

The perpendicular distance d_C from D to $\mathbf{F}_C$ is 20 ft, so the torque τ_C due to $\mathbf{F}_C$ is

$$\tau_C = F_C d_C = (8000 \text{ lb})(20 \text{ ft}) = 1.6 \times 10^5 \text{ lb} \cdot \text{ft}$$

This torque is also positive. Both torques exerted by the load

*τ is the Greek (lowercase) letter tau.

FIGURE 3.4

A truss supporting a 12,000-lb load at joint B and an 8000-lb load at joint C.

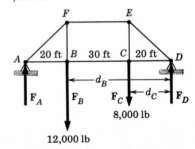

tend to rotate the truss counterclockwise about D. As we shall see, this tendency is opposed by the torque due to the support force F_A. ☐

Example 3.2 The hatchway door in Fig. 3.5a is counterweighted to aid someone lifting it. What torque does the cord exert on the door about the hinge O?

The tension in the cord is 500 N, and the cord exerts this force at a point 0.4 m from the hinge. Figure 3.5b shows that the perpendicular distance d from the hinge to the line of action of the force is the side of a right triangle whose hypotenuse is 0.4 m. Consequently, we have

$$\frac{d}{0.4 \text{ m}} = \sin 42° \qquad \text{or} \qquad d = (0.4 \text{ m}) \sin 42° = 0.27 \text{ m}$$

The torque exerted by the force, then, is

$$\tau = -Fd = -(500 \text{ N})(0.27 \text{ m}) = -135 \text{ N} \cdot \text{m}$$

The torque is negative because the force tends to rotate the door clockwise about the hinge.

> **REMARK** Whether a rotation is clockwise or counterclockwise depends on the position of the observer. Every drawing assumes a certain point of observation, and it is from this position that the sign of the torque is to be determined. ☐

An object which has no tendency to start rotating is said to be in *rotational equilibrium*. The necessary condition for rotational equilibrium is that the clockwise torques on an object equal the counterclockwise torques. Because we have assigned opposite signs to clockwise and counterclockwise torques, we can state the condition for rotational equilibrium as follows:

The Torque Condition *For an object to be in rotational equilibrium the sum of the torques produced by all the forces acting on the object must be zero.*

From Newton's first law we know that if the sum of the forces acting on an object is zero, the object will remain at rest. An object which remains at rest and does not tend to start rotating is said to be in *static equilibrium*. Thus the following conditions must be satisfied for an object to be in static equilibrium:

Conditions for Static Equilibrium *For an object to be in static equilibrium the vector sum of all the forces acting on the object must be zero (Newton's first law) and the sum of all the torques exerted on the object must be zero.* If the symbol Σ is used to indicate a sum, the equilibrium conditions can be written

$$\Sigma F = 0 \qquad\qquad\qquad 3.2a$$

$$\Sigma \tau = 0 \qquad\qquad\qquad 3.2b$$

In applying the torque condition, all the torques must be calculated about the same point. However, if the object is in static equilibrium, it does not matter where this point is located.

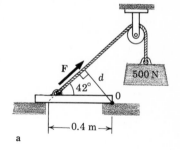

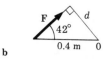

FIGURE 3.5

(a) A cord and weight attached to a hatchway door. (b) Perpendicular distance d from the hinge 0 to the line of action of the force **F**.

Example 3.3 Show that the conditions for static equilibrium are satisfied for the seesaw in Fig. 3.2. (Neglect the weight of the board.)

To apply the static-equilibrium conditions, one must first determine what the object is and what the forces on it are. In this case, the object is the board, and the forces are the normal forces exerted on it by the children and the pivot (Fig. 3.6). The torque about the pivot (point O) produced by the force F_1 (exerted by the boy on the board) is

$$\tau_1 = (-F_1)(6 \text{ ft}) = -(40 \text{ lb})(6 \text{ ft}) = -240 \text{ lb} \cdot \text{ft}$$

The torque about O produced by the force F_2 (exerted by the girl on the board) is

$$\tau_2 = (F_2)(4 \text{ ft}) = (60 \text{ lb})(4 \text{ ft}) = 240 \text{ lb} \cdot \text{ft}$$

The torque about O produced by the force F_n (exerted on the board by the pivot) is zero, since the line of action of this force passes through O. That is, the perpendicular distance from O to the line of action of F_n is zero, so

$$\tau_n = (F_n)(0) = 0$$

The sum of these three torques is clearly zero, as required for the board to be in equilibrium.

The only virtue in calculating the torques about O is that knowledge of F_n is not needed. However, from Newton's first law we know that $F_n = -(F_1 + F_2)$, so F_n is a 100-lb force directed upward. Therefore, it is just as easy to calculate the torques about any other point. For instance, about the point O', where the girl sits, the torques are

$$\tau_1' = -(40 \text{ lb})(10 \text{ ft}) = -400 \text{ lb} \cdot \text{ft}$$
$$\tau_2' = (60 \text{ lb})(0) = 0$$
$$\tau_n' = (100 \text{ lb})(4 \text{ ft}) = 400 \text{ lb} \cdot \text{ft}$$

and again their sum is zero. Thus, while the individual torques change when the point about which they are calculated changes, the sum of the torques is zero regardless of the point chosen.

REMARK To apply the torque condition in an equilibrium situation, all the torques must be calculated about the same point, but any point can be used. ☐

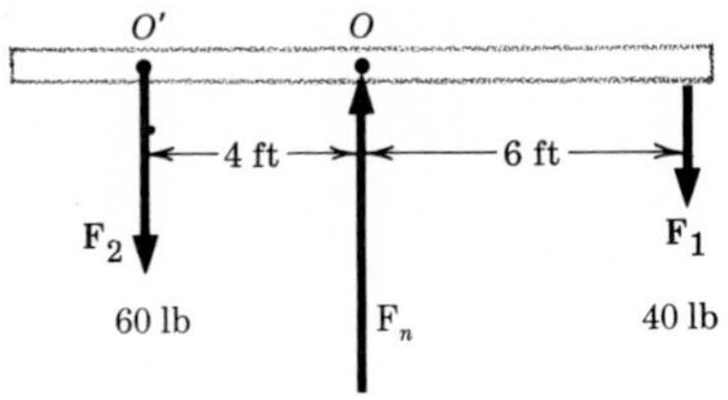

Example 3.4 Find the magnitudes of the forces F_A and F_D which are supporting the truss in Fig. 3.4.

The free-body diagram of the truss is shown in Fig. 3.7. (It is necessary to show only the external forces F_A, F_B, F_C, and F_D which act on the truss as a whole. The internal forces exerted by an individual member appear only in a free-body diagram in which the member was cut to isolate a particular joint.)

This is a statics problem with two unknown magnitudes. However, because the forces are in a line, the force equation (Eq. 3.2a) gives us only one equation:

$$F_A + F_D = 20{,}000 \text{ lb} \qquad\qquad 3.3$$

FIGURE 3.6

Free-body diagram of the seesaw in Fig. 3.2.

FIGURE 3.7

Free-body diagram of the truss in Fig. 3.4.

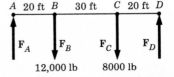

To find an equation for F_A alone, we must use the torque equation (Eq. 3.2b). In Example 3.1 we found that about point D the forces $\mathbf{F}_B$ and $\mathbf{F}_C$ exert the torques

$$\tau_B = 6 \times 10^5 \text{ lb} \cdot \text{ft}$$

$$\tau_C = 1.6 \times 10^5 \text{ lb} \cdot \text{ft}$$

About this same point $\mathbf{F}_D$ exerts zero torque, because its perpendicular distance from D is zero:

$$\tau_D = 0$$

Therefore, from Eq. 3.2b we can find the torque τ_A exerted by $\mathbf{F}_A$:

$$\tau_A + \tau_B + \tau_C + \tau_D = 0$$

$$\tau_A = -(\tau_B + \tau_C + \tau_D) = -7.6 \times 10^5 \text{ lb} \cdot \text{ft}$$

The perpendicular distance d_A from D to $\mathbf{F}_A$ is 70 ft, so

$$\tau_A = -F_A(70 \text{ ft}) = -7.6 \times 10^5 \text{ lb} \cdot \text{ft}$$

or

$$F_A = \frac{7.6 \times 10^5 \text{ lb} \cdot \text{ft}}{70 \text{ ft}} = 10,860 \text{ lb}$$

Once F_A is found, F_D is obtained from Eq. 3.3:

$$10,860 \text{ lb} + F_D = 20,000 \text{ lb} \qquad \text{or} \qquad F_D = 9,140 \text{ lb}$$

Although we could have taken the torques about any point, we chose D in order to eliminate one of the unknown forces from our torque equations. For this reason it is usually best to take torques about a point at which one of the unknown forces acts. For practice, repeat this problem taking torques about A instead of D. (The torques due to $\mathbf{F}_B$ and $\mathbf{F}_C$ will have to be recalculated.) ☐

3.2 Center of Gravity

The problem of calculating the torque τ_g produced by the weight of an extended object requires special consideration because gravity acts on every point in the object. For instance, there is a gravitational force acting on every section of the cantilevered beam in Fig. 3.8. Each of these forces has its own line of action and produces its own torque about O. The sum of all these forces is the total force of gravity $\mathbf{F}_g$ on the beam, and the sum of these torques is the total torque τ_g due to gravity.

The gravitational torque τ_g produced by the force of gravity $\mathbf{F}_g$ on an extended object is calculated in terms of $\mathbf{F}_g$ and the position of a special point in the object called the center of gravity.

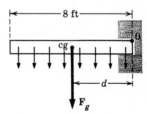

FIGURE 3.8

The force of gravity on a cantilevered beam.

Definition The *center of gravity* of an object is the point where, for the purposes of calculating the gravitational torque τ_g, the total force of gravity $\mathbf{F}_g$ can be considered to act.

In the case of a uniform beam, the center of gravity is located at the geometric center. Thus for the 8-ft beam in Fig. 3.8, the center of gravity (cg) is 4 ft from O. If the beam weighs 75 lb, the torque τ_g about O due to the weight of the beam is

$$\tau_g = F_g d = (75\,\text{lb})(4\,\text{ft}) = 300\,\text{lb} \cdot \text{ft}$$

Several properties of the center of gravity which follow directly from its definition are summarized here for convenience.

1 *The force of gravity on an object produces zero torque about the center of gravity of the object.* This is true because, by definition, the line of action of the force of gravity passes through the center of gravity, so the distance from the center of gravity to this line is zero. This property provides a method for locating the center of gravity of simple objects.

Example 3.5 Locate the center of gravity of the object in Fig. 3.9, which consists of two weights A and B connected by a bar of negligible weight.

Let the center of gravity be located a distance x from weight A. Then the torques about the center of gravity due to the individual forces of gravity on A and B are

$$\tau_A = F_A x \qquad \text{and} \qquad \tau_B = -F_B(d - x)$$

Since the total torque about the center of gravity due to the force of gravity is zero, we get

$$F_A x - F_B(d - x) = 0$$

or

$$x = \frac{F_B}{F_A + F_B} d \qquad\qquad 3.4$$

Thus if A is 50 lb, B is 25 lb, and d is 3 ft, the center of gravity is located the distance

$$x = \frac{25\,\text{lb}}{50\,\text{lb} + 25\,\text{lb}} (3\,\text{ft}) = 1\,\text{ft}$$

from A. $\qquad\qquad\qquad\qquad\qquad\qquad\qquad\qquad$ ☐

2 *The center of gravity of a rigid object is the balance point.* If a single support is placed directly under the center of gravity of an object (Fig. 3.10), the normal force $\mathbf{F}_n$ it exerts on the object equals $-\mathbf{F}_g$, so the total force on the object is zero. Furthermore, both $\mathbf{F}_g$ and $\mathbf{F}_n$ produce zero torques about the center of gravity since their lines of action pass through it. Consequently, the total torque about the center of gravity is zero, so the object is balanced (in equilibrium).

This property can also be used to locate the center of gravity of complex objects. For example, the center of gravity of a protractor can be found by first suspending it, together with a plumb line, from one point (Fig. 3.11a). The normal force $\mathbf{F}_n$ acts vertically upward at the point of suspension. Since it is the only force other than gravity on the object, it must produce zero torque about the center of gravity. Consequently, the center of gravity must lie along the line of action of $\mathbf{F}_n$, that is, along the plumb line. This line is marked on the protractor, and the protractor is then suspended from a second point not on the first

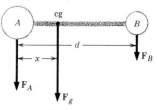

FIGURE 3.9

Locating the center of gravity of two weights connected by a bar of negligible weight.

FIGURE 3.10

An object balanced on a single support placed directly under its center of gravity.

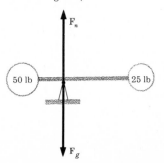

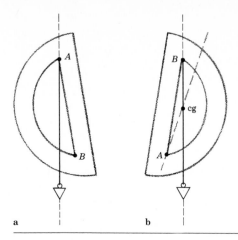

a   b

FIGURE 3.11

(a) A protractor suspended from
point *A* hangs with its center of
gravity somewhere along the vertical line defined by the plumb line.
(b) The protractor is suspended
from a second point (*B*) to determine a second line through the
center of gravity. The center of
gravity is located at the intersection of the two plumb lines.

plumb line (Fig. 3.11*b*). The center of gravity of the protractor is
located at the intersection of the two plumb lines.

3 *For a rigid object, the center of gravity is a fixed point relative to the
object, though it does not necessarily lie in the object itself.* The
center of gravity of the bar and weights is a fixed point on the bar, and
it does not change its position on the bar when the bar moves. The
center of gravity of a protractor lies in the hole, as shown in Fig. 3.11*b*.

4 *For a flexible object, such as the human body, the position of the
center of gravity relative to the object changes as the object changes
shape.* The center of gravity of a person standing erect is located at the
level of the second sacral vertebra on a vertical line touching the floor
about 3 cm in front of the ankle joint (Fig. 3.12*a*). When the woman
bends over (Fig. 3.12*b*), her center of gravity moves to a point outside
her body. During a Fosbury high jump (Fig. 3.13) the center of gravity
also lies outside the body. The ability to change the position of the
center of gravity of the body by moving body segments is of critical
importance in maintaining balance while walking and in the successful performance of many athletic feats.

3.3 Examples Involving Torque

The unknown forces acting on a mechanical system are calculated by the systematic application of the equilibrium conditions. In this section a number of problems are solved this way.

Example 3.6 Find the tensions T_A and T_B in the cables supporting the scaffold in Fig. 3.14*a*. A 180-lb worker is standing 4 ft
from *A*. The scaffold weighs 120 lb and its center of gravity is at
its geometric center.

   The free-body diagram of the scaffold is shown in Fig. 3.14*b*.
The torques about *A* due to the worker's weight and the scaffold's weight are

$$\tau_1 = -(180 \text{ lb})(4 \text{ ft}) = -720 \text{ lb} \cdot \text{ft}$$
$$\tau_2 = -(120 \text{ lb})(7.5 \text{ ft}) = -900 \text{ lb} \cdot \text{ft}$$

The tension T_A produces zero torque about *A* because the line of
action of this force passes through *A*. The tension T_B produces

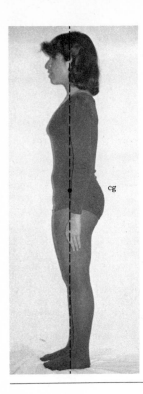
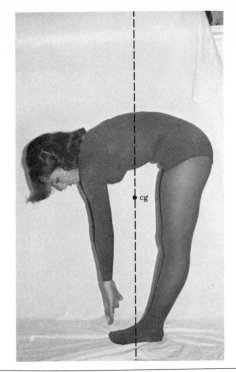

FIGURE 3.12

(a) The center of gravity of a person standing erect lies on a vertical line that touches the floor 3 cm in front of the ankle joint. (b) When a person bends over, the center of gravity lies outside the body.

FIGURE 3.13

Dick Fosbury doing a Fosbury high jump. As he passes over the bar, his center of gravity lies outside his body. (*Herb Scharfman for* Sports Illustrated © *Time, Inc.*)

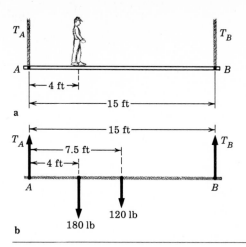

a

b

FIGURE 3.14

(a) A 180-lb worker standing on a scaffold. (b) Free-body diagram of the scaffold.

the torque

$$\tau_B = T_B(15 \text{ ft})$$

From the torque condition (Eq. 3.2b) we have

$$0 = \Sigma\tau = \tau_A + \tau_B + \tau_1 + \tau_2$$

or

$$\tau_B = -(\tau_A + \tau_1 + \tau_2) = 1620 \text{ lb} \cdot \text{ft}$$

Thus

$$T_B(15 \text{ ft}) = 1620 \text{ lb} \cdot \text{ft}$$

and

$$T_B = \frac{1620 \text{ lb} \cdot \text{ft}}{15 \text{ ft}} = 108 \text{ lb}$$

From Newton's first law of motion (Eq. 3.2a) we have

$$0 = \Sigma F = T_A + T_B - 180 \text{ lb} - 120 \text{ lb}$$

so

$$T_A = 300 \text{ lb} - T_B = 192 \text{ lb} \qquad \square$$

Example 3.7 An object rests on two scales 7 ft apart (Fig. 3.15a). Scale A reads 45 lb and scale B reads 25 lb. Find the weight of the object and the perpendicular distance d from A to the vertical line passing through the center of gravity (cg) of the object.

The free-body diagram of the object is shown in Fig. 3.15b. The scales exert upward normal forces on the object equal to their reading. The weight of the object acts at its center of gravity. From Newton's first law we have

$$F_g = F_A + F_B = 70 \text{ lb}$$

The torques about A due to $\mathbf{F}_A$, $\mathbf{F}_g$, and $\mathbf{F}_B$ are

$$\tau_A = 0$$
$$\tau_g = -F_g d = -(70 \text{ lb})d$$
$$\tau_B = (25 \text{ lb})(7 \text{ ft}) = 175 \text{ lb} \cdot \text{ft}$$

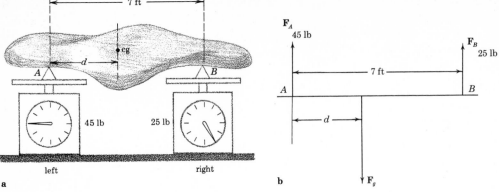

a

b

F_A
45 lb

F_B
25 lb

A

B

7 ft

d

F_g

FIGURE 3.15

(a) An object resting on two scales.
(b) Free-body diagram of the object.

Since the sum of these torques is zero, we have

$-(70\,\text{lb})d + 175\,\text{lb}\cdot\text{ft} = 0$

or

$d = \dfrac{175\,\text{lb}\cdot\text{ft}}{70\,\text{lb}} = 2.5\,\text{ft}$

The center of gravity of the object is on a line 2.5 ft from A.

REMARK Note the similarity and the difference between this example and Example 3.6. In both examples the torque condition was used to find the numerical value of an unknown torque. In Example 3.6 this value was used to determine an unknown force acting at a known distance, and in the present example it was used to determine the unknown distance at which a known force acted. □

Example 3.8 An 800-N sign is suspended from a horizontal beam as shown in Fig. 3.16a. The beam is pivoted at one end and supported near the other by a cable. Find the tension T in the cable and the magnitude and direction of the force $\mathbf{F}$ at the pivot. Neglect the weight of the beam.

The free-body diagram of the beam is shown in Fig. 3.16b. There are three unknowns in this problem: the tension T, the magnitude of the pivot force $\mathbf{F}$, and the angle θ that this force makes with the horizontal.

We know that $\mathbf{F}$ exerts zero torque about the pivot P, so the sum of the torques about P is

$\Sigma\tau = -(800\,\text{N})(2.5\,\text{m}) + Td = 0$ ⠀⠀⠀⠀3.5

where d is the perpendicular distance from P to the line of action of the tension. This line lies along the cable. From the right triangle in Fig. 3.16b we get

$\dfrac{d}{2.0\,\text{m}} = \sin 52°$ ⠀⠀or⠀⠀ $d = (2.0\,\text{m})\sin 52° = 1.58\,\text{m}$

Therefore, from Eq. 3.5 we find

$Td = (800\,\text{N})(2.5\,\text{m}) = 2000\,\text{N}\cdot\text{m}$

and

FIGURE 3.16

(a) A sign suspended from a horizontal beam. (b) Free-body diagram of the beam.

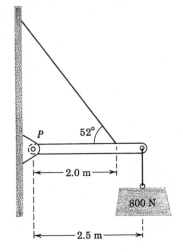

a

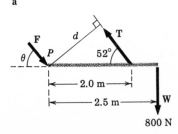

b

$$T = \frac{2000\text{ N} \cdot \text{m}}{d} = \frac{2000\text{ N} \cdot \text{m}}{1.58\text{ m}} = 1266\text{ N}$$

The force **F** is found by the methods of Chap. 2. We first calculate the x and y components of the known forces (Fig. 3.17):

$$T_x = -T\cos 52° = -(1266\text{ N})\cos 52° = -779\text{ N}$$
$$W_x = 0$$
$$T_y = T\sin 52° = (1266\text{ N})\sin 52° = 998\text{ N}$$
$$W_y = -800\text{ N}$$

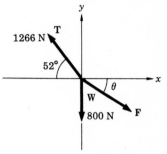

The sums of the x and y components of all the forces are zero, so

$$0 = F_x + T_x + W_x = F_x - 779\text{ N}$$
$$0 = F_y + T_y + W_y = F_y + 998\text{ N} - 800\text{ N}$$

or

$$F_x = 779\text{ N} \qquad \text{and} \qquad F_y = -198\text{ N}$$

Therefore, the magnitude of **F** is

FIGURE 3.17

The forces acting on the beam in Fig. 3.16.

$$F = \sqrt{F_x^2 + F_y^2} = \sqrt{(779\text{ N})^2 + (-198\text{ N})^2} = 804\text{ N}$$

and the angle θ is

$$\theta = \arctan \frac{F_y}{F_x} = \arctan \frac{-198\text{ N}}{779\text{ N}} = -14.3°$$

Since F_y is negative, **F** lies in the lower right-hand quadrant, as shown in Fig. 3.17. □

Example 3.9 An 18-ft ladder is leaning against a wall with its base 5 ft from the wall (Fig. 3.18a). A 110-lb painter stands 6 ft from the top of the ladder. The ladder weighs 25 lb and its center of gravity is 8 ft from the base. Find the normal force $\mathbf{F}_n$ which the wall exerts on the top of the ladder and the x and y components of the force **F** which the ground exerts on the base of the ladder. What is the minimum value of the coefficient of friction between the base and the ground required to prevent slipping?

The free-body diagram of the ladder is shown in Fig. 3.18b. The angle θ is given by

$$\sin \theta = \frac{5\text{ ft}}{18\text{ ft}} = 0.278 \qquad \text{or} \qquad \theta = \arctan 0.278 = 16.1°$$

and so

$$\cos \theta = \cos 16.1° = 0.961$$

The sum of the torques about O is

$$0 = -(25\text{ lb})d_1 - (110\text{ lb})d_2 + F_n d_n$$

where $d_1 = (8\text{ ft})\sin \theta = 2.22\text{ ft}$
$d_2 = (12\text{ ft})\sin \theta = 3.34\text{ ft}$
$d_n = (18\text{ ft})\cos \theta = 17.3\text{ ft}$

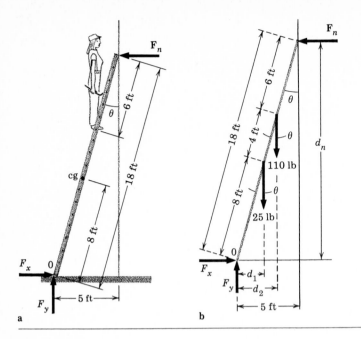

a   **b**

FIGURE 3.18

(a) A painter standing on a ladder. (b) Free-body diagram of the ladder.

Therefore, F_n is

$$F_n = \frac{(25 \text{ lb})(2.22 \text{ ft}) + (110 \text{ lb})(3.34 \text{ ft})}{17.3 \text{ ft}} = 24.4 \text{ lb}$$

The sums of the x and y components of the forces acting on the ladder are

$$0 = F_x - F_n \quad \text{or} \quad F_x = F_n = 24.4 \text{ lb}$$

and

$$0 = F_y - 25 \text{ lb} - 110 \text{ lb} \quad \text{or} \quad F_y = 135 \text{ lb}$$

F_y is the normal force on the base and F_x is the frictional force. The maximum force of friction

$$F_{f,\text{max}} = \mu_s F_n$$

must be greater than F_x in order that the ladder not slip. Therefore, we have

$$F_{f,\text{max}} = \mu_s F_y > F_x$$

or

$$\mu_s > \frac{F_x}{F_y} = \frac{24.4 \text{ lb}}{135 \text{ lb}} = 0.181$$

The coefficient of friction must be greater than 0.181 in order for the ladder not to slip. ☐

Example 3.10 Figure 3.19a shows a crane holding a 15,000-lb load. The boom is 40 ft long, weighs 4000 lb, and its center of gravity is at its midpoint. The hoisting cable with tension T_1 passes over a pulley at the top of the boom and attaches to an electric winch in the cab. The pendant cable with tension T_2 also

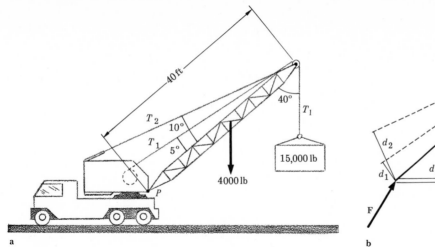

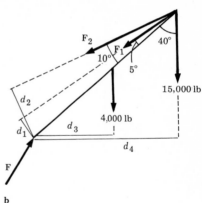

a   b

FIGURE 3.19

(a) A crane with a 40-ft boom
holding a 15,000-lb weight.
(b) Free-body diagram of the
boom.

is attached to a winch in the cab (not shown), but its other end
is fixed to the top of the boom. (The pendant cable raises and
lowers the boom and holds it in position.) Find the tension T_2 in
the pendant cable and the force **F** at the pivot P.

The free-body diagram of the boom is shown in Fig. 3.19b.
From this diagram the perpendicular distances from P to the
lines of action of the cables and weights are found to be

$d_1 = (40\ \text{ft}) \sin 5° = 3.49\ \text{ft}$
$d_2 = (40\ \text{ft}) \sin 10° = 6.95\ \text{ft}$
$d_3 = (20\ \text{ft}) \sin 40° = 12.86\ \text{ft}$
$d_4 = (40\ \text{ft}) \sin 40° = 25.71\ \text{ft}$

The tension T_1 in the hoisting cable is 15,000 lb, since it passes
over a pulley to the 15,000-lb load. The tension T_2 in the pendant
cable is unknown. The torques about P are

$\tau_1 = T_1 d_1 = (15{,}000\ \text{lb})(3.49\ \text{ft}) = 52{,}350\ \text{lb} \cdot \text{ft}$
$\tau_2 = T_2 d_2 = T_2(6.95\ \text{ft})$
$\tau_3 = -(4000\ \text{lb})d_3 = -51{,}440\ \text{lb} \cdot \text{ft}$
$\tau_4 = -(15{,}000\ \text{lb})d_4 = -385{,}650\ \text{lb} \cdot \text{ft}$

The sum of these torques is zero if the load is stationary, so
we have

$(52{,}350 - 51{,}440 - 385{,}650)\ \text{lb} \cdot \text{ft} + T_2(6.95\ \text{ft}) = 0$

or

$$T_2 = \frac{384{,}740\ \text{lb} \cdot \text{ft}}{6.95\ \text{ft}} = 55{,}360\ \text{lb}$$

The tension in the pendant cable is large because the perpendicular distance from P to the cable is small.

The forces $\mathbf{F}_1$ and $\mathbf{F}_2$ exerted by the hoisting and pendant
cables are shown in Fig. 3.20, together with the total downward
force of 19,000 lb due to the weights of the boom and load. The x

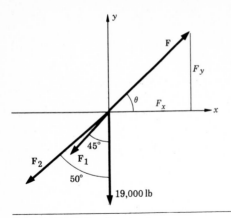

FIGURE 3.20

Forces on the boom in Fig. 3.19
(not drawn to scale).

and y components of $\mathbf{F}_1$ and $\mathbf{F}_2$ are

$F_{1x} = -F_1 \sin 45° = -(15{,}000 \text{ lb})(0.707) = -10{,}600 \text{ lb}$

$F_{1y} = -F_1 \cos 45° = -10{,}600 \text{ lb}$

$F_{2x} = -F_2 \sin 50° = -(55{,}360 \text{ lb})(0.766) = -42{,}400 \text{ lb}$

$F_{2y} = -F_2 \cos 50° = -35{,}600 \text{ lb}$

The sums of the x and y components of the forces due to the cables and weights are

$S_x = F_{1x} + F_{2x} = -53{,}000 \text{ lb}$

$S_y = F_{1y} + F_{2y} - 19{,}000 \text{ lb} = -65{,}200 \text{ lb}$

The total force on the boom is zero, so the x and y components of the force $\mathbf{F}$ at the pivot are

$F_x = -S_x = 53{,}000 \text{ lb}$

$F_y = -S_y = 65{,}200 \text{ lb}$

The magnitude of $\mathbf{F}$ is

$F = \sqrt{(53{,}000 \text{ lb})^2 + (65{,}200 \text{ lb})^2} = 84{,}000 \text{ lb}$

and the angle θ it makes with the x axis is

$\theta = \arctan \dfrac{F_y}{F_x} = \arctan \dfrac{65{,}200 \text{ lb}}{53{,}000 \text{ lb}} = 50.9°$

This force is shown in Fig. 3.20.

> **REMARK** The force at the pivot is much greater than the combined weight of the boom and the load. This is so because of the large force exerted on the boom by the pendant cable. Calculations like this one are necessary to design the size of the pin used in the pivot. $\square$

3.4 Elastic Deformation of Solids

A solid object has the ability to resist major deformation when subjected to external forces. However, a solid is not absolutely rigid; its size and shape are changed slightly when forces act on it. These small changes are measured with special instruments designed to test solids. From such measurements engineers learn about the deformation of solids.

NORMAL DEFORMATION

In the simplest test, the ends of a cylindrical sample of the material are connected to movable plates. The plates are pulled apart (or pushed together), subjecting the specimen to tension (or compression). Figure 3.21a shows a specimen in tension, and Fig. 3.21b shows it in compression. In each case the forces $\mathbf{F}_1$ and $\mathbf{F}_2$ have the same magnitude because the object is in equilibrium. The tension T is the magnitude of these forces (Sec. 2.2). Compression is distinguished from tension by the sign of T. The tension T is positive when the specimen is in tension (Fig. 3.21a) and negative when the specimen is in compression (Fig. 3.21b).

As a result of the tension, the length of the specimen is changed. Let L_0 be the length when the tension is zero, and let L be the length when the tension is nonzero, so that the change of length* ΔL is

$$\Delta L = L - L_0$$

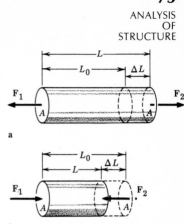

FIGURE 3.21

A cylindrical sample of material subjected to (a) tensile stress and (b) compressive stress.

Materials are tested by measuring the change of length produced by a given tension T. The change of length ΔL is positive when T is positive and negative when T is negative (Fig. 3.21).

The change of length depends on the size and composition of the specimen. For a cylindrical specimen, ΔL is proportional to its length L_0 and inversely proportional to its cross-sectional area A. That is, a long specimen is stretched more than a short one, and a thin specimen is stretched more than a thick one. Thus the effect of size on ΔL can be written

$$\Delta L = \frac{1}{E}\frac{L_0}{A}T \qquad\qquad 3.6$$

where E is a constant, called *Young's modulus*, that depends only on the composition of the specimen, not on its size. This equation also assumes that ΔL is proportional to T. This is true, as we shall see, provided T is not too large. Table 3.1 gives the Young's modulus of various solids.

The forces $\mathbf{F}_1$ and $\mathbf{F}_2$ which give rise to the tension or compression in a solid are perpendicular (normal) to the surfaces A on which they act (Fig. 3.21).

Definition The *normal stress s* in a solid is the ratio of the tension T in the solid to the cross-sectional area A.

$$s = \frac{T}{A}$$

Definition The *normal strain e* is the ratio of the change of length ΔL of a solid to the original length L_0.

$$e = \frac{\Delta L}{L_0}$$

In terms of these quantities, Eq. 3.6 can be written

$$e = \frac{s}{E} \quad \text{or} \quad s = Ee \qquad\qquad 3.7$$

*Δ is the Greek (capital) letter delta. It is often used to mean "change of."

TABLE **3.1**

Substance	Young's modulus, 10^9 N/m²	Elastic limit s_e, 10^7 N/m²	Tensile strength, 10^7 N/m²	Compressive strength, 10^7 N/m²
Aluminum	70	18	20	
Bone:				
Tensile	16		12	
Compressive	9			17
Brick	20			4
Copper	120	20	40	
Glass, fused				
quartz	70		5	110
Granite	50			20
Iron, wrought	190	17	33	
Marble	60			20
Polystyrene	3		5	10
Quartz	70			
Steel	200	30	50	
Wood	10			10

Young's modulus, elastic limit, and strength of some common solids

The values listed here are representative of each material; the actual values for a particular specimen can differ greatly from these because of different compositions and preparations.

REMARK The word "normal" indicates that the forces involved act perpendicular to a surface. It is used in distinction to the word "shear," which indicates that the forces act parallel to a surface. When stress and strain are used without qualification, they mean normal stress and normal strain.

UNITS Strain is the ratio of two lengths, so it is unitless. Stress is force divided by area, so in the English engineering system its unit is the pound per square foot (lb/ft²), and in the International System its unit is the newton per square meter (N/m²). From Eq. 3.7 we see that Young's modulus has the same unit as stress.

Figure 3.22 is a plot of stress against strain for a typical solid. Equation 3.7 is valid only for the straight-line portion of this curve, where the stress is proportional to the strain. The limits s_p and s_p' of this region are different for tensile and compressive stresses. The elastic limits s_e and s_e' are the maximum tensile and compressive stresses that can be applied to the solid such that it returns to its original size when the stress is removed. If the tensile stress exceeds s_e or the compressive stress exceeds s_e', the solid is permanently deformed and will not return to its unstressed condition when the stress is removed. The solid breaks when the stress exceeds the tensile strength s_u or the compressive strength s_u'. Table 3.1 gives the values of s_e, s_u, and s_u' for various solids.

If the stress on a supporting member of a mechanical structure exceeds the elastic limit, the member is permanently deformed and its physical properties are changed. In general, the deformed solid is weaker than the original material. For instance, a sheet of metal can be flexed back and forth indefinitely as long as it is not flexed beyond its elastic limit. But once bent beyond this limit, the sheet is easily broken by a few more twists. Thus the members of any mechanical structure are designed so that the greatest stress ever applied to them does not exceed their elastic limit.

Example 3.11 A 200-m-long aluminum transmission cable spans

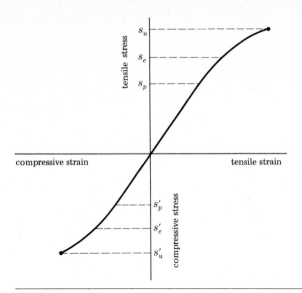

FIGURE 3.22

Stress-strain curve of a typical solid.

the space between two towers. The cable is 3 cm in diameter and is under a tension of 4×10^4 N. Find the stress and strain in the cable. Is the cable stretched beyond its elastic limit? How much is the cable stretched?

The radius of the cable is

$r = \frac{1}{2}d = 1.5\,\mathrm{cm} = 1.5 \times 10^{-2}\,\mathrm{m}$

and its cross-sectional area is

$A = \pi r^2 = (3.14)(1.5 \times 10^{-2}\,\mathrm{m})^2 = 7.07 \times 10^{-4}\,\mathrm{m}^2$

The stress in the cable therefore is

$$s = \frac{T}{A} = \frac{4 \times 10^4\,\mathrm{N}}{7.07 \times 10^{-4}\,\mathrm{m}^2} = 5.68 \times 10^7\,\mathrm{N/m^2}$$

This is less than one-third the elastic limit of $18 \times 10^7\,\mathrm{N/m^2}$ given in Table 3.1, so the cable is not stretched beyond its elastic limit.

Using Eq. 3.7 and the value of the Young's modulus for aluminum given in Table 3.1, we find that the strain is

$$e = \frac{s}{E} = \frac{5.68 \times 10^7\,\mathrm{N/m^2}}{70 \times 10^9\,\mathrm{N/m^2}} = 8.1 \times 10^{-4}$$

which means the cable is stretched the amount

$\Delta L = eL_0 = (8.1 \times 10^{-4})(200\,\mathrm{m}) = 0.16\,\mathrm{m} = 16\,\mathrm{cm}$ ☐

Example 3.12 What is the minimum diameter of the steel cable that can be used in a crane designed to lift a maximum weight of 20,000 lb?

The maximum tension in the cable will be

$T = 20{,}000\,\mathrm{lb} = 8.9 \times 10^4\,\mathrm{N}$

From Table 3.1, the elastic limit of steel is $30 \times 10^7\,\mathrm{N/m^2}$. If the stress is not to exceed this, the cross-sectional area of the cable must be greater than

$$A = \frac{T}{s_e} = \frac{8.9 \times 10^4 \, N}{30 \times 10^7 \, N/m^2} = 3.0 \times 10^{-4} \, m^2$$

and the diameter must be greater than

$$d = 2r = 2\sqrt{\frac{A}{\pi}} = 2\sqrt{\frac{3.0 \times 10^{-4} \, m^2}{3.14}}$$

$$= 1.95 \times 10^{-2} \, m = 1.95 \, cm$$

Of course, the diameter would be made larger than this to provide a margin of safety. □

The maximum elastic strain that can be produced in an object is found by substituting values of E and s_e from Table 3.1 in Eq. 3.7.* Steel, for instance, can have a maximum strain of

$$e = \frac{s_e}{E} = \frac{30 \times 10^7 \, N/m^2}{2 \times 10^{11} \, N/m^2} = 1.5 \times 10^{-3}$$

This means that if a steel rod is stretched by more than 0.15 percent, it will be stretched beyond its elastic limit.

> **REMARK** A normal force (Sec. 2.2) is the force a solid exerts as a result of its deformation. Because the Young's modulus of most solids is very large, the normal force exerted by a solid can vary over a wide range without a perceptible change in the shape of the solid. This is why in most statics problems the deformation involved in producing the normal force can be neglected.

SHEAR DEFORMATION

When a force acts parallel to a solid surface it is called a *shear*. Figure 3.23 shows four forces of equal magnitude applied parallel to four faces of a cubical solid. The forces are arranged so that the total force and torque on the solid are zero. Consequently, the solid is in equilibrium and remains at rest. However, it is deformed slightly into the shape shown by the dotted lines. The magnitude of the deformation is given by the *shear strain*, which is defined as the tangent of the angle θ. From Fig. 3.23 this is seen to be related to the size of the cube by

$$\text{Shear strain} = \tan \theta = \frac{\Delta L}{L_0}$$

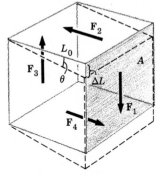

FIGURE 3.23

A solid object subjected to shear stress.

The forces apply a *shear stress* to the solid, which is defined as the magnitude of the parallel force on any face divided by the area of the face:

$$\text{Shear stress} = \frac{F_1}{A}$$

If the shear stress is not too large, it is related to the shear strain by

$$\frac{F_1}{A} = G \tan \theta$$

*Equation 3.7 is not strictly valid for stresses as large as s_e (Fig. 3.22). However, the stress-strain curve does not deviate much from a straight line at s_e, so Eq. 3.7 is a good approximation for stresses as large as s_e.

where G is a constant, called the *shear modulus,* that is charac- teristic of the solid. Table 3.2 gives the values of G for various solids. Generally, G is between $\frac{1}{2}E$ and $\frac{1}{3}E$.

> **REMARK** A fluid is distinguished from a solid by its inability to support a shear stress. Whenever one attempts to apply a shear stress to a fluid, the fluid flows; a solid, in contrast, responds to an applied shear stress by de- forming slightly.

3.5 Beams

A beam is a long, narrow structural member used to support both normal and shear stress. In the analysis of simple trusses given in Sec. 2.5, the members were assumed to be only in tension or compression; that is, they were assumed to be under normal but not shear stress. We now extend that analysis to the case of a horizontal beam supported at both ends and carrying a load in the middle (Fig. 3.24a). Such a structure is under both normal and shear stress. The purpose of this analysis is to under- stand the factors that determine the load-carrying capacity of such a beam.

Let L be the length of the beam and W the load the beam supports at its middle. If we neglect the weight of the beam itself, the upward support forces at either end will each be $\frac{1}{2}W$. Figure 3.24b shows the free-body diagram of the section of length x isolated in Fig. 3.24a. To balance the upward force $\frac{1}{2}W$ acting on the left-hand end of this section, there must be a downward force $\frac{1}{2}W$ acting on the right-hand end. This is a shear force because it is parallel to the cross section of the beam at which it acts. This force exerts a clockwise torque

$$\tau = -(\tfrac{1}{2}W)x = -\tfrac{1}{2}Wx$$

about point O. To balance this torque, there must be a counter- clockwise torque $\frac{1}{2}Wx$ exerted by some other force. The only possibility is that there is a force **F** pushing inward on the upper portion of the right-hand end. This force exerts the torque Fa

TABLE 3.2

Shear modulus of some common solids

The values listed here are repre- sentative of each substance; actual values for a particular specimen can differ greatly from these be- cause of different compositions and preparations.

Substance	Shear modulus G, 10^9 N/m²
Aluminum	25
Copper	40
Glass, fused quartz	30
Iron	50
Quartz	30
Steel	80
Tungsten	140
Wood	10

FIGURE 3.24

(a) Isolating a portion of a beam that is loaded at its center. (b) Free-body diagram of the iso- lated portion of the beam.

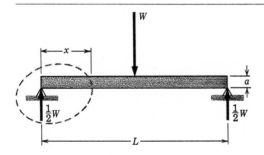

a

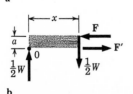

b

about O, where a is the thickness of the beam. The torque condition requires that Fa equal $\frac{1}{2}Wx$, so

$$Fa = \tfrac{1}{2}Wx \quad \text{or} \quad F = \frac{Wx}{2a} \tag{3.8}$$

Finally, to balance $\mathbf{F}$ there must be a force $\mathbf{F}'$ of equal magnitude pulling outward on the lower portion of the right-hand end. All the forces on the right-hand end are exerted by the portion of the beam to which the end is attached.

The force $\mathbf{F}$ subjects the upper portion of the beam to a compression, while the force $\mathbf{F}'$ subjects the lower portion to an equal tension. As a consequence, the top of the beam is shortened and the bottom is stretched. It is these deformations which cause the beam to bend.

Equation 3.8 shows that the tension (or compression) force is not equal throughout the beam, but increases with the distance x from the left-hand (or right-hand) end. The tension is zero at either end of the beam, and it has its maximum value at the position of the load ($x = \frac{1}{2}L$):

$$F_{max} = \frac{W(\frac{1}{2}L)}{2a} = \frac{WL}{4a}$$

The average tension (or compression) T is half the maximum tension, or

$$T = \tfrac{1}{2}F_{max} = \frac{WL}{8a}$$

Figure 3.25 shows a steel I beam commonly used in building construction. Under loading, the top flange is compressed, with an average compressive stress

$$s = \frac{T}{A} = \frac{WL}{8Aa} \tag{3.9}$$

where A is the cross-sectional area of the flange. The bottom flange has the same tensile stress.

From Eq. 3.7 the strain in each flange is

$$e = \frac{\Delta L}{L} = \frac{s}{E} \tag{3.10}$$

where ΔL is the amount a flange is stretched or compressed, and E is the Young's modulus of steel. From Eqs. 3.9 and 3.10 we find that the length of each flange is changed by the amount

$$\Delta L = \frac{sL}{E} = \frac{WL^2}{8AaE} \tag{3.11}$$

This equation is not as fearsome as it looks. It simply states that the change of length of a flange increases with the load W and the length squared L^2 of the beam and decreases with the cross-sectional area A of a flange, the distance a between the flanges, and the Young's modulus of steel.

Example 3.13 An I beam with the cross section shown in Fig. 3.25b carries a load of 5×10^4 N at its center. The beam is 7 m

a

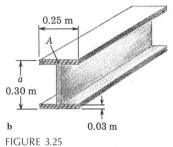

FIGURE 3.25

(a) A steel I beam being cut at the foundry. (*United States Steel Corp.*) (*b*) Dimensions of an I beam.

long. How much is the length of each flange changed? What is the maximum load the beam can support before it is deformed beyond its elastic limit?

In working problems involving many factors, it is advisable to list all the data. Here the data are

Load $W = 5 \times 10^4$ N

Flange area $A = (0.03 \text{ m})(0.25 \text{ m}) = 0.0075 \text{ m}^2$

Length $L = 7$ m

Separation distance $a = 0.30$ m

From Table 3.1 we get

Young's modulus of steel $E = 200 \times 10^9$ N/m^2

Elastic limit of steel $s_e = 30 \times 10^7$ N/m^2

From Eq. 3.9 the stress in each flange is

$$s = \frac{WL}{8Aa} = \frac{(5 \times 10^4 \text{ N})(7 \text{ m})}{8(0.0075 \text{ m}^2)(0.30 \text{ m})} = 1.94 \times 10^7 \text{ N/m}^2$$

This is less than the elastic limit of steel, so the load does stress the beam beyond its elastic limit. It is important to check this, because Eqs. 3.10 and 3.11 are valid only within this limit.

The change of length of each flange is found from Eq. 3.11 to be

$$\Delta L = \frac{sL}{E} = \frac{(1.94 \times 10^7 \text{ N/m}^2)(7 \text{ m})}{(200 \times 10^9 \text{ N/m}^2)}$$

$$= 6.8 \times 10^{-4} \text{ m} = 0.68 \text{ mm}$$

The length of each flange is changed by more than half a millimeter.

To find the maximum load the beam can support, we solve Eq. 3.9 for W:

$$W = \frac{8Aas}{L}$$

and substitute the elastic limit for s:

$$W_{max} = \frac{8(0.0075 \text{ m}^2)(0.30 \text{ m})(30 \times 10^7 \text{ N/m}^2)}{7 \text{ m}}$$

$$= 7.7 \times 10^5 \text{ N} \qquad \square$$

The stretching and compressing of the flanges of a beam causes the beam to bend (Fig. 3.26). In Appendix V we show that the amount h that the beam sags in the middle is given by the formula

$$h = \frac{WL^3}{24EAa^2} = \frac{(\Delta L)L}{3a} \qquad \qquad 3.12$$

For the beam in the last example, $\Delta L = 0.68$ mm, so the sag is

$$h = \frac{(6.8 \times 10^{-4} \text{ m})(7 \text{ m})}{3(0.30 \text{ m})} = 5.3 \times 10^{-3} \text{ m} = 5.3 \text{ mm}$$

Equations 3.9, 3.11, and 3.12 apply only to the case of an I beam loaded at its center. In an I beam the stress is all contained in narrow flanges separated by a known distance a. In a solid beam the stresses are distributed throughout the cross section, which complicates the details of the analysis. However, the principles remain the same: a horizontal beam under a vertical load is compressed along its upper side and stretched along its bottom side, causing the beam to bend. The factors which determine the ability of the beam to support its load are its length, cross-sectional area, and thickness, as shown in Eqs. 3.9, 3.11, and 3.12.

FIGURE 3.26

The bending of a beam loaded at its center.

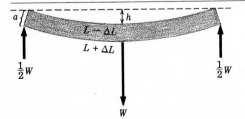

GUIDE TO MAJOR TOPICS

Topic	References	Problem
Torque	Sec. 3.1; Eq. 3.1; Examples 3.1 and 3.2	1 to 4
Static equilibrium	Secs. 3.1, 3.3; Eqs. 3.2 and 3.3; Examples 3.3 and 3.4, 3.6 to 3.10	5 to 10, 17 to 26
Center of gravity	Sec. 3.2; Eq. 3.4; Examples 3.5, 3.7	11 to 16
Stress and strain	Sec. 3.4; Eqs. 3.6 and 3.7; Examples 3.11 and 3.12; Tables 3.1 and 3.2	27 to 34
*Beams	Sec. 3.5; Eqs. 3.9 to 3.12; Example 3.13	35 to 38

*Supplemental material, which may be omitted.

PROBLEMS

1   A 25-lb force is applied to the handle of the wrench in Fig. 3.27. What is the torque which this force exerts about O?
Ans. -30 lb · ft.

2   Calculate the torque each of the forces in Fig. 3.28 exerts about A.

3   In the exercise situation shown in Fig. 3.29, the torque about the knee joint exerted by the 15-lb weight attached to the ankle varies with the elevation of the leg. (a) Calculate the torque due to this weight for the four positions shown. (b) Show that for any angle θ between the leg and the vertical, the torque is

$$\tau = -(20 \text{ lb} \cdot \text{ft}) \sin \theta$$

Ans. (a) 0, -10, -17.3, -20 lb · ft

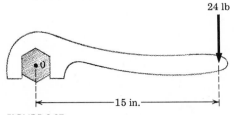

FIGURE 3.27

Problem 1.

FIGURE 3.28

Problem 2.

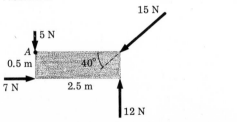

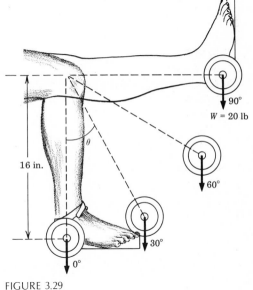

FIGURE 3.29

Problem 3.

4   The bracket in Fig. 3.30 is acted upon by a 200-N force. Calculate the torque this force exerts about points A and B. *Hint:* Make a scale drawing and measure the distances needed to find the torques.

5   Joy and Joe are carrying a 500-N weight which is sitting on a 3-m-long board (Fig. 3.31). What are the magnitudes of the forces Joy and Joe must exert on the board to support the load? (Neglect the weight of the board.)
Ans. 167 and 333 N

6   A 3-ft bar is used to lift a 300-lb stone, as shown in Fig. 3.32. What force **F** must be ap-

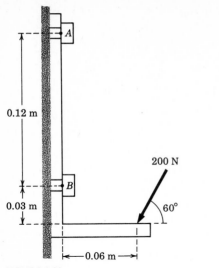

0.12 m

0.03 m

A

B

200 N

60°

0.06 m

FIGURE 3.30

Problem 4.

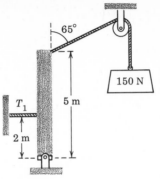

65°

150 N

T_1

5 m

2 m

FIGURE 3.33

Problem 7.

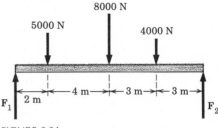

8000 N

5000 N

4000 N

F_1

2 m — 4 m — 3 m — 3 m

F_2

FIGURE 3.34

Problem 8.

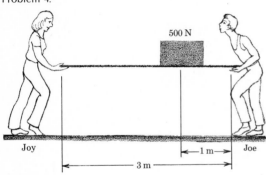

500 N

Joy

1 m

Joe

3 m

FIGURE 3.31

Problem 5.

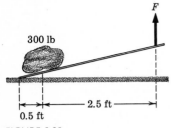

300 lb

F

2.5 ft

0.5 ft

FIGURE 3.32

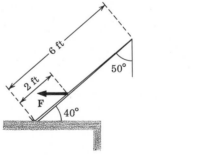

6 ft

2 ft

F

50°

40°

FIGURE 3.35

Problem 9.

Problem 6.

plied to the end of the bar to hold the stone in the position shown?

7   Figure 3.33 shows a vertical board pivoted at its base and held upright by two cords. Find the tension T_1 in the horizontal cord and the force acting at the pivot O. (Neglect the weight of the board.)
Ans. 340 and 214 N

8   Find the forces $\mathbf{F_1}$ and $\mathbf{F_2}$ which support the beam in Fig. 3.34. The beam is loaded as shown. (Neglect the beam's weight.)

9   A 10-lb fish hangs from the end of a 6-ft-long fishing rod (Fig. 3.35). (a) What force $\mathbf{F}$ must the fisherman exert at a point 2 ft from the base in order to hold the rod in equilibrium? (Neglect the weight of the rod.) (b) What are the horizontal and vertical components of the force which the ground exerts on the base?
Ans. (a) 35.9 lb; (b) 35.9 and 10 lb

REMARK Friction cannot hold the base of the rod in Prob. 9. The base must be pressed against a vertical stop.

10   A 750-N weight hangs from the post in Fig. 3.36. Find the tension T in the cord and the

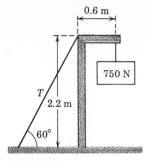

FIGURE 3.36

Problem 10.

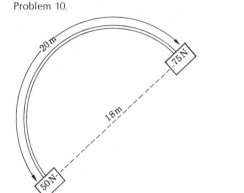

FIGURE 3.37

Problem 11.

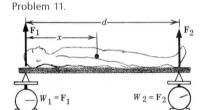

F_1 x d F_2

$W_1 = F_1$ $W_2 = F_2$

FIGURE 3.38

Problem 12.

force acting at the base of the post. (Neglect the weight of the post.)

11   Locate the common center of gravity of the two weights in Fig. 3.37. (Neglect the weight of the connecting rod.)
Ans. On the dotted line, 1.08 m from the 50-N weight.

12   The center of gravity of a person is measured by weighing the person on a board supported by wedges on two scales (Fig. 3.38). The scales are adjusted to read zero with the board alone, and the person is placed with his or her head and feet even with the wedges. Derive a formula for the distance x of the center of gravity from the head in terms of the scale readings W_1 and W_2 and the length d of the person.

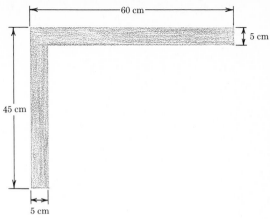

FIGURE 3.39

Problem 13.

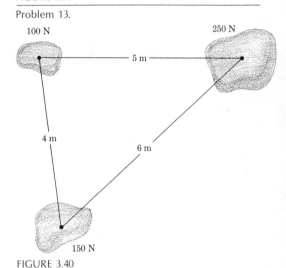

FIGURE 3.40

Problem 14.

13   Locate the center of gravity of the uniformly thick L-shaped plate in Fig. 3.39. (*Hint:* Divide the plate into two rectangles; the center of gravity of each rectangle is at its geometric center.)

14   Locate the position of the center of gravity of the three objects shown in Fig. 3.40.

15   Figure 3.41 shows a mobile consisting of four ornaments attached by threads to light rods. The distances (in centimeters) between the ornaments and the supporting threads and the weight of one ornament are indicated. If the mobile is to hang balanced, as shown, what are the weights of the other ornaments? Neglect the weight of the rods and thread. (*Hint:* Find the weight of *A* first.)
Ans. 1, 5, and 2.67 N

16   A 5-lb and a 10-lb weight are each attached to one end of a 3-ft-long rod. The other

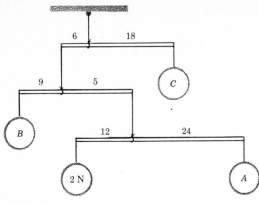

FIGURE 3.41

Problem 15.

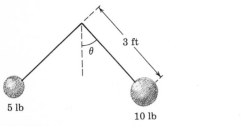

5 lb

10 lb

FIGURE 3.42

Problem 16.

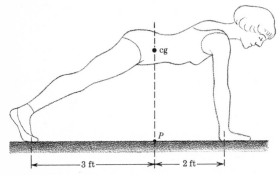

FIGURE 3.43

Problem 17.

ends of the two rods are welded together at right angles. When the assembly is freely suspended as shown in Fig. 3.42, the rod with the 10-lb weight makes an angle θ with the vertical. Find θ.

17 Figure 3.43 shows an athlete about to do a push-up. She weighs 125 lb, and her center of gravity is located above a point P on the ground 3 ft from her toes and 2 ft from her shoulders. What are the forces exerted by the floor on the athlete's hands and feet?
Ans. 75 and 50 lb

18 A 110-lb woman stands 5 ft from one end of a 20-ft scaffold (Fig. 3.44). The scaffold

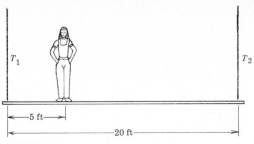

FIGURE 3.44

Problem 18.

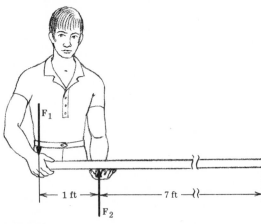

FIGURE 3.45

Problem 19.

weighs 150 lb. What are the tensions T_1 and T_2 in the ropes supporting the scaffold?

19 A man carries an 8-ft board with one hand pushing on it with a force $\mathbf{F}_1$ at one end and the other hand lifting it with a force $\mathbf{F}_2$ applied 1 ft from this end (Fig. 3.45). The board weighs 25 lb, and its center of gravity is at its center. Find F_1 and F_2.
Ans. 75 and 100 lb

20 Figure 3.46 shows a 9-m board hinged at one end (point O) and maintained at an angle of 30° to the horizontal by an 800-N weight suspended from a cord attached to the other end of the board. The center of gravity of the board is 4 m from the hinged end. (a) What is the weight of the board? (b) What is the magnitude of the normal force exerted by the hinge on the board?

21 The man in Fig 3.47 is in the process of putting his canoe up on his shoulders. The canoe is 18 ft long and weighs 84 lb; its center of gravity is at its midpoint. What is the magnitude of the force $\mathbf{F}_1$ that the man applies to the canoe while in the position shown? A friend, ignorant of physics, tries to help him by lifting the canoe at point A. Explain why this does not

86

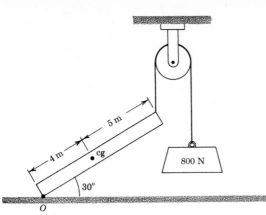

FIGURE 3.46

Problem 20.

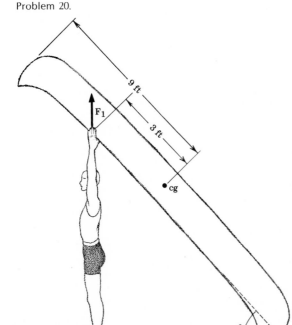

FIGURE 3.47

Problem 21.

help the man. Where should the friend lift to be of help? (Knowledge of the angle θ is not required for this problem, but if it will help, you may assume it to be 30°.)
Ans. 63 lb

22 In mountain climbing the standard method of descending a vertical wall is by *rappelling*. The climber attaches one end of a rope to a secure rock at the top of the cliff and the other end to a special buckle strapped to his or her waist, 0.5 ft above his or her center of gravity. (The buckle holds the rope and allows

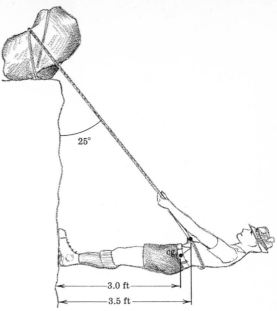

FIGURE 3.48

Problem 22.

the climber to increase the length of the rope by feeding the free end of the rope through the buckle. In this way the climber is able to walk down the wall with ease, as shown in Fig. 3.48.) The climber in the figure weighs 180 lb, his center of gravity is 3.0 ft from his feet, and his rope makes an angle of 25° with the wall. Find the tension in the rope and the magnitude and direction of the surface force exerted by the wall on his feet.

23 A 12-ft ladder leans against a wall at an angle $\theta = 34°$ (Fig. 3.49). A 200-lb worker is standing on the ladder, a distance $x = 3$ ft from the top. The ladder weighs 50 lb and its center of gravity is 5 ft from its base. (a) Assume the wall is frictionless so that the force $\mathbf{F}_1$ exerted on the ladder by the wall is perpendicular to the wall. What is the magnitude of $\mathbf{F}_1$? (b) Find the magnitude and direction of the force $\mathbf{F}_2$ exerted by the floor on the ladder at point O.
Ans. (a) 115 lb; (b) 275 lb, 65.3°

24 (a) Show that the magnitude of the horizontal component F_{2x} of the force $\mathbf{F}_2$ exerted by the ground on the ladder in Fig. 3.49 is given by

$$F_{2x} = W\left(1 - \frac{x}{L}\right)\tan\theta + W'\frac{d}{L}\tan\theta$$

where W is the weight of the worker, W' is the weight of the ladder, L is the length of the ladder, and d is the distance of the ladder's center of gravity from the base. Check this formula with the answer given in Prob. 23. (b) If the coefficient of static friction between the ground and the ladder is μ_s, show that the

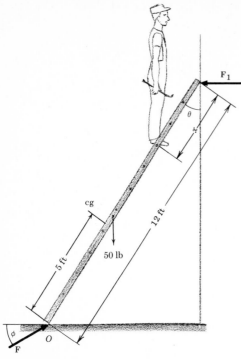

FIGURE 3.49

Problems 23 and 24.

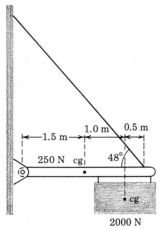

FIGURE 3.50

Problem 25.

largest angle θ at which the ladder can be placed such that the worker can safely stand anywhere on the ladder is given by

$$\tan \theta = \frac{(W + W') \mu_s}{W + W' \dfrac{d}{L}}$$

25 Find the tension in the cable and the force at the pivot of the sign in Fig. 3.50.
Ans. 2355 and 1653 N

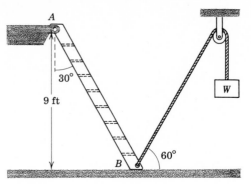

FIGURE 3.51

Problem 26.

26 The steel ladder in Fig. 3.51 is pivoted at A and is to be raised by the cable attached at B. The ladder weighs 500 lb and its center of gravity is at its midpoint. (a) Find the tension T in the cable required to lift B just off the ground. (b) What is the force $\mathbf{F}$ at the pivot in this case?

27 A wire 13.500 m long is stretched to a length of 13.507 m. (a) What is the strain of the stretched wire? (b) If the wire is copper, what is the stress required to produce this strain? (c) If the cross-sectional area of the wire is 4×10^{-5} m², what is the tension of the stretched wire?
Ans. (a) 5.18×10^{-4}; (b) 6.22×10^7 N/m²; (c) 2488 N

28 A marble column with a cross-sectional area of 25 cm² supports a weight of 7×10^4 N. (a) What is the stress in the column? (b) What is the strain in the column? (c) If the column is 2 m high, how much is its length changed by the weight? (d) What is the maximum weight the column can support?

29 An aluminum wire with a cross-sectional area of 7×10^{-5} m² is stretched to the elastic limit. (a) What is the tension in the wire? (b) What is the strain of the wire? (c) What tension is required to break the wire? (d) What tension is required to break an aluminum wire with twice the diameter?
Ans. (a) 1.26×10^4 N; (b) 2.57×10^{-3}; (c) 1.40×10^4 N; (d) 5.6×10^4 N

30 (a) What is the smallest diameter copper wire that can support a weight of 5000 N without exceeding its elastic limit? (b) What is the maximum weight such a wire can support without breaking?

88

31 A gold wire 50 cm long and 1 mm in diameter is stretched 0.2 mm when subjected to a tension of 25 N. What is the Young's modulus of gold?

Ans. $80 \times 10^{10} \, \text{N/m}^2$

32 The quantity $l = s_u/\rho g$ is used as a measure of the tensile strength of a material relative to its density. Calculate l for aluminum, bone, and steel using Tables 1.6 and 3.1.

33 Calculate the relative compressive strengths $l' = s'_u/\rho g$ of (a) bone, (b) glass, (c) granite, and (d) wood.

Ans. (a) 1.1×10^4 m; (b) 4.3×10^4 m; (c) 0.76×10^4 m; (d) 1.5×10^4 m

34 Show that the quantity l defined in Prob. 32 is equal to the maximum length of material that can hang together under its own weight. That is, show that if a length of material greater than l is hung from one end, it will break apart under its own weight.

35 The cross section of a 12-m-long steel I beam (Fig. 3.52) has the dimensions $a = 0.45$ m, $d = 0.06$ m, and $l = 0.30$ m. (a) What is the stress in each flange when the beam supports an 8×10^4 N load at its center? (b) How much is the upper flange shortened by this load? (c) What is the maximum load the beam can support at its center before it is deformed beyond its elastic limit?

Ans. (a) 1.48×10^7 N/m²; (b) 0.89 mm; (c) 1.62×10^6 N

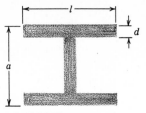

FIGURE 3.52

Problems 35, 36, 37, and 38.

36 A steel I beam 10 m long supports a 2×10^5 N load at its center. Find reasonable values for the dimensions d, l, and a of the beam's cross section (Fig. 3.52) so that the strain is 0.001. (There is no unique answer to this problem.)

37 A 2-m-long aluminum I beam with the dimensions (Fig. 3.52) $a = 0.03$ m, $l = 0.05$ m, and $d = 0.005$ m supports a 3×10^3 N load at its center. How much does the beam sag in the middle?

Ans. 6.4 cm

38 (a) Show that the maximum sag a centrally loaded I beam can have before it is stressed beyond its elastic limit is

$$h_{\text{max}} = \frac{s_e L^2}{3Ea}$$

where s_e is the elastic limit, and E is the Young's modulus of the beam material. (b) Calculate the maximum sag for the beam described in Prob. 37.

4

DYNAMICS

The last two chapters dealt with *statics,* the physics of objects at rest. In this chapter we discuss *dynamics,* the physics of objects in motion. We shall first generalize Newton's first law of motion to cover objects moving with constant velocity, and then we shall consider the motion of objects when the total force on them is not zero.

4.1 Frames of Reference

The last two chapters discussed fully the conditions for an object to remain at rest. The phrase "at rest," however, is very ambiguous, because an object can be at rest in one system while moving in another. For example, a person sitting in an airplane is at rest relative to the airplane but is moving at 600 mi/h relative to the ground. The airplane and the earth are two different reference systems, or *frames of reference,* with respect to which the person's motion can be referred. Because these two frames of reference are moving with respect to each other, the person can be at rest in one frame while moving in the other.

The phrase "at rest" is clearly meaningful only in a particular frame of reference. When the equilibrium conditions were applied in Chaps. 2 and 3, this frame was taken to be the earth. This seems reasonable, and in fact, up to the time of Galileo (1564–1642) it was thought that the earth was at rest at the center of the universe. However, since the discovery that the earth is in motion around the sun, it has been realized that there is nothing special about the earth's frame of reference. If the equilibrium conditions are true in the frame of the earth, they are probably true in some other frames as well.

Let us concentrate on Newton's first law of motion (Sec. 2.1). Briefly, it states that the total force on an object at rest is zero. For example, suppose we suspend a weight from two scales, as shown in Fig. 4.1. Newton's first law then asserts that the vector sum of the forces on the weight will be zero if the weight is at rest. We can verify this experimentally when the weight is at rest relative to the earth. But the angles and scale readings do not change when the experiment is done in a moving airplane, so the sum of the forces is also zero when the weight is at rest relative to a frame moving with respect to the earth. That is, Newton's first law of motion is true in the frame of reference of the airplane as well as in the frame of reference of the earth.

However, Newton's first law is not true in all frames. For example, it is not true in the frame of reference of a roller coaster going through a dip. If the apparatus in Fig. 4.1 were carried on a roller coaster, the angles and scale readings would change noticeably as the roller coaster went through a dip. The forces on the weight would no longer sum to zero, even though the weight was at rest in the frame of reference of the roller coaster. In this sense, we say that Newton's first law of motion is not true in the frame of reference of the roller coaster. This is just another way of saying that an object at rest relative to the dipping roller coaster is not "really" at rest. The problem remains of determining in which frames an object at rest has zero force on it.

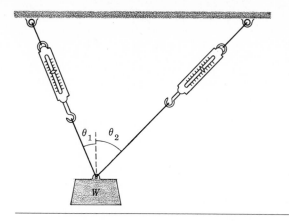

FIGURE 4.1

The vector sum of the forces on the weight W is zero if the weight is at rest in an inertial frame.

Definition An *inertial frame* is a frame of reference in which Newton's first law of motion is true. That is, the total force on an object at rest in an inertial frame is zero. To good approximation, both the earth and the airplane are inertial frames, but the dipping roller coaster is not.

Because of the special importance of inertial frames in physics, it is necessary to determine which frames are inertial. Galileo investigated this question for many years. He wanted to convince people that the earth could be moving without any apparent sensation of motion. The result of his investigations was the discovery of the following principle:

Galileo's Principle *Any frame that is moving with constant speed in a straight line relative to an inertial frame is itself an inertial frame.* Motion with constant speed in a straight line is called *uniform motion.*

The roller coaster is not an inertial frame because it moves in a curved path as it goes through a dip. The earth is not truly an inertial frame either, because objects at rest on its surface move through a large circle every day as the earth rotates on its axis. This motion has little effect, however, and for the purposes of this book, the earth is considered to be an inertial frame of reference.* Thus to the extent that the earth is an inertial frame, an airplane moving uniformly relative to the earth is also an inertial frame.

An immediate consequence of Galileo's principle is that the total force on an object is zero not only if the object is at rest, but also if it is moving uniformly relative to an inertial frame. This follows because an object moving uniformly relative to an inertial frame is itself an inertial frame. Since the object is clearly at rest in its own frame of reference, the total force on it is zero. The complete statement of Newton's first law then is:

Newton's First Law of Motion (Complete Statement) *For an object to remain at rest or move uniformly relative to an inertial*

---

*To understand certain large-scale motions, such as the circulation of winds around the earth or in cyclones and hurricanes, the earth's rotation must be taken into account. For these phenomena it is not a good approximation to consider the earth an inertial frame.

frame, it is necessary that the vector sum of all the forces acting on the object be zero.

Newton's first law of motion is true in all inertial frames, by definition. But what about the other laws of physics, such as the torque condition (Sec. 3.1)? Are they also true in all inertial frames? Nineteenth-century physicists thought that all the laws of mechanics were true in all inertial frames, but that other laws, especially laws concerning light and electromagnetism, were exactly true only in one special inertial frame, called the *ether*. The ether idea was abandoned after Einstein showed that the laws of light and electromagnetism, when properly understood, are exactly true in all inertial frames. This is the principle of relativity:

Principle of Relativity   *All the laws of physics are true in all inertial frames.*

Remember, an inertial frame is defined only as a frame of reference in which Newton's first law is true. Galileo's principle asserts further that any frame moving uniformly with respect to an inertial frame is itself an inertial frame. Then, according to the principle of relativity, all the laws of physics are true in all inertial frames. The principle of relativity thus establishes inertial frames as the proper frames of reference in which to describe the laws of physics.

4.2   Velocity and Acceleration

SPEED AND VELOCITY

The concept of uniform motion must now be made more precise. All motion must be described relative to some frame of reference, which we shall always take to be an inertial frame. However, it is not necessary to state which particular inertial frame we are using because the principle of relativity assures us that all inertial frames are equivalent for the purposes of physics.

Definition   An object moves with *constant speed v* if the distance *x* that it moves in time *t* is given by

$$x = vt \qquad 4.1$$

for all values of *t*. The constant *v* is the *speed* of the object.

UNITS   The dimensions of speed are

$$\frac{\text{Length}}{\text{Time}} = [l/t]$$

In the English engineering system the unit of speed is feet per second (ft/s), and in the SI the unit of speed is meters per second (m/s). Other common units of speed are kilometers per hour (km/h), miles per hour (mi/h), and kilometers per second (km/s).

Figure 4.2 is a graph of Eq. 4.1 for the case in which $v = 5$ m/s. In any 1-s interval, the object moves 5 m.

Example 4.1   An object moving with constant speed travels 48 m in 6 s. (a) What is the speed of the object? (b) How far does the

FIGURE 4.2

Graph of distance *x* against time *t* for an object moving with a constant speed of 5 m/s.

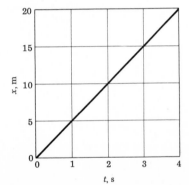

object travel in 5 s? (c) How long does it take the object to travel 100 m?

(a) Given the distance $x = 48$ m that the object travels in time $t = 6$ s, we can find the speed v by solving Eq. 4.1 for v:

$$v = \frac{x}{t} = \frac{48 \text{ m}}{6 \text{ s}} = 8 \text{ m/s}$$

(b) Knowing the speed $v = 8$ m/s and the travel time $t = 5$ s, Eq. 4.1 gives the distance traveled:

$$x = vt = (8 \text{ m/s})(5 \text{ s}) = 40 \text{ m}$$

(c) Knowing the speed $v = 8$ m/s and the distance $x = 100$ m, the travel time is found by solving Eq. 4.1 for t:

$$t = \frac{x}{v} = \frac{100 \text{ m}}{8 \text{ m/s}} = 12.5 \text{ s} \qquad \square$$

Definition The *velocity* **v** of a moving object is a vector quantity with magnitude (the speed v of the object) and direction (the direction of motion of the object).

Definition *Uniform motion* is motion with constant velocity. Since velocity is a vector, constant velocity implies two things: (1) the speed v is not changing (constant speed), and (2) the direction of motion is not changing (motion in a straight line). Thus uniform motion (constant velocity) is motion in a straight line with constant speed. An object at rest is a special case of uniform motion in which the speed is zero.

ACCELERATION

An object that is not moving with constant velocity is said to be *accelerating*. That is, an accelerating object either is not moving in a straight line or not moving with constant speed or both. Figure 4.3a shows a car moving with constant speed around a circular racetrack. The arrows, which represent the car's velocity at different positions on the track, have the same length but are pointing in different directions. This car is accelerating because its direction of motion is continuously changing. Figure 4.3b shows a car that is increasing its speed steadily while traveling in a straight line. Here the arrows, which represent the car's velocity at different positions, have different lengths but are pointing in the same direction. This car is accelerating because its speed is changing. Situations in which an object is simultaneously changing its speed and direction are also common, but we shall not consider them until Chap. 6.

Definition For an object with velocity $\mathbf{v}_1$ at time t_1 and velocity $\mathbf{v}_2$ at time t_2, the *acceleration* **a** during the time interval $\Delta t = t_2 - t_1$ is

$$\mathbf{a} = \frac{\mathbf{v}_2 - \mathbf{v}_1}{t_2 - t_1} = \frac{\Delta \mathbf{v}}{\Delta t} \qquad 4.2$$

where* $\Delta \mathbf{v} = \mathbf{v}_2 - \mathbf{v}_1$ and $\Delta t = t_2 - t_1$.

*As elsewhere in this book, the Greek (capital) letter delta (Δ) means "change of."

FIGURE 4.3

A car accelerating (a) because its direction of motion is changing and (b) because its speed is increasing.

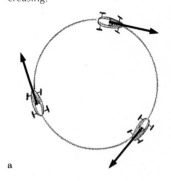

a

b

Like velocity, acceleration is also a vector quantity. Equation 4.2 gives only the average acceleration between t_1 and t_2, but if the time interval Δt is sufficiently small, the average acceleration is approximately equal to the instantaneous acceleration at the time midway between t_1 and t_2.

UNITS The dimensions of acceleration are $[v/t] = [l/t^2]$. In the English engineering system the units of acceleration are feet per second per second (ft/s^2), and in the SI the units are meters per second per second (m/s^2).

In this chapter we shall discuss only the special case of constant linear acceleration. Other types of accelerated motion, such as uniform circular motion and simple harmonic motion, are discussed in Chap. 6.

Definition *Constant linear acceleration* is accelerated motion in which an object moves in a straight line with an acceleration **a** that is constant in both magnitude and direction. In this motion, the velocity **v** of the object is constant in direction but variable in magnitude. At any time t, the magnitude v of the velocity is given by

$$v = v_0 + at \qquad\qquad 4.3$$

where v_0 is the speed at time $t = 0$ and a is the magnitude of the acceleration. The acceleration a is either positive or negative, depending on whether the speed of the object is increasing or decreasing.

REMARK To show that the constant a in Eq. 4.3 is the acceleration $\Delta v/\Delta t$, we substitute Eq. 4.3 into Eq. 4.2:

$$\frac{\Delta v}{\Delta t} = \frac{v_2 - v_1}{t_2 - t_1} = \frac{(v_0 + at_2) - (v_0 + at_1)}{t_2 - t_1}$$

$$= \frac{a(t_2 - t_1)}{t_2 - t_1} = a$$

This proves that the acceleration of an object with a speed v given by Eq. 4.3 is equal to the constant a in that equation.

The force of gravity is the total force on an unsupported object. The object therefore cannot be in equilibrium, so it falls. From his study of the motion of falling objects, Galileo found that all objects fall with constant linear acceleration and that (in vacuum) the magnitude of the acceleration is the same for all objects. This is not obvious: a feather certainly does not fall as fast as a rock. But this is so because the air exerts a relatively large upward force on the falling feather (air resistance) which partially supports it. In a vacuum, a feather and a rock do fall with the same acceleration.

The symbol g is used for the acceleration due to gravity. The value of g is the same for all objects at a given site, but it varies slightly with the latitude and to a lesser extent with the altitude and geology of the site. In doing problems it is sufficient to use the value of $32\ ft/s^2$ or $9.8\ m/s^2$ for g.

With $a = g = 9.8\ m/s^2$ in Eq.4.3, we see that an object falling from rest ($v_0 = 0$) has a speed of $4.9\ m/s$ at $t = 0.5\ s$, a speed of

9.8 m/s at $t = 1.0$ s, and a speed of 19.6 m/s at $t = 2.0$ s. How far does it fall in 1 s? At first you might be tempted to use Eq. 4.1 with the speed of 9.8 m/s that the object has after 1 s. The result would be

$$x = vt = (9.8 \text{ m/s})(1 \text{ s}) = 9.8 \text{ m}$$

This is wrong, however, because the speed of 9.8 m/s is obtained only at the end of the first second. Since the speed of the object was zero at the beginning of the first second, the average speed $\bar{v}$ during the first 1-s interval is

$$\bar{v} = \tfrac{1}{2}(0 \text{ m/s} + 9.8 \text{ m/s}) = 4.9 \text{ m/s}$$

Using the average speed in Eq. 4.1, we get the correct distance traveled:

$$x = \bar{v}t = (4.9 \text{ m/s})(1 \text{ s}) = 4.9 \text{ m}$$

The general formula for the distance x traveled in time t by an object that starts from rest and accelerates with constant acceleration a is thus

$$x = \bar{v}t = [\tfrac{1}{2}(0 + at)]t$$

or

$$x = \tfrac{1}{2}at^2 \qquad\qquad 4.4$$

Figure 4.4 is a plot of x against t for the case in which $a = 9.8$ m/s². (Compare this with Fig. 4.2.) If the object starts with an initial speed v_0, the distance traveled in time t is given by

$$x = v_0t + \tfrac{1}{2}at^2 \qquad\qquad 4.5$$

Figure 4.5 is a multiflash photograph of a ball dropped from rest. The time intervals between adjacent images are equal, so the distance between adjacent images is large when the speed is large and small when the speed is small. This photograph clearly shows that the speed of a falling object starts out small and increases with time, in accordance with Eq. 4.3. In addition, the photograph shows that the distance increases with the square of the time (in accordance with Eq. 4.4), since (for instance) image 8 is 4 times farther from the starting point than image 4*.

For your convenience the equations for constant speed and constant linear acceleration are listed together in Table 4.1.

Example 4.2 Calculate the position of the ball in Fig. 4.5 at $t = \tfrac{1}{30}$ s and $t = \tfrac{1}{3}$ s. Compare with Fig. 4.5.

*The time interval between successive images in Fig. 4.5 is $\tfrac{1}{30}$ s. However, the first image corresponds to both $t = 0$ and $t = \tfrac{1}{30}$ s, because the ball did not move enough to produce separate images.

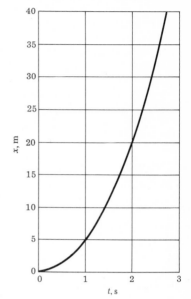

FIGURE 4.4

Graph of distance x against time t for an object moving with a constant acceleration of 9.8 m/s².

TABLE **4.1**

Equations for constant speed and constant linear acceleration

Constant speed:	
$x = vt$ v is constant	4.1

Constant linear acceleration:	
$v = v_0 + at$ a is constant; v_0 is speed at $t = 0$	4.3
$x = v_0t + \tfrac{1}{2}at^2$	4.5

Since the ball was released from rest, its position at $t = \frac{1}{30}$ s is given by Eq. 4.5 with $v_0 = 0$:

$$x = v_0 t + \tfrac{1}{2}at^2$$
$$= 0 + \tfrac{1}{2}(9.8 \text{ m/s}^2)(\tfrac{1}{30} \text{ s})^2$$
$$= 0.0054 \text{ m} = 0.54 \text{ cm}$$

The ball in Fig. 4.5 is about 5 cm in diameter, so at $t = \frac{1}{30}$ s it has moved only about one-tenth of its diameter. The image of the ball at this time is not distinguishable from the $t = 0$ image. At $t = \frac{1}{3}$ s the position of the ball is

$$x = 0 + \tfrac{1}{2}(9.8 \text{ m/s}^2)(\tfrac{1}{3} \text{ s})^2$$
$$= 0.54 \text{ m} = 54 \text{ cm}$$

Taking the second image in Fig. 5.4 to be at $t = \frac{2}{30}$ s, the tenth image is at $t = \frac{10}{30}$ s $= \frac{1}{3}$ s. The bottom of the tenth image coincides with the 54-cm mark on the meterstick. ☐

Example 4.3 An automobile starting from rest accelerates to 25 m/s (56 mi/h) in 12 s. (a) What is the acceleration of the automobile during this period? (b) How far does the automobile travel during this period?

(a) The initial speed is $v_0 = 0$, and the final speed is $v = 25$ m/s. Solving Eq. 4.3 for the acceleration, we get

$$a = \frac{v - v_0}{t} = \frac{25 \text{ m/s} - 0}{12 \text{ s}} = 2.08 \text{ m/s}^2$$

(b) With $v_0 = 0$ and $a = 2.08$ m/s^2, the distance traveled in 12 s is found from Eq. 4.5 to be

$$x = v_0 t + \tfrac{1}{2}at^2$$
$$= 0 + \tfrac{1}{2}(2.08 \text{ m/s}^2)(12 \text{ s})^2 = 150 \text{ m}$$ ☐

Example 4.4 The brakes are suddenly applied to an automobile traveling at 24 m/s (54 mi/h). While skidding to a stop, the deceleration (negative acceleration) is -6 m/s^2. How far does the automobile skid before stopping?

In this case, the initial speed is $v_0 = 24$ m/s, and the final speed is $v = 0$. Since the automobile is slowing down, the acceleration is negative. From Eq. 4.3 we find that the car skids for a time:

$$t = \frac{v - v_0}{a} = \frac{0 - 24 \text{ m/s}}{-6 \text{ m/s}^2} = 4 \text{ s}$$

The distance traveled in this time is given by Eq. 4.5:

$$x = v_0 t + \tfrac{1}{2}at^2$$
$$= (24 \text{ m/s})(4 \text{ s}) + \tfrac{1}{2}(-6 \text{ m/s}^2)(4 \text{ s})^2$$
$$= 96 \text{ m} - 48 \text{ m} = 48 \text{ m}$$

REMARK Note the importance of assigning a negative value to the acceleration in a case where the object is slowing down. ☐

Example 4.5 A ball is thrown straight up into the air with an initial speed of 40 ft/s. (a) How high does the ball go? (b) How

FIGURE 4.5

Multiflash photograph of a ball dropped from rest. (*From PSSC Physics, 4th ed., copyright © 1976 by Education Development Center, Inc.*)

long is the ball in the air? (c) What is the speed of the ball when it hits the ground?

(a) Figure 4.6 is a multiflash photograph of a ball thrown up in the air. As the ball rises, the distance between adjacent images decreases, indicating that the ball is slowing down. At its highest point the ball is momentarily at rest.

Thus for the upward portion of the ball's motion we have $v_0 = 40$ ft/s, $v = 0$, and $a = -g = -32$ ft/s^2. The acceleration is negative because the speed is decreasing. The problem now is just like the problem of the skidding car in Example 4.4. From Eq. 4.3 we find that the time for the upward journey is

$$t = \frac{v - v_0}{a} = \frac{0 - 40 \text{ ft/s}}{-32 \text{ ft/s}^2} = 1.25 \text{ s}$$

The distance traveled in this time is given by Eq. 4.5:

FIGURE 4.6

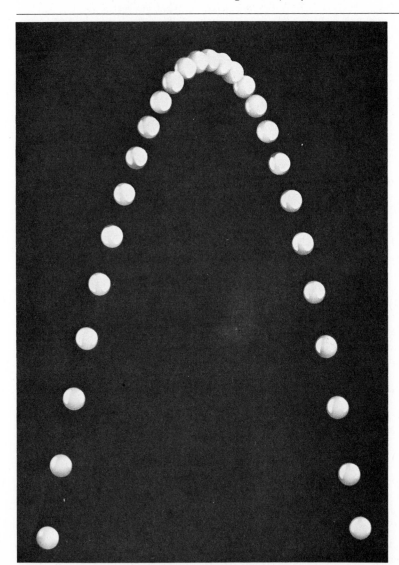

Multiflash photograph of a ball thrown into the air. (*From* PSSC Physics, *4th ed., copyright* © *1976 by Education Development Center, Inc.*)

$$x = v_0 t + \tfrac{1}{2}at^2$$
$$= (40 \text{ ft/s})(1.25 \text{ s}) + \tfrac{1}{2}(-32 \text{ ft/s}^2)(1.25 \text{ s})^2$$
$$= 50 \text{ ft} - 25 \text{ ft} = 25 \text{ ft}$$

The ball reaches its maximum height of 25 ft in 1.25 s.

(b) For the downward journey we have $v_0 = 0$ (since the ball is momentarily at rest at its highest point), $x = 25$ ft, and $a = g = 32$ ft/s^2. The time for the ball to fall to the ground from its highest point is found from Eq. 4.5:

$$t^2 = \frac{2x}{a} = \frac{2(25 \text{ ft})}{32 \text{ ft/s}^2} = 1.56 \text{ s}^2$$

or

$$t = \sqrt{1.56 \text{ s}^2} = 1.25 \text{ s}$$

The ball falls back to the ground in the same time it took to go up, and thus its total time in the air is $2t = 2.5$ s.

(c) The ball's speed when it reaches the ground is given by Eq. 4.3 with $v_0 = 0$:

$$v = v_0 + at = 0 + (32 \text{ ft/s}^2)(1.25 \text{ s}) = 40 \text{ ft/s}$$

Thus the ball returns to the ground with the same speed with which it was thrown. Its velocity is not the same, however, because initially its velocity had a magnitude of 40 ft/s and was directed upward, whereas at the end its velocity has a magnitude of 40 ft/s and is directed downward. Figure 4.6 shows that at any given point, the speed of the ball is the same for both upward and downward motion. ☐

4.3 Newton's Second Law of Motion

The basic problem of dynamics is to describe the motion of an object acted upon by a total force **F**. Newton's first law of motion states that when **F** is zero, the object remains at rest or moves with constant velocity. Newton's second law of motion describes what happens when **F** is not zero. It should be clear that in this case the object accelerates; i.e., the object does not move with constant velocity. The second law goes further, however, and states that the acceleration **a** of the object is proportional to **F**. This means that **a** has the same direction as **F** and that the magnitude of **a** is proportional to the magnitude of **F**.

Newton's Second Law of Motion *An object acted upon by a total force **F** has an acceleration **a** in the direction of **F**. The magnitude of **a** is F/m, where F is the magnitude of the force and m is an intrinsic property of the object, called its* mass. *Thus the second law can be written*

$$\mathbf{a} = \frac{\mathbf{F}}{m} \qquad \text{or} \qquad \mathbf{F} = m\mathbf{a} \qquad\qquad\qquad 4.6$$

UNITS In the International System the unit of mass is the *kilogram* (kg), which is the mass of a particular cylinder of platinum-iridium alloy kept in a vault in Sevres, France. The unit of force, the *newton* (N), is defined as the force which causes a mass of one kilogram to have an acceleration of one meter per second per second. From Eq. 4.6 we have

$$F = 1\,N = ma = (1\,kg)(1\,m/s^2) \qquad \text{or} \qquad 1\,N = 1\,kg \cdot m/s^2 \qquad\qquad 4.7$$

The kilogram, meter, and second are taken to be fundamental units in the SI*, and the units of force and all other quantities in mechanics are defined in terms of them.

In the English engineering system the unit of mass is the slug. In this system the pound, foot, and second are taken to be fundamental, and the *slug* is defined as the mass which has an acceleration of one foot per second per second when acted upon by a force of one pound. From Eq. 4.6 we have

$$m = 1\,slug = \frac{F}{a} = \frac{1\,lb}{1\,ft/s^2} \qquad \text{or} \qquad 1\,slug = 1\,lb \cdot s^2/ft \qquad\qquad 4.8$$

The total force on an unsupported object is the force of gravity F_g. Consequently, the object falls with an acceleration given by Eq. 4.6.

$$a = \frac{F_g}{m}$$

On the other hand, we know that all objects on the surface of the earth fall with the same acceleration g. Therefore, we have

$$g = \frac{F_g}{m} \qquad \text{or} \qquad F_g = mg \qquad\qquad 4.9$$

which shows that the force of gravity on an object is proportional to its mass.

Example 4.6 (a) What is the weight of a 15-kg mass? (b) What is the weight of a 1-slug mass?

(a) From Eq. 4.9 the force of gravity (weight) of a 15-kg mass is

$$F_g = mg = (15\,kg)(9.8\,m/s^2) = 147\,N$$

Note that since this problem is in SI units, g must be $9.8\,m/s^2$, and the unit of force must be newtons. Equation 4.7 shows that $1\,kg \cdot m/s^2$ is indeed 1 N.

(b) From Eq. 4.9 the weight of a 1-slug mass is

$$F_g = mg = (1\,slug)(32\,ft/s^2) = 32\,lb$$

Here the problem is in English engineering units, so g must be $32\,ft/s^2$, and the unit of force must be pounds. Equation 4.8 shows that $1\,slug \cdot ft/s^2$ is indeed 1 lb. $\square$

The distinction between mass and weight is not evident in everyday life because g is nearly the same everywhere on earth. The distinction becomes clearer when we transport an object, such as the Viking I lander, to Mars, where $g = 3.7\,m/s^2$. The mass of the lander (600 kg) is an intrinsic property that does not change with the lander's location. Thus on earth the lander weighs

$$F_g = mg_{earth} = (600\,kg)(9.8\,m/s^2) = 5880\,N$$

whereas on Mars it weighs

*Three other fundamental units are introduced later in this book in connection with temperature, electric charge, and luminosity.

$$F_g = mg_{\text{Mars}} = (600\,\text{kg})(3.7\,\text{m/s}^2) = 2220\,\text{N}$$

The weight of the lander is different on earth and Mars, but its mass is the same.

Example 4.7 A rock weighs 8 N on the moon, where $g = 1.67\,\text{m/s}^2$. How much does the rock weigh on earth?

We first use the rock's weight on the moon in Eq. 4.9 to find the rock's mass:

$$m = \frac{F_g(\text{moon})}{g_{\text{moon}}} = \frac{8\,\text{N}}{1.67\,\text{m/s}^2} = 4.8\,\text{kg}$$

Then we use the rock's mass to find its weight on earth:

$$F_g(\text{earth}) = mg_{\text{earth}} = (4.8\,\text{kg})(9.8\,\text{m/s}^2) = 47\,\text{N}$$

The rock weighs 6 times more on earth than on the moon, because the acceleration of gravity (g) is 6 times greater on the earth. $\qquad\qquad\square$

4.4 Examples Involving Newton's Second Law of Motion

Newton's second law of motion (Eq. 4.6) is the basis for the solution of all dynamical problems. When the forces on an object are constant, the law determines the constant acceleration to use in Eqs. 4.3 and 4.5 for the speed and position of the object. (When the forces are not constant, the second law still determines the acceleration, but Eqs. 4.3 and 4.5 no longer apply.) We shall restrict ourselves to constant-force problems, for which there are an abundant number of interesting and challenging examples.

Example 4.8 A constant horizontal force $\mathbf{F}_a$ of 150 N is applied to a 60-kg block which is initially resting on a frictionless surface (Fig. 4.7). After 3 s, what is the block's speed, and how far has it moved?

The force of gravity $\mathbf{F}_g$ and the normal force $\mathbf{F}_n$ also act on the block, so the total force $\mathbf{F}$ on it is

$$\mathbf{F} = \mathbf{F}_g + \mathbf{F}_n + \mathbf{F}_a$$

But $\mathbf{F}_g + \mathbf{F}_n = 0$, since the block does not move in the vertical direction. Therefore, the total force $\mathbf{F}$ equals $\mathbf{F}_a$, and the block accelerates in the direction of $\mathbf{F}_a$. Since the magnitude of $\mathbf{F}$ is 150 N, the acceleration of the block is found from Eq. 4.6 to be

$$a = \frac{F}{m} = \frac{150\,\text{N}}{60\,\text{kg}} = 2.5\,\text{m/s}^2$$

We know that a will be in the SI unit of acceleration because the force and mass are in the proper SI units. This illustrates the importance of keeping all the units in the same system.

The speed and position of the block are calculated using Eqs. 4.3 and 4.5. Initially, i.e., at time $t = 0$, the block was at rest, so $v_0 = 0$. Then at time $t = 3\,\text{s}$ its speed is

$$v = v_0 + at = 0 + (2.5\,\text{m/s}^2)(3\,\text{s}) = 7.5\,\text{m/s}$$

FIGURE 4.7

A constant horizontal force $\mathbf{F}_a$ applied to a block on a frictionless surface.

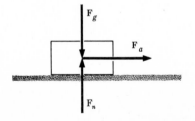

and the distance it has moved is

$$x = v_0 t + \tfrac{1}{2}at^2 = 0 + (\tfrac{1}{2})(2.5 \text{ m/s}^2)(3 \text{ s})^2 = 11.25 \text{ m} \qquad \square$$

Example 4.9 What is the acceleration of a 5-kg block sliding down a frictionless surface inclined 35° to the horizontal (Fig. 4.8)?

The magnitude of the force of gravity F_g on the block is

$$F_g = mg = (5 \text{ kg})(9.8 \text{ m/s}^2) = 49 \text{ N}$$

and F_g is directed vertically downward. The angle θ between F_g and the inclined surface is

$$\theta = 90° - 35° = 55°$$

because it is equal to the angle between the incline and the vertical.

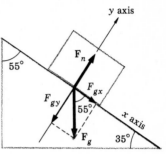

FIGURE 4.8

A block sliding down a frictionless surface inclined at 35° to the horizontal.

If we take the x axis along the incline and the y axis perpendicular to the incline, the components of F_g are

$$F_{gx} = F_g \cos 55° = (49 \text{ N})(0.574) = 28.1 \text{ N}$$
$$F_{gy} = -F_g \sin 55° = -(49 \text{ N})(0.819) = -40.1 \text{ N}$$

Since the incline is frictionless, the only other force on the block is the normal force F_n, with components

$$F_{nx} = 0 \qquad \text{and} \qquad F_{ny} = F_n$$

The block does not move perpendicular to the incline, so the sum of the y components is zero:

$$F_y = F_{ny} + F_{gy} = F_n - 40.1 \text{ N} = 0$$

Thus the normal force must be

$$F_n = 40.1 \text{ N}$$

The sum of the x components is

$$F_x = F_{nx} + F_{gx} = 0 + 28.1 \text{ N} = 28.1 \text{ N}$$

This is the net force on the block, so the block moves down the incline with the acceleration

$$a = \frac{F_x}{m} = \frac{28.1 \text{ N}}{5 \text{ kg}} = 5.6 \text{ m/s}^2$$

The acceleration is less than g because the full force of gravity does not act along the incline; only the component $F_{gx} = F_g \cos 55°$ acts along the incline. $\qquad \square$

FIGURE 4.9

(a) An accelerometer consisting of a mass and spring mounted on a shaft. (b) The displacement x of the mass is proportional to the acceleration of the device.

Example 4.10 An *accelerometer* is a device used to measure acceleration. It consists of a mass m attached to a spring, as shown in Fig. 4.9a. The mass rides on a frictionless rod, so that it can move only parallel to the spring. When the entire assembly has a forward acceleration a, the mass moves back a distance x (Fig. 4.9b). Find the relation between x, m, a, and the spring constant k.

When the spring is compressed a distance x from its unstretched position, it exerts the force

$$F_k = kx$$

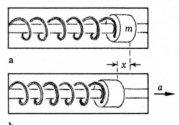

on the mass (Eq. 2.2). If the accelerometer is horizontal, this will be the net force on the mass. Thus the mass will have the acceleration given by Eq. 4.6:

$$kx = ma$$

If the accelerometer as a whole has the acceleration a, the last equation is still valid. The outside force that acts on the accelerometer to give it the acceleration a is transmitted to the mass by the spring. Thus the distance x is related to a by

$$x = \frac{ma}{k}$$

If m and k are known, the acceleration is determined by measuring x.

> **REMARK** The accelerometer and the mass are accelerating relative to the earth, which we take to be an inertial frame. However, as long as the acceleration is constant, the mass will not move relative to the body of the accelerometer. Thus the accelerometer is not an inertial frame, since the mass remains at rest in the frame of the accelerometer even though it has the net force $\mathbf{F}_k$ acting on it. □

Example 4.11 A mass m hangs from a string attached to the rearview mirror of an automobile. When the automobile accelerates forward, the string swings back, as shown in Fig. 4.10a. Find the angle θ that the string makes with the vertical while the automobile has a constant acceleration of 3 m/s².

The free-body diagram of the mass is shown in Fig. 4.10b. The forces on the mass are its weight $\mathbf{F}_g$ and the force $\mathbf{T}$ due to the tension in the string. The force $\mathbf{F}_g$ is directed downward and the force $\mathbf{T}$ is directed along the string.

The horizontal (x) and vertical (y) components of the forces (Fig. 4.10c) are

$$F_{gx} = 0 \quad \text{and} \quad F_{gy} = -mg$$
$$T_x = T \sin \theta \quad \text{and} \quad T_y = T \cos \theta$$

The mass has zero acceleration in the vertical direction, so the sum of the y components must be zero:

$$F_y = F_{gy} + T_y = -mg + T \cos \theta = 0$$

or

$$T = \frac{mg}{\cos \theta} \qquad \qquad 4.10$$

The mass has the acceleration a in the horizontal direction, so the sum of the x components is ma:

$$F_x = T \sin \theta = ma$$

We replace T in this last expression by Eq. 4.10 to get

$$\frac{mg \sin \theta}{\cos \theta} = ma$$

or

$$\tan \theta = \frac{a}{g}$$

FIGURE 4.10

(a) A mass suspended by a cord from the rearview mirror of an accelerating automobile. (b) Free-body diagram of the mass. (c) Vertical and horizontal components of the forces on the mass.

a

b

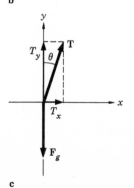

c

Here we have used the fact that

$$\frac{\sin\theta}{\cos\theta} = \tan\theta$$

Thus, if $a = 3\,\text{m/s}^2$, we have

$$\tan\theta = \frac{3\,\text{m/s}^2}{9.8\,\text{m/s}^2} = 0.306$$

so

$$\theta = \arctan 0.306 = 17.0° \qquad \square$$

Example 4.12 Figure 4.11a shows two masses, m_1 and m_2, connected by a flexible cord that passes over a frictionless pulley. If m_1 is greater than m_2, what will its acceleration be?

Since the masses are connected by the cord, they must move together, so that the magnitudes of their accelerations are equal. The force on each mass consists of the downward force of gravity and the upward force exerted by the cord. The difficulty of this problem is that both the acceleration and the tension of the cord are unknown.

Figure 4.11b shows the free-body diagram of each mass. The magnitude of the force exerted by the cord on each mass is equal to the tension T, which is the same throughout the cord (Sec. 2.2). We know that T is less than $m_1 g$, the force of gravity on m_1, because m_1 accelerates downward. Thus the total (downward) force on m_1 is $m_1 g - T$, and it is related to the (downward) acceleration a of m_1 by Eq. 4.6:

$$m_1 g - T = m_1 a \qquad\qquad 4.11$$

Likewise, T is greater than $m_2 g$ because m_2 accelerates upward. The total (upward) force $T - m_2 g$ on m_2 is related to the (upward) acceleration a of m_2 by

$$T - m_2 g = m_2 a \qquad\qquad 4.12$$

Here we have used the fact that the accelerations of m_1 and m_2 are equal.

Equations 4.11 and 4.12 are two linear equations in the two

FIGURE 4.11

(a) Two masses connected by a flexible cord passing over a frictionless pulley. (b) Free-body diagrams of the two masses.

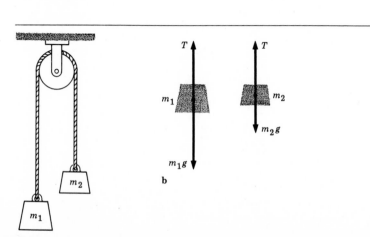

unknowns T and a. To solve them for a we must first eliminate T between them. This is easily done by just adding the equations together. The result is

$$(m_1g - T) + (T - m_2g) = m_1a + m_2a$$

We can remove the parentheses on the left and factor out a on the right, to give

$$m_1g - m_2g = (m_1 + m_2)a$$

or

$$a = \frac{m_1 - m_2}{m_1 + m_2} g$$

For instance, if $m_1 = 7.5$ kg and $m_2 = 4.5$ kg, the acceleration is

$$a = \frac{7.5 \text{ kg} - 4.5 \text{ kg}}{7.5 \text{ kg} + 4.5 \text{ kg}} (9.8 \text{ m/s}^2) = \frac{3 \text{ kg}}{12 \text{ kg}} (9.8 \text{ m/s}^2) = 2.45 \text{ m/s}^2 \square$$

4.5 Momentum

Definition The *momentum* **p** of a mass m is

$$\mathbf{p} = m\mathbf{v} \qquad\qquad 4.13$$

where **v** is the velocity of the mass. The vector **p** has the magnitude mv and the direction of **v**. For example, the momentum of a 3-kg mass with a velocity of 5 m/s to the west is 15 kg·m/s directed westward.

Suppose that in the time interval Δt the velocity of a mass changes from **v** to **v′**. Then the momentum of the mass changes from $\mathbf{p} = m\mathbf{v}$ to $\mathbf{p′} = m\mathbf{v′}$, and so the change of momentum $\Delta\mathbf{p}$ is

$$\Delta\mathbf{p} = \mathbf{p′} - \mathbf{p} = m\mathbf{v′} - m\mathbf{v}$$
$$= m(\mathbf{v′} - \mathbf{v}) = m\Delta\mathbf{v}$$

where $\Delta\mathbf{v} = \mathbf{v′} - \mathbf{v}$ is the change of velocity. When this expression for $\Delta\mathbf{p}$ is divided by Δt, we get

$$\frac{\Delta\mathbf{p}}{\Delta t} = m\frac{\Delta\mathbf{v}}{\Delta t}$$

But from Eq. 4.2 we know that when Δt is sufficiently small, $\Delta\mathbf{v}/\Delta t$ is equal to the acceleration **a** of the mass, so

$$\frac{\Delta\mathbf{p}}{\Delta t} = m\mathbf{a} \qquad \text{or} \qquad \frac{\Delta\mathbf{p}}{\Delta t} = \mathbf{F} \qquad\qquad 4.14$$

In the last expression we have used Newton's second law of motion (Eq. 4.6) to replace $m\mathbf{a}$ by the total force **F** acting on the mass.

Example 4.13 A pitched baseball ($m = 150$ g) passes over home plate with a speed of 30 m/s. After being hit by the bat, the ball moves away from home plate with a speed of 40 m/s. (a) What is the change of momentum of the baseball? (b) If the ball is in contact with the bat for 0.005 s, what force does the bat exert on the ball?

(a) The initial momentum of the ball has the magnitude

$$p = mv = (0.15\,\text{kg})(30\,\text{m/s}) = 4.5\,\text{kg}\cdot\text{m/s}$$

and its direction is toward home plate. The final momentum of the ball has the magnitude

$$p' = mv' = (0.15\,\text{kg})(40\,\text{m/s}) = 6.0\,\text{kg}\cdot\text{m/s}$$

and its direction is away from home plate. Figure 4.12 shows that because **p** and **p′** have opposite directions, the magnitude of their difference is equal to the sum of their magnitudes. That is, the magnitude of $\Delta\mathbf{p}$ is

$$\Delta p = (4.5 + 6.0)\,\text{kg}\cdot\text{m/s} = 10.5\,\text{kg}\cdot\text{m/s}$$

and its direction is the same as **p′**.

(b) From Eq. 4.14 the force exerted on the ball while the momentum is changing is

$$F = \frac{\Delta p}{\Delta t} = \frac{10.5\,\text{kg}\cdot\text{m/s}}{0.005\,\text{s}} = 2100\,\text{N} \qquad \square$$

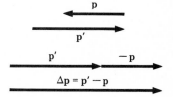

FIGURE 4.12

The difference $\Delta\mathbf{p}$ of two momenta **p** and **p′**. Since **p** and **p′** have opposite directions, the magnitude of their difference is equal to the sum of their magnitudes.

The total momentum of a system containing two or more objects is the vector sum of the momenta of the individual objects. Thus for a system of two masses m_1 and m_2 moving with velocities $\mathbf{v}_1$ and $\mathbf{v}_2$, the total momentum is

$$\mathbf{p} = \mathbf{p}_1 + \mathbf{p}_2 = m_1\mathbf{v}_1 + m_2\mathbf{v}_2$$

Suppose that as a result of colliding with each other the velocities of the two masses change from $\mathbf{v}_1$ and $\mathbf{v}_2$ to $\mathbf{v}'_1$ and $\mathbf{v}'_2$ (Fig. 4.13). Then the new momentum is

FIGURE 4.13

(a) Two masses heading toward each other with velocities $\mathbf{v}_1$ and $\mathbf{v}_2$. (b) During the collision of the masses, the forces on them are equal in magnitude and opposite direction. (c) After the collision, the masses move away from each other with velocities $\mathbf{v}'_1$ and $\mathbf{v}'_2$.

$$\mathbf{p}' = \mathbf{p}'_1 + \mathbf{p}'_2 = m_1\mathbf{v}'_1 + m_2\mathbf{v}'_2$$

and the total change of momentum is

$$\begin{aligned}\Delta\mathbf{p} = \mathbf{p}' - \mathbf{p} &= (\mathbf{p}'_1 + \mathbf{p}'_2) - (\mathbf{p}_1 + \mathbf{p}_2)\\ &= (\mathbf{p}'_1 - \mathbf{p}_1) + (\mathbf{p}'_2 - \mathbf{p}_2)\\ &= \Delta\mathbf{p}_1 + \Delta\mathbf{p}_2\end{aligned}$$

Thus the change of momentum of the system is equal to the sum of the changes of momentum of the individual masses.

If the masses are isolated from outside influences, they will not have any external forces acting on them. In this case, the only forces that occur are the forces that they exert on each other during the collision. According to Eq. 4.14, we have

$$\Delta\mathbf{p}_1 = \mathbf{F}_1\Delta t \qquad\text{and}\qquad \Delta\mathbf{p}_2 = \mathbf{F}_2\Delta t$$

where $\mathbf{F}_1$ is the force m_2 exerts on m_1, and $\mathbf{F}_2$ is the force m_1 exerts on m_2 (Fig. 4.13b). But these forces are an action-reaction pair, so according to Newton's third law of motion (Sec. 2.1),

$$\mathbf{F}_1 = -\mathbf{F}_2 \qquad\text{or}\qquad \mathbf{F}_1 + \mathbf{F}_2 = 0$$

Consequently, the total change of momentum is

$$\Delta\mathbf{p} = \Delta\mathbf{p}_1 + \Delta\mathbf{p}_2 = (\mathbf{F}_1 + \mathbf{F}_2)\Delta t = 0 \qquad\qquad 4.15$$

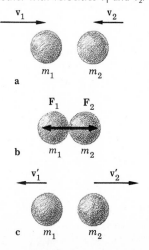

That is, when there are no external forces acting on the masses, the total momentum does not change. We say that the momentum is conserved. Although we have proven this important principle only in the case of two masses, it is true for any number of masses.

The Principle of Momentum Conservation *The total momentum of a system of masses remains unchanged as long as there are no external forces acting on the system.*

Example 4.14 A 2-kg rifle fires a 10-g bullet with a speed of 900 m/s. If the rifle is held loosely, what is its recoil speed?

Before the rifle is fired the speed of the rifle and the bullet are both zero:

$$v_r = 0 \quad \text{and} \quad v_b = 0$$

Therefore, the total momentum of the system is

$$p = m_r v_r + m_b v_b = 0$$

If the rifle is held loosely while it is fired, there will be no horizontal force on it, so the horizontal momentum will remain zero. Thus, after the bullet is fired, the momentum of the system is

$$0 = m_b v_b + m_r v_r = (0.010 \text{ kg})(900 \text{ m/s}) + (2 \text{ kg})v_r$$

where the bullet's speed is taken to be positive. The recoil speed of the rifle, then, is

$$v_r = -\frac{9 \text{ kg} \cdot \text{m/s}}{2 \text{ kg}} = -4.5 \text{ m/s}$$

The minus sign indicates that the recoil velocity of the rifle is in the direction opposite the bullet's velocity. ☐

Example 4.15 A 1500-kg automobile traveling north at 25 m/s collides with a 2000-kg automobile traveling east at 15 m/s. Immediately after the collision the two automobiles lock together and move with a common velocity **v**. Find **v**.

Over a period of time the total momentum of the two automobiles is not conserved because road friction is a large external force which acts on the system. However, if we just consider the momentum immediately before and after the collision, we can assume that it is conserved because the time interval is too short for the external forces to have much effect.

Before the collision the magnitudes of the momenta of the automobiles are

$$p_1 = m_1 v_1 = (1500 \text{ kg})(25 \text{ m/s}) = 3.75 \times 10^4 \text{ kg} \cdot \text{m/s}$$

and

$$p_2 = m_2 v_2 = (2000 \text{ kg})(15 \text{ m/s}) = 3.0 \times 10^4 \text{ kg} \cdot \text{m/s}$$

The momenta have the directions shown in Fig. 4.14, so the total momentum **p** = **p₁** + **p₂** is the vector sum of **p₁** and **p₂**. Since **p₁** and **p₂** are at right angles to each other in this case, **p** has the

FIGURE 4.14

The total momentum **p** of two automobiles with momenta **p₁** and **p₂**.

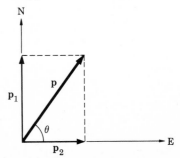

magnitude

$$p = \sqrt{p_1^2 + p_2^2} = 4.8 \times 10^4 \text{ kg} \cdot \text{m/s}$$

and makes the angle

$$\theta = \arctan \frac{p_1}{p_2} = 51.3°$$

with the east.

Immediately after the collision the total momentum is the same, but now the automobiles stay connected and move with a common velocity **v**. Thus we have

$$\mathbf{p} = m_1\mathbf{v} + m_2\mathbf{v} = (m_1 + m_2)\mathbf{v}$$

so the velocity has the magnitude

$$v = \frac{p}{m_1 + m_2} = \frac{4.8 \times 10^4 \text{ kg} \cdot \text{m/s}}{3500 \text{ kg}} = 13.7 \text{ m/s}$$

and the same direction as **p**.   ☐

GUIDE TO MAJOR TOPICS

Topic	References	Problems
Frames of reference	Sec. 4.1	
Constant velocity	Sec. 4.2; Eq. 4.1; Example 4.1	1 to 6
Constant linear acceleration	Sec. 4.2; Eqs. 4.3 to 4.5; Examples 4.2 to 4.5; Table 4.1	7 to 16
Newton's second law of motion	Secs. 4.3 and 4.4; Eq. 4.6	
Weight and mass	Eq. 4.9; Examples 4.6 and 4.7	17 to 20
Dynamics problems	Examples 4.8 to 4.12	21 to 30
Momentum	Sec. 4.5; Eq. 4.13	31 and 32
Change of momentum	Eq. 4.14; Example 4.13	33 and 34
Conservation of momentum	Eq. 4.15; Examples 4.14 and 4.15	35 to 38

PROBLEMS

1   An object is traveling at a constant speed of 25 m/s. (a) How far does it travel in 60 s? (b) How long does it take to travel 35 km ($1 \text{ km} = 10^3 \text{ m}$)?
Ans. (a) 1500 m; (b) 23.3 min

2   An earth satellite moving at constant speed travels 810 km in 3 min. (a) What is the speed of the satellite? (b) How far does the satellite travel in 1 h? (c) How long does it take the satellite to complete one orbit about the earth if the circumference of the orbit is 1.26×10^5 km?

3   The nearest star is about 4×10^{16} m away. At what speed would a rocket ship have to travel to get to this star in 10 years?
Ans. 1.27×10^5 km/s

4   (a) What is the average speed of a marathon runner who runs the 26.2-mi course in 2 h

15 min? (b) In what time (in minutes) does this runner run 1 mi?

5   Figure 4.15 is a graph of distance against time for an object moving with constant speed. (a) How far did the object travel between $t = 2$ s and $t = 5$ s? (b) What is the speed of the object?
Ans. (a) 22.5 m; (b) 7.5 m/s

6   Draw a graph of distance against time for an object moving at a constant speed of 12 m/s. Let $x = 0$ when $t = 0$.

7   An object dropped down a well hits the bottom in 3.2 s. How deep is the well?
Ans. 164 ft

8   (a) How long does it take an object starting from rest to fall 50 m? (b) How far would the object fall in twice that time?

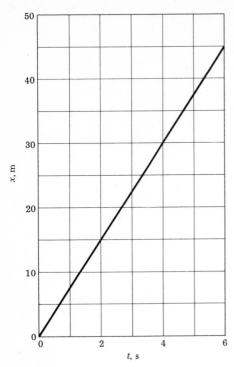

FIGURE 4.15

Problem 5.

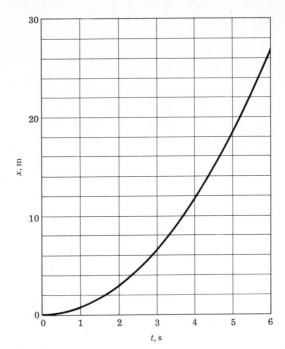

FIGURE 4.16

Problem 11.

9 An airplane accelerates down a 2000-m-long runway for 58 s before leaving the ground. (a) What is the acceleration of the airplane while moving down the runway? (b) What is the takeoff speed of the airplane?
Ans. (a) 1.19 m/s²; (b) 69 m/s (154 mi/h)

10 In a head-on collision, a car going 30 m/s is brought to rest in 0.1 s. What is the acceleration during the collision?

11 Figure 4.16 is a graph of distance against time for an object moving with constant acceleration. (a) What is the average speed of the object in the interval between $t = 0$ and $t = 4$ s? (b) What is the average speed of the object in the interval between $t = 4$ s and $t = 6$ s? (c) What is the acceleration of the object, assuming it starts from rest at $t = 0$?
Ans. (a) 3 m/s; (b) 7.5 m/s; (c) 1.5 m/s²

12 Draw a graph of distance against time for an object moving with a constant acceleration of 4 m/s². Take v_0 to be zero.

13 A ball is thrown straight up in the air with an initial speed of 12 m/s. (a) How long does it take the ball to reach its highest point? (b) How high does the ball go? (c) What is the time interval between the instant the ball leaves the hand until it is caught again?
Ans. (a) 1.22 s; (b) 7.29 m; (c) 2.44 s

14 A high jumper jumps 1.2 m straight up in the air. With what speed did the jumper leave the ground? (Hint: Remember that an object will hit the ground with the same speed that it left the ground.)

15 When making a vertical jump, a grasshopper extends its legs 2.5 cm in 0.025 s. (a) What is the acceleration of the grasshopper while extending its legs? (b) What is the speed of the grasshopper as it leaves the ground, i.e., at the instant its legs are fully extended? (c) How high does the grasshopper rise from the ground?
Ans. (a) 80 m/s²; (b) 2 m/s; (c) 0.20 m

16 The brakes are applied to an automobile that was traveling at a speed of 22 m/s. (a) If the automobile has a deceleration of −3 m/s², how long does it take to stop? (b) How far does the automobile travel during the braking period?

17 (a) What is the mass of a 500-lb weight? (b) What is the mass of a 500-N weight?
Ans. (a) 15.6 slug; (b) 51.0 kg

18 (a) How much does a 50-slug mass weigh on earth? (b) What does the mass weigh on the moon ($g_{moon} = 5.48$ ft/s²)?

19 A 75-kg mass weighs 660 N on Venus? (a) What is the acceleration due to gravity on Venus? (b) What is the weight on Venus of an

108

object that weighs 1000 N on earth?
Ans. (a) 8.8 m/s²; (b) 898 N

20   What is the weight on the moon of an astronaut who weighs 800 N on earth?

21   A 15-N force is the only force acting on a 4-kg mass. (a) What is the acceleration of the mass? (b) If the speed of the mass is zero at $t = 0$, what is its speed at $t = 10$ s? (c) How far did the mass travel during the 10-s interval?
Ans. (a) 3.75 m/s²; (b) 37.5 m/s; (c) 187.5 m

22   A 2000-kg car accelerates from rest to 35 m/s in 12 s. (a) What is the total force on the car during this acceleration? (b) How far does the car travel during the acceleration period?

23   The mass of the Saturn V rocket is 2.8×10^6 kg just before takeoff. During takeoff the rocket engines produce a thrust of 33×10^6 N. This means that an upward force of this magnitude is exerted on the rocket. (a) What is the acceleration of the rocket during takeoff? (Remember, the thrust is not the only force on the rocket.) (b) If this acceleration remains constant, what is the speed of the rocket after 60 s?
Ans. (a) 1.98 m/s²; (b) 119 m/s

24   The Viking I lander ($m = 600$ kg) used its rockets to brake its fall onto the surface of Mars. When the lander was 1200 m above the surface it was descending with a speed of 60 m/s. At this point its rockets were turned on to produce an upward acceleration of 1.5 m/s². (a) What force did the rocket engines exert on the lander ($g_{\text{Mars}} = 3.7$ m/s²)? (b) The lander hit the surface with a speed of 2 m/s. How long did the descent from 1200 m take?

25   The brakes are suddenly applied to a 1000-kg car going 25 m/s. The wheels lock, and the car skids to a stop in 5 s. (a) What is the force of friction on the car while it skids to a stop? (b) What is the coefficient of kinetic friction μ_k between the tires and the road? (c) How far does the car travel while skidding to a stop?
Ans. (a) 5000 N; (b) 0.51; (c) 62.5 m

26   A skier slides down a slope inclined 9° to the horizontal. (a) Neglecting friction, what is the acceleration of the skier? (b) If the slope is 50 m long, how long does it take the skier to reach the bottom? Assume she starts from rest at the top. (c) What is her speed when she reaches the bottom of the slope?

27   A 3-kg block on a horizontal surface is accelerated along the surface by a 10-N force applied at 40° to the horizontal (Fig. 4.17). Find

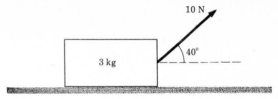

FIGURE 4.17

Problem 27.

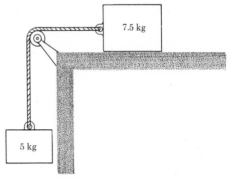

FIGURE 4.18

Problem 29.

the magnitudes of (a) the parallel and (b) the perpendicular components of the applied force. (c) What is the magnitude of the normal force acting on the block? (d) The coefficient of kinetic friction between the block and the surface is 0.25. What is the force of friction on the block? (e) What is the acceleration of the block?
Ans.   (a) 7.66 N;   (b) 6.43 N;   (c) 23.0 N; (d) 5.75 N; (e) 0.637 m/s²

28   A book sits on the dashboard of an automobile that has a forward acceleration of 3 m/s². What is the smallest value of the coefficient of friction between the book and the dashboard that prevents the book from sliding?

29   A 5-kg block is attached by a cord to a 7.5-kg block, as shown in Fig. 4.18. The 5-kg block hangs freely, while the 7.5-kg block is supported on a frictionless table. Initially both blocks are at rest, but as the 5-kg block falls, it pulls the 7.5-kg block along with it. Since the blocks are attached, they have the same speed and acceleration at any instant. Write equations similar to Eqs. 4.11 and 4.12 for these masses. (a) Eliminate the tension between these equations to find the acceleration of the masses. (b) What is the tension in the cord?
Ans. (a) 3.92 m/s²; (b) 29.4 N

30   A 2-kg block and a 3-kg block sitting on a frictionless surface are connected by a cord, as shown in Fig. 4.19. A horizontal 20-N force is applied to the 3-kg block. (a) What is the ac-

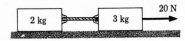

FIGURE 4.19

Problem 30.

celeration of the two-block system? (b) What is the tension in the connecting cord?

31 What is the momentum of a 20-g bullet moving with a speed of 800 m/s?
Ans. 16 kg · m/s

32 What is the total momentum of a system consisting of a 25-kg mass moving 7 m/s to the right and a 35-kg mass moving 3 m/s to the left?

33 A 1200-kg automobile going 30 m/s is involved in a head-on collision that brings it to a full stop in 0.1 s. What is the force exerted on the automobile during the collision?
Ans. 3.6×10^5 N

34 A 50-g golf ball is dropped from a height of 2 m and bounces back to a height of 1.5 m.

(a) What is the change of momentum of the ball? (b) If the ball is in contact with the ground for 0.02 s, what force does the ground exert on it?

35 A 50-kg Saint Bernard jumps from a stationary 40-kg canoe with a horizontal speed of 1.2 m/s. What is the recoil speed of the canoe?
Ans. 1.5 m/s

36 A 40-kg girl running at a speed of 3 m/s jumps onto an 8-kg sled. What is the speed of the girl-sled system immediately after the girl lands on the sled?

37 A 110-kg tackle running east at 3 m/s tackles an 80-kg quarterback running north at 4 m/s. Find the magnitude and direction of the velocity of the players immediately after the tackle.
Ans. 2.4 m/s, 44° north of east

38 A 20-g bullet with an initial speed of 600 m/s passes through a 500-g wooden block, emerging with a speed of 400 m/s. What is the speed of the block after the bullet emerges from it?

BIBLIOGRAPHY

BUTTERFIELD, HERBERT: *The Origins of Modern Science, 1300–1800,* rev. ed., Free Press, New York, 1965. A provocative account of the intellectual struggles that led to the scienctific revolution. The critical issues that confronted men like Copernicus and Galileo are placed in their historical context.

GALILEI, GALILEO: *Dialogue concerning the Two Chief Wold Systems,* trans. by Stillman Drake, rev. ed., University of California Press, Berkeley, 1967. Galileo's masterful account of the physical implications of Ptolemy's geocentric and Copernicus' heliocentric world systems. It is written in the form of a dialogue among three intelligent Venetian gentlemen, who meet to discuss the merits of these two systems. In the course of their witty and far-ranging discussion, most of Galileo's research on motion is expounded.

SHAMOS, MORRIS H.: *Great Experiments in Physics,* Holt, Rinehart and Winston, New York, 1959. A compilation of excerpts from the works of great physicists. Chapter 2 contains passages concerning accelerated motion from Galileo's *Dialogue.* Chapter 4 contains passages concerning the laws of motion from Newton's *Principia.*

M achines have been designed to perform a variety of functions, such as accelerating, lifting, bending, breaking, or abrading something. In spite of their different functions, however, all machines operate on the same basic physical principles. They all take energy in some form and transmit it to the thing on which they act. In the process of transmitting energy, the machine does work. The amount of work a machine can do is limited by the energy available to it, but within this limitation the machine can be made to apply a force or a torque of any magnitude. The principles governing this process are the subject of this chapter.

5

WORK, ENERGY, AND POWER

5.1   Simple Machines

Definition   A *simple machine* is a mechanical device, such as a lever or pulley system, that changes the direction or magnitude of an applied force. For example, when a relatively small downward force **F** is applied to the long end of the lever in Fig. 5.1, a relatively large upward force **F′** is applied to the block sitting on the short end.

It is worthwhile examining the lever in some detail, because it illustrates a number of important mechanical principles. Thus suppose the force **F** is applied at a point 2 m from the pivot, and a block of weight F_g is placed 0.5 m from the pivot. Figure 5.2 shows separately the forces on the lever and the block.

The block is in equilibrium under the actions of the downward force of gravity $\mathbf{F}_g$ and the upward force **F′** exerted on it by the lever. From the condition for equilibrium we get

$$\mathbf{F'} + \mathbf{F}_g = 0 \qquad \text{or} \qquad \mathbf{F'} = -\mathbf{F}_g \qquad\qquad 5.1$$

The reaction to **F′** is the downward force **R′** which the block exerts on the lever. From Newton's third law of motion we have

$$\mathbf{R'} = -\mathbf{F'}$$

while from Eq. 5.1 we have

$$-\mathbf{F'} = \mathbf{F}_g$$

so that

$$\mathbf{R'} = \mathbf{F}_g$$

Although **R′** and $\mathbf{F}_g$ have the same magnitude and direction, they are different forces: $\mathbf{F}_g$ is the force of gravity acting on the block, whereas **R′** is a normal force acting on the lever.

The lever is in equilibrium under the actions of the applied force **F**, the force **R′** exerted by the block, and the force **P** exerted by the pivot. The force **F** exerts the torque

$$\tau_1 = F(2 \text{ m})$$

about the pivot axis O. The force **R′** exerts the torque

$$\tau_2 = -R'(0.5 \text{ m})$$

about O. The force **P** exerts zero torque about O because it

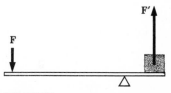

FIGURE 5.1

A relatively small downward force **F** applied to the long arm of a lever results in a relatively large upward force **F′** being applied by the short arm of the lever.

FIGURE 5.2

Free-body diagrams of (a) the lever and (b) the block in Fig. 5.1.

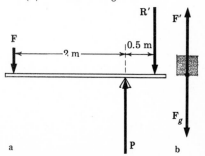

passes through O. Therefore, from the torque condition (*the sum of the torques on a body at rest is zero*) we have

$$0 = \tau_1 + \tau_2 = F(2\text{ m}) - R'(0.5\text{ m})$$

or

$$F = \frac{0.5\text{ m}}{2\text{ m}}R' = \tfrac{1}{4}R'$$

Since **F'** and **R'** are an action-reaction pair, they have the same magnitude ($R' = F'$). Consequently, we have

$$F = \tfrac{1}{4}F' \quad \text{or} \quad F' = 4F \qquad\qquad 5.2$$

That is, the downward force **F** applied to the long end of the lever results in the upward force $\mathbf{F'} = -4\mathbf{F}$ being applied to the block sitting on the short end.

Every simple machine can be thought of as a device which, when a force **F** is exerted on it, itself exerts a force **F'** on some other object. This is shown schematically in Fig. 5.3 as a *black box* whose input is **F** and whose output is **F'**. A class of devices, such as simple machines, is represented by a black box whenever we wish to emphasize the input-output characteristics of the class rather than the mechanism of a particular device. Figure 5.4 shows several particular devices which use the lever as the mechanism for transforming force.

Definition The *actual mechanical advantage* M_A of a simple machine is the ratio of its output force to its input force:

$$M_A = \frac{F'}{F}$$

From Eq. 5.2 the actual mechanical advantage of the lever in Fig. 5.2 is

$$M_A = \frac{F'}{F} = \frac{4F}{F} = 4$$

Example 5.1 What is the mechanical advantage of the crowbar in Fig. 5.4b?

When the crowbar exerts the force **F'** upward on the nail, the nail exerts the force $\mathbf{R'} = -\mathbf{F'}$ downward on the crowbar. Figure 5.5 shows the forces on the crowbar. Taking torques about the pivot axis O, we find:

Torque exerted by **F** $\tau_1 = -F(18\text{ in})$

Torque exerted by **R'** $\tau_2 = R'(0.5\text{ in})$

Torque exerted by **P**   zero

Then from the torque condition we have

$$0 = \tau_1 + \tau_2 = -F(18\text{ in}) + R'(0.5\text{ in})$$

so

$$R' = \frac{18\text{ in}}{0.5\text{ in}}F = 36F$$

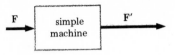

FIGURE 5.3

A simple machine is a device that transforms an input force **F** into an output force **F'**.

FIGURE 5.4

Examples of simple machines based on the lever: (a) bumper jack, (b) crowbar, and (c) pliers.

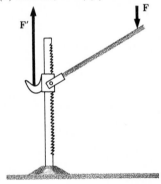

a

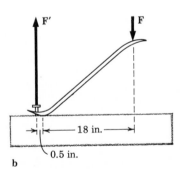

b

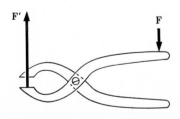

c

Since $R' = F'$, the mechanical advantage is

$$M_A = \frac{F'}{F} = \frac{R'}{F} = \frac{36F}{F} = 36 \qquad \square$$

Consider again the lever in Fig. 5.1, and imagine now that we start to lift the block (Fig. 5.6). This can be done by increasing F a very small amount so that the system is no longer in equilibrium. The increase is to be thought of as so small, however, that the system remains close to equilibrium at all times. Such a state of affairs is called *quasi-equilibrium*. A system in quasi-equilibrium is close enough to true equilibrium that the equilibrium conditions hold to good approximation. That is, in the case of the lever in Fig. 5.6, we can still assume that $F' = 4F$ even though the weight is slowly being lifted. In this way we see that by applying the force F to the long end of the lever, one can lift a block that weighs $4F$.

In spite of our familiarity with the force-multiplying properties of levers from everyday life, this phenomenon never ceases to amaze us. How is it possible to so drastically change the magnitude of a force by such a simple device? What is the trick?

There is no trick, but there is an important tradeoff. If you examine Fig. 5.6 carefully, you will notice that in order to lift the block a distance d', the applied force **F** has to move through the much greater distance d. In fact, the ratio d'/d of these distances is equal to the ratio of the short to the long arm of the lever:

$$\frac{d'}{d} = \frac{0.5\ \text{m}}{2\ \text{m}} = \tfrac{1}{4}$$

Thus we have

$$F' = 4F \qquad \text{and} \qquad d' = \tfrac{1}{4}d$$

so that

$$F'd' = (4F)(\tfrac{1}{4}d) = Fd$$

That is, the product of the force times the distance through which it moves is the same for both forces. We are able to make a large force move through a short distance by moving a small force through a large distance. This result, which we have demonstrated for the lever, is true for any simple machine *in the absence of friction*. A machine without friction is called an *ideal machine*. For such machines we have the following general principle.

The Principle of Simple Machines *In any ideal simple machine, the distance d through which the input force F moves and the distance d' through which the output force F' moves are related by*

$$Fd = F'd' \qquad \text{or} \qquad \frac{F'}{F} = \frac{d}{d'} \qquad \qquad 5.3$$

The *ideal mechanical advantage* M_I (also called the *theoretical mechanical advantage*) is equal to the ratio d/d':

$$M_I = \frac{d}{d'} \qquad \qquad 5.4$$

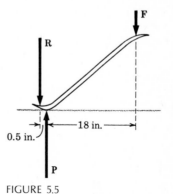

FIGURE 5.5

Free-body diagram of the crowbar in Fig. 5.4b.

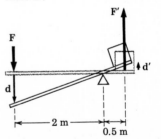

FIGURE 5.6

When the long arm of the lever in Fig. 5.1 is moved down a distance d, the short arm moves up a distance d'.

From Eq. 5.3 we see that for an ideal machine the ideal mechanical advantage d/d' is equal to the actual mechanical advantage:

$$M_A = \frac{F'}{F} \qquad\qquad 5.5$$

When friction is present, the product $F'd'$ is less than Fd. The *efficiency* e of a real machine is defined as the ratio of $F'd'$ to Fd:

$$e = \frac{F'd'}{Fd} \qquad\qquad 5.6a$$

This ratio is 1 for an ideal machine, and it is less than 1 for a real machine with friction. With a little algebraic manipulation we can rewrite Eq. 5.6a in the form

$$e = \frac{F'/F}{d/d'} = \frac{M_A}{M_I} \qquad\qquad 5.6b$$

where we have used Eqs. 5.4 and 5.5. This last equation shows that the efficiency is the ratio of the actual to the ideal mechanical advantage.

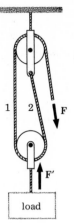

FIGURE 5.7

A pulley and cord arrangement.

Example 5.2 What is the ideal mechanical advantage of the pulley arrangement shown in Fig. 5.7?

When a downward force **F** is applied to the free end of the cord, an upward force **F'** is applied to the load. This system was analyzed in Example 2.4, where we found that (neglecting friction) $F' = 2F$. Thus the ideal mechanical advantage is 2.

To use the principle of simple machines, we note that when the free end of the cord is pulled down a distance d, sections 1 and 2 of the cord are each shortened by a distance d'. Since the total length of the cord is unchanged, we have

$$2d' = d$$

so

$$M_I = \frac{d}{d'} = 2 \qquad\qquad \square$$

FIGURE 5.8

A jackscrew.

Example 5.3 (a) What is the ideal mechanical advantage of the jackscrew in Fig. 5.8? (b) It is found in practice that a 65-lb force must be applied to the handle to lift a 5000-lb load. Find the actual mechanical advantage and the efficiency of the machine.

(a) The lead of the thread is 0.5 in, which means the load is lifted a vertical distance $d' = 0.5$ in when the screw is turned one complete revolution. In the course of this motion the force **F** applied to the handle moves a distance d equal to the circumference of a circle of radius 18 in:

$$d = 2\pi r = 2\pi(18 \text{ in}) = 113 \text{ in}$$

Thus the ideal mechanical advantage is

$$M_I = \frac{d}{d'} = \frac{113 \text{ in}}{0.5 \text{ in}} = 226$$

(b) The actual mechanical advantage is

$$M_A = \frac{F'}{F} = \frac{5,000 \text{ lb}}{65 \text{ lb}} = 77$$

and so the efficiency is

$$e = \frac{M_A}{M_I} = \frac{77}{226} = 0.34$$

The efficiency is much less than 1 in this case because of the large amount of friction between the threads of the screw and the frame.

> **REMARK**   A certain amount of friction is desirable in a jackscrew to prevent the load from running the screw back down when the force applied to the handle is removed.   ☐

THE HYDRAULIC PRESS

A hydraulic press is a device consisting of two cylinders connected by a pipe or flexible tube (Fig. 5.9). Each cylinder is fitted with a piston, and the space inside the cylinders is filled with oil or some other suitable fluid. When a force **F** is applied to the piston of radius r, the piston of radius r' applies the force **F'** to the load. This device finds many practical applications because the input and output cylinders can be connected by a flexible tube, eliminating the need for complicated mechanical linkage.

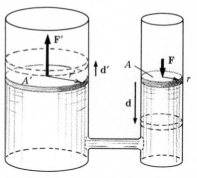

FIGURE 5.9

A hydraulic press. When the piston with cross-sectional area A moves down a distance d, the piston with cross-sectional area A' moves up a distance d'.

No special knowledge of fluid mechanics is needed to understand the hydraulic press, since it obeys the principle of simple machines. This is illustrated in the following example.

Example 5.4   What is the ideal mechanical advantage of the hydraulic press in Fig. 5.9?

The total volume of the fluid in the cylinders does not change as the pistons move. That is, as the small piston moves down a distance d displacing a volume V of oil, the large piston rises the distance d', increasing the volume of the large cylinder by the same amount.

Let A be the cross-sectional area of the small cylinder, and let A' be the cross-sectional area of the large cylinder. Then the volume V of oil displaced from the small cylinder is

$$V = Ad$$

and the volume by which the large cylinder is increased is

$$V = A'd'$$

Since these volumes are equal, we have

$$Ad = A'd'$$

or

$$M_I = \frac{d}{d'} = \frac{A'}{A}$$

Since the areas of the cylinders are $\pi r'^2$ and πr^2, respectively, the ideal mechanical advantage is

$$M_I = \frac{A'}{A} = \frac{\pi r'^2}{\pi r^2} = \frac{r'^2}{r^2} \qquad\qquad 5.7$$

FIGURE 5.10

Two hydraulic jacks lifting one end of a railroad car. (*Western-Cullen Division, Federal Signal Corp.*)

For instance, if the radius of the small cylinder is 0.5 in and the radius of the large cylinder is 3.0 in, the ideal mechanical advantage of the press is

$$M_I = \frac{r'^2}{r^2} = \frac{(3.0 \text{ in})^2}{(0.5 \text{ in})^2} = \frac{9 \text{ in}^2}{0.25 \text{ in}^2} = 36 \qquad \square$$

The *hydraulic jack* is a hydraulic press and pump combination used to lift heavy weights. Figure 5.10 shows two hydraulic jacks that have lifted one end of a railroad car. The ram cylinder

FIGURE 5.11

Construction of a hydraulic jack. Fluid in the reservoir between the inner and outer casings is pumped by the small piston to the underside of the large piston (ram).

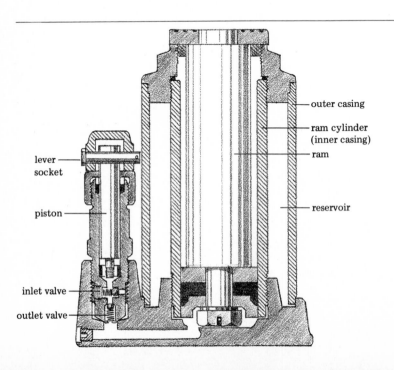

supporting the car is raised by inserting a lever bar (not shown) into one of the sockets on the right and pumping the small cylinder with it.

The construction of a hydraulic jack is shown in Fig. 5.11. The large cylinder (the *ram*) moves inside a double-walled casing. The space between the inner and outer casing is a reservoir filled with oil. The small cylinder (*piston*) is pumped up and down by means of a lever bar inserted into the lever socket (Fig. 5.12). When the piston is raised, oil from the reservoir flows through the inlet valve, and when the piston is lowered, this oil is pumped through the outlet valve to the underside of the ram. Each ball-and-spring valve prevents oil from flowing in the wrong direction.

The ideal mechanical advantage of a hydraulic jack is the product of the mechanical advantages of the pumping lever and the hydraulic press. For example, suppose the lever bar is 36 in long and is pivoted 1.5 in from the piston (Fig. 5.12). The mechanical advantage of the lever is

$$M_I(\text{lever}) = \frac{37.5 \text{ in}}{1.5 \text{ in}} = 25$$

If the diameters of the piston and ram are 0.7 and 2.5 in, respectively, the mechanical advantage of the press is

$$M_I(\text{press}) = \frac{(2.5 \text{ in})^2}{(0.7 \text{ in})^2} = 12.8$$

The overall ideal mechanical advantage of the jack is

$$M_I = M_I(\text{lever})M_I(\text{press}) = (25)(12.8) = 320$$

With this jack a 25-ton (50,000-lb) load can be lifted by applying 220 lb to the lever bar, so the actual mechanical advantage is

$$M_A = \frac{50,000 \text{ lb}}{220 \text{ lb}} = 227$$

FIGURE 5.12

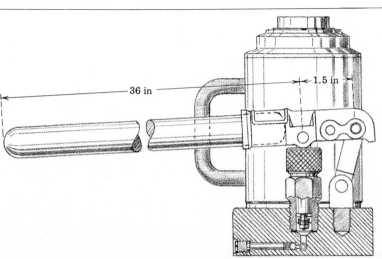

The small piston of a hydraulic jack is moved by a lever bar inserted into the lever socket.

36 in

1.5 in

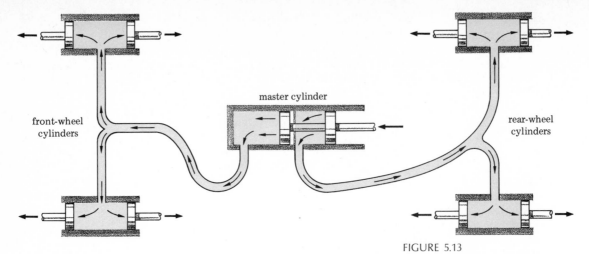

FIGURE 5.13

and the efficiency is

$$e = \frac{M_A}{M_I} = \frac{227}{320} = 0.71$$

The *hydraulic braking system* of an automobile uses the principle of the hydraulic press to transmit the force exerted on the brake pedal to the wheels. Figure 5.13 is a schematic diagram showing how fluid forced out of the master cylinder is used to expand the pistons in the wheel cylinders. In a dual braking system the master cylinder is divided into two compartments which separately work the front and rear brakes. This is a safety feature. If one half of the system should fail, the other half would still be available to stop the car.

Figure 5.14 shows the linkage between the foot pedal and the master cylinder. Note that the foot-pedal assembly is a lever. (What distances determine the mechanical advantage of the assembly?) As the pedal is depressed, fluid is forced from the master cylinder to the wheel cylinders. When the pedal is released, the fluid flows back.

In a *drum brake* (Fig. 5.15a), the fluid entering the wheel

FIGURE 5.14

Linkage between the brake pedal and the master cylinder of an automobile. (*After William H. Crouse, Automotive Mechanics, 7th ed., McGraw-Hill Book Co., New York, 1975.*)

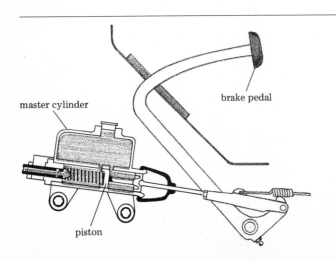

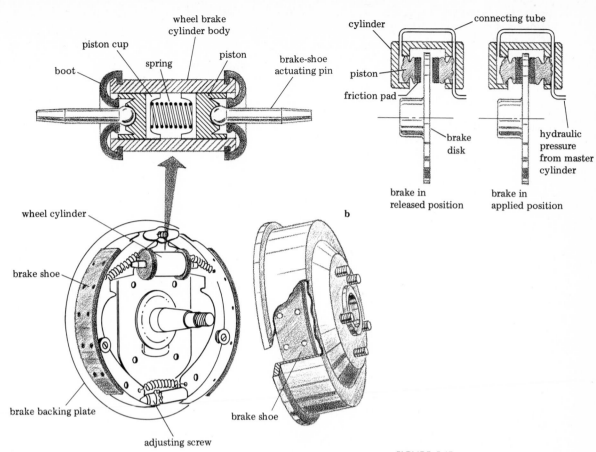

FIGURE 5.15

(a) Drum brake. Fluid from the master cylinder enters the wheel cylinder, forcing apart the two brake-shoe actuating pins. This action forces the brake shoe to press against the rotating brake drum. (b) Disk brake. Fluid entering the wheel cylinder forces the friction pads to press against the rotating disk. (After *William H. Crouse*, Automotive Mechanics, 7th ed., McGraw-Hill Book Co., New York, 1975.)

cylinder forces apart the pistons and actuating pins, causing the stationary brake shoe to press against the rotating brake drum. It is the frictional force between these surfaces that stops the rotating wheel. In a *disk brake* (Fig. 5.15b), the fluid entering the wheel cylinder causes two friction pads to press against opposite sides of a rotating disk. Again, it is the friction between the stationary and rotating surfaces that stops the rotation.

5.2 Work and Potential Energy

Our discussion of simple machines shows the special significance of the quantity Fd. In an ideal simple machine one gets out only as much Fd as one puts in. Thus Fd is something an ideal machine transmits without change, and something a real machine transmits with some loss. This something is called *work*.

Definition The *work* W done by a constant force **F** acting on an object which undergoes the displacement **d** is

$$W = Fd \cos \theta \qquad\qquad 5.8$$

where θ is the angle between **F** and **d** (Fig. 5.16).

FIGURE 5.16

A force **F** does work when it acts on an object undergoing a displacement **d**.

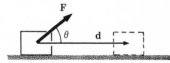

UNITS In SI units, force is in newtons and distance is in meters, so work is in newton-meters (N·m), or *joules* (J):

$$1 J = 1 N \cdot m$$

In the English engineering system, the unit of work is the foot-pound (ft · lb), which equals 1.35 J. Other equivalent units are the kilocalorie (kcal), the kilowatthour (kWh), and the British thermal unit (Btu). See Appendix II for the conversions among these units.

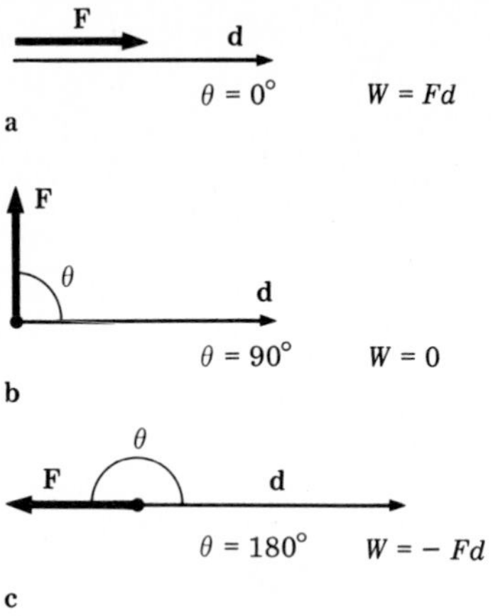

There are three special cases of Eq. 5.8 that should be noted.

1 **F** *is parallel to* **d** (Fig. 5.17a). In this case, $\theta = 0°$, so $\cos \theta = \cos 0° = 1$, and Eq. 5.8 reduces to

$$W = Fd$$

This is the form we used in our discussion of simple machines.

2 **F** *is perpendicular to* **d** (Fig. 5.17b). In this case, $\theta = 90°$, so $\cos \theta = \cos 90° = 0$, and Eq. 5.8 reduces to

$$W = 0$$

3 **F** *is antiparallel to* **d** (Fig. 5.17c). In this case, $\theta = 180°$, so $\cos \theta = \cos 180° = -1$, and Eq. 5.8 reduces to

$$W = -Fd$$

All three cases are illustrated in the following example.

Example 5.5 A block is pulled 3 m along a horizontal surface by a horizontal force of 15 N (Fig. 5.18). The force of friction on the block is 10 N. What is the work done by each force acting on the block?

The four forces acting on the block are shown in Fig. 5.18. The force of gravity F_g and the normal force F_n are each perpendicular to the direction of motion, so the work done by each force is zero:

$$W_g = 0 \quad \text{and} \quad W_n = 0$$

The applied force F_a is parallel to the direction of motion, so it does the work

$$W_a = F_a d = (15 \text{ N})(3 \text{ m}) = 45 \text{ N} \cdot \text{m} = 45 \text{ J}$$

The friction force F_f is antiparallel to the direction of motion, so it does the work

$$W_f = -F_f d = -(10 \text{ N})(3 \text{ m}) = -30 \text{ N} \cdot \text{m} = -30 \text{ J} \qquad \square$$

In the analysis of machinery one is especially concerned with the work done by the applied forces, i.e., by the forces actively exerted on or by the machine. There are other forces acting on all or parts of a machine, and we have to examine the work done by these forces as well.

Normal Force The *normal force* on an object sliding along a surface is the force that the surface exerts perpendicularly to itself. Thus, in this case, the normal force is perpendicular to the motion, so it does no work. However, when one object pushes another object in the direction of motion, the normal force

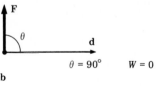

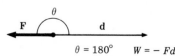

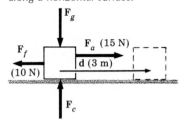

FIGURE 5.17

(a) When a force **F** is parallel to the displacement **d**, the work done is *Fd*. (b) When a force **F** is perpendicular to the displacement **d**, the work done is zero. (c) When a force **F** is antiparallel to the displacement **d**, the work done is *−Fd*.

FIGURE 5.18

The forces on a block moving along a horizontal surface.

which the first object exerts on the second does do work. The normal force in this case is usually referred to as an *applied force*.

A pair of pliers consists of two levers connected by a pin. The forces on one of these levers is shown in Fig. 5.19. As the pliers close to crimp a piece of metal, the applied force **F** does the work

$$W_{in} = Fd$$

and the force **F′** exerted by the jaw on the metal does the work

$$W_{out} = F'd'$$

If there is no friction in the pin, the force **P** of the pin on the lever will be a normal force; i.e., it will be perpendicular to the surface on which it acts. Thus, as the lever rotates around the pin, this force does no work. When there is no friction in the pin, the pliers is an ideal simple machine, so

$$W_{in} = W_{out}$$

In this case, the pliers is simply a device which transmits the work W_{in} done on the handle to the work W_{out} done on the metal.

Friction The frictional force on an object sliding along a surface is the force that the surface exerts parallel to itself. This force is usually antiparallel to the direction of motion, so the work done by a frictional force $\mathbf{F}_f$ acting on an object which moves a distance s is

$$W_f = -F_f s$$

Suppose that as the handle of the pliers in Fig. 5.19 moves the distance d, the lever slides a distance s around the pin. If there is a frictional force $\mathbf{F}_f$ exerted by the pin on the lever, the work done by the jaw on the metal will be

$$W_{out} = W_{in} + W_f = W_{in} - F_f s$$

From Eq. 5.6a we see that the efficiency of the device is thus

$$e = \frac{F'd'}{Fd} = \frac{W_{out}}{W_{in}} = \frac{W_{in} - F_f s}{W_{in}} = 1 - \frac{F_f s}{W_{in}}$$

which is less than 1. In the process of transmitting work to the jaws of the pliers, some of it was lost to friction. This is the meaning of the minus sign in the equation for the work done by friction. We shall see what becomes of this lost work in Sec. 5.4.

FIGURE 5.19

Forces on one arm of a pair of pliers.

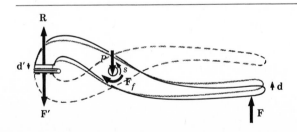

Gravity When the mass m in Fig. 5.20 is lowered vertically from point A to point C, the work done on it by the force of gravity $\mathbf{F}_g$ is

$$W_{AC} = F_g h = mgh$$

The sign is positive because $\mathbf{F}_g$ is directed from A to C; i.e., $\mathbf{F}_g$ is parallel to the displacement $\mathbf{h}$. The work done by gravity in moving the object horizontally from C to B is zero since $\mathbf{F}_g$ is perpendicular to the horizontal displacement. Thus

$$W_{CB} = 0$$

so the total work done on the object by gravity as it goes from A to C, and then from C to B, is

$$W_{ACB} = W_{AC} + W_{CB} = mgh + 0 = mgh$$

The object can also go from A to B by moving diagonally along the incline. According to Eq. 5.8, the work W_{AB} done by gravity in going along this path is

$$W_{AB} = mg \cos \theta \, d$$

where θ is the angle between $\mathbf{F}_g$ and the displacement $\mathbf{d}$. But from the right triangle ACB we have

$$\cos \theta = \frac{h}{d}$$

and so

$$W_{AB} = mg \frac{h}{d} d = mgh = W_{ACB}$$

Thus the work done by gravity on the object in going from A to B is the same whether it moves along the path ACB or along the diagonal path AB. In fact, it can be shown that the work done by gravity on the object in going from A to B is the same no matter what path is taken. The work is always mgh, where h is the separation in the vertical heights of A and B.

Definition A force which does the same work W_{AB} on an object going from A to B regardless of the path taken is called a *conservative force*. Gravitational and electric forces are examples of conservative forces.

Because the work W_{AB} done by a conservative force is independent of the path taken, W_{AB} must depend only on the positions of the points A and B themselves. Let h_A be the height of point A above some reference surface, and let h_B be the height of point B from this same surface (Fig. 5.21). Then the work done by gravity on an object going from A to B along any path can be written

$$W_{AB} = mgh_A - mgh_B = mg(h_A - h_B) = mgh \qquad 5.9$$

This expresses the work as the difference of two quantities (mgh_A and mgh_B) that depend only on the positions of the points A and B, respectively.

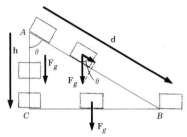

FIGURE 5.20

Comparison of the work W_{AB} done by the force of gravity on a mass moving from A to B along the incline with the work W_{ACB} done on the mass moving from A to B along the path ABC.

FIGURE 5.21

The work done by the force of gravity on a mass moving from A to B depends only on the difference h between the heights of the points and not on the path taken.

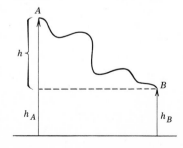

Definition For any conservative force it is possible to define at every point a quantity U, called the *potential energy*, such that the work done by the force in moving an object from A to B along any path is

$$W_{AB} = U_A - U_B \qquad\qquad 5.10$$

where U_A and U_B are the values of U at A and B. From Eq. 5.9 we see that the *potential energy for the gravitational force* is

$$U_A = mgh_A \qquad\qquad 5.11$$

where h_A is the vertical height of A above a reference surface.

> **REMARKS** (1) Since the reference surface is arbitrary, the absolute value of the potential energy at a point is arbitrary. If the reference surface is changed, the potential energy at all points will change by the same amount. This is not important, however, since in practice it is only the difference between potential energies at two points that is of interest. This difference is independent of the position of the reference surface. (2) Equation 5.11 is an approximation valid only for values of h small enough to permit the variation of the gravitational force with distance from the earth to be neglected. The expression for gravitational potential energy that is valid for all distances is given in Sec. 6.3.

The total work W_{total} done on an object is the sum of the works done by each force acting on the object. This can be written

$$W_{\text{total}} = W_a + W_f + W_g$$

where W_a is the work done by all the applied forces, W_f is the work done by all frictional forces, and W_g is the work done by gravity. If there is a single frictional force $\mathbf{F}_f$ which acts through the distance s, then

$$W_f = -F_f s$$

If the object moves from point A to point B, then

$$W_g = U_A - U_B$$

Thus the total work done on the object can be written

$$W_{\text{total}} = W_a - F_f s + (U_A - U_B) \qquad\qquad 5.12$$

Since the total force on an object in quasi-equilibrium is zero, the total work will be zero also. Therefore, for an object in equilibrium or quasi-equilibrium we have

$$W_{\text{total}} = W_a - F_f s + (U_A - U_B) = 0$$

or

$$W_a = F_f s + (U_B - U_A) \qquad \text{quasi-equilibrium} \qquad 5.13$$

Note that the term $(U_A - U_B)$ became $-(U_A - U_B) = (U_B - U_A)$ when it was transferred to the right-hand side of the equation.

> **REMARK** Equation 5.13 holds as long as there is no appreciable acceleration of the system and no deformation of the object. The modification of this equation required to treat accelerated motion is given in Sec. 5.3. The gener-

alization required to treat cases of deformation, such as the crimping of the metal in Fig. 5.19, is given in Sec. 5.4.

Example 5.6 A 200-kg block sits 0.4 m from the pivot of the lever in Fig. 5.22. (a) If the block is raised 0.05 m, what is its change of potential energy? (b) What is the applied work done on the block? (c) What is the work done by the force **F** which is applied at a point 1.8 m from the pivot? (d) What is the magnitude of **F**?

(a) If A is the position of the block before it is raised, and B is its position after it is raised, we have

$$h_A = 0 \quad \text{and} \quad h_B = 0.05 \text{ m}$$

so

$$U_A = mgh_A = 0$$

and

$$U_B = mgh_B = (200 \text{ kg})(9.8 \text{ m/s}^2)(0.05 \text{ m})$$
$$= 98 \text{ N} \cdot \text{m} = 98 \text{ J}$$

The change of potential energy is

$$\Delta U = U_B - U_A = 98 \text{ J} - 0 = 98 \text{ J}$$

(b) The only forces on the block itself are the applied force **F'** and the force of gravity (Fig. 5.22b). Therefore, applying Eq. 5.13 to the block alone, we find that the work W_a' done by the applied force **F'** is

$$W_a' = U_B - U_A = 98 \text{ J}$$

This is the same as the work W_{out} done by the machine on the block.

(c) Since there is no friction in the lever, the work W_{in} done by **F** is

$$W_{in} = W_{out} = W_a' = 98 \text{ J}$$

(d) The ratio d/d' of the distances the two ends of the lever move is equal to the ratio of the two arms of the lever:

$$\frac{d}{d'} = \frac{1.8 \text{ m}}{0.4 \text{ m}} = 4.5$$

Consequently, we have

$$d = 4.5 \, d' = (4.5)(0.05 \text{ m}) = 0.225 \text{ m}$$

FIGURE 5.22

(a) A lever raising a block.
(b) Free-body diagram of the block.

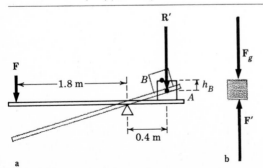

and so

$$F = \frac{W_{in}}{d} = \frac{98\,J}{0.225\,m} = 436\,N$$

Since we are assuming a weightless, frictionless lever, the only forces that do work on the lever itself are the applied force **F** exerted on the long arm, and the applied force **R′** that the block exerts on the short arm (the reaction to **F′**). The force **F** does the work Fd and the force $\mathbf{R'} = -\mathbf{F'}$ does the work

$$-R'd' = -F'd'$$

Consequently the total applied work done on the lever itself is

$$W_a = Fd - F'd' = 0$$

This is in agreement with Eq. 5.13, because for a weightless, frictionless lever, both F_f and $U_B - U_A$ are zero.

The lever is therefore just a device that transmits the work W_{in} to the block. The work W_{out} done on the block then goes to increase the block's potential energy. Work is a transient thing, which exists, so to speak, only while the lever is moving. Potential energy, on the other hand, exists as long as the block is in the elevated position. ☐

Example 5.7 How much applied work is required to push a 100-kg crate up the inclined plane shown in Fig. 5.23 if there is a 400-N force of friction between the crate and the plane?

The work done by the frictional force is

$$W_f = -F_f s = -(400\,N)(3\,m) = -1200\,J$$

and the change of the crate's potential energy is

$$\begin{aligned}U_B - U_A &= mgh_B - mgh_A \\ &= (100\,kg)(9.8\,m/s^2)(0.6\,m) - 0 \\ &= 588\,J\end{aligned}$$

Thus from Eq. 5.13 we have

$$\begin{aligned}W_a &= F_f s + (U_B - U_A) \\ &= 1200\,J + 588\,J = 1788\,J\end{aligned}$$

Since the applied force **F** moves the distance $d = 3\,m$, its magnitude must be

$$F = \frac{W_a}{d} = \frac{1788\,J}{3\,m} = 596\,N$$

FIGURE 5.23

A crate being moved up an incline.

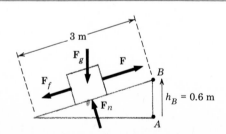

This force, in effect, lifts a block which weighs

$$F_g = mg = (100 \text{ kg})(9.8 \text{ m/s}^2) = 980 \text{ N}$$

so the inclined plane may be said to be a simple machine with the actual mechanical advantage of

$$M_A = \frac{F_g}{F} = \frac{980 \text{ N}}{596 \text{ N}} = 1.64$$

Its ideal mechanical advantage is the ratio of the distance d which **F** moves to the distance h which $\mathbf{F}_g$ moves:

$$M_I = \frac{d}{h} = \frac{3 \text{ m}}{0.6 \text{ m}} = 5$$

Thus the efficiency is only

$$e = \frac{M_A}{M_I} = \frac{1.64}{5} = 0.328$$

The efficiency is also equal to the fraction of the applied work that goes into potential energy:

$$e = \frac{U_B - U_A}{W_a} = \frac{588 \text{ J}}{1788 \text{ J}} = 0.328$$

The efficiency would be 1 if there were no friction. It is the friction which prevents all the applied work from going into potential energy. □

5.3 Kinetic Energy

Our discussion so far has been limited to equilibrium and quasi-equilibrium conditions in which the sum of the forces, and consequently the sum of the works done by these forces, is zero. When the sum of the forces acting on an object is not zero, neither is the total work. The effect of this work (when there is no deformation) is given by the following theorem:

Work-Energy Theorem *The total work W_{total} done on an object moving from some initial position A to some final position B is equal to the change in the kinetic energy of the object*

$$W_{\text{total}} = K_B - K_A \qquad\qquad 5.14$$

where, by definition, the kinetic energy K of an object of mass m moving with speed v is

$$K = \tfrac{1}{2}mv^2 \qquad\qquad 5.15$$

REMARK It is instructive to prove this theorem for the special case in which a constant force **F** acts on an object of mass m in the direction of the object's displacement **d**. In this case, the work done is

$$W_{\text{total}} = Fd$$

and the object has the constant acceleration

$$a = \frac{F}{m} = \frac{W_{\text{total}}}{md} \qquad\qquad 5.16$$

From Eq. 4.3 the initial and final speeds v_A and v_B are related to a by

$$v_B = v_A + at \qquad\qquad 5.17$$

while from Eq. 4.5 the distance d is related to v_A and a by

$$d = v_A t + \tfrac{1}{2}at^2 \qquad\qquad 5.18$$

We eliminate t between Eqs. 5.17 and 5.18 by solving Eq. 5.17 for t:

$$t = \frac{v_B - v_A}{a}$$

and substituting the result in Eq. 5.18:

$$d = v_A \frac{v_B - v_A}{a} + \tfrac{1}{2}a\left(\frac{v_B - v_A}{a}\right)^2 = \frac{(v_A v_B - v_A{}^2) + \tfrac{1}{2}(v_B - v_A)^2}{a}$$

$$= \frac{v_A v_B - v_A{}^2 + \tfrac{1}{2}v_B{}^2 - v_A v_B + \tfrac{1}{2}v_A{}^2}{a} = \frac{1}{a}(\tfrac{1}{2}v_B{}^2 - \tfrac{1}{2}v_A{}^2)$$

or

$$ad = \tfrac{1}{2}v_B{}^2 - \tfrac{1}{2}v_A{}^2$$

But from Eq. 5.16, $ad = W_{\text{total}}/m$, so

$$W_{\text{total}} = \tfrac{1}{2}mv_B{}^2 - \tfrac{1}{2}mv_A{}^2 = K_B - K_A$$

This proves the work-energy theorem for the special case considered. However, the theorem is true in general, even when $\mathbf{F}$ is not constant or not parallel to the displacement of the object.

UNITS The SI units of kinetic energy are

$$\text{kg} \cdot (\text{m/s})^2 = \text{kg} \cdot \text{m}^2/\text{s}^2 = \text{N} \cdot \text{m} = J$$

(Remember that $1\,\text{N} = 1\,\text{kg} \cdot \text{m/s}^2$.) Thus the units of work and kinetic energy are the same, which in fact they must be according to Eq. 5.14.

When we combine Eq. 5.12 for total work with Eq. 5.14 we get

$$W_a + W_f + (U_A - U_B) = K_B - K_A$$

or

$$W_a = -W_f + (U_B - U_A) + (K_B - K_A)$$

Although this looks rather awesome, it is easy to apply in most cases. But before giving some examples, let us simplify the notation a bit by introducing the symbol Δ to mean "change of":

$$\Delta U = U_B - U_A$$
$$\Delta K = K_B - K_A$$

Thus we have

$$W_a = -W_f + \Delta U + \Delta K \qquad\qquad 5.19$$

which says that the applied work on an object goes to change the kinetic and potential energies of the object and to overcome friction.

REMARK By definition Δ means the final value of something minus its initial value. Either ΔK or ΔU will be negative if the final value of K or U is less than its initial value.

Example 5.8 A 2000-kg automobile starting from rest accelerates to 25 m/s. During the acceleration period the automobile travels

200 m and has a frictional force of 1200 N acting on it. How much applied work is done on the automobile? What is the magnitude of the applied force?

The applied force **F** on an automobile is actually applied by the road to the rear tires (Fig. 5.24). This is technically a frictional force, but as long as the tire rolls without slipping, it does positive work. This work is delivered to the tires from the engine by means of the transmission, which is a not-so-simple simple machine (see Fig. 5.30).

The 1200-N frictional force F_f is a combination of wind resistance and rolling friction. It does the work

$$W_f = -F_f s = -(1200 \text{ N})(200 \text{ m}) = -2.4 \times 10^5 \text{ J}$$

Since the automobile is moving horizontally

$$\Delta U = 0$$

And since it starts from rest

$$v_A = 0$$

so

$$\Delta K = K_B - K_A = \tfrac{1}{2}mv_B^2 - 0$$
$$= \tfrac{1}{2}(2000 \text{ kg})(25 \text{ m/s})^2 = 6.25 \times 10^5 \text{ J}$$

Therefore, from Eq. 5.19, the work done by the applied force is

$$W_a = -W_f + \Delta U + \Delta K$$
$$= -(-2.4 \times 10^5 \text{ J}) + 0 + 6.25 \times 10^5 \text{ J}$$
$$= 8.65 \times 10^5 \text{ J}$$

and the magnitude of the applied force is

$$F = \frac{W_a}{d} = \frac{8.65 \times 10^5 \text{ J}}{200 \text{ m}} = 4325 \text{ N}$$

The work done by the automobile engine equals W_a plus the work lost to friction in the transmission. Thus the engine does work to increase the kinetic energy of the automobile and to overcome internal and external frictional forces. Once the automobile is moving at constant speed, the engine just has to overcome friction. ☐

Example 5.9 A pile hammer drops a 400-kg iron hammer from a height of 8 m onto the top of a pole (pile) being driven into the ground (Fig. 5.25). (a) What is the kinetic energy of the hammer just before it strikes the pole? (b) What is the speed of the hammer just before it strikes the pole? (c) How much work does the hammer do on the pole? (d) If the pole is driven 5 cm into the ground, what force did the hammer exert on the pole?

This problem illustrates the versatility of Eq. 5.19 because it has to be applied separately to the hammer while it is in free fall and to the hammer while it is impacting the pole.

(a) During free fall there are no frictional or applied forces on the hammer, so Eq. 5.19 reduces to

$$0 = \Delta U + \Delta K = (U_B - U_A) + (K_B - K_A)$$

FIGURE 5.24

The applied force **F** on an automobile is exerted by the road on the wheels.

FIGURE 5.25

A pile hammer.

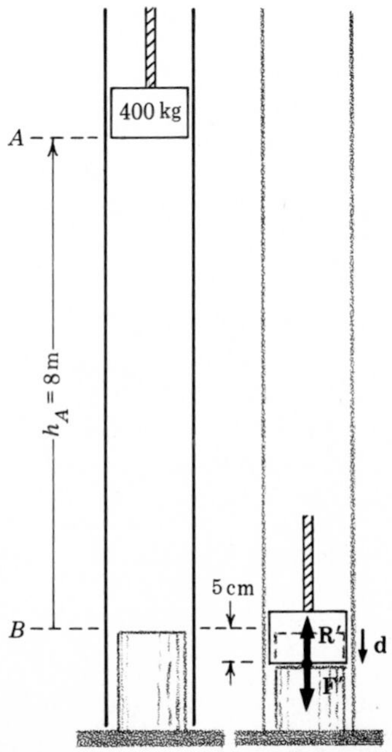

or

$$U_B + K_B = U_A + K_A \qquad\qquad 5.20$$

This equation, although only a special case of Eq. 5.19, is important enough to have its own number. It says that when gravity is the only force doing work on an object, the sum of the kinetic and potential energies of the object is constant. That is, while U and K may separately change as the object moves from point to point, $U + K$ has the same value at any two points.

REMARK Equation 5.20 applies also to an object, such as an ideal roller coaster, that moves along a curved frictionless surface, because the normal force exerted by the surface does no work.

In the present case, if A is the upper position of the hammer and B is its position just before it strikes the pole, we have

$$h_B = 0 \qquad \text{and} \qquad v_A = 0$$

so

$$U_B = mgh_B = 0 \qquad \text{and} \qquad K_A = \tfrac{1}{2}mv_A{}^2 = 0$$

and Eq. 5.20 reduces to

$$0 + K_B = U_A + 0$$

or

$$K_B = mgh_A = (400\text{ kg})(9.8\text{ m/s}^2)(8\text{ m}) = 3.14 \times 10^4\text{ J}$$

(b) Since

$$K_B = \tfrac{1}{2}mv_B{}^2$$

the speed v_B of the hammer just before impact is

$$v_B = \sqrt{\frac{2K_B}{m}} = \sqrt{\frac{2(3.14 \times 10^4\text{ J})}{400\text{ kg}}} = 12.5\text{ m/s}$$

(c) To find how much work the hammer does on the pole, we have to apply Eq. 5.19 to the hammer while it is impacting the pole. When the hammer first comes into contact with the pole, its kinetic energy is 3.14×10^4 J,. and when it comes to rest on top of the pole, its kinetic energy is zero. Thus the change of the hammer's kinetic energy is

$$\Delta K = 0 - 3.14 \times 10^4\text{ J} = -3.14 \times 10^4\text{ J}$$

During impact the hammer drops an additional 5 cm = 0.05 m, so its change of potential energy is

$$\Delta U = 0 - mgh = -(400\text{ kg})(9.8\text{ m/s}^2)(0.05\text{ m})$$
$$= -2.0 \times 10^2\text{ J}$$

The frictional work done on the hammer is negligible, so from Eq. 5.19 the applied work done on the hammer is

$$W_a = \Delta U + \Delta K = -3.14 \times 10^4\text{ J} - 0.020 \times 10^4\text{ J}$$
$$= -3.16 \times 10^4\text{ J}$$

The negative sign indicates that the applied work is decreasing the kinetic and potential energies of the hammer. We also can see from Fig. 5.25 that W_a must be negative, because the force $\mathbf{R}'$ which the pole exerts on the hammer is antiparallel to the direction of the hammer's motion.

The reaction to $\mathbf{R}'$ is the force $\mathbf{F}'$ which the hammer exerts on the pole. It has the same magnitude as $\mathbf{R}'$ and moves through the same distance, but it is parallel to its direction of motion. Consequently, the work W_a' done on the pole is

$$W_a' = -W_a = 3.16 \times 10^4 \text{ J}$$

(d) Since the force $\mathbf{F}'$ moves through the distance $d = 0.05$ m, its magnitude is

$$F' = \frac{W_a'}{d} = \frac{3.16 \times 10^4 \text{ J}}{0.05 \text{ m}} = 6.3 \times 10^5 \text{ N} = 1.4 \times 10^5 \text{ lb}$$

The pile hammer thus develops a momentary force of 140,000 lb while driving down the pole. But since the magnitude of this force depends entirely on how much the pole moves in a single blow, the more significant quantity is the work done on the pole.

In this example we see how the potential energy of the hammer is transformed into work done on the pole, whereas in Example 5.8 we saw how the work done by an automobile's engine is transformed into kinetic energy. The back-and-forth transformation of work and energy is a fundamental feature of all machinery. We shall continue our study of these transformations in the next two sections. ☐

5.4 Conservation of Energy

We see from Eq. 5.19 that the applied work not lost to friction comes from or goes into potential and kinetic energy. We now want to consider what happens to the work lost to friction, and where the applied work comes from in the first place.

When you rub the palms of your hands together vigorously, you notice a definite warming sensation. Similar warming is noticed during sanding, drilling, and milling operations. In all these cases, the frictional work results in an increase in the temperature of the surfaces being acted upon.

Temperature is a manifestation of the motions of the atoms and molecules which compose all matter. The atoms and molecules of an object are in constant motion, so each particle has a kinetic energy. The sum of these energies is called the *thermal energy* I_{th} of the object. In addition to their kinetic energy, the particles also have electric potential energy that results from the atomic forces that hold the molecules together. The sum of these energies is called the *chemical energy* I_{ch}. The sum of the chemical and thermal energies of an object is called its *internal energy* I:

$$I = I_{\text{th}} + I_{\text{ch}} \qquad\qquad 5.21$$

A rise in the temperature of an object is associated with an increase in the object's thermal energy, and thus with an increase in internal energy. The question then is: How much of the frictional work done on an object goes into internal energy? The answer is simple: *All of it.* That is, all the work lost to friction can be accounted for by an increase in internal energy:

$$-W_f = F_f s = \Delta I \qquad\qquad 5.22$$

REMARK Although Eq. 5.22 is easily stated, it was not so easily discovered. It took many scientists many decades to accumulate the experimental evidence to support this result.

When Eq. 5.22 is substituted into Eq. 5.19, the result is

$$W_a = \Delta I + \Delta U + \Delta K \qquad\qquad 5.23$$

All the applied work done on a system goes into one or another form of energy. Let E be the total energy of the system:

$$E = I + U + K$$

Then Eq. 5.23 can be written

$$\begin{aligned} W_a &= (I_B - I_A) + (U_B - U_A) + (K_B - K_A) \\ &= (I_B + U_B + K_B) - (I_A + U_A + K_A) \\ &= E_B - E_A = \Delta E \qquad\qquad 5.24 \end{aligned}$$

If no applied work is done on the system, then $\Delta E = 0$. That is, if the system is isolated from applied forces, its total energy does not change, though energy of one form may be transformed into another form. This is a consequence of the following fundamental law of physics.

Conservation of Energy *Energy can neither be created nor destroyed, but only transformed from one form to another.* In an isolated system, in which energy neither enters nor leaves, the total energy is constant. Work is a means by which energy is transferred into or out of a system. (Heat is another such means.) Thus work (and heat) can be thought of as *energy in transit.*

REMARK According to the conservation of energy, energy is not lost when work is transformed into internal energy, although in engineering practice the work is said to be lost. The reason for this is that internal energy, unlike potential and kinetic energy, cannot be readily transformed back into work. Thus there is a loss, not of energy, but of energy that can be transformed into work. From a practical point of view, internal energy is not as useful as kinetic and potential energy. The usefulness of energy is the subject of thermodynamics (Chap. 10).

We know that work must be done to deform an object, i.e., to break, bend, or abrade it. Since there is no change of kinetic or potential energy in the process, Eq. 5.23 implies that the work done goes to increase the internal energy of the object. The purpose of every machine is to apply work to change the kinetic, potential, or internal energy of something, i.e., to accelerate, lift, or deform something. Where does this work come from?

According to Eq. 5.24, the applied work W_a must be produced by a change of energy. A device which converts internal energy into work is called an *engine.* Common examples are batteries, automobile engines, steam turbines, and animal muscles. The details of the operation of an engine do not concern us here, only the general principle, which is summarized by the equation

$$-\Delta I_{fuel} = W_a + \Delta I_{env} \qquad 5.25$$

where ΔI_{fuel} is the change in the internal energy of the fuel, and ΔI_{env} is the change in the internal energy of the environment. The quantity ΔI_{fuel} is negative because the fuel has less internal energy after it is burned than before; consequently, $-\Delta I_{fuel}$ is a positive quantity. Equation 5.25 says, therefore, that for a given amount of internal energy released when fuel is burned, a certain amount of applied work W_a is generated and the differ- ence ΔI_{env} is wasted, increasing the internal energy of the envi- ronment. The wasted energy ΔI_{env} produced by an automobile engine, for example, appears in the form of high-temperature exhaust gas and the thermal energy that is continuously re- moved from the engine by the cooling system.

The *efficiency* e of an engine is the ratio of the applied work produced to the internal energy used to produce it

$$e = \frac{W_a}{-\Delta I_{fuel}} \qquad 5.26$$

Efficiency is often quoted as a percentage. For engines that use chemical energy, e.g., gasoline engines and animal muscles, the efficiency is only about 25 percent; i.e., for every 100 J of internal energy used, only 25 J of applied work is produced. The remain- ing 75 J is released into the environment in the form of heat. The adequate removal of this waste heat is a problem for any engine, whether it is an automobile engine, the steam turbine of an electric generating station, or a human athlete.

Example 5.10 A steam turbine generates 3.75×10^8 J of work per second to run an electric generator. The efficiency of the turbine is 0.35. How much energy (per second) is needed to run the turbine, and how much energy (per second) is transferred to the environment?

From Eq. 5.26 the fuel energy is

$$-\Delta I_{fuel} = \frac{W_a}{e} = \frac{3.75 \times 10^8 \text{ J}}{0.35} = 10.7 \times 10^8 \text{ J}$$

Therefore, from Eq. 5.25, the energy transferred to the environ- ment is

$$\Delta I_{env} = -\Delta I_{fuel} - W_a$$
$$= 10.7 \times 10^8 \text{ J} - 3.75 \times 10^8 \text{ J}$$
$$= 7.0 \times 10^8 \text{ J}$$

Sixty-five percent of the fuel energy is unavailable for work, and is thus wasted. To remove his enormous quantity of energy, power plants require extensive cooling systems which have considerable environmental impact. Figure 5.26 shows part of the spray cooling system used at the Brayton Point Power Station in Massachusetts. □

We can get an overall view of the energy flow between an

FIGURE 5.26

Spray cooling system at the Brayton Point Power Station in Massachusetts. This system cools the water used to remove waste energy from the condenser of the steam turbine. (*New England Electric System.*)

engine and a working system by using Eq. 5.24 to replace W_a in Eq. 5.25 with the energy change ΔE of the system on which the engine does work:

$$-\Delta I_{\text{fuel}} = \Delta E + \Delta I_{\text{env}} \quad \text{or} \quad \Delta E + \Delta I_{\text{fuel}} + \Delta I_{\text{env}} = 0$$

This states that the total energy of the environment, the fuel, and the system on which the engine works is conserved. Figure 5.27 shows schematically that the component systems taken together form a single closed system in which none of the energies is destroyed but only transformed from one component to another. The applied work W_a is the means by which energy is transferred from the engine to the system.

5.5 Power

Definition The *power P* of an engine is the work per second done by the engine. Thus if the engine does the work W_a in time t, the power is

$$P = \frac{W_a}{t} \qquad\qquad 5.27$$

FIGURE 5.27

Relation between the energy changes of an engine's fuel, the system on which the engine does work, and the environment.

UNITS Since the SI units of work and time are the joule (J) and the second (s), the SI unit of power is a joule per second (J/s), or watt (W):

$$1 \text{ W} = 1 \text{ J/s}$$

Other units of power are the foot-pound per second (ft·lb/s) and the horsepower (hp). See Appendix II for the conversions among these units.

The total rate R at which fuel energy is used by an engine is

$$R = \frac{-\Delta I_{\text{fuel}}}{t} \qquad\qquad 5.28$$

which from Eqs. 5.26 and 5.27 is

$$R = \frac{W_a/e}{t} = \frac{P}{e} \qquad\qquad 5.29$$

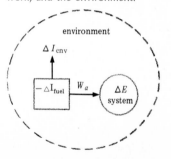

An electric generator produces electric power nearly equal to the rate at which work is done on it. That is, a generator can convert work into electric energy with an efficiency close to 1. The electricity is not a new source of energy, but only a convenient means by which energy transformed at the central generating plant can be distributed to customers. The primary energy source is usually fossil fuel (coal, gas, or oil), nuclear fuel, or potential energy in the form of elevated water (hydroelectric power). Potential energy can be transformed into work with an efficiency close to 1, but fossil and nuclear energy can be transformed into work with an efficiency of only 0.3 to 0.4.

Example 5.11 An oil-burning power plant generates 250 MW of power with an efficiency of 0.38. (a) What is the rate at which fuel energy is used? (b) If 1 gal of oil releases 1.4×10^5 Btu when burned, how many gallons of oil per hour are burned in the plant?

(a) From Eq. 5.29 the rate of fuel-energy consumption is

$$R = \frac{P}{e} = \frac{250 \text{ MW}}{0.38} = 658 \text{ MW}$$

(b) From, Eq. 5.28 the fuel energy $-\Delta I_{\text{fuel}}$ used in 1 h is

$$-\Delta I_{\text{fuel}} = Rt = (658 \times 10^6 \text{ J/s})(3600 \text{ s})$$
$$= 2.37 \times 10^{12} \text{ J}$$

The energy content c per gallon of fuel is

$$c = 1.4 \times 10^5 \text{ Btu/gal}$$
$$= (1.4 \times 10^5)(1055 \text{ J/gal})$$
$$= 1.48 \times 10^8 \text{ J/gal}$$

so in 1 h the power plant burns

$$\frac{-\Delta I_{\text{fuel}}}{c} = \frac{2.37 \times 10^{12} \text{ J}}{1.48 \times 10^8 \text{ J/gal}} = 1.6 \times 10^4 \text{ gal}$$

(Appendix II was used to find the conversion between joules and Btus.) ☐

Figure 5.28 shows an engine applying a constant force $\mathbf{F}$ to a block. There is a frictional force $\mathbf{F}_f$ equal to $-\mathbf{F}$ acting on the block, so the block slides at constant speed v. In time t the block moves the distance

$$d = vt$$

so the work done by $\mathbf{F}$ is

$$W_a = Fd = Fvt$$

and the power is

$$P = \frac{W_a}{t} = Fv \qquad\qquad 5.30$$

That is, the power of an engine is equal to the force it exerts times the speed with which the force is moved. For a given

FIGURE 5.28

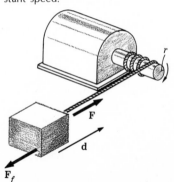

An engine applying a constant force to a block moving at constant speed.

power rating, an engine can move a large force slowly or a small force quickly. Gears are often used to change the speed and force of an engine.

Instead of writing power in terms of force and linear speed, engineers write it in terms of torque τ and rotational speed. In one revolution of the engine shaft, the cord in Fig. 5.28 moves the distance $2\pi r$ equal to the circumference of the shaft. If the shaft makes n revolutions in time t, the cord moves the distance

$$d = n2\pi r$$

at the speed

$$v = \frac{d}{t} = \frac{n2\pi r}{t} = 2\pi r\omega$$

where* $\omega = n/t$ and is the rotational speed of the shaft in revolutions per second. The power of the engine can thus be written

$$P = Fv = F(2\pi r)\omega$$
$$= 2\pi\tau\omega \qquad\qquad 5.31$$

where $\tau = Fr$ and is the torque exerted by the engine. Equation 5.31 is equivalent to Eq. 5.30. It shows that for a given power the engine can exert a large torque at small rotational speed or a small torque at large speed.

A simple machine can be thought of as a device which transmits power. In an ideal simple machine there is no power lost in the process, so the input power Fv is equal to the output power $F'v'$:

$$Fv = F'v' \qquad\qquad 5.32$$

In terms of torque, the equivalent equation is

$$2\pi\tau\omega = 2\pi\tau'\omega'$$

or

$$\tau\omega = \tau'\omega' \qquad\qquad 5.33$$

Recall that for an ideal simple machine, the principle of simple machines (Eq. 5.3) is

$$Fd = F'd'$$

Equation 5.32 follows at once from this by dividing by the time t in which the distances d and d' were traveled.

Figure 5.29 shows a simplified picture of the planetary-gear system used in automatic transmissions to change the torque and speed applied to the wheels of an automobile. The sun-gear shaft is driven by the engine and the planet-pinion shaft carries the power to the rear wheels. In low gear the ring, or internal gear,† is held stationary so that as the sun gear rotates, the planetary gears rotate around the ring gear. If the radius of the ring gear is twice the radius of the sun gear, the sun gear must make two revolutions to rotate the planetary gears once around

FIGURE 5.29

Planetary-gear system. The sun-gear shaft is connected to the engine and the planet-pinion shaft is connected to the drive wheels. (*After William H. Crouse,* Automotive Mechanics, *7th ed., McGraw-Hill Book Co., New York, 1975.*)

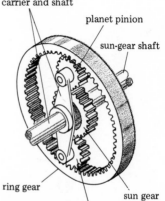

planet-pinion
carrier and shaft

planet pinion

sun-gear shaft

ring gear

sun gear

planet pinion

*ω is the Greek (lowercase) letter omega.

†It is called an internal gear because the teeth are on the inside.

the ring gear. Therefore the planetary-gear shaft will rotate at half the rotational speed of the sun-gear shaft. If no power is lost to friction, Eq. 5.33 shows that the torque delivered by the planetary-gear shaft will be twice the torque of the sun-gear shaft.

This increased torque is required while the automobile is accelerating. Once the automobile reaches cruising speed, a hydraulic control system locks the planetary gears to the ring gear. This locks the entire assembly, so that the two shafts rotate together. This is direct drive.

Figure 5.30 shows where the planetary-gear system fits in an automatic transmission. The entire transmission is a simple machine for transmitting power from the engine to the wheels. The word "simple" clearly does not mean that the device is not complex. It means only that the device does not have its own source of energy.

Example 5.12 (a) What is the power (in horsepower) of an automobile engine that develops a torque of 200 N · m at an engine speed of 3000 r/min (revolutions per minute)? (b) At this engine speed, find the torque transmitted to the wheels in low gear if the speed-reduction ratio is 2.2 to 1 and no energy is lost to work against friction.

(a) The engine speed in revolutions per second (r/s) is

$$\omega = \frac{3000 \text{ r/min}}{60 \text{ s/min}} = 50 \text{ r/s}$$

So from Eq. 5.31 the power is

$$P = 2\pi\tau\omega = 2\pi(200 \text{ N} \cdot \text{m})(50 \text{ r/s})$$
$$= 6.28 \times 10^4 \text{ W} = 62.8 \text{ kW}$$
$$= (62.8)(1.34 \text{ hp}) = 84.2 \text{ hp}$$

(b) A speed reduction of 2.2 to 1 means that the rotational speed of the shaft going to the wheels is $\frac{1}{2.2}$ times the rotational speed of the engine. In this case, the reduced speed is

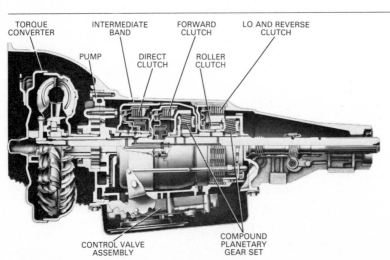

TORQUE CONVERTER   INTERMEDIATE BAND   FORWARD CLUTCH   LO AND REVERSE CLUTCH

PUMP   DIRECT CLUTCH   ROLLER CLUTCH

CONTROL VALVE ASSEMBLY   COMPOUND PLANETARY GEAR SET

FIGURE 5.30

Automatic transmission. (*General Motors Corporation.*)

$$\omega' = \frac{50\,\text{r/s}}{2.2} = 22.7\,\text{r/s}$$

So from Eq. 5.33 the transmitted torque is

$$\tau' = \frac{\omega}{\omega'}\tau = \frac{50\,\text{r/s}}{22.7\,\text{r/s}}\,(200\,\text{N}\cdot\text{m})$$

$$= (2.2)(200\,\text{N}\cdot\text{m}) = 440\,\text{N}\cdot\text{m}$$

The transmitted torque is 2.2 times the engine torque. ☐

The power output at the shaft of an automobile engine is called the *brake power* because at one time it was measured by applying a steel brake to the shaft, as shown in Fig. 5.31. The forces on the brake are the tensions T_1 and T_2 in the cords pulling the brake against the shaft, the normal force exerted by the shaft against the brake, and the frictional force of the shaft against the brake. Since the brake is at rest, the torque exerted on it by the cords

$$\tau = (T_2 - T_1)r$$

must equal in magnitude the torque exerted on the brake by the shaft. By measuring T_1, T_2, and the rotational speed of the engine, the brake power can be calculated from Eq. 5.31. The engine speed depends on how tightly the brake is pulled against the shaft. By varying this, the brake power of the engine can be measured at different speeds.

Today brake power is measured by a dynamometer (Fig. 5.32), which is an electric generator connected to the drive shaft of the engine. The power of the engine is measured by the power output of the generator. Figure 5.33 shows a typical curve of brake power against engine speed. Note that power increases with speed up to a certain point and then begins to decrease.

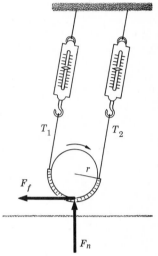

FIGURE 5.31

Method for measuring the brake power of an engine.

FIGURE 5.32

Engineers measuring the brake power of an engine on a modern dynamometer. (*General Motors Corporation.*)

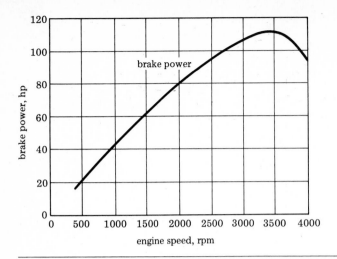

engine speed, rpm

FIGURE 5.33

Variation of brake power with engine speed. (*After William H.
Crouse,* Automotive Mechanics, *7th
ed., McGraw-Hill Book Co., New
York, 1975.*)

This is so because at the higher speed air cannot be mixed fast
enough with the fuel in the carburetor to sustain power.

Brake power is less than the power developed in the cylinders
because of friction in the engine, especially between the pistons
and the cylinders. The power developed in the cylinders is
called *indicated power.*

The efficiency for converting fuel energy into indicated power
is about 38 percent. The efficiency for converting indicated
power into brake power is about 66 percent. Thus the overall
efficiency for converting fuel energy to brake power is

$$e_{bp} = (0.38)(0.66) = 0.25$$

or 25 percent. Finally, there is further loss in transmitting the
brake power to the wheels. During an average driving cycle, only
about 48 percent of the brake power reaches the wheels. This
means that only

$$(0.48)e_{bp} = (0.48)(0.25) = 0.12$$

or 12 percent of the fuel energy reaches the wheels. It is this 12
percent which is available to accelerate the automobile and
keep it moving at constant speed against wind resistance and
rolling friction. Figure 5.34 summarizes the allocation of fuel
energy in an automobile.

Example 5.13 The engine of an automobile traveling at 55 mi/h
has a brake power of 50 hp. Assuming 48 percent of the power
reaches the wheels, what is the total external frictional force on
the automobile?

The power reaching the wheels is

$$P = (0.48)(50 \text{ hp}) = 24 \text{ hp}$$
$$= (24)(746 \text{ W}) = 1.8 \times 10^4 \text{ W}$$

and the speed is

$$v = 55 \text{ mi/h} = (55)(0.447 \text{ m/s}) = 24.6 \text{ m/s}$$

So from Eq.5.30 the applied force on the wheels is

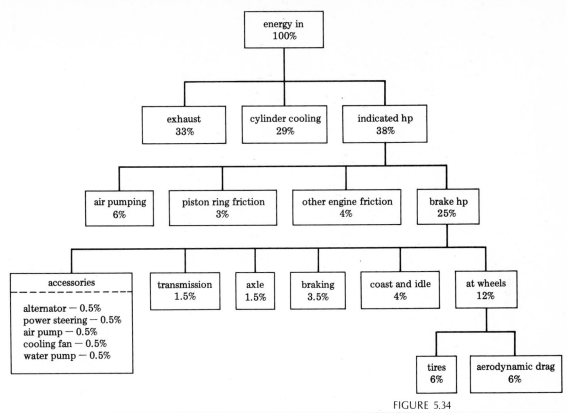

FIGURE 5.34

Allocation of fuel energy in an automobile. (*O. Pinkus and D. F. Wilcock*, Strategy for Efficient Energy Utilization through Tribology, *American Society of Mechanical Engineers, New York, 1977.*)

$$F = \frac{P}{v} = \frac{2.24 \times 10^4 \, \text{W}}{24.6 \, \text{m/s}} = 730 \, \text{N}$$

Since the automobile is not accelerating, the total force on it is zero. This means there is an opposing frictional force of magnitude 730 N.

PUMPED-STORAGE HYDROELECTRIC POWER

As a final application and review of the principles discussed in this chapter, we shall consider the energetics of pumped-storage hydroelectric power. The energy for a conventional hydroelectric-power-generating station comes from the potential energy of the water stored behind a tall dam (Fig. 5.35). As the water is brought from the higher to the lower elevation, its potential energy is converted into electric energy.

The water got to the higher elevation in the form of rain which fell on the highlands above the dam and then ran off into brooks and streams that ultimately emptied into the river on which the dam was built. The rain, in turn, came from clouds formed when water in the oceans and lowlands was evaporated by the heat of the sun. The sun is thus the ultimate source of conventional hydroelectric power. Conventional hydroelectric dams are the best way we have today to harness the sun's energy for power production.

Figure 5.36 is an aerial view of the Bear Swamp Project, a pumped-storage hydroelectric generating facility on the Deer-

FIGURE 5.35

Hydroelectric dam. (*Power Authority of the State of New York.*)

field River in northwestern Massachusetts. It was built by the New England Electric System in 1974 at a cost of $125 million. The upper reservoir, shown in the upper right of Fig. 5.36, was built by constructing four dikes around a natural depression (Bear Swamp) in the mountain top. A total of 2.17×10^6 m^3 of land fill went into these dikes. The lower reservoir, directly below the upper reservoir, was built by constructing the dam shown in the lower left of Fig. 5.36. (There is a small 10-MW conventional hydroelectric generator in the dam, but this is incidental to the main project.) In the normal course of operation, the water levels in the upper and lower reservoirs can rise

FIGURE 5.36

Aerial view of the Bear Swamp Pumped-Storage Hydroelectric Generating Facility on the Deerfield River in Massachusetts. (*New England Electric System.*)

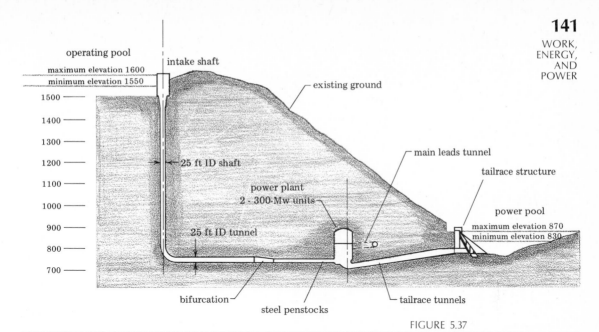

operating pool
intake shaft
maximum elevation 1600
minimum elevation 1550
existing ground
1500
1400
1300
1200 — 25 ft ID shaft
main leads tunnel
tailrace structure
1100
power plant
2 - 300-Mw units
1000
power pool
900 25 ft ID tunnel
maximum elevation 870
minimum elevation 830
800
700
bifurcation
tailrace tunnels
steel penstocks

FIGURE 5.37

Profile of the Bear Swamp Pumped-Storage Hydroelectric Facility, showing the flow of water between the upper and lower reservoirs. (*New England Electric System.*)

and fall 50 ft (15.2 m) in 5 h. For the public's safety, both reservoirs are fenced.

Power is generated when water from the upper reservoir flows into the lower reservoir through two 300-MW generators built into the base of the mountain. Water from the upper reservoir passes vertically down through a single 25-ft-diameter cement-lined shaft (Fig. 5.37), which then branches into two 17.5-ft-diameter steel penstocks that carry the water to the generator turbine (Fig. 5.38). During generation, water flows at a 150 m³/s rate through each turbine. Vertical distance between the upper and lower reservoirs is (on the average) 220 m.

A cubic meter of water has a mass of 1000 kg, so the change in the potential energy of 1 m³ of water passing from the upper to the lower reservoir is $\Delta U = 0 - mgh = -(1000\ \text{kg/m}^3)$ $(9.8\ \text{m/s}^2)(220\ \text{m}) = -2.16 \times 10^6\ \text{J/m}^3$. According to Eq. 5.19, if there is no change in the kinetic and internal energies of the water, the work done on the water by the turbine through which it passes is ΔU, and the work the water does on the turbine is $-\Delta U$. Thus, if 150-m³/s flow through the turbine, the available power is $P = (150\ \text{m}^3/\text{s})(2.16 \times 10^6\ \text{J/m}^3) = 3.24 \times 10^8\ \text{W} = 324\ \text{MW}$.

Of this, 300 MW is converted into electric energy by the generator, and 24 MW is lost. The efficiency is

$$e = \frac{300\ \text{MW}}{324\ \text{MW}} = 0.926$$

The power loss in this case is mainly in the form of the kinetic energy of the discharged water.

If the turbine was not in the way, the potential energy of the water would be converted entirely into kinetic energy. With a kinetic energy of 2.16×10^6 J, a cubic meter (1000 kg) of water would have a speed v given by $\frac{1}{2}mv^2 = 2.16 \times 10^6$ J or $v =$

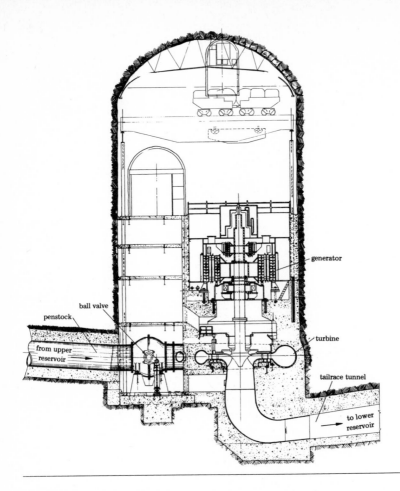

penstock

ball valve

from upper
reservoir

generator

turbine

tailrace tunnel

to lower
reservoir

FIGURE 5.38

Cross section of one of the two
turbine generators at the Bear
Swamp Pumped-Storage Hydroelectric Facility. (*New England Electric
System.*)

$\sqrt{[2(2.16 \times 10^6 \text{ J})]/1000 \text{ kg}} = 64.5$ m/s. If the water were discharged directly into the lower reservoir, no work at all would be done, and the kinetic energy would soon be converted into thermal energy. Even if the high-speed water was directed against a turbine, only a small fraction of the kinetic energy would be transformed into work. This is so because as the high-speed water hit the blades of the turbine, most of the energy would be converted into thermal energy.

In the Bear Swamp Project, as in any conventional hydroelectric generating facility, most of the available potential energy of the water is converted directly into work and is not allowed to become kinetic energy. When the generator is at full power, it takes enormous force to turn the turbine blades. This force is produced by the weight of the water coming from the upper reservoir. A column of water 220 m high exerts a force of 2.16×10^6 N on each square meter of blade surface (see Sec. 7.2). It is this force which pushes the blade forward. Most of the potential energy of the water goes directly into the work done by this force in moving the blades of the turbine, and this work is then converted by the generator into electric energy.

However, since the water must flow through the turbine at some speed, it does carry away some kinetic energy. At Bear

Swamp the water discharged from each turbine has a speed of 17 m/s. Since each turbine discharges 150 m³ or 150,000 kg of water per second, the kinetic energy per second discharged is

$$\frac{\frac{1}{2}mv^2}{t} = \frac{1}{2}(1.5 \times 10^5 \text{ kg/s})(17 \text{ m/s})^2 = 21.7 \text{ MW}$$

Of the 324 MW of available power, 300 MW of electric power is generated, 21.7 MW of power is lost in the form of kinetic energy, and the rest is lost to friction in the turbine.

> **REMARK** These fundamental concepts have important consequences for the design of hydroelectric and steam turbines. Because water has much more mass per cubic meter than steam, it is important to keep the speed of the water small to prevent the discharge of a large amount of kinetic energy. According to Eq. 5.30, the power produced is Fv, where F is the total force applied by the water to the turbine blades. In a hydroelectric turbine, the blades are very large, so that F will be large and v can be small. In a steam turbine, much smaller blades are used, because the same power can be efficiently generated with smaller F and larger v.

The generating mode of the Bear Swamp Project is the same as that of any conventional hydroelectric installation. However, unlike a conventional hydroelectric facility, there is no natural source of water for the upper reservoir at Bear Swamp, since the upper reservoir is *above* the river. All the water in the upper reservoir must be *pumped* there from the lower reservoir.

In the pumping mode, electricity from fossil and nuclear plants flows into the Bear Swamp generators, which then act as electric motors, turning the turbines in the reverse direction. The turbines in reverse act as pumps, pumping water from the lower to the upper reservoir. The purpose of this operation is to store the excess energy which is available late at night from large fossil and nuclear power plants.

When the demand for electricity is high, say at 5 P.M. in the wintertime, the water stored in the upper reservoir is used to generate power. The Bear Swamp Project is not itself a prime source of energy, but is only an energy storage system. The energy is stored in the form of the potential energy of the water in the upper reservoir.

For every 100 J of electric energy used during the pumping, about 77 J is recovered in the generating mode. The difference goes into the kinetic energy of the discharged water, frictional losses in the generator and turbine, and transmission losses. In spite of these losses, the Bear Swamp Project is economical because it uses the low-cost electric energy generated at night to meet high daytime demand. The alternative would be to build a large fossil-fuel generating station which operated only a few hours a day.

The unit of energy used in the electric-power industry is the *kilowatt-hour* (kWh), which is equal to one kilowatt times one hour:

$$1 \text{ kWh} = (1 \text{ kW})(1 \text{ h}) = (1 \times 10^3 \text{ W})(3.6 \times 10^3 \text{ s}) = 3.6 \times 10^6 \text{ J}$$

The upper reservoir at Bear Swamp contains enough water to run the two 300-MW generators at full power for 5 h, so the total available energy stored is

$$E = Pt = 2(300\,\text{MW})(5\,\text{h})$$
$$= (6 \times 10^5\,\text{kW})(5\,\text{h}) = 3.0 \times 10^6\,\text{kWh}$$

The Bear Swamp Project uses the principle of the conservation of energy to store excess electric energy available at night for use during peak daytime demand. The massive size of the project demonstrates that practical methods of energy storage are as important for a sound energy policy as are new energy sources.

GUIDE TO MAJOR TOPICS

Topic	References	Problems
Simple machines	Sec. 5.1	
Principle of simple machines	Eq. 5.3	
Mechanical advantage and efficiency	Eqs. 5.4 to 5.6; Examples 5.1 to 5.4	1 to 10
Work·	Sec. 5.2; Eq. 5.8; Example 5.5	11 to 14
Potential energy	Sec. 5.2; Eqs. 5.10 to 5.13; Examples 5.6 and 5.7	15 to 18
Kinetic energy	Sec. 5.3; Eqs. 5.14 and 5.15, 5.19 and 5.20; Examples 5.8 to 5.9	19 to 32
Conservation of energy	Sec. 5.4; Eqs. 5.21 to 5.24	33 to 36
Engines	Eqs. 5.25 and 5.26; Example 5.10	37 to 40
Power	Sec. 5.5; Eqs. 5.27 to 5.33; Examples 5.11 to 5.13	41 to 48

PROBLEMS

1   (a) What is the ideal mechanical advantage of the nut cracker in Fig. 5.39? (b) If a force of 40 N is applied to the handle, what is the force applied to the nut?
Ans. (a) 3; (b) 120 N

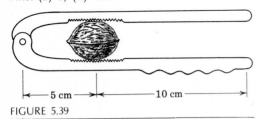

FIGURE 5.39

Problem 1.

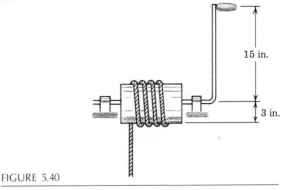

FIGURE 5.40

Problem 2.

2   (a) What is the ideal mechanical advantage of the winch in Fig. 5.40? (b) If a 100-N force must be applied to the handle to lift a 440-N weight, what is the actual mechanical advantage? (c) What is the efficiency of the winch?

3   (a) A mechanical jackscrew, similar to the one in Fig. 5.8, has a thread lead of 12 mm. How long is the handle if the ideal mechanical advantage is 250?
Ans. 47.7 cm

4   A loaded skid is pushed up an inclined plane with a force applied parallel to the incline. Show that the ideal mechanical advantage of this arrangement is $1/\sin\theta$, where θ is the angle between the incline and the horizontal.

5   A crate is pushed up a plane inclined 15° to the horizontal. The applied force is parallel to the incline and the coefficient of kinetic friction between the crate and the plane is 0.3.

What is the actual mechanical advantage of the system?

Ans. 1.82

6 What force F is required to pull a 1200-N block up the frictionless incline in Fig. 5.41?

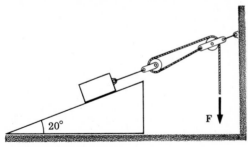

FIGURE 5.41

Problem 6.

7 (a) What is the ideal mechanical advantage of the block-and-tackle system in Fig. 5.42? (b) If the efficiency is 0.65, what force must be applied to lift a 5000-N load?

Ans. (a) 5; (b) 1538 N

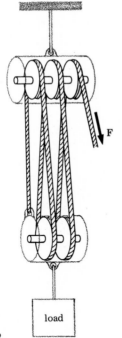

load

FIGURE 5.42

Problems 7 and 17.

8 (a) What is the mechanical advantage of the wedge in Fig. 5.43? (b) When the top of the wedge is struck with a sledgehammer, a momentary force of 1000 lb is applied to the wedge. What is the force F' exerted by one face of the wedge?

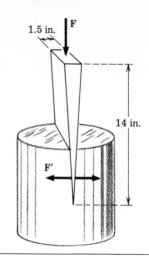

FIGURE 5.43

Problem 8.

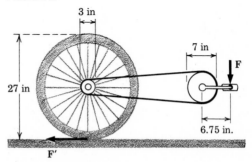

FIGURE 5.44

Problems 9, 45, and 46.

9 Figure 5.44 shows the mechanical connections to the rear wheel of a bicycle. (a) What is the ideal mechanical advantage of this arrangement; i.e., what is the ratio F/F' of the force **F** applied by the foot on the pedal to the force **F'** applied by the wheel to the road? (The reaction to **F'** is the force **R'** which the road applies to the wheel. It is **R'** which accelerates the bicycle forward.) (b) If the chain is shifted to a rear-wheel sprocket that is 4 in in diameter, will the bicycle be easier or harder to pedal? Will the bicycle move faster or slower for the same pedaling speed?

Ans. (a) $\frac{3}{14}$; (b) easier and slower

10 The *chain hoist,* or *differential pulley,* consists of a continuous chain that passes over two upper pulleys and one lower pulley, as shown in Fig. 5.45. The radii of the upper pulleys are unequal $(R > r)$, so as these pulleys complete one revolution, a length of chain $2\pi R$ long is raised onto the large pulley while a length of the chain $2\pi r$ long is lowered off of the small pulley. The chain around the lower pulley is thereby shortened the net amount $2\pi(R - r)$, and the lower pulley itself is raised the vertical distance $\pi(R - r)$. (a) Show that

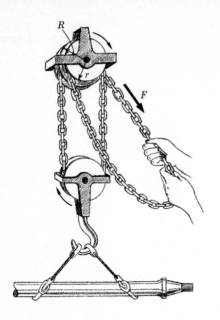

FIGURE 5.45

Problems 10 and 43.

the ideal mechanical advantage of a differential pulley is $M_I = 2R/(R - r)$.
(b) Find the ideal mechanical advantage of a chain hoist with $R = 6$ in and $r = 5.75$ in.

11 A horizontal force of 24 N is applied to a 4-kg mass that moves a distance of 3 m on a horizontal frictionless surface. What is the work done by each of the forces which act on the mass?
Ans. 0, 0, and 72 J

12 Repeat Prob. 11 when the coefficient of kinetic friction between the mass and the horizontal surface is 0.45.

13 A girl pulls a 30-kg sled with a force of 60 N directed at 40° to the horizontal (Fig. 5.46). What is the work done by each of the forces on the sled in moving it 15 m along a frictionless horizontal surface?
Ans. 0, 0, and 689.4 J

14 Repeat Prob. 13 when the coefficient of kinetic friction between the sled and the snow is 0.1. (*Hint:* The normal force is not equal to *mg* in this case.)

15 Calculate the total work done on a 75-kg skier who goes 40 m down a frictionless 15° slope.
Ans. 7.61 × 10³ J

16 Repeat Prob. 15 when the coefficient of kinetic friction between the skier and the snow is 0.08.

17 A block-and-tackle device, like the one in Fig. 5.42, raises a 750-kg load a vertical distance of 12 m. (a) How much is the potential energy of the load increased in the process? (b) How much work does the machine do on the load? (c) In lifting the load, 11 × 10⁴ J of work is done on the machine. How much is the work done by friction? (d) What is the efficiency of the machine?
Ans. (a) 8.82 × 10⁴ J; (b) 8.82 × 10⁴ J; (c) −2.18 × 10⁴ J; (d) 0.80

18 A 50-kg rock falls from a cliff to the floor of the valley 300 m below. During the fall there is a constant frictional force of 74 N acting on the rock. What is the total work done on the rock during its fall?

19 A resultant force of 100 N acts on a 25-kg mass. (a) If the mass starts from rest, what is its kinetic energy after moving 5 m? (b) What is its speed after moving 5 m?
Ans. (a) 500 J; (b) 6.32 m/s

20 A force of 180 N is applied on a 30-kg mass parallel to its initial velocity of 5 m/s. What is the speed of the mass after moving 25 m?

21 The brakes are suddenly applied to a 1000-kg car going 25 m/s. The wheels lock, and the car skids 62 m before stopping. (a) What is the force of friction on the car? (b) What is the coefficient of kinetic friction μ_k between the tires and the road?
Ans. (a) 5040 N; (b) 0.514

FIGURE 5.46

Problems 13 and 14.

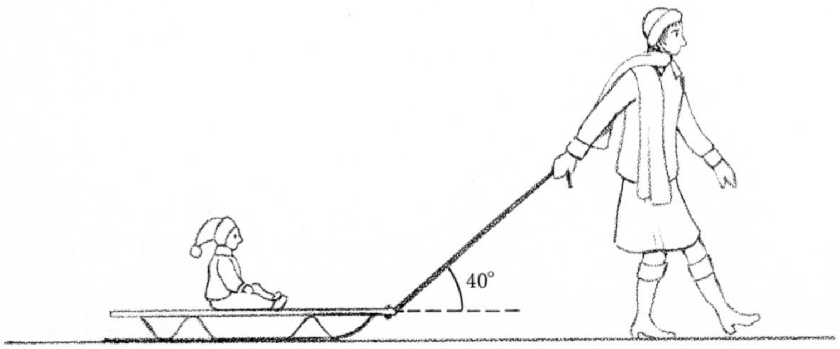

40°

22 Show that when the brakes are suddenly applied to an automobile moving with speed v, the mininum stopping distance d is $d = v^2/2\mu_k g$, where μ_k is the coefficient of kinetic friction between the tires and the road.

23 A 1000-kg car traveling 25 m/s collides with a brick wall. The car moves 0.5 m before stopping. What is the average force exerted on the car by the wall during the collision?
Ans. 6.25×10^5 N

24 In shuffleboard, a long stick is used to give a shove to a 0.3-kg wooden puck and start it moving across the floor (Fig. 5.47). A player pushes down with a force of 20 N along the stick, which is inclined 45° to the ground. (a) How much work is done on the puck by the stick as the stick pushes the puck from point A to B, a distance of 0.5 m? (b) If the puck starts from rest at A, what is its speed at B (neglect friction)? (c) From point B, where the stick stops pushing, the puck slides a distance d before stopping. If there is a force of friction of 1 N from point B onward, what is d?

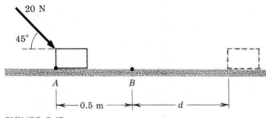

FIGURE 5.47

Problem 24.

25 A 10-N force at an angle of 35° to the horizontal is applied to the 5-kg block in Fig. 5.48. In addition, there is a 7-N frictional force acting on the block as shown. (a) Find the work done by each of the four forces acting on the block in moving the block 12 m. (b) What is the total work done on the block? (c) If the block starts from rest, what is its speed after moving the 12 m?
Ans. (b) 14.3 J; (c) 2.39 m/s

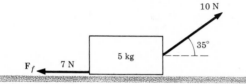

FIGURE 5.48

Problem 25.

26 A 0.2-kg hockey puck is given an initial speed of 12 m/s. If the coefficient of friction between the puck and the ice is 0.1, what is the puck's speed after traveling 15 m?

27 A bullet of mass 5×10^{-3} kg is moving horizontally with a speed of 400 m/s. (Neglect gravity in this problem.) (a) What is the kinetic energy of the bullet? (b) The bullet penetrates a 5-cm-thick piece of wood, emerging with a speed of 200 m/s. What is the work done on the bullet by the wood? (c) What is the average force exerted on the bullet by the wood?
Ans. (a) 400 J; (b) -300 J; (c) 6000 N

28 A 100-kg skier, skiing along level ground at a speed of 7 m/s, comes to a hill (Fig. 5.49). (a) If the skier coasts up the hill, what is his speed at point B, assuming the slope exerts no friction on the skis? (b) What is his speed at point B, assuming the slope exerts a constant 75-N force of friction on the skis?

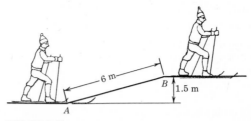

FIGURE 5.49

Problem 28.

29 A 6-kg block starting from rest slides 4 m down the incline in Fig. 5.50. (a) What is the potential energy of the block (relative to the bottom of the incline) when it is at the top? (b) If the incline is frictionless, what is the speed of the block when it reaches the bottom of the incline? (c) If there is a constant 8-N force of friction on the block while it slides down the incline, what is its speed at the bottom?
Ans. (a) 176 J; (b) 7.66 m/s; (c) 6.93 m/s

FIGURE 5.50

Problem 29.

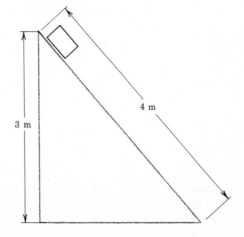

30 A boy takes a running jump off the high-dive board at the local pool. His speed as he leaves the board is 3 m/s, and the board is 5 m above the water. What is his speed as he hits the water?

31 A girl throws a 0.2-kg ball a distance of 6 m straight up in the air. (a) What is the kinetic energy of the ball as it leaves the girl's hand? (b) How much work does the girl do in throwing the ball? (c) If the girl's arm muscle contracted a distance of 0.05 m while throwing the ball, what was the average force exerted by the muscle?
Ans. (a) 11.8 J; (b) 11.8 J; (c) 236 N

32 A 7-g bullet shot straight up in the air with an initial speed of 200 m/s reaches a vertical height of 900 m. What is the average frictional force on the bullet?

33 A 10-kg block, starting from rest, slides 3 m down a plane inclined 22° to the horizontal. The block has a speed of 4 m/s when it reaches the bottom. Find the change of (a) potential energy, (b) kinetic energy, and (c) internal energy.
Ans. (a) −110 J; (b) 80 J; (c) 30 J

34 The brakes are applied to a 1200-kg automobile going 30 m/s just as it starts up a hill with a 10° inclination. The automobile comes to a stop after moving 125 m along the incline. (a) What is the increase in internal energy? (b) What is the frictional force on the automobile during the braking period?

35 A 1500-kg automobile accelerates from rest to 25 m/s while moving a distance of 200 m along a horizontal road. During this period there is a constant frictional force of 1100 N opposing the motion. (a) What is the magnitude of the applied force exerted on the automobile? (The applied force due to the engine is exerted on the automobile by the road.) (b) What is the work done by the applied force during the acceleration process?
Ans. (a) 3444 N; (b) 6.89 × 10⁵ J

36 A ball is thrown up in the air with an initial speed v_A. If the air exerts a constant frictional force of magnitude F_f on the ball, show that the height h reached by the ball is

$$h = \frac{v_A{}^2}{2(g + F_f/m)}$$

and that the speed v_B of the ball as it hits the ground is

$$v_B = v_A\sqrt{\frac{mg - F_f}{mg + F_f}}$$

37 An engine produces 750 J of work for every

2000 J of fuel energy used. (a) What is the efficiency of the engine? (b) How much energy is released to the environment?
Ans. (a) 0.375; (b) 1250 J

38 An engine with an efficiency of 0.28 produces 5000 J of work. (a) How much fuel energy is used? (b) How much energy goes into the environment?

39 Human muscles have an efficiency of about 0.22 for converting chemical energy into work. During normal daily activity the body uses about 3 × 10⁵ J of chemical energy per hour. (a) How much additional chemical energy is used when a 70-kg person climbs a 900-m-high mountain? (b) The *kilocalorie* (kcal) is the unit of energy referred to by nutritionists as the *calorie*. Given that 1 kcal = 4185 J, calculate how many extra kilocalories are used in the mountain climb.
Ans. (a) 2.81 × 10⁶ J; (b) 671 kcal

40 What is the efficiency of an engine that releases 75 percent of its fuel energy into the environment?

41 What is the power of a motor required to raise a 2000-kg elevator cab to the top of a 150-m-tall building in 65 s? Because the cab is counterweighted, the motor has to exert a vertical force only $\frac{1}{10}$ the weight of the cab.
Ans. 4.5 kW

42 How fast can a 50-hp motor raise a 1500-kg mass?

43 A mechanic can pull the chain of the hoist in Prob. 10 at a speed of 0.5 m/s. () How fast is the load lifted? (b) What is the power required to lift a 1000-kg load?
Ans. (a) 0.0104 m/s; (b) 102 W

44 An automobile engine exerts a torque of 300 N · m when it is rotating at the rate of 1500 r/min. What is the brake power at this speed?

45 A cyclist on the bicycle in Fig. 5.44 rotates the pedals at the rate of 0.6 r/s. (a) If she exerts an average force of 15 N on the pedals, what is her power output? (b) What is the rate of rotation of the rear wheel? (c) What is the forward speed of the bicycle? (d) What is the total frictional force opposing the bicycle's motion? (Since the bicycle is moving at constant speed, the power exerted by the cyclist goes entirely into overcoming frictional resistance.)
Ans. (a) 9.7 W; (b) 1.4 r/s; (c) 3.0 m/s; (d) 3.2 N

46 The cyclist in Prob. 45 shifts the chain to a rearwheel sprocket that has a diameter of 4 in.

If the frictional force on the bicycle remains the same (3.2 N), and she continues to rotate the pedals at the same rate (0.6 rps), calculate (a) the forward speed of the bicycle, (b) the power input, and (c) the force exerted on the pedals.

REMARK The 4-in sprocket is a lower gear than the 3-in sprocket because for the same pedaling speed, the bicycle moves slower.

47 Assume that the total frictional force F_f opposing the motion of an automobile is given by

$$F_f = (100 \text{ N} \cdot \text{s/m})v$$

where v is the speed in meters per second. (a) Calculate the frictional force on the automobile at 10, 20, and 30 m/s (22.4, 44.7, and 67.1 mi/h). (b) Calculate the power P required to move the automobile at each speed. (c) Assume that gasoline consumption C is related to P by

$$C = [0.05 \text{ gal/(h} \cdot \text{hp)}]P$$

where C is in gallons per hour (gal/h), and P is in horsepower (hp). Calculate the gasoline used in 1 h at each speed. (d) Calculate the fuel usage in miles per gallon at each speed. *Ans.* (a) 1000, 2000, and 3000 N; (b) 13.4, 53.6, and 120.6 hp; (c) 0.67, 2.68, and 6.03 gal/h; (d) 33.4, 16.7, and 11.1 mi/gal.

48 The water for a 200-MW hydroelectric generator comes from a height of 300 m. (a) What is the mass of water per second that passes through the generator? (b) What is the volume of water per second passing through the generator?

BIBLIOGRAPHY

CROUSE, WILLIAM H.: *Automotive Mechanics,* 7th ed., McGraw-Hill, New York, 1975. This comprehensive and authoritative book covers all aspects of the automobile from engine to battery in a thorough yet easy-to-understand manner. Chapter 16 on engine power and its measurement and Chap. 56 on automotive brakes are especially relevant to this chapter.

GREY, JERRY, GEORGE W. SUTTON, and MARTIN ZLOTNICK: "Fuel Conservation and Applied Research," *Science* **200**:135 (1978). This article discusses the various sources of energy loss in an automobile and how these losses may be decreased by further research.

Newton's second law of motion completely determines the motion of an object in terms of the forces acting on it. If the total force is zero, then the acceleration is zero, and the object moves with constant velocity. If the total force is constant, then the acceleration is constant, and the object moves with constant linear acceleration. These two types of motion were treated in Sec. 4.2. Other possible forces lead to other motions. In this chapter we shall consider some of the other motions that commonly arise in engineering practice.

6

COMPLEX MOTIONS

6.1 Parabolic Motion

When a projectile is thrown at some angle to the vertical, it does not move straight up and down, but rather it moves in a curved path called a *parabola* (Fig. 6.1; also see Fig. 4.6). To study such motion, we first establish a coordinate system. The origin of the axes is placed at the initial, or launch, position of the projectile, and the x and y axes are aligned with the horizontal and vertical directions, respectively.

Once the projectile is launched, the only force on it is the force of gravity $\mathbf{F}_g$, and so its acceleration is

$$\mathbf{a} = \frac{\mathbf{F}_g}{m} = \mathbf{g}$$

where $\mathbf{g}$ is the constant acceleration of gravity. Since $\mathbf{g}$ is directed vertically downward, the x and y components of $\mathbf{a}$ are

$$a_x = 0 \quad \text{and} \quad a_y = -g = -9.8\,\text{m/s}^2$$

The x component is zero because we have chosen the x axis to be perpendicular to the force of gravity; the y component is negative because we have chosen the positive y direction to be up, whereas the direction of the acceleration of gravity is down.

If the initial velocity $\mathbf{v}_0$ of the projectile makes an angle θ with the x axis, the x and y components of $\mathbf{v}_0$ are

$$v_{0x} = v_0 \cos\theta \quad \text{and} \quad v_{0y} = v_0 \sin\theta \qquad 6.1$$

Since there is no force and no acceleration in the x direction, the x component of velocity does not change. The projectile moves with a constant horizontal speed, so at any time t its x position is

$$x = v_{0x}t \qquad 6.2$$

where v_{0x} has been substituted for v in Eq. 4.1.

Likewise, since the acceleration is constant in the y direction, the projectile moves with constant linear acceleration along the y axis. At any time t the projectile's vertical speed is

$$v_y = v_{0y} - gt \qquad 6.3$$

and its y position is

$$y = v_{0y}t - \tfrac{1}{2}gt^2 \qquad 6.4$$

These last equations are the equations for constant linear accel-

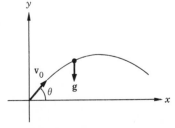

FIGURE 6.1

The parabolic path of a projectile.

eration (Eqs. 4.3 and 4.5), with y, $-g$, v_y, and v_{0y} substituted for x, a, v, and v_0.

Projectile motion is thus a combination of uniform motion in the x direction and constant acceleration in the y direction. Equations 6.2 and 6.4 determine the x and y coordinates of the projectile at any time.

Example 6.1 A ball is thrown with an initial speed of 21 m/s at 34° to the horizontal. (a) Plot the position of the ball at 0.4-s intervals. (b) Calculate the speed of the ball at 0.4-s intervals.

(a) From Eq. 6.1 the x and y components of v_0 are

$$v_{0x} = v_0 \cos \theta = (21 \text{ m/s}) \cos 34° = 17.4 \text{ m/s}$$

$$v_{0y} = v_0 \sin \theta = (21 \text{ m/s}) \sin 34° = 11.7 \text{ m/s}$$

and so Eqs. 6.2 and 6.4 for the x and y coordinates become

$$x = v_{0x}t = (17.4 \text{ m/s})t$$

$$y = v_{0y}t - \tfrac{1}{2}gt^2 = (11.7 \text{ m/s})t - (4.9 \text{ m/s}^2)t^2$$

At $t = 0$ these equations give $x = 0$ and $y = 0$, which just means that the ball started from the origin at $t = 0$. At $t = 0.4$ s these equations give

$$x = (17.4 \text{ m/s})(0.4 \text{ s}) = 7.0 \text{ m}$$

$$y = (11.7 \text{ m/s})(0.4 \text{ s}) - (4.9 \text{ m/s}^2)(0.4 \text{ s})^2 = 3.9 \text{ m}$$

This means that the ball is at the point with coordinates $x = 7.0$ m and $y = 3.9$ m at $t = 0.4$ s. The calculations are the same for $t = 0.8$, 1.2, 1.6, 3.0, and 2.4 s. The results are given in Table 6.1 and plotted in Fig. 6.2. The smooth curve through these points is a parabola.

From these results we see that the ball was in the air for 2.4 s,* traveled 41.8 m along the x axis, and reached a maximum height of 7.0 m.

(b) The speed of the ball at any time is

$$v = \sqrt{v_x^2 + v_y^2}$$

where v_x and v_y are the x and y components of the velocity. The x component remains constant, equal to v_{0x}, while the y component is given by Eq. 6.3. Thus at $t = 0.8$ s, we have

$$v_x = v_{0x} = 17.4 \text{ m/s}$$

*The small negative value of y at 2.4 s means that the ball was at $y = 0$ slightly before 2.4 s.

TABLE **6.1**

t, s	x, m	y, m	v_y, m/s	v, m/s	The x and y coordinates of the ball in Example 6.1 at 0.4-s intervals
0.0	0.0	0.0	11.7	21.0	
0.4	7.0	3.9	7.8	19.1	Also the y component of the velocity v_y and the speed $v = \sqrt{v_x^2 + v_y^2}$ are tabulated.
0.8	13.9	6.2	3.9	17.8	
1.2	20.9	7.0	0.0	17.4	
1.6	27.8	6.2	-4.0	17.8	
2.0	34.8	3.8	-7.9	19.1	
2.4	41.8	-0.1	-11.8	21.0	

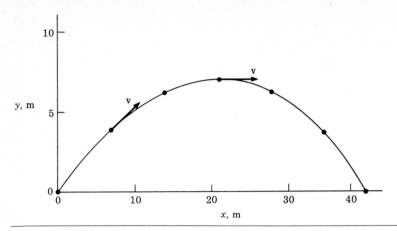

FIGURE 6.2

Plot of the positions of the ball in Example 6.1. The dots are the x and y coordinates of the ball given in Table 6.1.

$$v_y = v_{0y} - gt = (11.7 \text{ m/s}) - (9.8 \text{ m/s}^2)(0.8 \text{ s})$$
$$= 3.86 \text{ m/s}$$

so the speed is

$$v = \sqrt{v_x^2 + v_y^2}$$
$$= \sqrt{(17.4 \text{ m/s})^2 + (3.86 \text{ m/s})^2} = 17.8 \text{ m/s}$$

The results for other times are given in Table 6.1.

The velocity vector is always tangent to the parabola, as shown in Fig. 6.2. At the top of the parabola, for instance, the velocity is horizontal. This is so because v_y is zero at this point. □

Example 6.2 A rifle fires a bullet with an initial speed of 600 m/s at a target 200 m away. The rifle is held horizontally and aimed directly at the bullseye. How far below the bullseye will the bullet hit the target?

The situation is shown in Fig. 6.3. The bullet, aimed directly at the bullseye, does not hit it because in the time required to reach the target, the bullet will fall some distance.

The x and y components of the initial velocity are

$$v_{0x} = 600 \text{ m/s} \quad \text{and} \quad v_{0y} = 0$$

So the equations for x and y are

$$x = v_{0x}t = (600 \text{ m/s})t$$
$$y = v_{0y}t - \tfrac{1}{2}gt^2 = -(4.9 \text{ m/s}^2)t^2$$

The time t required to travel the 200 m to the target is found from the x equation:

FIGURE 6.3

A rifle shooting directly at the center of a target. Because the bullet follows a parabolic path, it strikes the target below the center.

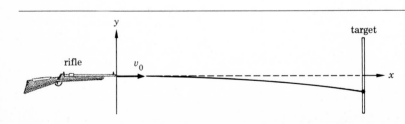

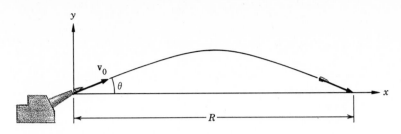

FIGURE 6.4

The range R of a projectile is the horizontal distance it travels along level ground.

$$t = \frac{x}{v_{0x}} = \frac{200 \text{ m}}{600 \text{ m/s}} = 0.33 \text{ s}$$

In this time the bullet will fall the vertical distance given by the y equation:

$$y = -(4.9 \text{ m/s}^2)(0.33 \text{ s})^2 = -0.53 \text{ m}$$

The minus sign indicates that the bullet hits the target 0.53 m below the bullet's initial vertical position. ☐

Example 6.3 The range R of a cannon is the horizontal distance the shell travels along level ground. Find the formula for the range of a cannon in terms of the initial speed v_0 of the shell and the angle θ the velocity vector makes with the horizontal.

The y coordinate of the shell is zero when the shell leaves the cannon (at time $t = 0$) and is zero again when the shell hits the ground (Fig. 6.4). (We neglect the height of the cannon above the ground.) The times at which $y = 0$ are found by setting y to zero in Eq. 6.4:

$$0 = v_{0y}t - \tfrac{1}{2}gt^2$$

This is an equation for t, the two solutions of which are

$$t = 0 \quad \text{and} \quad t = \frac{2v_{0y}}{g}$$

The first solution is the time when the shell leaves the cannon, and the second solution is the time when the shell hits the ground.

To find the horizontal distance the shell traveled, we insert the second value of t into Eq. 6.2:

$$x = v_{0x}t = \frac{2v_{0x}v_{0y}}{g}$$

This is the range R of the shell. From Eq. 6.1 we can write it in the form

$$R = \frac{2v_0{}^2 \sin\theta \cos\theta}{g}$$

This formula shows that the range is zero when $\theta = 0°$ and $90°$. The maximum range occurs when $\theta = 45°$. ☐

6.2 Uniform Circular Motion

Uniform circular motion is motion in a circle of radius r with constant speed v (Fig. 6.5). Although the speed is constant, the acceleration is not zero, because the direction of the velocity vector is continuously changing. According to Eq. 4.2 there is acceleration whenever there is a change in either the magnitude or the direction of the velocity vector.

In the case of uniform circular motion, the acceleration is called *centripetal acceleration* because at any instant it is directed toward the center of the circle. Thus as the object moves, the direction of the acceleration changes, as shown in Fig. 6.5. The magnitude of this acceleration is related to the speed v and the radius r by

$$a = \frac{v^2}{r} \tag{6.5}$$

Notice that v^2/r has the usual units of acceleration:

$$\left[\frac{v^2}{r} \right] = \frac{(m/s)^2}{m} = m/s^2$$

Since the speed v is constant, the distance s that the object moves around the circle in time t is given by

$$s = vt \tag{6.6}$$

REMARK It is instructive to derive Eq. 6.5 from the definition of acceleration given by Eq. 4.2. Thus consider an object moving at constant speed v in a circle of radius r (Fig. 6.6). At time t_1 the object is at point A on the circle. The velocity $\mathbf{v}_1$ of the object at time t_1 is tangent to the circle at A. At the later time t_2, the object is at point B, having moved through an angle θ. The velocity $\mathbf{v}_2$ at this time is tangent to the circle at B. We shall assume that $t_2 - t_1$ is small enough for Eq. 4.2 to be used to calculate the instantaneous acceleration.

Since the object is moving with constant speed v, the time interval $\Delta t = t_2 - t_1$ is related to the distance s traveled along the circle by

$$\Delta t = \frac{s}{v}$$

But if θ is in radians, we have (Appendix IV)

$$s = r\theta$$

and so

$$\Delta t = \frac{r\theta}{v}$$

To find $\Delta \mathbf{v} = \mathbf{v}_2 - \mathbf{v}_1$ we note that the velocity vectors are perpendicular to the two radii which form the sides of θ, so (by Theorem 3 in Appendix IV) the angle between $\mathbf{v}_1$ and $\mathbf{v}_2$ is also θ. In Fig. 6.7 the vector $\mathbf{v}_1$ is replaced by $-\mathbf{v}_1$, and both it and $\mathbf{v}_2$ have been moved to the point C midway between A and B for the purpose of vector addition. The sum of $\mathbf{v}_2$ and $-\mathbf{v}_1$ is, of course, $\mathbf{v}_2 - \mathbf{v}_1$, and from Fig. 6.7 this sum is seen to be in the direction of the line from C to the center of the circle. This shows that the acceleration is directed toward the center of the circle. (Since we are interested only in the case in which θ is very small, the three points A, B, and C are nearly coincident.)

Figure 6.8 shows that Δv, the magnitude of $\Delta \mathbf{v}$, is the chord of a circle of radius v which subtends the angle θ. If θ is very small, the length of the chord is nearly equal to the arc length s', so

$$\Delta v = s' = v\theta$$

Thus from Eq. 4.2 the magnitude of the acceleration is

$$a = \frac{\Delta v}{\Delta t} = \frac{v\theta}{r\theta/v} = \frac{v^2}{r}$$

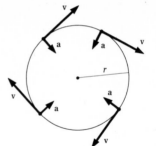

FIGURE 6.5

An object moving at constant speed v in a circle of radius r. At each instant the velocity $\mathbf{v}$ is tangent to the circle and the centripetal acceleration $\mathbf{a}$ is directed toward the center of the circle.

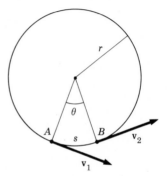

FIGURE 6.6

Two velocities $\mathbf{v}_1$ and $\mathbf{v}_2$ of an object moving with uniform circular motion. At time t_1 the object is at A and at time t_2 the object is at B.

FIGURE 6.7

The vectors $-\mathbf{v}_1$ and $\mathbf{v}_2$ drawn from a common point midway between A and B.

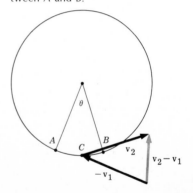

This result does not depend on θ provided that θ is so small that the chord in Fig. 6.8 is equal to the arc length. But it is only in this case that Eq. 4.2 gives the instantaneous acceleration. Therefore, Eq. 6.5 gives the magnitude of the acceleration exactly. We have already shown that under these same conditions, the acceleration is directed, at each instant, from the position of the object to the center of the circle.

In order for an object to maintain uniform circular motion, Newton's second law requires that the object be acted upon by a force of magnitude

$$F = ma = m\frac{v^2}{r} \qquad\qquad 6.7$$

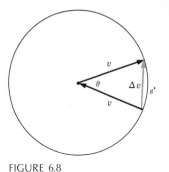

directed toward the center of the circle. A force directed toward a center of motion is called a *centripetal force*. For example, when a mass at the end of a string is twirled in a circle, the tension in the string exerts the centripetal force on the mass. This same tension pulls outward on the hand holding the string, but this outward force is not the force on the mass. The force on the mass is always directed toward the center of the circle.

FIGURE 6.8

Δv is the chord of an arc that subtends an angle θ on a circle of radius v.

Example 6.4 The exit ramp of an expressway is curved in a circular arc 70 m in radius. If the coefficient of friction between the tires of an automobile and the road is 0.6, what is the maximum speed the automobile can have on the ramp before it starts to skid?

The centripetal force that keeps the automobile moving in a circle of radius 70 m is the sideways frictional force exerted by the road on the tires. The maximum value of this force is

$$F_{f,\max} = \mu_s mg$$

where m is the mass of the car. Consequently, the maximum centripetal acceleration the car can have is given by

$$ma_{\max} = F_{f,\max} = \mu_s mg$$

so

$$a_{\max} = \mu_s g$$

The maximum acceleration occurs at the speed v given by

$$\frac{v^2}{r} = \mu_s g$$

or

$$v = \sqrt{\mu_s gr}$$
$$= \sqrt{(0.6)(9.8\text{ m/s}^2)(70\text{ m})} = 20.3\text{ m/s}$$

If the automobile's speed is greater than this, the frictional force cannot keep it in a circle of radius 70 m, so the car will skid outward, moving in a circle with a larger radius. ◻

Example 6.5 The *period* τ of a satellite is the time the satellite takes to complete one revolution about the earth. Find the period of a satellite that moves in a circular orbit close to the earth's surface.

The centripetal force that keeps the satellite in a circular orbit is the force of gravity $\mathbf{F}_g$ exerted by the earth. For a satellite close to the earth's surface, F_g is approximately mg, the same as at the surface. Thus the speed v of the satellite is given by the relation

$$mg = m\frac{v^2}{r}$$

or

$$v = \sqrt{gr}$$

where r is the radius of the satellite's orbit.

For a satellite orbiting near the earth's surface, r is just slightly greater than the radius R_e ($= 6.4 \times 10^6$ m) of the earth. Using R_e for r, the satellite's speed is

$$v = \sqrt{gr} = \sqrt{(9.8 \text{ m/s}^2)(6.4 \times 10^6 \text{ m})} = 7.9 \times 10^3 \text{ m/s}$$

The distance s that the satellite travels in one revolution is the circumference of a circle of radius R_e:

$$s = 2\pi R_e = 2(3.14)(6.4 \times 10^6 \text{ m/s}) = 40 \times 10^6 \text{ m}$$

From Eq. 6.6 the time τ required to move this distance is

$$\tau = \frac{s}{v} = \frac{40 \times 10^6 \text{ m}}{7.9 \times 10^3 \text{ m/s}} = 5.1 \times 10^3 \text{ s}$$

or about 85 min. The period of a low-orbiting satellite is actually about 90 min, a familiar number since the beginning of the space age. Our calculation gives a slightly smaller value because it neglected the height of the satellite above the earth's surface. ☐

6.3 Motion in the Presence of a Gravitational Force

UNIVERSAL LAW OF GRAVITATION

The gravitational force the earth exerts on a mass m is approximately the same everywhere on the earth's surface, but it decreases with distance from the earth. For problems involving objects that move far from the earth's surface, such as space stations, meteors, and the moon, account must be taken of the variation of the gravitational force with distance.

The law that describes this variation was discovered around 1664 by Isaac Newton, who was then twenty-three years old. He arrived at it from theoretical considerations of the moon's motion around the earth. Newton knew that a centripetal force was required to keep the moon in a circular orbit around the earth, but he did not know the nature of this force. He was thinking about the problem while sitting in his garden one day, when a falling apple started him wondering whether the centripetal force on the moon could be the same as the force on the apple.

Newton calculated the centripetal acceleration of the moon (see Prob. 9) and found that it was only 2.7×10^{-3} m/s^2, much less than the acceleration of gravity at the earth's surface. However, he thought it was reasonable that the force should decrease with distance from the earth, just like the loudness of a sound decreases with distance from the source. If, in fact, the force decreased as the square of the distance from the earth's center, it would have just the right magnitude to give the moon its correct centripetal acceleration.

Newton went on to consider the planets, which revolve in nearly circular orbits around the sun. Again their motion could be explained if one assumed that the sun exerted a force on them that decreased as the square of the distance from the sun's center. But then if both the sun and the earth exert the same kind of force, probably other objects do so too. On the basis of these considerations Newton stated the following law:

Universal Law of Gravitation   *Between any two objects of mass* m_1 *and* m_2 *there exists an attractive force proportional to the product of the masses and inversely proportional to the square of the distance between them.* In symbols this force is written

$$F = G\frac{m_1 m_2}{r^2} \qquad\qquad 6.8$$

where r is the distance between the objects, and G is a universal constant of nature. In SI units

$$G = (6.673 \pm 0.003) \times 10^{-11}\,\text{N} \cdot \text{m}^2/\text{kg}^2$$

REMARKS   (1) Forces always come in pairs, of course, and Eq. 6.8 gives the magnitude of either member of the pair. Since the force is attractive, the force F_1 on m_1 is directed toward m_2, and the (reaction) force F_2 on m_2 is directed toward m_1 (Fig. 6.9). (2) Equation 6.8 holds exactly only for point objects (particles). It holds for extended objects only if r is very much larger than the objects themselves. The one very important exception to this occurs when the object is a sphere. For spherical objects it is a remarkable fact, first proved by Newton, that Eq. 6.8 holds if r is the distance between the centers of the spheres.

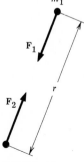

FIGURE 6.9

The forces on two masses attracted to each other by gravity.

Since the earth is nearly spherical, Eq. 6.8 can be used for objects on its surface. The force of gravity on an object of mass m on the earth's surface, then, is

$$F_g = G\frac{m_e m}{R_e^{\,2}}$$

where m_e is the mass of the earth, and $R_e = 6.37 \times 10^6$ m is the radius of the earth. But since all objects have the same acceleration g near the earth's surface, we also have $F_g = mg$, so

$$G\frac{m_e m}{R_e^{\,2}} = mg$$

After canceling the mass m on both sides of this equation, we get a very useful relation between the acceleration of gravity g and the mass and radius of the earth:

$$g = \frac{Gm_e}{R_e^{\,2}} \qquad \text{or} \qquad Gm_e = gR_e^{\,2} \qquad\qquad 6.9$$

At any distance r from the earth's center (provided $r > R_e$), the gravitational force on an object of mass m is

$$F_g = G \frac{m_e m}{r^2}$$

Using Eq. 6.9 to replace Gm_e by gR_e^2, the force can be written

$$F_g = mg\left(\frac{R_e^2}{r^2}\right) = mg\left(\frac{R_e}{r}\right)^2 \qquad \qquad 6.10$$

Example 6.6 What is the period of a satellite that moves in a circular orbit of radius $r = 8 \times 10^6$ m? (See Example 6.5.)

Figure 6.10 shows that such a satellite orbits at a height

$$h = r - R_e = (8 \times 10^6 \text{ m}) - (6.37 \times 10^6 \text{ m})$$
$$= 1.63 \times 10^6 \text{ m} = 1630 \text{ km}$$

above the earth's surface. From Eq. 6.10 the gravitational force on the satellite is

$$F_g = mg\left(\frac{R_e}{r}\right)^2 = mg\left(\frac{6.37 \times 10^6 \text{ m}}{8 \times 10^6 \text{ m}}\right)^2$$
$$= mg\left(\frac{6.37}{8}\right)^2 = (0.634)mg$$

And so from Newton's second law we get

$$(0.634)mg = m\frac{v^2}{r}$$

or

$$v^2 = (0.634)gr$$
$$v = \sqrt{(0.634)gr} = \sqrt{(0.634)(9.8 \text{ m/s}^2)(8 \times 10^6 \text{ m})}$$
$$= 7.05 \times 10^3 \text{ m/s}$$

In one revolution the satellite moves the distance

$$s = 2\pi r = 2(3.14)(8 \times 10^6 \text{ m}) = 5.02 \times 10^7 \text{ m}$$

so its period is

$$\tau = \frac{s}{v} = \frac{5.02 \times 10^7 \text{ m}}{7.05 \times 10^3 \text{ m/s}} = 7.12 \times 10^3 \text{ s} = 119 \text{ min}$$

Note that this problem is essentially the same as Example 6.5, except that the gravitational force on the satellite was reduced by the factor $(R_e/r)^2 = 0.634$. ☐

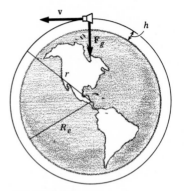

FIGURE 6.10

A satellite in circular orbit about the earth.

Example 6.7 A communication satellite moves in a synchronous orbit of radius r about the earth's equator. A synchronous orbit is a circular orbit with a period τ of 24 h. Since the earth and the satellite each complete a revolution in the same time, the satellite remains above the same point on the earth's surface. What is the radius of the orbit?

The speed of the satellite is related to the radius r and the period τ by

$$v = \frac{s}{\tau} = \frac{2\pi r}{\tau}$$

so the centripetal acceleration is

$$a = \frac{v^2}{r} = \frac{1}{r}\left(\frac{2\pi r}{\tau}\right)^2 = \frac{4\pi^2 r}{\tau^2}$$

The force on the satellite is given by Eq. 6.10, so Newton's second law shows that

$$mg\left(\frac{R_e^{\,2}}{r^2}\right) = m\frac{4\pi^2 r}{\tau^2} \qquad \text{or} \qquad r^3 = \frac{gR_e^{\,2}\tau^2}{4\pi^2}$$

This equation, true for any orbit, shows that the cube of the radius is proportional to the square of the period. In the present case,

$$\tau = 24\,h = (24\,h)(60\,min/h)(60\,s/min) = 8.64 \times 10^4\,s$$

so

$$r^3 = \frac{(9.8\,m/s^2)(6.37 \times 10^6\,m)^2(8.64 \times 10^4\,s)^2}{4\pi^2}$$

$$= 7.52 \times 10^{22}\,m^3$$

and

$$r = 4.22 \times 10^7\,m = 4.22 \times 10^4\,km = 26{,}200\,mi$$

Thus a satellite moving about the earth's equator in a circular orbit with a radius of 26,200 mi will remain above a fixed point on the earth's surface. □

POTENTIAL ENERGY

When a mass m stays near the earth's surface, the gravitational force on it is mg, and its potential energy is

$$U = mgh$$

where h is the height of the mass above a reference surface. In using this form of the potential energy, it is assumed that h is small compared with the radius of the earth.

When a mass m moves far from the earth, the gravitational force on it is given by Eq. 6.10, and the potential energy is

$$U = mgR_e\left(1 - \frac{R_e}{r}\right) \qquad\qquad 6.11$$

At the earth's surface, $r = R_e$, and the potential energy is

$$U_e = mgR_e\left(1 - \frac{R_e}{R_e}\right) = mgR_e(1 - 1) = 0$$

At great distance from the earth, $r = \infty$ and $R_e/r = 0$, so

$$U_\infty = mgR_e$$

Thus the change in the potential energy of a mass in going from the earth's surface to infinity is

$$\Delta U = U_\infty - U_e = mgR_e$$

Example 6.8 A rocket is launched from the earth's surface with an initial speed v. This means that the engine accelerates the rocket to this speed and then shuts off. Thereafter the rocket is acted upon only by the earth's force of gravity. What is the minimum value of v which will allow the rocket to move infinitely far from the earth?

The kinetic energy of the rocket at the earth's surface is

$$K_e = \tfrac{1}{2}mv^2$$

If K_e is the smallest kinetic energy which will allow the rocket to escape to infinity, the rocket's kinetic energy when it gets to infinity will be zero. Therefore, the change in kinetic energy is

$$\Delta K = K_\infty - K_e = 0 - \tfrac{1}{2}mv^2 = -\tfrac{1}{2}mv^2$$

Once the rocket engine shuts off, there is no applied work done on the rocket. Consequently, the total energy

$$E = K + U$$

of the rocket is constant; i.e., the change of energy ΔE is zero. Thus we have

$$\Delta E = \Delta K + \Delta U = 0$$

so

$$-\tfrac{1}{2}mv^2 + mgR_e = 0$$

or

$$\begin{aligned} v &= \sqrt{2gR_e} \\ &= \sqrt{2(9.8 \text{ m/s}^2)(6.37 \times 10^6 \text{ m})} \\ &= 1.12 \times 10^4 \text{ m/s} = 11.2 \text{ km/s} \end{aligned}$$

This is called the *escape speed* of the rocket, since it is the smallest speed that will enable the rocket to completely escape the earth. It is approximately the speed a rocket needs to reach the moon. ☐

Example 6.9 A rocket is given an initial speed of 9000 m/s. To what distance r from the earth's center will the rocket rise before falling back to earth?

The rocket's initial speed is less than the escape speed (Example 6.8), so the rocket will not escape from the earth. It will move out a distance r from the earth's center and then fall back to earth. At the maximum distance r its kinetic energy will momentarily be zero, so the change in kinetic energy in going from the earth's surface to r is

$$\Delta K = K_r - K_e = 0 - \tfrac{1}{2}mv^2 = -\tfrac{1}{2}mv^2$$

The change in potential energy is

$$\Delta U = U_r - U_e = mgR_e\left(1 - \frac{R_e}{r}\right) - 0$$

where we have used Eq. 6.11 for U_r.

Since the total energy of the rocket does not change, we have

$$\Delta E = \Delta K + \Delta U = 0$$

or

$$-\tfrac{1}{2}mv^2 + mgR_e\left(1 - \frac{R_e}{r}\right) = 0$$

This can be written

$$1 - \frac{R_e}{r} = \frac{v^2}{2gR_e}$$

Since the initial speed v is 9000 m/s, we have

$$1 - \frac{R_e}{r} = \frac{(9000 \text{ m/s})^2}{2(9.8 \text{ m/s}^2)(6.37 \times 10^6 \text{ m})} = 0.649$$

or

$$\frac{R_e}{r} = 1 - 0.649 = 0.351$$

Thus the distance r is

$$r = \frac{1}{0.351}R_e = 2.85\,R_e$$

This is the distance from the earth's center. The height of the rocket above the earth's surface is

$$h = r - R_e = 1.85\,R_e \qquad \square$$

6.4 Simple Harmonic Motion

Figure 6.11 shows a mass m connected to two springs which exert the forces $\mathbf{F}_1$ and $\mathbf{F}_2$ on it. At one point, which we can take to be $x = 0$, the total force $\mathbf{F} = \mathbf{F}_1 + \mathbf{F}_2$ is zero. This point is called the *equilibrium position* of the mass, because the mass can remain at rest there.

When the mass is displaced to some other point, say $x = A$ (Fig. 6.11b), the magnitude of $\mathbf{F}_2$ increases and the magnitude of $\mathbf{F}_1$ decreases, so $\mathbf{F} = \mathbf{F}_1 + \mathbf{F}_2$ is no longer zero, but is a force

FIGURE 6.11

A mass connected to two springs. (a) In the equilibrium position of the mass, the sum $\mathbf{F}$ of the forces $\mathbf{F}_1$ and $\mathbf{F}_2$ exerted by the springs is zero. (b) When the mass is displaced to the right, the resultant force $\mathbf{F}$ is directed to the left. (c) When the mass is displaced to the left, the resultant force $\mathbf{F}$ is directed to the right.

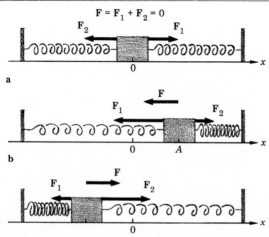

directed from m toward the equilibrium position. If the mass is released from A, the force $\mathbf{F}$ will accelerate it back toward its equilibrium position. However, when it arrives at $x = 0$, the mass will be moving to the left, so it will move out along the negative axis. Once the mass is to the left of its equilibrium position (Fig. 6.11c), the force $\mathbf{F}$ will be directed to the right, and it will slow down the mass and eventually accelerate it back to the right. If the mass is sliding on a frictionless surface, it will continue to oscillate back and forth about its equilibrium position.

In the simplest case, the total force on the mass is the *harmonic-oscillator force*

$$F = -kx \qquad\qquad 6.12$$

Here k is a constant, and x is the distance of the mass from its equilibrium position. The minus sign indicates that the force is opposite the displacement. That is, when the mass is on the right (x positive), the force is directed to the left (F negative), and when the mass is on the left (x negative), the force is directed to the right (F positive). Whatever the position of the mass, the force tends to push it back toward the equilibrium position. Any force with this property is called a *restoring force*, because it tends to restore the mass to its equilibrium position. A mass subjected to a restoring force will oscillate about its equilibrium position.

When the restoring force is the harmonic-oscillator force (Eq. 6.12), the resulting motion is called *simple harmonic motion*. The position x at any time t of a mass executing simple harmonic motion is given by

$$x = A \cos\left(\frac{2\pi}{\tau}t\right) \qquad\qquad 6.13$$

where

$$\tau = 2\pi\sqrt{\frac{m}{k}} \qquad\qquad 6.14$$

is the *period* and A is the *amplitude* of the motion.

To see how Eq. 6.13 describes the motion, we calculate x at times $t = 0, \frac{1}{4}\tau, \frac{1}{2}\tau, \frac{3}{4}\tau$, and τ.

At $t = 0$,

$$x = A \cos\left(\frac{2\pi}{\tau}0\right) = A \cos 0 = A(1) = A$$

At $t = \frac{1}{4}\tau$,

$$x = A \cos\left(\frac{2\pi}{\tau}\frac{1}{4}\tau\right) = A \cos \frac{1}{2}\pi = A(0) = 0$$

(Here we have used the fact that $\frac{1}{2}\pi$ is the radian equivalent of 90°.) At $t = \frac{1}{2}\tau$,

$$x = A \cos\left(\frac{2\pi}{\tau}\frac{1}{2}\tau\right) = A \cos \pi = A(-1) = -A$$

since π is the radian equivalent of 180°. At $t = \frac{3}{4}\tau$,

$$x = A \cos\left(\frac{2\pi}{\tau}\tfrac{3}{4}\tau\right) = A \cos\left(\tfrac{3}{2}\pi\right) = A(0) = 0$$

since $\tfrac{3}{2}\pi$ is the radian equivalent of 270°. At $t = \tau$,

$$x = A \cos\left(\frac{2\pi}{\tau}\tau\right) = A \cos 2\pi = A(1) = A$$

since $\cos 2\pi = \cos 360° = \cos 0° = 1$.

These calculations show that in a time interval equal to the period τ, the mass moves from its maximum positive position A to 0, to its maximum negative position $-A$, back to 0, and finally back to A. Such a complete oscillation is called a *cycle*. The period τ is the time required to execute one cycle, and the amplitude A is the maximum displacement during a cycle. Figure 6.12a is a graph of the position of the mass against time.

The speed and acceleration of the mass at any time are given by

$$v = -\frac{2\pi A}{\tau} \sin\left(\frac{2\pi}{\tau}t\right) \qquad\qquad 6.15$$

and

$$a = -\left(\frac{2\pi}{\tau}\right)^2 A \cos\left(\frac{2\pi}{\tau}t\right) \qquad\qquad 6.16$$

The speed and acceleration also complete one full cycle in the period τ. Graphs of these quantities are shown in Fig. 6.12b.

Example 6.10 An object is executing simple harmonic motion with an amplitude of 0.08 m and a period of 2 s. What are the maximum speed and acceleration of the mass?

Since the sine and cosine functions are always between -1

FIGURE 6.12

(a) Position x of a mass executing simple harmonic motion plotted against time t. (b) The speed v and acceleration a of the mass plotted against time t.

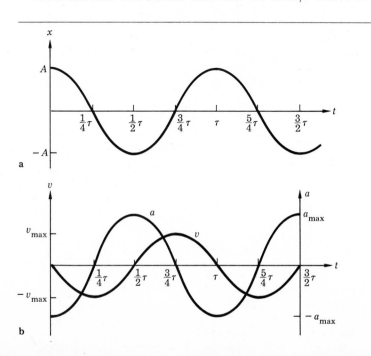

and 1, the maximum values of the speed and acceleration occur when the sine and cosine have these values. From Eq. 6.15 the maximum speed is

$$v_{max} = \frac{2\pi A}{\tau} = \frac{2\pi(0.08 \text{ m})}{2 \text{ s}} = 0.25 \text{ m/s}$$

From Eq. 6.16 the maximum acceleration is

$$a_{max} = \left(\frac{2\pi}{\tau}\right)^2 A = \left(\frac{2\pi}{2 \text{ s}}\right)^2 (0.08 \text{ m}) = 0.79 \text{ m/s}^2$$

From Fig. 6.12 we see that the maximum velocity occurs at $x = 0$, i.e., as the mass passes through its equilibrium position. The maximum acceleration occurs at $x = \pm A$, i.e., when the mass is farthest from its equilibrium position. ☐

Example 6.11 Show that the simple-harmonic-motion equations are consistent with Newton's second law of motion.
  For the harmonic-oscillator force (Eq. 6.12), Newton's second law ($F = ma$) becomes

$$-kx = ma$$

The values of x and a at any time t are given by Eqs. 6.13 and 6.16. When these equations are substituted into the last equation, we get

$$-kA \cos\left(\frac{2\pi}{\tau} t\right) = -m\left(\frac{2\pi}{\tau}\right)^2 A \cos\left(\frac{2\pi}{\tau} t\right)$$

The common factor $-A \cos\left(\frac{2\pi}{\tau} t\right)$ cancels on both sides, leaving

$$k = m\left(\frac{2\pi}{\tau}\right)^2 \qquad \text{or} \qquad \tau = 2\pi\sqrt{\frac{m}{k}}$$

Newton's second law will be satisfied if this last equation is true. But this is the condition already stated in Eq. 6.14. Thus the simple-harmonic-motion equations are consistent with Newton's second law. ☐

POTENTIAL ENERGY

The potential energy of a mass subjected to the harmonic-oscillator force (Eq. 6.12) is

$$U = \tfrac{1}{2}kx^2 \qquad\qquad 6.17$$

As the mass oscillates back and forth about its equilibrium position, the sum of its kinetic and potential energies remains constant:

$$K + U = \tfrac{1}{2}mv^2 + \tfrac{1}{2}kx^2 = E \qquad\qquad 6.18$$

This relation is illustrated in Fig. 6.13, which shows a graph of U against x. A horizontal line is drawn to represent the constant energy E of a mass executing simple harmonic motion with amplitude A. At any point along the line, the potential and kinetic energies of the mass are given by the vertical lines U and

FIGURE 6.13

Harmonic-oscillator potential energy U plotted against the displacement x of the mass from equilibrium. The sum E of the kinetic energy K and the potential energy U of the mass remains constant as the mass moves between $-A$ and A.

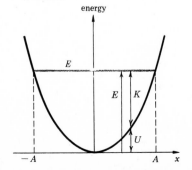

K drawn from the x axis to the curve and from the curve to the horizontal line. This diagram shows that at maximum displacement (A and $-A$) the kinetic energy is zero, and all the energy is in the form of potential energy. At $x = 0$ the potential energy is zero, and all the energy is in the form of kinetic energy. At intermediate positions the energy is a mixture of potential and kinetic energy.

Simple harmonic motion is important because it is exhibited, at least approximately, by many mechanical systems. For example, consider a simple pendulum, which is just a mass m suspended from a cord of length L (Fig. 6.14). When the mass is pulled to one side and released, it oscillates in an arc about its lowest (equilibrium) position.

When the mass is pulled a horizontal distance x from its equilibrium position, it is raised a vertical distance h, so its gravitational potential energy is

$$U = mgh$$

Figure 6.14 shows that L is the hypotenuse of a right triangle with sides x and y, so from the pythagorean theorem

$$L^2 = x^2 + y^2$$

But $y = L - h$, so this can be written

$$L^2 = x^2 + (L - h)^2$$
$$= x^2 + L^2 - 2Lh + h^2$$

or

$$2Lh = x^2 + h^2$$

For small displacements from equilibrium, h will be much smaller than L, so we can drop the term h^2 in comparison with $2Lh$ in the last equation. To this approximation we get

$$2Lh = x^2 \qquad \text{or} \qquad h = \frac{1}{2L}x^2$$

Thus the potential energy is

$$U = mgh = \tfrac{1}{2}\frac{mg}{L}x^2$$

This is identical to the harmonic-oscillator potential energy (Eq. 6.17) with $k = mg/L$. Consequently, from Eq. 6.14 the period of the pendulum is

$$\tau = 2\pi\sqrt{\frac{m}{mg/L}} = 2\pi\sqrt{\frac{L}{g}} \qquad\qquad 6.19$$

provided the displacement is sufficiently small to justify the approximation.

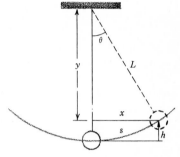

FIGURE 6.14

A simple pendulum. The height h of the mass is $h = L - L\sin\theta$, where L is the length of the pendulum, and θ is the angular displacement.

Example 6.12 What is the length of a pendulum with a period of 1 s?

From Eq. 6.19 we have

$$\tau^2 = (2\pi)^2\frac{L}{g}$$

or

$$L = \frac{g\tau^2}{(2\pi)^2} = \frac{(9.8 \text{ m/s}^2)(1 \text{ s})^2}{(6.28)^2} = 0.248 \text{ m}$$

REMARK The pendulum has been used for centuries as the basic timing mechanism for clocks because its period is independent of its amplitude, provided only that the amplitude is small. Modern technology has replaced the pendulum with atomic and crystal oscillators of much greater accuracy. □

6.5 Rigid-Body Motion

A *rigid body* is an extended object that does not change its shape as it moves. We are interested in studying the motion of such an object in the special case in which the object rotates about a fixed axis. This case is of great practical importance because it includes the motion of wheels, turbines, and propellers.

Figure 6.15 shows a rigid body rotating about an axis through A. As the body rotates, each point in the body moves in a circle about A. The radius of the circle is equal to the distance r from A to the point. Because the body is rigid, all points rotate through the same angle in the same time t.

FIGURE 6.15

A rigid body rotating about an axis through A.

Definition The *angular speed* ω of a point moving in a circle is

$$\omega = \frac{\theta}{t} \qquad\qquad 6.20$$

where θ is the angle through which the object rotates in time t. All points of a rigid body have the same angular speed, since they all rotate through the same angle in the same time.

UNITS Angles are measured in degrees, radians, and revolutions. A complete circle is 360°, 2π rad, or 1 r, so these units are related by

$$360° = 2\pi \text{ rad} = 1 \text{ r}$$

The SI unit of angular speed is radians per second (rad/s). Other common units of angular speed are revolutions per second (r/s) and revolutions per minute (r/min). The conversion between radians per second and revolutions per minute is

$$1 \text{ r/min} = \frac{2\pi \text{ rad}}{60 \text{ s}} = 0.105 \text{ rad/s}$$

FIGURE 6.16

The arc s of a circle is equal to $r\theta$, where r is the radius of the circle, and θ is the angle (in radians) subtended by the arc.

Figure 6.16 shows a point on a circle of radius r rotating through an angle θ. When θ is measured in radians, the distance s that the point moves along the circle is related to θ by

$$\theta = \frac{s}{r} \qquad \text{or} \qquad s = \theta r$$

Dividing both sides of this equation by the time t of the motion, we get

$$\frac{s}{t} = \frac{\theta r}{t}$$

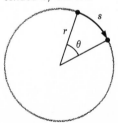

*ω is the Greek (lowercase) letter omega.

or

$$v = \omega r \qquad\qquad 6.21$$

where $\omega = \theta/t$ is the angular speed in radians per second, and $v = s/t$ is the linear speed.

Example 6.13 What is the linear speed of a point on the edge of a 12-in record rotating at $33\frac{1}{3}$ r/min?
  The radius of the record is

$$r = 6 \text{ in} = 0.5 \text{ ft} = 0.152 \text{ m}$$

and its angular speed is

$$\omega = 33.33 \text{ r/min} = (33.33)(0.105 \text{ rad/s})$$
$$= 3.5 \text{ rad/s}$$

So from Eq. 6.21 the linear speed at the edge is

$$v = \omega r = (3.50 \text{ rad/s})(0.152 \text{ m}) = 0.53 \text{ m/s}$$

> **REMARK** Notice that we dropped rad in the final unit expression. This is because the radian, being the ratio of two lengths, is a dimensionless quantity. We keep rad in the expression for the angular speed, however, in order to remind us of the angular measure we are using. □

All points in a rigid body have the same angular speed, but the linear speed of a point depends on its distance r from the axis of rotation. The rigid body in Fig. 6.17 has been divided into a number of small areas, which are labeled by the mass of the area. Thus m_5 is the mass of area 5, and m_i is the mass of any area i. Each area is small enough that all points in it are approximately the same distance r_i from the axis of rotation A.
  The linear speed of area i is

$$v_i = \omega r_i$$

and so the kinetic energy of the area is

$$K_i = \tfrac{1}{2}m_i v_i{}^2 = \tfrac{1}{2}m_i (\omega r_i)^2$$
$$= \tfrac{1}{2}m_i \omega^2 r_i{}^2$$

The total kinetic energy K of the rigid body is the sum of the kinetic energies of all its areas, so

$$K = \Sigma K_i$$

where Σ indicates the sum over all areas. From the equation for K_i we get

$$K = \Sigma K_i = \Sigma \tfrac{1}{2}m_i \omega^2 r_i{}^2$$

But $\tfrac{1}{2}\omega^2$ is a factor common to all the terms in the sum, so it can be factored outside the summation sign, giving us

$$K = \tfrac{1}{2}\omega^2 \Sigma m_i r_i{}^2 \qquad \text{or} \qquad K = \tfrac{1}{2}I\omega^2 \qquad\qquad 6.22$$

where

$$I = \Sigma m_i r_i{}^2 \qquad\qquad 6.23$$

is a constant called the *moment of inertia* of the rigid body.
  Equation 6.22 expresses the rotational kinetic energy of a rigid

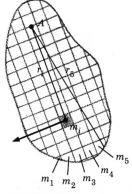

FIGURE 6.17

A rigid body divided into a number of small areas. The ith area has a mass m_i and is a distance r_i from the axis of rotation.

body rotating with angular speed ω in a form analogous to the normal kinetic energy of a mass moving with linear speed v. The moment of inertia I plays the role of the mass m, and the angular speed plays the role of the linear speed.

The moment of inertia depends on the geometry of the body and on the axis of rotation, but it does not depend on the speed or position of the body. Table 6.2 gives the moment of inertia of some common geometric shapes. In each, mass m of the body is assumed to be distributed uniformly throughout the body.

Example 6.14 What is the rotational kinetic energy of a 500-kg solid cylinder of radius 0.8 m rotating at 750 r/min?

From Table 6.2 moment of inertia of the cylinder is $I = \frac{1}{2}mr^2 = \frac{1}{2}(500 \text{ kg})(0.8 \text{ m})^2 = 160 \text{ kg} \cdot \text{m}^2$. And angular speed is $\omega = 750 \text{ r/min} = (750)(0.105 \text{ rad/s}) = 78.8 \text{ rad/s}$. Therefore the rotational kinetic energy is $K = \frac{1}{2}I\omega = \frac{1}{2}(600 \text{ kg} \cdot \text{m}^2)(78.8 \text{ rad/s})^2 = 4.97 \times 10^5 \text{ kg} \cdot \text{m}^2/\text{s}^2 = 4.97 \times 10^5 \text{ J}$. ☐

TABLE **6.2**

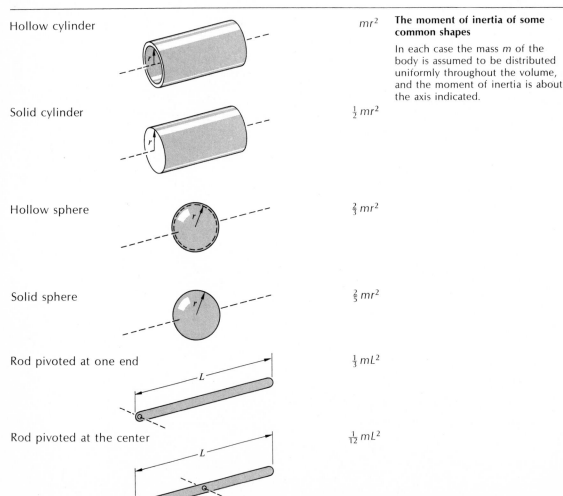

Hollow cylinder	mr^2
Solid cylinder	$\frac{1}{2}mr^2$
Hollow sphere	$\frac{2}{3}mr^2$
Solid sphere	$\frac{2}{5}mr^2$
Rod pivoted at one end	$\frac{1}{3}mL^2$
Rod pivoted at the center	$\frac{1}{12}mL^2$

The moment of inertia of some common shapes

In each case the mass m of the body is assumed to be distributed uniformly throughout the volume, and the moment of inertia is about the axis indicated.

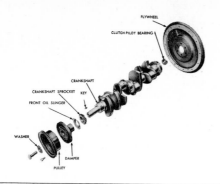

FIGURE 6.18

Exploded view of an automobile
crankshaft, showing the flywheel.
(*General Motors Corporation.*)

THE FLYWHEEL

A *flywheel* is a rotating wheel used to store energy. Automobiles
have a flywheel attached to the crankshaft (Fig. 6.18) to smooth
out the energy delivered by the engine cylinders. As each cylinder fires, it delivers an impulse of energy to the crankshaft. The
purpose of the flywheel is to increase the total moment of
inertia of the crankshaft, so that the angular speed of the shaft is
not changed much by the firing of a single cylinder. The flywheel thus keeps the crankshaft rotating at a nearly constant
speed between the power strokes of the cylinders.

The desirability of using high-speed flywheels to store large
quantities of energy has long been recognized. The main difficulty is that as the speed of the flywheel is increased, so are the
internal centripetal forces that keep each part of the wheel
moving in a circle. At very high speeds these forces will exceed
the tensile strength of the material, and the flywheel will break
apart. The development in recent years of new synthetic fibers
which are much stronger and lighter than steel may make possible some exciting new uses of the flywheel.

One idea is to power an automobile entirely by the energy
stored in a large flywheel. A 100-kg flywheel with a radius of
0.8 m would have a rotational kinetic energy of 1.1×10^8 J at a
rotational speed of 25,000 r/min. At such a high speed the flywheel would have to be sealed inside a vacuum chamber to
eliminate air resistance. Estimates are that in a vacuum the
flywheel could remain spinning for several months.

The flywheel automobile would be driven by electric motors
attached to the wheels. Electricity to run these motors would
come from an electric generator connected to the flywheel. Like
an electric car, the flywheel car would have to be energized
periodically from an external source of electricity. In an electric
car, the electricity is used to charge a bank of storage batteries, a
process that takes several hours. In the flywheel car, the electricity runs a motor which spins the flywheel up to full speed, a
process that takes only 5 min.

Electric utilities are studying the possibility of using flywheels
to store large quantities of energy for use during periods of peak
load. The system would work like the pumped hydroelectric
storage system described in Sec. 5.5. Electricity generated during
off-peak periods would be used to speed up the flywheels, and

the stored energy would later be used to run electric generators
to meet peak demand.

170

MECHANICS

Example 6.15 The flywheel of a proposed electrical storage system has a mass of 2×10^5 kg and a radius of 2.5 m. If the flywheel is to store 7.2×10^{10} J (2×10^4 kWh), what is its angular speed? What is the linear speed of a point on the outer rim of the flywheel? Assume the wheel is a solid cylinder.

The moment of inertia of the flywheel is

$$I = \tfrac{1}{2}mr^2 = \tfrac{1}{2}(2 \times 10^5 \text{ kg})(2.5 \text{ m})^2$$
$$= 6.25 \times 10^5 \text{ kg} \cdot \text{m}^2$$

so the kinetic energy of the flywheel is

$$K = \tfrac{1}{2}I\omega^2 = (3.12 \times 10^5 \text{ kg} \cdot \text{m}^2)\omega^2$$

or

$$\omega^2 = \frac{K}{3.12 \times 10^5 \text{ kg} \cdot \text{m}^2} = \frac{7.2 \times 10^{10} \text{ J}}{3.12 \times 10^5 \text{ kg} \cdot \text{m}^2}$$
$$= 2.3 \times 10^5 \text{ rad}^2/\text{s}^2$$

Thus the angular speed of the flywheel is

$$\omega = 480 \text{ rad/s} = 4.58 \times 10^3 \text{ r/min}$$

and the linear speed of the rim is

$$v = \omega r = (480 \text{ rad/s})(2.5 \text{ m}) = 1.2 \times 10^3 \text{ m/s}$$
$$= 2680 \text{ mi/h} \qquad \square$$

GUIDE TO MAJOR TOPICS

Topic	References	Problems
Parabolic motion	Sec. 6.1; Eqs. 6.1 to 6.4; Examples 6.1 to 6.3	1 to 6
Uniform circular motion	Sec. 6.2; Eqs. 6.5 to 6.7; Examples 6.4 to 6.5	7 to 12
Law of universal gravitation	Sec. 6.3; Eqs. 6.8 to 6.10	13 to 16
Circular motion	Examples 6.6 to 6.7	17 to 20
Potential energy	Eq. 6.11; Examples 6.8 to 6.9	21 to 24
Simple harmonic motion	Sec. 6.4; Eqs. 6.12 to 6.16; Examples 6.10 to 6.11	25 to 30
Energy conservation	Eqs. 6.17 to 6.18	31 to 34
Pendulum	Eq. 6.19; Example 6.12	35 to 36
Rigid-body motion	Sec. 6.5; Eqs. 6.20 to 6.23; Examples 6.13 to 6.15; Table 6.2	37 to 42

PROBLEMS

1 A girl throws a stone with an initial speed of 12 m/s. As the stone leaves her hand it is moving horizontally at a height of 1.5 m above the ground. (a) How long does it take the stone to hit the ground? (b) How far from the girl does the stone hit the ground? (c) What is the speed of the stone when it hits the ground? *Ans.* (a) 0.553 s; (b) 6.64 m; (c) 13.2 m/s

2 An artillery shell has an initial velocity with components $v_{0x} = 50$ m/s and $v_{0y} = 40$ m/s. (a) Find the horizontal coordinate x of the shell at times 0.5, 1.0, 1.5, 2.0, 2.5, 3.0, 3.5, and 4.0 s. (b) Find the vertical coordinate y of the shell at these same times. (c) Plot y against x and draw a smooth curve through the points.

3 The initial velocity of a golf ball has a magnitude of 30 m/s and makes an angle of 25° with the horizontal. (a) How long does it take the ball to reach its maximum height? (*Hint:* At its maximum height $v_y = 0$.) (b) What is the maximum height reached by the ball? (c) What was the range of the ball?
Ans. (a) 1.30 s; (b) 16.5 m; (c) 70.7 m

4 A motorcycle runs off a ramp that is inclined 30° to the horizontal. The motorcycle travels a horizontal distance of 20 m before landing on a second ramp of the same height. What was the speed of the motorcycle as it left the first ramp?

5 A batter strikes a baseball, giving it an initial velocity with an x component of 30 m/s. The ball just clears a 5-m-high fence that is 75 m from the batter. (a) What is the y component of the ball's initial velocity? (b) What angle does the initial velocity make with the horizontal?
Ans. (a) 14.25 m/s; (b) 25.4°

6 At time $t = 0$ the velocity of a missile has a magnitude of 400 m/s and makes an angle of 70° with the horizontal. (a) Find the magnitude and direction of the velocity at time $t = 5$ s. (b) What is the position of the missile at $t = 5$ s?

7 A 2000-kg automobile goes around a curve of radius 120 m at a speed of 25 m/s. (a) What is the centripetal acceleration of the automobile? (b) What is the centripetal force on the automobile? (c) What is the minimum value of the coefficient of friction between the tires and the road required to prevent the automobile from skidding?
Ans. (a) 5.21 m/s²; (b) 1.04×10^4 N; (c) 0.53

8 A merry-go-round 4 m in radius makes a complete revolution every 10 s. (a) What is the speed of a point on the outer edge of the merry-go-round? (b) A 30-kg child is sitting on the edge of the merry-go-round. What force is required to prevent the child from sliding off?

9 The moon travels in an approximately circular orbit of radius 3.8×10^8 m about the earth, completing one revolution every 27.3 days. (a) What is the speed of the moon? (b) What is the moon's centripetal acceleration?
Ans. (a) 1.01×10^3 m/s; (b) 2.68×10^{-3} m/s²

10 (a) What is the speed of a Viking satellite that orbits Mars near its surface. (b) What is the period of the satellite? (Use data from Appendix VI.)

11 A phonograph record is rotating at 45 r/min. (a) What is the speed of a point 8 cm from the center? (b) What is the centripetal acceleration of this point? (c) If a penny is placed on this point, what is the minimum value of the coefficient of friction between the penny and the record that prevents the penny from sliding?
Ans. (a) 0.377 m/s; (b) 1.78 m/s²; (c) 0.18

12 A man swings a bucket of water in a vertical circle of radius r (Fig. 6.19). Show that the water will not spill out of the bucket if the speed of the bucket at its highest point is equal to or greater than $\sqrt{gr}$.

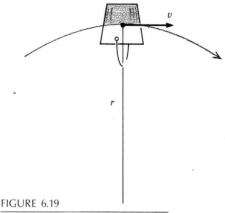

FIGURE 6.19

Problem 12.

13 What is the force of gravity that a 2×10^{15} kg asteroid exerts on a 4×10^4 kg spacecraft at a distance of 10^4 m?
Ans. 53.4 N

14 Find the force of gravity on an 80-kg astronaut (a) on the surface of the earth, (b) in orbit 200 km above the earth's surface, and (c) halfway to the moon, a distance of 1.9×10^5 km from the earth's center.

15 What is the acceleration due to the earth's gravity at a distance of 3.8×10^8 m from the earth's center?
Ans. 2.75×10^{-3} m/s²

REMARK This is the distance to the moon. Compare this answer to the answer to Prob. 9b. It was the similarity of the moon's centripetal acceleration and the acceleration due to gravity at the distance of the moon that led Newton to associate the centripetal force required to keep the moon in orbit about the earth with the earth's gravitational force.

16 Calculate the acceleration of gravity on the surface of Mercury using the data in Appendix VI.

17 Find the period of a satellite that orbits the earth at a distance of 2×10^7 m from the earth's center.
Ans. 7.83 h

18 Calculate the earth's orbital speed about the sun, using Eqs. 6.7, 6.8, and Appendix VI.

19 When viewed through a telescope, many bright stars are seen to consist of two stars in close proximity. If one star is much less massive than the other, it will orbit the more massive star just as the earth orbits the sun. Suppose the distance between two such stars is 4×10^{12} m and the period is 60 years. (a) What is the velocity of the less massive star? (b) What is its centripetal acceleration? (c) What is the mass of the more massive star? *Ans.* (a) 1.33×10^4 m/s; (b) 4.4×10^{-5} m/s^2; (c) 1.06×10^{31} kg

20 The length of a day on Mars is 24.6 h. Calculate the radius of a satellite in synchronous orbit about Mars.

21 What must be the initial speed of a rocket launched from earth, if it is to reach a maximum height of 1.2×10^7 m from the earth's center?
Ans. 7.65×10^3 m/s

22 With what speed must a rocket be launched from the earth's surface in order to have a speed of 5 km/s when it is infinitely far from earth?

23 What is the escape speed from Mars?
Ans. 5.15×10^3 m/s

24 A meteorite heading toward earth has a speed of 3 km/s when it is 2.5×10^7 m from the earth's center. What is the meteorite's speed when it reaches the earth's surface?

25 Figure 6.20 is a plot of position x against time t for an object executing simple harmonic motion. (a) What are the amplitude and period of the oscillation? (b) What are the maximum speed and acceleration of the object? (c) Where is the object when it has its maxi-

FIGURE 6.20

Problem 25.

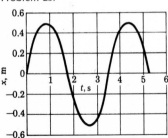

mum positive acceleration?
Ans. (a) 0.5 m, 3.5 s; (b) 0.898 m/s, 1.61 m/s^2; (c) -0.5 m

26 (a) Write the equations for the position, speed, and acceleration of an object executing simple harmonic motion with a period of 5 s and an amplitude of 0.03 m. (b) Draw a graph of position x against time t for the object.

27 The position (in meters) of an object executing simple harmonic motion is given by $x = 0.25 \cos 5t$. (a) What are the amplitude and period of the oscillation? (b) What are the maximum speed and acceleration of the object? (c) What are the position, speed, and acceleration of the object at time $t = 0.15$ s?
Ans. (a) 0.25 m, 1.26 s; (b) 1.25 m/s, 6.25 m/s^2; (c) 0.183 m, -0.852 m/s, -4.57 m/s^2

28 An object starting from rest moves a distance A in time $\frac{1}{4}\tau$ with constant acceleration. Compare its acceleration with the maximum acceleration of an object executing simple harmonic motion with amplitude A and period τ.

29 When a 0.5-kg mass is suspended from a spring, the spring stretches 4 cm. The mass is then displaced 2 cm from its equilibrium position and released. (a) What is the spring constant? (b) What is the period of the subsequent oscillation? (c) What is the speed of the mass as it passes through its equilibrium position?
Ans. (a) 122.5 N/m; (b) 0.40 s; (c) 0.313 m/s

30 A 2.5-kg mass suspended from a spring is executing simple harmonic motion with a period of 1.5 s. The mass has a speed of 0.75 m/s as it passes through its equilibrium position. (a) What is the spring constant? (b) What is the amplitude of the oscillation?

31 (a) Plot the potential energy of a harmonic-oscillator force with force constant $k = 250$ N/m for values of x between -10 and $+10$ cm. (b) On your graph draw a horizontal line to represent an object executing simple harmonic motion with a total energy E of 0.75 J. (c) What is the amplitude of the object's oscillation? (d) At what positions is its potential energy 0.50 J? (e) At what positions is its kinetic energy 0.50 J?
Ans. (c) 0.0775 m; (d) ±0.0632 m; (e) ±0.0447 m

32 (a) Show that the speed v of an object of mass m executing simple harmonic motion with amplitude A is given by

$$v = \sqrt{\frac{k}{m}(A^2 - x^2)}$$

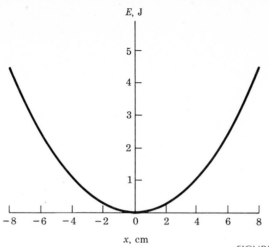

E, J

x, cm

FIGURE 6.21

Problem 34.

where k is the spring constant. (b) Show that at $x = \frac{1}{2}A$, the speed of an object is $\frac{1}{2}\sqrt{3}$ times its maximum speed. (c) At what displacement, expressed as a fraction of A, is the kinetic energy equal to the potential energy?

33 An object is subject to a harmonic-oscillator force with force constant $k = 300$ N/m. At time $t = 0$ the object is executing simple harmonic motion with an amplitude of 7 cm, but due to air resistance the amplitude is reduced to 4 cm at time $t = 5$ min. What is the change of mechanical energy of the system during this time interval?
Ans. −0.495 J

34 Figure 6.21 is a plot of the potential energy of a certain harmonic-oscillator force. Consider a mass moving in this force with a total energy of 2.5 J. (a) What is the amplitude of the motion? (b) What is the kinetic energy of the mass at $x = 4$ cm. (c) What is the force constant k?

35 What is the period of a pendulum of length 2 m?
Ans. 2.84 s

36 The length of a clock's pendulum increases by 0.1 percent as a result of a change in room temperature. By what percentage does the period of the pendulum change?

37 (a) Find the earth's angular speed in radians per second. (b) What is the linear speed of a point on the earth's equator?
Ans. (a) 7.27×10^{-5} rad/s; (b) 463 m/s

38 The 0.5-m-long rod in Fig. 6.22 has a mass of 0.25 kg. A 0.5-kg weight is attached to the rod at a point 0.4 m from the rotational axis A. Find the total moment of inertia of the rod and weight.

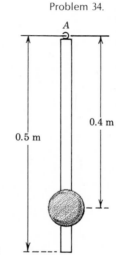

FIGURE 6.22

Problem 38.

39 (a) Calculate the earth's moment of inertia, assuming the earth is a sphere of uniform density. (b) What is the earth's rotational kinetic energy (see Prob. 37)?
Ans. (a) 9.7×10^{37} kg · m²; (b) 2.6×10^{29} J

40 The wheels of a bicycle are 68.5 cm (27 in) in diameter. (a) What is the angular speed of a wheel when the linear speed of the bicycle is 5 m/s. (b) The mass of a wheel is 5 m/s. Assuming that this mass is all concentrated on the rim, what is the rotational kinetic energy of a wheel?

41 A flywheel with a moment of inertia of 5000 kg · m² is used to store 8×10^7 J of energy. What is the angular speed of the flywheel?
Ans. 1704 r/min

42 (a) Calculate the centripetal acceleration

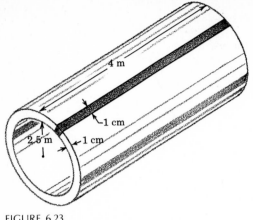

of a point on the outer edge of the flywheel described in Example 6.15. (*b*) The density of the material of which the flywheel is made is 2500 kg/m³ and the wheel is 4 m thick. Calculate the centripetal force on a narrow strip 1 cm wide, 1 cm deep, and 4 m long located on the outer edge of the wheel (Fig. 6.23). (*c*) Calculate the stress on this strip. Compare your answer with the tensile strength of steel.

FIGURE 6.23

Problem 42.

BIBLIOGRAPHY

FEYNMAN, RICHARD P.: *The Character of Physical Law,* M.I.T. Press, Cambridge, Mass., 1965. Chapter 1 gives a very readable account of the origin and development of the law of gravitation.

POST, RICHARD F., and STEPHEN F. POST: "Flywheels," *Scientific American* **229**:17 (December 1973). Descriptions of possible flywheels designed to store energy for electric power and to propel trucks and automobiles.

A solid is a rigid substance that maintains its shape against external distorting forces, whereas a *fluid* is a nonrigid substance (gas or liquid) that does not maintain its shape against such forces. Instead, a fluid will flow when acted upon by distorting forces. *Fluid mechanics* is the study of fluids at rest and in motion. It has application to such diverse fields as oceanography, meteorology, hydraulic engineering, and aerodynamics. In this chapter we shall obtain the basic laws of fluid mechanics by applying Newton's laws of motion to a deformable substance, and then we shall apply these laws to problems in deep-sea diving, fluid flow, and aerodynamics.

7
FLUID MECHANICS

7.1 Hydrostatics

Hydrostatics is the branch of fluid mechanics which deals with fluids at rest. Because a fluid is nonrigid, it can remain at rest only in the absence of distorting forces. For example, consider the cube in Fig. 7.1a, which is being acted upon by four equal forces directed parallel to four of the cube's sides. The forces are directed so that the total force and total torque on the cube are zero. If the cube is a solid, it will be deformed slightly by such a system of forces (Sec. 3.4), but it will resist major deformation. Thus a solid cube will be in equilibrium under these forces. However, if the cube is a fluid, its deformation (Fig. 7.1b) will increase as long as there are any forces acting parallel to a fluid surface. That is, a fluid always flows under the action of parallel forces, so consequently, if a fluid is at rest, there can be no parallel forces acting on it. This important fact can be stated as follows:

Fundamental Property of Fluids *Any force exerted on or by a fluid at rest must be perpendicular to the surface on which it acts.*

To prove this, consider the force **F** in Fig. 7.2 which is acting on the fluid surface shown. If the force is not perpendicular to the surface, it will have a component $\mathbf{F}_\parallel$ parallel to the surface. This parallel component will cause the fluid to flow. So if the fluid is to remain at rest, the parallel component must be zero; i.e., the force **F** must be perpendicular to the surface. Likewise, the fluid cannot itself exert a parallel force on a solid surface, because the surface would exert a parallel reaction force back on the fluid (Newton's third law). Thus, as long as the fluid is at rest, no parallel force can be exerted on or by the fluid.

Another way to say this is that a *fluid has no static coefficient of friction.* Imagine a boat floating on water. If a force **F** is applied to the boat parallel to the water, the boat will not remain at rest no matter how small **F** is because the water cannot apply a parallel force to balance **F**. Once the boat starts to move, however, the situation changes, since the fluid is now moving relative to the boat. A moving fluid does exert a force parallel to a surface, the magnitude of which increases with speed (Sec. 7.5). Consequently, the boat accelerates under the

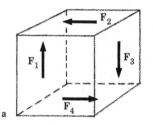

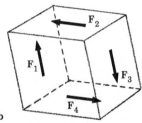

FIGURE 7.1

(a) A cube acted upon by four forces that are parallel to four of its faces. (b) The deformation caused by the forces.

FIGURE 7.2

If a force **F** is not perpendicular to a surface, it will have a component $\mathbf{F}_\parallel$ parallel to the surface.

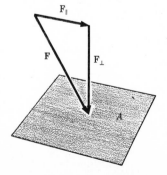

action of **F** until it reaches the speed at which the magnitude of the frictional force of the water on it equals the magnitude of **F**. Similarly, the air (which is a fluid) offers little resistance to slowly moving objects, but air resistance becomes large for high-speed objects.

REMARK A lubricant reduces the friction between two solid objects by introducing a thin layer of fluid, such as oil, between their surfaces. Since the fluid itself can exert no static friction, the friction between the surfaces is greatly reduced.

Definition *Pressure p* is the force per unit area exerted perpendicularly on a surface. Thus the pressure exerted by the force **F** in Fig. 7.2 on the surface of area A is

$$p = \frac{F_\perp}{A}$$

where $F_\perp$ is the component of **F** perpendicular to the surface. Since all the forces in a fluid at rest are perpendicular to the surfaces on which they act, we can drop the symbol $\perp$ in the definition of p, and simply write

$$p = \frac{F}{A} \qquad\qquad 7.1$$

UNITS The dimensions of pressure are force per unit area. The SI unit of pressure is the *pascal* (Pa), which is one newton per square meter (N/m^2):

$$1\,Pa = 1\,N/m^2$$

The English engineering system unit of pressure is the pound per square foot (lb/ft^2). Other common units of pressure are atmosphere (atm), pounds per square inch (lb/in^2), millimeters of mercury (mmHg or Torr), and bars (bar). Table 7.1 gives the conversion factors from customary units to pascals. Appendix II gives the conversion between other pairs of units.

REMARK The definition of pressure is the same as the definition of normal stress given in Sec. 3.4. *Shear stress* is defined as the force per unit area

TABLE **7.1**

Unit	Symbol or alternate name	Value in pascals
Atmosphere	atm	1.013250×10^5
Bar	bar	10^5
Dyne per square centimeter	dyn/cm^2	10^{-1}
Inch of mercury	inHg	3386.388
Inch of water	inH_2O	249.082
Millimeter of mercury	mmHg or Torr	133.3224
Pascal	Pa or N/m^2	1
Pound per square foot	lb/ft^2	47.88026
Pound per square inch	lb/in^2 or psi	6894.757
Torr	Torr or mmHg	133.3224

Factors for converting from customary units of pressure to pascals

To convert between two customary units, convert from the first unit to pascals, and then from pascals to the second unit. For example, to convert from atmospheres to millimeters of mercury, write

$$1\,atm = 1.013250 \times 10^5\,Pa$$

$$1\,Pa = \frac{1}{133.3224}\,mmHg$$

so

$$1\,atm = \frac{1.013250 \times 10^5}{133.3224}\,mmHg$$

$$= 760\,mmHg$$

See Appendix II for other conversion factors.

exerted parallel to a surface. Therefore, the fundamental property of fluids is equivalent to saying that there are no shear stresses in a fluid at rest. Furthermore, in a solid the normal stress varies with the position and orientation of the surface on which the normal force acts, whereas in a fluid the pressure is the same everywhere. This remarkable fact is the content of the following law.

Pascal's Law *In the absence of gravity*, i.e., neglecting the weight of the fluid itself, *the pressure in a fluid at rest is the same everywhere.*

This law is proved by showing that the pressure is the same at any two points P and Q in the fluid. Thus choose any two points P and Q in a fluid at rest and consider the fluid inside the cylindrical region shown in Fig. 7.3. Since the fluid is at rest everywhere, the total force on this cylinder of fluid, as on every other region of fluid, must be zero. Furthermore, from the fundamental property of fluids, the forces on this region are perpendicular to its surface. Therefore, there is a force of magnitude F_P perpendicular to the cylinder at P and a force of magnitude F_Q perpendicular to the cylinder at Q. Since these forces are parallel to the long axis of the cylinder, whereas all the other forces are perpendicular to this axis, the forces $\mathbf{F}_P$ and $\mathbf{F}_Q$ must have the same magnitude if the total force parallel to the axis is to be zero. Thus we have

$$F_P = F_Q$$

Dividing both sides of this equation by the area A of either end of the cylinder, we get

$$\frac{F_P}{A} = \frac{F_Q}{A} \qquad \text{or} \qquad p_P = p_Q$$

where p_P and p_Q are the pressures at P and Q. Since P and Q are any two points in the fluid, this proves that the pressure is the same everywhere in the fluid.

> **REMARK** The fundamental property of fluids is essential in the proof of Pascal's law because it assures us that the forces on the body of the cylinder have no components parallel to the axis of the cylinder.

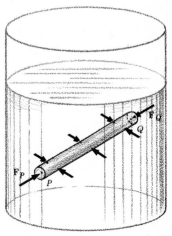

FIGURE 7.3

The forces on a cylindrical region in a fluid. The surface of the cylinder is an imaginary boundary that defines the region of interest.

FIGURE 7.4

A fluid confined to a cylinder with a movable piston of area A.

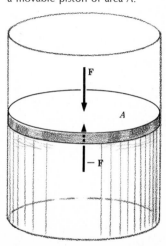

To see how these fluid properties are applied in practice, consider a fluid contained in a cylinder of cross-sectional area A (Fig. 7.4). If a downward force $\mathbf{F}$ is applied to the movable piston that closes the top of the cylinder, the fluid must apply the opposite force $-\mathbf{F}$ on the piston when the piston is at rest. Therefore, at equilibrium the pressure exerted by the fluid on the piston is $p = F/A$, which by Pascal's law is also the pressure everywhere else in the fluid. (Remember, we are neglecting gravity.)

Suppose now this cylinder is connected by a pipe to a smaller cylinder of cross-sectional area A' (Fig. 7.5). What is the magnitude F' of the force that must be applied to the smaller piston to maintain equilibrium? Since the pressure is the same everywhere in the fluid, the pressure exerted by the fluid on the small cylinder must also be $p = F/A$. On the other hand, the force exerted by the fluid on the smaller piston must have a magnitude of F' to balance the applied force, so $p = F'/A'$. Equating these two evaluations of the pressure, we have

$$p = \frac{F}{A} = \frac{F'}{A'} \quad \text{or} \quad F' = pA' = \frac{A'}{A}F$$

Example 7.1 If the cross-sectional areas of the two cylinders in Fig. 7.5 are $A = 0.1 \text{ m}^2$ and $A' = 0.02 \text{ m}^2$, what force F' must be applied to the smaller piston to balance a force $F = 900 \text{ N}$ applied to the larger piston?

The pressure in the fluid is

$$p = \frac{F}{A} = \frac{900 \text{ N}}{0.1 \text{ m}^2} = 9000 \text{ N}/\text{m}^2$$

so the upward force applied by the fluid to the smaller piston is

$$F' = pA' = (9000 \text{ N}/\text{m}^2)(0.02 \text{ m}^2) = 180 \text{ N}$$

Therefore, to keep the system in equilibrium, a downward force F' of only 180 N has to be applied to the smaller piston. This is the principle of the hydraulic press commonly used to raise large weights. In Sec. 5.1 this same principle was derived from the principle of simple machines. □

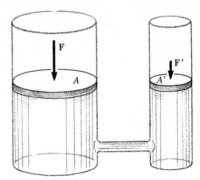

FIGURE 7.5

Two cylinders, each with a movable piston, connected by a tube.

7.2 Hydrostatic Pressure

Pascal's law is true only insofar as the force of gravity on a fluid can be neglected, in which case the pressure in the fluid can be considered to be produced entirely by applied forces, e.g., the pistons in Figs. 7.4 and 7.5.

To study the effect of gravity on fluid pressure, we consider the fluid inside the cylinder in Fig. 7.6. A force $\mathbf{F}$ is applied perpendicular to the piston, which has a cross-sectional area A, so the pressure directly under the piston is

$$p_0 = \frac{F}{A}$$

FIGURE 7.6

The forces on the fluid in a cylinder. $\mathbf{F}_g$ is the force of gravity on the fluid, $\mathbf{F}$ is the force exerted by the piston, and $\mathbf{F}_n$ is the normal force exerted by the base of the cylinder.

The subscript 0 indicates that this is the pressure at the top of the fluid. From Pascal's law, the pressure p_h at the bottom of the fluid would equal p_0 if gravity were neglected. However, because of gravity, the total downward force on the fluid is $\mathbf{F} + \mathbf{F}_g$, where $\mathbf{F}_g$ is the force of gravity on the fluid. Since the fluid is in equilibrium, there must be an upward normal force $\mathbf{F}_n = -(\mathbf{F} + \mathbf{F}_g)$ exerted on the fluid by the bottom of the cylinder. The reaction to $\mathbf{F}_n$ is the force $\mathbf{R}_n = -\mathbf{F}_n = \mathbf{F} + \mathbf{F}_g$ that the fluid exerts downward on the bottom of the cylinder. Thus the pressure p_h at the bottom is

$$p_h = \frac{F + F_g}{A} = \frac{F}{A} + \frac{F_g}{A} = p_0 + \frac{F_g}{A} \qquad 7.2$$

The pressure at the bottom of the fluid is greater than at the top because of the weight of the fluid itself.

This increase in pressure with depth is related to the density ρ of the fluid. Recall (Sec. 1.5) that the mass density ρ of a substance is the ratio of its mass m to its volume V:

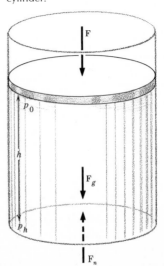

$$\rho = \frac{m}{V}$$

Thus, since the volume of the fluid in the cylinder is $V = Ah$, where h is the height of the fluid, the mass of the fluid is

$$m = \rho V = \rho Ah$$

and the weight of the fluid is

$$F_g = mg = \rho Ahg$$

Consequently, the increase in pressure at the bottom of the cylinder due to the weight of the fluid is

$$\frac{F_g}{A} = \rho gh$$

This increase is called the *hydrostatic pressure*. From Eq. 7.2 the total pressure at the bottom of the cylinder is

$$p_h = p_0 + \rho gh \qquad\qquad 7.3$$

There is, of course, nothing special about the bottom of a container. The pressure at any point inside the fluid is also given by Eq. 7.3, where h is the distance of the point from the surface. The pressure in a fluid thus increases steadily with depth below the surface.

Consider two points A and B in a fluid (Fig. 7.7). Let A be a distance h_A below the surface, and let B be a distance h_B below the surface. According to Eq. 7.3, the pressures p_A and p_B at these points are

$$p_A = p_0 + \rho gh_A \qquad \text{and} \qquad p_B = p_0 + \rho gh_B$$

We subtract the second equation from the first to get

$$p_A - p_B = (p_0 + \rho gh_A) - (p_0 + \rho gh_B)$$
$$= \rho gh_A - \rho gh_B = \rho g(h_A - h_B)$$

where $h_A - h_B$ is the vertical distance between points A and B. This is the generalization of Pascal's law needed to include the effect of gravity. It can be stated formally as follows:

The Law of Hydrostatic Pressure *The pressure in a fluid at rest is the same at all points at equal depth, and the difference in pressure between two points A and B at depths h_A and h_B is*

$$p_A - p_B = \rho gh_A - \rho gh_B = \rho g(h_A - h_B) \qquad 7.4$$

Here h_A and h_B are positive when measured downward from the surface of the fluid (Fig. 7.7).

REMARKS (1) In the English engineering system of units, the weight density d is used in place of the mass density. The weight density of a substance is the ratio of its weight (in pounds) to its volume (in cubic feet):

$$d = \frac{W}{V}$$

In terms of the weight density, Eq. 7.4 becomes

$$p_A - p_B = d(h_A - h_B) \qquad 7.5$$

where h is in feet and p is in pounds per square foot.

FIGURE 7.7

Two points A and B in a fluid. The vertical separation h of these points is equal to the difference $h_A - h_B$ between their distances from the surface of the fluid.

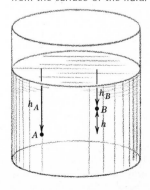

(2) Equations 7.4 and 7.5 assume that the density of the fluid is the same at all points between A and B. The density of a liquid varies only slightly with changes in temperature and pressure, so in most practical situations this assumption is valid. The density of a gas, on the other hand, does change appreciably with changes in temperature and pressure, so the density may not be the same at all points. In this case, the average density between A and B should be used.

Example 7.2 What is the pressure on a scuba diver 30 m below the surface of the ocean?

Let p_B be the pressure p_0 at the top of the ocean (atmospheric pressure), and let p_A be the pressure 30 m below the surface. Then with $p_B = p_0$ and $h_B = 0$, Eq. 7.4 can be written

$$p_A - p_0 = \rho g h_A$$

From Table 1.6 in Sec. 1.5 we find that the mass density of seawater is $1.025 \times 10^3 \, kg/m^3$. Thus the pressure p_A at depth $h_A = 30 \, m$ is

$$p_A = p_0 + \rho g h_A$$
$$= p_0 + (1.025 \times 10^3 \, kg/m^3)(9.8 \, m/s^2)(30 \, m)$$
$$= p_0 + 3.01 \times 10^5 \frac{kg \cdot m/s^2}{m^2}$$
$$= p_0 + 3.01 \times 10^5 \, N/m^2$$

Here we have used the fact that $1 \, N = 1 \, kg \cdot m/s^2$ to reduce the units to N/m^2.

The pressure p_0 at the surface is atmospheric pressure. From Table 7.1 we see that

$$1 \, atm = 1.01 \times 10^5 \, N/m^2$$

so the pressure on the diver (in atmospheres) is

$$p_A = p_0 + \frac{3.01 \times 10^5}{1.01 \times 10^5} \, atm$$
$$= p_0 + 2.98 \, atm = 3.98 \, atm$$

where we have taken $p_0 = 1 \, atm$. We see from this calculation that the pressure on a deep-sea diver increases by 1 atm for each 10 m of depth below the surface. $\square$

The atmosphere is a layer of air surrounding the earth. The weight of this air exerts a pressure at the surface of the earth which we call *atmospheric pressure*. The unit of pressure called the *atmosphere* is defined by the relation

$$1 \, atm = 760 \, mmHg = 1.0133 \times 10^5 \, Pa$$

This unit is equal to the average pressure of the atmosphere at sea level, although the actual pressure varies by about 5 percent depending on weather conditions. The average pressure at a site above sea level is less than at sea level because there is less air above it.

Equation 7.4 can be used to calculate the (average) air pressure p_A at a site above sea level, but a little care is required. The distances in Eq. 7.4 are measured from the top of the fluid, whereas the elevation of a geographic site is measured from sea level, which is at the bottom of the fluid. However, since only the difference in the distances below the surface is used in Eq. 7.4, the distances can be measured downward from any convenient reference level.

Example 7.3 What is the atmospheric pressure p_A at Mexico City, which is 1500 m above sea level?

For convenience we measure distances from Mexico City. That is, we take h_A to be zero, so that the depth at sea level is $h_B = 1500$ m. (Remember, distances are measured positively downward.) From Table 1.6 the density of air at sea level is 1.2 kg/m³. At Mexico City the density is about 1.0 kg/m³, so the average density of 1.1 kg/m³ should be used. Then Eq. 7.4 gives

$$p_A - p_B = \rho g h_A - \rho g h_B = -\rho g h_B$$
$$= -(1.1 \text{ kg/m}^3)(9.8 \text{ m/s}^2)(1.5 \times 10^3 \text{ m})$$
$$= -0.16 \times 10^5 \text{ Pa}$$
$$= -0.16 \text{ atm}$$

The pressure at Mexico City is less than the pressure at sea level, as expected. With $p_B = 1$ atm, the pressure at Mexico City is

$$p_A = p_B - 0.16 \text{ atm} = 0.84 \text{ atm} \qquad \square$$

Example 7.4 What is the total force exerted on the face of the dam in Fig. 7.8? The water is 80 ft deep and the dam is 200 ft wide.

We shall work this problem in English units, using Eq. 7.5. From Table 1.6 the weight density of water (at 20°C) is 62.4 lb/ft³, so the pressure p_A at the base of the dam is

$$p_A = p_B + dh_A$$
$$= p_B + (62.4 \text{ lb/ft}^3)(80 \text{ ft})$$
$$= p_0 + 4.99 \times 10^3 \text{ lb/ft}^2$$

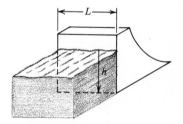

FIGURE 7.8

A dam of width L holding back water of depth h.

Here we have taken the pressure p_B at the top to be atmospheric pressure p_0. Since the same atmospheric pressure is exerted on both sides of the dam, only the hydrostatic pressure dh_A exerts a net force on the dam. The hydrostatic pressure is zero at the top of the dam and increases linearly to its maximum value of 4.99×10^3 lb/ft² at the base, so the average hydrostatic pressure is

$$p_{av} = \tfrac{1}{2}(0 + dh_A) = 2.5 \times 10^3 \text{ lb/ft}^2$$

The total submerged area of the dam is

$$A = hL = (80 \text{ ft})(200 \text{ ft}) = 16 \times 10^3 \text{ ft}$$

so the net force exerted on the dam by the hydrostatic pressure of the water is

$$F = p_{av}A$$
$$= (2.5 \times 10^3 \text{ lb/ft}^2)(16 \times 10^3 \text{ ft}^2)$$
$$= 40 \times 10^6 \text{ lb} \qquad \square$$

7.3 Pressure-Measuring Instruments

Almost every shop and laboratory has some sort of pressure-measuring device. Such instruments vary from a simple pocket gauge used to measure tire pressure to a large, mercury-filled manometer used for accurate pressure measurements. The basic operation of some of these devices is described here.

THE BAROMETER

The *barometer* is an instrument used to measure atmospheric pressure. It consists of a straight glass tube, over 76 cm long, that is closed at one end. The tube is filled with mercury and then inverted into a dish of mercury (Fig. 7.9). After the tube is inverted, the mercury column falls away from the closed end, leaving a vacuum there.

The pressure p_0 at point 0 on the surface of the mercury in the dish is the prevailing atmospheric pressure. From the principle of hydrostatic pressure (Eq. 7.4), the pressure p_A at point A inside the tube is

$$p_A = p_B + \rho g h$$

where p_B is the pressure at the top of the mercury column, ρ is the density of mercury, and h is the height of the mercury column. But the pressure p_B is zero because the space above the mercury column is a vacuum, and the pressure p_A is equal to p_0 because points A and 0 are at the same level in the same fluid. Thus the atmospheric pressure p_0 is given by

$$p_0 = \rho g h \qquad \qquad 7.6$$

Example 7.5 The height of the mercury column in a barometer is 73.2 cm. What is the atmospheric pressure in atmospheres, bars, and millimeters of mercury?

From Table 1.5 the density of mercury is found to be $13.6 \times 10^3 \ \mathrm{kg/m^3}$, so Eq. 7.6 gives

$$p_0 = \rho g h = (13.6 \times 10^3 \ \mathrm{kg/m^3})(9.8 \ \mathrm{m/s^2})(0.732 \ \mathrm{m})$$
$$= 0.976 \times 10^5 \ \mathrm{N/m^2}$$

This is easily converted to the other units using the conversion factors in Table 7.1:

$$p_0 = \frac{0.976 \times 10^5}{10^5} \ \mathrm{bar} = 0.976 \ \mathrm{bar}$$

$$= \frac{0.976 \times 10^5}{1.01 \times 10^5} \ \mathrm{atm} = 0.966 \ \mathrm{atm}$$

$$= \frac{0.976 \times 10^5}{133.3} \ \mathrm{mmHg} = 732 \ \mathrm{mmHg}$$

Note that the pressure in millimeters of mercury is just the height h of the mercury column in millimeters. ☐

The average atmospheric pressure at sea level (1 atm = 760 mmHg) can support a column of mercury 760 mm high (about 30 in or 2.5 ft). This same pressure can support a column of

FIGURE 7.9

Barometer: (a) diagram and (b) wall-mounted. (*Central Scientific Co.*)

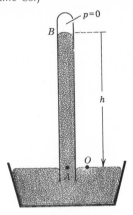

a

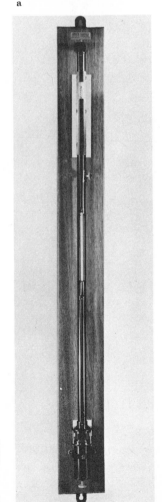

b

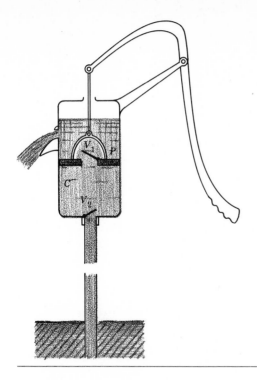

FIGURE 7.10

Lift pump.

water 34 ft high, the ratio of 34 ft to 2.5 ft being equal to the ratio of the density of mercury to the density of water (13.6).

A simple lift pump (Fig. 7.10) operates by lowering the air pressure at the upper end of a pipe. When the piston P is raised by pushing down on the pump handle, the pressure in chamber C is reduced below atmospheric pressure. As a consequence, valve V_1 closes, valve V_2 opens, and water is forced up the pipe by the atmospheric pressure. When the piston is lowered, valve V_2 closes, valve V_1 opens, so the water remains in the chamber. Since the smallest the pressure in the chamber can be is zero, the maximum pressure difference between the top and bottom of the pipe is 1 atm. This pressure can support a column of water 34 ft high, so the maximum height that water can be raised with a lift pump is 34 ft.

> **REMARK** It is a remarkable fact that water in trees can be raised a vertical height of 200 ft and more. The water travels from the roots to the leaves through a system of narrow capillaries called the *xylem*. The pressure in the xylem is actually negative—it can be -30 atm or less. This negative pressure means that the water in the xylem is in a state of tension, like a stretched cord. Negative pressure in a liquid is difficult to obtain in the laboratory, so it has so far found little industrial application. We shall assume that in ordinary practice pressures are always positive or zero.

THE MANOMETER
The *manometer* consists of a U-shaped tube partially filled with a liquid, usually either mercury or water. The tube is mounted in a vertical position with a measuring stick in back of it (Fig. 7.11). One end of the tube is connected to the vessel whose pressure p is to be measured, and the other end is open to the atmosphere.

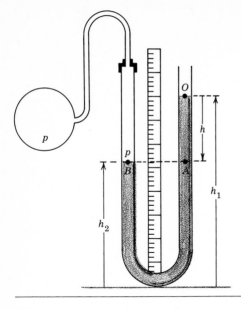

FIGURE 7.11

A manometer measuring the pressure p in a vessel.

Since the pressure at the top of the open tube is atmospheric pressure p_0, the pressure p_A at a point A in the manometer liquid that is a distance h from the top is given by

$$p_A = p_0 + \rho g h$$

Here ρ is the density of the liquid.

The pressure p_A at A is equal to the pressure p_B at B because these two points are at the same elevation in the same fluid. But p_B is equal to the pressure p in the vessel, so we have

$$p = p_0 + \rho g h \qquad 7.7$$

where $h = h_1 - h_2$ is the difference in the heights of the liquid in the two arms of the tube.

The *absolute pressure* p in the vessel is obtained by measuring h and adding to the term $\rho g h$ the atmospheric pressure p_0. A barometer must be used to find p_0. The manometer alone can measure only the difference $p - p_0$ between the absolute pressure in the vessel and atmospheric pressure.

Definition *Gauge pressure* $\bar{p}$ is the difference between the absolute pressure p in a fluid and the prevailing atmospheric pressure p_0:

$$\bar{p} = p - p_0 \qquad 7.8$$

Thus a manometer alone measures only gauge pressure.

In certain situations the pressure of interest is the gauge pressure. Tire pressures, for instance, are always gauge pressures. That is, when a tire is said to have a pressure of 24 lb/in², this means that the pressure inside the tire is 24 lb/in² greater than the pressure outside the tire. The absolute pressure in the tire is thus 24 lb/in² plus atmospheric pressure (about 14.7 lb/in²). Likewise, the gauge pressure of the blood is given when discussing blood pressure. A blood pressure of 120 mmHg means the pressure in the blood is 120 mmHg above the pressure outside the body.

Medical manometers, which are used to measure blood pressure, are mercury-filled manometers consisting of a long, narrow column connected to a short, wide column, as shown in Fig. 7.12. The long column is calibrated by the manufacturer to read the true difference between the heights of the mercury in the two columns. For instance, at a pressure of 100 mmHg the mercury in the short column might be 9 mm below the zero-pressure mark, while the mercury in the long column is 91 mm above this mark. In this case, the point 91 mm above the zero-pressure mark would be marked 100 mmHg. The advantage of this arrangement is that the practitioner can read the pressure directly on one column. But because no two glass tubes have the same bore, each instrument must be individually calibrated. This is not true of manometers in which the heights of both mercury columns are read.

THE BOURDON GAUGE

The *Bourdon gauge* is a convenient mechanical gauge that can be made to measure either low or high pressure. It consists of a metal tube, closed at one end, and bent into a figure C (Fig. 7.13). The open end of the tube is rigidly connected to a fitting, and the closed end, or tip, is connected through a mechanical linkage to an indicator needle.

The Bourdon tube has an oval cross section. As the pressure inside the tube increases, this cross section becomes more circular, causing the tube to straighten slightly. The mechanical linkage transforms the small movement of the tube's tip into a large movement of the needle. The scale in back of the needle is calibrated by the manufacturer so that the position of the needle indicates the gauge pressure in the tube. The Bourdon gauge measures gauge pressure, because the extent to which the tube is straightened depends on the difference between the pressure inside the tube and the pressure outside the tube (atmospheric pressure).

The tube in a sensitive Bourdon gauge has a very flat cross section, which causes the tube to bend easily with relatively small changes in pressure. However, the tube in a Bourdon gauge designed to measure very large pressures (10,000 lb/in^2 or more) has a nearly circular cross section, so that large pressure changes are needed to move the tip.

THE ANEROID BAROMETER

The *aneroid barometer* is a compact mechanical barometer consisting of a flexible chamber that has been evacuated of air and sealed (Fig. 7.14). The chamber is compressed by the outside pressure, and the extent of this compression changes with changes in the prevailing air pressure. As the chamber contracts or expands, it moves a mechanical linkage connected to a needle. The scale in back of the needle is calibrated to read the pressure.

This device is used to measure the air pressure outside an

FIGURE 7.12

Medical manometer with cuff bag and bulb pump. The cuff is wrapped around the patient's upper arm and inflated with the bulb until the pressure in the bag equals the blood pressure. The manometer is connected to the bag and measures the pressure in it. (*W. A. Baum Co., Inc.*)

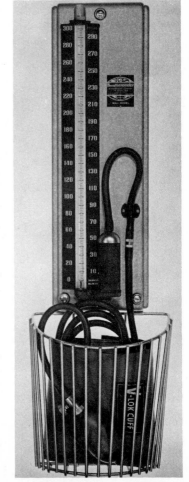

airplane. Since atmospheric pressure decreases with increasing elevation in a known way (Example 7.3), the scale of the barometer can be marked to directly show the height of the airplane above sea level.

ELECTRICAL GAUGES

A number of devices exist which produce an electric signal that is proportional to the pressure applied to them. The electric signal is then processed electronically to produce a readable signal in the form of a needle deflection or a digital display. Collectively known as *pressure transducers,* these devices find many industrial applications because of their great variety and versatility.

7.4 Buoyancy

A fluid exerts an upward force $\mathbf{F}_b$ on an object immersed in it. This is called the *buoyant force.* The magnitude of this force is given by one of the oldest laws in physics.

Archimedes' Principle *The buoyant force exerted by a fluid on an object is equal to the weight of the fluid displaced by the object.*

If the object is totally submerged, then the volume of the displaced fluid is equal to the volume of the object itself. If the object is partially submerged, the volume displaced is equal to the volume of the submerged portion of the object.

Example 7.6 Figure 7.15 shows a block of wood floating on water. (a) What is the buoyant force on the block? (b) What is the density of the wood?

(a) The volume V_s of the portion of the block submerged in the water is

$$V_s = (0.5\ \text{m})(0.3\ \text{m})(0.15\ \text{m}) = 0.0225\ \text{m}^3$$

Since the density of water is 1000 kg/m³ (Table 1.6), the mass of

FIGURE 7.13

Bourdon gauge. (*Dresser Industries, Industrial Valve and Instrument Division, Heise Bourdon Gauge.*)

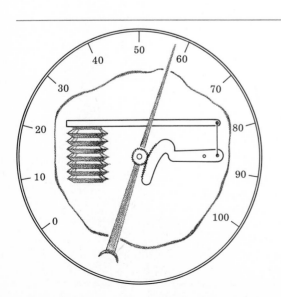

FIGURE 7.14

Aneroid barometer. The shape of the sealed evacuated chamber on the left changes with changes in atmospheric pressure. The mechanical linkage converts the change of shape of the chamber into a change in the position of the pointer.

the water which would occupy this volume is

$$m_f = \rho_f V_s = (1000 \text{ kg/m}^3)(0.0225 \text{ m}^3) = 22.5 \text{ kg}$$

where the subscript f stands for *fluid*. The buoyant force is equal to the weight of the displaced water, so

$$F_b = m_f g = (22.5 \text{ kg}(9.8 \text{ m/s}^2) = 220.5 \text{ N}$$

(b) Since the block is in static equilibrium, the upward buoyant force equals the weight W_o of the block. Thus we have

$$W_o = F_b \quad \text{or} \quad m_o g = m_f g$$

so

$$m_o = m_f = 22.5 \text{ kg}$$

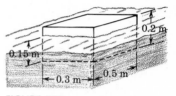

FIGURE 7.15

A block of wood floating on water.

where the subscript o stands for *object*. The mass m_o of the block is equal to the mass m_f of the displaced fluid. The volume V_o of the block is

$$V_o = (0.5 \text{ m})(0.3 \text{ m})(0.2 \text{ m}) = 0.030 \text{ m}^3$$

so its density ρ_o is

$$\rho_o = \frac{m_o}{V_o} = \frac{22.5 \text{ kg}}{0.030 \text{ m}^3} = 750 \text{ kg/m}^3$$

An object will float when its density is less than the density of the fluid in which it is placed. □

Archimedes' principle is not an independent law of physics. It can be shown to follow from the principles discussed in Secs. 7.2 and 7.3 Thus consider the forces exerted on the block immersed in the fluid in Fig. 7.16. The force F_B exerted by the fluid on the top face is directed downward, and its magnitude is $p_B A$, where A is the area of this face and p_B is the pressure in the fluid at this depth. Likewise, the force F_A on the bottom face is directed upward, and its magnitude is $p_A A$. The sum of these forces has the magnitude

$$F_A - F_B = p_A A - p_B A$$

and is directed upward, since $F_A > F_B$. From Eq. 7.5 this magnitude can be written

$$F_A - F_B = A\rho_f g(h_A - h_B) = A\rho_f g h \qquad 7.9$$

where ρ_f is the density of the fluid, and h is the height of the block.

The sum of the forces on the other faces of the block is zero, because for any region on one vertical face, such as S in Fig. 7.16, there is another region S' on the opposite face with the same area and pressure. Since the forces on these two regions have equal magnitude but opposite direction, their sum is zero. All other regions on the vertical faces can be similarly paired, so that the total force on these faces is zero. Equation 7.9 thus gives the magnitude of the total force exerted by the fluid on the block. This is the buoyant force F_b.

The buoyant force can be written somewhat differently by noting that Ah is the volume V of the block, so that $\rho_f Ah = \rho_f V$

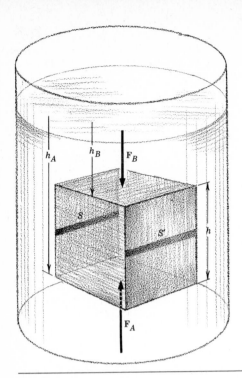

FIGURE 7.16

The forces on a block immersed in a fluid.

is the mass m_f of an equal volume of fluid. The buoyant-force on an object of volume V, then, is

$$F_b = F_A - F_B = \rho_f V g \qquad 7.10$$
$$= m_f g = \text{weight of an equal volume of fluid}$$

This is Archimedes' principle. Although it has been derived for the special case of a rectangular block, it is true in general.

> **REMARK** It is not difficult to see why Archimedes' principle is true even for irregularly shaped objects. The thing to remember is that since the fluid is in contact only with the surface of the object, the force it exerts on the object depends only on the object's shape and not on the material of which the object is made. Thus the buoyant force on the irregularly shaped object in Fig. 7.17 is the same as the buoyant force on any other object of the same shape. In Fig. 7.18 the object has been removed, but a region of fluid of the same shape as the object has been outlined. The fluid inside this region is in equilibrium under the actions of the force of gravity and the buoyant force due to the rest of the fluid, so the buoyant force on the region is equal to the weight of the fluid inside the region. This same buoyant force acts on any other object of the same shape.

FIGURE 7.17

An irregularly shaped object immersed in a fluid.

SPECIFIC GRAVITY

Definition The *specific gravity*, or *relative density*, s of an object is the ratio of the density ρ_o of the object to the density ρ_w of water:

$$s = \frac{\rho_o}{\rho_w} \qquad 7.11$$

For instance, the specific gravity of the block in Example 7.6 ($\rho_o = 750 \text{ kg/m}^3$) is

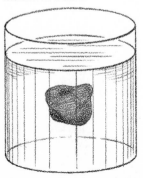

$$s = \frac{\rho_o}{\rho_w} = \frac{750 \text{ kg/m}^3}{1000 \text{ kg/m}^3} = 0.75$$

Table 7.2 gives the specific gravities of some common substances in order of increasing magnitude.

The specific gravity of a solid object denser than water is readily measured by weighing the object in air and when suspended in water, as shown in Fig. 7.19. The scale measures the tension in the cord. When the object is in air, we have

$$T_a = mg = \rho_o V g$$

and when the object is in water, we have

$$T_w = mg - F_b = T_a - F_b$$

Thus the ratio of the two scale readings is

$$\frac{T_w}{T_a} = \frac{T_a - F_b}{T_a} = 1 - \frac{F_b}{T_a}$$

$$= 1 - \frac{\rho_w V g}{\rho_o V g} = 1 - \frac{\rho_w}{\rho_o}$$

This can be written

$$\frac{\rho_w}{\rho_o} = 1 - \frac{T_w}{T_a}$$

or

$$s = \frac{\rho_o}{\rho_w} = \frac{1}{1 - \dfrac{T_w}{T_a}} = \frac{T_a}{T_a - T_w} \qquad 7.12$$

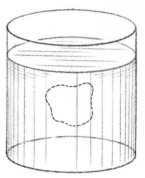

FIGURE 7.18

A region of fluid of the same shape as the object in Fig. 7.17.

Example 7.7 The weight of an object is 500 N in air and 435 N in water. (a) What is the specific gravity of the object? (b) What is the volume of the object?

(a) With $T_a = 500$ N and $T_w = 435$ N, Eq. 7.12 gives

$$s = \frac{T_a}{T_a - T_w} = \frac{500 \text{ N}}{500 \text{ N} - 435 \text{ N}} = 7.7$$

From Table 7.2 we see that the specific gravity of the object is a little less than the specific gravity of iron. The object might be an iron alloy, such as stainless steel, which has a specific gravity of 7.7.

(b) The buoyant force on the object is

$$F_b = T_a - T_w = 500 \text{ N} - 435 \text{ N} = 65 \text{ N}$$

But from Eq. 7.10 we have

$$F_b = \rho_w V g$$

so the volume V of the object is

$$V = \frac{F_b}{\rho_w g} = \frac{65 \text{ N}}{(1000 \text{ kg/m}^3)(9.8 \text{ m/s}^2)}$$

$$= 6.6 \times 10^{-3} \text{ m}^3 \qquad \square$$

TABLE **7.2**

The specific gravities of some common substances in order of increasing magnitude

Substance	Specific gravity
Air	0.00122
Cork	0.24
Wood, maple	0.70
Ethanol	0.791
Ice	0.917
Water, pure (4°C)	1.000
Water, sea	1.025
Glycerin	1.260
Sugar	1.59
Aluminum	2.7
Granite	2.7
Iron	7.86
Copper	8.96
Lead	11.35
Mercury	13.6
Uranium	18.9
Tungsten	19.3
Gold	19.3

The density of a liquid is conveniently measured with a *hydrometer*. This device, several examples of which are shown in Fig. 7.20, consists of a weighted bulb connected to a long stem. The density of the hydrometer is less than that of the liquid it is measuring, so when it is placed in the liquid it floats with its stem partially out of the water.

Let V_s be the submerged volume of the hydrometer, and let W be the total weight of the hydrometer. At equilibrium the buoyant force equals the weight so

$$F_b = W \qquad \text{or} \qquad \rho_f V_s g = W$$

Thus the density of the liquid is inversely proportional to the submerged volume:

$$\rho_f = \frac{W}{V_s g}$$

The greater the density of the liquid, the smaller the submerged volume and the higher the stem rides out of the water. A scale etched on the stem is calibrated to read the specific gravity of the liquid at the point where the surface of the liquid intersects the stem.

Special-purpose hydrometers are used to measure the composition of various solutions. For instance, the density of a water and antifreeze mixture depends on the percentage of antifreeze in the mixture. The calibration on a hydrometer used to test the antifreeze in an automobile radiator indicates the temperature at which the mixture freezes. Likewise, the calibration on a winemaker's hydrometer used to measure the density of a sugar and water mixture indicates the percentage of alcohol the mixture will yield after fermentation.

7.5   Hydrodynamics

Hydrodynamics is the study of fluid motion. Such motion can be very complex, because each small region of a fluid can move more or less independently of every other region. Even though all motion, no matter how complex, is governed by Newton's laws of motion, it is difficult or impossible to obtain useful mathematical results in the case of very complex fluid motion. For this reason we shall restrict the types of motion we consider to those for which simple and useful results are available. These results, restricted as they are, nevertheless find many important applications in hydraulic and aerodynamic engineering.

STREAMLINE FLOW

In the most complex type of fluid motion, the velocity and pressure of the fluid at each point changes in a rapid and unpredictable way. It is such motion, called *turbulence*, that cannot be treated by simple mathematics. However, there are many cases in which the fluid flows in a more regular way, so that the velocity at each point is constant. Such motion is called *streamline flow*. We shall restrict our discussion of fluid motion to the case of streamline flow.

In streamline flow the fluid can be thought of as moving

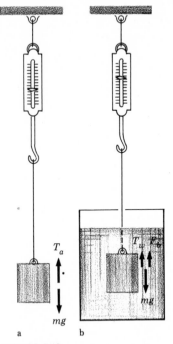

FIGURE 7.19

An object suspended from a scale. (a) When the object is in air, the scale measures the weight *mg* of the object. (b) When the object is suspended in water, the scale measures $mg - F_b$, where F_b is the buoyant force on the object.

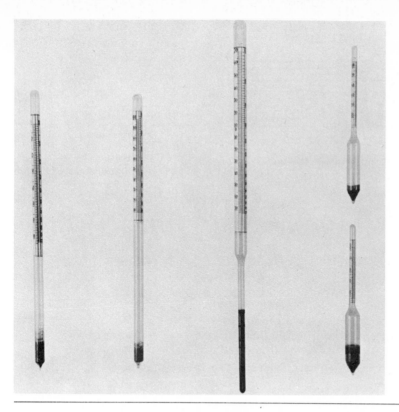

FIGURE 7.20

Hydrometers of various types.
(*Central Scientific Co.*)

along fixed lines, called *streamlines* (Fig. 7.21). At each point, the direction of the fluid velocity is tangent to the streamline at that point. Two streamlines can never cross, because if they did, the fluid would have two velocities at the same point.

The magnitude of the velocity can change along a streamline, and it can change from one streamline to another. To see how the velocity changes along a streamline, consider the fluid flow shown in Fig. 7.22. The fluid is flowing from one pipe of cross-sectional area A to another pipe of cross-sectional area A'. The connection between the pipes is tapered to maintain the streamline flow and avoid turbulence. Because A' is less than A, the streamlines move closer together as they go from pipe 1 to pipe 2.

Let v be the velocity of each streamline as it passes through the surface S, and let v' be the velocity of each streamline as it passes through surface S'. In time t all the fluid a distance $d = vt$

FIGURE 7.21

Streamline flow of a fluid around an obstacle. At each point the velocity of the fluid is parallel to the streamline passing through the point.

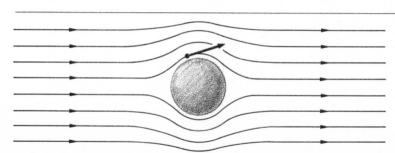

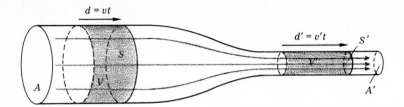

FIGURE 7.22

Fluid flowing from a pipe of cross-sectional area A to a pipe of cross-sectional area A'. The transition region between the two pipes is tapered to preserve the streamline flow.

upstream from S will pass through this surface, so the total volume V of fluid that passes through S in time t is

$$V = dA = vtA$$

Likewise, the volume V' of fluid that passes through S' in the same time t is

$$V' = d'A' = v'tA'$$

If the total amount of fluid between S and S' remains unchanged, the volume V of fluid that flowed into this region in time t must equal the volume V' that flowed out. This is called the *continuity condition,* and it is exactly true for incompressible fluids and approximately true for compressible fluids under most circumstances. We shall assume its validity from now on. That is, we assume

$$V = V' \quad \text{or} \quad vtA = v'tA'$$

so

$$vA = v'A' \qquad\qquad 7.13$$

Equation 7.13 shows that the product of velocity and area remains constant. Thus, as the fluid moves into a pipe of smaller cross-sectional area, its velocity increases in order to maintain the same rate of fluid flow. The same reasoning shows that in general whenever two streamlines come closer together, the fluid speed increases.

Definition The *rate of fluid flow* Q is the volume per second of fluid that passes through a given surface. The rate of flow across the surface S in Fig. 7.22 is

$$Q = \frac{V}{t} = \frac{vtA}{t} \quad \text{or} \quad Q = vA \qquad\qquad 7.14$$

From Eq. 7.13 we see that this is the same as the rate of flow across surface S'.

Example 7.8 Water at the rate of 1500 gal/min passes through a 0.10-m-diameter pipe. (a) What is the velocity of the fluid in the pipe? (b) What is the fluid velocity when the pipe narrows to a diameter of 0.06 m?

(a) The rate of fluid flow must be expressed in SI units. From Appendix II we have

$$1 \text{ gal} = 3.79 \times 10^{-3} \text{ m}^3$$

so

$$Q = 1500\,\text{gal/min} = \frac{(1500)(3.79 \times 10^{-3}\,\text{m}^3)}{60\,\text{s}}$$

$$= 0.0948\,\text{m}^3/\text{s}$$

The cross-sectional area of the pipe is

$$A = \pi r^2 = \pi(0.05\,\text{m})^2 = 7.85 \times 10^{-3}\,\text{m}^2$$

so from Eq. 7.14 the fluid velocity is

$$v = \frac{Q}{A} = \frac{0.0948\,\text{m}^3/\text{s}}{7.85 \times 10^{-3}\,\text{m}^2} = 12.1\,\text{m/s}$$

(b) When the diameter narrows to 0.06 m, the area becomes

$$A' = \pi(0.03\,\text{m})^2 = 2.83 \times 10^{-3}\,\text{m}^2$$

and from Eq. 7.13 the velocity is

$$v' = \frac{vA}{A'} = (12.1\,\text{m/s})\frac{7.85 \times 10^{-3}\,\text{m}^2}{2.83 \times 10^{-3}\,\text{m}^2}$$

$$= 33.6\,\text{m/s} \qquad \qquad \square$$

BERNOULLI'S EQUATION

The mass m of the fluid in the volume V in Fig. 7.22 is

$$m = \rho V$$

where ρ is the density of the fluid. Since the fluid in this volume has the velocity v, the kinetic energy of this mass is

$$K = \tfrac{1}{2}mv^2 = \tfrac{1}{2}\rho V v^2$$

When this mass moves into V', its kinetic energy becomes

$$K' = \tfrac{1}{2}mv'^2 = \tfrac{1}{2}\rho V v'^2$$

Since v' is greater than v, the kinetic energy of the mass has increased in going from V to V'. (Remember $V = V'$.) From the work-energy theorem (Eq. 5.14), the change in kinetic energy $K - K'$ is equal to the total work done on this mass.

The forces which do work on the fluid are friction, gravity, and the force due to the pressure of the fluid itself. We shall neglect friction altogether for now and consider only the work done by gravity and fluid pressure. Furthermore, in a horizontal pipe, gravity has no effect on the fluid, so we can neglect it for the moment and concentrate on the work done by the fluid pressure. To understand how this work comes about, let the fluid behind volume V be replaced by a piston which exerts the fluid pressure

$$p = \frac{F}{A}$$

on the fluid to the right (Fig. 7.23). Similarly, let the fluid in front of volume V' be replaced by a piston that exerts the fluid pressure

$$p' = \frac{F'}{A'}$$

on the fluid to the left. These pressures are not equal to each

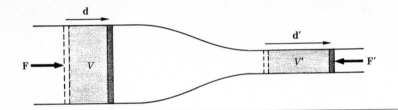

FIGURE 7.23

The pressure p on the left of the fluid in the large pipe does the work pAd, while the pressure p' on the right of the fluid in the small pipe does the work $-p'A'd'$.

other in this case because the fluid is not at rest: *Pascal's law does not apply here.*

As the fluid in volume V moves the distance d to the right, the force $F = pA$ does the work

$$W = Fd = pAd = pV$$

on the fluid; at the same time, the fluid in volume V' moves the distance d' to the right, and the force $F' = p'A'$ does the work

$$W' = -F'd' = -p'A'd' = -p'V' = -p'V$$

on the fluid. Here we have used the fact that $V' = V$. Thus the net work W_p done by the fluid pressure is

$$W_p = W + W' = (p - p')V$$

From the work-energy theorem this work is equal to the change in the kinetic energy of the mass, so

$$(p - p')V = \tfrac{1}{2}\rho V v'^2 - \tfrac{1}{2}\rho V v^2 \qquad \text{or} \qquad \text{7.15}$$
$$p - p' = \tfrac{1}{2}\rho v'^2 - \tfrac{1}{2}\rho v^2$$

Rearranging terms we get

$$\tfrac{1}{2}\rho v'^2 + p' = \tfrac{1}{2}\rho v^2 + p \qquad \text{7.16}$$

This is Bernoulli's equation for a level pipe. It shows that the pressure in a fluid must decrease whenever the velocity of the fluid increases.

If a pipe is not level, as in Fig. 7.24, work is done by gravity as well as by the fluid pressure. The gravitational work W_g done on the mass ρV in going from V to V' is

$$W_g = mgh - mgh' = V\rho gh - V\rho gh'$$

and this must be added to the left-hand side of Eq. 7.15.

After carrying out the necessary algebra, Bernoulli's equation becomes

FIGURE 7.24

Gravity does the work $W_g = V\rho g\,(h - h')$ on a volume V of fluid moving from a height h to a height h'.

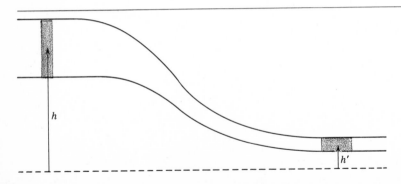

$$\tfrac{1}{2}\rho v'^2 + p' + \rho gh' = \tfrac{1}{2}\rho v^2 + p + \rho gh \qquad 7.17$$

Although this equation has been derived for fluid flow in a pipe, it is true in all cases of incompressible streamline flow in which friction can be neglected. The primed and unprimed quantities in Eq. 7.17 represent the values of these quantities at any two points on the same streamline, so the quantity

$$\tfrac{1}{2}\rho v^2 + p + \rho gh$$

has the same value at every point on a given streamline.

If the fluid is not flowing at all ($v = v' = 0$), Eq. 7.17 becomes

$$p' - p = \rho g(h - h')$$

which is the same as the hydrostatic equation given in Eq. 7.4.

Bernoulli's equation finds application in many different technical areas. The examples that follow will demonstrate the importance of Bernoulli's equation in engineering practice.

Torricelli's theorem The rate of discharge of a liquid from a hole in a tank can be calculated from Bernoulli's equation. Thus consider the situation in Fig. 7.25. We apply Bernoulli's equation to a streamline from the top of the liquid to a point just outside the orifice. At the top of the liquid the pressure p is equal to the atmospheric pressure p_0. If the tank is large, the water level will drop slowly, so the velocity v of the water at the top is very small; for simplicity we take $v = 0$.

At the orifice the pressure p' is also the atmospheric pressure p_0. Inserting these values into Eq. 7.17 we get

$$\tfrac{1}{2}\rho v'^2 + p_0 + \rho gh' = \tfrac{1}{2}\rho(0)^2 + p_0 + \rho gh$$

or

$$v'^2 = 2g(h - h') \qquad 7.18$$

This is called *Torricelli's theorem*. It shows that the discharge velocity is the same as that of a mass that falls from the vertical height $\Delta h = h - h'$.

The rate of discharge Q of the liquid is

$$Q = A_v v' = A_v \sqrt{2g\,\Delta h} \qquad 7.19$$

where Δh is the distance of the orifice from the surface of the liquid, and A_v is the *vena contracta*, the smallest cross-sectional area the stream obtains after it emerges from the orifice. The *vena contracta* is located a short distance beyond the orifice, because the streamlines continue to converge briefly after they leave the orifice. For a sharp circular orifice of cross-sectional area A, A_v is approximately $0.65\,A$.

Example 7.9 The diameter of a bathtub drain is 4 cm. What is the rate of discharge when the depth of water in the tub is 18 cm?

The *vena contracta* is

$$A_v = 0.65\,A = (0.65)\pi(0.02\ \text{m})^2 = 0.82 \times 10^{-4}\ \text{m}^2$$

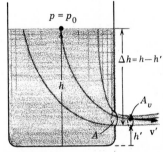

FIGURE 7.25

The speed v' of fluid discharge from an orifice is given by Bernoulli's equation. The smallest cross-sectional area A_v of the discharge stream occurs a short distance beyond the orifice.

so from Eq. 7.19 the rate of discharge is

$$Q = A_v \sqrt{2g\,\Delta h} = (0.82 \times 10^{-4}\ \text{m}^2)\ \sqrt{2(9.8\ \text{m/s}^2)(0.18\ \text{m})}$$
$$= 1.53 \times 10^{-3}\ \text{m}^3/\text{s} = 24.2\ \text{gal/min} \qquad \square$$

The Venturi Meter   One way to measure the rate of flow of a fluid through a pipe is to insert into the pipe a constriction, or throat, as shown in Fig. 7.26. Such a constriction, with its ends tapered to maintain streamline flow, is called a *Venturi tube*. The smaller cross-sectional area of the Venturi tube causes the velocity of the fluid to increase, in accordance with Eq. 7.13. The pressure in this faster-moving fluid is less than the pressure in the main pipe, in accordance with Bernoulli's equation (Eq. 7.16). The drop in pressure as a fluid flows through a Venturi tube is called the *Venturi effect*. It has many applications in engineering.

In a Venturi meter, the pressures p and p' in the main pipe and in the Venturi tube are measured by suitable devices, such as the manometers in Fig. 7.26. In this case, the pressure difference $p - p'$ is related to the height difference h of the fluid in the manometers by

$$p - p' = \rho g h \qquad\qquad\qquad 7.20$$

The rate of flow Q is the same through both the main pipe and the Venturi tube, and it is given by Eqs. 7.13 and 7.14:

$$Q = Av = A'v'$$

where A and A' are the cross-sectional areas of the main pipe and the Venturi tube, respectively. From this equation we get

$$v' = \frac{A}{A'} v$$

which when inserted into Eq. 7.16 yields

$$\tfrac{1}{2}\rho \left(\frac{A}{A'}\right)^2 v^2 + p' = \tfrac{1}{2}\rho v^2 + p$$

or

$$v^2 = \frac{p - p'}{\tfrac{1}{2}\rho \left(\dfrac{A^2}{A'^2} - 1\right)}$$

Thus the velocity of the fluid in the main pipe is

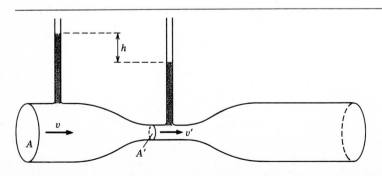

$$v = \sqrt{\frac{p - p'}{\frac{1}{2}\rho\left(\frac{A^2}{A'^2} - 1\right)}}$$ 7.21a

and the rate of flow is

$$Q = Av = A\sqrt{\frac{p - p'}{\frac{1}{2}\rho\left(\frac{A^2}{A'^2} - 1\right)}}$$ 7.21b

The rate of flow is determined by measuring the pressure difference between the main pipe and the Venturi tube.

Example 7.10 A Venturi meter of cross-sectional area $A' = 5$ cm^2 is inserted into a water main of cross-sectional area $A = 50$ cm^2. The pressure difference between the main and the Venturi tube is found to be 2×10^4 Pa. What is the velocity of the water in the main?
    We have

$$p - p' = 2 \times 10^4 \text{ Pa}$$

and

$$\frac{A^2}{A'^2} = \frac{(50 \text{ cm})^2}{(5 \text{ cm})^2} = (10)^2 = 100$$

so from Eq. 7.21a

$$v = \sqrt{\frac{p - p'}{\frac{1}{2}\rho\left(\frac{A^2}{A'^2} - 1\right)}} = \sqrt{\frac{2 \times 10^4 \text{ Pa}}{\frac{1}{2}(1000 \text{ kg/m}^3)(100 - 1)}}$$

$$= 0.63 \text{ m/s}$$

The rate of flow is

$$Q = Av = (50 \times 10^{-4} \text{ m}^2)(0.63 \text{ m/s}) = 3.18 \times 10^{-3} \text{ m}^3/\text{s} \qquad \square$$

The Automobile Carburetor An automobile carburetor uses the Venturi effect to mix air with vaporized gasoline. Air on its way to the engine passes through a Venturi tube into which the fuel nozzle projects (Fig. 7.27). The smaller cross section of the Venturi tube causes the velocity of the air to increase and the air pressure to drop. The reduced pressure in the tube draws tiny droplets of gasoline from the nozzle. These droplets quickly vaporize, and the air-gasoline mixture passes on to the engine through the throttle valve. This valve (which is *not* the choke) is connected by mechanical linkage to the accelerator pedal. As the pedal is depressed, the valve opens, allowing more air and gasoline to reach the engine.
    An *atomizer* is any device, such as a carburetor or perfume spray bottle, that uses the passage of air through a Venturi tube to draw out droplets of liquid. The liquid is said to be atomized, but in fact it is only broken into tiny droplets. Because each droplet is so small, it very quickly vaporizes.

Lift on an Airplane Wing The lowering of air pressure with increased speed is the principal cause of lift on an airplane

FIGURE 7.27

Venturi tube in an automobile carburetor.

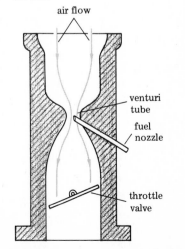

wing. The wing, or airfoil, has a curved cross section which causes the air to flow more rapidly around the upper surface than the lower surface (Fig. 7.28). Consequently, the pressure is less along the upper surface than along the lower surface, and so there is a net upward force, called the *lift,* on the wing.

> **REMARK** If the frictional (viscous) force between the flowing air and the airfoil is neglected, it is possible to calculate the net force on the airfoil. The result, which was first obtained toward the end of the nineteenth century, is that the force is exactly zero! In the absence of friction, the reduced pressure on the larger upper surface exerts a downward force exactly equal to the upward force exerted by the higher pressure on the smaller lower surface; consequently, there is no lift. Physicists at the time were unable to explain heavier-than-air flight, and in fact the first successful airplane was built in 1903 by two bicycle mechanics [Orville (1871–1948) and Wilbur (1867–1912) Wright], not by physicists. Only later was the critical role of friction understood sufficiently to enable scientists to make significant contributions to airplane design. (Because of friction, a circulation is established around an airfoil that increases the pressure difference between the upper and lower surfaces, resulting in a net upward force on the airfoil.)

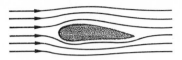

FIGURE 7.28

Streamlines around an airfoil.

VISCOSITY

Bernoulli's equation neglects the frictional forces that act on a moving fluid. This is valid provided the total work done by fluid pressure and gravity is much greater than the work done by friction. When this is not the case, the frictional force must be taken into account. For instance, in the case of the steady flow of a fluid through a horizontal pipe of constant cross section, the frictional force is equal in magnitude to the force due to the fluid pressure. All the mechanical work done on the fluid goes to maintain the constant flow against the retarding frictional force. Thus for all problems involving the transport of water, oil, or gas by pipelines, the effect of friction is important.

The fundamental property of fluids (Sec. 7.1) states that a fluid at rest exerts no force parallel to a surface. However, a fluid flowing along a surface exerts a force $\mathbf{F}_\parallel$ parallel to the surface in the direction of flow. The reaction $\mathbf{F}_v$ to $\mathbf{F}_\parallel$ is a force exerted by the surface on the fluid directed opposite to the direction of flow. This force, called the *viscous force,* plays a role in fluid flow similar to that of friction in the motion of one solid upon another. That is, the viscous force opposes the motion. In order to maintain a steady flow, an external driving force must be applied to the fluid to balance the viscous force.

Consider a fluid flowing along a flat surface S_1, as shown in Fig. 7.29. To maintain the flow, a second surface S_2 placed on top of the fluid is moved in the direction of fluid flow with constant speed v. (S_1 and S_2 are actually parts of two larger surfaces, such as the solid cylinder and U-shaped frame in Fig. 7.30.) Since the fluid exerts the force $\mathbf{F}_\parallel$ parallel to S_1 an external force $\mathbf{F}_a = -\mathbf{F}_\parallel$ must be applied to S_1 to keep it at rest.* The viscous force $\mathbf{F}_v$, which is the force the surface S_1 applies to the fluid, is the reaction to $\mathbf{F}_\parallel$, so that

$$\mathbf{F}_v = -\mathbf{F}_\parallel = -(-\mathbf{F}_a) = \mathbf{F}_a$$

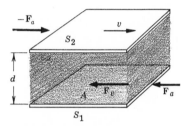

FIGURE 7.29

Fluid flowing along a flat surface S_1. The upper surface S_2 moves with constant speed v to maintain the flow, while S_1 is held stationary.

FIGURE 7.30

Viscosimeter. (*Brookfield Engineering Laboratories, Inc.*)

*The fluid may also apply a force perpendicular to S_1, but this is just the fluid pressure, which has already been studied.

Thus $\mathbf{F}_v$ is determined by measuring the external force $\mathbf{F}_a$ needed to keep S_1 from moving, or alternatively, the force $-\mathbf{F}_a$ needed to move S_2 with constant speed.

REMARK $\mathbf{F}_a$ and $\mathbf{F}_v$ are two different forces that happen to be equal. $\mathbf{F}_a$ is the force applied to S_1 by some outside agent, whereas $\mathbf{F}_v$ is the viscous force applied by S_1 to the fluid.

From studies of $\mathbf{F}_v$ made using an arrangement of surfaces similar to Fig. 7.29, it is found that the magnitude of $\mathbf{F}_v$ is directly proportional to the speed v of S_2 and to the area A of S_1 and inversely proportional to the distance d between the surfaces. In symbols this is expressed as

$$F_v = \frac{\eta A v}{d} \qquad\qquad 7.22$$

where η^* is a constant, called the *viscosity*, that is characteristic of the fluid.

UNITS From Eq. 7.22 the dimensions of viscosity are found to be

$$[\eta] = \frac{[f][l]}{[l^2][v]} = \frac{[f]}{[l][l/t]} = \frac{[f][t]}{[l^2]}$$

The SI units of viscosity are $N \cdot s/m^2$; this is called a *poiseuille* (Pl). Another common unit of viscosity, the *poise* (P), is equal to 0.1 Pl:

1P = 0.1 Pl

1 cP = (10^{-3}) Pl = 1 mPl

Before the adoption of SI units, the centipoise (cP) was commonly used to express viscosity; the equivalent SI unit is the millipoiseuille (mPl).

Viscosity is measured in a device called a *viscosimeter*. One type of viscosimeter, shown in Fig. 7.30, consists of a cylinder that rotates on a shaft inside a U-shaped frame. The device is placed in the liquid being studied, and the torque required to rotate the cylinder at constant speed is measured. The viscosity is determined from the torque, speed of rotation, and dimensions of the cylinder and frame.

Table 7.3 gives the viscosities of some common liquids and gases. Since the viscosity of a fluid varies rapidly with temperature, the temperature at which the viscosity is measured must be given. Note that the viscosity of a gas is much less than the viscosity of a liquid.

When a fluid moves between two surfaces, as in Figs. 7.29 and 7.31, it is found that the streamlines are parallel to the surfaces and to each other. This is called *laminar flow*. Furthermore, it is found that the layers of fluid adjacent to each surface adhere to these surfaces and move with them. For example, the layer of fluid adjacent to S_2 does not slip along the surface: it moves with the same speed v as S_2 (Fig. 7.31). The layer of fluid immediately below it is dragged along by this first layer, but with a slightly smaller speed because one fluid layer slips over the other. This second fluid layer in turn drags the next layer with a still smaller speed, and so on, until the layer of fluid adjacent to S_1 is

*η is the Greek (lowercase) letter eta.

FIGURE 7.31

The speed of flow of the fluid between a moving surface and a stationary surface varies from zero, for the fluid in contact with the stationary surface, to the speed v of the moving surface, for the fluid in contact with the moving surface.

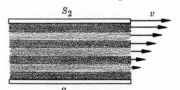

TABLE **7.3**

Fluid	Temperature, °C	Viscosity, mPl	Viscosities of some common liquids and gases
Liquids:			
Acetone	25	0.316	
Ethanol	20	1.2	
Ether	20	0.233	
Glycerin	20	1490	
Mercury	20	1.55	
Oil, light machine	16	113	
	38	34	
	100	4.9	
Water	0	1.79	
	20	1.00	
	60	0.467	
	100	0.282	
Gases:			
Air	0	0.0171	
	18	0.0183	
	229	0.0264	
Methane	20	0.0109	
Water vapor	100	0.0125	

reached. Since this layer adheres to the stationary surface S_1, it has zero speed. This is shown schematically in Fig. 7.31, where the speed in each layer is indicated by an arrow whose length is proportional to the speed. The fluid moves with different speeds at different distances from S_1, but all the layers of fluid move parallel to each other.

In reality, therefore, the viscous force does not have its origin in the slipping of the fluid along a surface (as with friction), but rather in the slipping of one layer of fluid along another. The viscous force is large when the force needed to produce this slippage is large. The viscosity η is a measure of how much force is required to slip one layer of fluid over another. A large value of η corresponds to a viscous fluid like glycerin or oil, whereas a small value corresponds to a nonviscous fluid like water or ether. The viscosity is an intrinsic property of a fluid and does not depend on the nature of the surface along which the fluid is moving.

FLUID FLOW THROUGH PIPES

We now consider the problem of a fluid flowing though a pipe of radius r and length L (Fig. 7.32). The speed of the layer of fluid adjacent to the wall of the pipe is zero, and the fluid moves with the maximum speed v_m along the central axis of the pipe. In going outward from the central axis the speed in each concentric layer of fluid varies steadily from v_m to zero in such a way that the average speed of the fluid is $\overline{v} = \frac{1}{2}v_m$.

The motion of the fluid through the pipe is opposed by the viscous force exerted on the fluid by the walls of the pipe. Equation 7.22 applies only to the case of two flat surfaces, but it can be used to estimate F_v in the present case. For the area in Eq. 7.22 we take the area of the pipe wall, because this is the area in

FIGURE 7.32

Fluid flowing through a pipe of radius r and length L. The speed of the fluid varies from zero for fluid in contact with the pipe to v_m for fluid in the center of the pipe.

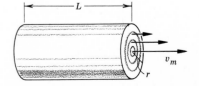

contact with the fluid. The area of the pipe wall is equal to the circumference of the pipe times its length, or

$$A = 2\pi r L$$

For the distance d in Eq. 7.22 we take the radius r of the pipe because this is the distance over which the fluid speed varies from v_m to zero. Finally, for v we take the maximum fluid speed v_m. The result is

$$F_v = \eta \frac{2\pi r L v_m}{r} = 2\pi \eta L v_m$$

This estimate gives the correct dependence of F_v on η, L, and v_m, but it is too small by a factor of 2. A more careful analysis shows that

$$F_v = 4\pi \eta L v_m \qquad\qquad 7.23$$

   This viscous force opposes the fluid flow, so to maintain constant flow, there must be a driving force of magnitude F_v. If gravity is neglected, the only other force on the fluid is caused by the fluid pressure. Fluid entering the pipe on the left at pressure p_1 exerts a force $p_1 A$ to the right on the fluid inside the pipe, where now $A = \pi r^2$ is the cross-sectional area of the pipe. The fluid leaving the pipe on the right at pressure p_2 exerts a force $p_2 A$ to the left on the fluid inside the pipe. If p_2 is less than p_1, there will be a net driving force of magnitude

$$p_1 A - p_2 A = (p_1 - p_2)A = (p_1 - p_2)\pi r^2$$

to the right on the fluid. The condition for steady flow is that this driving force just equal F_v,

$$(p_1 - p_2)\pi r^2 = 4\pi \eta L v_m$$

so that

$$v_m = \frac{(p_1 - p_2)r^2}{4\eta L} \qquad\qquad 7.24$$

This equation gives the speed in the center of a pipe in terms of the pressure difference between the ends of the pipe, the radius and length of the pipe, and the viscosity of the fluid. For a given pressure difference, v_m increases with r^2 and decreases with η and L, which is reasonable. Solving Eq. 7.24 for $p_1 - p_2$, we have

$$p_1 - p_2 = \frac{4\eta L v_m}{r^2} \qquad\qquad 7.25$$

which gives the pressure difference produced in a fluid as it moves through a pipe.

Example 7.11 (a) What is the pressure drop per mile in a 12-in water main carrying water at a maximum speed of 5 m/s? (b) What is the rate of flow through the pipe?
   (a) The radius of the main is

$$r = 6 \text{ in} = 0.152 \text{ m}$$

and its length is

$$L = 1 \text{ mi} = 1610 \text{ m}$$

From Table 7.3 the viscosity of water (at 20°C) is 1.00 mPl, so from Eq. 7.25 the pressure drop is

$$p_1 - p_2 = \frac{4\eta L v_m}{r^2} = \frac{4(1.0 \times 10^{-3} \text{ Pl})(1610 \text{ m})(5 \text{ m/s})}{(0.152 \text{ m})^2}$$

$$= 1.39 \times 10^3 \text{ Pa} = 0.20 \text{ lb/in}^2$$

If the gauge pressure of the water at the pumping station is 20 lb/in², the water would flow through 100 mi of pipeline before its gauge pressure was reduced to zero by the viscous forces acting on it.

(b) The average speed of the water in the pipe is

$$\bar{v} = \tfrac{1}{2}v_m = 2.5 \text{ m/s}$$

and the pipe's cross-sectional area is

$$A = \pi r^2 = \pi(0.152 \text{ m})^2 = 0.0726 \text{ m}^2$$

so from Eq. 7.14 the flow rate is

$$Q = \bar{v}A = (2.5 \text{ m/s})(0.0726 \text{ m}^2)$$

$$= 0.181 \text{ m}^3/\text{s} = 2865 \text{ gal/min} \qquad \square$$

Equations 7.14 and 7.24 can be combined to give an important equation relating the rate of fluid flow to the pressure difference in a pipe. To do this we use $\bar{v} = \tfrac{1}{2}v_m$ in Eq. 7.14, obtaining

$$Q = \tfrac{1}{2}\pi r^2 v_m$$

Substituting the right-hand side of Eq. 7.24 for v_m in this last expression, we get

$$Q = \frac{\pi r^4(p_1 - p_2)}{8\eta L} \qquad\qquad 7.26$$

which is known as *Poiseuille's law*. This law says that *the amount of fluid that flows through a pipe is proportional to the pressure drop along the pipe and to the fourth power of the radius of the pipe*. That is, for the same pressure difference, 16 times as much fluid will flow through a pipe of radius 2 cm as will flow through a pipe of radius 1 cm. Equation 7.26 is the basic equation describing the flow of fluids in all pipelines from city water mains to the trans-Alaska pipeline (Fig. 7.33).

REMARK Poiseuille's law is really only an approximation, valid when the flow speed $\bar{v}$ is sufficiently small. When the flow speed is large, the flow no longer consists of concentric layers of fluid all moving parallel to each other at different speeds (laminar flow). Instead turbulence develops, in which some of the fluid moves in small circles inside the pipe (eddies). The conditions for laminar flow are discussed in Prob. 39.

Example 7.12 A power plant uses cooling water at the rate of 30 m³/s. This water is brought from the ocean through a pipe that is 10 km long and 2 m in diameter. Taking the viscosity of seawater to be 1 mPl, calculate the pumping pressure required

FIGURE 7.33

A section of the 48-in-diameter trans-Alaska pipeline. (*Alyeska Pipeline Service Co.*)

to deliver this water to the plant at a gauge pressure of 2 lb/in².

In this problem we have $Q = 30 \text{ m}^3/\text{s}$, $r = 1 \text{ m}$, $L = 10^4 \text{ m}$, and $\eta = 1 \times 10^{-3} \text{ Pl}$, so Eq. 7.26 yields

$$p_1 - p_2 = \frac{8Q\eta L}{\pi r^4} = \frac{8(30 \text{ m}^3/\text{s})(1 \times 10^{-3} \text{ Pl})(10^4 \text{ m})}{\pi (1 \text{ m})^4}$$

$$= 764 \text{ Pa} = 0.11 \text{ lb/in}^2$$

If the pressure p_2 at the power plant is to be 2 lb/in², the original pressure p_1 of the water at the inlet end of the pipe must be

$$p_1 = p_2 + 0.11 \text{ lb/in}^2 = 2.11 \text{ lb/in}^2$$

The total pressure drop along the 10-km pipe is low because the diameter of the pipe is large (over 6 ft). If a pipe with half the diameter were used, the pressure drop would be 16 times as great, or 1.76 lb/in². ◻

A pump takes a fluid at the inlet pressure p_{in} and discharges it at the outlet pressure p_{out}. To do this the pump exerts the net force

$$F = (p_{out} - p_{in})A \qquad 7.27$$

on the fluid as it flows into an outlet pipe of cross-sectional area A. If the fluid moves a distance d in time t, the work W done by the pump is

$$W = Fd$$

and the power P of the pump is

$$P = \frac{W}{t} = \frac{Fd}{t} = F\bar{v}$$

where $\bar{v} = d/t$ is the average speed of the fluid as it flows through the pipe. Using Eq. 7.27 we see that the power of a pump is related to the pressure increase it provides and the rate at which it discharges the fluid by

$$P = F\bar{v} = (p_{out} - p_{in})\bar{v}A$$

or

$$P = (p_{out} - p_{in})Q \qquad\qquad 7.28$$

where Eq. 7.14 hs been used for Q.

Example 7.13 What is the power required to pump the water through the pipe in Example 7.12?

The pressure drop along the pipe is 764 Pa, so

$$p_1 - p_2 = 764 \text{ Pa}$$

The smallest possible value for the pressure p_2 of the water as it reaches the power plant is atmospheric pressure p_0, so the smallest value for the pressure $p_1 = p_{out}$ of the water as it leaves the pump is

$$p_{out} = p_1 = p_2 + 764 \text{ Pa} = p_0 + 764 \text{ Pa}$$

The pump takes in water at atmospheric pressure, so

$$p_{out} - p_{in} = p_1 - p_0 = 764 \text{ Pa}$$

Thus the minimum power required to pump the water is

$$P = (p_{out} - p_{in})Q = (764 \text{ Pa})(30 \text{ m}^3/\text{s}) = 2.29 \times 10^4 \text{ W}$$
$$= 22.9 \text{ kW}$$

Of course, greater power is required if the water is to reach the plant at a pressure p_2 greater than p_0.

REMARK If the diameter of the pipe were 1 m instead of 2 m, the pressure drop along the pipe would be 16 times as great, or 1.22×10^4 Pa, and the power requirement would be 16 times as great, or 366 kW The lower construction cost involved in building a 1-m pipe instead of a 2-m pipe has to be weighed against the higher operating cost involved in pumping the water through the smaller pipe. □

PROBLEMS

1 The edge of a chisel has an area of 0.12 in². When struck with a hammer, the chisel exerts a momentary force of 20 lb on a brick. What is the pressure exerted directly under the edge of the chisel?
Ans. 167 lb/in²

2 An explosion creates a momentary increase in the pressure of the surrounding air (overpressure). Calculate the total force exerted on the wall of a building 20 m high and 30 m wide by an overpressure of 0.03 atm.

3 In an automobile lift, compressed air is fed into a cylinder fitted with a long piston. The air raises the piston, on which the car sits. If the air pressure is 8×10^5 Pa and the total mass of the automobile and piston assembly is 2500 kg, what is the radius r of the piston?
Ans. 9.9 cm

4 The pressure in an automobile tire is 32 lb/in². Convert this to (a) pascals, (b) torrs, and (c) atmospheres.

GUIDE TO MAJOR TOPICS

5 The two cylinders of a hydraulic lift (Fig. 7.5) have cross-sectional areas of 12 and 65 cm². What force must be applied to the smaller cylinder to lift a 1500-N load sitting on the larger cylinder?
Ans. 277 N

6 A hydraulic lift raises a 3000-lb weight when a 200-lb force is applied to the smaller piston. If the smaller piston has a cross-sectional area of 5 in², what is the cross-sectional area of the larger piston?

7 A force of 200 N must be applied to the smaller piston of a hydraulic lift in order to raise a 3500-N weight sitting on the larger piston. If the diameter of the smaller piston is 0.05 m, what is the diameter of the larger piston?
Ans. 0.21 m

8 Choose suitable diameters for the two cylinders of a hydraulic lift so that a 5000-N weight can be lifted with an applied force of 600 N.

9 A hole forms in the side of a water storage tank at a point 6 m below the surface of the water. (a) What is the hydrostatic pressure of the water near the hole? (b) A plug is inserted into the hole to stop the leak. If the area of the hole is 0.5 cm², with what force must the plug be secured to prevent it from popping out?
Ans. (a) 5.88×10^4 Pa; (b) 2.94 N

10 The top deck of a submarine has an area of 500 m². Calculate the net force on this deck when the submarine is submerged to a depth of 60 m.

11 (a) The density of blood plasma is 1030 kg/m³. What is the pressure of the plasma entering a patient's vein, if it flows from a bag held 1.5 m above the patient's arm? (b) If the blood pressure in the vein is 12 mmHg, what is the minimum height at which the bag must be held for the plasma to flow into the vein? (c) Suppose an astronaut needed a transfusion on the moon. What would the minimum height of the bag have to be in this case? On the moon g is 1.63 m/s².
Ans. (a) 114 mmHg; (b) 0.16 m (because of the viscosity of the plasma, the bag must be held much higher than this to get any appreciable flow); (c) 0.95 m

12 Around 1646 Pascal performed the experiment shown in Fig. 7.34. A very long tube, with cross-sectional area $A = 3 \times 10^{-5}$ m², was connected to a wine barrel which had a lid of area $A' = 0.12$ m². The barrel was first filled with water, and then water was added to the tube until the barrel burst. This happened when the column of water was 12 m high. Just before the barrel burst, what was (a) the weight of the water in the tube, (b) the pressure of the water on the lid of the barrel, and (c) the net force exerted on the lid?

REMARK Note that the water in the tube, although it weighed only a few newtons, was able to exert a force of thousands of newtons on the barrel lid.

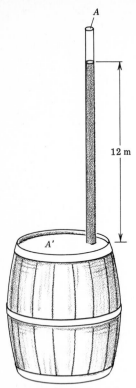

A

12 m

A′

FIGURE 7.34

Problem 12.

13 Some people experience ear trouble going up in an elevator because of the pressure change. If the pressure in back of the eardrum does not change during the ascent, the decrease in the outside pressure causes a net outward force on the eardrum. (a) What is the change in air pressure in going up 100 m in an elevator? (b) What is the net force on an eardrum with an area of 0.6 cm²?
Ans. (a) −1176 Pa; (b) 0.0706 N

14 The density of air decreases steadily with increasing height from its sea-level value of 1.2 kg/m³. But for the fun of it, suppose the density of air remained constant at 1.2 kg/m³ up to a height h and was zero above this height. Calculate the height h of this imaginary atmosphere, given that the air pressure at sea level is 1.01×10^5 Pa.

15 How tall would a barometer filled with glycerin ($\rho = 1260$ kg/m³) have to be to measure a pressure of 1 atm?
Ans. 8.18 m

16 On an imaginary planet the atmospheric pressure is 2.5 bar, and the acceleration of gravity is 15 m/s². How tall would a mercury barometer be on this planet?

17 A mercury manometer is connected to a vessel as shown in Fig. 7.35. (a) What is the gauge pressure in the vessel? (b) What is the absolute pressure in the vessel, assuming atmospheric pressure is 1.01×10^5 N/m²? (c) If the absolute pressure in the vessel is doubled, what is the gauge pressure?
Ans.   (a) 0.33×10^5 Pa;   (b) 1.34×10^5 Pa; (c) 1.67×10^5 Pa

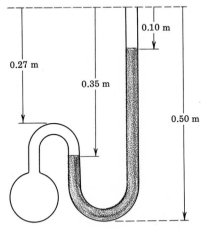

0.10 m

0.27 m

0.35 m

0.50 m

FIGURE 7.35

Problem 17.

18 A mercury manometer is connected to a vessel as shown in Fig. 7.36. (a) If the height d_A of the left-hand column is 0.22 m, what is the height d_B of the right-hand column when the gauge pressure inside the vessel is 0.16×10^5 Pa? (b) What are the heights d_A and d_B when the gauge pressure is 0.32×10^5 Pa?

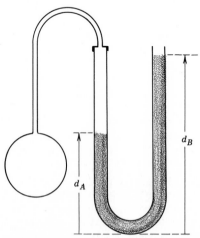

d_A

d_B

FIGURE 7.36

Problem 18.

19 A cylinder with cross-sectional area $A = 4 \times 10^{-4}$ m² is connected by a tube to

206

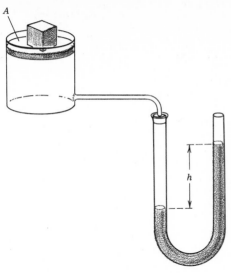

A

FIGURE 7.37

Problem 19.

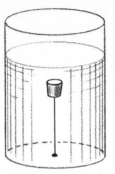

FIGURE 7.39

Problem 21.

one side of a mercury manometer (Fig. 7.37). What is the difference in the heights of the two columns when a 3-kg mass is placed on the piston of the cylinder?
Ans. 0.55 m

20 Oil ($\rho = 920$ kg/m³) is poured into one arm of a U-tube that is partially filled with water. The oil forms a 5-cm high column in the left arm (Fig. 7.38). Find the height difference h of the columns in the two arms.

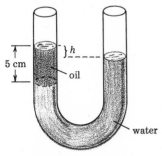

5 cm

$\}h$

oil

water

FIGURE 7.38

Problem 20.

21 Figure 7.39 shows a 20-g cork ($\rho = 250$ kg/m³) attached by a cord to the bottom of a water-filled bowl. (a) What is the bouyant force on the cork? (b) What is the tension in the cord? (c) If the cord is cut, the cork will accelerate upward. What will be its initial acceleration?
Ans. (a) 0.784 N; (b) 0.588 N; (c) 29.4 m/s²

22 (a) An object floating on water has the fraction f of its total volume below water.

Show that the density ρ_0 of the object is

$$\rho_0 = f\rho_w$$

where ρ_w is the density of water. (b) What fraction of an iceberg is below the surface? See Table 1.6.

23 A hot-air balloon consists of a gondola suspended from a nearly spherical balloon 17 m in diameter (Fig. 7.40). The total mass of the balloon and gondola is 2500 kg. (a) What is the bouyant force on the balloon in sea-level air at 20°C? See Table 1.6. (b) The balloon ascends until the bouyant force equals the weight of the balloon. What is the density of air at the highest point reached by this balloon? (Neglect the volume of the gondola and treat the balloon as though it were a perfect sphere.)
Ans. (a) 3.03×10^4 N; (b) 0.97 kg/m³

24 A 2-kg block of aluminum is suspended in water by a cord attached to a scale (Fig. 7.19b). What does the scale read?

25 A 5-kg object has a volume of 2.2×10^{-3} m³. What is its specific gravity?
Ans. 2.27

26 An object with a specific gravity of 3.2 has a mass of 40 kg. What is its volume?

27 An object weighs 75 lb in air and 50 lb in water. (a) What is the specific gravity of the object? (b) What is the volume of the object?
Ans. (a) 3; (b) 0.40 ft³

28 A hydrometer consists of a weighted bulb attached to a cylindrical stem of cross-sectional area 0.5 cm². The total mass of the hydrometer is 12 g. When placed in water, the stem extends 8 cm above the surface. When placed in a liquid of specific gravity 0.8, how much does the stem extend above the surface?

29 Water flows through a 1-in-diameter

207

FIGURE 7.40

pipe at the rate of 5 gal/min (3.15×10^{-4} m³/s). What is the average speed of the water in this pipe?
Ans. 0.62 m/s

30 Water is flowing through a 10-cm-diameter pipe at the rate of 0.025 m³/s. A clog at one point in the pipe narrows the cross-sectional area of the pipe to 30 cm². What is the speed of the water in this constricted region?

31 A cylindrical water tank 15 m in diameter is being filled at the rate of 5×10^4 gal/h. How fast is the water level rising in the tank?
Ans. 0.3 mm/s

32 Eight 4-in-diameter sewer pipes drain into a single 12-in-diameter pipe. If the average speed of the liquid in each 4-in pipe is 10 ft/s, what is the average speed in the 12-in pipe?

33 Calculate the rate at which water flows out of the hole in Prob. 9.
Ans. 1.1×10^{-3} m³/s

34 Gasoline pours from a hole in the side of a gasoline storage tank with a speed of 22 m/s. How far is the hole below the surface of the gasoline?

35 The intake tube of an automobile carburetor (Fig. 7.27) has a diameter of 5 cm. This narrows to a diameter of 1 cm in the Venturi tube. (*a*) Calculate the speed at which air must flow into the intake tube to produce a pressure drop of 150 Pa in the Venturi tube. (*b*) What is the rate (in grams per second) at which air flows through the carburetor?
Ans. (*a*) 3.23 m/s; (*b*) 7.6 g/s

36 Figure 7.41 shows a device, called a *Pitometer log* or *Pitot tube*, which is used to measure the speed of a ship or airplane. Water flows past the opening A of the Pitometer log with a speed v equal to the ship's speed, whereas the speed v' of the water at point B is zero. (*a*) Show that the pressure difference $p' - p$ between points

208

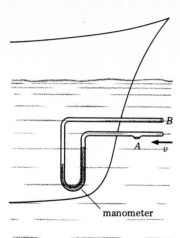

manometer

FIGURE 7.41

Problem 36.

A and B is given by

$$p' - p = \tfrac{1}{2}\rho v^2$$

(b) What is the ship's speed when the manometer connected to the device measures a pressure difference of 1.2 bar?

37 A steel cube 10 cm on a side slides at a constant speed of 15 cm/s along a flat steel plate. (a) What is the force required to keep the cube moving when the surface is lubricated with a 0.2-mm-thick layer of oil ($\eta = 0.4$ Pl)? (b) What would be the required force in the absence of lubrication, assuming a coefficient of kinetic friction of 0.5? Use Table 1.6.
Ans. (a) 3 N; (b) 37.7 N

38 A fluid exerts a viscous force $\mathbf{F}_v$ on any object moving through it. For a small sphere of radius r moving slowly with speed v, the force is given by Stokes' law

$$F_v = 6\pi\eta rv$$

(a) What is the viscous force on a water droplet of radius $r = 0.02$ cm moving through air with the speed $v = 2$ m/s? (b) A falling droplet increases its speed until the viscous force balances the weight of the droplet. Thereafter the droplet falls at constant speed v_t, called the *terminal speed*. Show that the terminal speed is given by

$$v_t = \frac{2\rho r^2 g}{9\eta}$$

where ρ is the density of the droplet, and η is the viscosity of air. (c) What is the terminal speed of the droplet in part (a)?
Ans. (a) 1.38×10^{-7} N; (c) 4.76 m/s

39 It is found experimentally that the flow of a fluid of density ρ and viscosity η through a pipe of radius r is laminar as long as the *Reynolds number*

$$Re = \frac{\bar{v}r\rho}{\eta}$$

is less than 1000. Here $\bar{v}$ is the average speed of the fluid in the pipe. (a) At what speed $\bar{v}$ does the Reynolds number for water flowing in a 12-in-diameter pipe equal 1000? (b) Calculate the Reynolds number of the air flowing through the intake tube and the Venturi tube of the carburetor described in Prob. 35.
Ans. (a) 0.66 cm/s; (b) 5400, 27,000

REMARK When the Reynolds number is much greater than 1000, the flow is no longer laminar, and so Bernoulli's and Poiseuille's equations give only approximate results.

40 (a) Show that the Reynolds number defined in Prob. 39 can also be written

$$Re = \frac{Q\rho}{\pi r\eta}$$

(b) Show that the flow of water through a capillary tube is laminar as long as Q (in cubic centimeters per second) is less than the diameter of the tube (in millimeters).

41 A glass capillary tube is 20 cm long and 0.12 cm in diameter. What pressure difference must be maintained between the ends of this tube to force water (at 20°C) through it at the rate 0.5 cm³/s?
Ans. 123 Pa

42 A hypodermic needle of length 8 cm and internal radius 0.04 cm is attached to a syringe with a plunger of area 3.5 cm². What force must be applied to the plunger to get water to flow from the syringe into a patient's vein at the rate $Q = 2$ cm³/s? Assume the blood in the vein has a gauge pressure of 9 Torr.

43 Oil, with a viscosity of 0.15 Pl, flows through a 20-mi section of pipeline at an average speed of 0.8 m/s. The pipe is 1.5 m in diameter. (a) What is the pressure difference between the ends of the pipe? (b) What is the rate of flow through the pipe? (c) How much power is required to pump the oil?
Ans. (a) 0.55×10^5 Pa; (b) 1.4 m³/s; (c) 77 kW

44 How much power is required to pump 5000 gallons of water per minute through a 10-mi-long pipe that is 10 in in diameter?

45 The human heart pumps blood at the rate of 80 cm³/s in the average resting adult. In the process it raises the average gauge pressure of

the blood from 0 to 100 Torr. What is the power
of the heart?
Ans. 1.06 W

BIBLIOGRAPHY

HAYWARD, ALAN T.: "Negative Pressure in Liquids: Can It Be Harnessed to Serve Man?" *American Scientist* **59**:434 (1971). An account of the present and future applications of negative pressure, i.e., of liquids under tensive stress.

WALKER, JEARL: "The Amateur Scientist: Serious Fun with Polyox, Silly Putty, Slime and Other Non-Newtonian Fluids," *Scientific American* **239**:186 (November 1978). A newtonian fluid is a fluid whose viscosity is independent of the viscous force. This article describes interesting non-newtonian fluids whose viscosity depends on the magnitude or the rate of change of the shear stress applied on them.

PART II

THERMAL PHYSICS

The laws of mechanics are stated in terms of a few general properties of matter, such as force, mass, and energy. In the next three chapters we shall study phenomena that depend on another general property—*temperature*. These phenomena include the transition of matter from solid to liquid and liquid to gas, the flow of heat from one object to another, and the transformation of heat into work in an engine. The laws we shall study find application in many areas of chemical and mechanical engineering, such as engine design, heating and air-conditioning, and energy conservation.

First steam car (Verbiest, 1678). (*National Museum of Science and Technology, Milan.*)

M echanical phenomena, such as the action of a pulley, imply nothing about the ultimate structure of matter. However, many other phenomena, such as the melting of ice or the vaporization of water, can be properly understood only in terms of the atomic theory of matter. This theory explains the observed macroscopic properties of a substance in terms of the motions and interactions of the invisible atoms of which it is composed. In this chapter we shall review the basic concepts of atoms and molecules and then show how these ideas can be used to understand temperature, the gas laws, and phase transitions.

8

THE STRUCTURE OF MATTER

8.1 Atoms and Molecules

Definition A *chemical element* is any one of the 104 known substances that cannot be decomposed into a simpler substance by chemical means. A table of the elements is given inside the back cover of this book. This table includes many familiar substances, like oxygen, carbon, and iron, as well as less familiar (and much rarer) substances like ruthenium and astatine. The table also gives the symbol used for each element, such as O for oxygen, C for carbon, and Fe for iron.

An element is composed of an immense number of identical units called *atoms*. Each atom is very small. For example, the diameter of a carbon atom is about 1.5×10^{-10} m, and the mass m_C of a carbon atom is

$$m_C = 19.9 \times 10^{-27} \text{ kg}$$

We use the term *microscopic* to refer to quantities of atomic size, even though an atom is too small to be seen under a microscope. The term *macroscopic* refers to quantities of normal size, such as a gram or a cubic centimeter. Because atoms are so small, there are always an immense number of them in anything of macroscopic size.

Example 8.1 How many atoms are in 1 mg of carbon?
The mass m of N carbon atoms is N times the mass of one atom:

$$m = Nm_C$$

In this case, the mass m is

$$m = 1 \text{ mg} = 1 \times 10^{-3} \text{ g} = 1 \times 10^{-6} \text{ kg}$$

so

$$N = \frac{m}{m_C} = \frac{1 \times 10^{-6} \text{ kg}}{19.9 \times 10^{-27} \text{ kg}} = 5 \times 10^{19}$$

Thus even in 1 mg of carbon, the number of carbon atoms is immense. ☐

Definition A *molecule* is a microscopic entity composed of two or more atoms bound together by atomic forces. The atoms may be of the same or different elements. For instance, two oxygen atoms can combine to form a molecule of oxygen (O_2), and an atom of oxygen and an atom of carbon can combine to form a molecule of carbon monoxide (CO). The symbol, or formula, for a molecule is formed from the symbols of the atoms of which it is composed. A subscript is used to indicate when two or more atoms of the same element are present. Thus the formula for water, H_2O, indicates that a molecule of water is composed of two hydrogen atoms and one oxygen atom. Examples of the formulas of other molecules are given in Table 8.1.

Each atom in a molecule has a definite position. In a water molecule, for instance, the two hydrogen atoms make a 105° angle with the oxygen atom (Fig. 8.1a), whereas in a deoxyribonucleic acid (DNA) molecule, the thousands of precisely arranged atoms form a complex double spiral (Fig. 8.1b).

Definition A *compound* is a substance composed of identical molecules made of atoms from two or more different elements. Thus carbon monoxide (CO) is a compound, whereas molecular oxygen (O_2) is an element. *Chemistry* is the study of the ways in which compounds and elements can be combined to form other compounds.

ATOMIC MASSES

The mass of an individual atom can be measured very precisely with an instrument called a *mass spectrometer* (Sec. 17.3). When a sample of carbon is placed in a mass spectrometer it is found that 1 out of every 100 carbon atoms is a little heavier than the others. This heavier atom is identical with the more abundant carbon atoms in all respects except for its mass, which is about 8 percent greater. Atoms of the same element which differ from one another in mass are called *isotopes*.

Because atomic masses are so small by macroscopic standards, it is convenient to measure these masses in *atomic mass units* (u). This unit is defined in terms of the abundant carbon isotope, which is assigned a mass of exactly 12.00000 u. Since the rarer isotope is 8 percent heavier, its mass is about 13 u. Similarly, since the mass of the common hydrogen isotope is one-twelfth the mass of carbon, its mass is about 1 u.

TABLE **8.1**

Chemical name	Common name	Formula	Examples of compounds and their chemical formulas
Carbon monoxide		CO	
Carbon dioxide	Dry ice (when solid)	CO_2	
Ethanol	Ethyl alcohol	C_2H_6O	
Hydrogen oxide	Water	H_2O	
Hydrogen peroxide		H_2O_2	
Methane	Marsh gas	CH_4	
Oxygen		O_2	
Silicone dioxide	Quartz, Sand	SiO_2	
Sodium chloride	Common salt	NaCl	
Sucrose	Cane sugar	$C_{12}H_{22}O_{11}$	
Tetrachloromethane	Carbon tetrachloride	CCl_4	

When we want to specify a particular isotope of an element, we put a superscript equal to its mass (rounded to the nearest whole number) to the left of the chemical symbol. Thus the two isotopes of carbon are denoted ^{12}C and ^{13}C. The table of the elements gives the average mass of the atoms of an element. For example, 99 percent of the atoms in natural carbon are ^{12}C, and 1 percent are ^{13}C; therefore the average mass of a carbon atom in nature is

$$m_C = \frac{(99)(m_{^{12}C}) + (1)(m_{^{13}C})}{100}$$

$$= \frac{(99)(12\ u) + (1)(13\ u)}{100} = 12.01\ u$$

In this chapter we use the average masses given in the table of the elements unless we specify a particular isotope by using a superscript.

The mass of a molecule, often called its *molecular weight*, is equal to the sum of the masses of its atoms. For instance, since a molecule of water (H_2O) contains two hydrogen atoms and one oxygen atom, its molecular weight is

$$m_{H_2O} = 2m_H + m_O$$
$$= (2)(1.008\ u) + (1)(16.00\ u)$$
$$= 18.01\ u \qquad\qquad 8.1$$

Here we have rounded off the atomic masses given in the table of the elements.

The mass of a ^{12}C atom in grams is found experimentally to be

$$m_{^{12}C} = 19.92637 \times 10^{-24}\ g$$

Since the definition of the atomic mass unit is

$$m_{^{12}C} = 12.00000\ u$$

the conversion between grams and atomic mass units is

$$1\ g = \frac{12\ u}{19.92637 \times 10^{-24}} = 6.022 \times 10^{23}\ u$$

or

$$1\ u = 1.66 \times 10^{-24}\ g = 1.66 \times 10^{-27}\ kg$$

The number

$$N_A = 6.022 \times 10^{23}$$

is called *Avogadro's number.* In terms of this number, the conversion between grams and atomic mass units is

$$1\ g = N_A\ u \qquad \text{or} \qquad 1\ u = \frac{1}{N_A} g \qquad\qquad 8.2$$

Definition   A *mole* (abbreviated mol) is a quantity of a substance that contains N_A molecules. It is often convenient to measure substances in moles rather than in grams, because certain properties of a substance depend on the number of molecules present rather than on the mass.

FIGURE 8.1

Models of molecules: (a) water (H_2O), and (b) deoxyribonucleic acid (DNA). (*Ealing Scientific Co.*)

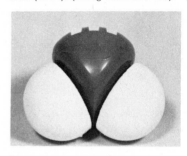

a

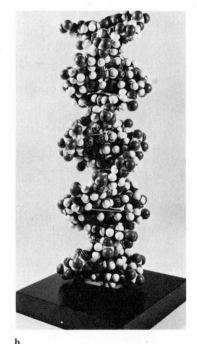

b

Example 8.2 What is the mass in grams of 1 mol of water?

From Eqs. 8.1 and 8.2, the molecular mass of one water molecule is

$$m_{H_2O} = 18\,u = (18)\left(\frac{1}{N_A}g\right) = \frac{18}{N_A}g$$

so the mass of N_A water molecules is

$$N_A m_{H_2O} = N_A\left(\frac{18}{N_A}g\right) = 18\,g$$

In general, the mass in grams of 1 mol of a substance is equal to its molecular mass in atomic mass units. $\square$

Example 8.3 What is the number n of moles in 50 g of oxygen?

A molecule of oxygen consists of two oxygen atoms, so the mass of a molecule of oxygen is

$$m_{O_2} = (2)m_O = (2)(16.0\,u) = 32\,u$$

and the mass of 1 mol of oxygen is 32 g. The number of moles in 50 g therefore is

$$n = \frac{50\,g}{32\,g} = 1.56 \qquad\qquad \square$$

8.2 The Three Phases of Matter

Matter can exist normally in only one of three phases: gas, liquid, or solid. Many substances are transformed from one phase to another by changes in temperature and pressure. The most familiar example of this is water, which is transformed from the liquid phase to the solid phase at temperatures below 0°C. When a wet towel is hung up to dry, we observe the transformation of water from the liquid to the gaseous phase at room temperature. Other familiar examples of phase changes are the melting of solder when touched with a hot iron, the vaporization of gasoline in the carburetor of a car, and the condensation of water on a cold window pane.

A *solid* is characterized by having a definite *volume* and *shape*. Its shape can be changed only by the application of considerable force, such as that required to bend a steel bar. This rigidity of form is the result of the strong forces between the molecules of the solid, which are packed tightly together in fixed positions. To bend a solid, this very stable molecular arrangement must be altered, which requires the application of a strong force.

A *liquid* is characterized by having a definite *volume* but not a definite shape. A liquid flows to conform to the shape of the container in which it is placed. It has a definite volume, however, which it maintains in spite of changes in shape. The molecules of a liquid are almost as close together as in a solid,* but they do not have fixed positions. It is because the molecules of a liquid are free to move relative to each other that a liquid has no

---

*In a few substances, most notably water, the molecules are actually closer in the liquid phase than in the solid phase.

rigidity. On the other hand, it is because of the strong attraction between its molecules that a given quantity of liquid occupies a definite volume, regardless of its shape.

A *gas* is characterized by having neither a definite volume nor a definite shape. A gas will expand to fill any closed container in which it is placed, and if the container is opened, the gas will leak out the opening. (Only the gravitational attraction of the earth keeps the gaseous atmosphere from expanding into space. The moon, which exerts only one-sixth the gravitational force of the earth, cannot hold a gaseous atmosphere.) In a dilute gas the molecules are so far apart that they exert forces on each other only when they collide. Consequently, every molecule travels freely in a straight line until it hits another molecule or the walls of its container. It is this unconstrained molecular motion that gives a gas its inherent expandability. Furthermore, all very dilute gases tend to have the same properties because the frequency of molecular collisions is so small that the behavior of different gases is unaffected by differences in the strength of the forces between their molecules.

8.3 Temperature

The concept of temperature is rather subtle. Although everyone has a rough idea that temperature is a measure of the relative hotness or coldness of a substance, this does not take one very far without an understanding of the physical meaning of hot and cold. It turns out that it is best to begin by describing the procedure used to measure temperature. The fundamental physical meaning of temperature will then emerge as the properties of an ideal gas are developed in the next section.

An instrument that measures temperature is called a *thermometer*. A normal laboratory thermometer consists of a sealed glass tube with a bulb at one end (Fig. 8.2). The bulb and part of the tube are filled with a liquid (usually mercury), and the rest of the tube is vacuum. The temperature of a water and ice mixture is defined to be 0°C on the Celsius scale. Therefore, to calibrate the thermometer, it is first put into an ice-water mixture, and a line is etched on the outside of the tube opposite the position of the top of the mercury column. The thermometer is then placed in boiling water, the temperature of which (at atmospheric pressure) is defined to be 100°C. When heated, mercury expands more than glass, and so the mercury rises in the glass tube. A second line is etched on the tube opposite the position of the top of the mercury column after it stops rising. Intermediate temperature marks are made on the thermometer by dividing the distance between the 0°C mark and the 100°C mark into 100 equal intervals. The temperature of a substance can then be measured with this thermometer by placing its bulb in contact with the substance and noting the position of the top of the mercury column. If it is at the twentieth division between the 0°C line and the 100°C line, its temperature is 20°C. Temperatures above 100°C and below 0°C are measured by extending the scale above and below the 100 and 0°C marks on the tube.

Unfortunately, the temperature measured this way depends

FIGURE 8.2

Mercury thermometer. (*Central Scientific Co.*)

to some extent on the liquid used in the thermometer. For instance, a mercury thermometer and an alcohol thermometer, both calibrated in the same way, give slightly different measurements for the temperature of a substance because these liquids expand differently with temperature. One could *define* temperature by specifying a particular substance to be used in the standard thermometer, but such a definition lacks fundamental physical significance. To get a more meaningful definition of temperature, it is desirable to have a thermometer that does not depend on the particular properties of any one substance. Since all gases behave the same at low density, a dilute-gas thermometer is used for the standard definition of temperature.

In a standard gas thermometer, a dilute gas is contained in a glass bulb that is connected through a narrow tube to a mercury-filled manometer (Fig. 8.3). The left-hand side of the manometer is sealed at zero pressure, so that the difference in the heights of the two mercury columns is equal to the absolute pressure of the gas. The two sides of the manometer are connected by a flexible rubber tube to allow the left-hand side to be raised or lowered relative to the right-hand side. During the operation of this instrument, the left-hand side is moved up and down until the mercury in the right-hand side just reaches the top of the manometer tube. In this way the volume occupied by the gas is kept constant.

This instrument is calibrated by first measuring the absolute pressure p_i in the bulb when the bulb is in a water and ice mixture (ice temperature). When the bulb is then placed in boiling water (steam temperature), the pressure of the gas increases, so that some of the mercury is pushed out of the right-hand tube. By raising the left-hand tube, the mercury level in the right-hand tube is brought back to the top of the tube. After this adjustment is made, the pressure p_s of the gas at the steam temperature is measured. The values of p_i and p_s depend, of course, on the amount of gas in the bulb. However, if the gas is sufficiently dilute, the ratio p_s/p_i depends neither on the density nor the chemical composition of the gas. It is a universal constant, independent of the properties of any one substance. Careful measurements of this ratio using very dilute gases have determined it to be

$$\frac{p_s}{p_i} = 1.3661$$

The only thing special about the temperatures of ice and steam is that they are easily reproduced in the laboratory, so that they are convenient calibration points. However, the ratio p/p_i, where p is the pressure in the bulb at any other temperature, is also independent of the nature of the gas. Thus it is natural to define the fundamental temperature scale in terms of this ratio.

Definition   The *fundamental*, or *absolute*, *temperature T* is defined by

$$T = a\frac{p}{p_i} \qquad\qquad 8.3$$

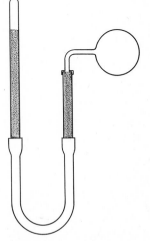

FIGURE 8.3

Gas thermometer. A glass bulb containing a dilute gas is connected to a mercury-filled flexible-tube manometer.

where p is the pressure in the bulb at this temperature, and a is an arbitrary constant.

This constant is chosen to make the difference between the steam temperature T_s and the ice temperature T_i exactly 100 temperature units, or kelvins (K). By definition, these temperatures are given by Eq. 8.3, with p replaced by p_s and p_i, respectively. That is,

$$T_s = a\frac{p_s}{p_i} = 1.3661a \qquad \text{and} \qquad T_i = a\frac{p_i}{p_i} = a$$

And so their difference is

$$T_s - T_i = 1.3661a - a = 0.3661a$$

Equating this to 100 K we get

$$T_s - T_i = 100 \text{ K} = 0.3661a$$

or

$$a = \frac{100 \text{ K}}{0.3661} = 273.15 \text{ K} \qquad\qquad\qquad 8.4$$

Thus the absolute temperature T of the bulb when its pressure is p is

$$T = (273.15 \text{ K})\frac{p}{p_i}$$

Since the gas pressure in the bulb cannot be less than zero, temperatures below 0 K cannot be defined this way. In fact, temperatures below absolute zero do not exist in the usual sense, and so the zero point on the absolute scale corresponds to the coldest temperature possible.

> **UNITS** Temperature is a fundamental dimension, like mass, length, and time, and its unit is the kelvin (K) or degree Celsius (°C). On the Kelvin scale, absolute zero is 0 K, the ice point is 273.15 K, and the steam point is 373.15 K. The units of temperature on the Kelvin and Celsius scales are equal,
>
> $$1 \text{ K} = 1°\text{C}$$
>
> but their zero points are shifted by 273.15 K. A temperature t on the Celsius scale is related to a temperature T on the Kelvin scale by
>
> $$t = T - 273.15 \text{ K} \qquad\qquad\qquad 8.5$$
>
> so that on the Celsius scale absolute zero is $-273.15°\text{C}$, the ice point is 0°C, and the steam point is 100°C. Note that the degree sign (°) is still used for Celsius temperatures, but is no longer used for Kelvin temperatures.
> Since the gas thermometer is not convenient for everyday laboratory use, other devices, such as the mercury thermometer and thermal transducers (Sec. 19.2), are used instead. These instruments are calibrated to measure the temperature defined by the gas thermometer.

8.4 The Ideal Gas

IDEAL-GAS LAW

Definition An *ideal gas* is a gas in which the molecules are so far apart that they spend relatively little time in direct contact with one another. The distance between molecules in a gas increases as the density of molecules decreases. This density is the ratio N/V of the number N of molecules in the gas to the volume V occupied by the gas. Thus every real gas is an ideal gas when N/V is sufficiently small. This fact was used in the last section to define the absolute temperature T in terms of the pressure of a dilute gas.

The pressure of a gas is caused by the gas molecules colliding with the walls of the container. For an ideal gas at a given temperature, the pressure is proportional to N/V, since as N/V increases, so does the number of molecules that can strike the wall. Thus the pressure p_i of an ideal gas at the ice point can be written

$$p_i = b\frac{N}{V}$$

where b is a constant independent of the chemical composition of the gas. When this expression is substituted into Eq. 8.3, we get

$$T = a\frac{p}{p_i} = \frac{ap}{bN/V} = \frac{a}{bN}pV$$

or

$$pV = \frac{bN}{a}T \qquad\qquad 8.6$$

The number N of molecules in a gas is

$$N = nN_A$$

where n is the number of moles of gas, and N_A is Avogadro's number (the number of molecules in a mole). Thus Eq. 8.6 can be written

$$pV = n\frac{bN_A}{a}T \qquad\qquad 8.7$$

The quantity bN_A/a is the *gas constant* R. Its dimensions are [energy/kelvin], and its magnitude is

$$R = \frac{bN_A}{a} = 8.314\,\text{J/K}$$

In terms of R, Eq. 8.7 becomes

$$pV = nRT \qquad\qquad 8.8$$

which is the ideal-gas law. This law gives the relation between the pressure p, volume V, and temperature T of n mol of an ideal gas. Care and thought must be exercised in using this equation because the quantities that appear in it must be of the proper form and in the correct units. The pressure is the absolute pressure in pascals (newtons per square meter), the volume is in cubic meters, and the temperature is the absolute temperature in kelvins.

Example 8.4 What is the mass density of oxygen at *standard temperature and pressure* (STP), i.e., at 0°C and 1 atm?

The ideal-gas law is used to determine the number of moles n

of an ideal gas in unit volume $V = 1\,m^3$ at a pressure of $p = 1\,atm = 1.01 \times 10^5\,Pa$ and a temperature of $T = 273\,K$. Equation 8.8 first is solved for n, and then these values are substituted into it. The result is

$$n = \frac{pV}{RT} = \frac{(1.01 \times 10^5\,Pa)(1\,m^3)}{(8.31\,J/K)(273\,K)} = 44.5\,mol$$

Since the mass of a mole of oxygen, O_2, is $32\,g$, the mass of $44.5\,mol$ of oxygen is $(44.5)(32\,g) = 1425\,g = 1.425\,kg$. Therefore, the STP density of oxygen is $1.43\,kg/m^3$. $\square$

Example 8.5 An automobile tire is filled with air at a temperature of 20°C to a (gauge) pressure of 24 lb/in.^{2} After driving 100 mi on a highway, the temperature of the tire is 50°C. What is the tire pressure at this temperature?

The best way to proceed is to first rewrite the ideal-gas law so that the quantities that change are on one side of the equation and the quantities that remain constant are on the other side. In this case p and T change, but the volume V and, of course, n and R do not. Therefore Eq. 8.8 is rewritten in the form

$$\frac{p}{T} = \frac{nR}{V} \qquad\qquad 8.9$$

Next, the given quantities must be put into the proper forms. The Celsius temperatures must be expressed as absolute temperatures, and the gauge pressures must be expressed as absolute pressures. From Eq. 8.5 we have that the initial temperature is

$$T_I = 273 + t_I = 273 + 20 = 293\,K$$

and the final temperature is

$$T_F = 273 + t_F = 273 + 50 = 323\,K$$

Assuming that atmospheric pressure is

$$p_0 = 1\,atm = 14.7\,lb/in^2$$

the initial absolute pressure is

$$\begin{aligned} p_I &= p_0 + \bar{p}_I = 14.7\,lb/in^2 + 24.0\,lb/in^2 \\ &= 38.7\,lb/in^2 \end{aligned}$$

At the initial temperature, Eq. 8.9 is

$$\frac{p_I}{T_I} = \frac{nR}{V}$$

and at the final temperature it is

$$\frac{p_F}{T_F} = \frac{nR}{V}$$

so we have

$$\frac{p_I}{T_I} = \frac{p_F}{T_F}$$

Finally, we solve this for p_F to get

$$p_F = \frac{T_F}{T_I} p_I = \frac{323\text{ K}}{293\text{ K}}(38.7\text{ lb/in}^2) = 42.6\text{ lb/in}^2$$

The gauge pressure in the tire at the higher temperature is

$$\bar{p}_F = p_F - p_0 = 42.6\text{ lb/in}^2 - 14.7\text{ lb/in}^2$$
$$= 27.9\text{ lb/in}^2$$

which is 3.9 lb/in² higher than the initial gauge pressure. ☐

REMARK In this last example the gas constant R is not needed because only the change in pressure is calculated. By using proportions to find p_F, the pressure did not have to be converted into SI units. In the first example, on the other hand, R is needed to find the number of moles in a gas under specified conditions. Since the value of R is given in SI units, all other quantities had to be in SI units also. It is sometimes convenient to measure pressure in atmospheres, torrs, or pounds per square inch and to measure volume in liters, cubic centimeters, or cubic feet. Instead of converting these units into SI units, the value of R appropriate to these units is used. Table 8.2 gives the value of R to be used in Eq. 8.8 when p and V are measured in these other units.

TABLE **8.2**

Volume	Pressure	R
m³	pascal	8.314 J/K
liter	atm	8.205×10^{-2} atm $\cdot$ L/K
liter	torr	62.4 torr $\cdot$ L/K
cm³	atm	82.05 atm $\cdot$ cm³/K
cm³	torr	6.24×10^4 torr $\cdot$ cm³/K
ft³	atm	2.90×10^{-3} atm $\cdot$ ft³/K
ft³	lb/in²	4.26×10^{-2} lb $\cdot$ ft³/in² $\cdot$ K

Values of the gas constant R appropriate for various units of pressure and volume

The temperature is always in kelvins.

PRESSURE, TEMPERATURE, AND KINETIC ENERGY
The pressure a gas exerts on the walls of its container is the result of the collisions of the molecules of the gas with the walls. It is possible to calculate this pressure in terms of the average kinetic energy of the molecules in an ideal gas and to relate this kinetic energy to temperature.

A molecule of mass m moving with speed v perpendicularly toward a wall will rebound with a speed $-v$ in the opposite direction (Fig. 8.4). The initial and final momenta of the molecule are

$$p_I = mv \quad \text{and} \quad p_F = m(-v) = -mv$$

so the change in momentum is

$$\Delta p = p_F - p_I = (-mv) - (mv) = -2mv$$

From Eq. 4.15 the force F_m exerted on the molecule by the wall is

$$F_m = \frac{\Delta p}{\Delta t} = \frac{-2mv}{\Delta t}$$

where Δt is the time interval in which F_m acts, i.e., the time taken by the molecule to reverse direction. The reaction force

FIGURE 8.4

A molecule striking a wall perpendicularly with velocity **v** rebounds in the opposite direction with velocity $-$**v**.

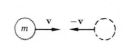

$$F_w = -F_m = \frac{2mv}{\Delta t} \qquad\qquad 8.10$$

is the force exerted on the wall by the molecule.

From Fig. 8.5 we see that all the molecules which strike an area A of the wall in time Δt must be contained in a cylindrical region of volume $V' = Ad$, where $d = v\Delta t$ is the distance a molecule with speed v moves in time Δt. The number N' of molecules in this volume is

$$N' = \frac{V'}{V}N$$

where N is the total number of molecules in the total volume V occupied by the gas. The number N of molecules is related to the number n of moles by

$$N = nN_A$$

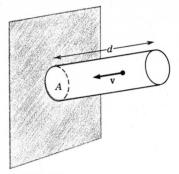

FIGURE 8.5

A region of cross-sectional area A and height d on the wall of a container of gas.

where N_A is Avogadros's number. The volume V' is related to v and Δt by

$$V' = Ad = Av\Delta t$$

so

$$N' = \frac{Av\Delta t}{V}nN_A$$

Of course, not all the molecules in the volume V' are moving toward the wall. We can assume that on the average, one-third of the molecules are moving in each perpendicular direction, and that of the one-third moving perpendicularly to the wall, one-half are moving toward the wall and one-half are moving away. Thus at any instant only $\frac{1}{3} \times \frac{1}{2} = \frac{1}{6}$ of the molecules in V' are moving perpendicularly toward the wall. Each of these molecules exerts the force F_w on the wall, so from Eq. 8.10, the total force F exerted on the area A is

$$F = \tfrac{1}{6}N'F_w = \tfrac{1}{6}\frac{AvnN_A\Delta t}{V}\frac{2mv}{\Delta t} = \tfrac{1}{3}\frac{nN_A A}{V}mv^2$$

and the pressure exerted is

$$p = \frac{F}{A} = \tfrac{1}{3}\left(\frac{nN_A}{V}\right)mv^2 \qquad\qquad 8.11$$

Actually the molecules in a gas have a wide distribution of speeds, so v^2 in Eq. 8.11 must be replaced by $\overline{v^2}$, the average of the squared speed. In terms of the average kinetic energy $\overline{K} = \tfrac{1}{2}m\overline{v^2}$ of a molecule, the pressure is

$$p = \tfrac{2}{3}\frac{nN_A\overline{K}}{V}$$

The pressure is also related to the temperature by Eq. 8.8:

$$p = \frac{nRT}{V}$$

By equating these two expressions for p we obtain the following relation between the temperature and the average kinetic energy of a gas molecule:

$$\frac{nRT}{V} = \frac{2}{3}\frac{nN_A\bar{K}}{V}$$

or

$$\bar{K} = \frac{3}{2}\frac{RT}{N_A} = \frac{3}{2}kT \qquad \qquad 8.12$$

where $k = R/N_A$ is called the *Boltzmann constant*. Since the constant R is the gas constant for 1 mol of gas and N_A is the number of molecules in a mole, the Boltzmann constant k is the gas constant per molecule. Its magnitude is

$$k = \frac{R}{N_A} = \frac{8.314 \text{ J/K}}{6.02 \times 10^{23}} = 1.38 \times 10^{-23} \text{ J/K}$$

The Boltzmann constant relates the absolute temperature T to the average kinetic energy of the molecules of an ideal gas. In fact, temperature is just a measure of this kinetic energy. This is the fundamental physical meaning of temperature. In solids, liquids, and nonideal gases, the temperature is still a measure of the average kinetic energy of the molecules, but the relation is not as simple as in Eq. 8.12.

8.5 Real Gases

The pressure and volume of a real gas are related by the ideal-gas law (Eq. 8.8) only when the molecular density is small. As the pressure of a gas is increased, the decrease in volume produces an increase in density. At high pressure, therefore, the pressure-volume relation of a real gas can be expected to deviate from that of an ideal gas. For most gases these deviations become appreciable at molecular densities above 10^{21} molecules/cm^3, or about 1 mol/L. That is, at a pressure at which 1 mol of gas occupies a volume of 1 L or so, the ideal-gas law does not apply.

pV DIAGRAMS

In Fig. 8.6 the pressure p is plotted against the volume V occupied by 1 mol of gaseous H_2O. Each curve gives the variation of pressure with volume at a fixed temperature, so these curves are called *isotherms* (meaning "equal temperature"). The light curves are the pV relations for an ideal gas, whereas the dark curves are the actual pV relations of H_2O. As expected, the real curve deviates appreciably from the ideal curve for values of V less than 1 or 2 L.

Definition The *critical temperature* of a gas is the temperature above which all the isotherms are smooth unbroken curves. Table 8.3 gives the critical temperatures of some common substances.

The isotherms in Fig. 8.6 are typical of any real gas. Above a certain critical temperature, the real isotherms are smooth curves which lie somewhat above or below the corresponding ideal isotherm. The real isotherm lies below the ideal isotherm

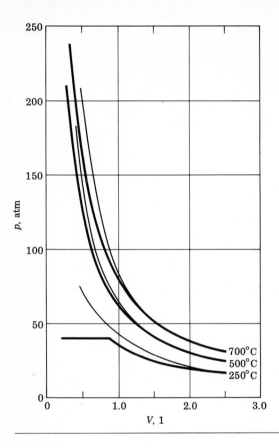

FIGURE 8.6

Plot of the pressure p against the volume V occupied by 1 mol of gaseous H_2O. The dark lines are the real isotherms at the temperature indicated, while the light lines are the isotherms of an ideal gas at the same temperatures.

when the molecules of the gas are so strongly attracted to each other that they tend to stick together briefly after colliding, and consequently they strike the walls less frequently than in an ideal gas. The real isotherm lies above the ideal isotherm when the molecules are not strongly attracted to each other. In this case they tend to rebound rapidly from each other and strike the walls more frequently than in an ideal gas. Below the critical temperature, the isotherms are no longer simple smooth curves, but like the 250°C isotherm in Fig. 8.6, consist of a horizontal section joined abruptly to a smoothly varying section.

A horizontal section in an isotherm indicates that the volume occupied by the gas can be decreased without an increase in pressure. This is possible because as the gas is compressed, some of it condenses into the liquid phase. However, a gas can be liquefied by compression only if its temperature is below the critical temperature of the gas. Note that two of the isotherms in Fig. 8.6 are at temperatures above 374.1°C, the critical temperature of H_2O, and one is at a temperature below the critical temperature.

The isotherm of a substance at its critical temperature, the *critical isotherm*, is horizontal only at a single point, the *critical point*. The isotherms of H_2O in the vicinity of its critical point are shown in Fig. 8.7. The scale in this figure is enlarged over that in Fig. 8.6 in order to display the necessary detail. An isotherm

TABLE **8.3**

Substance	Melting temperature, °C	Boiling temperature, °C	Critical temperature °C
Aluminum	659	2467	
Ammonia	−77.7	−33.3	132.4
Argon	−189.2	−185.7	
Carbon dioxide	−56.6*	*	31.0
Chlorine	−101	−34.6	144
Copper	1083	2595	
Ethanol	−114.5	78.5	126.9
Helium	−272.2*	*	
Iodine	113.5	184.3	512
Iron	1530	3000	
Lead	327	1744	
Mercury	−38.9	357	
Methane	−182.5	−164	−82.1
Neon	−248.7	−245.9	
Nitrogen	−209.9	−195.8	−146.9
Oxygen	−218.8	−183	−118.5
Propane	−189.7	−42.1	96.8
Silicon dioxide	1470	2230	
Sodium chloride	800	1413	
Sulfur	119	445	1040
Tetrachloro- methane	−23	76.5	283.2
Tungsten	3387	5927	
Water	0	100	374.1

The melting, boiling, and critical temperatures of some common substances

The melting and boiling temperatures are at a pressure of 1 atm, except for carbon dioxide and helium.

*This is the triple-point temperature. The triple-point pressures of CO_2 and He are 5.11 and 26 atm, respectively, so these substances do not exist as liquids at atmospheric pressure.

just above the critical isotherm has a kink in it near the critical point (point C in Fig. 8.7), but the isotherm is never horizontal. An isotherm below the critical isotherm is split into two distinct sections connected by a horizontal line. The critical isotherm, which is the transition between these two types of isotherms, is horizontal only at the critical point.

In Fig. 8.7 the ends of the horizontal lines are connected by a curve, called the *saturation curve*. This curve touches the critical isotherm at the critical point, so together these two curves divide the pV diagram into four regions. In the region above the critical isotherm, the substance remains a gas for all values of the pressure; this is the *gas region*. In the region below the critical isotherm and to the right of the saturation curve, the substance is still a gas, but it can be liquefied by compression; this is the *vapor region*. Below the critical isotherm and to the left of the saturation curve is the *liquid region*. The area under the saturation curve is the *liquid-vapor region*.

Consider point i in Fig. 8.7. This represents 1 mol of gaseous H_2O at a temperature of 350°C and a pressure of 155 atm. If this gas is compressed while its temperature is held constant, its pressure will increase. For example, decreasing the volume from 0.180 to 0.165 L increases the pressure from 155 to 160 atm. The substance is now represented by point h. As the substance is further compressed, its corresponding point in the pV diagram

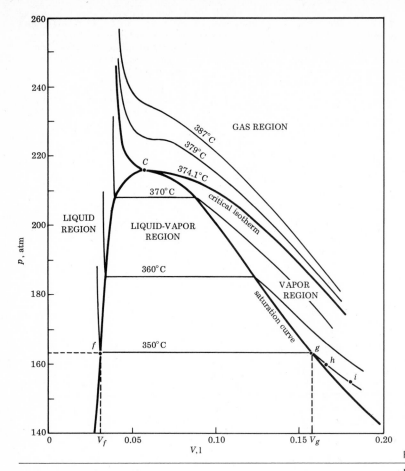

FIGURE 8.7

Isotherms of H_2O in the vicinity of
the critical point.

moves along the isotherm. When the substance reaches point g, it begins to condense into the liquid phase, so the point moves along the horizontal line fg. At the pressure $p_v = 163$ atm of this line, the substance can occupy any volume between V_f and V_g. The point g corresponds to the substance's being all in the gaseous phase, whereas the point f corresponds to the substance's being all in the liquid phase. Between these two points part of the substance is liquid and part is gas. When all the substance is in the liquid phase, a further decrease in its volume requires an enormous increase in pressure; hence the isotherm rises steeply to the left of f.

VAPOR PRESSURE

Definition   *Vapor pressure* p_v is the pressure of the horizontal section of an isotherm. It is the pressure at which the vapor and liquid can coexist at the temperature of the isotherm.

Table 8.4 gives the vapor pressure of H_2O for various temperatures between 0 and 374.1°C. (The vapor pressure obviously does not exist above the critical temperature.) Also listed in Table 8.4 are the volumes V_f and V_g occupied by 1 mol of H_2O at

TABLE **8.4**

Temperature, °C	Pressure, atm	V_f, L	V_g, L
374.1	218.3	0.057	0.057
300	84.8	0.0252	0.390
250	39.2	0.0225	0.901
200	15.3	0.0208	2.04
150	4.69	0.0196	7.01
120	1.96	0.0190	16.0
100	1.00	0.0188	30.1
80	0.466	0.0185	61.4
60	0.196	0.0183	138
40	0.0728	0.0182	352
37	0.0621	0.0181	410
30	0.0418	0.0181	593
20	0.0230	0.01804	1042
15	0.0168	0.01803	1405
10	0.0121	0.01802	1917
5	0.00856	0.01801	2651
0	0.00626	0.01800	3717

Vapor pressure of water at temperatures from 0 to 374.1°C

The table also lists the volumes V_f and V_g occupied by 1 mol of H_2O in the liquid and gaseous phases, respectively

the pressure p_v when it is completely in the liquid and gaseous phases, respectively.

Table 8.4 shows that 1 mol of water (liquid H_2O) at 20°C has a volume of 0.018 L. If it is placed inside an evacuated 2-L container, some of the liquid will evaporate. That is, some of the H_2O molecules leave the liquid, forming a gas in the empty space above the liquid. As more molecules evaporate, the pressure of this vapor increases. At the same time, some of the molecules in the vapor collide with the surface of the liquid, reentering the liquid phase. The number of molecules which recondense increases as the pressure of the vapor increases, until eventually a point of *dynamic equilibrium* is reached, where the rate at which molecules evaporate from the liquid equals the rate at which molecules condense back into the liquid. The pressure of the vapor at this equilibrium point is the vapor pressure given in Table 8.4, which at 20°C is 0.0230 atm. When the pressure reaches this value, the number of molecules in the liquid and gaseous phases remains constant. That is, there is no net evaporation or condensation, so the system is in equilibrium. However, there is still a constant flow of molecules from the liquid to the gas and an equal flow of molecules from the gas to the liquid, which is why the equilibrium is termed *dynamic*. The system is represented by a point in the liquid-vapor region of the H_2O pV diagram.

At equilibrium, the vapor occupies a volume of 2 L minus the volume occupied by the liquid, or

$$V = 2\,L - 0.018\,L = 1.982\,L$$

From Table 8.4 we see that 1 mol of gaseous H_2O occupies 1042 L at this temperature and pressure. Thus the 1.982 L of vapor is only

$$\frac{1.982\,L}{1042\,L} = 0.00190\ \text{mol}$$

That is, only 0.0190 mol of H_2O is in the gaseous phase. There is still 0.99810 mol of H_2O in the liquid phase. If the volume of the container is increased while keeping the temperature constant at 20°C, a larger fraction of the H_2O will go into the gaseous phase in order to maintain the pressure of H_2O at 0.0230 atm. When the volume of the container reaches 1042 L, all the H_2O is in the gaseous phase. A further increase in the volume beyond this produces a decrease in pressure because the number of molecules in the gaseous phase cannot be increased. The system is now represented by a point in the vapor region of the pV diagram. The vapor pressure is the maximum pressure the vapor can have at a given temperature.

Definition The *boiling point* of a liquid is the temperature at which the vapor pressure equals atmospheric pressure. At this temperature bubbles of vapor are formed in the interior of the liquid. Vapor bubbles cannot be formed at a lower temperature because the pressure inside them, which is the vapor pressure, is less than atmospheric pressure, so that the bubbles collapse. From Table 8.4 we see that the vapor pressure of H_2O equals 1.0 atm at 100°C. This, of course, is not a coincidence, since 100°C was first defined as the boiling point of water. However, this is the boiling point only at a pressure of 1 atm. At elevations above sea level the air pressure is less than 1.0 atm, so water boils at a lower temperature. In Mexico City, for instance, where the air pressure is only 0.85 atm, water boils at 95°C. On the other hand, in a pressure cooker the absolute pressure is about 2 atm, so water boils at a temperature of about 120°C.

Table 8.3 gives the boiling points (at $p_0 = 1$ atm) of some common substances.

THE COMPOSITION OF AIR

Air contains gaseous H_2O in concentrations that vary from 0 to 6 percent, depending on the weather conditions. Dry air, i.e., air with no H_2O in it, is 78.1 percent nitrogen, 20.9 percent oxygen, 0.9 percent argon, and 0.03 percent carbon dioxide. Each component of a gas mixture exerts a partial pressure independent of the other components. In a mixture of ideal gases, the total pressure is the sum of the partial pressures exerted by the components. This fact is usually expressed as follows.

Dalton's Law of Partial Pressure *In a mixture of ideal gases, each component gas exerts a partial pressure proportional to its molecular concentration.* The total pressure of the mixture is equal, therefore, to the sum of the partial pressures of all the component gases.

Example 8.6 (a) What is the partial pressure of H_2O in air which contains 2 percent water vapor? (b) What is the partial pressure of oxygen in this air?

(a) One mole of this air contains 2 percent H_2O, or 0.02 mol. Since the total pressure exerted by 1 mol of air is $p_0 = 1$ atm, the partial pressure p_{H_2O} of H_2O is

$$p_{H_2O} = (0.02)p_0 = (0.02)(1\ atm) = 0.02\ atm$$

(b) One mole of this air contains 0.98 mol of dry air, or

$$(0.209)(0.98\ mol) = 0.205\ mol$$

of oxygen (since dry air is 20.9 percent oxygen). Thus the partial pressure p_{O_2} of oxygen is 0.205 atm. □

The maximum partial pressure of H_2O possible at a particular air temperature is the vapor pressure of H_2O at that temperature. For instance, at a temperature of 20°C the partial pressure of H_2O can be at most 0.023 atm (Table 8.4).

Definition The *relative humidity* of air is the ratio of the actual partial pressure of H_2O in the air to the maximum partial pressure of H_2O possible at the existing air temperature. It is often expressed as a percentage.

Example 8.7 (a) What is the relative humidity at 20°C of air which has a partial H_2O pressure of 0.0085 atm? (b) What is the relative humidity of this air at 10°C?

(a) From Table 8.4 the vapor pressure at 20°C is 0.0230 atm, so the relative humidity at 20°C is

$$\frac{p_{H_2O}}{p_v}(100) = \frac{0.0085\ atm}{0.0230\ atm}(100) = 37\%$$

(b) Since at 10°C the vapor pressure is 0.0121 atm, the relative humidity at 10°C is

$$\frac{0.0085\ atm}{0.0121\ atm}(100) = 70\%$$ □

Definition The *dew point* of the air is the temperature at which the air would have a relative humidity of 100 percent.

Example 8.8 What is the dew point of air in which the partial pressure of H_2O is 0.0085 atm?

From Table 8.4 we see the the vapor pressure of H_2O is 0.0085 atm at a temperature of 5°C. Thus the dew point of this air is 5°C. During the day, with an air temperature of 15°C, the relative humidity of this air is 50 percent. But at night, if the temperature falls to 5°C, the relative humidity will reach 100 percent. At this point moisture will start to condense on the ground. Moisture will also condense on any surface that is 5°C or less, such as a cold windowpane or the outside of a glass containing an iced drink. □

8.6 Solids

The molecules of a solid, like those of a liquid, are so close together that they exert strong forces on each other. However, in a liquid the molecules are free to move about, whereas in a solid the molecules have fixed positions. It is the fixed positions of the molecules that give a solid its rigidity.

Definition A *crystal* is a solid in which the molecules are arranged in a regular three-dimensional pattern that persists throughout the solid. The arrangement of molecules determines the external shape of the crystal, which has plane surfaces meeting at definite angles characteristic of the molecular pattern. Figure 8.8 shows one possible pattern, a simple cubic lattice.

Figure 8.9 shows some naturally occurring crystals of halite (NaCl) and quartz (SiO_2).The regularity and symmetry of a crystal are the visible manifestation of the regularity and symmetry of the arrangement of the molecules in the solid.

If identical molecules occupy every point in this lattice, the molecule at any one lattice point has the same surroundings as the molecule at any other point (except, of course, for molecules located near the surface of the crystal). Thus nearly all the molecules in the interior of a crystal are subject to identical conditions, and so they behave identically. It is the identical behavior of the molecules in a crystal that gives these solids their special characteristics.

A solid is often formed under conditions that do not favor the formation of a single crystal. Instead, as the solid forms, thousands of minute crystals grow together. Within each microcrystal the molecules have an ordered arrangement, but the microcrystals are randomly arranged with respect to each other. Metals generally have this kind of structure. Because each microcrystal contains 10^{16} or more molecules, the vast majority of molecules are in the interior of a large ordered array of molecules. Only the relatively few molecules on the boundary between two microcrystals are out of order (Fig. 8.10). Thus a solid composed of microcrystals behaves in many ways like a single crystal.

Definition A *crystalline solid* is a solid consisting either of a single crystal or of many microcrystals fused together. This section discusses properties common to all crystalline solids.

> **REMARK** In a noncrystalline, or *amorphous,* solid the molecules are randomly arranged, so that the forces on one molecule are different from the forces on another. Glass is perhaps the most common example of a noncrystalline solid. Chemically it is predominantly SiO_2, the same as quartz. Its physical properties differ from those of quartz, however, because its molecules are arranged differently.

The most characteristic property of a crystalline solid is the existence of a definite temperature T_m, called the *melting point,* at which the solid is transformed into a liquid. The temperature of a solid increases as its internal energy is increased, until the melting point is reached. Thereafter the addition of more energy transforms the solid into a liquid at the temperature T_m. Table 8.3 gives the melting points of some common substances.

At the melting point a substance can coexist in the solid and liquid phases. For example, a mixture of ice and water at 0°C is stable if no energy is added or removed. However, if energy is added, the ice will melt, and if energy is removed, the water will freeze. At temperatures above T_m the substance cannot exist as a

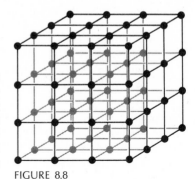

FIGURE 8.8

Position of molecules in a simple cubic lattice.

FIGURE 8.9

Crystals: (a) halite, and (b) quartz, (*The American Museum of Natural History.*)

a

b

solid, and at temperatures below T_m the substance cannot normally exist as a liquid.*

REMARK A noncrystalline solid, such as glass, has no definite melting point. Since the molecules of the solid are randomly arranged, some molecules can break away from their positions more easily than others. As the temperature of glass is increased, the glass becomes softer and softer, until eventually it flows freely, but there is no sharp transition from the solid to the liquid. Conversely, as molten glass cools, its viscosity increases steadily, until no perceptible flow occurs. Butter and margarine are also examples of noncrystalline solids that have no definite melting point.

The melting point of a crystalline solid depends slightly on pressure. For example, the melting point of H_2O, which is 0°C at a pressure of 1 atm, decreases to −1.5°C at a pressure of 200 atm. The conventional melting point T_m is the melting point at a pressure of 1 atm. A plot of melting point against pressure gives the melting curve shown in Fig. 8.11. The curve is nearly vertical because p has to change by many atmospheres to change the melting point by a fraction of a degree. (The melting curve in Fig. 8.11 slopes to the right because the melting point of most substances, unlike that of water, increases with increasing pressure.) A substance is a solid for values of p and T to the left of the melting curve and a liquid for values of p and T to the

*Under certain circumstances a liquid can be cooled below the melting point without solidifying. This is an unstable situation, however, and the entire liquid immediately solidifies once any part of it starts to solidify.

FIGURE 8.10

Only the molecules on the boundary between two microcrystals are out of order.

FIGURE 8.11

pT diagram of a crystalline substance. The melting curve separates the solid and liquid phases, the vaporization curve separates the liquid and gas phases, and the sublimation curve separates the gas and solid phases.

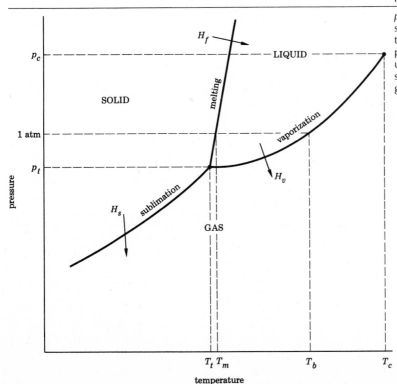

right of the curve. Points on the melting curve give the values of p and T for which the liquid and solid can coexist.

Similarly, a vaporization curve is a plot of the vapor pressure of a liquid against temperature. As shown in Fig. 8.11, a substance is a liquid for values of p and T above this curve and a gas for values of p and T below the curve. A point on the vaporization curve gives the pressure p of the vapor that is in equilibrium with the liquid at the temperature T. The vaporization curve terminates at the critical temperature T_c, which is the highest temperature at which the substance can exist as a liquid (Sec. 8.5). The position of the boiling point T_b, which is the temperature of a liquid at which the vapor pressure is 1 atm, is also indicated in Fig. 8.11.

The vaporization curve intersects the melting curve at a point called the *triple point*. The temperature T_t and pressure p_t at this point are characteristic of the substance. For instance, at the triple point of H_2O they are 0.01°C and 0.0060 atm. At this temperature and vapor pressure, the solid, liquid, and gaseous phases of a substance all coexist. For many substances the temperatures T_m and T_t differ by only a few hundredths of a degree, which is inconsequential for most purposes.*

A third curve, called the *sublimation curve,* also intersects the vaporization and melting curve at the triple point (Fig. 8.11). This curve gives the vapor pressure of the gas that is in equilibrium with the solid at various temperatures. A substance is a solid for values of p and T above the curve and a gas for values of p and T below the curve. For example, the vapor pressure of ice at $-10°C$ is 0.0026 atm. On a cold, dry winter day, when the temperature is $-10°C$ (14°F) and the partial pressure of H_2O in the air is less than 0.0026 atm, the values of p and T correspond to a point below the sublimation curve. Under these conditions the only stable phase of H_2O is the gaseous phase. This means that any snow or ice on the ground is not in equilibrium with the vapor but is slowly being transformed directly into vapor, a process called *sublimation*. No liquid H_2O is formed because it does not exist below the triple point.

*One important exception is carbon dioxide; see Table 8.3.

PROBLEMS

1 What is the molecular weight of sucrose?
Ans. 342

2 Ketene, a compound with a molecular weight of 42.04, is known to be composed of carbon, oxygen, and hydrogen. Determine the formula for this compound.

3 How many molecules are there in 5 g of sugar? See Prob. 1.
Ans. 8.8×10^{21}

4 How many moles are in 250 g of ethanol?

5 (a) How many molecules are there in a cubic centimeter of oxygen at STP (0°C and 1 atm)? The density of oxygen at STP is 1.25 kg/m^3. (b) How many molecules are there in a cubic centimeter of water?
Ans. (a) 2.7×10^{19}; (b) 3.3×10^{22}

> **REMARK** The last problem shows that the density of molecules is about one thousand times greater in a liquid than in a gas at STP.

6 What is the distance between neighboring silicon dioxide molecules in quartz? The specific gravity of quartz is 2.6.

7 Air in a sealed bulb has a pressure of 1 atm at 20°C. (a) What is the pressure in the bulb at

GUIDE TO MAJOR TOPICS

150°C? (b) At what temperature is the pressure in the bulb 2 atm? (c) At what temperature is the pressure in the bulb 0.5 atm?
Ans. (a) 1.44 atm; (b) 313°C; (c) −126.5°C

8   Gas in a sealed bulb has a pressure of 0.15 atm at the ice point. What is the temperature of the bulb when the gas pressure is 0.18 atm?

9   In a student experiment the ratio p_s/p_i of a gas is found to be 1.41. What value does this give for absolute zero on the Celsius scale?
Ans. −244°C

10   On the Fahrenheit temperature scale, the temperature of the ice point is 32°F, and the temperature of the steam point is 212°F. (a) How many Fahrenheit degrees equal one Celsius degree? (b) Show that the conversion from a temperature t_F on the Fahrenheit scale to a temperature t_C on the Celsius scale is given by

$$t_C = \tfrac{5}{9}(t_F - 32°)$$

and the conversion from the Celsius scale to the Fahrenheit scale is given by

$$t_F = \tfrac{9}{5}t_C + 32°$$

(c) Plot t_F against t_C. At what temperature does $t_F = t_C$? (d) What is absolute zero on the Fahrenheit scale?

11   A gas occupies a volume of 22 L when its gauge pressure is 1.5 atm. (a) What is the absolute pressure of the gas? (b) The gas is compressed until it occupies 10 L at the same temperature. What is the new absolute pressure of

the gas? (c) What is the new gauge pressure?
Ans. (a) 2.5 atm; (b) 5.5 atm; (c) 4.5 atm

12   A bubble of volume $V = 15$ cm³ is formed at the bottom of a lake, 12 m below the surface. What is the volume of the bubble when it reaches the surface of the water? (Assume the temperature does not change.)

13   A cylinder contains 1.8 ft³ of oxygen at a temperature of 25°C and a pressure of 2200 lb/in². (a) What volume does this gas occupy at 25°C and atmospheric pressure? (b) A welder uses this oxygen at the rate of 0.8 ft³/min. How long will one cylinder of oxygen last?
Ans. (a) 269 ft³; (b) 5.61 h

14   A cylinder of compressed air will supply a scuba diver with enough air to last for 90 min on the surface of the water. How long will the same tank last when the diver is swimming 20 m below the surface? (*Hint:* The volume of air inhaled per minute does not change with depth.)

15   From the ideal-gas law, calculate the mass density of helium at STP.
Ans. 0.178 kg/m³

16   From the ideal-gas law, calculate the mass density of steam at 100°C and 1 atm.

17   What is the mass of 5 L of CO_2 at a temperature of 35°C and a pressure of 2.5 atm? (*Hint:* Use Table 8.2.)
Ans. 21.8 g

233

18 What is the mass of the oxygen in the tank described in Prob. 13?

19 What is the volume of air that must escape from a room when the temperature of the room is raised from 15 to 25°C? The dimensions of the room are 10 by 7 by 4 m, and the pressure remains unchanged while the room is being heated.
Ans. 9.72 m³

20 Calculate the volume occupied by 1 mol of an ideal gas at STP.

21 Calculate the volume occupied by 1 g of oxygen, O_2, at STP.
Ans. 0.700 L

22 Show that at atmospheric pressure the density of air (average molecular mass = 29 u) at temperature T is given by

$$\rho = \frac{353 \text{ K}}{T} \text{ kg/m}^3$$

23 What is the average kinetic energy of the molecules of a gas at 37°C?
Ans. 6.42×10^{-21} J

24 What is the total kinetic energy of the molecules in 1 mol of gas at 0°C?

25 (a) What is the average velocity $\overline{v^2}$ of an oxygen molecule at 27°C? (b) The square root of $\overline{v^2}$ is called the root-mean-square (rms) speed. Calculate the rms speed of an oxygen molecule at 27°C.
Ans. (a) 23.4×10^4 m²/s²; (b) 484 m/s

26 An ideal-gas mixture contains molecules of mass m and m'. Show that the ratio of the rms speeds v and v' (Prob. 25) of these molecules is

$$\frac{v'}{v} = \sqrt{\frac{m}{m'}}$$

27 The density of a liquid approaches the density of its vapor as its temperature approaches the critical temperature. What is the ratio of the density of liquid H_2O to the density of its vapor at 350°C and 163 atm? Use Fig. 8.7.
Ans. 5.1

28 Plot the 387°C isotherm for 1 mol of an ideal gas in the range $V = 0.05$ L to $V = 0.20$ L. Compare it to the 387°C isotherm of H_2O in Fig. 8.7.

29 Figure 8.12 is a plot of the vapor pressure p_v of trichloromethane (chloroform) against temperature. Ten grams of chloroform are injected into a sealed 2-L bulb which initially contained air at 20°C and 500 Torr. (a) What is

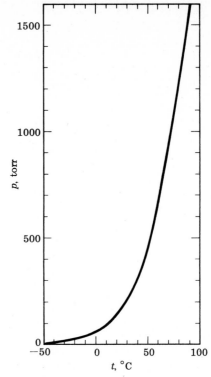

FIGURE 8.12

Plot of the vapor pressure of trichloromethane against temperature (Probs. 29 and 30).

the total pressure of the gas (air + chloroform) in the bulb at 20°C? (b) What is the total pressure of the gas when it is heated to 30°C?
Ans. (a) 650 Torr; (b) 737 Torr

30 What is the boiling point of chloroform (see Fig. 8.12)?

31 One mole of H_2O occupies a volume of 0.10 L at 300°C. (a) What fraction of the H_2O is in the liquid phase? (b) What is the volume occupied by the liquid H_2O? Use Table 8.4.
Ans. (a) 79.5%; (b) 0.020 L

32 Five grams of liquid tetrachloromethane, CCl_4, in injected into a sealed 500-cm³ vessel filled initially with air at 22°C and 1 atm. The pressure in the vessel increases to 1.114 atm. (a) What is the vapor pressure of CCl_4 at 22°C? (b) What percent of the CCl_4 is in the form of vapor?

33 How many moles of oxygen are there in 1 L of air at STP?
Ans. 9.33×10^{-3}

34 How many grams of argon, Ar, are there in 1 kg of air? (Argon does not form molecules; it exists as individual atoms.)

234

35 Deep-sea divers breathe air at the pressure of the surrounding water. Oxygen, however, becomes toxic at a partial pressure of 2 atm, so special gas mixtures must be used below certain depths. (a) At what depth does the partial pressure of oxygen in air equal 2 atm? (b) A gas mixture of 3 percent oxygen and 97 percent helium is used for very deep work. What is the partial pressure of oxygen in this gas at a depth of 200 m?
Ans. (a) 86 m; (b) 0.63 atm

36 A gas composed of 3.5 g of oxygen and 1.5 g of helium occupies a volume of 3 L at a temperature of 25°C. (a) What are the partial pressures of oxygen and helium in this gas? (b) What is the density of the gas mixture?

37 On a winter day, outside air at 0°C and relative humidity of 30 percent is taken into a house, where it is warmed to 20°C. What is the relative humidity of the air in the house?
Ans. 8.16%

38 Air with a relative humidity of 65 percent at 20°C is placed in a cylinder and compressed until moisture starts to condense on the inside surface of the cylinder. What is the pressure of the air when moisture first appears? Assume that the temperature remains 20°C.

39 The relative humidity is 37.2 percent on a day when the temperature is 20°C. What is the dew point?
Ans. 5°C

40 The dew point is 30°C on a day when the temperature is 37°C. What is the relative humidity?

41 (a) Define the frost point in analogy to the dew point. (b) What is the relative humidity of air at 10°C if it has a frost point of −5°C? The vapor pressure of ice at −5°C is 0.004 atm.
Ans. (b) 33%

42 In what phase is a substance with $T > T_t$ and $p > p_t$?

9

HEAT

Heat is energy that spontaneously flows from a hotter object to a colder one. Much of engineering practice is devoted to either promoting or reducing this flow. For instance, the cooling system of an automobile is designed to increase the rate of heat flow from the engine to the atmosphere, whereas home insulation is designed to decrease the rate of heat flow from the interior of a house. Refrigerators, air-conditioners, and heat pumps are mechanical devices which reverse the normal flow of heat by pumping heat from a colder region to a hotter one. These devices make use of the energy changes that occur when a substance undergoes a phase change.

9.1 The First Law of Thermodynamics

Consider a fixed amount of gas in a cylinder fitted with a piston and a thermometer, as shown in Fig. 9.1. By moving the piston and by heating or cooling the cylinder, the pressure p, volume V, and temperature T of the gas can be varied. The *thermodynamic state* of the gas is specified by giving the values of the *thermodynamic variables p, V,* and *T.* When these variables are changed, the state of the gas is changed.

A system is in a definite thermodynamic state only if its temperature and pressure are the same throughout. In the case of the gas in the cylinder, we know from Pascal's law (Sec. 7.1) that the pressure is the same everywhere if the gas is at rest, but that the pressure can differ from point to point if the gas is in motion. For instance, suppose the volume of the gas is suddenly increased by moving the piston. The gas will rush into the new volume, and for a time the system will not be in *mechanical equilibrium* because of the turbulent motion of the gas. Similarly, if the gas is heated from below, the system will not be in *thermal equilibrium* because different parts of the gas will be at different temperatures. A system that is in both mechanical and thermal equilibrium is said to be in *thermodynamic equilibrium;* only systems in thermodynamic equilibrium are in a definite thermodynamic state. A thermodynamic state is thus the same as an equilibrium state.

Thermodynamics considers the relation between a *system* $\mathcal{S}$, such as the gas in the cylinder in Fig. 9.1, and the *environment* $\mathcal{E}$ surrounding it. The environment is everything outside the system that can affect the system, which in most cases includes only the immediate surroundings of the system. The system and the environment together constitute the *universe* $\mathcal{U}$.

The internal energy $E_\mathcal{S}$ of the system is the sum of the kinetic energies of the molecules of the system (thermal energy) and the potential energy of the atoms in the molecules (chemical energy). The energy $E_\mathcal{S}$ depends on the state of the system, changing when the state changes. However, conservation of energy (Sec. 5.4) implies that the energy of a system can change only by energy flowing into or out of the system.

For instance, if the energy of a system changes from an initial value $E_\mathcal{S}$ to a final value $E'_\mathcal{S}$, the energy change

FIGURE 9.1

A fixed amount of gas contained in a cylinder with a piston. With this arrangement the temperature, pressure, and volume of the gas can be varied.

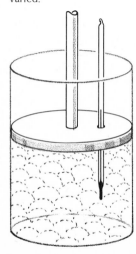

$$\Delta E_8 = E'_8 - E_8$$

must either be energy that flowed into the system (if ΔE_8 is positive) or energy that flowed out of the system (if ΔE_8 is negative). The place where this energy comes from or goes to is the environment. Therefore, the energy change ΔE_ε of the environment

$$\Delta E_\varepsilon = E'_\varepsilon - E_\varepsilon$$

must be the opposite of the energy change of the system:

$$\Delta E_\varepsilon = -\Delta E_8 \qquad\qquad 9.1$$

This is just a statement of the law of conservation of energy, because it says that the energy $E_\mathfrak{u}$ of the universe

$$E_\mathfrak{u} = E_\varepsilon + E_8$$

remains unchanged:

$$\Delta E_\mathfrak{u} = \Delta E_\varepsilon + \Delta E_8 = 0$$

HEAT AND WORK
Energy can be transferred between the system and the environment in two essentially different ways.

Definition *Heat* is energy that flows from one object to another as the result of a temperature difference between the objects. The molecules of an object with a temperature T_1 have, on the average, greater kinetic energy than the molecules of an object with a lower temperature T_2. If two such objects are in contact, their molecules collide with each other where the objects touch. In each collision the more energetic molecule of the hotter object loses energy, while the less energetic molecule of the colder object gains energy. Through a succession of billions upon billions of such collisions, energy is transferred from the hotter object to the colder object. Energy can also be transferred from a hotter to a colder object by radiation without the objects being in contact (Sec. 9.5).

Definition *Work* is energy that is transferred from one object to another as the result of a force **F** moving through a distance d. For instance, the expansion of the gas in Fig. 9.2 does work in moving the piston. The gas exerts a force of magnitude $F = pA$ on the piston, where p is the gas pressure and A is the cross-sectional area of the piston. Therefore, if the piston is moved a distance d, the work done is

$$W = Fd = pAd = p\,\Delta V \qquad\qquad 9.2$$

where

$$\Delta V = V_{\text{final}} - V_{\text{initial}} = Ad$$

is the change of volume of the gas.* Work W done by the gas is

FIGURE 9.2

The expansion of a gas does work.

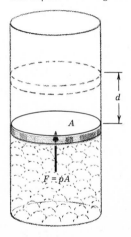

*Equation 9.2 is exact if the pressure of the gas is constant during the expansion. If the pressure is not constant, this equation is only an approximation, valid for small volume changes.

produced at the expense of the internal energy of the system and goes to increase the energy of the environment. Thus by doing work, energy is transferred from the system to the environment. Likewise, if the gas is compressed, work is done *on* the system and the energy of the system is increased at the expense of the energy of the environment.

The First Law of Thermodynamics *The change of energy ΔE_s of a system is equal to the heat Q that enters the system minus the work W done by the system:*

$$\Delta E_s = Q - W \qquad \text{first law} \qquad 9.3$$

This is just another statement of conservation of energy. Note that a different sign convention is used for heat and work. Heat Q is *positive* when it *enters* the system and *negative* when it *leaves*, whereas work W is *positive* when it *decreases* the energy of the system (i.e., when work is done *by* the system) and *negative* when it *increases* the energy of the system (i.e., when work is done *on* the system).

9.2 Specific Heat

When energy is added to a system, the temperature of the system may increase. For a given energy increase ΔE_s, the temperature change ΔT depends on how the process occurs. For instance, the volume of the system could be kept constant (an *isochoric* process), or the pressure of the system could be kept constant (an *isobaric* process). In an isochoric (constant-volume) process no work is done ($W = 0$), so Q is equal to ΔE_s. The temperature change is related to ΔE_s by

$$Q = \Delta E_s = C_v \Delta T \qquad 9.4$$

where C_v is the *heat capacity* of the system at constant volume.

Definition The *specific heat c_v* of a substance is its heat capacity divided by its mass:

$$c_v = \frac{C_v}{m} \qquad 9.5$$

The specific heat is a characteristic property of a substance. It depends on temperature, but over a narrow temperature range it can be treated as a constant. Combining Eqs. 9.4 and 9.5, we get

$$\Delta E_s = Q = mc_v \Delta T \qquad \text{isochoric} \qquad 9.6$$

Many processes occur at constant pressure rather than constant volume. In an isobaric (constant-pressure) process the temperature change is related to ΔE_s by

$$\Delta E_s + p \Delta V = mc_p \Delta T \qquad 9.7$$

where c_p is the specific heat at constant pressure. From Eqs. 9.2, 9.3, and 9.7 the heat absorbed in an isobaric transformation is

$$Q = \Delta E_s + p \Delta V = mc_p \Delta T \qquad \text{isobaric} \qquad 9.8$$

UNITS The SI units of specific heat are joules per kilogram per kelvin [J/(kg·K)], or joules per kilogram per degree Celsius [J/(kg·°C)]. The *kilocalorie* (kcal) is a unit of energy often used for heat. It is defined as the quantity of heat required to raise the temperature of one kilogram of water from 14.5 to 15.5°C. The conversion between joules and kilocalories is

1 kcal = 4184 J = 4.184 kJ

In terms of the kilocalorie, the units of specific heat are kilocalories per kilogram per kelvin [kcal/(kg·K)], or kilocalories per kilogram per degree Celsius [kcal/(kg·°C)]. Table 9.1 gives the specific heats at constant pressure of some common substances in both units.

The unit of heat in the English engineering system is the British thermal unit (Btu), equal to 1/180 of the heat required to raise the temperature of one pound of water from 32 to 212°F. The units of specific heat in the engineering system are British thermal units per pound per degree Fahrenheit [Btu/(lb·°F)]. These units will not be used in this book, but the conversions between them and SI units are given in Appendix II.

Example 9.1 How much heat is required to raise the temperature of 5 kg of water from 20 to 100°C?

From Table 9.1 the specific heat of water in this range is approximately 1 kcal/(kg·°C), so from Eq. 9.8 the heat required is

$$Q = mc_p \Delta T = (5 \text{ kg})[1 \text{ kcal/(kg·°C)}](80°C)$$
$$= 400 \text{ kcal} = (400)(4.184 \text{ kJ}) = 1.67 \times 10^3 \text{ kJ} \qquad \square$$

Example 9.2 (a) What is the heat capacity of a 75-g flint-glass beaker that contains 250 g of water? (b) How much heat must

TABLE **9.1**

Substance	Temperature, °C	c_p kcal/(kg·°C)	J/(kg·°C)	Specific heat at constant pressure of 1 atm of some common substances
Gases:				
Air	0–100	0.240	1000	
Carbon dioxide	15	0.199	833	
Oxygen	15	0.218	913	
Nitrogen	15	0.248	1040	
Water (stream)	100	0.482	2020	
Liquids:				
Ethanol	25	0.581	2430	
Mercury	20	0.0332	139	
Water	0	1.0076	4217.7	
	15	1.0000	4185.8	
	30	0.9983	4178.5	
	50	0.9988	4180.7	
	100	1.0072	4216.0	
Solids:				
Aluminum	20	0.214	899	
Brass	20	0.0917	384	
Copper	20	0.0921	386	
Glass, crown	20	0.161	674	
Flint	20	0.117	490	
Granite	20	0.192	804	
Human body (avg.)	37	0.83	3500	
Iron	20	0.115	481	
Water (ice)	0	0.492	2060	
Wood	20	0.42	1760	

be removed from this system in order to lower its temperature from 50 to 30°C?

(a) From Eq. 9.5 the heat capacity (at constant pressure) of the beaker and water are

$$C_{beaker} = m_{glass}c_{glass} = (0.075 \text{ kg})[490 \text{ J}/(\text{kg} \cdot °\text{C})] = 37 \text{ J}/°\text{C}$$
$$C_{water} = m_{water}c_{water} = (0.250 \text{ kg})[4180 \text{ J}/(\text{kg} \cdot °\text{C})] = 1045 \text{ J}/°\text{C}$$

The total heat capacity C of the system is the sum of the heat capacities of the individual components, so

$$C = C_{beaker} + C_{glass} = 1082 \text{ J}/°\text{C}$$

(b) The heat required to produce a temperature change of

$$\Delta T = 20°\text{C} - 50°\text{C} = -30°\text{C}$$

is

$$Q = C \Delta T = (1082 \text{ J}/°\text{C})(-30°\text{C}) = -3.25 \times 10^4 \text{ J}$$

The temperature change ΔT is negative because the final temperature (20°C) is less than the initial temperature (50°C). The negative value of Q means that the heat flows *out* of the system. ☐

CALORIMETRY

The heat transferred to or from a system is measured in a device called a *calorimeter*. One type of calorimeter (Fig. 9.3) consists of a small sample container immersed in a large outer vessel of water. The outer vessel is heavily insulated on the outside to prevent room heat from reaching the water, whereas the inner container is made of copper or some other heat-conducting material to permit heat to be readily exchanged between it and the water. The inner container encloses the system $\mathcal{S}$ to be measured, and the water surrounding the container is the environment $\mathcal{E}$.

No work is done by either the system or the environment; consequently, the temperature change of the environment (water) is just due to the heat exchanged between it and the system. This temperature change is measured with a thermometer, and the heat exchanged is calculated from the known mass and specific heat of the water. From conservation of energy (Eq. 9.1), the heat gained by the system is the negative of the heat lost by the environment, and vice versa. The calorimeter thus measures the heat exchanged by the system under specified conditions.

Example 9.3 Fifty grams of ethanol at a temperature of 30°C is placed in a calorimeter containing 2.50 kg of water at an initial temperature of 15.00°C. As the ethanol cools, the temperature of the water increases until the water and the ethanol reach the same temperature, which is found experimentally to be 15.17°C. What is the specific heat of ethanol?

The temperature change of the environment (water) is

$$\Delta T = 15.17°\text{C} - 15.00°\text{C} = 0.17°\text{C}$$

so from Eq. 9.8 and the specific heat of water (at 15°C) given in Table 9.1 we find that the heat absorbed by the environment is

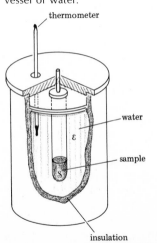

FIGURE 9.3

A calorimeter. The small sample container is immersed in a large vessel of water.

thermometer

water

$\mathcal{E}$

sample

$\mathcal{S}$

insulation

$$Q_\varepsilon = mc_p \Delta T = (2.50 \text{ kg})(4186 \text{ J/kg} \cdot {}^\circ\text{C})(0.17{}^\circ\text{C})$$
$$= 1780 \text{ J}$$

The temperature change of the system (alcohol) is

$$\Delta T = 15.17{}^\circ\text{C} - 30{}^\circ\text{C} = -14.83{}^\circ\text{C}$$

so the heat absorbed by the system is

$$Q_S = mc_p \Delta T = (0.050 \text{ kg})(c_p)(-14.83{}^\circ\text{C})$$
$$= (-0.74 \text{ kg} \cdot {}^\circ\text{C})c_p$$

where c_p is now the specific heat of the alcohol. (The minus sign indicates that the system lost heat.) Since the heat absorbed by the environment is equal to the heat lost by the system, we have

$$(0.74 \text{ kg} \cdot {}^\circ\text{C})c_p = 1780 \text{ J}$$

From this the value of c_p is found to be

$$c_p = \frac{1780 \text{ J}}{0.74 \text{ kg} \cdot {}^\circ\text{C}} = 2400 \text{ J/(kg} \cdot {}^\circ\text{C)} \qquad \square$$

9.3 Temperature Coefficients

Tables 1.6, 7.3, and 9.1 show that the density, viscosity, and specific heat of water change with temperature. Most physical properties of a substance show such variation. Some properties, like viscosity, are strongly temperature dependent, whereas others, like density and specific heat, vary only slightly with small temperature changes. The degree of temperature variation of a particular property is given by the *temperature coefficient* of that property.

Let Z_1 and Z_2 be the values of a physical property of a substance at two temperatures T_1 and T_2. We can write

$$Z_2 = Z_1 + (Z_2 - Z_1)$$
$$= Z_1 + \Delta Z$$

where $\Delta Z = Z_2 - Z_1$ is the change in Z. Likewise, we have

$$T_2 = T_1 + (T_2 - T_1)$$
$$= T_1 + \Delta T$$

where $\Delta T = T_2 - T_1$ is the temperature change. When ΔT is small, the relative change $\Delta Z/Z_1$ in the quantity Z will be proportional to ΔT, so it can be written

$$\frac{\Delta Z}{Z_1} = \alpha_1 \Delta T \qquad\qquad 9.9$$

The constant of proportionality α_1 is called the *temperature coefficient* of Z. Since this coefficient itself may depend on temperature, the subscript 1 is used to indicate that α_1 is the value of the coefficient at temperature T_1.

UNIT The relative change $\Delta Z/Z_1$ is the ratio of two quantities which have the same units, so it is unitless. Equation 9.9 then shows that α has the dimension $[1/T]$, so its unit is $1/{}^\circ\text{C}$, or ${}^\circ\text{C}^{-1}$.

To understand Eq. 9.9 better, consider Fig. 9.4, which shows a

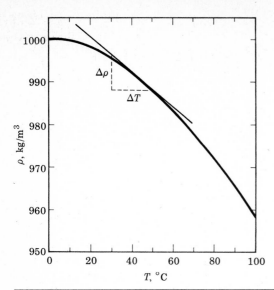

FIGURE 9.4

Plot of the density of water against
temperature. The tangent to the
curve at 40°C approximates the
curve between 30 and 50°C.

graph of the density ρ of water against temperature T. The curve is not a straight line, which means that the linear relation given by Eq. 9.9 is not valid over the entire range from 0 to 100°C. However, over a small range, such as 30 to 50°C, the curve can be approximated by a straight line. The straight line shown in Fig. 9.4 is tangent to the curve at 40°C. The slope s of this line is

$$s = \frac{\Delta\rho}{\Delta T} = \frac{\rho_{50} - \rho_{30}}{(50 - 30)°C}$$

$$= \frac{(988 - 996)\,\text{kg/m}^3}{20°C} = -0.40\,\text{kg/(m}^3 \cdot °C)$$

The density ρ of water at a temperature T in the range from 30 to 50°C is then given approximately by the formula for this line:

$$\rho = \rho_{40} + s(T - 40°C)$$

Thus we have

$$\Delta\rho = \rho - \rho_{40} = s\,\Delta T$$

or

$$\frac{\Delta\rho}{\rho_{40}} = \frac{s}{\rho_{40}}\,\Delta T = \alpha_{40}\,\Delta T$$

where $\Delta T = T - 40°C$ and

$$\alpha_{40} = \frac{s}{\rho_{40}} = \frac{-0.40\,\text{kg/(m}^3 \cdot °C)}{992\,\text{kg/m}^3} = -0.40 \times 10^{-4}°C^{-1}$$

is the temperature coefficient at 40°C. The minus sign indicates that the density is decreasing with increasing temperature.

Example 9.4 At 20°C the density of mercury is 13,546 kg/m³, and the temperature coefficient is $-1.8 \times 10^{-4}°C^{-1}$. What is the density of mercury at 35°C?

We use Eq. 9.9, with $Z = \rho$ and

$$\Delta T = (35 - 20)°C = 15°C$$

$$Z_1 = \rho_{20} = 13{,}546 \text{ kg/m}^3$$

$$\alpha_1 = \alpha_{20} = -1.8 \times 10^{-4}°C^{-1}$$

The change of density $\Delta\rho$ then is

$$\begin{aligned}\Delta\rho &= \rho_{20}\alpha_{20}\,\Delta T \\ &= (13{,}546 \text{ kg/m}^3)(-1.8 \times 10^{-4}°C^{-1})(15°C) \\ &= -37 \text{ kg/m}^3\end{aligned}$$

and the density at 35°C is

$$\begin{aligned}\rho_{35} &= \rho_{20} + \Delta\rho \\ &= (13{,}546 - 37) \text{ kg/m}^3 = 13{,}509 \text{ kg/m}^3\end{aligned} \qquad \square$$

A temperature coefficient can be defined for any physical property that varies with temperature. Table 9.2 gives the temperature coefficients for some of the properties of water.

Example 9.5 Use the data in Table 9.2 to determine the temperature at which the speed of sound in water is 1500 m/s.

From Table 9.2 the speed at 20°C is

$$v_{20} = 1486 \text{ m/s}$$

so

$$\Delta v = v_T - v_{20} = (1500 - 1486) \text{ m/s} = 14 \text{ m/s}$$

The temperature change ΔT required to produce this change in speed is given by Eq. 9.9:

$$\begin{aligned}\Delta T &= \frac{1}{\alpha_{20}}\frac{\Delta v}{v_{20}} \\ &= \frac{1}{(1.61 \times 10^{-3}°C^{-1})} \times \frac{14 \text{ m/s}}{1486 \text{ m/s}} \\ &= 5.9°C\end{aligned}$$

Thus the temperature T at which the speed of sound is 1500 m/s is

$$T = 20°C + \Delta T = 25.9°C \qquad \square$$

THERMAL EXPANSION

We saw in Sec. 8.4 that temperature is a measure of the average kinetic energy of the molecules of a substance. This means that

TABLE **9.2**

Property	Magnitude	Temperature coefficient, $°C^{-1}$	Some properties of water and their temperature coefficients at 20°C
Density	998.2 kg/m³	-2.06×10^{-4}	
Heat of vaporization	44.16 kJ/mol	-9.70×10^{-4}	
Specific heat	4.182 kJ/(kg · °C)	-1.58×10^{-4}	
Speed of sound	1486 m/s	1.61×10^{-3}	
Thermal conductivity	0.597 W/(m · °C)	2.90×10^{-3}	
Viscosity	1.02 mPl	-2.30×10^{-2}	

as the temperature of an object increases, the molecules in it move faster and faster. As a consequence, the molecules move farther apart and so the object expands. This is most noticeable in a gas, where the volume V is directly proportional to the absolute temperature T (Eq. 8.8). A similar but much smaller effect occurs in solids and liquids as well.

Let L_1 and L_2 be the length of an object such as a metal rod at temperatures T_1 and T_2, respectively. The change in length

$$\Delta L = L_2 - L_1$$

is related to the change in temperature

$$\Delta T = T_2 - T_1$$

by the equation

$$\frac{\Delta L}{L_1} = \alpha \, \Delta T \qquad\qquad 9.10$$

where α is the *coefficient of thermal expansion*. Equations 9.9 and 9.10 have the same form; the coefficient of thermal expansion is just the temperature coefficient of length. Table 9.3 gives the coefficients of thermal expansion of various substances.

Example 9.6 The steel boiler of an electric-power station is 30 m tall. When it is fired up its temperature changes from 20 to 500°C. By how much does the length of the boiler change in this process?

Table 9.3 shows that the value of α for steel varies with temperature. Between 20 and 100°C the increase in length is

$$\Delta L_1 = L_1\alpha \, \Delta T = (30 \text{ m})(10.5 \times 10^{-6}\text{°C}^{-1})(80\text{°C})$$
$$= 2.52 \times 10^{-2} \text{ m}$$

Between 100 and 200°C the increase is

$$\Delta L_2 = (30 \text{ m})(11.5 \times 10^{-6}\text{°C}^{-1})(100\text{°C})$$
$$= 3.45 \times 10^{-2} \text{ m}$$

Similarly, the increase from 200 to 300°C is found to be $\Delta L_3 = 3.9 \times 10^{-2}$ m, and the increase from 300 to 500°C is $\Delta L_4 = 9.0 \times 10^{-2}$ m. Thus the total change in the length of the

TABLE **9.3**

Substance	Temperature range, °C	Coefficient of thermal expansion, 10^{-6}°C^{-1}	The coefficients of thermal expansion for various substance for the temperature ranges indicated
Aluminum	20 to 100	24.0	
Cement and concrete	—	10 to 14	
Glass, plate	0 to 100	8.9	
Iron, soft	0 to 100	12.1	
Quartz, fused	0 to 1200	0.585	
Steel	0 to 100	10.5	
	100 to 200	11.5	
	200 to 300	13	
	300 to 600	15	
Wood, pine			
Parallel to fiber	2 to 34	5.4	
Across fiber	2 to 34	34.1	

boiler is

$$\Delta L = \Delta L_1 + \Delta L_2 + \Delta L_3 + \Delta L_4$$
$$= 0.189 \text{ m} = 7.4 \text{ in}$$

Large boilers must be specially designed to accommodate such a large change in size. They are usually suspended from above, rather than supported from below, to give them the freedom to expand (Fig. 9.5). □

Example 9.7 (a) What is the strain produced in a steel girder when its temperature changes from 20 to 35°C? (b) If the girder were not allowed to expand, what would be the stress that developed in it?

(a) We recall from Sec. 3.4 that strain e is the ratio $\Delta L/L$ of an object, so from Eq. 9.10 and Table 9.3 the strain in the girder is

$$e = \frac{\Delta L}{L} = \alpha \, \Delta T$$
$$= (10.5 \times 10^{-6} °\text{C}^{-1})(15°\text{C})$$
$$= 0.16 \times 10^{-3}$$

(b) If the girder is held rigidly at both ends, so that it is not free to expand, a large stress will develop. This stress is equal to the stress required to produce a strain of -0.16×10^{-3}. To see

SUPERHEATER SECTION

FIGURE 9.5

Large power-plant boilers are suspended from above to allow for thermal expansion. (*Foster Wheeler Energy Corp.*)

this, imagine that the girder is first allowed to thermally expand, and then it is compressed back to its original length. The compressional stress s required to produce a strain e is given by Eq. 3.7

$$s = eE$$

where E is the Young's modulus of steel (Table 3.1). Thus the stress developed is

$$s = eE = (-0.16 \times 10^{-3})(200 \times 10^9 \text{ N/m}^2)$$
$$= -3.2 \times 10^7 \text{ N/m}^2$$

The minus sign indicates that the stress is compressional. Since this stress is less than the elastic limit of steel (Table 3.1), the girder can support it, provided the girder does not have to support any other very large compressional loads. □

9.4 Latent Heat

Definitions *Latent heat* is the energy required to transform one kilogram of a substance from one phase to another at constant temperature. The *heat of fusion* H_f is the latent heat for the transformation from solid to liquid; the *heat of vaporization* H_v is the latent heat for the transformation from liquid to gas; the *heat of sublimation* H_s is the latent heat for the transformation from solid to gas. Table 9.4 gives the heats of fusion and vaporization for various substances.

Let us consider what happens as we heat 1 kg of ice that is initially at $-10°C$ (Fig. 9.6). At first the temperature of the ice increases. According to Eq. 9.8 and Table 9.1, the temperature of the ice will reach 0°C when the quantity of heat

$$Q = mc_p \Delta T = (1 \text{ kg})[0.492 \text{ kcal/(kg} \cdot °C)](10°C)$$
$$= 4.92 \text{ kcal}$$

has been added. Once the melting point is reached, the addition of more heat does not cause a further temperature increase. Instead, the ice starts to melt. According to Table 9.4, it requires

TABLE **9.4**

Substance	Temperature, °C	Heat of fusion, kcal/kg	Temperature, °C	Heat of vaporization, kcal/kg
Ethanol	−114.5	24.9	78.3	204
Freon-12 (CCl_2F_2)			0	37.0
			20	34.5
Carbon dioxide	−56.2	45.3	20	35.1*
Iron	1530	69.1	2735	1515
Mercury	−39	2.82	357	65
Oxygen	−219	3.30	−183	50.9
Sulfur	119	13.2	445	78.1
Tungsten	3387	45.8	3410	978
Water	0	79.7	0	596
			20	585
			50	569
			100	540

Heats of fusion and vaporization for various substances

*Heat of sublimation.

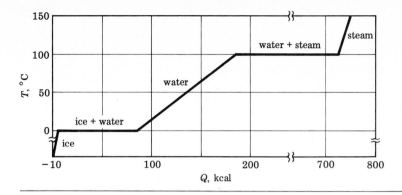

FIGURE 9.6

Plot of the temperature change against absorbed heat of 1 kg of H_2O. During a phase change, the temperature remains constant because the absorbed heat all goes into latent heat.

79.7 kcal to transform 1 kg of ice into water at 0°C. Figure 9.6 shows that while this heat is being absorbed, the temperature remains constant.

After all the ice has melted, the temperature will again start to increase with the addition of more heat. Although there is some evaporation as well, we shall neglect it. That is, we shall assume that all the heat absorbed goes to increase the temperature. To increase the temperature of water from 0 to 100°C requires the absorption of

$$Q = mc_p \, \Delta T = (1 \text{ kg})[1 \text{ kcal}/(\text{kg} \cdot °C)](100°C)$$
$$= 100 \text{ kcal}$$

Once the boiling point is reached, the addition of more heat does not cause a further temperature rise. Instead, the water is vaporized. According to Table 9.4, it required 540 kcal to transform 1 kg of water into vapor at 100°C.

The following is a tabulation of the energy required to transform 1 kg of ice at $-10°C$ into 1 kg of vapor at 100°C:

To raise the temperature of ice from -10 to 0°C	4.92 kcal
To melt the ice at 0°C	79.70 kcal
To raise the temperature of water from 0 to 100°C	100.00 kcal
To vaporize the water at 100°C	540.00 kcal
Total	724.60 kcal

Figure 9.6 summarizes the same data in graphical form.

Example 9.8 How much heat must be removed from 200 g of water at 15°C to produce 200 g of ice at $-15°C$?

To cool the water to 0°C, the quantity of heat

$$Q_1 = mc_p \, \Delta T = (0.2 \text{ kg})[1 \text{ kcal}/(\text{kg} \cdot °C)](0°C - 15°C)$$
$$= -3 \text{ kcal}$$

must be removed. To freeze the water at 0°C, the quantity of heat

$$Q_2 = -mH_f = -(0.2 \text{ kg})(79.7 \text{ kcal}/\text{kg}) = -16 \text{ kcal}$$

must be removed. Finally, to cool the ice to $-15°C$, the quantity of heat

$$Q_3 = mc_p \, \Delta T = (0.2 \text{ kg})[0.492 \text{ kcal}/(\text{kg} \cdot {}^\circ\text{C})](-15{}^\circ\text{C} - 0{}^\circ\text{C})$$
$$= -1.5 \text{ kcal}$$

must be removed. Thus the total heat that must be removed is

$$Q = Q_1 + Q_2 + Q_3 = -20.5 \text{ kcal} \qquad \square$$

REFRIGERATION

Heat spontaneously flows from a region of high temperature to a region of low temperature by a number of processes described in the next section. Heat is thus somewhat analogous to a fluid, which spontaneously flows from a region of high pressure to a region of low pressure. A *refrigerator* is a mechanical device that lowers the temperature of an insulated compartment below that of the environment by transporting heat from the low-temperature compartment to the high-temperature environment. A refrigerator is thus analogous to a pump which transports a fluid from a region of low pressure to a region of high pressure. Like a pump, a refrigerator does work in moving the heat opposite to its natural direction of flow.

A commercial refrigerator uses a substance called a *refrigerant,* which is a gas at room temperature and pressure (20°C and 1 atm), but which is a liquid at room temperature and high pressure (10 atm or so). Fluorocarbons, such as Freon-12 (CCl_2F_2), are commonly used today as refrigerants. The refrigerant is stored in liquid form at high pressure and room temperature in a *storage tank* (Fig. 9.7).

To cool the low-temperature compartment, the liquid refrigerant is allowed to vaporize at low pressure in a network of coils

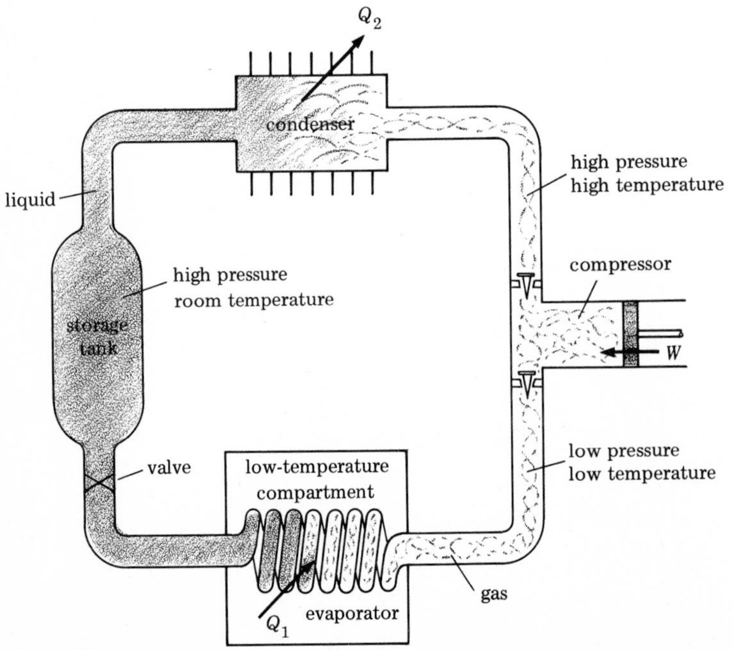

FIGURE 9.7

Schematic diagram of a refrigerator. Liquefied refrigerant is stored at high pressure and room temperature in the storage tank. When the refrigerant vaporizes in the evaporator, heat Q_1 is extracted from the low-temperature compartment. The gaseous refrigerant is then compressed and the heat Q_2 removed from it in the condenser. Liquefied refrigerant is returned to the storage tank.

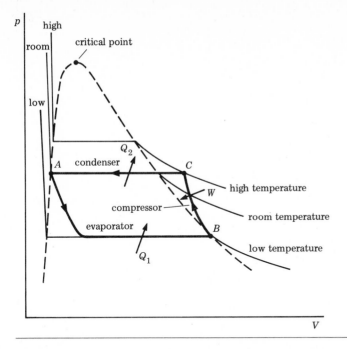

FIGURE 9.8

Plot of the refrigeration cycle on a *pV* diagram. The vaporization of the refrigerant occurs along *AB*, the compression of the gas occurs along *BC*, and the condensation of the gas occurs along *CA*.

called the *evaporator.* These coils are in close contact with the compartment, and the heat of vaporization required to vaporize the liquid is drawn from the compartment. This lowers the temperature of the compartment and the gas.

The low-pressure gas then passes out of the evaporator and into the *compressor,* where it is compressed into a high-pressure (10-atm) gas. The compression process raises the temperature of the gas above room temperature. The hot high-pressure gas is then cooled back to room temperature in the *condenser,* which is just a chamber that promotes the normal transfer of heat from the hot gas to the cooler environment. As the gas is cooled, it liquefies and returns to the storage tank.

There are several points to note about this process. (1) It is a complete cycle; i.e., it starts and ends with a high-pressure liquid at room temperature. (2) In one complete cycle, heat is taken from a low-temperature region and transferred to a high-tempera-ture region. This exhaust heat can be felt coming from vents on the bottom of a modern kitchen refrigerator. The refrigerator must always be located so that these vents are not blocked. (3) The work required to transfer the heat from a low- to a high-temperature region is done by the compressor when it compresses the gas.

The states through which the refrigerant passes in one cycle are shown on the *pV* diagram in Fig. 9.8. At *A* the refrigerant is a liquid at high pressure and room temperature. In the evaporator the refrigerant vaporizes to a gas at low temperature and pres-sure (*B*). In the process the heat Q_1 is absorbed by the gas. The compressor raises the pressure and temperature of the gas (*C*). No heat is exchanged in this process, but the work *W* is done on the gas. Finally, the gas is liquefied by removing the heat Q_2 in the condenser.

Example 9.9 How many grams of Freon-12 must be vaporized to transform 200 g of water at 15°C to 200 g of ice at −15°C?

In Example 9.8 we found that 20.5 kcal of heat must be removed from the H_2O. The average heat of vaporization H_v of Freon-12 between 15 and −15°C is seen from Table 9.4 to be about 37 kcal/kg, so the mass m of Freon-12 that must be evaporated is

$$m = \frac{Q_1}{H_v} = \frac{20.5 \text{ kcal}}{37.0 \text{ kcal/kg}} = 0.55 \text{ kg} = 550 \text{ g}$$

It is not necessary for the refrigerator to contain this much refrigerant, however, since the same refrigerant can be cycled through the evaporator as many times as is required. ☐

The change ΔE_8 of the internal energy of the refrigerant is given by the first law of thermodynamics (Eq. 9.3)

$$\Delta E_8 = Q - W$$

where Q is the net heat absorbed by the refrigerant, and W is the work done by it. In one complete cycle the refrigerant returns to its original state, so $\Delta E_8 = 0$. That is, the liquid is returned to storage with the same internal energy it had when it left. Thus in one complete cycle we have

$$Q = W \qquad\qquad 9.11$$

According to the sign convention used in the first law, W is negative because work is done on the refrigerant by the compressor. To avoid confusion we write

$$W = -|W|$$

where $|W|$ is the positive value of W. The net heat absorbed is

$$Q = Q_1 + Q_2$$

where Q_1 is the heat absorbed by the refrigerant in the evaporator, and Q_2 is the heat lost in the condenser. According to the sign convention for heat, we have

$$Q_1 = |Q_1| \qquad \text{and} \qquad Q_2 = -|Q_2|$$

so

$$Q = Q_1 + Q_2 = |Q_1| - |Q_2|$$

Then Eq. 9.11 can be written

$$-|W| = |Q_1| - |Q_2| \qquad \text{or} \qquad |Q_2| = |Q_1| + |W| \qquad 9.12$$

The heat lost in the condenser is equal to the heat absorbed in the evaporator plus the work done by the compressor.

Definition The *coefficient of performance* e of a refrigerator is the ratio of the heat $|Q_1|$ removed from the compartment to the work $|W|$ done by the compressor:

$$e = \frac{|Q_1|}{|W|} = \frac{|Q_1|}{|Q_2| - |Q_1|} \qquad\qquad 9.13$$

Equation 9.12 was used to write W in terms of Q_1 and Q_2 in the last expression. Typically e is about 3.

Example 9.10 (a) If the refrigerator in Example 9.9 has a coefficient of performance of 3.5, how much work is done by the compressor in removing 20.5 kcal of heat from the compartment? (b) How much heat is exhausted in the condenser?

(a) From Eq. 9.13 the work done is

$$|W| = \frac{|Q_1|}{e} = \frac{20.5 \text{ kcal}}{3.5} = 5.86 \text{ kcal} = 24.5 \text{ kJ}$$

(b) From Eq. 9.12 the heat Q_2 vented by the condenser is

$$|Q_2| = |Q_1| + |W|$$
$$= 20.5 \text{ kcal} + 5.9 \text{ kcal} = 26.4 \text{ kcal} \qquad \square$$

Air-conditioners The last example shows why you cannot cool a room by opening the refrigerator door. The refrigerator vents more heat into the room than it removes from the compartment. To cool a room it is necessary to vent the exhaust heat $|Q_2|$ to the outside. An *air-conditioner* is a refrigerator that does this. The compartment in this case is the room and the environment is the outside air.

Heat Pumps A *heat pump* is a device that uses the principle of refrigeration to warm a house. The condenser is located inside the house and the evaporator is outside. When the outside air is colder than the inside air, the heat pump extracts heat from the cold outside air (making it even colder) and vents this heat inside the house. The coefficient of performance e_{hp} of a heat pump is the ratio of the heat $|Q_2|$ delivered to the house to the work $|W|$ required to deliver it:

$$e_{hp} = \frac{|Q_2|}{|W|}$$

From Eqs. 9.12 and 9.13 this can be written

$$e_{hp} = \frac{|Q_1| + |W|}{|W|} = \frac{|Q_1|}{|W|} + 1$$
$$= e + 1$$

where e is the coefficient of performance of the same device acting as a refrigerator.

A heat pump can be an efficient way to heat a house, especially in temperate climates. With a coefficient of performance of 2.5, a heat pump requires only 1 kcal of electric energy to deliver 2.5 kcal of heat. This is to be compared with electric heating, which requires 2.5 kcal of electric energy to produce 2.5 kcal of heat. Furthermore, heat pumps are designed to operate as air-conditioning units in summer time, saving the cost of separate installation. Figure 9.9 shows the inside of a heat pump.

9.5   Heat Transfer

Heat is energy that flows from one object to another as a consequence of the temperature difference between them. The natural direction of spontaneous heat flow is always from the higher-temperature object to the lower-temperature object. For

FIGURE 9.9

A heat pump designed for rooftop installation. (*General Electric.*)

instance, if a hot object is placed in contact with a cold object, heat will flow from the hot object to the cold object, making the hot object cooler and the cold object warmer. The reverse process, in which heat flows from the cold object to the hot object, making the hot object hotter and the cold object colder, never happens by itself. That is, to make heat flow from a lower-temperature object to a higher-temperature object some outside agent, such as a refrigerator, must be used. These observations concerning the direction of heat flow are the basis of the *second law of thermodynamics*, discussed in more detail in Chap. 10.

There are three basic mechanisms by which heat flows spontaneously from a higher-temperature region to a lower-temperature region: conduction, convection, and radiation. The rate of heat flow can be controlled to some extent by increasing or decreasing the effectiveness of one or another of these mechanisms.

CONDUCTION

Conduction is the transfer of energy through a material medium by the successive collisions of neighboring molecules. In each collision, on the average, the faster-moving molecules in the higher-temperature part of the medium transfer some of their kinetic energy to their slower-moving neighbors in the lower-temperature part. These neighboring molecules in turn collide with still slower nearby molecules, transferring energy to them. In this way energy is transferred from the higher- to the lower-temperature part of the medium while the molecules remain near their original positions.

> **REMARK** In metals, which are good heat conductors, most of the energy is transferred by collisions between free electrons and fixed molecules. A molecule vibrating energetically about a fixed position transfers some of its vibrational energy to a free electron, which in turn passes it on to a less energetic molecule. The existence of free electrons in metals increases the rate at which the energy can be transferred.

Figure 9.10 shows a sheet of material of thickness d and area A. If the two sides of the sheet are maintained at temperatures T_1 and T_2, heat will flow by conduction from the higher- to the lower-temperature side.

Definition The *rate of flow P* is the heat that flows from one side to the other in unit time. Thus if the quantity of heat Q flows in time t, the rate P is

$$P = \frac{Q}{t} \qquad\qquad 9.14$$

The magnitude of P is proportional to the temperature difference $\Delta T = T_1 - T_2$ between the sides and to the area A of the sheet and is inversely proportional to the thickness d of the sheet. Thus we have

$$P \propto \frac{A\,\Delta T}{d}$$

or

FIGURE 9.10

Heat being conducted through a sheet of material. The bottom surface of the sheet is at temperature T_1, and the top surface is at the cooler temperature T_2.

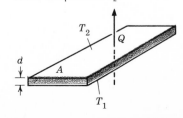

$$P = K\frac{A\,\Delta T}{d} \qquad\qquad 9.15$$

where K is a constant characteristic of the material, called the *thermal conductivity*.

> **UNITS** Equation 9.14 shows that the unit of P is joules per second (J/s), or watts (W). Then from Eq. 9.15 the units of K are watts per meter per degree Celsius [W/(m · °C)]. Table 9.5 gives the thermal conductivity of some common substances.

From Table 9.5 it is seen that substances differ greatly in the magnitude of their thermal conductivity. Substances with large conductivities, like metals, are called *conductors*; materials with small conductivities, like air and wood, are called *insulators*. However, even the best insulator conducts some heat, though at a slower rate than a conductor.

Example 9.11 What is the rate of heat flow through a glass window pane 0.5 cm thick when the outside surface is 6°C and the inside surface is 6.8°C? The dimensions of the window are 2 by 3 m.

The area of the window is $A = (2\,\text{m})(3\,\text{m}) = 6\,\text{m}^2$, and the temperature difference is $\Delta T = 0.8°C$. From Table 9.5 the thermal conductivity of glass is $K = 0.80\,\text{W/m} \cdot °C$, and so from Eq. 9.15 the rate of heat flow is

$$P = K\frac{A\,\Delta T}{d} = \frac{[0.8\,\text{W/(m} \cdot °\text{C)}](6\,\text{m}^2)(0.8°\text{C})}{0.5 \times 10^{-2}\,\text{m}} = 768\,\text{W}$$

The small temperature difference between the inner and outer surfaces of the window in this example is typical of the situation on a winter day when the outside temperature is $-4°C$ and the inside temperature is 20°C. Most of the resistance to heat flow comes not from the glass itself, but from thin layers of air that stick to both sides of the glass. This is discussed more fully following Example 9.12. ☐

Definitions The *conductance* C of a sheet of material with thermal conductivity K, thickness d, and area A is

$$C = \frac{KA}{d} \qquad\qquad 9.16$$

The *resistance* R of a sheet of material with conductance C is

$$R = \frac{1}{C} = \frac{d}{KA} \qquad\qquad 9.17$$

From Eqs. 9.15 and 9.17 we see that the rate of flow P through a sheet with conductance C (or resistance R) can be written

$$P = C\,\Delta T \qquad\qquad 9.18$$

or

$$P = \frac{\Delta T}{R} \qquad\qquad 9.19$$

REMARK The *R-value*, commonly used in the United States to grade insulat-

TABLE 9.5

Thermal conductivity of some common substances

Substance	Thermal conductivity, W/(m · °C)
Silver	430
Copper	400
Aluminum	240
Iron	80
Concrete	1.2
Glass	0.80
Water	0.60
Body fat	0.20
Wood, pine	0.12
Insulating material	0.040
Air	0.025

ing material, is the resistance (in English units) of one square foot of insulating material:

$$R\text{-value} = \frac{d}{K(1 \text{ ft}^2)} \qquad \text{English units} \qquad 19.20a$$

Here K is the conductivity of the material in Btu-inches per hour per square foot per degree Fahrenheit [Btu·in/(h·ft²·°F)], and d is the thickness in inches. If K and d are in SI units [W/(m·°C) and m], the R-value is given by

$$R\text{-value} = (5.68)\frac{d}{K} \qquad \text{SI units} \qquad 19.20b$$

For example, the R-value of a 5-cm-thick sheet of insulating material with a conductivity of 0.06 W/(m·°C) is

$$R\text{-value} = (5.68)\frac{0.05}{0.06} = 4.7$$

Units are usually omitted when giving an R-value. The R-value is used to compare different insulations, the insulation with the higher R-value giving the higher resistance to heat flow.

Often one deals with layers of material, such as the two sheets of insulation shown in Fig. 9.11. The material with resistance R_1 is in contact with the warm temperature T_1, and the material with resistance R_2 is in contact with the cold temperature T_2. The temperature T' between the layers is between T_1 and T_2.

From Eq. 9.19 the rate of flow P_1 through R_1 is

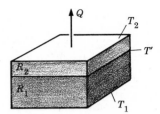

FIGURE 9.11

Two sheets of insulating material laid on top of each other. The temperature T' of the inner surfaces is in between the temperatures T_1 and T_2 of the outer surfaces.

$$P_1 = \frac{T_1 - T'}{R_1} \qquad \text{or} \qquad T_1 - T' = R_1 P_1$$

and the rate of flow P_2 through R_2 is

$$P_2 = \frac{T' - T_2}{R_2} \qquad \text{or} \qquad T' - T_2 = R_2 P_2$$

When we add these last two expressions we get

$$(T_1 - T') + (T' - T_2) = R_1 P_1 + R_2 P_2$$

or

$$\Delta T = T_1 - T_2 = R_1 P_1 + R_2 P_2$$

But all the heat that flows through R_1 also flows through R_2, so $P_1 = P_2 = P$, where P is the rate of flow of heat through both layers. Thus we have

$$\Delta T = R_1 P + R_2 P = P(R_1 + R_2) = PR$$

where $R = R_1 + R_2$ is the total resistance of the two sheets. In general, the total resistance R of n sheets with resistances R_1, R_2, ..., R_n is

$$R = R_1 + R_2 + \cdots + R_n \qquad 19.21$$
$$= \Sigma R_i$$

The rate of heat flow through several layers of material is calculated using Eqs. 9.17, 9.19, and 9.21.

Example 9.12 (a) Calculate the resistance of one square meter

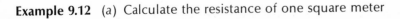

of window glass that is 0.5 cm thick. (b) Special insulated window pane consists of two 0.5-cm-thick sheets of glass separated by a 0.15-cm-thick layer of dry air (Fig. 9.12). Calculate the resistance of one square meter of this pane.

(a) From Table 9.5 and Eq. 9.17 we have that the resistance R_g of a single sheet of glass is

$$R_g = \frac{d}{KA} = \frac{0.5 \times 10^{-2}\,m}{[0.60\,W/(m \cdot °C)](1\,m^2)} = 0.0083°C/W$$

(b) The resistance R_a of the layer of air between the two sheets of glass is

$$R_a = \frac{d}{KA} = \frac{0.15 \times 10^{-2}\,m}{[0.025\,W/(m \cdot °C)](1\,m^2)} = 0.060°C/W$$

Since the pane consists of two layers of glass and one layer of air, its total resistance R is

$$
\begin{aligned}
R &= R_g + R_a + R_g \\
&= 2(0.0083°C/W) + 0.060°C/W \\
&= 0.077°C/W
\end{aligned}
$$

The resistance of the double-sheeted pane is more than 9 times greater than the resistance of the single-sheeted pane, and most of this increased resistance comes from the layer of air between the glass sheets. ☐

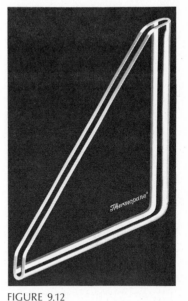

FIGURE 9.12

Two-sheeted window pane. (*Libbey-Owens-Ford.*)

A single-sheeted pane by itself offers very little resistance to heat flow. Its main function is to provide a barrier that prevents the mixing of the cold outside air with the warm inside air. Because air does not flow through the barrier, thin layers of air adhere to both its surfaces. On the interior surface of a vertical barrier, the air layer is typically 3 mm thick. This layer provides an internal resistance R_i. For each square meter of surface area, the internal resistance is

$$R_i = \frac{d}{KA} = \frac{0.003\,m}{[0.025\,W/(m \cdot °C)](1\,m^2)} = 0.12°C/W$$

A similar layer forms on the exterior surface, but its thickness depends on external conditions, especially the wind speed. If a typical value of 1.5 mm is used for the thickness of the exterior layer, the exterior resistance R_e of one square meter of surface is 0.06°C/W. Together these two air layers provide a *surface resistance* $R_s = R_i + R_e = 0.18°C/W$ that is independent of the insulating properties of the surface itself.

Definition The *effective resistance* R_{eff} of a barrier is the sum of the barrier resistance R and the surface resistance R_s:

$$R_{eff} = R + R_s \qquad\qquad 9.22$$

Example 9.13 What are the effective resistances of one square meter of the single-sheeted and double-sheeted window panes described in Example 9.12?

The resistance R of one square meter of a single sheet of glass is 0.0083°C/W, so its effective resistance is

$$R_{eff} = R + R_s = 0.0083°C/W + 0.18°C/W = 0.19°C/W$$

The glass itself contributes very little to this resistance, which is why the temperature difference between its inner and outer surfaces is small (Example 9.11). The effective resistance of one square meter of double-sheeted pane is

$$R_{eff} = R + R_s = 0.077°C/m + 0.18°C/W = 0.26°C/W$$

The effective resistance of the double-sheeted pane is only 37 percent greater than the effective resistance of the single-sheeted pane, because both panes have a surface resistance of 0.18°C/W. ☐

Definition The *transmission coefficient U*, or *U-value*, of a barrier of area A is

$$U = \frac{1}{AR_{eff}} = \frac{1}{A(R + R_s)} \qquad 9.23$$

The rate of heat flow P through the barrier is

$$P = \frac{\Delta T}{R_{eff}} = AU \, \Delta T \qquad 9.24$$

where ΔT is the temperature difference between the interior of the room and the outside air. Table 9.6 gives the U-values of some barriers used in housing construction.

Example 9.14 (a) Calculate the U-values of the single-sheeted and double-sheeted window panes described in Examples 9.12 and 9.13. (b) Calculate the heat flow through a 2- by 3-m window made of each type of pane. Take the interior temperature to be 20°C and the exterior temperature to be −4°C.

(a) From Eq. 9.23 and the results of Example 9.13 we find that the U-value of the single-sheeted pane is

TABLE **9.6**

Barrier	$U,$ W/(m² · °C)	Transmission coefficients, or U-values, of some barriers used in housing construction
Brick wall		
0.105 m thick	3.3	
0.335 m thick	1.7	
Frame wall, with 0.5-cm asbestos sheet		
On bare frame	5.3	
On frame closed with aluminum-backed plaster-board	1.8	
On closed frame with 2.5 cm of fiber glass insulation	1.1	
Flat roof, wood with asphalt shingles		
Over aluminum-backed plasterboard ceiling	0.9	
Over ceiling with 2.5 cm of fiber glass insulation	0.6	
Window, wood frame		
Single-sheeted	5.3	
Double-sheeted	3.8	

$$U_1 = \frac{1}{AR_{\text{eff}}} = \frac{1}{(1\,\text{m}^2)(0.19°\text{C/W})} = 5.3\,\text{W/(m}^2 \cdot °\text{C})$$

and the U-value of the double-sheeted pane is

$$U_2 = \frac{1}{AR_{\text{eff}}} = \frac{1}{(1\,\text{m}^2)(0.26°\text{C/W})} = 3.8\,\text{W/(m}^2 \cdot °\text{C})$$

(b) The area of the window is 6 m², and the temperature difference is 24°C, so from Eq. 9.24 the rate of heat flow through the single-sheeted pane is

$$P_1 = AU_1\,\Delta T = (6\,\text{m}^2)[5.3\,\text{W/(m}^2 \cdot °\text{C})](24°\text{C}) = 763\,\text{W}$$

This is similar to the value we found in Example 9.11. In that example we just considered the flow through the glass, so we had to use the temperature difference between the inner and outer surfaces of the glass. In this example, by using the transmission coefficient, we are taking into account the flow through the air layers on either side of the glass, so we can use the full difference between the interior and exterior air temperatures.

The rate of flow through the double-sheeted pane is

$$P = AU_2\,\Delta T = (6\,\text{m}^2)[3.8\,\text{W/(m}^2 \cdot °\text{C})](24°\text{C}) = 547\,\text{W}$$

which is 72 percent of the heat flow through the single-sheeted pane. $\square$

CONVECTION

Convection is the transfer of energy in a liquid or gas by the actual transfer of higher-temperature fluid from a higher-temperature region to a lower-temperature region. The higher-temperature fluid has greater internal energy than the lower-temperature fluid it displaces, so energy is transferred to the lower-temperature region along with the fluid.

The most familiar example of convective heat transfer is the air circulation which is established by a room radiator (Fig. 9.13). The temperature of the air near the radiator is increased by conduction, making it less dense than the cooler room air. As a consequence, this warmer air rises and is replaced by cooler air from near the windows, establishing an air circulation that transfers heat from the radiator to the other parts of the room and ultimately, by conduction through the windows, to the outside environment.

Much of the resistance to heat conduction through a closed window comes from the thin layers of air that adhere to the inside and outside surfaces of the window. Winds on the outside and convection currents on the inside tend to reduce the thickness of these air layers and thus increase the rate at which heat is conducted through the window. Drapes hung in front of a window prevent convection currents from reaching the window. This reduces heat conduction through the window by allowing a relatively thick layer of air to adhere to the inside surface of the window. Convection is a rapid mode of heat transfer, so its control is essential in promoting or reducing heat loss. In the cooling system of an automobile, the heat is first

FIGURE 9.13

The convection current in a room heated by a radiator.

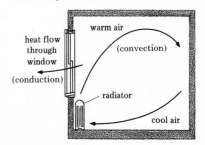

transferred from the engine block to the cooling water by conduction, but the heated water is then pumped (forced convection) to the radiator. In the radiator, heat from the water is conducted to the outside air through the radiator fins. Finally, this heated air is blown away by the engine fan (more forced convection).

RADIATION

Radiation is electromagnetic energy that travels through empty space with the speed of light (3×10^8 m/s). Light is one form of radiation, but there are others, such as infrared and ultraviolet radiation, which differ from light only by the size of their wavelengths (Sec. 13.1). All objects emit radiation: objects at room temperature emit mostly infrared radiation, whereas objects at high temperature, such as a light-bulb filament, emit visible as well as infrared radiation.

The rate P_e at which radiant energy is emitted by an object with surface area A and absolute temperature T_e is

$$P_e = \epsilon \sigma A T_e^4 \qquad \text{emission} \qquad 9.25$$

where $\sigma = 5.67 \times 10^{-8}$ W/m² · K⁴ is a universal constant called the *Stefan-Boltzmann constant*, and ϵ is a dimensionless parameter called the *emissivity*, which varies between 0 and 1 depending on the nature of the surface. The same object placed in an enclosure with walls at absolute temperature T_a will absorb radiation from the walls at the rate

$$P_a = \epsilon \sigma A T_a^4 \qquad \text{absorption} \qquad 9.26$$

Thus if the object is hotter than the walls of the enclosure ($T_e > T_a$), there will be a net flow of energy from the object to the walls at the rate

$$P = P_e - P_a = \epsilon \sigma A (T_e^4 - T_a^4) \qquad \text{net loss} \qquad 9.27$$

If the object is colder than the walls ($T_e < T_a$), the object will gain energy at the rate given by Eq. 9.27.

An object with the maximum emissivity of 1 is called a *blackbody* because it absorbs all the radiation incident on it. An object with an emissivity of zero is a perfect reflector, which absorbs none of the radiation incident on it. The same emissivity appears in both Eqs. 9.25 and 9.26, so a good absorber is also a good emitter and a poor absorber is a poor emitter. Table 9.7 gives the emissivity of various substances.

Example 9.15 The tungsten filament of a 60-W light bulb has a temperature of 1500°C. What is the surface area of the filament?

The power rating of a light bulb is the total rate at which radiant energy is transferred. If we neglect the heat conducted through the wire leads connected to the filament, all the power is emitted as radiation. Thus the data for this problem are

$P = 60$ W

$T = 273°C + 1500°C = 1773$ K

$\epsilon = 0.23$

TABLE **9.7**

Substance	Temperature, °C	Emissivity
Aluminum, oxidized	100	0.10
	600	0.19
Brick, fire	1000	0.75
Carbon	500	0.81
Iron, oxidized	100	0.74
	500	0.84
	1000	0.89
Unoxidized	100	0.05
Tungsten, unoxidized	500	0.071
	1000	0.15
	1500	0.23
	2000	0.28

where we have used Table 9.7 for ϵ. Then from Eq. 9.25 the area of the filament is

$$A = \frac{P_e}{\epsilon \sigma T^4} = \frac{60 \text{ W}}{(0.23)[5.67 \times 10^{-8} \text{ W/(m}^2 \cdot \text{K}^4)](1773 \text{ K})^4}$$
$$= 4.66 \times 10^{-4} \text{ m}^2$$
$$= 4.66 \text{ cm}^2$$

The filament is a long, thin wire, twisted into a helix so that it will fit into a compact space. ☐

The infrared radiation emitted by objects at room temperature cannot be seen, but it can be detected by special equipment. Figure 9.14 is a picture of the infrared radiation emitted by a house. The warmer surfaces of the house emit more radiation than the colder surfaces, so this picture shows where the house loses most of its heat. As can be seen, the doors and windows of the house are major sources of loss.

FIGURE 9.14

Infrared scan of a house. (*Inframetrics, Inc.*)

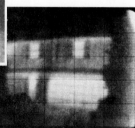

GUIDE TO MAJOR TOPICS

PROBLEMS

1 A gas absorbs 1500 J of heat while expanding and doing 400 J of work. What is the change of energy of the gas?
Ans. 1100 J

2 In compressing a gas, 1200 J of work is done. In the process the gas loses 300 J of heat. What is the change of energy of the gas?

3 The internal energy of a system increases by 3200 J when it absorbs 1400 J of heat. How much work is done by (or on) the system during the process?
Ans. −1800 J

4 The internal energy of a system decreases by 1700 J when it does 2200 J of work. How much heat is absorbed (or lost) by the system in this process?

5 Find the quantity of heat, in joules and kilocalories, required to raise the temperature of 650 g of water from 22 to 85°C.
Ans. 1.71×10^5 J, 40.9 kcal

6 (a) By how much is the temperature of 1.25 kg of water raised when 2×10^5 J of heat is added to it? (b) By how much is the temperature of 750 g of ethanol raised when 35 kcal of heat is added to it?

7 A 0.4-kg rock falls 1200 m into a bucket containing 2.5 kg of water. How much does the temperature of water rise?
Ans. 0.45°C

8 A 20-g iron nail is being pounded by a 0.45-kg (1-lb) hammer. The speed of the hammer is 9 m/s when it strikes the nail. If half the kinetic energy of the hammer is converted into internal energy of the nail, how many times must the nail be struck to raise its temperature 25°C?

9 (a) What is the heat capacity of a 350-g brass bar? (b) If the bar absorbs 75 cal of heat, how much is its temperature raised?
Ans. (a) 0.032 kcal/°C; (b) 2.3°C

10 (a) What is the heat capacity of a system consisting of 2.5 kg of water in a 0.75-kg aluminum bucket? (b) How much heat is required to raise the temperature of the system from 16 to 85°C?

11 It takes 880 J to raise the temperature of 350 g of lead from 0 to 20°C. What is the specific heat of lead?
Ans. 126 J/(kg · °C)

12 The inner vessel of a calorimeter contains 100 g of trichloromethane ($CHCl_3$) at 35°C. The vessel is surrounded by 1.75 kg of water at 18°C. After a time the trichloromethane and the water reach the common temperature of 18.22°C. What is the specific heat of trichloromethane?

13 A 500-g aluminum cylinder is heated to 100°C and then plunged into 300 g of water at 15°C. If no heat is gained or lost from the aluminum-water system, what final temperature does the system reach?
Ans. 23.2°C

14 The molar heat capacity is the heat capacity of 1 mole of a substance. (a) Calculate the molar heat capacities (at constant pressure) of carbon dioxide (CO_2), oxygen (O_2), and nitrogen (N_2). (b) The molar heat capacity (at constant pressure) of an ideal diatomic (two-atoms per molecule) gas can be shown to be $\frac{7}{2}R$, where $R = 8.32$ J/K is the gas constant. Check this for oxygen and nitrogen.

260

15 Use Table 9.2 to find the thermal conductivity of water at 35°C.
Ans. 0.64 W/(m · °C)

16 From the data in Table 9.4 calculate the temperature coefficient of the heat of vaporization of water at 20°C.

17 How much does an 8-m-long aluminum rod expand when its temperature is raised from 20 to 50°C?
Ans. 0.58 cm

18 How much does a 40-m-tall steel boiler contract when its temperature falls from 600 to 20°C?

19 An aluminum rod at 100°C is bolted at both ends and then cooled. (a) What is the stress in the rod when the temperature is 20°C? (b) At what temperature will the rod break? Use Table 3.1.
Ans. (a) 13.4×10^7 N/m²; (b) -19°C

20 A steel rod and a fused quartz rod, each 1 cm² in cross section, are bolted at both ends when their temperatures are both 20°C. What is the compressive force in each rod when its temperature is increased to 45°C? Use Table 3.1.

21 (a) How much heat is required to melt 250 kg of iron at its melting point (1530°C)? (b) How much heat does it take to melt 250 kg of iron starting with iron at 20°C? (Assume the specific heat of iron remains constant throughout this temperature range.)
Ans. (a) 1.73×10^4 kcal; (b) 6.07×10^4 kcal

22 How much heat is released when 125 g of Freon-12 is condensed from a gas to a liquid at 20°C?

23 A 50-g ice cube at 0°C is dropped into 250 g of water at 18°C. (a) Assuming all the heat required to melt the ice comes from the water, calculate the temperature of the 250 g of water when the ice has melted. Neglect the mixing of the melted ice with the original water. (b) What will be the final temperature of the 300 g of water after mixing?
Ans. (a) 2.06°C; (b) 1.72°C

24 How much heat must be removed from 50 g of oxygen at 20°C to liquefy it at -183°C? (Assume its specific heat remains constant throughout this temperature range.)

25 The evaporation of 50 g of trichloromethane ($CHCl_3$) from the inner vessel of a calorimeter lowers the temperature of the 1.6 kg of water surrounding the vessel by 1.9°C. What is the heat of vaporization of trichloromethane?
Ans. 60.7 kcal/kg

26 By how many degrees does the evaporation of 65 g of ethanol from the inner vessel of a calorimeter lower the temperature of the 2.3 kg of water surrounding the vessel?

27 A refrigerator removes 50 kcal of heat from the cold compartment and exhausts 65 kcal of heat through the condenser. (a) How much work is done by the compressor? (b) What is the coefficient of performance of this refrigerator?
Ans. (a) 15 kcal; (b) 3.33

28 A refrigerator with a coefficient of performance of 2.5 removes 400 kcal of heat from the cold compartment. (a) How much work is done by the compressor? (b) How much heat is exhausted through the condenser?

29 A refrigerator contains 200 g of Freon-12. How many times must this refrigerant be cycled through the cold compartment in order to remove 350 kcal of heat?
Ans. 48

30 A heat pump with a coefficient of performance of 2.8 delivers heat at the rate of 2.5×10^4 kcal/h. (a) What is the rate at which the compressor does work? (b) How much heat per hour is removed from the outside air? (c) If the outside air is cooled from 5 to -5°C, how many kilograms of air must be cooled per second?

REMARK The large quantity of air that must be blown across the evaporator limits the effectiveness of air-cooled heat pumps. More often the heat is extracted from a stream or underground water supply.

31 One end of an aluminum rod is maintained at 220°C, while the other end is maintained at 0°C. The rod is 2 m long and 1 cm in diameter. What is the rate of conduction of heat along the rod?
Ans. 2.07 W

32 What is the rate of heat transfer through a 1.2 by 2.4 m sheet of insulating material 8 cm thick when one side is at 22°C and the other side is at 4°C? The conductivity of the material is 0.035 W/(m · °C).

33 A house is sheeted with 250 m² of 1.6-cm-thick plywood. (a) What is the conductance of this sheeting? (b) What is the resistance of this sheeting? Use Table 9.5.
Ans. (a) 1875 W/°C; (b) 5.3×10^{-4}°C/W

34 Calculate the R-value of the air layer that adheres to both sides of a vertical barrier. Use the typical values of the thicknesses of these layers given in Sec. 9.5.

35 A 10-cm-thick layer of insulation with thermal conductivity 0.035 W/(m · °C) is laid over a 5-cm-thick layer of insulation with thermal conductivity 0.050 W/(m · °C). (a) What is the resistance of one square meter of this combination? (b) What is the R-value of each layer and of the two layers combined?
Ans. (a) 3.86°C/W; (b) 16.2, 5.68, 21.9

36 The wall of a house consists of a 5-cm-thick layer of fiber glass insulation [$K = 0.04$ W/(m · °C)] sandwiched between a 1.6-cm-thick layer of plywood and a 0.5-cm-thick layer of wooden paneling. (a) What is the resistance of one square meter of this wall? (b) What is the U-value of the wall?

> **REMARK** The purpose of the insulating material is to trap air inside the wall. Without the insulation, air would leak through the wall and heat would be lost by convection. The insulation traps the air, so that heat can be transferred only by conduction.

37 A room has 22 m² of outside wall, of which 6 m² is window. The non-window portion of the wall has a U-value of 2.8 W/(m² · °C), and the window portion has a U-value of 4.8 W/(m² · °C). Calculate the total rate of heat loss through the outside wall when the inside temperature is 20°C and the outside temperature is −8°C.
Ans. 2.06 kW

38 Consider the two insulating sheets with resistances R_1 and R_2 shown in Fig. 9.11. The outside surface of insulator R_1 is at temperature T_1 and the outside surface of insulator R_2 is at temperature T_2. (a) Show that the temperature T' of the inner surfaces is given by

$$T' = \frac{R_2 T_1 + R_1 T_2}{R_1 + R_2}$$

(b) The interior air layer on the inside surface of a 0.5-cm-thick single-sheeted window pane is 3 mm thick. The inside temperature is 20°C and the outside temperature is −4°C. Neglecting the exterior air layer, calculate the temperature T' on the inside surface of the pane.

39 The surface of a radiator has an emissivity of 0.55 and an area of 1.5 m². (a) At what rate is radiation emitted by the radiator when its temperature is 50°C? (b) At what rate is radiation absorbed by the radiator when the walls of the room are 22°C? (c) What is the net rate of radiation from the radiator?
Ans. (a) 509 W; (b) 354 W; (c) 155 W

> **REMARK** This last problem shows that radiation is not an important mechanism for heat transfer at these temperatures. The radiator, in spite of its name, transfers most of its heat to the room by convection.

40 From measurements made on earth it is known that the sun has a surface area of 6.1×10^{18} m² and radiates energy at the rate of 3.9×10^{26} W. Assuming the emissivity of the sun's surface is 1, calculate the temperature of the surface.

41 Show that when the difference $\Delta T = T_e - T_a$ between the temperature T_e of an object and the temperature T_a of the walls of its container is small compared with T_a, the net rate of heat transfer by radiation from the object to the walls can be written $P = 4\epsilon\sigma T_a^3 \Delta T$.

> **REMARK** Newton's law of cooling states that the rate of heat loss of an object is proportional to the difference ΔT between the temperatures of the object and its surroundings. Equation 9.15 shows that conduction is proportional to ΔT. Problem 39 shows that radiation is also proportional to ΔT, provided ΔT is small compared with the temperature of the surroundings.

42 An aluminum sphere 5 cm in diameter is suspended by a fine thread inside an evacuated jar, so that it can lose heat only by radiation. The sphere's initial temperature is 100°C, and the wall of the jar is always 22°C. (a) What is the initial net rate of heat loss of the sphere? (b) At the rate in part (a), how long will it take the sphere to go from 100 to 90°C? (c) What is the net rate of heat loss at 90°C? (d) At the rate in part (c), how long will it take the sphere to go from 90 to 80°C? (Hint: Use Tables 1.6 and 9.1 to calculate the heat capacity of the sphere.)
Ans. (a) 0.524 W; (b) 50.6 min; (c) 0.436 W; (d) 60.8 min

BIBLIOGRAPHY

FABER, OSCAR, and J. R. KELL: *Heat and Air-Conditioning of Buildings*, 5th ed., Architectural Press, London, 1971. A relatively modern treatment of heating theory and practice, written in SI units.

T

hermodynamics is the study of the relation of heat, work, and energy, and, in particular, of the transformation of energy into work. It originated in the nineteenth century when engineers were interested in designing more efficient steam engines, i.e., engines that could do more work with less fuel. The first law of thermodynamics (Sec. 9.1) states (in effect) that an engine cannot do more work than the energy released by the fuel. The second law of thermodynamics (Sec. 10.2) states (in effect) that not all the energy released can be converted into work, and it goes on to specify the maximum amount of work obtainable from a given amount of energy under ideal conditions. From the point of view of the first law, all energy is the same, but from the point of view of the second law, energy at high temperature is more valuable for producing work than energy at low temperature.

10 THERMO-DYNAMICS

10.1 Thermodynamic Transformations

The thermodynamic state of a system is specified by a few thermodynamic variables, such as pressure p, volume V, and temperature T. In an ideal gas these variables are related by the equation of state (Eq. 8.8)

$$pV = nRT \qquad 10.1$$

where n is the number of moles of gas, and $R = 8.314\,J/K$ is the gas constant. Consequently, if any two of these three variables are given, the third variable is determined. This means that only two variables are required to specify the state, because the third variable can be found from Eq. 10.1. Even if the gas is not ideal, only two variables are required, because there always exists an equation of state relating these variables. Of course, the equation of state of a nonideal gas is more complicated than Eq. 10.1.

If p and V are chosen to specify the state, the state may be represented by a point on a graph of p against V (Fig. 10.1). For instance, the state A, with pressure p_A and volume V_A, is represented by the point A in Fig. 10.1. The temperature of this state is determined from the equation of state.

All states with the same temperature lie on a curve called the *isotherm*.* Figure 10.1 shows several isotherms for different temperatures; the temperature of a particular isotherm is greater than the temperature of all isotherms lying below it and less than the temperature of all isotherms lying above it. At high temperature the isotherms are smooth curves given by Eq. 10.1, but at lower temperature the shape of the isotherm is more complex because the gas is no longer ideal, as discussed in Sec. 8.5.

When the state of a system is suddenly changed, the system may not be in equilibrium for a time. For instance, when the volume of the gas in the cylinder in Fig. 10.2 is suddenly increased from V_A to V_B by moving the piston, the motion of the gas as it rushes into the new volume causes the pressure to vary

*Iso- is a common prefix meaning "equal"; thus, *isotherm* means "equal temperature."

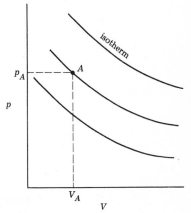

FIGURE 10.1

The state A of a gas is represented by a point on a pV diagram. The curves, called *isotherms*, connect states with the same temperature.

FIGURE 10.2

Change in the volume of a gas. For a time after the volume is suddenly changed, the gas may not be in thermodynamic equilibrium.

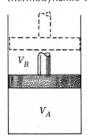

throughout the gas. The initial state of the gas is represented by point A in Fig. 10.3, and after a time the gas settles into a new equilibrium state represented by point B. But during the transformation from A to B the gas was not in equilibrium; i.e., it was not in any definite thermodynamic state. Consequently, the conditions of the gas during the transformation cannot be represented by points on a pV diagram. Transformations like this, in which a system goes from one thermodynamic state to another without passing through intermediate states, are called *irreversible* and are indicated by broken lines on a pV diagram (Fig. 10.3).

On the other hand, the volume or temperature of the gas can be changed so slowly that the gas is always in equilibrium. Then, at every step of the transformation from one state to another the system is in a thermodynamic state, and the entire process can be represented by a solid line connecting the initial and final states and passing through all the intermediate states. This is called a *reversible* transformation because the system can be transformed from the final state back to the initial state through the same intermediate states. Each of the infinite number of lines that can be drawn between two states represents a different reversible transformation. A few special cases of particular importance are given special names.

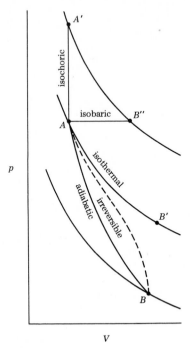

Definition An *adiabatic* transformation is a transformation in which no heat is allowed to enter or leave the system. This is accomplished by surrounding the cylinder with insulating material (Fig. 10.4a). From the first law of thermodynamics (Eq. 9.3), the change in energy ΔE_{S} of a system is

$$\Delta E_{\mathrm{S}} = Q - W \qquad 10.2$$

But in an adiabatic transformation $Q = 0$, so

$$\Delta E_{\mathrm{S}} = -W$$

In the adiabatic expansion of a gas, the work done is

$$W = p\,\Delta V$$

where ΔV is the change of volume of the gas. By convention, W is positive when the system does work on its environment. In the adiabatic expansion of a gas, W is positive and ΔE_{S} is negative. This means that the internal energy of the gas decreases, and hence its temperature decreases also. The system follows a path like the solid path from A to B in Fig. 10.3. Conversely, in the adiabatic compression of a gas, work is done on the gas, so W is negative and ΔE_{S} is positive. This means that the internal energy of the gas increases, and hence its temperature increases also. The system follows a path like the solid path from B to A in Fig. 10.3.

FIGURE 10.3

A gas initially in state A is transformed to states A', B, B', and B" by isochoric, adiabatic, isothermal, and isobaric transformations, respectively. An irreversible transformation from A to B is indicated by a dotted line because the system does not pass through any of the intermediate states between A and B.

Definition An *isothermal* transformation is a transformation in which the temperature is held constant. This can be accomplished by placing the cylinder in contact with a large reservoir of water that is at the desired temperature (Fig. 10.4b). The cylinder has thin metallic walls through which heat can readily flow back and forth between the reservoir and the gas. The reservoir is large enough for its own temperature to be unaffected by the amount of heat exchanged with the gas. During an isothermal expansion heat flows into the gas to keep its temperature constant. (Remember, the temperature of a gas decreases if heat is prevented from flowing into it while it expands.) The system follows a path along an isotherm, such as the path from A to B' in Fig. 10.3.

Definition An *isochoric* transformation is a transformation in which the volume of the system is held constant. This can be accomplished by clamping the piston in a fixed position. The state of the gas is changed by heating the gas (Fig. 10.4c). Since the piston is held fixed, no work is done by the system during the transformation. The system follows a vertical path, such as the line AA' in Fig. 10.3.

Definition An *isobaric* transformation is a transformation in which the pressure of the system is held fixed. This can be accomplished by applying a constant external pressure to the piston (Fig. 10.4d). The state of the gas is changed by heating the gas. The system follows a horizontal path, such as the line AB'' in Fig. 10.3.

There are countless other transformations in which nothing is held constant, but it is sufficient for our purposes to consider only these special cases. There is no need to learn these transformations now, but you should review this discussion whenever a particular transformation is mentioned later in this chapter.

10.2 Engines and Refrigerators

An *engine* is a device that converts heat into work. A *refrigerator* is a device that uses work to remove heat. Both engines and refrigerators involve a process in which a substance is made to undergo a sequence of thermodynamic transformations that eventually brings the substance back to its original state. Such a sequence of transformations is called a *cycle*. In this section we shall examine the cycles used by a number of engines and refrigerators. We already discussed the refrigeration cycle in Sec. 9.4 (see Fig. 9.8), but we shall repeat some of the discussion here to show the close relation between refrigerators and engines.

STEAM ENGINE

The first engines were crude devices that used steam to create a vacuum inside a cylinder, allowing atmospheric pressure to do work on the piston. The first widely used engines of this type were built by Thomas Newcomen (1663–1729) to pump water from the mines of England. When the piston was in the uppermost position (Fig. 10.5), steam at atmospheric pressure was let into the cylinder, forcing air and water out through the eduction pipe. Next, cold water was injected into the cylinder to condense the steam and create a low pressure inside the cylinder. Atmospheric pressure then drove the piston down, raising the pump rod on the other side of the rocker. When the pressure

FIGURE 10.4

Schematic diagrams representing (a) an adiabatic, (b) an isothermal, (c) an isochoric, and (d) an isobaric transformation.

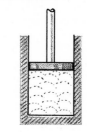
a

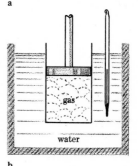
b

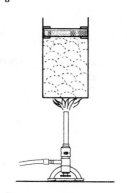
c

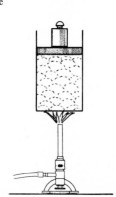
d

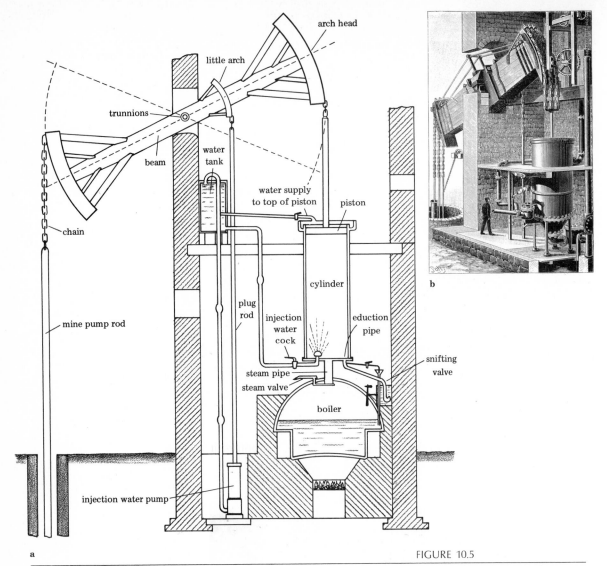

arch head

little arch

trunnions

beam

water tank

water supply to top of piston

piston

chain

cylinder

mine pump rod

plug rod

injection water cock

eduction pipe

steam pipe

steam valve

snifting valve

boiler

injection water pump

a

b

FIGURE 10.5

Newcomen engine. (a) Schematic diagram and (b) engraving of an engine installed in Bristol, England in the eighteenth century. (*The Bettmann Archive, Inc.*)

inside the cylinder reached atmospheric pressure, the weight of the pump rod pulled the piston back to its starting position. Crude as they were, the Newcomen engines played an essential role in the exploitation of the mineral resources of England in the eighteenth century.

James Watt (1736–1819) greatly improved the efficiency of the steam engine by keeping the cylinder always at high temperature. This was accomplished by exhausting the steam from the cylinder to a separate tank, the *condenser*, where it was condensed, creating a low pressure in the cylinder. He also sealed the top of the cylinder, and allowed steam to enter on this side and aid in pushing the piston down. Watt made many other improvements as well, but his basic invention was the condenser, on which he held a patent until 1800. He did not invent the principle of using steam to generate work, nor did he use steam at much above atmospheric pressure. His engines, like Newcomen's, used the pressure of the atmosphere, not the steam, as the principal driving force. Although his engines had an efficiency of only 3 percent, they were 3 times more efficient than the Newcomen engines.

The early development of the steam engine was entirely the work of technicians. Newcomen was a dealer in iron tools and Watt was an instrument maker. But the need to develop more efficient engines in the nineteenth century stimulated scientists to look at the problem more generally, and resulted in the development of the science of thermodynamics. One of the results of thermodynamics is the realization that the efficiency of an engine increases as its operating temperature increases. Watt intuitively saw this when he decided to maintain the cylinder at high temperature. But thermodynamics showed that to achieve efficiencies much above 3 percent, it would be necessary to operate at temperatures well above 100°C. This meant working with superheated steam at high pressure.

The operating temperature and pressures of steam engines have steadily increased as technology has developed the metals and fittings that can reliably withstand them. All modern coal, gas, oil, and nuclear power plants use high-pressure steam to drive large turbine blades, like the one shown in Fig. 10.6. A sequence of blades of increasing size is used, so that as the steam pressure drops, the lower-pressure steam acting on the large blades exerts the same force as the higher-pressure steam on the small blades.

The steam in a modern power plant passes continuously through a sequence of transformations, as shown in Fig. 10.7. A feed pump raises the pressure of the water to 170 atm before it

FIGURE 10.6

Steam-turbine blades from a power plant. (*New England Electric System.*)

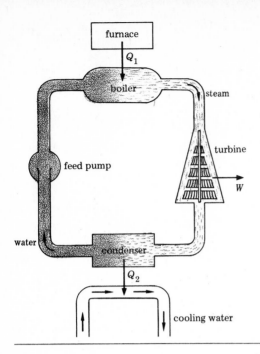

FIGURE 10.7

Schematic diagram of the transformations of water in a steam turbine. The feed pump delivers water at high pressure to the boiler, where it is converted into high-pressure steam by the absorption of the heat Q_1. This steam drives the turbine, producing the work W. The low-temperature, low-pressure steam is converted to water in the condenser by the removal of the heat Q_2.

reaches the boiler. In the boiler, the temperature of this high-pressure water is raised to 600°C, which transforms the water into steam at constant pressure. This high-temperature, high-pressure steam expands against the turbine blades, doing work on the blades. The temperature and pressure of the steam is reduced in the process, and the low-pressure steam passes into the condenser, where it is condensed into water before returning to the feed pump.

The thermodynamic transformations of the water-steam system are shown on the pV diagram in Fig. 10.8. From A to B the water pressure is increased by the feed pump. This is an isothermal and isochoric transformation, since the temperature and volume of the water are virtually unchanged by this process. The water enters the boiler at B, where it undergoes an isobaric transformation from B to C as it is heated at constant pressure and transformed into steam. From C to D the high-temperature, high-pressure steam does work against the turbine blades. This is an adiabatic transformation since there is little heat exchanged during the process. As the steam expands against the turbine blades, its temperature and pressure are reduced. At D the low-pressure, low-temperature steam enters the condenser, where it undergoes isobaric condensation to water at A.

In a steam engine water is the substance that undergoes a cycle of thermodynamic transformations. This substance is the system $\S$, and the change in energy $\Delta E_\S$ of the system is given by the first law of thermodynamics (Eq. 10.2). In one complete cycle, however, the system returns to its initial state, so $\Delta E_\S = 0$. Then from Eq. 10.2 we have that in one cycle

$$Q = W$$

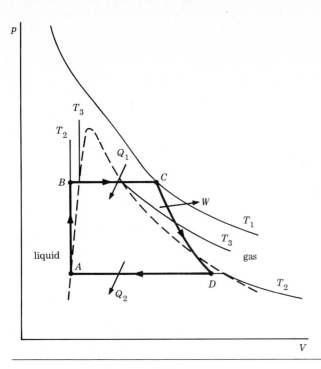

FIGURE 10.8

Plot of the steam-engine cycle on a
pV diagram. Water pressurized by
the feed pump (AB) is vaporized
(BC) and then expanded (CD)
against the turbine blades. The exhausted steam is then condensed
(DA) before returning to the
feed pump.

That is, the work done by the system is equal to the heat absorbed. The system absorbs the heat Q_1 in the boiler during the
isobaric transformation from B to C and loses the heat
$Q_2 = -|Q_2|$ in the condenser during the isobaric transformation
from D to A. (Q_2 is negative because it is heat lost by the system;
$|Q_2|$ is the positive value of Q_2.) The work done is equal to the
net heat absorbed, so

$$W = Q_1 + Q_2 = |Q_1| - |Q_2| \qquad\qquad 10.3$$

The efficiency e of an engine is the ratio of the work done to
the heat Q_1 absorbed in the boiler:

$$e = \frac{W}{Q_1} \qquad\qquad 10.4$$

Efficiency is defined this way because the work W is the benefit
and Q_1 is the cost of the process. From Eq. 10.3 the efficiency can
also be written

$$e = \frac{|Q_1| - |Q_2|}{|Q_1|} = 1 - \frac{|Q_2|}{|Q_1|} \qquad\qquad 10.5$$

This shows that the maximum efficiency allowed by the first law
of thermodynamics is 1, and that this occurs only if $|Q_2| = 0$.
That is, one cannot get more work out of an engine than the
heat one puts in. However, the second law of thermodynamics
(Sec. 10.3) says that there must always be some heat Q_2 lost
during part of the cycle, so e is always less than 1. In fact, for a
modern steam turbine, the maximum possible efficiency is about
55 percent, and actual efficiencies are about 40 percent.

REMARK A perpetual-motion machine of the first kind is an engine with an efficiency greater than 1. Such a machine is impossible according to the first law of thermodynamics. (The first law says you can never win.) A perpetual-motion machine of the second kind is an engine with an efficiency of 1. Such a machine is impossible according to the second law of thermodynamics. (The second law says you cannot even break even!)

Assuming an efficiency of $\frac{1}{3}$, Eq. 10.5 shows that

$$|Q_2| = \tfrac{2}{3}|Q_1|$$

That is, two-thirds of the heat generated in the boiler is lost through the condenser; only one-third is converted into work.

Example 10.1 A 850-MW nuclear power plant has an efficiency of 28 percent. It cools its condenser by taking in seawater at 10°C and discharging it at 18°C. How much water is required?

The plant produces electric energy (work) at the rate

$$P = 850 \, \text{MW} = 850 \times 10^6 \, \text{W} = 8.5 \times 10^8 \, \text{J/s}$$

so the work produced in 1 s is

$$W = 8.5 \times 10^8 \, \text{J}$$

From Eq. 10.4 the heat absorbed in the boiler in 1 s is

$$Q_1 = \frac{W}{e} = \frac{8.5 \times 10^8 \, \text{J}}{0.28} = 30.4 \times 10^8 \, \text{J}$$

and from Eq. 10.3 the heat lost in the condenser (in 1 s) is

$$|Q_2| = |Q_1| - W = 21.9 \times 10^8 \, \text{J}$$

The specific heat of water (Table 9.1) is $c_p = 4.18 \, \text{kJ}/(\text{kg} \cdot °\text{C})$, so with a temperature change of $\Delta T = 8°\text{C}$, Eq. 9.8 shows that the mass m of water required to absorb this heat is

$$m = \frac{Q_2}{c_p \, \Delta T} = \frac{21.9 \times 10^5 \, \text{kJ}}{[4.18 \, \text{kJ}/(\text{kg} \cdot °\text{C})](8°\text{C})} = 6.5 \times 10^4 \, \text{kg}$$

and the volume V of this water is

$$V = \frac{m}{\rho} = \frac{6.5 \times 10^4 \, \text{kg}}{1000 \, \text{kg/m}^3} = 65 \, \text{m}^3 = 25 \times 10^3 \, \text{gal}$$

Thus water at the rate of 25,000 gal/s or 1.5 million gal/min is required.

It is important to remember that waste heat is a necessary by-product of any work-producing system. It is not peculiar to nuclear plants, although it is somewhat more severe for them because their efficiency is less than that of a fossil-fuel plant. But every athlete faces the same problem of heat removal during strenuous exercise; sweating is the way the human body removes heat from its power plant. ☐

GASOLINE ENGINE

In an internal combustion engine, such as a gasoline engine, the fuel is burned inside the cylinder rather than in a separate boiler. This mixing of the fuel with the working substance confuses the picture somewhat. The substance which undergoes the thermodynamic cycle in an internal combustion engine is air. Because of the sudden heating of the air that occurs when the gasoline is ignited, and because of the acceleration of the piston, the air is not in definite thermodynamic states during most of the cycle. This makes an accurate plot of the cycle on a pV diagram impossible.

However, if we simplify the process drastically, we can imagine an idealized set of transformations that approximate the true conditions. Such an idealized cycle is called an *Otto cycle*. It is shown on the pV diagram in Fig. 10.9.

On the *intake stroke* of the piston (O to A), a quantity V_1 of air (mixed with gasoline) is drawn into the cylinder. This mixture is compressed adiabatically during the *compression stroke* (A to B). At B the mixture is ignited, which causes an isochoric increase in the temperature and pressure of the gas (B to C). There is relatively little motion of the piston during this part of the cycle, but the gas absorbs the heat Q_1 from the burning of the fuel. The gas then expands adiabatically on the *power stroke* (C to D). At D the exhaust valve on the piston is opened, causing a sudden drop in temperature and pressure of the gas and a loss of heat Q_2 (D to A). Finally, on the *exhaust stroke* (A to O) the combustion gas is exhausted at atmospheric pressure.

In this idealized cycle, the exhaust gas has the same temperature as the intake air, so the transformations O to A and A to O cancel each other. The heat Q_1 is absorbed during ignition (B to C), and the heat Q_2 is all lost when the valve is opened. In a real engine, the waste heat is expelled along with the exhaust gas through the exhaust system of the engine. The engine's cooling

FIGURE 10.9

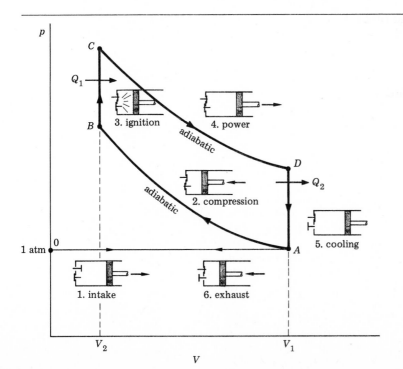

water is used to remove the additional heat generated by the friction between the pistons and the walls of the cylinders.

Equations 10.3, 10.4, and 10.5 give the relations between the work W done by the engines and the heat absorbed and lost. Furthermore, for the idealized Otto cycle it can be shown that the efficiency of the engine is related to the maximum and minimum volumes V_1 and V_2 of the cylinder by

$$e = 1 - \left(\frac{V_1}{V_2}\right)^{-0.4} = 1 - r^{-0.4}$$

where $r = V_1/V_2$ is the *compression ratio*. With $r = 8$, this equation gives an efficiency of 56.5 percent. The efficiency of a real engine is less than this, but it too increases with r.

When the compression ratio is too large, the rise in temperature of the air during the nearly adiabatic compression stroke is great enough to cause some of the gasoline in the cylinder to ignite prematurely (knocking). This can be prevented by mixing additives, such as tetraethyl lead, to the fuel. But since the use of leaded fuel has been banned, automobile manufacturers have had to reduce the compression ratio of their engines, thereby reducing their efficiency. This is just one instance of the tradeoff that must often be made between energy conservation and pollution control.

REFRIGERATOR

An engine is a device that converts heat into work. A refrigerator is a device that uses work to remove heat from a cold compartment. In both devices a substance, called the *system*, undergoes a closed cycle of thermodynamic transformations, so that its internal energy E_8 is unchanged. Consequently, the net work W done by (or on) the system is equal to the net heat Q absorbed (or lost) by the system.

Figure 10.10a shows that an engine absorbs a quantity of heat Q_1 from a hot reservoir (e.g., the boiler of a steam engine or the burning fuel of a gasoline or diesel engine) and loses a quantity of heat Q_2 to a cold reservoir (e.g., the cooling water for the condenser of a steam engine). The difference $W = |Q_1| - |Q_2|$ is the work done by the engine.

Figure 10.10b shows that a refrigerator absorbs a quantity of heat Q_1 from a cold reservoir (e.g., the cold compartment of the refrigerator) and loses a quantity of heat Q_2 to a hot reservoir (e.g., the room). The work $|W| = |Q_2| - |Q_1|$ must be done on the system in order to move the heat from the cold to the hot reservoir. A comparison of Fig. 10.10a and b shows that a refrigerator is the reverse of an engine.

The thermodynamic cycle of the refrigerant of a refrigerator is shown in Fig. 10.11. (Refrigeration is discussed in more detail in Sec. 9.4.) The heat Q_1 is absorbed from the cold reservoir when the refrigerant vaporizes (A to B), the work W is done on the refrigerant when it is compressed by the compressor (B to C), and the heat Q_2 is lost to the hot reservoir when the refrigerant condenses (C to A). Compare this cycle with the cycles for a steam engine (Fig. 10.8) and a gasoline engine (Fig. 10.9). The

FIGURE 10.10

(a) An engine absorbs heat Q_1 at a high temperature and exhausts heat Q_2 at a lower temperature. In the process, the engine does the work $W = |Q_1| - |Q_2|$. (b) A refrigerator does the work $W = |Q_1| - |Q_2|$ to remove the heat Q_1 from a low-temperature reservoir and exhaust the heat Q_2 to a higher-temperature reservoir.

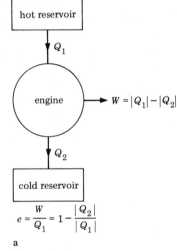

a

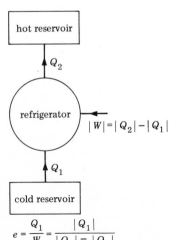

b

refrigeration cycle is similar to the cycle of an engine, except that the transformations proceed counterclockwise around the cycle, whereas in an engine the transformations proceed clockwise. In a clockwise cycle work is done on the environment, whereas in a counterclockwise cycle work is done on the system.

In a refrigerator the benefit is the heat Q_1 removed from the cold reservoir and the cost is the work W done on the system, so the *coefficient of performance e* of a refrigerator is

$$e = \frac{Q_1}{W} = \frac{|Q_1|}{|Q_2| - |Q_1|} \qquad 10.6$$

CARNOT CYCLE

The founders of thermodynamics were interested in finding the limits, if there were any, to the maximum efficiency of an engine. They knew that for practical reasons it was necessary to make the temperature of the hot reservoir greater than the temperature of the system, so that when the system came in contact with the hot reservoir heat would flow rapidly from the reservoir to the system. This large temperature difference made it possible to extract a large amount of heat in a short time, but it produced irreversible effects that lowered the efficiency of the engine.

In 1824 the French engineer Sadi Carnot (1796–1832) proposed a thermodynamic cycle that operates between two fixed temperatures T_1 and T_2. These can be thought of as the temperatures of the boiler and condenser, respectively, of a steam engine. However, in the Carnot cycle there is to be almost no temperature difference between the working substance (system) and the reservoir in which it is in contact. This means that the rate at which heat flows from the reservoir to the system is too slow to make the cycle the basis of a practical engine. The importance of the Carnot engine (i.e., an engine that operates on a Carnot cycle) is that it is the most efficient engine that operates between T_1 and T_2.

The Carnot cycle is shown in Fig. 10.12. At point A the working substance at temperature T_1 is brought into contact with the hot reservoir at T_1. If the system and the reservoir were at exactly the same temperature, there would be no heat flow between them, so we must imagine that the temperature of the hot reservoir is slightly greater than the temperature of the system. This allows us to imagine that the heat Q_1 flows from the hot reservoir to the system while both remain at essentially the same temperature T_1. Thus the system undergoes an isothermal expansion from A to B, absorbing the heat Q_1 from the hot reservoir.

At B the system is insulated and allowed to expand adiabatically from B to C. There is no heat flow during this transformation, so the temperature of the system falls to the temperature T_2 of the cold reservoir. From A to B to C we can imagine the substance expanding in a cylinder, and thus doing the work W_1.

At C the system is brought in contact with the cold reservoir. The temperature of the cold reservoir is slightly less than the temperature of the system, so the heat Q_2 flows from the system

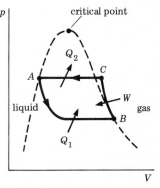

FIGURE 10.11

Plot of the refrigerator cycle on a pV diagram. The refrigerant is vaporized (AB), absorbing the heat Q_1 from the cold reservoir. The work W is done to compress the vapor (BC), and the compressed vapor is then liquefied by the removal of the heat Q_2 (CA).

FIGURE 10.12

Plot of the Carnot cycle on a pV diagram. Heat Q_1 is absorbed during the isothermal expansion AB, and the temperature is lowered from T_1 to T_2 during the adiabatic expansion BC. The lower-temperature gas is then compressed isothermally (CD), exhausting the heat Q_2, and compressed adiabatically (DA), raising its temperature from T_2 to T_1.

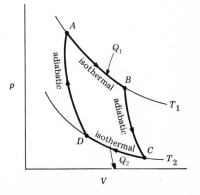

to the reservoir. The system is thus isothermally compressed from C to D. At D the system is again insulated and compressed adiabatically. This raises its temperature back to T_1 and returns the system to A. From C to D to A the system is being compressed by the return stroke of the piston, so the work $W_2 = -|W_2|$ is done on the system.

The system does the work W_1 during expansion and has the work W_2 done on it during compression. Since the expansion is done at higher pressure than the compression, $|W_1|$ is greater than $|W_2|$, so the net work done by the system

$$W = W_1 + W_2 = |W_1| - |W_2|$$

is positive. It is equal to the net heat

$$Q = Q_1 + Q_2 = |Q_1| - |Q_2|$$

absorbed by the system. The efficiency is defined the same way as any other engine, namely,

$$e = \frac{W}{Q_1} = \frac{|Q_1| - |Q_2|}{|Q_1|} = 1 - \frac{|Q_2|}{|Q_1|}$$

On the power stroke of the Carnot cycle (ABC) the heat Q_1 could be completely converted to work W_1. But because the engine operates on a cycle, the piston must be returned to its starting position. In order to do this with less work than was obtained on the power stroke, the system's temperature is lowered by extracting heat from it. The smaller this lost heat Q_2 is, the greater will be the efficiency e.

The following properties can be proved for the Carnot cycle.

1 A Carnot engine is the most efficient engine that operates between two temperatures T_1 and T_2.

2 The efficiency of a Carnot engine is independent of the working substance. That is, the efficiency is the same whether the system is an ideal gas, a real gas, or even a liquid.

3 The efficiency of a Carnot engine operating between the absolute temperatures T_1 and T_2 is

$$e = 1 - \frac{T_2}{T_1} \qquad\qquad 10.7$$

4 A Carnot engine can be operated in reverse, becoming a Carnot refrigerator. It is then the most efficient refrigerator which can extract the heat Q_1 from the cold reservoir T_1 and transfer the heat Q_2 to the hot reservoir T_2. The coefficient of performance of a Carnot refrigerator is

$$e = \frac{Q_1}{W} = \frac{T_1}{T_2 - T_1} \qquad\qquad 10.8$$

Example 10.2 The temperature of the boiler of a steam turbine is 550°C, and the temperature of the condenser is 10°C. What is the efficiency of a Carnot engine operating between these temperatures?

The absolute temperatures of the hot and cold reservoirs are

$$T_1 = 273°C + 550°C = 823 \text{ K}$$

and
$$T_2 = 273°C + 10°C = 283 \text{ K}$$

so from Eq. 10.7 the efficiency of a Carnot engine operating between these temperatures is

$$e = 1 - \frac{T_2}{T_1} = 1 - \frac{283}{823} = 0.66$$

The efficiency of any real engine is less than that of a Carnot engine. A modern fossil-fuel-fired steam turbine has an efficiency of about 0.40, while a nuclear-powered turbine has an efficiency of 0.30. The efficiency of a nuclear generating plant is less than that of a fossil-fuel plant because, for safety reasons, the nuclear plant is operated at lower maximum temperature.

10.3 The Second Law of Thermodynamics

A thermodynamic system S and the environment $\mathcal{E}$ with which it exchanges energy constitute the universe $\mathcal{U}$. The first law of thermodynamics (Sec. 9.1) states that the energy $E_\mathcal{U}$ of the universe, which is the sum of the energy E_S of the system and the energy $E_\mathcal{E}$ of the environment

$$E_\mathcal{U} = E_S + E_\mathcal{E}$$

is conserved; i.e., in any thermodynamic transformation

$$\Delta E_\mathcal{U} = 0$$

or

$$\Delta E_\mathcal{E} = -\Delta E_S \qquad \text{first law} \qquad\qquad 10.9$$

Alternatively, since the energy E_S of the system changes only as a consequence of the heat Q that enters the system from the environment or the work W done by the system on the environment, the first law is usually written

$$\Delta E_S = Q - W \qquad \text{first law}$$

Although no transformation is possible which violates the first law, many transformations which do not violate the first law do not occur. The second law of thermodynamics characterizes these impossible transformations.

An object of mass m released from a height h spontaneously falls to the ground, where it comes to rest. In this situation the energy of the universe is the sum of the internal energies of the object and the ground and the mechanical energy of the object. Before it is released, the object has mechanical energy just equal to its potential energy $U = mgh$, and after it comes to rest on the ground, its mechanical energy is zero. In this process, therefore, the mechanical energy of the universe decreases from mgh to zero. Since the total energy of the universe does not change (the first law of thermodynamics), the internal energy of the universe must increase by mgh. This increase in internal energy manifests itself as a slight increase in the temperature of the object and the ground.

It is well known from everyday experience that an object

initially at rest on the ground never spontaneously jumps into the air. On first thought such an event seems to be impossible because it violates the first law of thermodynamics. If an object were to jump into the air, there would clearly be an increase in the mechanical energy of the universe. This would not be a violation of the first law, however, if there were a corresponding decrease in the internal energy of the universe. The first law, therefore, does not explain why objects never spontaneously jump into the air.

The process "object spontaneously jumps into the air" is the reverse of the process "object spontaneously falls to the ground." The one process occurs with ease, whereas the reverse process never occurs at all. There are many other irreversible processes that can proceed in only one direction. For example, when a cold object and a hot object are placed in contact, heat always flows from the hot object to the cold object, never from the cold object to the hot object. Consequently, the temperature of the hot object decreases, while the temperature of the cold object increases. If the reverse process were to occur, the cold object would get colder while the hot object got hotter. As another example, ink dropped into a glass of water diffuses until it is evenly distributed throughout the water. The reverse process, in which a uniform mixture of water and ink spontaneously separates into pure water and pure ink, never occurs.

KELVIN AND CLAUSIUS STATEMENTS OF THE SECOND LAW OF THERMODYNAMICS

The second law of thermodynamics specifies those processes which cannot occur spontaneously. There are many equivalent statements of the second law, and we shall consider a number of them in this chapter. We begin with the form of the second law stated by Lord Kelvin (William Thomson, 1824–1907).

Kelvin's Statement of the Second Law of Thermodynamics *It is impossible to construct a device that can without other effect lift one object by extracting internal energy from another.* This form of the second law is essentially the assertion that objects do not spontaneously jump into the air. But it goes beyond this by asserting that even with the help of special machinery, no matter how complex, an object cannot be lifted solely by extracting internal energy. In other words, internal energy cannot be completely converted into potential energy without other effect, i.e., without something else happening.

There is another form of the second law attributed to Rudolf Clausius (1822–1888).

Clausius's Statement of the Second Law of Thermodynamics *It is impossible to construct a device that can without other effect transfer heat from a cold object to a hotter object.* This form of the second law is essentially the assertion that heat does not spontaneously flow from a cold object to a hotter object. But it goes beyond this by asserting that even with the help of special machinery, no matter how complex, heat cannot be transferred from a cold object to a hot object without other effect.

The Kelvin and Clausius statements are equivalent. This means that if it were possible to construct a device that violated the Kelvin statement, it would be possible to construct a device that violated the Clausius statement, and vice versa. For example, suppose a device existed that could lift a mass solely by extracting internal energy from object A, in violation of the Kelvin statement. Then, with suitable mechanical connections, this device could be made to lift the mass repeatedly and drop it onto another object B (Fig. 10.13). Each time the mass was lifted, the internal energy and temperature of object A would decrease, and each time the mass fell, the internal energy and temperature of object B would increase. Consequently, internal energy would be transferred from A, which would become colder, to B, which would become hotter, in violation of the Clausius statement. This proves that a device that violates the Kelvin statement can be used to construct a device that violates the Clausius statement. The reverse is also true, but we shall not prove it here.

CONSEQUENCES OF THE SECOND LAW

The Kelvin statement is about engines. We have seen that an engine extracts the heat Q_1 from a hot reservoir and transfers the heat Q_2 to a cold reservoir. The Kelvin statement asserts that it is impossible to build an engine that does not transfer some heat Q_2 to the cold reservoir.

For example, consider the general thermodynamic cycle shown in Fig. 10.14. Along path 1 the system absorbs the heat Q_1 and does the work W_1. In the process the state of the system changes from A to B, so there has been an "effect" other than the extraction of the heat Q_1 and the performance of the work W_1. The Kelvin statement thus does not apply to a part of a cycle. However, on the return stroke of the piston, the system is brought back to its initial state A along path 2. In the process the work $W_2 = -|W_2|$ is done on the system. Since the return stroke is at lower pressure (on the average) than the power stroke, $|W_2|$ is less than $|W_1|$. So the net work done in one cycle

$$W = W_1 + W_2 = |W_1| - |W_2|$$

is positive. If no heat is lost along path 2, we would have succeeded in extracting the heat Q_1 from the hot reservoir and doing the work W without other effect. Since this violates the Kelvin statement, some heat must be lost along path 2. The Kelvin statement is equivalent to saying that all the heat extracted from the hot reservoir cannot be converted into work: the efficiency of an engine must be less than 1.

An engine that violates the second law of thermodynamics is called a perpetual-motion machine of the second kind. An example would be an engine that could do work by cooling the temperature of the environment. If such a machine were possible, a ship could propel itself across the ocean by cooling seawater. When the temperature of 1 kg of water is lowered from 18 to 10°C, 3.35×10^4 J of heat is given off. If some of this heat could be converted into work, we would have a perpetual-motion machine of the second kind. Such a machine does not

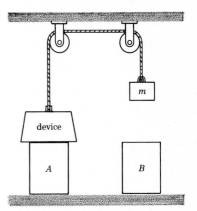

FIGURE 10.13

A device that violates the Kelvin statement by lifting a mass m using internal energy extracted from A.

FIGURE 10.14

A cycle for converting heat into work.

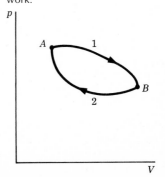

violate the first law of thermodynamics. It does not create energy. But it transfers energy in a way forbidden by the second law.

There is a futuristic proposal to generate electric energy (work) from seawater that does not violate the second law. In this proposal, the power plant is submerged hundreds of meters below the surface of a tropical sea (Fig. 10.15). The surface water, warmed by the sun, has a temperature of 30°C, whereas the deep water has a temperature of 4°C. The warm surface water is the hot reservoir. This water is pumped into the plant, where heat is extracted at 30°C and the temperature of the water lowered to (say) 8°C. In the process work is done and waste heat is produced. The waste heat is removed by using the deep 4°C water as the coolant.* Although an engine that operated between reservoirs at 30 and 4°C would have a very low efficiency, it would not violate the second law. Someday such a scheme might be an effective way to use solar energy to generate electricity.

The heat from the sun does work in creating the winds and in

*The working substance of such an engine would have to have a boiling point around 6°C (e.g., hexafluoro-1, 2-butadiene, C_4F_6).

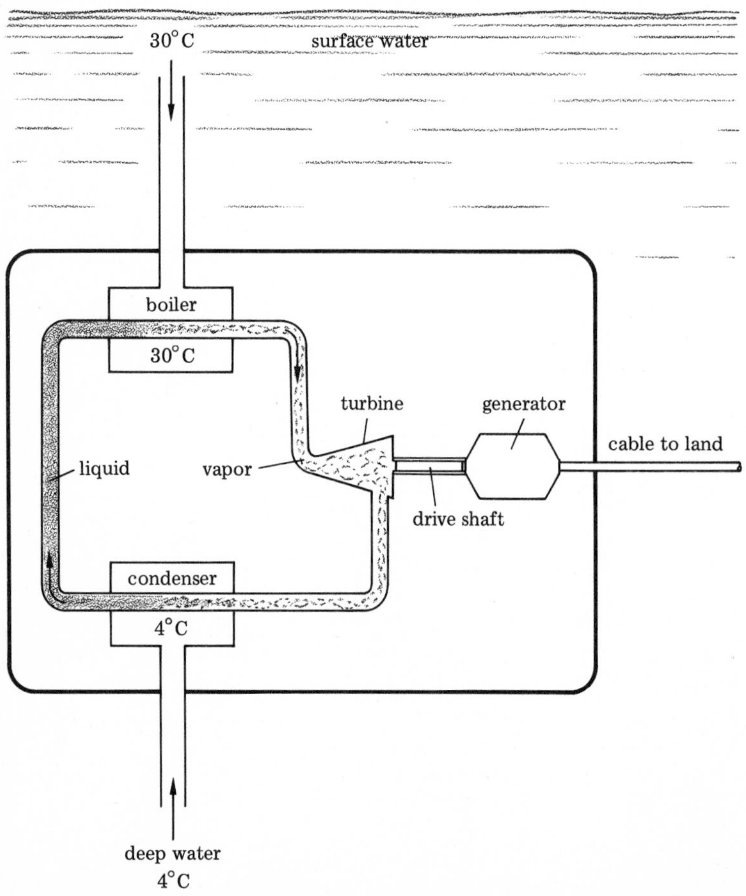

FIGURE 10.15

Futuristic power plant that operates on the temperature difference between the upper and lower regions of a tropical ocean. Heat extracted from the higher-temperature water vaporizes a volatile liquid. The vapor is condensed using the lower-temperature water.

raising water vapor into the clouds. (This work is harnessed when the wind turns a windmill or the rain, falling behind a dam, turns a hydroelectric generator.) It is possible to obtain work from the sun because some of the heat absorbed from the sun at high temperature during the day is lost at low temperature into outer space at night. If the earth did not rotate, there would be no cycle of day and night, and consequently no wind or rain from which to obtain work from the sun's heat.

The Clausius statement of the second law is about refrigerators. We have seen that a refrigerator requires a certain amount of work to move heat from a hot reservoir to a cold reservoir. This work is the other effect mentioned in the Clausius statement. The Clausius statement thus says that while heat will spontaneously flow from a hot to a cold reservoir, work is required to move heat in the opposite direction.

The two statements of the second law of thermodynamics are equivalent, although the Clausius statement may seem more obviously true. It is the logical equivalence of these two statements that makes it as impossible to build a perpetual-motion machine of the second kind as it is to freeze a kettle of water by placing it on a hot stove.

10.4   Statistical Formulation of the Second Law

The Kelvin and Clausius statements of the second law are useful because they apply directly to practical engineering problems. However, they sound more like the consequences of a fundamental law rather than the law itself. The second law of thermodynamics is fundamentally a law of statistics, which can be stated as follows:

Statistical Statement of the Second Law of Thermodynamics
The total disorder of the universe never decreases. The concept of disorder will be clarified shortly, but for now it is sufficient to use the ordinary meaning of the word.

For example, when ink is dropped into water, it diffuses until it is evenly distributed throughout the water. Water and ink are in a more disordered state when mixed than when separated, because their molecules are randomly mixed together in one state and not in the other. Consequently, the disorder of the universe increases during the diffusion process. The reverse process, in which a uniform mixture of water and ink spontaneously separates into pure water and pure ink, is forbidden by the second law because it would decrease the disorder of the universe.

As another example, consider again an object of mass m held at a height h above the ground. Its potential energy mgh is distributed equally among all its molecules, since each molecule has the same potential energy $m_{mol}gh$, where m_{mol} is the mass of an individual molecule. When the object is dropped, its potential energy is ultimately converted into kinetic energy of the molecules of the ground. This energy is distributed randomly among these molecules, because each molecule can have a different energy. Consequently, the disorder of the universe is increased when potential energy is converted into internal energy. The reverse process, in which internal energy is completely converted into potential energy, is forbidden by the second law because it would decrease the disorder of the universe. This shows how the statistical statement of the second law is related to the Kelvin statement, though we shall not give a complete proof of their equivalence.

The concept of disorder can be made quantitative by considering the relation between a thermodynamic state and a molecular state of a system. A molecular state (*microstate*) is specified by giving the position and velocity of every molecule in the system, whereas a thermodynamic state (*macrostate*) is specified by giving a few average molecular properties, such as temperature and pressure. Different microstates with the same average properties correspond to the same macrostate. The number D of distinct microstates corresponding to a macrostate is the *disorder* of the macrostate.

As the molecules of the system change their positions and velocities, the system changes from one microstate to another. The equilibrium macrostate of the system is the macrostate of maximum disorder, i.e., the state with the greatest number of microstates. At equilibrium the system does not change macrostates, even though its microstates are constantly changing, because almost all the microstates correspond to the equilibrium macrostate. This abstract discussion will become clearer after considering the following example.

A tray is filled with 100 plastic chips. Each chip is white on one side and black on the other. When the tray is viewed from a large distance, the individual chips cannot be seen. All that is seen is the average color of the tray, which can be any shade of gray or pure white or pure black, depending on how many chips are turned white side up and how many are turned black side up. The color of the tray is thus an average property of the system, analogous to a thermodynamic variable. A quantitative description of the color is given by the *grayness g*, which is defined as the fraction of chips turned black side up. The grayness varies from 0 (pure white) to 1 (pure black). The value of g specifies the macrostate of the tray, just as the values of p and T specify the macrostate of a thermodynamic system.

A microstate of the tray is specified by giving the color of each chip. For instance, one microstate may have the first chip black side up, the second chip white side up, the third white, the fourth black, and so on. This microstate is represented by the sequence

bwwb $\cdots$

Another microstate may have the first chip black, the second black, the third white, the fourth white, and so on. It is represented by the sequence

bbww $\cdots$

Since each chip can have either one of two colors (black or white), the total number N of distinct microstates is

$$N = (2)(2)(2)(2) \cdots = 2^{100}$$

From the definition of logarithm (Appendix III) we can write

$$2 = 10^{\log 2} = 10^{0.30}$$

and so

$$N = 2^{100} = (10^{0.30})^{100} = 10^{30}$$

This shows that even a simple system with only 100 particles in it possesses an immense number of microstates.

Of even greater importance is the number of microstates $D(g)$ corresponding to a given macrostate g. The all-white macrostate ($g = 0$) has only one microstate because every chip must be white side up; hence $D(0) = 1$. The $g = 0.01$ macrostate has 99 chips white side up and 1 chip black side up. Since the macrostate is the same regardless of which one of the 100 chips is black, there are 100 different microstates corresponding to $g = 0.01$; hence $D(0.01) = 100$. The $g = 0.02$ macrostate has 98 white chips and 2 black. There are 4950 different microstates in this case; hence $D(0.02) = 4950$. Table 10.1 gives the disorder $D(g)$ of various macrostates between 0 and 1. This shows that $D(g)$ is very different for different macrostates and that it is a maximum for the state $g = 0.50$, in which 50 chips are white and 50 black.

When the tray is shaken up and down, the chips randomly flip from one side to another. With each shake, the tray makes a random transition from one microstate to another. Although any one microstate is as likely to occur as any other, some macrostates are much more probable than others. The probability $P(g)$ of the tray's being in macrostate g after any one shake is equal to the number of microstates $D(g)$ corresponding to g divided by the total number N of microstates of the system:

$$P(g) = \frac{D(g)}{N} \qquad 10.10$$

For example, of the 10^{30} microstates of the tray, only one corresponds to $g = 0$. Therefore, the probability of the all-white macrostate occurring after any one shake of the tray is

$$P(g) = \frac{D(0)}{N} = \frac{1}{10^{30}} = 10^{-30}$$

This is an incredibly small probability; the tray could be shaken a billion times a second for 100 billion years without the all-white macrostate's occurring. An event with such a low probability of occurrence is, for all practical purposes, impossible. That is, we can safely say that the all-white macrostate cannot be produced by randomly shaking the tray. There is no mechanical law that prevents its occurrence, but its probability is so small that it can be considered zero.

The macrostate of maximum disorder ($g = 0.50$) occurs with the greatest probability:

$$P(0.50) = \frac{1.0 \times 10^{29}}{10^{30}} = 0.10$$

This means that on an average the tray is in the $g = 0.50$ state

TABLE **10.1**

Number of microstates $D(g)$ corresponding to various macrostates g of a tray containing 100 black and white chips

Macrostate g	Disorder $D(g)$
0.00	1
0.02	4.9×10^3
0.05	7.5×10^7
0.10	1.7×10^{13}
0.20	5.5×10^{20}
0.40	1.4×10^{28}
0.50	1.0×10^{29}
0.60	1.4×10^{28}
0.80	5.5×10^{20}
0.90	1.7×10^{13}
0.95	7.5×10^7
0.98	4.9×10^3
1.00	1

once in every 10 shakes of the tray. The macrostates with g close to 0.50 also occur frequently. In fact, it can be shown that 98 percent of all the microstates of the tray correspond to macrostates between $g = 0.40$ and $g = 0.60$. Therefore, although the microstate changes with each shake of the tray, the macrostate is always close to 0.50.

If the number of chips in the tray is increased, the probability of the tray's being in a macrostate other than 0.50 decreases. For instance, in a tray with a million chips, over 99.99 percent of the microstates correspond to macrostates between $g = 0.499$ and $g = 0.501$. Suppose that initially these chips are all turned white side up, so that the tray is started in the $g = 0$ state. When the tray is shaken, it randomly changes its microstate and will, with overwhelming probability, make a transition to a microstate corresponding to a macrostate close to $g = 0.50$. Thereafter, with each shake of the tray, the microstate changes while the macrostate remains close to 0.50.

In a thermodynamic system, which consists of billions upon billions of molecules, the number of microstates corresponding to a state of greater disorder is overwhelmingly larger than the number of microstates corresponding to a macrostate of lesser disorder. Consequently, the system never spontaneously changes to a state of lesser disorder. The second law of thermodynamics is fundamentally a statistical law. There is no physical law that prevents a uniform mixture of ink and water from spontaneously separating into pure ink and pure water. However, there are so many more ways to arrange ink and water molecules in the mixed system than there are in the separated system that the probability of the molecules of the mixture ever being, by pure chance, in a microstate of the separated system is essentially zero.

> **REMARK** It is important to realize that the second law says that the disorder of the *universe* never decreases. The disorder of a part of the universe can decrease if there is a corresponding increase in the disorder of another part. Such a decrease is not spontaneous, however, since it is produced by an interaction with another part of the universe. For instance, the heat Q_1 removed from the hot environment of a Carnot engine decreases the disorder of the hot environment. But the heat Q_2 added to the cold environment increases the disorder of the cold environment. The second law requires that the increase of disorder of the cold environment be at least as large as the decrease of disorder of the hot environment.

It is sometimes said that evolution violates the second law because in the course of hundreds of millions of years more and more complex forms of life have developed. The more complex forms are clearly more ordered than the lower forms. Thus evolution, the argument goes, is a process in which life progresses from a state of less order to a state of more order, in contradiction to the second law. Some people even maintain that this proves that life has unique properties that are beyond the laws of physics.

It is true that evolution has, on the whole, produced forms of increased order. Even the growth of an individual organism, from egg to maturity, is a process in which order increases. However, these are not violations of the second law because the earth is not an isolated system. The earth receives energy from the sun during the day and radiates energy into outer space at night. The energy-releasing reactions that occur on the sun increase the disorder of the sun. Thus the energy used on earth to produce ordered life from disordered matter is produced at the expense of increased disorder of the sun. The total disorder of the earth and sun increases, in compliance with the second law.

10.5 Entropy

Definition *Entropy S* is a thermodynamic variable that measures the disorder of a thermodynamic state. This means that every state has a definite entropy and that the entropy of one state is greater than the entropy of another if the disorder of the first state is greater than the disorder of the second. But unlike disorder, the entropy of a composite system is simply the sum of the entropies of the component subsystems.* In particular, the entropy $S_\mathfrak{u}$ of the universe is the sum of the entropies of a system and its environment:

$$S_\mathfrak{u} = S_\mathfrak{s} + S_\mathfrak{E} \qquad\qquad 10.11$$

Let $S_\mathfrak{s}$ and $S_\mathfrak{E}$ be the entropies of the system and the environment when the system is in one state, and let $S'_\mathfrak{s}$ and $S'_\mathfrak{E}$ be the entropies after the system has transformed to another state. According to the statistical form of the second law, the entropy of the universe never decreases, so $S'_\mathfrak{u}$ must be greater than or equal to $S_\mathfrak{u}$:

$$S'_\mathfrak{u} \ge S_\mathfrak{u}$$

From Eq. 10.11 we have

$$S'_\mathfrak{s} + S'_\mathfrak{E} \ge S_\mathfrak{s} + S_\mathfrak{E} \qquad \text{or} \qquad (S'_\mathfrak{s} - S_\mathfrak{s}) + (S'_\mathfrak{E} - S_\mathfrak{E}) \ge 0$$

Using ΔS to represent these entropy changes, we can write this as

---

*The disorder D of a system composed of two subsystems is the product $D_1 D_2$ of the disorders of the two subsystems. This is because for every microstate of subsystem 1, the composite system has D_2 microstates. Since subsystem 1 has D_1 different microstates, the composite system has a total of $D_1 D_2$ microstates.

The entropy S of a state is proportional to the logarithm of its disorder. Thus the individual entropies of subsystems 1 and 2 are

$$S_1 = k \log D_1 \qquad \text{and} \qquad S_2 = k \log D_2$$

and the entropy of the composite system is

$$S = k \log D = k \log D_1 D_2$$

where k is the constant of proportionality. From the rules of logarithms we have

$$\log D_1 D_2 = \log D_1 + \log D_2$$

and so

$$S = k \log D_1 + k \log D_2 = S_1 + S_2$$

Thus the entropy of a composite system is equal to the sum of the entropies of its subsystems, while the disorder of the composite system is equal to the product of the disorders of its subsystems.

$$\Delta S_{\mathfrak{u}} = \Delta S_{\mathfrak{s}} + \Delta S_{\mathfrak{e}} \geq 0$$

or

$$\Delta S_{\mathfrak{e}} \geq -\Delta S_{\mathfrak{s}} \qquad \text{second law} \qquad\qquad 10.12$$

This last equation expresses the second law of thermodynamics in a form analogous to the first law in Eq. 10.9.

It is always possible, in principle, to transform a system from one state A to another state B without changing the entropy of the universe ($\Delta S_{\mathfrak{u}} = 0$). Such a transformation is said to be reversible because the system can be transformed from B back to A through the same intermediate states. Normally, however, $\Delta S_{\mathfrak{u}}$ increases during a transformation, so that the reverse process is impossible. For example, $\Delta S_{\mathfrak{u}}$ increases when ink spontaneously diffuses in water. This is an irreversible process:* the spontaneous separation of a mixture of ink and water cannot occur.

Transformations are classified as irreversible, reversible, or impossible depending on whether $\Delta S_{\mathfrak{u}}$ is greater than, equal to, or less than zero:

$$\Delta S_{\mathfrak{u}} \begin{cases} > 0 & \text{irreversible} \\ = 0 & \text{reversible} \\ < 0 & \text{impossible} \end{cases}$$

Reversible transformations are idealized processes in which there is no increase in disorder. They are like frictionless processes in mechanics, which are idealized processes with no decrease in mechanical energy. Only a reversible transformation can be represented by a solid line on a pV diagram.

Entropy, being a thermodynamic variable like temperature and pressure, has a definite value for every state of a system. Consequently, the entropy difference $\Delta S_{\mathfrak{s}}$ between two states of a system depends only on these states, and not on the particular process used to transform the system from one state to another. The entropy difference between states can be calculated from the following rules:

1 If a system can be transformed from one state to another by means of a reversible adiabatic transformation, then the entropy difference between these states is zero:

$$\Delta S_{\mathfrak{s}} = 0 \qquad \text{adiabatic} \qquad\qquad 10.13$$

2 If a system can be transformed from one state to another by means of a reversible isothermal transformation, then the entropy difference between these states is

$$\Delta S_{\mathfrak{s}} = \frac{Q}{T} \qquad\qquad 10.14$$

where Q is the heat absorbed in the transformation at the absolute temperature T.

UNITS From Eq. 10.14 we see that the units of entropy are kilocalories per kelvin (kcal/K) or joules per kelvin (J/K).

*Ink and water can be mixed reversibly using a piston and a semipermeable membrane.

As an example, we can calculate the entropy change of a system in the course of a Carnot cycle (Fig. 10.12). The system, starting in state A, is transformed isothermally to state B at the temperature T_1. During the transformation the system absorbs the heat Q_1, so by rule 2 the entropy change is

$$S_B - S_A = \frac{Q_1}{T_1}$$

The system is then transformed adiabatically from state B to state C, so by rule 1 the entropy change is zero:

$$S_C - S_B = 0$$

During the isothermal transformation from state C to state D, the system loses the heat $Q_2 = -|Q_2|$ at the temperature T_2, so the entropy change is

$$S_D - S_C = \frac{-|Q_2|}{T_2}$$

Finally, the system is transformed adiabatically from D to A, so there is again no entropy change:

$$S_A - S_D = 0$$

The sum of these entropy changes is

$$\Delta S_8 = \frac{|Q_1|}{T_1} - \frac{|Q_2|}{T_2}$$

which is the total entropy change of the system in one cycle. But since the final state is the same as the initial state, ΔS_8 is zero. Consequently, we have

$$\frac{|Q_1|}{T_1} - \frac{|Q_2|}{T_2} = 0 \quad \text{or} \quad \frac{|Q_2|}{|Q_1|} = \frac{T_2}{T_1} \qquad 10.15$$

This equation shows that the efficiency e of a Carnot engine is

$$e = 1 - \frac{|Q_2|}{|Q_1|} = 1 - \frac{T_2}{T_1}$$

which is the result given in Eq. 10.7.

The environment loses the heat $-|Q_1|$ at temperature T_1 and gains the heat $|Q_2|$ at temperature T_2, so the entropy change of the environment is

$$\Delta S_\varepsilon = \frac{-|Q_1|}{T_1} + \frac{|Q_2|}{T_2}$$

But in any one cycle, whether reversible or irreversible, the entropy change ΔS_8 of the system is zero, since the system returns to its initial state. Therefore, from Eq. 10.12 we have that in one cycle

$$\Delta S_\varepsilon \geq 0 \quad \text{or} \quad \frac{|Q_2|}{|Q_1|} \geq \frac{T_2}{T_1}$$

If the cycle is reversible, the equal sign holds, and these results reduce to Eq. 10.15. If the cycle is irreversible, the inequality holds, and $|Q_2|$ is greater than it would be for a reversible cycle. This means that an irreversible cycle is always less efficient than a reversible one. All real engines have some irreversible processes due to turbulence and finite heat differences, but in large steam-generated power plants great effort is made to reduce these to a minimum.

GUIDE TO MAJOR TOPICS

Topic	Reference	Problems
Thermodynamic transformations	Sec. 10.1; Eq. 10.2	1 to 6
Engines and refrigerators	Sec. 10.2	
Efficiency	Eqs. 10.3 to 10.5; Example 10.1	7 to 10
Coefficient of performance	Eq. 10.6	11 and 12
Carnot cycle	Eqs. 10.7 and 10.8; Example 10.2	13 to 16
Second law of thermodynamics	Sec. 10.3	17 and 18
Statistical formulation of the second law	Sec. 10.4; Eq. 10.10; Table 10.1	19 and 20
Entropy	Sec. 10.5; Eqs. 10.11 to 10.14	21 to 24

PROBLEMS

1 During a transformation a system absorbs 1500 J of heat and does 900 J of work. What is the change of energy ΔE_s of the system?
Ans. 600 J

2 When a system goes from state A to state B its energy changes by 5000 J. If, in the process, the system absorbs 6000 J of heat, how much work is done by (or on) the system?

3 During an adiabatic transformation the change of energy ΔE_s of a system is $+950$ J. How much work is done by (or on) the system during the transformation?
Ans. -950 J

4 During an adiabatic transformation the work $|W| = 75$ J is done *on* a system. What is the change of energy ΔE_s of the system?

5 During an isobaric transformation at 1 atm the volume of a gas changes from 1.0×10^{-3} to 1.5×10^{-3} m³ and the gas absorbs 30 J of heat. What is the change of energy ΔE_s of the system?
Ans. -20.5 J

6 During an isochoric transformation a system absorbs 35 J of heat. What is the change of energy ΔE_s of the system?

7 A gas absorbs 800 J of heat and does 500 J of work as it is transformed from state A to state B along path 1 in Fig. 10.14. (a) What is the change of energy ΔE_s of the system? (b) As the gas is transformed from state B back to state A along path 2, 300 J of work is done on it. What is the efficiency e of the cycle ($A \rightarrow B \rightarrow A$)? (c) How much heat is released from the gas along path 2?
Ans. (a) 300 J; (b) 0.25; (c) 600 J

8 A gas absorbs 7000 J of heat as it is transformed from state A to state B along path 1 in Fig. 10.14, and it releases 5000 J of heat as it is transformed from state B back to state A along path 2. (a) What is the net amount of work done in one cycle? (b) What is the efficiency of the cycle?

9 The efficiency of an engine is 0.21. For every 1000 J of heat absorbed by the engine, how much (a) net work is done by it and (b) heat is released by it?
Ans. (a) 210 J; (b) 790 J

10 The United States generates about 2×10^{16} J of electric energy a day. Electric energy is equivalent to work, because it can be converted into work with nearly 100 percent efficiency by an electric motor. (a) If this electric energy is generated by power plants with an average efficiency of 0.30, how much heat is dumped into the environment each day? (b) How much water is required to absorb this heat if the water temperature is increased by 5°C?

11 A refrigerator does 150 J of work to remove 500 J of heat from the cold compartment. (a) What is the coefficient of performance of the refrigerator? (b) How much heat is exhausted into the room?
Ans. (a) 3.33; (b) 650 J

12 A refrigerator with a coefficient of performance of 2.5 is required to remove 950 kcal of heat from the cold compartment. (a) How much work does this require? (b) How much heat is exhausted into the room?

13 (a) Calculate the efficiency of a Carnot engine that operates between the temperatures 600 and 30°C. (b) If the engine absorbs 100 kcal of heat at the high temperature, how much heat does it release at the low temperature?
Ans. (a) 0.64; (b) 36 kcal

14 A Carnot engine with an efficiency of 0.30 operates between a high temperature T_1 and a low temperature $T_2 = 310$ K. (a) What is T_1? (b) How much heat is released at T_2 for every 100 J of work done?

15 What is the coefficient of performance of a Carnot refrigerator that operates between the temperatures −15 and 20°C?
Ans. 7.4

16 Show that in a Carnot engine the heat Q_2 released at the lower temperature T_2 is related to the work W done by the engine by

$$Q_2 = \frac{T_2}{T_1 - T_2} W$$

17 The work $W_1 = -|W_1|$ is done on a refrigerator to remove a quantity of heat Q_1 from a cold reservoir and to transfer the heat Q_2 to a hot reservoir. The heat Q_2 is then absorbed by an engine operating between the same two reservoirs. Show that the work W_2 done by the engine cannot be greater than $|W_1|$, i.e., prove that $W_2 \le |W_1|$.

REMARK This last problem shows that you cannot build a perpetual-motion machine of the second kind by using a refrigerator to supply the heat to an engine.

18 Prove that a device that violates the Clausius statement of the second law of thermodynamics can be used to build a device that violates the Kelvin statement of the second law.

19 Consider a tray with six chips, each of which is white on one side and black on the other. (a) How many microstates does this system have? (b) How many macrostates does the system have? (c) List all the distinct microstates of the system and determine the disorder D of each macrostate. (d) After shaking the tray, what is the probability that the system will be in the $g = 0.5$ macrostate?
Ans. (a) 64; (b) 7; (d) 0.31

20 What is the probability that a tray with 100 black and white chips will be in the $g = 0.10$ macrostate after shaking?

21 (a) What is the entropy change of 1 kg of H_2O when it changes from ice to water at 0°C (see Table 9.4)? (b) If the ice is in contact with an environment at a temperature of 10°C, what is ΔS_u when the ice melts? (c) If the temperature of the environment is −10°C, what would ΔS_u be if the ice melted? (d) What must be the temperature of the environment for ΔS_u to be zero?
Ans.   (a) 0.292 kcal/K;   (b) 0.010 kcal/K; (c) −0.011 kcal/K; (d) 0°C

22 The temperature of the sun's surface is 5800 K, and the temperature of the earth's surface is 300 K. What is the entropy change of the universe when 10^4 J of heat is transferred from the sun to the earth?

23 What is the entropy change of 850 g of water that is heated from 20 to 50°C?
Ans. 11.6 J/K

REMARK Equation 10.14 is exact only for an isothermal process. However, it can be used to good approximation whenever the temperature change ΔT is small compared to the initial temperature T_1 of the system. In such a case, the average temperature of the system

$$T_{av} = \tfrac{1}{2}(T_1 + T_2)$$

should be used in the denominator of Eq. 10.14.

24 A 500-g block of iron at a temperature of 60°C is placed in contact with a 500-g block of iron at a temperature of 20°C. (a) The blocks soon come to a common temperature of 40°C. What is the entropy change in this case? (b) Calculate the entropy change, assuming the temperature of the hotter block increased to 80°C while the temperature of the colder block decreased to 0°C. (Since the entropy of the universe decreases in this case, this process is impossible.)

BIBLIOGRAPHY

BENT, HENRY A.: *The Second Law: An Introduction to Classical and Statistical Thermodynamics,* Oxford University Press, New York, 1965. An excellent introduction to thermodynamics that uses no advanced mathematics. The book contains extensive quotations from the works of the founders of thermodynamics, which provide fascinating glimpses of the history of the subject.

WILSON, DAVID GORDON: "Alternative Automobile Engines," *Scientific American* **239**:39 (July 1978). Comparison of the advantages and disadvantages of the various engines suitable for automobiles.

PART III

WAVES

Light and sound are fundamentally different physical phenomena which nevertheless have many common properties because they are both waves. Other examples of waves are vibrating cords, x-rays, and ocean waves. In each case the phenomenon involves a disturbance that propagates through a continuous medium. For a vibrating cord, the medium is the cord; for an ocean wave, the medium is the surface of the water; for sound, the medium is air; and for light and x-rays (as well as for radar, infrared and ultraviolet radiation, and gamma rays), the medium is space itself.

Our ears and eyes are natural instruments for the detection of sound and light, and

First telescopes
(Galileo, 1609).
(*History of Science Museum, Florence.*)

our mouth is an instrument for the production of sound. Technology has developed many specialized instruments for the detection and production of sound and light; such as photographic film, microphones, light bulbs, and musical instruments. All such instruments play a vital role in the communication of ideas, information, and entertainment.

W

aves on a vibrating cord are easily visualized, while at the same time they exhibit most of the general properties common to all waves. For this reason they serve as a useful introduction to the study of waves. The theory of vibrating cords has direct application to musical instruments, such as the guitar, piano, and violin, and to overhead cables, such as power lines, telephone lines, and bridge suspensions. The theory has indirect application to the study of atomic structure (Chap. 19) because of the close analogy between the normal modes of a cord and the energy states of an atom.

11
VIBRATING CORDS

11.1 Waves on a Stretched Cord

Imagine a long cord fixed at one end.* If the other end is given a sharp flip up and down, a pulse is generated which moves along the cord at constant speed (Fig. 11.1). The pulse is a limited region of the cord which is temporarily disturbed from its normal (equilibrium) position. It is this region of disturbance that moves along the cord.

If the free end of the cord is shaken steadily, a steady wave pattern is generated on the cord (Fig. 11.2). We shall often come across such wavy patterns, but the word "wave" is used generally for any disturbance that propogates in this manner, regardless of its shape.

Definition A *wave* is a disturbance pattern in a medium that propagates through the medium at a constant speed v characteristic of the medium. In the present example, the cord is the medium and the disturbance is the displacement of points of the cord from their undisturbed, or equilibrium, position.

Every medium has an equilibrium configuration in which all the points of the medium are at rest. Figure 11.3a shows the equilibrium configuration of a stretched cord. Coordinate axes are drawn in back of the cord, and each point of the cord is labeled by its equilibrium position. For instance, point A on the cord is the point that is at $x = 4.0$ m, $y = 0$ when the cord is in its equilibrium position.

Figure 11.3b shows the cord with a wave on it. Point A is no longer on the axis but is displaced from the axis by the distance $y_{4.0} = 0.3$ cm. The subscript 4.0 indicates that $y_{4.0}$ is the displacement of the point A whose x coordinate at equilibrium is 4.0 m. This displacement is the measure of how much the medium is disturbed at this point. As the wave moves to the right, the displacement of point A changes. In Fig. 11.3c the wave has moved so that now point A has the displacement $y_{4.0} = 0.6$ cm, and point B has the displacement $y_{5.0} = 0.3$ cm. Note that each point of the cord moves only up and down while the disturbance itself moves with constant speed v along the cord. This speed is not the same as the up-and-down v speed of a particular point on the cord. In fact, the up-and-down speed is not constant at all, as we shall see.

*For demonstration purposes a long spring is sometimes used.

FIGURE 11.1

A pulse moving along a cord at constant speed. (*From PSSC Physics, 4th ed., copyright © 1976 by Educational Development Center, Inc., used with permission.*)

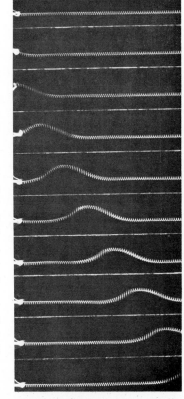

To be specific, suppose the time interval between Fig. 11.3*b* and *c* is 0.2 s. In this time the point on the cord with the displacement 0.3 cm has changed from *A* (at $x = 4.0$ m) to *B* (at $x = 5.0$ m), so the speed v of the wave is

$$v = \frac{5.0 \text{ m} - 4.0 \text{ m}}{0.2 \text{ s}} = 5 \text{ m/s}$$

On the other hand, during this same time the displacement of point *A* changed from $y_{4.0} = 0.3$ cm to $y_{4.0} = 0.6$ cm, so the (average) up-and-down, or *transverse*, speed of this point during this time interval is

$$v = \frac{0.6 \text{ cm} - 0.3 \text{ cm}}{0.2 \text{ s}} = 1.5 \text{ cm/s}$$

The transverse speed v is related to the energy carried by the wave (Sec. 11.4), whereas the wave speed v is the speed with which this energy is transferred along the cord. For a given medium the wave speed is a constant, independent of the shape of the wave, whereas the transverse speed varies from wave to wave, and even from point to point and moment to moment within a given wave.

Example 11.1 (*a*) Draw the displacement of the cord 0.2 s after the cord is in the position shown in Fig. 11.3*c*. (*b*) What is the displacement of point *A* at this instant? (*c*) What is the (average) transverse speed of point *A* during this 0.2-s interval?

(*a*) Since the speed v of the wave is 5 m/s, in 0.2 s the wave moves the distance

FIGURE 11.2

A steady wave pattern moving along a cord.

FIGURE 11.3

Portion of a stretched cord: (*a*) equilibrium position of the cord, (*b*) displacement at $t = 0$ of the cord produced by a wave moving along it, and (*c*) displacement at $t = 0.2$ s of the cord.

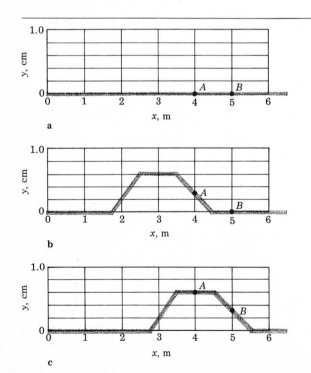

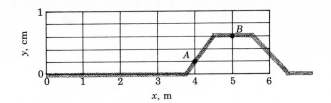

FIGURE 11.4

Displacement at $t = 0.4$ s of the
cord in Fig. 11.3.

$d = vt = (5\text{ m/s})(0.2\text{ s}) = 1\text{ m}$

along the x axis. This places the center of the peak at $x = 5$ m, as
shown in Fig. 11.4. The rest of the wave is drawn accordingly.

(b) From Fig. 11.4 we see that A now has the displacement
$y_{4.0} = 0.2$ cm.

(c) During this interval the position of A changes from
$y_{4.0} = 0.6$ cm to $y_{4.0} = 0.2$ cm, so its average speed is

$$v = \frac{0.2\text{ cm} - 0.6\text{ cm}}{0.2\text{ s}} = -2.0\text{ cm/s}$$

The minus sign indicates that A is moving in the negative y
direction during this time interval. $\square$

Definition A *transverse wave* is a wave in which the points of
the medium move perpendicular to the direction of propagation
of the wave. The wave on a cord is transverse, since the points of
the cord move perpendicular to the cord while the wave propagates along the cord.

Definition A *longitudinal wave* is a wave in which the points of
the medium move back and forth parallel to the direction of
propagation of the wave. A longitudinal wave can be established in a long spring by squeezing together a few of the coils at
one end and then releasing them. As these coils move back to
their equilibrium position, neighboring coils are compressed,
which in turn return to their equilibrium positions, compressing
coils still further along the spring. The result is a longitudinal
compression wave that moves along the spring with constant
speed.

Figure 11.5a shows a spring in its equilibrium position, and
Fig. 11.5b shows the spring when points A and B are squeezed
together. The displacements $y_{2.0}$ and $y_{4.0}$ of these points are
measured from their respective equilibrium positions. Even
though these points move parallel to the motion of the longitu-

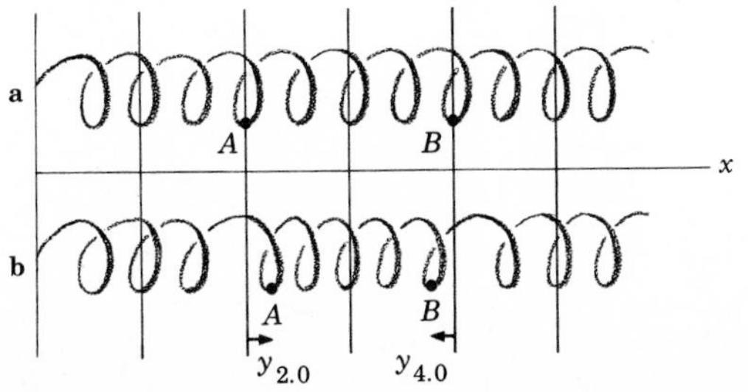

FIGURE 11.5

Portion of a spring: (a) equilibrium
configuration of the spring and
(b) longitudinal displacements of
points of the spring.

dinal wave, there is still no net flow of matter along the spring because the points move back and forth about their equilibrium positions only.

Both transverse and longitudinal waves can propagate through solid media. For example, after an earthquake (or a moonquake) both transverse and longitudinal *seismic* waves travel outward from the quake site through the earth's (or moon's) crust. These two types of waves have different speeds, which depend on the properties of the rocks through which they travel. Measurements of the speed of seismic waves provide valuable information about the nature of the rocks inside the earth and moon.

Transverse waves can also propagate along the surface of a liquid, but only longitudinal waves can propagate through the interior of a fluid, because a fluid cannot exert forces parallel to a surface (Sec. 7.1). This is discussed more fully in Sec. 12.1, where it is shown that sound is a longitudinal compression wave in air. Waves that are disturbances in material media, whether solid or fluid, are called *mechanical* waves, to distinguish them from electromagnetic waves, such as light.

11.2   The Physics of Mechanical Waves

The propagation of a mechanical wave through a medium can be understood by considering the forces that act on a small bit of the medium. When this bit is displaced slightly from its equilibrium position, it is pulled back to equilibrium by forces exerted on it by neighboring bits of matter. At the same time, by Newton's third law, the displaced matter exerts forces on its neigh-

FIGURE 11.6

Successive positions of a pulse traveling to the right along a cord. A small region *R* of the cord is marked so that its motion can be studied.

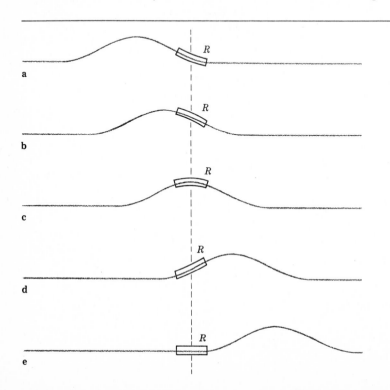

bors, so that as the displaced matter is pulled back to equilibrium, it pulls its neighbors away from equilibrium. These neighboring bits of matter are in turn acted upon by their neighbors, and so the process repeats, propagating the original disturbance through the medium.

> **REMARK** The displacements must be so small that the medium is not permanently deformed. For example, a piece of metal that is flexed slightly will spring back to its original shape. However, if it is flexed beyond a certain amount, it will be permanently deformed. Mechanical waves are limited to displacements that do not exceed the material's ability to spring back to its original shape.

Figure 11.6 shows the successive positions of a pulse traveling to the right along a long cord. A small region R of the cord is isolated so that the forces on it can be studied as the pulse passes by.

In Fig. 11.6a the pulse is just reaching R. Since R has not yet started to move, its vertical speed is zero. At the same time, the cord exerts the forces $\mathbf{F}_1$ and $\mathbf{F}_2$ on either end of R. The magnitudes of $\mathbf{F}_1$ and $\mathbf{F}_2$ are both T (the tension in the cord), but because the region R is curved, the sum $\mathbf{S} = \mathbf{F}_1 + \mathbf{F}_2$ of these forces is not zero. Figure 11.7a shows that since the angle θ is small, $\mathbf{S}$ is directed upward, perpendicular to the cord. Therefore, by Newton's second law, the region R has an upward acceleration $a = S/m$, where m is the mass of R. It is this acceleration that causes R to start to move upward as the pulse moves along the cord.

A short time later R has acquired a considerable upward speed and is in the position shown in Fig. 11.6b. But now the total force $\mathbf{S}$ on R is directed downward, as shown in Fig. 11.7b. This means that R will continue to move upward, but with decreasing speed.

When R reaches its maximum displacement (Fig. 11.6c), it again has zero speed. In this position the tension in the cord exerts a downward force $\mathbf{S}$ on R (Fig. 11.7c), and so R starts to move downward with increasing speed. In Fig. 11.6d R has a large downward speed, while Fig. 11.7d shows that the total force on it is now upward. Thus R's downward speed now decreases until R returns to its equilibrium position (Fig. 11.6e), where it has zero speed and zero acceleration (Fig. 11.7e).

This discussion shows that while the pulse as a whole moves to the right with constant speed v, each point on the cord moves with continually changing speed in the vertical (transverse) direction. Figure 11.8 shows the speed of different points of the pulse. The arrows point in the direction of motion, and their lengths are proportional to the speed at that point. The points in the leading half of the pulse (the right-hand side) are moving upward, the transverse speed of the peak is zero, and points in the trailing half of the pulse (the left-hand side) are moving downward.

This qualitative discussion indicates how the displacement itself generates the forces that keep it moving. A wave of any shape propagates in a similar fashion.

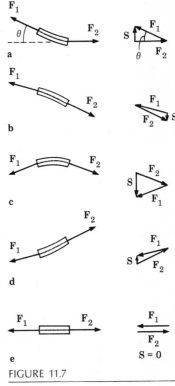

FIGURE 11.7

The forces $\mathbf{F}_1$ and $\mathbf{F}_2$ that act on the ends of the region R in Fig. 11.6, and the sum $\mathbf{S} = \mathbf{F}_1 + \mathbf{F}_2$.

FIGURE 11.8

Speed of different points of a pulse.

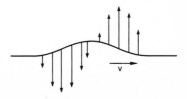

SPEED OF A WAVE ON A CORD

It can be shown that the speed v of a wave on a cord of mass m and length L is

$$v = \sqrt{\frac{T}{m/L}}$$

where T is the tension. The quantity m/L is the mass per unit length, or *linear density*, of the cord. Thus, if we let

$$\mu = \frac{m}{L} \qquad\qquad 11.1$$

the speed of the wave is

$$v = \sqrt{\frac{T}{\mu}} \qquad\qquad 11.2$$

This equation shows that the speed of a wave on a stretched cord increases with the square root of the tension and decreases with the square root of the mass per unit length.

Example 11.2 What is the speed of a wave on a guitar string which has a tension of 30 N and a mass per unit length of 0.015 kg/m?

From Eq. 11.2 the speed of transverse waves on this string is

$$v = \sqrt{\frac{T}{\mu}} = \sqrt{\frac{30\text{ N}}{0.015\text{ kg/m}}}$$
$$= \sqrt{2000\text{ N}\cdot\text{m/kg}} = \sqrt{2000\text{ m}^2/\text{s}^2} = 44.7\text{ m/s}$$

If the tension were quadrupled, the speed would be doubled, and if the mass per unit length were quadrupled, the speed would be halved. □

> **REMARK** It is not possible to give an elementary derivation of Eq. 11.2. However, in Appendix VII the speed of a pulse is estimated by considering the forces acting on it. This calculation is worth studying in order to appreciate how the principles of mechanics are applied to a mechanical wave.

11.3   Superposition

More than one wave can exist at the same point in a medium at the same time. This property of waves clearly distinguishes them from material objects. For example, the two pulses that are heading toward each other on the cord in Fig. 11.9a and b pass through each other and continue on their way undisturbed (Fig. 11.9c). Material objects do not interpenetrate this way. A wave, however, is not a material object. It is a self-propagating pattern of displacements of points in a material medium.

> **REMARK** A wave is a real physical entity, even though it is not a material object. It moves, carries energy, and interacts with material objects. In fact, waves and material objects are the two basic physical entities in terms of which the physical world is described.

What is the displacement of the cord while the two pulses in Fig. 11.9 are interpenetrating? This is answered by the following remarkable principle.

FIGURE 11.9

Two pulses moving on the same cord: (a) and (b) show successive positions of the pulses as they approach each other, and (c) shows the pulses after they have passed through each other.

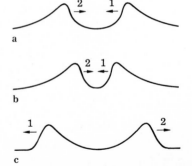

Superposition Principle   *If at any instant two or more waves simultaneously exist at a point, the displacement of the point is the sum of the displacements the point would have with each wave separately.* In applying this principle, displacements to one side of the equilibrium position are taken to be positive, and displacements to the other side are taken to be negative.

Figure 11.10 shows how this principle applies to the pulses in Fig. 11.9 as they are interpenetrating. In each part of the figure, the light lines represent the displacements the cord would have with each pulse separately. Pulse 1 moves steadily from right to left, while pulse 2 moves steadily from left to right. The actual displacements of the cord at each instant are represented by the dark lines. These displacements are found by adding the displacements produced by each pulse separately, in accordance with the superposition principle.

For instance, consider the point A on the cord. In Fig. 11.10a the peak of pulse 2 and the leading edge of pulse 1 are simultaneously at point A. Since pulse 1 alone would have displaced A by 5 cm and pulse 2 alone would have displaced it by 10 cm, the actual displacement of point A at this instant is 15 cm, as shown. The rest of the black curve in Fig. 11.10a is calculated similarly.

FIGURE 11.10

The superposition of the two pulses in Fig. 11.9 as they pass through each other.

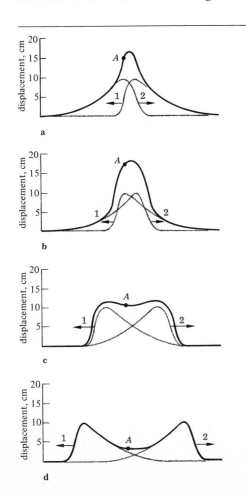

In Fig. 11.10*b* the peak of pulse 1 and the trailing side of pulse 2 are simultaneously at *A*. Again the actual displacement of *A* is the sum of the displacements produced by each of these pulses separately. All the points on the black curves are calculated in the same way. These dark curves represent the actual displacement of the cord, which can be quite complicated, even when it is just the sum of two simple pulses.

Interpenetration of a positive and a negative pulse is shown in Fig. 11.11. Figure 11.11c shows that at the instant when these pulses completely overlap, the cord has zero displacement everywhere. This is because, by the superposition principle, the sum of a positive and a negative displacement of equal magnitude is zero. (Study Figs. 11.10 and 11.11 carefully until you are sure you understand how the dark waves are constructed from the two light pulses.)

It is surprising that the cord can evolve from the apparent equilibrium configuration shown in Fig. 11.11c into the two traveling pulses shown in Figs. 11.10*d* and *e*. The point is that the cord is not really in equilibrium in Fig. 11.11c, because the points in the region of superposition are not at rest. Figure 11.12 shows that the points on the leading side of the positive pulse are moving away from equilibrium, while the points on the trailing side are moving toward equilibrium; the points of the negative pulse behave similarly. Furthermore, the superposition principle applies to speed as well as to displacement. That is, the actual speed of each point of the cord is the sum of the speeds it would have with each pulse separately. Thus when the two pulses overlap, they combine momentarily to give a wave with zero displacement for each point but not zero speed. Each point in the wave is moving through its equilibrium position with the sum of the speeds it would have if each pulse were present alone (Fig. 11.12).

The superposition principle makes it possible to express a complex wave as the sum of several simpler waves. It is, in fact, possible to express any wave, no matter how complex, as the sum of waves of an especially simple form, called *sine waves.* These are discussed in the next section.

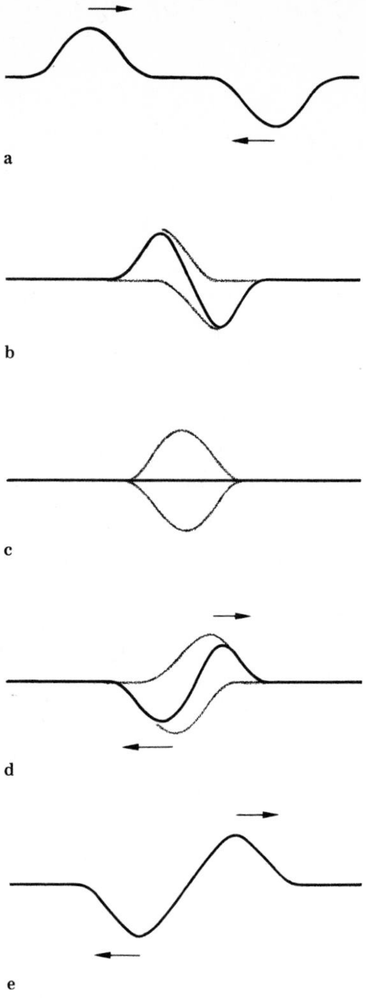

11.4 Sine Waves
WAVELENGTH AND FREQUENCY

A *sine wave* is a particular wave pattern in which positive and negative pulses alternate successively (Fig. 11.13).

Definition The *wavelength* λ of a sine wave is the distance between any two neighboring positive peaks.* For a given sine wave any two neighboring positive peaks have the same separation λ, and any two points in the wave that are one wavelength apart (such as *a* and *b* in Fig. 11.13) have the same displacement. This kind of wave pattern is said to be *periodic* because the pattern repeats itself exactly at intervals of λ.

*λ is the Greek (lowercase) letter lambda.

a

b

c

d

e

FIGURE 11.11

Superposition of a positive and a negative pulse as they pass through each other.

FIGURE 11.12

Superposition of the transverse speeds of the two pulses in Fig. 11.11c.

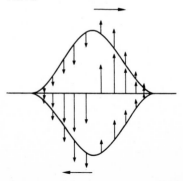

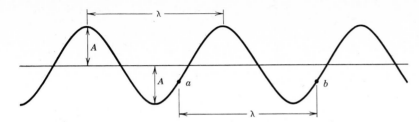

FIGURE 11.13

A sine wave of amplitude A and wavelength λ.

Definition The *amplitude A* of a sine wave is the maximum displacement of a wave. It is the same for both positive and negative displacements.

The shape of a sine wave is related to the trigonometric sine. To understand this, consider one wavelength of a sine wave of amplitude A, as shown in Fig. 11.14. The interval between $x = 0$ and $x = \lambda$ has been divided into 360 equal divisions, corresponding to the 360° in a circle. Therefore, any point x on the x axis can be related to an angle θ by the proportion

$$\frac{x}{\lambda} = \frac{\theta}{360°} \quad \text{or} \quad \theta = \frac{x}{\lambda} 360°$$

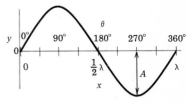

FIGURE 11.14

One wavelength of a sine wave divided into 360 degrees.

The displacement y of a sine wave at point x is given in terms of the sine of θ by

$$y = A \sin \theta = A \sin \left(\frac{x}{\lambda} 360°\right) \qquad 11.3a$$

Alternatively, if angles are measured in radians rather than degrees, the displacement is written

$$y = A \sin \left(2\pi \frac{x}{\lambda}\right) \qquad 11.3b$$

since 2π rad equals 360° (Appendix IV).

Example 11.3 A sine wave has an amplitude $A = 0.5$ cm and a wavelength $\lambda = 30$ cm. What is its displacement at $x = 6$ cm?
From Eq. 11.3a the displacement is

$$y = A \sin \left(\frac{x}{\lambda} 360°\right) = (0.5 \text{ cm}) \sin \left(\frac{6 \text{ cm}}{30 \text{ cm}} 360°\right)$$

$$= (0.5 \text{ cm})(\sin 72°) = (0.5 \text{ cm})(0.951) = 0.475 \text{ cm} \qquad \square$$

> **REMARK** It is natural to associate the points in one wavelength of a sine wave with points on a circle because a sine wave repeats after one wavelength just as a circle repeats after 360° or 2π rad.

Figure 11.13 shows a sine wave at one particular instant of time. If the wave is traveling to the right with speed v, the whole pattern moves with time, as shown in Fig. 11.15. The open circles in this figure show how a characteristic feature of the wave moves, whereas the solid circles show how a fixed material point in the medium moves. The peak of the wave moves to the right with constant speed v, whereas a point in the medium moves down and up with simple harmonic motion.

After the wave has moved a distance λ, the pattern looks the same again. In the time $\tau = \lambda/v$ that the wave takes to move the

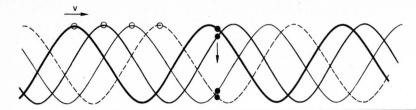

FIGURE 11.15

A sine wave at four successive instants of time. The open circles mark a characteristic feature of the wave (a peak), whereas the solid circles mark a physical point in the medium.

distance λ, a fixed point in the medium returns to its initial position, completing one *cycle* of motion.

Definition The *period τ* of a sine wave is the time required for a point in the medium to complete one cycle. It is related to the speed and wavelength of the wave by

$$\tau = \frac{\lambda}{v} \qquad \text{or} \qquad v = \frac{\lambda}{\tau} \qquad\qquad 11.4$$

This relation between the speed, wavelength, and period of a sine wave is of fundamental importance in the study of wave motion.

The period τ is also the time it takes for one whole wavelength of a wave to pass a given point in space. If $\tau = 0.25$ s, 1 wavelength passes a given point every 0.25 s, so that $1/0.25 = 1/\tau = 4$ wavelengths pass each second.

Definition The *frequency f* of a sine wave is

$$f = \frac{1}{\tau} \qquad\qquad 11.5$$

and is equal to the number of wavelengths that pass a point in 1 s.

> **UNITS** The unit of frequency is $1/s = s^{-1}$, called the *hertz* (Hz). The frequency is also the number of cycles that a point in the medium executes in 1 s, so *f* is often quoted as so many *cycles per second* (c/s). Thus the frequency of a wave with the period $\tau = 0.25$ s is
>
> $$f = \frac{1}{\tau} = 4\,s^{-1} = 4\,Hz = 4\,c/s$$

An alternative form of Eq. 11.4 is obtained by combining it with Eq. 11.5 to get

$$v = \lambda f \qquad\qquad 11.6$$

Only Eqs. 11.5 and 11.6 need to be remembered because Eq. 11.4 is easily derived from them.

For a particular wave in a particular medium, the wave speed *v* is a constant. For example, the speed of a wave on a cord with a certain tension and linear mass density is independent of the wavelength or frequency of the wave. The same is true for the speed of sound in air and the speed of light in vacuum,* although the speeds of these different kinds of waves are different

*The speed of light in materials such as glass and water does depend slightly on its wavelength, but this can be neglected for most purposes.

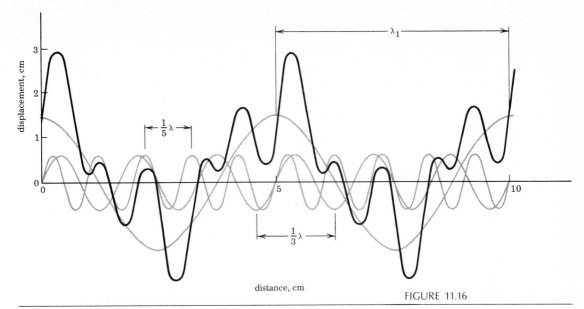

distance, cm

FIGURE 11.16

A complex periodic wave produced by the superposition of three sine waves of different amplitude and wavelength.

from each other. For each kind of wave in a given medium, Eq. 11.6 states the relation between wavelength and frequency of sine waves in that medium.

Frequency is important because the tone or pitch of a sound depends on the frequency of the sound wave, and the color of an object depends on the frequency of the light wave reflected from the object. For example, the musical tone A is a sine wave (of sound) with a frequency of about 440 Hz, and the blue color in a rainbow is a sine wave (of light) with a frequency of about 7.5×10^{14} Hz.

FOURIER'S THEOREM

The connection between the frequency of sine waves and our perception of tone and color is certainly remarkable. However, sine waves have an even more remarkable mathematical property that jusitifies giving them special attention.

Fourier's Theorem *Any wave, of whatever shape, can be uniquely expressed as the superposition (sum) of sine waves of definite wavelengths and amplitudes.*

For example, according to this theorem, the rather complex wave pattern (dark curve) in Fig. 11.16 is equal to the superposition of sine waves. These sine waves, which are called the *Fourier components* of the complex pattern, are shown as gray curves in Fig. 11.16. These components are unique; i.e., only sine waves with these wavelengths and amplitudes combine to equal the given pattern. Any pattern, no matter how complex, is similarly equal to a sum of sine waves, though many sine waves may have to be superimposed to equal a very complex pattern.

The longest-wavelength Fourier component in Fig. 11.16 has a wavelength $\lambda_1 = 5$ cm and an amplitude of 1.5 cm. If the wave speed is 12 m/s, the frequency of this component is

$$f_1 = \frac{v}{\lambda_1} = \frac{12 \text{ m/s}}{0.05 \text{ m}} = 240 \text{ s}^{-1} = 240 \text{ Hz}$$

This component is represented on a graph of amplitude against frequency (Fig. 11.17) by a vertical line 1.5 cm high, positioned at 240 Hz on the horizontal axis. The wavelength of the next-longest wavelength component is

$$\lambda_2 = \tfrac{1}{3}\lambda_1 = 1.67 \text{ cm}$$

and the frequency of this component is

$$f = \frac{12 \text{ m/s}}{0.0167 \text{ m}} = 720 \text{ s}^{-1} = 720 \text{ Hz}$$

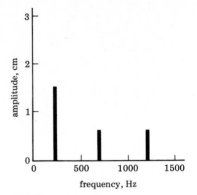

FIGURE 11.17

Spectrum of the complex wave in Fig. 11.16.

This component has an amplitude of 0.6 cm, so it is represented by a vertical line 0.6 cm high, positioned at 720 Hz on the horizontal axis. Likewise, the third component is represented by a line 0.6 cm high, positioned at 1200 Hz.

A representation of the amplitudes and frequencies of the components of a particular wave, such as Fig. 11.17, is called a *spectrum*. It is a convenient way to specify a wave uniquely. Figure 11.18 shows several complex waves with their corresponding spectra.

Complex wave patterns are either periodic or aperiodic. A *periodic wave,* such as a single sine wave, repeats its pattern exactly at intervals of length λ_1. An *aperiodic wave* never repeats its pattern. The wave in Fig. 11.18c is aperiodic, and the waves in Figs. 11.16, 11.18a, and 11.18b are periodic. The spectrum of an aperiodic wave contains a continuous distribution of component frequencies, whereas the spectrum of a periodic wave is discrete.

The spectrum of a periodic wave is discrete because in order for a wave to be periodic with an interval length λ_1, the wavelengths of its Fourier components must be members of the sequence

$$\lambda_1, \tfrac{1}{2}\lambda_1, \tfrac{1}{3}\lambda_1, \ldots, \tfrac{1}{n}\lambda_1, \ldots$$

where n is an integer ($n = 1, 2, 3, \ldots$). For example, the components of the wave in Fig. 11.16 are the first, third, and fifth members of such a sequence. In the interval λ_1 there is exactly 1 wavelength of the component equal to λ_1, exactly 3 wavelengths of the component equal to $\tfrac{1}{3}\lambda_1$, and exactly 5 wavelengths of the component equal to $\tfrac{1}{5}\lambda_1$. Thus, since all the component waves are the same at the end of the interval as they are at the beginning, their sum repeats after this interval. Clearly there are exactly n wavelengths of the component equal to $(1/n)\lambda_1$, so that a sum of such components repeats in the interval λ_1.

Let f_1 be frequency of component with wavelength λ_1:

$$f_1 = \frac{v}{\lambda_1}$$

The frequencies of the components of a periodic wave then must be members of the sequence

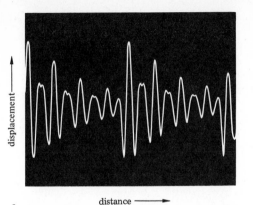

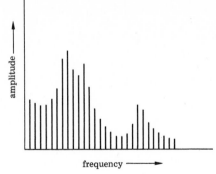

a

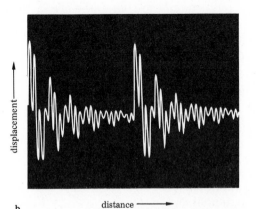

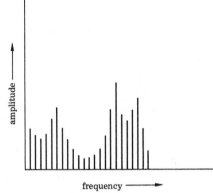

b

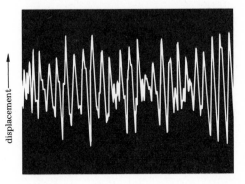

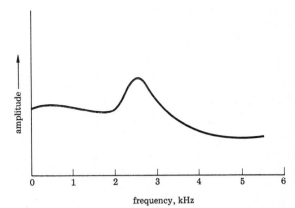

c

$$f_1 = \frac{v}{\lambda_1}$$

$$f_2 = \frac{v}{\frac{1}{2}\lambda_1} = \frac{2v}{\lambda_1} = 2f_1$$

..................................⋮

$$f_n = \frac{v}{(1/n)\lambda_1} = \frac{nv}{\lambda_1} = nf_1$$

FIGURE 11.18

Some complex waves and their spectra: (a) and (b) periodic waves, (c) aperiodic wave. (*After P. B. Denes and E. N. Pinson,* The Speech Chain, *Bell Telephone Laboratories, Inc., Copyright © 1963, used with permission.*)

or

$f_1, 2f_1, 3f_1, \ldots, nf_1, \ldots$

Such a sequence of evenly spaced frequencies is called a *harmonic sequence*. The spectrum of a periodic wave consists only of members of a harmonic sequence, although not all the members of the sequence have to be present, not even f_1. For instance, a wave whose three Fourier components have the frequencies 120, 180, and 300 Hz is periodic because each of these frequencies is equal to an integer times 60 Hz. Such a wave repeats in an interval of length

$$\lambda_1 = \frac{v}{f_1} = \frac{v}{60 \text{ Hz}}$$

even though f_1 is not one of its components.

ENERGY IN A SINE WAVE

When a sine wave of frequency f and amplitude A moves through a medium, each point of the medium moves back and forth with simple harmonic motion (Sec. 6.4). The energy E of a point mass m is (*Eq. 6.18*)

$$E = \tfrac{1}{2}mv^2 + \tfrac{1}{2}ky^2$$

where v is the speed of the point when it is a distance y from its equilibrium position. The total energy E is the sum of the instantaneous kinetic energy $\tfrac{1}{2}mv^2$ and the instantaneous potential energy $\tfrac{1}{2}ky^2$ of the point. Although the kinetic and potential energies of the point continually change as the point moves back and forth about its equilibrium position, the total energy E remains constant. When the point has its maximum displacement ($y = A$), the kinetic energy is zero, so that the total energy is equal to the potential energy at this instant:

$$E = \tfrac{1}{2}kA^2$$

The constant k is a measure of the strength of the force which the medium exerts on a displaced point (Eq. 6.12). It is related to the period τ of oscillation of the point by Eq. 6.14:

$$\tau = 2\pi\sqrt{\frac{m}{k}} \quad \text{or} \quad k = \frac{4\pi^2 m}{\tau^2} = 4\pi^2 f^2 m$$

Thus the energy E of a point in the medium is

$$E = \tfrac{1}{2}kA^2 = 2\pi^2 mf^2 A^2$$

On a cord with linear mass density μ, the mass of a one-wavelength-long section of cord is $m = \lambda\mu$, so the energy of this section is

$$E = 2\pi^2 \mu\lambda f^2 A^2$$
$$= 2\pi^2 \mu v f A^2 \qquad\qquad 11.7$$

where $v = \lambda f$ is the speed of the wave on the cord.

The energy in the wave is passed from point to point as the wave moves through the medium. The number of wavelengths of the wave that pass a given point in 1 s is equal to the frequency f of the wave, so the energy per second P that passes this point is

$$P = fE = 2\pi^2 \mu v f^2 A^2 \qquad\qquad 11.8$$

The symbol P is used for this because it has the same dimension as power [energy/time]. The importance of this result is that it shows that a sine wave traveling in a given direction carries energy and that the rate P of energy transport is proportional to the square of the amplitude A of the wave.

11.5 Standing Waves

A wave pattern of particular interest is formed by the superposition of two sine waves with the same wavelength and amplitude traveling in opposite directions. At the instant ($t = 0$) when two such sine waves are in the positions shown in Fig. 11.19a, their superposition (dark curve) results in zero displacement of the medium everywhere. A quarter of a period later ($t = \frac{1}{4}\tau$) one sine wave has moved one-fourth wavelength to the right while the other has moved one-fourth wavelength to the left. This causes the two waves to coincide, as shown in Fig. 11.19c, so that their superposition is a sine wave of amplitude 2A.

Figure 11.19b shows that the superposition of the waves at an intermediate time ($t = \frac{1}{8}\tau$) is also a sine wave, but with an amplitude less than 2A. Furthermore, the positions where the resulting displacement of the medium is zero are the same at both times. Figure 11.19d and e shows that the situation is similar at $t = \frac{1}{2}\tau$ and $t = \frac{3}{4}\tau$. Thus, as the two sine waves move through each other, their superposition remains a sine wave with changing amplitude and fixed points where the displacement is always zero.

The resulting displacement of the medium at various times is shown in Fig. 11.20. This kind of wave pattern is called a *standing wave* because although the pattern changes with time, it does not move along the cord. The points of zero displacement, or *nodes*, of the pattern are fixed points that are evenly spaced at intervals of one-half wavelength. A point of the medium located at a node has no vertical oscillation, while a point midway between two nodes (called an *antinode*) executes simple harmonic motion with amplitude 2A. A point between a node and an antinode has an amplitude between 0 and 2A.

All points in a standing wave complete one oscillation in the time τ equal to the period of the sine waves of which it is composed. This is seen in Fig. 11.19, where the point O in the medium is shown going through one oscillation in the time τ. (The pattern at time $t = \tau$ is not shown because it is identical to the pattern at time $t = 0$.)

Standing waves are important because they can be confined in a limited space. They are thus the natural waves to consider when studying the wave motion of a finite medium, e.g., a guitar string, which is just a cord of length L fixed at both ends. A

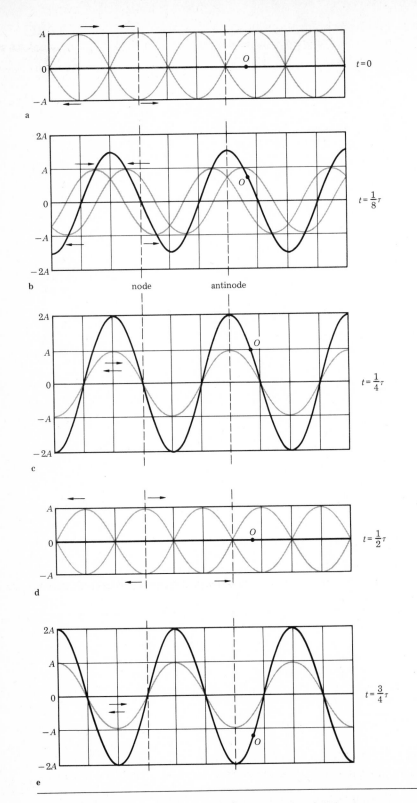

FIGURE 11.19

Superposition of two sine waves of
equal amplitude and wavelength
traveling in opposite directions.

single sine wave moving in one direction cannot exist on the cord, because such a wave requires the end points to oscillate. However a standing wave can exist on the cord if the ends of the cord coincide with nodes of the wave. Figure 11.21 shows several possible standing waves that can exist on a cord of length L.

The condition that the end points be at nodes puts a severe restriction on the wavelengths of the possible standing waves. Since the nodes occur at intervals of $\frac{1}{2}\lambda$, and since there must be exactly a whole number of these half-wavelengths on the cord, the wavelength of the standing wave is related to the length L of the cord by

$$n\tfrac{1}{2}\lambda = L$$

where n is an integer. Thus only wavelengths that satisfy the condition

$$\lambda_n = \frac{2L}{n} \qquad n = 1, 2, 3, \ldots \tag{11.9}$$

can exist as standing waves on the cord. Figure 11.21 shows standing waves with n equal to 1, 2, and 5.

The frequency of oscillation of the cord is related to the wavelength and the speed of the wave by Eq. 11.6. Thus a cord fixed at both ends can oscillate only with the frequencies f_n given by

$$f_n = \frac{v}{\lambda_n} = \frac{nv}{2L} \tag{11.10}$$

The lowest frequency, which occurs when $n = 1$, is

$$f_1 = \frac{v}{2L}$$

It is called the *fundamental* and corresponds to $L = \frac{1}{2}\lambda$ (Fig. 11.21a). Equation 11.10 shows that all the other frequencies are integer multiples of f_1:

$$f_n = nf_1 \tag{11.11}$$

The possible frequencies of an oscillating cord thus form a harmonic sequence.

Example 11.4 A 120-m-long high-voltage transmission cable is suspended between two towers. The linear mass density of the cable is 1.6 kg/m, and the tension in the cable is 3.6×10^4 N. (a) What is the lowest frequency with which this cable will oscillate? (b) What is the frequency of the standing wave on this cable when the nodes are 1.50 m apart?

(a) From Eq. 11.2 the speed of a transverse wave on the cable is

$$v = \sqrt{\frac{T}{\mu}} = \sqrt{\frac{2.6 \times 10^4 \text{ N}}{1.6 \text{ kg/m}}} = 150 \text{ m/s}$$

The lowest frequency oscillation is the fundamental (Fig. 11.21a), which from Eq. 11.9 has the wavelength

$$\lambda_1 = 2L = 2(120 \text{ m}) = 240 \text{ m}$$

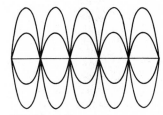

FIGURE 11.20

Displacements of a standing wave at several instants of time.

FIGURE 11.21

Some possible standing waves that can exist on a cord of length L held fixed at both ends.

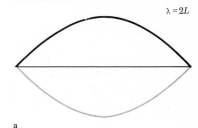

$\lambda = 2L$

a

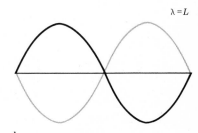

$\lambda = L$

b

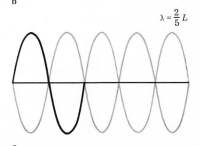

$\lambda = \frac{2}{5}L$

c

and the frequency

$$f_n = \frac{v}{\lambda_n} = \frac{150 \text{ m/s}}{3.0 \text{ m}} = 50.0 \text{ Hz}$$

(b) The nodes of a standing wave are one-half wavelength apart, so the wavelength λ_n of the standing waves with nodes 1.50 m apart is

$$\lambda_n = 2(1.50 \text{ m}) = 3.00 \text{ m}$$

and the corresponding frequency is

$$f_n = \frac{v}{\lambda_n} = \frac{150 \text{ m/s}}{3.0 \text{ m}} = 50.0 \text{ Hz}$$

From Eq. 11.11, f_n is n times the fundamental, where

$$n = \frac{f_n}{f_1} = \frac{50.0 \text{ Hz}}{0.625 \text{ Hz}} = 80$$

is an interger. □

STRINGED INSTRUMENTS

Instruments such as the guitar, violin, and piano use vibrating strings to produce the musical notes. Since each string is fixed at both ends, it can oscillate only at one or more of the harmonic frequencies given by Eq. 11.10. For example, when a guitar string is plucked, the resulting oscillation is the superposition of many different standing waves, with the frequency of each standing-wave component satisfying Eq. 11.10. Usually the fundamental frequency dominates; i.e., it has an amplitude much larger than the other components. The oscillating string causes the air to oscillate, which generates a sound wave with the same frequencies. The ear hears these waves as a musical tone of frequency f_1. The other frequencies (called *overtones*) contribute to the quality of the tone.

All stringed musical instruments produce tones in this way. The differences in the quality of the sound produced by different instruments when playing the same note, i.e., oscillating with the same fundamental frequency, is caused by the differences in the amplitudes of the overtones. Figure 11.22 shows the spectrum of sound produced by a violin and a piano playing the same note ($f_1 = 440$ Hz). Each instrument produces the same sequence of harmonic frequencies but with different amplitudes.

A stringed instrument is tuned by adjusting the tension in its strings. From Eq. 11.2 this is seen to change the speed v of waves on the string, and so, from Eq. 11.10, this changes the oscillation frequencies of the string as well. Once in tune, a violin or guitar is made to play different notes by pressing its strings at different points, thus changing the length L that is free to oscillate. Other instruments, such as a harp and a piano, have a large number of strings of different lengths and densities, each of which is adjusted to produce a different note.

Example 11.5 A guitar string 60 cm long has a linear mass den-

FIGURE 11.22

Spectra of the musical note A played by a piano and a violin.

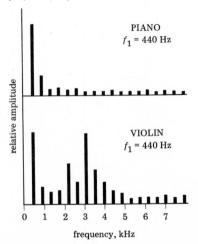

sity of 8 g/m. (a) What tension is required to produce a fundamental frequency of 120 Hz? (b) Where should the string be
pressed to produce a fundamental of 160 Hz?

(a) The wavelength of the fundamental is

$$\lambda_1 = 2L = 2(0.60 \text{ m}) = 1.2 \text{ m}$$

so the wave speed must be

$$v = \lambda_1 f_1 = (1.2 \text{ m})(120 \text{ Hz}) = 144 \text{ m/s}$$

But from Eq. 11.2

$$v = \sqrt{\frac{T}{\mu}}$$

so

$$T = \mu v^2 = (0.008 \text{ kg/m})(144 \text{ m/s})^2 = 166 \text{ N}$$

(b) With $v = 144$ m/s and $f_1 = 160$ Hz, the wavelength of the
fundamental now is

$$\lambda_1 = \frac{v}{f_1} = \frac{144 \text{ m/s}}{160 \text{ Hz}} = 0.9 \text{ m}$$

so the length of the vibrating portion of the string is

$$L = \tfrac{1}{2}\lambda_1 = 0.45 \text{ m}$$

Thus the string must be pressed at a point 0.15 m from one end,
leaving 0.45 free to vibrate. ☐

11.6 Resonance

Definition The frequencies with which a system can vibrate in
the absence of an external driving force are called the *normal,* or
characteristic, frequencies of the system. For example, when a
guitar string is plucked, it is given a momentary displacement
from its equilibrium position and then allowed to vibrate freely
about this position. The string will vibrate with one or more of
its normal frequencies, which are given by Eq. 11.10. A simple
pendulum, on the other hand, has only one period (Eq. 6.19),
and hence only one normal frequency. When a pendulum is
displaced from equilibrium and released, it vibrates freely with
this one frequency.

What happens when we try to force a system to vibrate at a
frequency other than a normal frequency? To be specific, consider a pendulum with a normal frequency f_{nor}. What happens
when we repeatedly push back and forth on the pendulum bob
with a driving frequency f_d? The bob is now forced to vibrate
with the frequency f_d rather than with its normal frequency f_{nor},
and as a consequence the amplitude A of the motion will generally be small. The actual value of A depends on the magnitude
of the driving force and on the ratio f_d/f_{nor} of the driving frequency to the normal frequency.

Definition The value of the driving frequency f_d that produces
the maximum amplitude of vibration is called the *resonant
frequency.* A system driven at its resonant frequency is said to be

in *resonance*. If there is little or no friction in the system, the resonant frequency is nearly equal to the normal frequency,

$$f_{res} \approx f_{nor} \qquad 11.12$$

and the amplitude will be very large. In fact, in the ideal case of zero friction, the amplitude at resonance becomes infinite. What this means in practice is that if there is not enough friction in the system, the amplitude at resonance will exceed the elastic limit of the vibrating material, causing permanent damage.

The magnitude of a driving force can be characterized by the maximum displacement A_0 it produces on the system when the driving frequency is zero. The actual amplitude A of the system when it is driven at a frequency f_d is proportional to A_0. Thus we can write

$$A = rA_0 \qquad \text{or} \qquad r = \frac{A}{A_0}$$

where r is the *response factor* of the system. Near resonance, r is large, so a small driving force can produce a large amplitude.

The response factor is a function of the ratio f_d/f_{nor}, as shown in Figure 11.23. This figure gives curves of r against f_d/f_{nor} for different amounts of friction in the system. These curves peak near $f_d/f_{nor} = 1$, i.e., when the driving frequency f_d is nearly equal to the normal frequency (Eq. 11.12). The quantity* Γ is the *width* of a response curve at $1/\sqrt{2}$ ($= 0.707$) of its peak value. That is, r is within 70 percent of its peak value when f_d is between $f_{nor} - \frac{1}{2}\Gamma$ and $f_{nor} + \frac{1}{2}\Gamma$.

The width Γ (measured in hertz) depends on the amount of friction in the system. A system with zero friction has Γ equal to zero, and Γ increases as the amount of friction increases. The peak of the response factor is exactly at $f_d/f_{nor} = 1$ when $\Gamma = 0$, and it is displaced slightly from this value as Γ increases. Furthermore, the peak value of r is equal to f_{nor}/Γ, so that the peak increases as the width Γ decreases. These facts are all displayed in Fig. 11.12. The important point to remember is that a system with little friction has a tall, narrow response curve and a system with much friction has a short, broad response curve.

CONTROL OF VIBRATIONS IN TRANSMISSION CABLES

An electric transmission cable suspended between two towers has the same normal frequencies as a string fixed at both ends (Example 11.4). When a steady wind blows across such a cable, little air vortices are formed which detach themselves alternately from the top and bottom of the cable. Each detachment produces a minute vertical force on the cable, alternately directed up and down. Thus the cable is subjected to a small driving force. The frequency f_d of this force is related to the wind velocity v (in meters per second) and the radius r of the cable (in meters) by the equation

$$f_d = c\frac{v}{r} \qquad 11.13$$

*Γ is the Greek (capital) letter gamma.

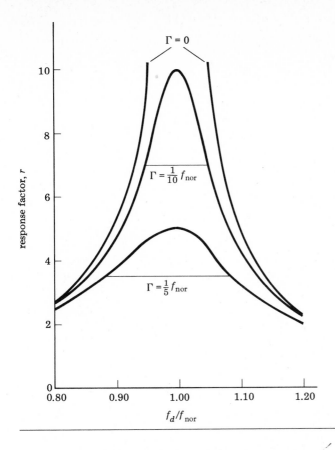

FIGURE 11.23

Plot of the response factor r
against f_d/f_{nor} for different values
of the width Γ.

where $c = 0.0923 \text{ s}^{-2}$. The constant c has been determined by experimental studies on real cables.

Example 11.6 The cable in Example 11.4 has a diameter of 2.8 cm. (a) What is the frequency of the driving force exerted on this cable by a 25-km/h wind? (b) What wind speed would cause the $n = 80$ normal frequency of the cable to resonate?
(a) The wind speed is

$$v = 25 \text{ km/h} = (25)\frac{1000 \text{ m}}{3600 \text{ s}} = 6.94 \text{ m/s}$$

and the cable radius is

$$r = \tfrac{1}{2}d = (0.5)(0.028 \text{ m}) = 0.014 \text{ m}$$

so from Eq. 11.13 the frequency of the driving force is

$$f_d = c\frac{v}{r} = (0.0923 \text{ s}^{-2})\frac{6.94 \text{ m/s}}{0.014 \text{ m}}$$

$$= 45.8 \text{ Hz}$$

(b) From Example 11.4 we know that the $n = 80$ normal frequency of the cable is $f_{80} = 50.0$ Hz. Resonance occurs when the driving frequency f_d is equal to the normal frequency. Thus the wind velocity v at which this occurs is given by

$$f_d = c\frac{v}{r} = f_{80} = 50.0\,\text{Hz}$$

or

$$v = (50.0\,\text{Hz})\frac{r}{c} = (50.0\,\text{Hz})\frac{0.014\,\text{m}}{0.0923\,\text{s}^{-2}}$$

$$= 7.58\,\text{m/s} = 27.3\,\text{km/h}$$

If the wind velocity stayed steady at exactly this value long enough, a standing wave of very large amplitude would be established on the cable. For an unprotected cable, the amplitude of the standing wave can become large enough to damage and even rupture the cable. ☐

FIGURE 11.24

Vibration damper attached to a power cable. (*Alcoa Conductor Products Co.*)

The magnitude of the amplitude of a vibrating cable at resonance depends on the amount of friction in the cable. Most large overhead cables consist of 20 or 30 strands of aluminum and steel wire twisted together. As the cable vibrates, these strands rub against each other, providing some internal friction. When this is not sufficient to protect the cable, a device called a *vibration damper* is clamped to one end of the cable (Fig. 11.24). This is just a pair of weights connected to each other by a flexible rod.

To understand how the damper increases the frictional force on the cable, we must remember that a standing wave is actually the superposition of two sine waves traveling in opposite directions. As one of these waves reaches an end of the cable that is held fixed, it is reflected and travels back in the direction from which it came. When a damper is fixed near the end of a cable, the wave traveling toward that end causes the damper to vibrate. Some of the energy of the wave goes into vibrational energy of the damper, and the wave that passes on to the end of the cable is reduced in amplitude. The damper in turn loses its vibrational energy through air friction. If the damper absorbs the energy of a wave as fast as the driving force adds energy to the wave, the amplitude of the standing wave will not become dangerously large. In effect, the peak value of the cable's response factor has been decreased by increasing the frictional force on the cable.

GUIDE TO MAJOR TOPICS

Topic	Reference	Problems
Graphical description of wave motion	Sec. 11.1; Example 11.1	1 to 4
Speed of a wave	Sec. 11.2; Eqs. 11.1 and 11.2; Example 11.2	5 to 8
Superposition principle	Sec. 11.3	9 and 10, 30, 34
Sine waves	Sec. 11.4; Eqs. 11.3 to 11.6; Example 11.3	11 to 20
Fourier's theorem		21 to 24
Energy in a sine wave	Eqs. 11.7 and 11.8	25 to 28, 36 to 38
Standing waves	Sec. 11.5; Eqs. 11.9 to 11.11; Examples 11.4 and 11.5	29 to 38
Resonance	Sec. 11.6; Eqs. 11.12 and 11.13; Example 11.6; Fig. 11.23	39 to 42

PROBLEMS

1 Figure 11.25 shows a pulse on a cord at times $t = 0$ and $t = 0.2$ s. (Note that the vertical and horizontal scales are different.) (a) What is the speed of the pulse? (b) What is the vertical speed of point A on the cord during this time?
Ans. (a) 1.5 m/s; (b) 30 cm/s

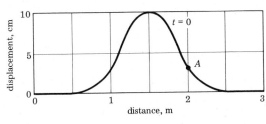

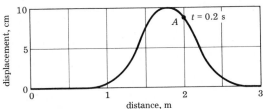

FIGURE 11.25

Problems 1 and 3.

2 Figure 11.26 shows a pulse on a cord at time $t = 0$. The pulse is moving to the right with a speed of 2.5 m/s. (a) Draw the shape of the cord at the time $t = 0.6$ s. (b) What is the vertical distance that point A moves between $t = 0$ and $t = 0.6$ s? (c) What is the (average) vertical speed of point A during this time?

3 (a) What is the position of the peak of the pulse in Fig. 11.25 at time $t = 3.0$ s (see Prob. 1)? (b) At what time will the peak of the pulse arrive at the 4-m position?
Ans. (a) 6.0 m; (b) 1.66 s

4 Figure 11.27 shows a complex wave moving

FIGURE 11.26

Problem 2.

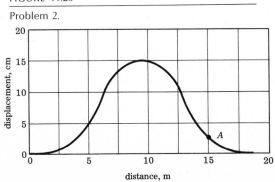

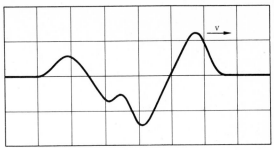

FIGURE 11.27

Problem 4.

to the right along a cord. Draw the shape of the cord an instant later, and determine which parts of the cord are moving upward and which parts are moving downward.

5 A metal guitar string has a linear density $\mu = 3.2 \times 10^{-3}$ kg/m. What is the speed of transverse waves on this string when its tension is 90 N?
Ans. 168 m/s

6 The speed of transverse waves on a cord is 200 m/s. If the linear density of the cord is 7×10^{-3} kg/m, what is the tension of the cord?

7 When the tension in a cord is 75 N, the wave speed is 140 m/s. What is the linear density of the cord?
Ans. 3.83×10^{-3} kg/m

8 When the tension in a cord is 100 N, the wave speed is 120 m/s. What is the wave speed when the tension is 200 N?

9 Figure 11.28 shows two pulses on a cord at time $t = 0$. The pulses are moving toward each other, and the speed of each pulse is 40 cm/s. Sketch the shape of the cord at 0.15, 0.25, and 0.30 s.

10 Figure 11.29 shows two pulses on a cord at time $t = 0$. The pulses are moving toward each other, and the speed of each pulse is 2.5 m/s.

FIGURE 11.28

Problem 9.

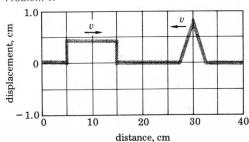

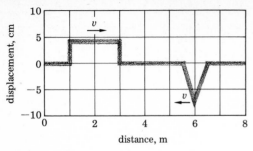

FIGURE 11.29

Problem 10.

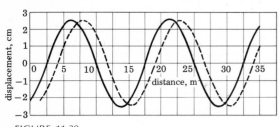

FIGURE 11.30

Problem 11.

Sketch the shape of the cord at 0.6, 0.8, and 0.9 s.

11   Figure 11.30 shows a sine wave traveling to the right on a cord. The solid curve represents the shape of the cord at time $t = 0$, and the dotted curve represents the shape of the cord at time $t = 0.12$ s. (Note that the horizontal and vertical scales are different.) Find (a) the amplitude, (b) the wavelength, (c) the speed, (d) the frequency, and (e) the period of the wave.
Ans.   (a) 2.6 cm;   (b) 15 m;   (c) 12.5 m/s; (d) 0.833 Hz; (e) 1.2 s

12   Draw a sine wave with a wavelength of 10 cm and an amplitude of 6 cm. Plot at least 20 points accurately on graph paper, and connect them with a smooth curve.

13   The equation of a sine wave is

$$y = (3.5 \text{ cm}) \sin (60° \text{ cm}^{-1} x)$$

where the argument of the sine is in degrees. Find (a) the amplitude and (b) the wavelength of this wave.
Ans. (a) 3.5 cm; (b) 6.0 cm

14   Write the equation of a sine wave whose amplitude is 12 cm and whose wavelength is 30 cm.

15   At $t = 0$ the equation of a sine wave is

$$y = (0.21 \text{ cm}) \sin (0.3\pi \text{ cm}^{-1} x)$$

where the argument of the sine is in radians. Find (a) the amplitude and (b) the wavelength of this wave. (c) What is the displacement at $x = 0.5$ cm?
Ans. (a) 0.21 cm; (b) 6.67 cm; (c) 0.0953 cm

16   If the wave described in Prob. 15 is moving with a speed of 66.7 cm/s to the right, what is its displacement at $x = 0.5$ cm and $t = 0.005$ s?

17   What is the speed of a sine wave whose frequency and wavelength are 500 Hz and 0.5 m, respectively?
Ans. 250 m/s

18   What is the wavelength of a sine wave whose speed and period are 75 m/s and 0.005 s, respectively?

19   What is the frequency of a sine wave whose speed and wavelength are 120 m/s and 30 cm, respectively?
Ans. 400 Hz

20   A sine wave with an amplitude of 5 cm and a wavelength of 1.6 m is traveling with a speed of 80 m/s. (a) If at time $t = 0$ a point in the medium has a displacement of zero, what are the possible displacements of this point at time $t = 0.003$ s? (b) If at time $t = 0$ a point in the medium has a displacement of 2 cm, what are the possible displacements of this point at time $t = 0.002$ s?

21   If 256 and 320 Hz are two consecutive members of a harmonic sequence, what is the fundamental frequency of this sequence?
Ans. 64 Hz

22   The spectrum of a periodic wave contains Fourier components with frequencies 12, 24, 48, and 96 Hz. Which harmonics are missing?

23   A periodic wave is composed of three sine waves whose frequencies are 36, 60, and 84 Hz. If the speed of the wave is 180 m/s, what is the shortest distance in which the wave pattern repeats?
Ans. 15 m

24   A complex wave is traveling at a speed of 126 m/s. The shortest distance in which the wave pattern repeats is 7 m. Which of the following frequencies are not possible Fourier components of this wave: 9, 18, 27, 36, 54, 64, 72 Hz?

25   What is the energy per second carried by a sine wave of amplitude 0.2 cm, frequency 120 Hz, and speed 200 m/s that is traveling on a cord with a linear mass density of 0.015 kg/m?
Ans. 3.41 W

26   Show that the energy per second carried

by a sine wave of amplitude A and wavelength λ moving on a cord with tension T is

$$P = 2\pi^2 Tv\left(\frac{A}{\lambda}\right)^2$$

27 (a) Show that the energy per wavelength of a sine wave of amplitude A and wavelength λ traveling on a cord with tension T is

$$E = 2\pi^2\frac{T}{\lambda}A^2$$

(b) What is the energy per wavelength of a wave of amplitude 5 cm and wavelength 2.0 m that is traveling on an overhead transmission cable with a tension of 6000 N? (c) If the cable is 150 m long, what is the total wave energy on the cable?
Ans. (b) 148 J; (c) 1.11×10^4 J

28 (a) What are the ratios of the energies (per wavelength) of the Fourier components of the wave in Fig. 11.16? (b) What are the ratios of the energies (per centimeter) of these components?

29 A standing wave is established on a 175-m-long overhead transmission cable suspended from two towers. Which of the following are *not* possible values for the distance between adjacent nodes: 0.70, 1.4, 2.0, 2.8, 3.5, and 4.2 m?
Ans. 2.0, 2.8, and 4.2 m

30 Draw the standing wave given by the superposition of the two sine waves in Fig. 11.19 at times $t = 0.1\tau$ and $t = 0.2\tau$.

31 A cord of length 1.5 m is fixed at both ends. Its mass per unit length is 1.2×10^{-3} kg/m. (a) If the cord has a tension of 12 N, what is the frequency of the fundamental oscillation? (b) What tension is required if the third harmonic f_3 is to have a frequency of 500 Hz?
Ans. (a) 33.3 Hz; (b) 300 N

32 A string of length 55 cm is held fixed at both ends. When the tension is 25 N, the fundamental frequency is 40 Hz. (a) What is the wavelength of the fundamental? (b) What is the speed of a wave on this string? (c) What is the mass per unit length of the string?

33 The lowest note on a piano has a frequency of 27.5 Hz. The piano wire is 2.0 m long and has a tension of 300 N. What is the total mass of the wire?
Ans. 49.6 g

34 A 20-cm cord fixed at both ends is oscillating simultaneously with the frequencies of its

fundamental f_1 and second harmonic $2f_1$. The amplitude of the fundamental is 5 cm, and the amplitude of the second harmonic is 3 cm. At time $t = 0$ both harmonics have maximum displacement. (a) Draw the displacement produced by each wave separately at $t = 0$, and then draw the shape of the cord with both waves present simultaneously. (b) Draw the shape of the cord at time $t = \frac{1}{2}\tau_1$.

35 A guitar string 0.75 m long has a fundamental frequency of 440 Hz. (a) What is the speed of a wave on this string? (b) To produce other frequencies, the effective length L of the string is shortened by pressing it at a point below the end of the string. What length is needed to produce a fundamental frequency of 660 Hz?
Ans. (a) 660 m/s; (b) 0.50 m

36 Show that the total energy in a standing wave of amplitude $2A$ is $4\pi^2 L\mu f^2 A^2$, where L is the total length of the cord. (Hint: The energy of a standing wave is the sum of the energies of its two sine-wave components.)

37 A guitar string 0.75 m long has a fundamental frequency of 440 Hz. The linear density of the string is 2.2×10^{-3} kg/m. (a) When the string is plucked hard, the fundamental harmonic vibrates with a maximum displacement of 0.2 cm. What is the energy of the fundamental (see Prob. 36)? (b) What is the energy of the third harmonic if its amplitude is 0.05 cm? (Hint: The amplitude of a standing wave is $2A$ where A is the amplitude of its sine-wave components.)
Ans. (a) 1.26×10^{-2} J; (b) 7.09×10^{-3} J

38 A standing wave with an amplitude $2A$ of 5 cm is established on an overhead transmission cable. The nodes are 1.2 m apart, the cable is 240 m long, its linear mass density is 1.3 kg/m, and the tension is 45,000 N. What is the total energy of this vibrating cable? See Probs. 36 and 37.

39 A pendulum with a normal frequency of 5.0 Hz is subjected to a driving force of frequency 5.4 Hz. (a) What is the response factor if there is no friction in the system? (b) What is the response factor if the width Γ of the response curve is 0.5 Hz? (Use Fig. 11.23.)
Ans. (a) 6.1; (b) 4.9

40 An oscillating system has a normal frequency of 60 Hz and a response width Γ of 2 Hz. What is the response factor at resonance?

41 A 120-m-long transmission cable has a diameter of 3.2 cm, a linear mass density of 2.1 kg/m, and a tension of 4.2×10^4 N.

(a) What is the driving frequency produced by a 9-m/s wind? (b) What is the normal frequency of the cable closest to the driving frequency? (c) What is the response factor if the damping produces a width Γ of 2.5 Hz?
Ans. (a) 51.92 Hz; (b) 51.85 Hz; (c) 20.7

42   Show that for wind velocities above 6 m/s, the ratio f_d/f_{nor} of the driving frequency to the closest normal frequency of the cable in Prob. 41 is between 0.992 and 1.008. Thus the cable is in resonance regardless of the exact wind velocity.

S ound is a longitudinal mechanical wave that propagates through air, water, and other material media. It is produced by a vibrating object transferring some of its vibrational motion to the molecules of the surrounding medium, and it is detected by the vibrating molecules of the medium transferring some of their vibrational motion to a movable diaphragm, such as the diaphragm of a microphone or the eardrum of a person.

Some vibrating objects, such as the string of a violin or the vocal cords of a singer, produce pleasing sounds, while others, such as a motorcycle engine or a jackhammer, produce displeasing sounds, called *noise*. Sound engineering, or *acoustics*, is concerned with the problems both of sound transmission and reproduction and of noise control and suppression.

12
SOUND

12.1 The Physics of Sound

The loudspeaker in Fig. 12.1 consists of a plastic cone connected to an electromagnet. When an alternating electric current of frequency f is supplied to the electromagnet, the cone is forced to vibrate back and forth with the driving frequency f. The cone in turn forces the air molecules adjacent to it to vibrate with the same frequency. If the cone is executing simple harmonic motion, the displacement y of these molecules from their equilibrium position is given by Eq. 6.13:

$$y = A \cos (2\pi f t) \tag{12.1}$$

Here A is the amplitude of the cone's motion, and the relation (Eq. 11.5)

$$f = \frac{1}{\tau} \tag{12.2}$$

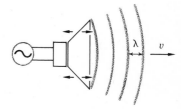

FIGURE 12.1

A loudspeaker emitting a sound wave.

between the frequency f and the period τ of an oscillator has been used to replace τ by f in Eq. 6.13.

As the cone moves forward, the air immediately in front of it is compressed, and the pressure p of this air rises above the pressure p_0 of the undisturbed surrounding air. As the cone moves back, this region of high-pressure air expands, compressing the air ahead of it. This new region of high-pressure air then expands, compressing the air ahead of it. In this manner, a region of high pressure propagates through the air. Likewise, as the cone moves backwards, the air in front of it is expanded, and the pressure p of this air falls below the pressure p_0 of the surrounding air. As the cone moves forward, this region of low-pressure air contracts, allowing the air ahead of it to expand. Thus as the cone vibrates back and forth, alternating regions of high- and low-pressure air are formed that propagate away from the speaker. This propagating pattern of pressure variations is the sound wave.

The wave travels with the wave speed v characteristic of the medium. (In air the speed of sound v is about 340 m/s. See Table 12.1.) The wavelength λ of a sinusoidal wave of frequency f is given by Eq. 11.6:

$$f\lambda = v \tag{12.3}$$

TABLE **12.1**

Substance	Temperature, °C	Density, kg/m³	Speed of sound, m/s	Acoustic impedance, kg/(m²·s)
Gases:				
Air, dry	0	1.293	331	428
	20	1.20	343	412
Argon	0	1.783	308	549
Carbon dioxide	0	1.977	259	512
Helium	0	0.178	965	172
Hydrogen	0	0.0899	1284	115
Methane	0	0.7168	430	308
Nitrogen	0	1.251	334	418
Oxygen	0	1.429	316	452
Liquids:				
Ethanol	25	790	1207	0.95×10^6
Mercury	25	13,500	1450	19.6×10^6
Tetrachloromethane	25	1595	926	1.48×10^6
Water, distilled	25	998	1498	1.50×10^6
Water, sea	25	1025	1531	1.57×10^6

Speed of sound in various gases and liquids

This table also gives the density ρ and acoustic impedance ρv of each substance

Example 12.1 What is the wavelength in air of the sound wave generated by a speaker oscillating with a frequency of 1000 Hz?
From Eq. 12.3 we have

$$\lambda = \frac{v}{f} = \frac{340 \text{ m/s}}{1000 \text{ Hz}} = 0.34 \text{ m}$$

Typically the wavelength of sound is between 2 cm and 20 m, depending on the frequency. ☐

Sound is a longitudinal wave because the displacements of the air molecules are in the direction in which the wave propagates. Figure 12.2a is a graph of the displacements in a sinusoidal sound wave of amplitude A at one instant of time. Figure 12.2b shows the actual displacements of the air molecules at this instant. A positive value of the displacement y corresponds to a longitudinal displacement to the right, and a negative value of y corresponds to a displacement to the left.

Notice that molecules on either side of the displacement node P in Fig. 12.2 are displaced toward the node, so the density and pressure of the air at this point are above normal. Likewise, the molecules on either side of the displacement node Q are displaced away from the node, so the density and pressure at this point are below normal. Figure 12.2c is a graph of the pressure difference

$$\bar{p} = p - p_0$$

in the wave. If at any instant the displacement y in the wave is given by

$$y = A \cos\left(2\pi \frac{x}{\lambda}\right) \qquad 12.4$$

then at this same instant the pressure difference $\bar{p}$ is given by

FIGURE 12.2

Displacements in a sinusoidal sound wave. (a) Plot of the longitudinal displacements of points in the medium against their equilibrium positions. (b) Displacements of points in the medium. (c) Plot of the pressure variation of the medium against position.

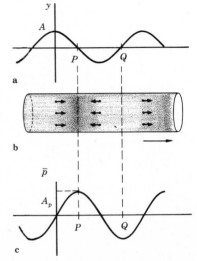

$$\bar{p} = A_p \sin\left(2\pi\frac{x}{\lambda}\right) \qquad\qquad 12.5$$

where the pressure amplitude A_p is the maximum value of $\bar{p}$.

The pressure amplitude A_p is related to the displacement amplitude A by the equation

$$A_p = 2\pi f\rho vA \qquad\qquad 12.6$$

where ρ is the density of the medium, and v is the speed of sound in the medium. The quantity ρv is called the *acoustic impedance* of the medium. Values of ρ and ρv are given in Table 12.1 for various media.

Example 12.2 The pressure amplitude of the sound in a normal conversation is about 0.02 Pa. What is the displacement amplitude of this sound if the frequency is 1000 Hz?

From Table 12.1 the acoustic impedance of air at 20°C is 411.6 kg/(m² · s), so from Eq. 12.6 the displacement amplitude A is

$$A = \frac{A_p}{2\pi f\rho v} = \frac{0.02\ \mathrm{N/m^2}}{2\pi(1000\ \mathrm{Hz})[411.6\ \mathrm{kg/(m^2 \cdot s)}]}$$

$$= 7.7 \times 10^{-9}\ \mathrm{m}$$

This minute displacement is too small to be detected directly. Most sound detectors detect the pressure variation rather than the displacement of a sound wave.

REMARK Using the relations

1 Pa = 1 N/m² = 1 kg/(m · s²)

and

1 Hz = 1 s⁻¹

you can show that $A_p/(2\pi f\rho v)$ has the unit of meters. But by always using SI units, the complications of units in complex equations like Eq. 12.6 are largely avoided. ☐

SPEED OF SOUND IN GASES AND LIQUIDS

The speed of sound in an ideal gas is given by

$$v = \sqrt{\frac{\gamma p_0}{\rho}} \qquad \text{gas} \qquad\qquad 12.7$$

where ρ is the density of the gas, p_0 is the undisturbed pressure, and γ is equal to c_p/c_v, the ratio of the specific heat of the gas at constant pressure to the specific heat at constant volume (Sec. 9.2). For diatomic gases, such as oxygen, nitrogen, and air, γ equals $\frac{7}{5}$, or 1.4.

The speed of sound in a liquid is given by

$$v = \sqrt{\frac{B}{\rho}} \qquad \text{liquid} \qquad\qquad 12.8$$

where B, the bulk modulus, is defined as follows:

Definition If a liquid occupies a volume V at a pressure p_0, the volume will decrease to $V - \Delta V$ when the pressure is increased to $p_0 + \Delta p$. The *bulk modulus B* of the liquid is

$$B = \Delta p \, \frac{V}{\Delta V} \qquad\qquad 12.9$$

The bulk modulus is a measure of the resistance to compression of the liquid. The relative volume change $\Delta V/V$ produced by a pressure change Δp is related to the bulk modulus by

$$\frac{\Delta V}{V} = \frac{\Delta p}{B} \qquad\qquad 12.10$$

Like the Young's modulus E of a solid (Sec. 3.4), the bulk modulus is a characteristic property of a substance.

The density ρ, speed of sound v, and acoustic impedance ρv of various gases and liquids are given in Table 12.1.

Example 12.3 (a) Calculate the bulk modulus of water from the data in Table 12.1. (b) How much is the volume of 10^4 cm^3 of water decreased when the water is subjected to a pressure of 200 atm?

(a) From Eq. 12.8 we have

$$B = v^2\rho$$

so from the values v and ρ given for water in Table 12.1, we get

$$B = (1498 \text{ m/s})^2 (998 \text{ kg/m}^3)$$
$$= 2.2 \times 10^9 \text{ N/m}^2$$

(b) From Eq. 12.10 the decrease in volume ΔV produced by the pressure change

$$\Delta p = 200 \text{ atm} = 2.02 \times 10^7 \text{ N/m}^2$$

is

$$\Delta V = \Delta p \frac{V}{B} = (2.02 \times 10^7 \text{ N/m}^2) \frac{10^4 \text{ cm}^3}{2.2 \times 10^9 \text{ N/m}^2}$$
$$= 91.8 \text{ cm}^3 \qquad\qquad \square$$

SPEED OF SOUND IN SOLIDS*

Only a longitudinal wave can propagate in a gas or liquid because a fluid lacks the rigidity necessary to support a transverse wave. However, both longitudinal and transverse waves propagate through solids. Furthermore, two types of longitudinal waves are possible in a solid, depending on the dimensions of the solid. Thus there are three different types of sound waves in solids and three corresponding speeds.

Extensional Waves An *extensional wave* is a longitudinal wave that propagates along the axis of a long rod whose diameter d is small compared with the wavelength λ. The minute pressure changes that propagate along the rod cause minute changes in the diameter of the rod. This is shown in Fig. 12.3a. That is,

*This is optional material, which may be omitted.

FIGURE 12.3

Types of waves in a sound: (a) extensional wave, (b) compressional (longitudinal) wave, and (c) shear (transverse) wave.

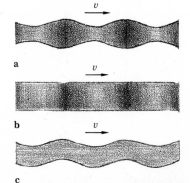

although the molecules of the solid move mainly in the longitudinal direction, there is a small transverse motion that results from the small diameter of the rod.

The speed v_{ex} of an extensional wave is given by

$$v_{ex} = \sqrt{\frac{E}{\rho}} \qquad 2.11$$

where E is the Young's modulus of the solid (Sec. 3.4). This equation is often stated in elementary textbooks, but it is seldom pointed out that it applies only for the speed along a narrow rod. It is nevertheless an important equation because the extensional speed can be measured very precisely. Equation 12.11 can then be used to determine the Young's modulus of a solid from a measurement of the extensional speed. Table 12.2 gives the extensional speeds of sound in various solids.

Example 12.4 Calculate the Young's modulus of lead using the data in Table 12.2.

From Eq. 12.11 we have

$$E = v_{ex}^2\rho$$

so from the values of v_{ex} and ρ given for lead in Table 12.2 we get

$$E = (1210 \text{ m/s})^2(11,400 \text{ kg/m}^3)$$
$$= 17 \times 10^9 \text{ N/m}^2$$

This value agrees with the value given in Table 3.1. ☐

Compressional or Pure Longitudinal Waves The highest wave speed is that for a pure longitudinal wave traveling through an extended medium. A *pure longitudinal wave* differs from an extensional wave in that there is no transverse motion of the molecules (Fig. 12.3b). It is thus the same as a longitudinal wave in a gas or liquid. The wave speed v of a pure longitudinal wave is related to the extensional speed v_{ex} by the equation

$$v = \sqrt{\frac{1 - \mu}{(1 + \mu)(1 - 2\mu)}}\, v_{ex} \qquad 12.12$$

where μ is a characteristic property of a solid called the *Poisson*

TABLE **12.2**

Extensional, longitudinal, and transverse speeds of sound in various solids

Substance	Density, kg/m³	Extensional speed, m/s	Longitudinal speed, m/s	Transverse speed, m/s
Aluminum, rolled	2700	5000	6420	3040
Brick	1800	3650		
Copper, rolled	8930	3750	5010	2270
Cork	250	500		
Glass:				
Flint	3880	3720	3980	2380
Pyrex	2320	5170	5640	3280
Lead, rolled	11,400	1210	1960	690
Lucite	1180	1840	2680	1100
Rubber, gum	950	40	1550	23
Steel, hardened	7840	5070	5854	3150
Wood, along fiber	600 to 900	3400 to 4700		

ratio. For any substance μ is between 0 and 0.5, so the square-root factor in Eq. 12.12 is always greater than 1.

The Poisson ratio measures the extent to which the diameter of a long rod contracts when the rod is stretched. If the diameter of the rod does not change when its length is increased, then μ is zero. If the diameter of the rod contracts sufficiently to keep the volume of the rod unchanged when its length is increased, then μ has its maximum value of 0.5. For most solids μ is around 0.3, so the square-root factor in Eq. 12.12 is about 1.2. That is, for most solids the longitudinal speed is about 20 percent greater than the extensional speed. But for some substances, notably rubber, μ is close to 0.5. In this case, the factor $1/(1 - 2\mu)$ in Eq. 12.12 is very large, so the longitudinal speed is much greater than the extensional speed. This is shown in Table 12.2, which compares the extensional and longitudinal speeds of sound in various substances.

Shear or Transverse Waves In a *transverse wave*, the molecules of the medium move perpendicularly to the direction of propagation of the wave (Fig. 12.3c). Transverse waves can propagate through a solid, but not through a gas or liquid, because a fluid cannot support a shear stress. The transverse speed v_t of a wave is given by

$$v_t = \sqrt{\frac{G}{\rho}} \qquad\qquad 12.13$$

where G is the shear modulus of the solid. Table 12.2 gives the transverse speeds of sound in various solids.

An earthquake is a sudden breakage of rock inside the earth's surface. An earthquake generates both longitudinal and transverse waves which travel to the surface of the earth through the intervening rock. Waves produced this way are called *seismic waves* and are detected on an instrument called a *seismograph* (Fig. 12.4). When an earthquake occurs at a point inside the earth, seismographs at different locations record the time at which the longitudinal and transverse waves arrive. From these data the location of the earthquake and the speed of the waves can be determined. Since the speeds depend on the mechanical properties of the rock through which the waves travel, seismologists are able to learn about the nature of these rocks from studies of earthquakes.

Seismic waves can also be produced by setting off explosions on the surface of the earth. These waves then travel into the earth and are reflected back to the surface by abrupt changes in the rock structure inside the earth. Seismographs stationed on the earth's surface detect these reflected waves and use the data collected to study the underlying rocks. This technique is especially important in oil exploration.

REMARK The Young's modulus E, shear modulus G, and Poisson ratio μ of a solid are related by the equation

$$G = \frac{E}{2(1 + \mu)} \qquad\qquad 12.14$$

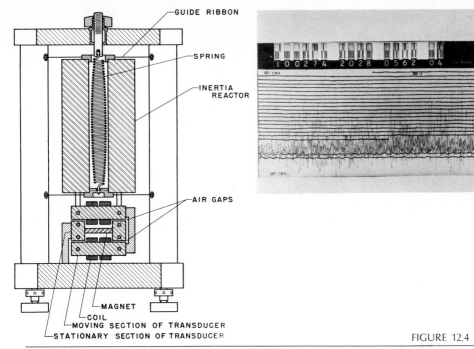

GUIDE RIBBON

SPRING

INERTIA REACTOR

AIR GAPS

MAGNET
COIL
MOVING SECTION OF TRANSDUCER
STATIONARY SECTION OF TRANSDUCER

FIGURE 12.4

(a) Schematic diagram of a seismograph (*Environmental Science Services Administration*). (b) Seismogram showing the oscillations recorded by a seismograph (*National Oceanographic and Atmospheric Administration*).

As a consequence, the longitudinal speed v, the extensional speed v_{ex}, and the transverse speed v_t are related by the equation

$$v = v_t \sqrt{1 + \frac{1}{3 - v_{ex}^2/v_t^2}} \qquad 12.15$$

You can check this last equation with the data in Table 12.2.

12.2 Intensity and Intensity Level

Figure 12.5 shows a sound wave traveling through a column of air of length d and cross-sectional area a. The total mass m of the air in this column is

$$m = \rho V = \rho ad$$

where ρ is the density of the air, and $V = ad$ is the volume of the column. The linear mass density μ is the mass per unit length of the column:

$$\mu = \frac{m}{d}$$

For the column in Fig. 12.5, the linear mass density is

$$\mu = \frac{\rho ad}{d} = \rho a \qquad 12.16$$

As a wave travels through a medium, the kinetic and potential energies of the moving molecules are transported from one region of space to another with the speed v of the wave. For a wave on a string of linear mass density μ, the rate of energy transport, or power P, is given by Eq. 11.8:

$$P = 2\pi^2 \mu v f^2 A^2$$

FIGURE 12.5

Sound wave traveling through a column of air.

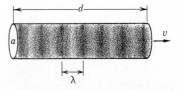

This equation also holds for the power of a sound wave, if we use Eq. 12.16 for the linear mass density of air. That is, the rate at which sound energy is transported through a column of air of cross-sectional area a is

$$P = 2\pi^2\rho avf^2A^2 \qquad\qquad 12.17$$

where A is the displacement amplitude, f is the frequency, and v is the speed of the wave.

Definition The *intensity* I of a wave is the rate at which the wave transports energy across a unit area. Thus, if P is the rate at which the wave transports energy across an area a, the intensity of the wave is

$$I = \frac{P}{a} \qquad\qquad 12.18$$

Using Eq. 12.17 for the power of a sound wave, we find that the intensity is

$$I = 2\pi^2\rho vf^2A^2 \qquad\qquad 12.19a$$

Using the relation between the pressure amplitude A_p and the displacement amplitude A given in Eq. 12.6, the intensity of a sound wave can also be written

$$I = \frac{A_p{}^2}{2\rho v} \qquad\qquad 12.19b$$

UNITS The unit of power P is joules per second (J/s), or watts (W). The unit of intensity is thus watts per square meter (W/m²).

Example 12.5 (a) The diaphragm of a microphone has an area of 3 cm². If the power P of the sound incident on the diaphragm is 6×10^{-10} W, what is the intensity I of the sound? (b) What is the pressure amplitude A_p of this sound?

(a) From Eq. 12.18 the intensity I is

$$I = \frac{P}{a} = \frac{6 \times 10^{-10}\,\text{W}}{3 \times 10^{-4}\,\text{m}^2} = 2 \times 10^{-6}\,\text{W/m}^2$$

(b) From Eq. 12.19b and the value of ρv for air given in Table 12.1, the pressure amplitude A_p is found to be

$$\begin{aligned}
A_p &= \sqrt{2\rho vI} \\
&= \sqrt{2[412\,\text{kg/(m}^2 \cdot \text{s)}](2 \times 10^{-6}\,\text{W/m}^2)} \\
&= 4.06 \times 10^{-2}\,\text{Pa} \qquad\qquad \square
\end{aligned}$$

DECIBEL SCALE

The human ear can detect sounds that vary in intensity from 10^{-12} W/m² to 1 W/m². Below 10^{-12} W/m² the sound is inaudible, and above 1 W/m² the sound produces discomfort and pain. Because of the wide range of audible intensities it is convenient to use a logarithmic intensity-level scale, called the *decibel* (dB) scale.*

*See Appendix III for a review of logarithms.

Definition The *intensity level β* (in decibels) of a sound of intensity I is

$$\beta = (10 \text{ dB}) \log \frac{I}{I_0} \qquad\qquad 12.20$$

where $I_0 = 10^{-12} \text{ W/m}^2$ is the *reference intensity*.

On the decibel scale the audible range is from

$$\beta = (10 \text{ dB}) \log \frac{10^{-12} \text{ W/m}^2}{10^{-12} \text{ W/m}^2}$$

$$= (10 \text{ dB}) \log (1) = 0 \text{ dB}$$

to

$$\beta = (10 \text{ dB}) \log \frac{1 \text{ W/m}^2}{10^{-12} \text{ W/m}^2}$$

$$= (10 \text{ dB}) \log (10^{12}) = 120 \text{ dB}$$

Table 12.3 gives the intensities I and intensity levels β of some common sounds.

Since the decibel scale is logarithmic, a constant addition to a decibel value corresponds to a constant multiple of the intensity. For instance, an increase of 10 dB in the intensity level corresponds to a factor of 10 increase in intensity; an increase of 20 dB corresponds to a factor of $10 \times 10 = 100$ increase in intensity. A factor of 2 increase in intensity corresponds to a 3 dB increase in the intensity level.

Example 12.6 (*a*) What is the intensity level of the sound in Example 12.5? (*b*) What would be the intensity level of a sound with twice the intensity of the sound in Example 12.5?

(*a*) The intensity of the sound in Example 12.5 is $I = 2 \times 10^{-6} \text{ W/m}^2$. So from Eq. 12.20 we have

$$\beta_I = (10 \text{ dB}) \log \frac{I}{I_0} = (10 \text{ dB}) \log \frac{2 \times 10^{-6} \text{ W/m}^2}{10^{-12} \text{ W/m}^2}$$

TABLE **12.3**

Sound level, dB	Intensity, W/m²	Sound	Sound levels and intensities of some common sounds
0	10^{-12}	Threshold of hearing	
10	10^{-11}	Rustle of leaves	
20	10^{-10}	Whisper (1 m away)	
30	10^{-9}	Quiet home	
40	10^{-8}	Average home, quiet office	
50	10^{-7}	Average office	
60	10^{-6}	Normal conversation, average traffic	
70	10^{-5}	Noisy office	
80	10^{-4}	Busy traffic, inside car in traffic	
90	10^{-3}	Inside subway train	
100	10^{-2}	Machine shop	
120	10^{0}	Pneumatic chipper (2 m away), threshold of pain	
140	10^{2}	Jet airplane (10 m away)	

$$= (10\,\text{dB})\,\log\,(2 \times 10^6)$$
$$= (10\,\text{dB})(\log 2 + \log 10^6)$$

The logarithm of 10^6 is 6, and from Table III.1 the logarithm of 2 is found to be 0.30. Thus we have

$$\beta_I = (10\,\text{dB})(0.30 + 6) = 63\,\text{dB}$$

The logarithm of 2×10^6 can also be obtained directly on a calculator with a log key.

(b) The intensity level β_{2I} of a sound with an intensity of $2I$ is

$$\beta_{2I} = (10\,\text{dB})\,\log\left(\frac{2I}{I_0}\right) = (10\,\text{dB})(\log 2 + \log I/I_0)$$

$$= (10\,\text{dB})\,\log 2 + (10\,\text{dB})\,\log\frac{I}{I_0} = 3\,\text{dB} + \beta_I = 66\,\text{dB}$$

or 3 dB more than the intensity level β_I of a sound with intensity I. As mentioned before, an increase of 3 dB in the intensity level corresponds to a doubling of the intensity. ☐

12.3 Hearing

Human beings, like most vertebrates and arthropods, have the capacity to detect sound waves and to distinguish between waves of different intensity and frequency. The ear detects a sound wave by transforming it into a sequence of electric impulses. These impulses then travel through neurons to the brain, where they are interpreted.

ANATOMY OF THE EAR

Figure 12.6 shows the anatomy of the human ear. The outer ear, consisting of the *pinna* and *auditory canal,* channels sound waves to the eardrum (*tympanic membrane*). The eardrum is a delicate membrane that separates the outer ear from the middle ear. Sound waves impinging on the eardrum force it to vibrate. These vibrations are picked up and amplified by three tiny bones (the *auditory ossicles*) in the middle ear. The bone in contact with the eardrum is called the *malleus* (hammer); it connects to the *incus* (anvil) and *stapes* (stirrup), as shown in Fig. 12.6. The stapes terminates at the *oval window,* which is the membrane separating the middle and inner ears. The malleus, incus, and stapes provide a mechanical linkage which transmits vibrations from the eardrum to the oval window; vibrations of the oval window pass into the *cochlea.*

REMARK The eardrum is sensitive enough to detect pressure variations as small as $10^{-5}\,\text{Pa} = 10^{-10}$ atm. Because it is so sensitive, the static air pressure must be the same on both sides. Pressure equalization is maintained by the *eustachian tube,* which provides a restricted air passage between the middle ear and the mouth. This tube opens whenever a person yawns or swallows, allowing the pressure in the inner ear to adjust to any change in the atmospheric pressure. The effect of unequal pressures can be felt in a rapidly ascending elevator. In this case, the outside air pressure changes faster than the pressure in the middle ear, causing a net force to be exerted on the eardrum. This force produces a painful sensation which can be easily relieved by yawning or swallowing.

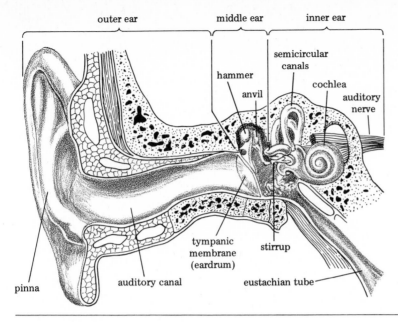

outer ear   middle ear   inner ear

semicircular canals

hammer

anvil

cochlea

auditory nerve

tympanic membrane (eardrum)

stirrup

pinna

auditory canal

eustachian tube

FIGURE 12.6

Anatomy of the ear.

The cochlea is a spiral-shaped organ filled with a viscous fluid (*perilymph*). The interior of the cochlea is divided into two chambers by the *basilar membrane,* a flexible structure covered with more than 20,000 *basilar fibers.* Figure 12.7a shows the cochlea uncoiled. Figure 12.7b shows a magnified view of a basilar fiber. One end of the fiber is free to vibrate, while the other end is connected to sensory hair cells. When the fiber vibrates, it stimulates the hair cells to transmit nerve impulses to the brain.

Vibrations transmitted by the stapes to the oval window produce a wave in the perilymph. The wave could pass all the way around the end of the basilar membrane and exit through the *round window,* but the thickness of the basilar membrane varies in such a way that only low-frequency waves reach the end of the membrane. Higher-frequency waves pass through the membrane at some point before reaching the end. The higher the frequency, the shorter the distance the wave travels along the membrane before passing through. This means that the sensory cells at the far end of the membrane can be stimulated only by low-frequency waves.

The length and thickness of the basilar fibers vary with their positions on the basilar membrane. Near the oval window the fibers are short and fat, and consequently they vibrate with a high characteristic frequency. Near the opposite end of the basilar membrane, the fibers are long and thin, and consequently they vibrate with a low characteristic frequency. A low-frequency wave causes the fibers near the oval window to vibrate with a small amplitude, because the frequency of the wave is not near the resonant frequency of these fibers. On the other hand, a low-frequency wave causes the fibers near the end of the membrane to resonate with a large amplitude, because the frequency of the wave is equal to the characteristic frequency of these

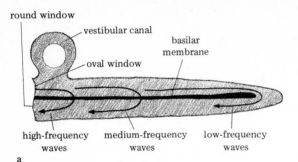

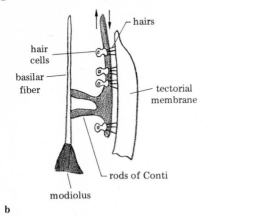

a

b

FIGURE 12.7

(a) Sagittal view of uncoiled cochlea. (b) Magnified view of a basilar fiber.

fibers. Likewise, a high-frequency wave causes the fibers near the oval window to resonate, and a medium-frequency wave causes the fibers midway down the basilar membrane to resonate.

The structure of the cochlea is thus designed so that each frequency component of a complex wave produces resonant vibrations in fibers located at different positions along the basilar membrane. These fibers in turn stimulate impulses on the different nerves that go to the brain. This enables the ear to respond differently to waves with different frequency components.

PSYCHOPHYSICS

It is important to distinguish between the physical and the psychological properties of a sound. Intensity is a physical property; it can be measured objectively with a suitable instrument. Loudness is a psychological property; it is measured subjectively by asking a listener to judge a particular sound. Likewise, frequency is a physical property, whereas pitch is a psychological property. *Psychophysics* is a branch of psychology concerned with measuring the relation between the physical and psychological properties of a stimulus.

As an example, consider the following psychophysical experiment. A subject compares a standard sound, which has (say) a frequency of 1000 Hz and an intensity level of 60 dB, to a test sound, which has a frequency f. The intensity of the test sound is varied until the subject judges it to have the same loudness as

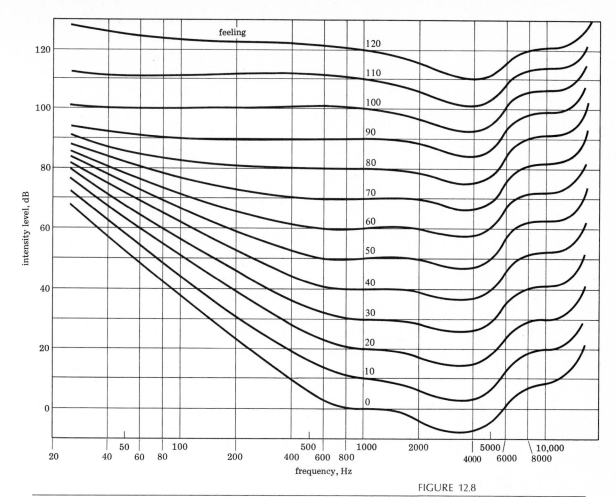

feeling

120
110
100
90
80
70
60
50
40
30
20
10
0

intensity level, dB

frequency, Hz

20   40 50 60 80 100   200   400 500 600 800 1000   2000   4000 5000 6000 8000 10,000

FIGURE 12.8

Psychophysical relation of loudness to intensity and frequency. Each curve gives the intensities at which sounds of various frequencies have the same apparent loudness. The loudness level is arbitrarily taken to equal the intensity level of the 1000-Hz signal.

the standard 1000-Hz sound. This is repeated for a range of frequencies until a curve, such as the one labeled 60 in Fig. 12.8, is obtained. Each point on the curve gives the intensity level at which a signal of the corresponding frequency is judged to have the same loudness as the standard signal (1000 Hz at 60 dB). For instance, a 100-Hz signal must have an intensity level of 72 dB to be judged as loud as a 1000-Hz signal at 60 dB. Figure 12.8 also shows curves produced when the 1000-Hz standard signal is given at different intensity levels. From these curves it is seen that hearing is most acute for sounds with frequencies between 3000 and 4000 Hz.

In another test, the intensity of a test signal of frequency f is decreased until the subject no longer hears it. By repeating this for a range of frequencies, a curve of the individual's hearing acuity is obtained. Figure 12.9 summarizes the results of thousands of these tests given by the U.S. Public Health Service. Each curve is labeled by the percentage of people in the population whose hearing acuity was below the curve. Thus 99 percent of the population could hear signals below the 99 percent curve, whereas only 10 percent could hear signals below the 10 percent

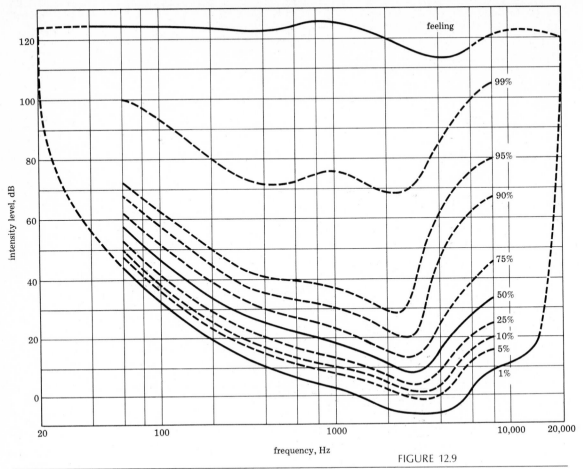

FIGURE 12.9

Hearing-acuity profiles of the United States population. The number to the right of each curve is the percentage of the population that can hear all signals below the curve.

curve. (The 1 percent of the population that cannot hear below the 99 percent level require hearing aids, since they cannot hear a normal conversation at 60 dB.) The 1 percent curve is often taken as the standard of unimpaired, or "normal," hearing. Only children and young adults can hear at this level, because acuity decreases with age. These curves have a dip between 3000 and 4000 Hz, which again indicates that hearing is most acute in this frequency range.

12.4 Standing Waves

The air inside a cavity, like a cord fixed at both ends, can vibrate freely only at certain discrete frequencies that are characteristic of the cavity. In Sec. 11.5 it was shown that the characteristic frequencies of a cord fixed at both ends correspond to standing waves on the cord. Likewise, the characteristic frequencies of a cavity correspond to the standing waves that can exist in the cavity. The frequencies of these waves are determined by the size and shape of the cavity, but they can be easily calculated only for cavities of very simple geometry.

An important special case is that of a cylindrical cavity of length L open at both ends (open pipe). By blowing air across one end of the cylinder, waves are produced that travel down the cylinder. If the diameter of the cylinder is small compared to the wavelength of the waves, they will be reflected back into the cylinder when they reach the other end. The superposition of waves of equal amplitude and frequency traveling in opposite directions produces a standing-wave pattern (Sec. 11.5). The pressure at each end is fixed at atmospheric pressure because the ends are open to the atmosphere. Therefore, only standing waves that have pressure nodes at each end can exist in the cylinder (Fig. 12.10). The situation is identical to that of a cord fixed at both ends: only standing waves are possible that have wavelengths given by

$$L = n\frac{\lambda_n}{2} \qquad \text{or} \qquad \lambda_n = \frac{2L}{n}$$

where n is an integer. The frequencies of these waves are

$$f_n = \frac{v}{\lambda_n} = \frac{nv}{2L} \qquad n = 1, 2, 3, \ldots \qquad\qquad 12.21$$

Like the frequencies of a vibrating cord, this set includes the fundamental, $f_1 = v/2L$, and all the harmonics $f_n = nf_1$.

Organ pipes and other wind instruments, including the human voice, produce tones by establishing standing waves in cavities. However, the characteristic frequencies are a harmonic sequence only if the cavity is a perfect cylinder that is open at both ends (open pipe). For cavities of different shape, the frequencies do not form a harmonic sequence.

For example, some organ pipes are closed at one end and open at the other (closed pipe). Such a pipe has a pressure antinode at the closed end because the air has zero displacement there. (Figure 12.2 shows that the pressure has an antinode wherever the displacement has a node.) The characteristic standing waves, then, have a node at the open end and an antinode at the closed end. Figure 12.11 shows the first three standing waves that satisfy this condition. These waves have $\frac{1}{4}, \frac{3}{4}, \frac{5}{4}, \ldots$ wavelength in the pipe, so that

$$L = m\frac{\lambda_m}{4} \qquad \text{or} \qquad \lambda_m = \frac{4L}{m}$$

where m is an odd integer ($m = 1, 3, 5, 7, \ldots$). The characteristic frequencies are

$$f_m = \frac{mv}{4L} \qquad m = 1, 3, 5, \ldots \qquad\qquad 12.22$$

The fundamental frequency is $f_1 = v/4L$, which is half the fundamental frequency of an open pipe of the same length. The overtones of the closed pipe consist only of the odd harmonics: $f_1, 3f_1, 5f_1, \ldots$.

Example 12.7 What is the length of an open organ pipe whose fundamental frequency is 440 Hz?

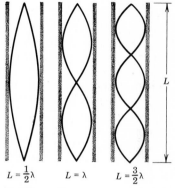

$$L = \tfrac{1}{2}\lambda \qquad L = \lambda \qquad L = \tfrac{3}{2}\lambda$$

FIGURE 12.10

Some standing waves in an open pipe. The curves show the pressure variations inside the pipe.

FIGURE 12.11

Some standing waves in a closed pipe. The curves show the pressure variation inside the pipe.

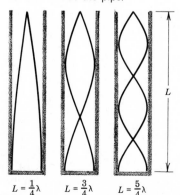

$$L = \tfrac{1}{4}\lambda \qquad L = \tfrac{3}{4}\lambda \qquad L = \tfrac{5}{4}\lambda$$

The length L given by Eq. 12.21 is

$$L = \frac{nv}{2f_n} = \frac{(1)(340 \text{ m/s})}{2(440 \text{ Hz})} = 0.386 \text{ m} \qquad \square$$

12.5 Doppler Effect and Beats

DOPPLER EFFECT

Consider a sound source that is vibrating at a frequency f_s. The sound waves emitted by this source travel to a detector and cause it to vibrate with a frequency f. The detected frequency f depends on the source frequency f_s and on the motions of the source and detector. If the speed v_s of the source and the speed v_d of the detector are both zero, then $f = f_s$. If either of these speeds is not zero, then f is not equal to f_s; there is a change in the detected frequency due to the motion of the source or detector. This phenomenon is known as the *Doppler effect*. We shall consider the effect of each motion separately.

Moving Source Suppose the source is moving with speed v_s toward the detector (Fig. 12.12). From time $t = 0$ to time $t = T$ the source makes $f_s T$ oscillations and moves the distance $v_s T$ toward the detector. The wave emitted by the source moves with the wave speed v of the medium. At $t = T$ the portion of the wave emitted at $t = 0$ will be the distance vT from the original position of the source and the distance

$$vT - v_s T$$

from the final position of the source (Fig. 12.12). The number $f_s T$ of wavelengths emitted by the source are all contained within this distance, so the wavelength of the wave is

$$\lambda = \frac{(v - v_s)T}{f_s T} = \frac{v - v_s}{f_s}$$

and its frequency is

$$f = \frac{v}{\lambda} = \frac{v}{v - v_s} f_s \qquad \text{moving source} \qquad 12.23$$

The frequency f of the wave is greater than the frequency f_s of the source when the source is moving toward the detector. When the source is moving away from the detector, v_s is negative, and f is less than f_s.

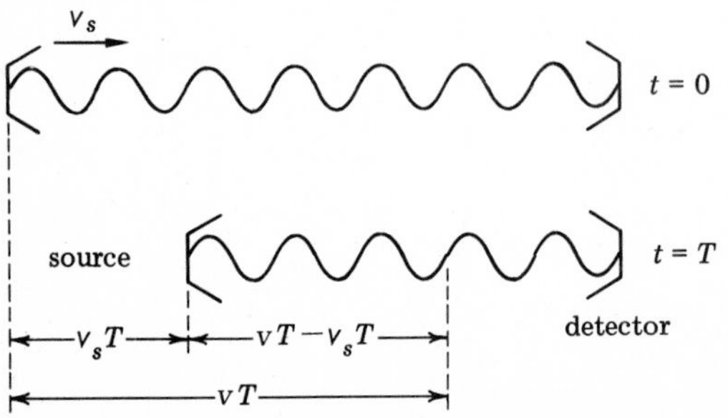

FIGURE 12.12

Sound wave emitted by a source moving toward a stationary detector.

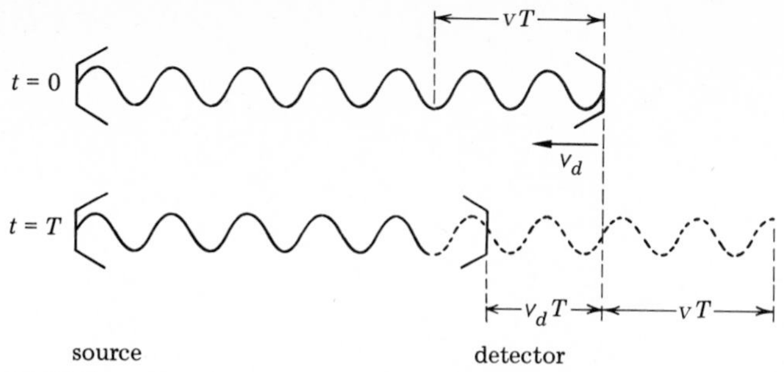

FIGURE 12.13

Sound wave detected by a detector
moving toward a stationary source.

Example 12.8 An ambulance siren oscillates at a frequency of 1200 Hz. What is the frequency of the sound reaching a stationary detector when the ambulance is moving toward the detector at 25 m/s?

Taking the speed of sound in air to be 340 m/s, Eq. 12.23 yields

$$f = \frac{v}{v - v_s} f_s$$

$$= \frac{340 \text{ m/s}}{340 \text{ m/s} - 25 \text{ m/s}} (1200 \text{ Hz}) = 1295 \text{ Hz}$$

If the ambulance is moving away from the detector, $v_s = -25$ m/s, so the frequency is

$$f = \frac{340 \text{ m/s}}{340 \text{ m/s} - (-25 \text{ m/s})} (1200 \text{ Hz})$$

$$= \frac{340 \text{ m/s}}{365 \text{ m/s}} (1200 \text{ Hz}) = 1118 \text{ Hz}$$

Thus, as the ambulance passes a stationary observer, the frequency of the sound will drop from 1295 to 1118 Hz. This is the origin of the characteristic change of pitch observed whenever a moving sound source passes by. ☐

Moving Detector Suppose the detector is moving with speed v_d toward the source (Fig. 12.13). Since the source is stationary, the wave in the air has the same frequency f_s as the source, and its wavelength is $\lambda_s = v/f_s$. However, this is not the frequency f detected by the moving detector. Figure 12.13 shows that in the time interval from $t = 0$ to $t = T$ a wave of length

$$vT + v_d T$$

will pass into the detector. The number of wavelengths in this train is

$$\frac{vT + v_d T}{\lambda_s} = \frac{vT + v_d T}{v/f_s} = \frac{v + v_d}{v} Tf_s$$

and is equal to the number fT of oscillations produced in the detector. Thus the detected frequency f is given by

$$fT = \frac{v + v_d}{v} Tf_s$$

or

$$f = \frac{v + v_d}{v} f_s \qquad \text{moving detector} \qquad 12.24$$

Although the form of Eq. 12.24 is different from Eq. 12.23, both equations give similar numerical results when v_d and v_s are much smaller than v.

DETECTION OF MOTION

The Doppler effect causes the frequency of a wave reflected from a moving object to be shifted. This phenomenon is the basis of a number of devices designed to detect the speed of an object. To understand how these devices work, let us consider what happens when a sound wave impinges on a solid object.

An incident sound wave of frequency f_s impinging on a stationary object causes the molecules of the object to oscillate with frequency f_s. This oscillation produces a *refracted* wave of frequency f_s that travels into the object and a *reflected* wave of frequency f_s that moves back into the air (Fig. 12.14). The object thus acts both as a detector and as a source of sound.

If the object is moving with speed v_0 toward the incident wave, it will act as a moving detector. That is, the molecules of the object will oscillate with the frequency detected by a detector moving with speed v_0 (Eq. 12.24, with $v_d = v_0$). These oscillating molecules then act as a source of frequency

$$f'_s = \frac{v + v_0}{v} f_s$$

But since the object is moving, the frequency of the reflected wave will be the frequency f generated by a source of frequency f'_s moving with speed v_0. That is, f will be given by Eq. 12.23, with $v_s = v_0$ and $f_s = f'_s$:

$$f = \frac{v}{v - v_0} f'_s = \frac{v + v_0}{v - v_0} f_s \qquad \text{moving reflector} \qquad 12.25$$

Some burglar alarm systems use the Doppler effect to detect the motion of a prowler. The system consists of a source and a detector (Fig. 12.15). The source fills the room with sound, and the detector detects the sound that is reflected from the walls and objects in the room. As long as the objects in the room are stationary, the detected sound waves will have the frequency f_s of the source. But if an object, such as a prowler, is moving about the room, the waves reflected from it will have the frequency given by Eq. 12.25. The detector sounds an alarm whenever it detects a wave with such a frequency.

Example 12.9 The source of a burglar alarm generates sound waves with a frequency of 2500 Hz. What will be the frequency of the waves reflected from an object moving with a speed of 1 m/s toward the source?

Equation 12.25 applies to an object moving toward the source. Thus with $v = 340$ m/s and $v_0 = 1$ m/s this equation gives

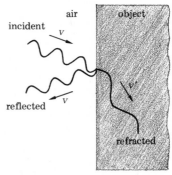

FIGURE 12.14

When a sound wave is incident on a surface, some of the sound is reflected from the surface and some is refracted into the surface.

FIGURE 12.15

A burglar alarm that uses sound to detect a moving object. (*Sonaguard.*)

$$f = \frac{v + v_0}{v - v_0} f_s = \frac{341 \text{ Hz}}{339 \text{ Hz}} (2500 \text{ Hz}) = 2515 \text{ Hz}$$

for the frequency of the reflected wave. Although the difference $f - f_s = 15$ Hz is only 0.6 percent of the source frequency f_s, it is easily measured using the phenomenon of *beats*. $\quad\square$

BEATS

The superposition of two sine waves with nearly equal frequencies f_1 and f_2 results in a pulsating wave (Fig. 12.16c). The frequency $\bar{f}$ of this wave is the average of the frequencies of its components

$$\bar{f} = \tfrac{1}{2}(f_1 + f_2) \tag{12.26}$$

and the amplitude of the wave oscillates with the *beat frequency* Δf, which is equal to the difference between f_1 and f_2:

$$\Delta f = f_2 - f_1 \tag{12.27}$$

To understand this, consider what happens in a time interval T equal to $1/\Delta f$. In this time the wave f_1 completes $f_1 T$ cycles, and wave f_2 completes

$$f_2 T = (f_1 + \Delta f) T$$
$$= f_1 T + \Delta f T = f_1 T + 1$$

FIGURE 12.16

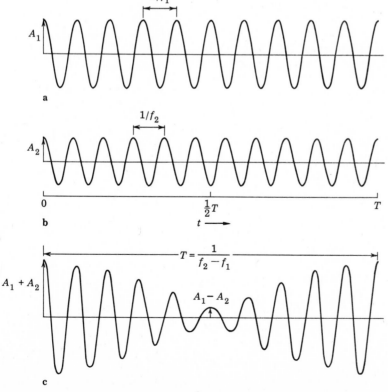

(a) and (b) Two waves with nearly the same frequency. (c) Superposition of the waves in (a) and (b).

displacement

cycles. That is, in the time $T = 1/\Delta f$, wave f_2 completes exactly one more cycle than wave f_1.

Figure 12.16 shows the case in which $f_1 T = 10$ and $f_2 T = 11$. At time $t = 0$ both waves have their maximum positive displacements A_1 and A_2, so the sum of these waves has the displacement $A_1 + A_2$. At time $t = T$, wave f_2 has completed exactly one more cycle than f_1, so again their displacements are A_1 and A_2, and their sum has the displacement $A_1 + A_2$. However, at time $t = \frac{1}{2}T$, wave f_2 has completed one-half cycle more than wave f_1, so wave f_2 has the displacement $-A_2$ while wave f_1 has the displacement A_1, and their sum has the displacement $A_1 - A_2$. Figure 12.16c shows the sum of waves f_1 and f_2 resembles a sine wave of frequency $\bar{f} = \frac{1}{2}(f_1 + f_2)$ whose amplitude varies between $A_1 + A_2$ and $A_1 - A_2$. The time interval $T = 1/\Delta f$ between amplitude maxima is called the *beat period*, and the number of maxima per second Δf is the *beat frequency*.

The Doppler effect is used to measure the speed of a distant object by superimposing a transmitted wave and a wave reflected from the object. The resulting mixture beats with a frequency equal to the difference between the frequencies of these waves (Eq. 12.27). The frequency of the reflected wave is determined by electronically measuring the beat frequency, and the speed of the object is then determined from the frequency of the reflected wave by Eq. 12.25. The entire calculation is done electronically by a microprocessor inside the instrument.

The highway police use radar to monitor the speed of automobiles. The radar signal, which is an electromagnetic wave similar to a radio wave (Sec. 13.1), is reflected from an oncoming vehicle and detected by an instrument which mixes it with a sample of the transmitted wave. The beat frequency is then converted electronically into the vehicle's speed, which is displayed on an instrument panel. Figure 12.17 shows a new speed detector which contains the transmitter, detector, and data-processing system in a small handheld unit.

FIGURE 12.17

Handheld speed detector used by highway police. (*Kustom Signals Division, Kustom Electronics, Inc.*)

Example 12.10 Show that when the speed v of the wave used to measure the speed v_0 of an object is much greater than v_0, the speed v_0 is related to the beat frequency Δf by

$$v_0 = \frac{1}{2}v\frac{\Delta f}{f_s} \qquad\qquad 12.28$$

The speed v of radar is equal to the speed of light (3×10^8 m/s), whereas the speed of an automobile is only around 30 m/s. It is difficult to obtain a precise result by substituting these numbers directly into Eq. 12.25. A better formula can be derived by making use of the fact that $v_0/v \ll 1$. First we write

$$v + v_0 = v\left(1 + \frac{v_0}{v}\right) \qquad \text{and} \qquad v - v_0 = v\left(1 - \frac{v_0}{v}\right)$$

so that Eq. 12.5 can be written

$$f = \frac{v(1 + v_0/v)}{v(1 - v_0/v)}f_s = \frac{1 + v_0/v}{1 - v_0/v}f_s$$

Then we use the formula

$$\frac{1}{1-x} \approx 1 + x$$

(which is valid when $x \ll 1$) to write

$$f = (1 + v_0/v)(1 + v_0/v)f_s$$
$$= [1 + 2v_0/v + (v_0/v)^2]f_s$$

Since $v_0/v \ll 1$, the expression $(v_0/v)^2$ is much less than v_0/v, so it can be dropped. Thus to good approximation we have

$$f = \left(1 + \frac{2v_0}{v}\right)f_s = f_s + 2\frac{v_0}{v}f_s$$

When the reflected wave of frequency f is mixed with a wave of frequency f_s, the resulting beat frequency is

$$\Delta f = f - f_s = 2\frac{v_0}{v}f_s$$

or

$$v_0 = \tfrac{1}{2}v\frac{f}{f_s}$$

For instance, the beat frequency produced when a radar wave of frequency $f_s = 10^{10}$ Hz is reflected from a vehicle moving at 30 m/s is

$$\Delta f = \frac{2v_0 f}{v} = \frac{2(30 \text{ m/s})(10^{10} \text{ Hz})}{3 \times 10^8 \text{ m/s}} = 2000 \text{ Hz} \qquad \square$$

12.6   Noise Control

Noise is unwanted sound. It may be the roar of a jet airplane, the clatter of a business machine, or the music from someone else's phonograph. Technology has so greatly increased the number of sound sources in our environment that noise control has become a major field of engineering.

The best method of noise control is to suppress it at the source. By taking the problem of noise suppression seriously, engineers can decrease the noise output of engines and machines. For instance, the noise from a modern jet engine (Fig. 12.18) is reduced by bypassing some of the intake air around the

FIGURE 12.18

High-bypass-ratio jet engine designed to reduce noise. (*General Electric.*)

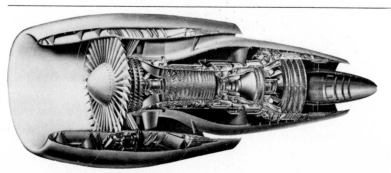

primary jet plume. Beyond the engine the bypass air mixes with the primary exhaust gases to produce an air stream of large cross section and reduced velocity. This leads to a lower noise level for the same engine thrust.

In this section we shall consider the major factors that affect noise levels once the noise is produced. These factors are distance, absorption in air, and transmission and reflection by a barrier.

VARIATION OF INTENSITY WITH DISTANCE

Distance is the most obvious way to reduce noise, since the intensity of the sound produced by a source decreases with the distance from the source. If the source is small (a *point source*), the sound spreads out from it in spherical waves (Fig. 12.19) in which the intensity I_1 is the same at all points that are the same distance r_1 from the source. From Eq. 12.19 the energy per second, or power P_1, that passes through the sphere of radius r_1 centered at the source is the product of the intensity I_1 at this distance and the area $a_1 = 4\pi r_1^2$ of the sphere:

$$P_1 = a_1 I_1 = 4\pi r_1^2 I_1$$

Likewise, the power P_2 that passes through a sphere of radius r_2 is

$$P_2 = a_2 I_2 = 4\pi r_2^2 I_2$$

In going from r_1 to r_2, some of the energy of the wave is absorbed by the air, but this energy is small if the distance from r_1 to r_2 is not too large. If we neglect this energy loss entirely, all the energy that passes through sphere 1 must also pass through sphere 2. Therefore, the energy per second P_1 that passes through sphere 1 is equal to the energy per second P_2 that passes through sphere 2. So from the last two equations we get

$$4\pi r_1^2 I_1 = 4\pi r_2^2 I_2$$

or

$$I_2 = \frac{r_1^2 I_1}{r_2^2} \qquad\qquad 12.29$$

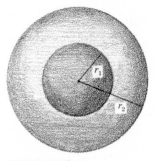

FIGURE 12.19

Sound waves spreading out from a point source.

The intensity I_2 decreases as the inverse square of the distance r_2 from the source. This is similar to the inverse-square decrease of the gravitational (Sec. 6.3) and electric (Sec. 15.2) forces.

Example 12.11 Table 12.3 shows that the sound intensity of a jet airplane is $10^2\,\text{W/m}^2$ at a distance of 10 m. Find the intensity and intensity level 1000 m from the airplane.

From Eq. 12.29 the intensity I_2 at a distance $r_2 = 1000$ m is

$$I_2 = \frac{r_1^2 I_1}{r_2^2} = \frac{(10\,\text{m})^2(10^2\,\text{W/m}^2)}{(1000\,\text{m})^2}$$

$$= 10^{-2}\,\text{W/m}^2$$

Table 12.3 shows that this intensity corresponds to an intensity level of 100 dB. We get the same result using Eq. 12.20:

$$\beta = (10 \text{ dB}) \log \frac{I_2}{I_0} = (10 \text{ dB}) \log \frac{10^{-2} \text{ W/m}^2}{10^{-12} \text{ W/m}^2}$$

$$= (10 \text{ dB}) \log (10^{10}) = 100 \text{ dB} \qquad \qquad \square$$

The difference between the intensity levels β_1 and β_2 corresponding to the intensities I_1 and I_2 is

$$\beta_1 - \beta_2 = (10 \text{ dB}) \log \frac{I_1}{I_0} - (10 \text{ dB}) \log \frac{I_2}{I_0}$$

$$= (10 \text{ dB}) \log \frac{I_1/I_0}{I_2/I_0} = (10 \text{ dB}) \log \frac{I_1}{I_2}$$

That is, the *difference* in intensity levels is proportional to the logarithm of the *ratio* of the intensities. If I_1 and I_2 are the intensities at the distances r_1 and r_2 from a point source, then from Eq. 12.29 we have

$$\frac{I_1}{I_2} = \frac{r_2^2}{r_1^2} = \left(\frac{r_2}{r_1}\right)^2$$

and so the difference in the intensity levels at r_1 and r_2 is

$$\beta_1 - \beta_2 = (10 \text{ dB}) \log \frac{I_1}{I_2} = (10 \text{ dB}) \log \left(\frac{r_2}{r_1}\right)^2$$

or

$$\beta_1 - \beta_2 = (20 \text{ dB}) \log \frac{r_2}{r_1} \qquad \qquad 12.30$$

For instance, if r_2 is twice as far from a source as r_1, we have

$$\beta_1 - \beta_2 = (20 \text{ dB}) \log 2 = 6 \text{ dB} \qquad \text{or} \qquad \beta_2 = \beta_1 - 6 \text{ dB}$$

Thus a doubling of the distance from a point source produces a 6-dB decrease in the intensity level.

ABSORPTION IN AIR

As the air in a sound wave is compressed, the energy of the wave is transformed into an increase in the translational and rotational kinetic energies of the gas molecules. As the air expands, the translational kinetic energy is fully recovered, but some of the rotational energy is not. The air in effect absorbs energy from the sound wave, transforming it into internal energy. The absorption of sound energy in air is an important mechanism in the attenuation of sound intensity.

The decrease $\Delta\beta$ in the intensity level of a sound wave that travels a distance x in air is

$$\Delta\beta = -(4.34 \text{ dB}) \, \alpha x \qquad \qquad 12.31$$

where α is the attenuation coefficient. The decrease in intensity level given by Eq. 12.31 is in addition to the inverse-square decrease given by Eq. 12.30.

The value of the attenuation coefficient depends on the frequency of the sound and the humidity and temperature of the air. Figure 12.20 shows curves of the attenuation coefficient for

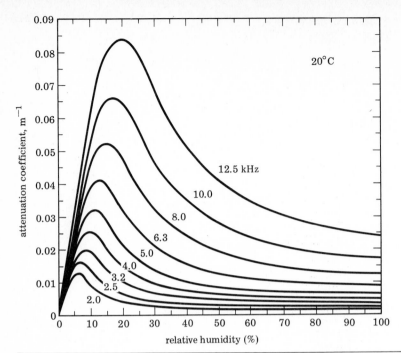

FIGURE 12.20

Plot of the attenuation coefficient against relative humidity for different frequencies. (*From C. M. Harris, Journal of the Acoustical Society of America 40:148 (1966)*).

different frequencies as a function of relative humidity. For frequencies below 2000 Hz, α is 0.002 m^{-1} or less, but it increases rapidly with frequencies above 2000 Hz. This means that high-frequency sound is absorbd more quickly than low-frequency sound. Fog horns emit a low-frequency sound because it will travel farther than a higher-frequency sound. The high-frequency whine of a jet engine does not travel more than a few hundred meters, but the low-frequency roar of the exhaust gas can be heard several kilometers away.

Example 12.12 Assume that 10 m from a jet engine half the noise is low-frequency sound with an attenuation coefficient of 0.002 m^{-1} and half is high-frequency sound with an attenuation coefficient of 0.02 m^{-1}. If the total intensity level at 10 m is 140 dB, find the intensity levels of each frequency component at 10 and 1000 m.

A 3-dB reduction in intensity level corresponds to a factor of 2 decrease in intensity. Thus at 10 m each frequency component has an intensity level of 137 dB. Example 12.11 showed that the intensity level at 1000 m is 40 dB less than its value at 10 m just because of the inverse-square law. Thus, neglecting absorption, the intensity levels of the two frequency components would both be 97 dB at 1000 m.

Equation 12.31 shows that the intensity level of the low-frequency component is further reduced by

$$\Delta\beta = -(4.34 \text{ dB})\, \alpha x = -(4.34 \text{ dB})(0.002 \text{ m}^{-1})(990 \text{ m})$$
$$= -8.7 \text{ dB}$$

because of absorption. Therefore, the intensity level of the low-frequency component at 1000 m is $(97 - 8.7)$ dB $= 88.3$ dB. Likewise, air absorption reduces the intensity level of the high-frequency component by

$$\Delta\beta = -(4.34 \text{ dB})\, \alpha x = -(4.34 \text{ dB})(0.02 \text{ m}^{-1})(990 \text{ m})$$
$$= -87 \text{ dB}$$

This reduces the intensity level of this component to $(97 - 87)$ dB $= 10$ dB, which is negligible compared to the intensity level of the low-frequency component. $\square$

Sound absorption in water is much less than in air. For this reason, underwater sound (sonar) can be used for communication and echolocation. Whale songs can be heard a thousand kilometers away, and hours after an underwater atomic bomb explosion in the Pacific, low-frequency signals were still being detected in California. These late signals were the result of multiple reflections from the islands of the Pacific Ocean. Presumably, if it were not for the relatively large absorption coefficient of air, our environment would be even noisier than it is.

TRANSMISSION AND REFLECTION BY A BARRIER

Barriers (interior and exterior walls, floors, and ceilings) are the major architectural elements used in the control of noise. They are most frequently used to isolate sound, i.e., to prevent sound from the noisy side of the barrier from reaching the quiet side. But they also can be used to reduce the noise already present in a room by reducing the level of reflected sound.

When sound is incident on a barrier, some of it will be reflected back into the original space, some will be transmitted through the barrier, and some will be absorbed by the barrier (Fig. 12.21). The absorbed sound is converted into internal energy of the barrier. From conservation of energy, we see that the incident intensity I_i is equal to the sum of the reflected intensity I_r, the transmitted intensity I_t, and the absorbed intensity I_a:

$$I_i = I_r + I_t + I_a \qquad \qquad 12.32$$

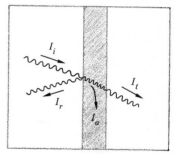

FIGURE 12.21

Sound reflected from and transmitted through a barrier.

Definition The *noise-reduction rating* $\Delta\beta_{nr}$ of a barrier is the difference between the intensity levels of the incident and transmitted sound:

$$\Delta\beta_{nr} = \beta_i - \beta_t = (10 \text{ dB}) \log \frac{I_i}{I_t} \qquad \qquad 12.33$$

The intensity level of the sound transmitted through a wall with a noise-reduction rating of 30 dB will be 30 dB less than the intensity level of the sound incident on the wall. That is, if the intensity level on the noisy side of the wall is 65 dB, the intensity level of the transmitted sound will be 65 dB $-$ 30 dB $=$ 35 dB. Remember, a reduction of 30 dB in intensity level corresponds to a factor of 1000 decrease in intensity.

Sound incident on a barrier causes the barrier to vibrate. The more massive the barrier, the smaller is the amplitude of the resulting vibration. For this reason, barriers made of concrete or brick have large noise-reduction ratings and thus provide good sound isolation. Double walls consisting of gypsum panels mounted on studs also have good sound isolation properties. In this case, the layer of air between the panels greatly reduces the intensity of the transmitted sound.

The noise-reduction rating of a wall depends critically on how the wall is mounted. Any cracks or holes in the wall will significantly impair its sound-isolation quality. For instance, it is wise not to place electric outlets on the double wall between private offices, because the holes will greatly decrease the noise-reduction rating of an otherwise well-constructed wall.

Definition The *reflection coefficient r* of a barrier is the ratio of the reflected to the incident intensity:

$$r = \frac{I_r}{I_i} \hspace{4cm} 12.34$$

This ratio depends on the frequency of the incident wave. Table 12.4 gives the reflection coefficients of some common building materials at 1000 Hz. Unglazed brick, for instance, reflects 96 percent of the sound intensity incident on it. Of the remaining 4 percent of the sound energy, some is transmitted through the brick, and the rest is converted to internal energy in the brick. An open window has a reflection coefficient of zero, because none of the sound impinging on the opening is reflected back into the room.

> **REMARK** Acoustical engineers prefer to work with the *absorption coefficient a*, which is equal to $1 - r$. We shall not use this term, however, because we are restricting the meaning of absorption to sound energy that is converted into internal energy.

Materials with low reflection coefficients (carpets, acoustic tiles) reduce the noise level in a room by preventing sound generated in the room from being reflected back into the room. Such materials usually have a light porous structure with many interconnected air spaces. A sound wave easily penetrates these spaces, and very little of the wave is reflected. The vibration of the air in the pores absorbs some of the sound energy, and the rest passes through.

It is interesting to note that the characteristics of a good nonreflecting barrier (light, porous) are the opposite of the characteristics of a good noise-isolating barrier (heavy, dense). A nonreflecting barrier, by its nature, must transmit a large fraction

TABLE **12.4**

Material	Reflection coefficient
Open window	0
Acoustic tile, mounted on wood furring	0.01 to 0.30
Carpet on foam rubber	0.31
Glass window	0.88
Brick, unglazed	0.96
Water	0.985
Marble, glazed tile	0.99

Reflection coefficients of some common building materials

These coefficients are for sound with a frequency of 1000 Hz. Acoustical engineers use the absorption coefficient a, which is equal to $1 - r$.

of the sound incident on it. It does little good, for instance, to stick acoustic tiles onto a concrete wall, because sound will penetrate through the tile, reflect off the concrete, and pass back into the room. Engineers must often compromise between the goal of reducing the noise level in a noisy room and the goal of preventing this noise from passing into another room.

Example 12.13 A 4-in brick wall has a noise-reduction rating of 40 dB. What fraction of the nonreflected intensity is absorbed in the brick, and what fraction is transmitted through the brick?

From Table 12.4 we see that the reflection coefficient of brick is 0.96, so from Eq. 12.34 we have

$$r = \frac{I_r}{I_i} = 0.96 \quad \text{or} \quad I_r = (0.96)I_i$$

From Eq. 12.32 the nonreflected intensity is

$$I_t + I_a = I_i - I_r = I_i - (0.96)I_i = (0.04)I_i$$

A noise-reduction rating of 40 dB means that the transmitted intensity I_t is 10^{-4} times the incident intensity I_i. That is,

$$I_t = 10^{-4}I_i$$

so the absorbed intensity I_a is

$$I_a = (0.04)I_i - I_t = (0.04)I_i - 10^{-4}I_i$$
$$= (0.0399)I_i$$

The ratio of the absorbed intensity to the nonreflected intensity is

$$\frac{I_a}{I_a + I_t} = \frac{(0.0399)I_i}{(0.04)I_i} = 0.9975$$

and the ratio of the transmitted intensity to the nonreflected intensity is

$$\frac{I_t}{I_a + I_t} = \frac{10^{-4}I_i}{(0.04)I_i} = 0.0025$$

This shows that only 0.25 percent of the nonreflected intensity is transmitted through the wall; the rest is absorbed. ☐

PROBLEMS

REMARK Use 340 m/s for the speed of sound in air.

1 Ultrasonic waves are sound waves with a frequency above the upper limit of human hearing (20,000 Hz). (a) What is the wavelength of an ultrasonic wave of frequency 8×10^4 Hz? (b) What is the period of the wave?
Ans. (a) 0.425 cm; (b) 12.5 μs

2 The displacement y of the air molecules produced by a sound wave is given by

$$y = (0.4) \, \mu m) \cos (500 \, \pi t)$$

What are the amplitude, frequency, and wavelength of the wave?

3 (a) What is the wavelength in air of a sound wave with a frequency of 125 Hz? (b) What is the wavelength of a wave with the same frequency in seawater?
Ans. (a) 2.72 m; (b) 12.0 m

4 What frequency must a sound wave have in seawater to have the same wavelength as a 500-Hz sound wave in air?

GUIDE TO MAJOR TOPICS

5 The lowest sound that can be heard by a young adult has a frequency of 3000 Hz and a pressure amplitude of 30 μPa. What is the displacement amplitude of this sound?
Ans. 3.9×10^{-12} m

6 What is the pressure amplitude of the wave described in Prob. 2?

7 Find the bulk modulus of ethanol.
Ans. 1.15×10^9 N/m^2

8 Use Eq. 12.7 and the ideal-gas law (Eq. 8.8) to show that the speed of sound in an ideal gas can be written

$$v = \sqrt{\frac{\gamma k T}{m}}$$

where k is the Boltzmann constant (Sec. 8.4), m is the mass of a single molecule, and T is the absolute temperature.

9 Find (a) the Young's modulus and (b) the Poisson ratio of lucite.
Ans. (a) 4.0×10^9 N/m^2; (b) 0.4

10 Find the shear modulus of copper.

11 A sound wave with an intensity of 3×10^{-4} W/m^2 is incident on an eardrum of area 0.60 cm^2. How much energy is incident on the eardrum in 10 min?
Ans. 1.08×10^{-5} J

12 A loudspeaker emits sound energy at a rate of 0.7 W uniformly in all directions. What is the intensity of the sound 10 m from the speaker?

13 Ultrasonic waves (see Prob. 1) have many applications in science and technology. One of their advantages is that ultrasonic waves of large intensity can be used without danger to the ear. Consider an ultrasonic wave of intensity $I = 2 \times 10^5$ W/m^2. (a) What is the pressure amplitude of this wave? (b) What is the intensity in water of an ultrasonic wave with this same pressure amplitude?
Ans. (a) 1.28×10^4 Pa; (b) 5.5×10^4 W/m^2

14 (a) What is the displacement amplitude in air of an ultrasonic wave of intensity 2×10^5 W/m^2 and frequency 5×10^4 Hz? (b) What is the displacement amplitude of its wave in water?

15 What is the intensity level of a sound with an intensity of 7.5×10^{-8} W/m^2?
Ans. 48.8 dB

16 What is the intensity of a sound with an intensity level of 53 dB?

344

17 What is the intensity level of a sound with a pressure amplitude of 0.01 Pa?
Ans. 50.8 dB

18 Show that the intensity level of a sound with pressure amplitude A_p is given by

$$\beta = (20\ \text{dB}) \log \frac{A_p}{A_{p0}}$$

where A_{p0} is the reference pressure. What is the appropriate value of A_{p0}?

19 (a) What percentage of the population cannot hear a 200-Hz sound at 40 dB? (b) What is the lowest intensity level of a 1000-Hz sound that can be heard by 90 percent of the population?
Ans. (a) 25%; (b) 30 dB

20 What is the intensity level of a 60-Hz signal that sounds as loud as a 600-Hz signal with an intensity level of 24 dB?

21 (a) What are the first three characteristic frequencies of a closed pipe of length 1.5 m? (b) What are the first three characteristic frequencies if the pipe is filled with carbon dioxide at 0°C instead of air?
Ans. (a) 56.7, 170, and 283 Hz; (b) 43.2, 129, and 216 Hz

22 The outer ear consists of the externally visible parts of the ear and the *ear canal,* which is the passageway from the outside to the eardrum. (a) Find the fundamental frequency of this passageway if it is 2 cm long. (b) Suggest a role that this cavity might play in hearing.

23 A test vehicle has a loudspeaker mounted on it that oscillates at a frequency of 3000 Hz. A ground-based monitor measures the frequency of the sound coming from the speaker. When the vehicle is moving directly toward the monitor, the measured frequency is 3250 Hz. What is the speed of the vehicle?
Ans. 26.1 m/s

24 Astronomers studying the light coming from distant galaxies find that the frequency of this light is less than it would be if it were emitted by a stationary object. For instance, light that would have a frequency of 6.0×10^{14} Hz if it came from a stationary object might have a frequency of 5.8×10^{14} Hz coming from a particular galaxy. This is interpreted as a Doppler shift due to the motion of the galaxy *away* from the earth. Find the speed of the galaxy in this case. (The speed of light is 3×10^8 m/s.)

25 A submarine moving at 15 m/s detects a sonar (underwater sound) signal coming directly at it. If the detected frequency is 3.0×10^4 Hz, what is the frequency of the wave in the water?
Ans. 2.97×10^4 Hz

26 Ultrasonic waves can be used to measure the speed of a piston inside an engine by aiming a beam at the piston and measuring the frequency of the reflected wave. If the maximum frequency of the reflected wave is 7 percent greater than the frequency of the incident wave, what is the maximum speed of the piston?

27 A burglar alarm system uses an ultrasonic wave with a frequency of 35,000 Hz to detect motion. When a signal reflected from a moving object is mixed with the original wave, a beat frequency of 200 Hz is produced. What is the speed of the object? (Assume the object is moving directly toward the detector.)
Ans. 0.97 m/s

28 A radar beam with a frequency of 2.5×10^{10} Hz is reflected from an oncoming vehicle. The beat frequency produced when the reflected signal is mixed with the original signal is 3500 Hz. What is the speed of the vehicle?

REMARK Neglect the absorption of sound in air in Probs. 29 through 34.

29 The intensity level 2 m from a motorcycle is 90 dB. At what distance is the intensity level 60 dB?
Ans. 63.2 m

30 The sound from an organ pipe has a sound level of 120 dB at a point 1 m from the open end of the pipe. (a) What is the sound level at a distance of 32 m from the end of the pipe? (b) What is the rate at which sound energy emerges from the open end of the pipe?

31 A firework rocket explodes at an altitude of 400 m, producing an average sound intensity of 6.7×10^{-2} W/m² for 0.2 s at a point on the ground directly below it. (a) What is the average intensity of the sound at a distance of 10 m from the rocket? (b) What is the sound level (in decibels) of the sound 10 m from the rocket? (c) What is the total sound energy radiated in the explosion?
Ans. (a) 107 W/m²; (b) 140 dB; (c) 2.69×10^4 J

32 The intensity level 30 m from a truck is 84 dB. What is the intensity level 5 m from the truck?

33 The pressure amplitude of a sound wave is 0.04 N/m² at a distance of 12 m from its source.

345

What is the pressure amplitude of the wave 150 m from the source?
Ans. 0.0032 N/m²

34 A line source, such as a busy highway, emits a semicylindrical sound wave (Fig. 12.22). The intensity I_1 is the same at all points on a cylinder of radius r_1 centered on the highway. Show that the intensity I_2 on a cylinder of radius r_2 is given by

$$I_2 = \frac{r_1 I_1}{r_2}$$

and the difference in the intensity levels at r_1 and r_2 is

$$\beta_1 - \beta_2 = (10 \text{ dB}) \log \frac{r_2}{r_1}$$

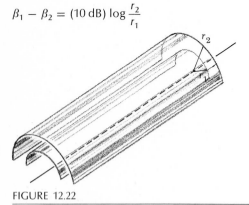

FIGURE 12.22

Waves spreading out from a line source (Prob. 34).

35 A loudspeaker is emitting low-frequency sound uniformly in all directions at the rate of 0.8 W. The attenuation coefficient for this sound is 0.0075 m⁻¹. What is the intensity level 60 m from the speaker?
Ans. 70.5 dB

36 A loudspeaker is emitting high-frequency sound uniformly in all directions. Assume the attenuation coefficient for this sound is 0.03. What must be the power of the emitted sound in order to produce an intensity level of 70 dB at a distance of 50 m from the speaker?

37 A loudspeaker is emitting sound uniformly in all directions at the rate of 0.2 W. The intensity level 100 m from the speaker is 60 dB. What is the attenuation coefficient for this sound?
Ans. 0.0047 m⁻¹

38 Show that the intensity level at a distance r from a point source that is emitting sound at the rate P is

$$\beta = 109 \text{ dB} + (10 \text{ dB}) \log \frac{P}{r^2} - (4.34 \text{ dB})\alpha r$$

where α is the attenuation coefficient for this sound.

39 The noise-reduction rating of a barrier is 20 dB and its reflection coefficient is 0.2. What percentage of the incident intensity is absorbed?
Ans. 79 percent

40 A barrier with a reflection coefficient of 0.85 absorbs 99 percent of the nonreflected intensity. What is the noise-reduction rating of the barrier?

BIBLIOGRAPHY

DENES, PETER B., and ELLIOT N. PINSON: *The Speech Chain,* Bell Telephone Laboratories, Murray Hill, N.J., 1963. A clear and concise description of the physics of speech and hearing. Especially interesting is the account of how the position of the tongue controls the resonant frequencies of the mouth cavity.

MEYER, ERWIN, and ERNST-GEORG NEUMANN: *Physical and Applied Acoustics,* Academic Press, New York, 1972. Chapter 1 gives a detailed description of the different types of acoustic waves in solid bodies.

RIGDEN, JOHN S.: *Physics and the Sound of Music,* John Wiley & Sons, Inc., New York, 1977. This book describes the production, propagation, perception, and reproduction of musical sounds. Besides explaining the physics of musical instruments, it gives useful information about sound-reproduction equipment.

SUNDBERG, JOHAN: "The Acoustics of the Singing Voice," *Scientific American* **236**:82 (March 1977). How the resonant cavity of the mouth and throat is used to produce sounds in singing and speaking.

WHITE, FREDERICK A.: *Our Acoustic Environment,* John Wiley & Sons, Inc., New York, 1975. Broad treatment of diverse topics in environmental acoustics for nonspecialists. Topics include the character of sound; production, propagation, and perception of sound; noise control; speech; music; auditorium acoustics; and the psychophysics of hearing.

L

ight, like sound, is a wave that can transport energy without transporting mass. Unlike sound, however, light does not have a basis in mechanics. Sound is the vibration of the molecules of a material medium produced by mechanical forces, so its properties can be derived from the Newtonian laws of mechanics. A light wave, on the other hand, is not the vibration of a material substance. Light is fundamentally different from sound, and its properties cannot be derived from the laws of mechanics.

Nevertheless, a wave is a wave, and the general properties of waves discussed in Chapter 11 apply as well to light as to sound. In this chapter we shall discuss some important properties of light that can be understood without a detailed knowledge of the structure of a light wave. These properties find application in optics, photometry, and stress analysis. Further application to spectroscopy will be discussed in Chapter 20.

13

LIGHT

13.1 The Nature of Light

WAVE NATURE OF LIGHT

It was known to Aristotle (384–322 B.C.) that sound is caused by vibrations in the air. This knowledge was probably based on the observation that music is produced by vibrating strings. In fact, the study of the relation of musical tones to the length of the vibrating string was well developed in the ancient world. Of course, an adequate explanation of sound waves was not possible until the time of Newton.

Though nothing was known about the fundamental nature of light in Newton's time, it was natural to speculate that light was a wave, similar to sound. Christian Huygens (1629–1695), a contemporary of Newton, developed a wave theory of light, but Newton himself favored a theory according to which light is composed of massless particles (corpuscles). Newton hypothesized that these corpuscles travel through space at constant speed and that there is a different type of corpuscle for each color. His main objection to the wave theory was that light, unlike sound, does not appear to bend around corners.

Newton's great reputation and the absence of any definitive evidence one way or another led to the general acceptance of the corpuscular theory during the eighteenth century. This did not hinder developments in optics, however, because these developments were based on empirical laws that are consistent with either a wave or corpuscular theory.

The wave nature of light was finally established by a series of experiments that demonstrated that light obeys the superposition principle (Sec. 11.3). These experiments were first performed by Thomas Young (1773–1829), the great Egyptologist, and later and more definitively by Augustin Fresnel (1788–1827). Young's experiment is discussed in the next section.

ELECTROMAGNETIC NATURE OF LIGHT

At the time of these advances in the study of light, important discoveries were being made in the fields of electricity and magnetism (Sec. 15.1). The basic laws of electricity and magnet-

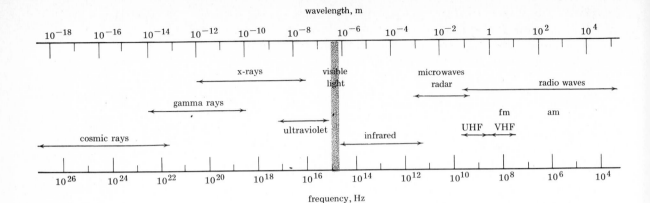

FIGURE 13.1

Electromagnetic radiation between 10^4 and 10^{26} Hz.

ism, which had been discovered in the first half of the nineteenth century, were formulated into a comprehensive mathematical theory by James Clerk Maxwell (1831–1879). From this theory Maxwell deduced that there should exist electromagnetic waves consisting essentially of oscillating electric and magnetic fields that propagate through space with a definite speed. According to Maxwell's theory, this speed is given in terms of certain well-known electric constants that enter into the theory. When Maxwell calculated the speed of electromagnetic waves from these constants, he found it equal to the speed of light. This great success of Maxwell's theory established that light is a form of electromagnetic radiation.

Today physicists are familiar with electromagnetic waves with wavelengths ranging from less than 10^{-17} m to more than 10^4 m. Only waves with wavelengths in the narrow range between 4×10^{-7} and 7×10^{-7} m are detected by the human eye and thus constitute visible light. Waves with longer and shorter wavelengths have special names, such as radio waves, microwaves, infrared waves, ultraviolet waves, x-rays, etc. Figure 13.1 shows the full range of electromagnetic radiation that has been studied, together with the names given to different regions. Except for the visible region, the boundaries between the regions are not sharply defined.

REMARK In analogy with all other waves, electromagnetic waves were originally thought to be displacements in a medium. It would have been no ordinary medium, of course, since light and other forms of electromagnetic radiation travel to us from distant stars, through outer space. During the nineteenth century it was thought that there was a special substance, called the *ether*, permeating all space. Electromagnetic waves were thought to be vibrations in the ether.

The ether would have had to be an extraordinary substance, however, because it would offer no resistance to the motion of the planets about the sun. Any ordinary material, even a very dilute gas, would apply frictional forces to the planets that would produce observable effects on their motion. A substance, like the hypothetical ether, which produced no effect on ordinary matter would seem to be indistinguishable from space itself. That is, it had no physical reality other than being the medium for the propagation of electromagnetic waves. The failure of attempts made toward the end of the nineteenth century to detect the ether led physicists to abandon the idea of the ether altogether and to accept that there is no mechanical foundation to electromagnetic waves.

SPEED OF LIGHT

In addition to the question of the nature of light, there is the question of its speed. Indeed, at one time it was not universally believed that light had a finite speed. In about the year 1600, Galileo tried and failed to measure the speed of light with a crude experiment. In 1676, Ole Römer (1644–1710) obtained the first finite value for the speed of light, based on discrepancies in the motions of the moons of Jupiter that depended on the distance between Jupiter and earth. However, his result was not generally accepted.

It was not until the middle of the nineteenth century that direct experimental measurements of the speed of light were obtained. The first measurement was made in 1849 by Armand Hippolyte Louis Fizeau (1819–1896), who used an ingenious modification of Galileo's unsuccessful measurement. Galileo had stationed two assistants, each with a lantern, some distance apart (Fig. 13.2a). Assistant A would open his lantern, and assistant B was instructed to open his lantern when he saw the light from A's lantern. Galileo tried to measure the time between the opening of A's lantern and the arrival of the light from B's lantern back to A. Even with a distance of 5 mi between A and B, however, the time is only 5×10^{-5} s, which is much too small for Galileo to have detected.

Fizeau replaced assistant B by a mirror (Fig. 13.2b), so that the light would be instantaneously returned to A without any delay caused by B's reaction time. Assistant A was replaced by a rotating toothed wheel in front of a light source, so that as the wheel rotated, pulses of light were automatically transmitted to the mirror. The rate of rotation of the wheel was increased until an observer in back of the wheel did *not* see the reflected light.

FIGURE 13.2

Measurements of the speed of light. (a) Galileo tried unsuccessfully to measure the time interval between when assistant A opens his lantern and when A first sees B open his lantern. (b) Fizeau replaced assistant A by a toothed wheel rotating in front of a lantern and assistant B by a mirror that instantaneously reflects light pulses back toward the mirror.

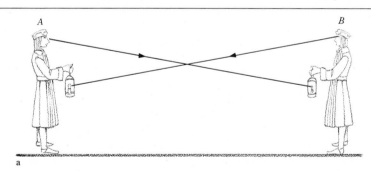

a

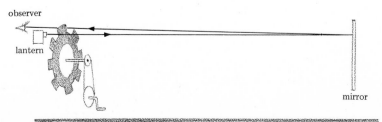

b

This happens when the time required for the light to travel from the wheel to the mirror and back again is equal to the time required for the wheel to move from a space to the next tooth, so that the tooth blocks the returning light from the observer's vision. From knowledge of the distance between the wheel and the mirror and the rate of rotation of the wheel, Fizeau calculated the speed of light.

Since Fizeau's first measurement, the speed of light has been measured repeatedly by different methods with increasing accuracy. Table 1.3 gives the results of some of these measurements. For our purposes the speed of light (in vacuum) is taken to be

$$c = 3.00 \times 10^8 \text{ m/s}$$

The symbol c is always used for the speed of light in vacuum.

The speed of light in transparent substances, e.g., water and glass, is found from measurements to be always less than the speed of light in vacuum. (The measurement of Fizeau and other early investigators was in air, of course. The speed of light in air is only slightly different from its speed in vacuum and this difference can be neglected in most cases.)

Definition The *index of refraction n* of a substance is the ratio of the speed of light c in vacuum to the speed of light v in the particular substance:

$$n = \frac{c}{v} \qquad 13.1$$

Example 13.1 If the speed of light in water is 2.25×10^8 m/s, what is the index of refraction of water?

From Eq. 13.1 the index of refraction is

$$n = \frac{c}{v} = \frac{3.00 \times 10^8 \text{ m/s}}{2.25 \times 10^8 \text{ m/s}} = 1.33 \qquad \square$$

Since the speed of light in any substance is always less than c, the index of refraction of every substance is greater than 1. For a particular substance, the value of n depends somewhat on the wavelength of light. Table 13.1 gives the index of refraction of some common substances at the wavelength of yellow light, $\lambda = 5.89 \times 10^{-7}$ m = 589 nm.*

TABLE **13.1**

Index of refraction of some common substances

These values are for yellow light.

Substance	Index of refraction
Acetone at 20°C	1.3584
Air at STP	1.0002926
Canada balsam	1.530
Diamond	2.4168
Ethanol at 20°C	1.36008
Glass:	
Fused quartz	1.458
Heavy flint	1.650
Zinc crown	1.517
Quartz, crystalline	1.553
Sodium chloride	1.544
Water at 20°C	1.33335

13.2 Reflection and Refraction

Definition A *plane wave* is a wave in which the displacement is the same at all points in a plane perpendicular to the direction of propagation. These planes, called *wavefronts,* move with the speed v of the wave. Figure 13.3 shows a transverse plane wave traveling along the z axis. Figure 13.4 is a two-dimensional representation of a similar wave. The solid lines are the edge-on view of the planes in which the displacement has its maximum positive value (the *crests*), and the broken lines are the edge-on

*1 nm = 1 nanometer = 10^{-9} m.

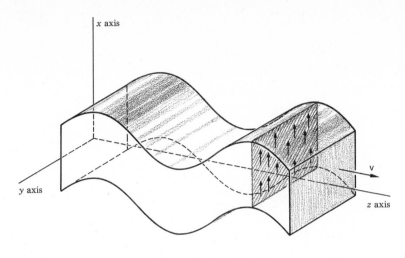

x axis

y axis

z axis

v

FIGURE 13.3

A transverse-plane wave traveling along the z axis.

view of the planes in which the displacement has its maximum negative value (the *troughs*). The distance between solid (or broken) lines is the wavelength λ of the wave. As long as the wave remains planar, it can be represented by a single line, called a *ray*, drawn perpendicular to the wavefronts. The ray represents the direction of propagation of the wave.

Figure 13.5a shows a plane wave propagating with speed v_1 in medium 1. When this wave arrives at the boundary separating medium 1 from medium 2, part of the wave is *reflected* back into medium 1, and part is *refracted* into medium 2. The intensity of the reflected wave is rI_0, where I_0 is the intensity of the incident wave, and r is the fraction of the wave reflected. If there is no absorption, the intensity of the refracted wave is $I_0 - rI_0 = (1 - r)I_0$. The value of r varies between 0 and 1, depending on the circumstances. It may be exactly 1.00, in which case there is no refracted wave and the incident wave is totally reflected. However, r is never exactly zero: there is always some reflection from the boundary separating two media.

Figure 13.5a shows both the wavefronts and the rays of the incident, reflected, and refracted waves. The rays correspond physically to narrow beams of light, as seen in Fig. 13.5b. This figure shows an incident beam of light reflected and refracted at the boundary between air and glass. Of particular interest are the angles that the incident, reflected, and refracted rays make with a boundary. These angles are customarily measured with respect to a line (the normal) drawn perpendicular to the boundary. The angles of incidence θ_1, of reflection θ_1', and of refraction θ_2 are shown in Fig. 13.5a. Note that these are also the angles between the respective wavefronts and the boundary itself.

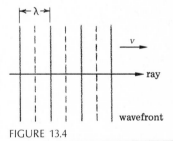

λ

v

ray

wavefront

FIGURE 13.4

Two-dimensional representation of a plane wave.

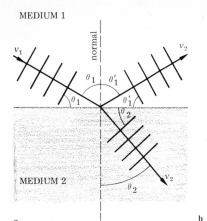

MEDIUM 1

MEDIUM 2

a

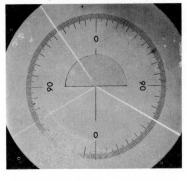

b

FIGURE 13.5

Reflection and refraction of light at the boundary between two media. (a) The angles of incidence θ_1, reflection θ_1', and refraction θ_2 are measured with respect to the normal. (b) Photograph of a light ray reflected and refracted at the boundary of air and glass. Note that at the curved glass-air boundary the angles of incidence and refraction are both zero. (*From* PSSC Physics, 4th ed., copyright © 1976, by Education Development Center, Inc., used with permission.)

REFLECTION

Law of Reflection *The angle of reflection θ_1' is equal to the angle of incidence θ_1:*

$$\theta_1' = \theta_1 \qquad\qquad 13.2$$

REMARK The law of reflection follows from the fact that the speed v_1 of the reflected wave is the same as the speed v_1 of the incident wave. Figure 13.6 shows the position of a wavefront $A'A$ of the incident wave at time $t = 0$ and the position of the reflected portion BB' of this wavefront at time $t = T$. In time T part of the wave is reflected from A' to B', while part is still moving from A to B. But since the incident and reflected waves have the same speed, the distance d that the incident wave travels in time T is equal to the distance d' that the reflected wave travels in time T. With $d = d'$, it follows that the right triangles $A'AB$ and $BB'A'$ are congruent, which proves that $\theta_1 = \theta_1'$.

Example 13.2 Figure 13.7 shows two mirrors at right angles to each other. A ray of light incident on one mirror at an angle $\theta = 62°$ is reflected toward the other mirror. (a) What is the angle of incidence of the ray on the second mirror? (b) Show that after being reflected from the second mirror, the ray emerges parallel to the incoming ray.

(a) Let θ' be the angles of reflection from the first mirror, and let ϕ be the angles of incidence on the second mirror (Fig. 13.8a). From the law of reflection applied to the first mirror we have $\theta' = \theta = 62°$. The normals OP and OP' to the two perpendicular mirrors are perpendicular to each other, so POP' is a right triangle. Consequently,

$$\theta' + \phi = 90° \qquad \text{or} \qquad \phi = 90° - \theta = 90° - 62° = 28°$$

(b) The angle of reflection from the second mirror is $\phi' = \phi = 28°$, as shown in Fig. 13.8b. Both the incident ray SP and the emerging ray $S'P'$ make the angle $2 \times 62° = 124°$ with the line PP', so these rays are parallel to each other. This is true for any angle of incidence θ. ☐

REMARK Two or three mirrors arranged at right angles to each other is called a *corner mirror*, or a *corner reflector*. It has the property that a ray incident on it from any direction will be reflected back on itself. Astronauts placed a corner mirror on the surface of the moon during the first moon landing. Later a laser beam was aimed at this mirror through a telescope on

FIGURE 13.6

A plane wavefront reflected from a plane boundary.

MEDIUM 1

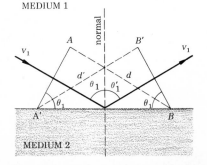

MEDIUM 2

earth, and the reflected light was collected by the same telescope. By timing the interval between the transmission of a light pulse and its return, the distance to the moon was determined with great precision.

When a plane wave is incident on a smooth, flat surface, such as a mirror, the angle of incidence is the same at all points on the surface, so all the reflected light has the same angle of reflection. This is called *specular,* or *regular,* reflection. But most surfaces, such as the paper of this book, are not smooth. Under high magnification, the surface is seen to consist of millions of microsurfaces, each oriented in a different direction (Fig. 13.9), so that the angle of incidence of the light illuminating the paper at any point depends on the orientation of the microsurface at that point. Consequently, light that is incident from a single direction is reflected in different directions from different points on the surface. This is called *diffuse reflection*. It is not a contradiction of the law of reflection because the law is still valid for each microsurface.

Figure 13.10 shows the distribution of light from various surfaces. For a mirrorlike surface, all light incident at angle θ is reflected at angle θ' (Fig. 13.10a). The reflection is 100 percent specular. For a rough-mat surface, such as a plaster wall, light incident at one angle is reflected uniformly at all angles (Fig. 13.10b). The reflection is nearly 100 percent diffuse. For a glossy surface, such as porcelain-enameled steel, some of the incident light is specularly reflected and some is diffusely reflected (Fig. 13.10c). It is important to realize that the law of reflection applies rigorously only to mirrorlike surfaces or to microsurfaces. Most macrosurfaces reflect some light in all directions.

REFRACTION

Law of Refraction *The angle of incidence θ_1 is related to the angle of refraction θ_2 by*

$$n_1 \sin \theta_1 = n_2 \sin \theta_2 \qquad 13.3$$

where n_1 and n_2 are the indexes of refraction of the first and second media. Equation 13.3 is also known as Snell's law.

REMARK The law of refraction follows from the fact that the speed v_2 of the refracted wave is different from the speed v_1 of the incident wave. Figure 13.11 shows the position of a wavefront $A'A$ of the incident wave at time $t = 0$ and the position of the refracted portion $C'C$ of this wavefront at time $t = T$. In time T part of the wave travels from A to C in medium 1, while part travels from A' to C' in medium 2. Thus the distances d and d' are

$$d = v_1T \quad \text{and} \quad d' = v_2T$$

so

$$\frac{d}{d'} = \frac{v_1}{v_2}$$

In terms of the indexes of refraction n_1 and n_2 of the two media, we have

$$v_1 = \frac{c}{n_1} \quad \text{and} \quad v_2 = \frac{c}{n_2}$$

so

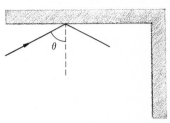

FIGURE 13.7

Two mirrors at right angles to each other. A ray is incident on the horizontal mirror.

FIGURE 13.8

(a) The ray incident on the horizontal mirror in Fig. 13.7 at angle θ is incident on the vertical mirror at angle $\phi = 90° - \theta$. (b) The final ray $S'P'$ emerges parallel to the incident ray SP.

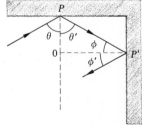

a

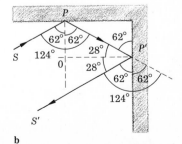

b

$$\frac{d}{d'} = \frac{n_2}{n_1}$$

From the right triangles $A'AC$ and $CC'A'$ in Fig. 13.11 we see that

$$\sin \theta_1 = \frac{d}{A'C} \quad \text{and} \quad \sin \theta_2 = \frac{d'}{A'C}$$

so

$$\frac{\sin \theta_1}{\sin \theta_2} = \frac{d}{d'} = \frac{n_2}{n_1}$$

or

$$n_1 \sin \theta_1 = n_2 \sin \theta_2$$

This proves the law of refraction

Example 13.3 A beam of light is incident at an angle θ_1 on one side of a sheet of glass with parallel sides (Fig. 13.12). What is the angle ϕ_1 at which the beam emerges from the other side of the glass?

According to Eq. 13.3, the angle of refraction θ_2 of the light in the glass is given by

$$\sin \theta_2 = \frac{n_1}{n_2} \sin \theta_1$$

Since the sides of the sheet of glass are parallel, θ_2 is equal to the angle ϕ_2 at which the light is incident on the second surface. As the light passes back into air, it is again refracted. This angle of refraction ϕ_1 is given by

$$\sin \phi_1 = \frac{n_2}{n_1} \sin \phi_2 = \frac{n_2}{n_1} \sin \theta_2 = \frac{n_2}{n_1} \frac{n_1}{n_2} \sin \theta_1 = \sin \theta_1$$

Therefore, $\phi_1 = \theta_1$, which means that the light emerges from the glass parallel to the incident light ray. That is, a sheet of glass with parallel sides produces no net deflection of a ray of light, although the emerging ray is displaced relative to the incident ray. $\square$

Example 13.4 A beam of light is incident at an angle $\theta_1 = 40°$ on one side of a glass prism with *apex angle* $A = 60°$ and index of refraction $n_2 = 1.5$ (Fig. 13.13). What is the *angle of deviation* θ_D between the incident and emergent beam?

Because the sides of the prism are not parallel, the beam does not emerge parallel to the incident beam. To find the angle of deviation, we must trace the beam through the prism. Thus the angle θ_2 at which the incident beam is refracted into the prism is

$$\sin \theta_2 = \frac{n_1}{n_2} \sin \theta_1 = \frac{1}{1.5} \sin 40° = 0.429$$

or

$$\theta_2 = 25.4°$$

The angle of incidence ϕ_2 of this ray on the other side of the prism is found by simple geometry. The sum of the interior angles of the triangle OST is 180°:

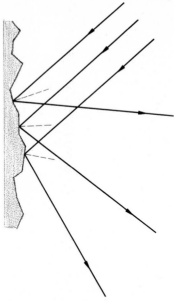

FIGURE 13.9

Enlarged view of light reflected from the microsurfaces of a rough surface.

FIGURE 13.10

Light reflected from various surfaces: (a) specular reflection from a mirrorlike surface, (b) diffuse reflection from a mat surface, and (c) mixed specular and diffuse reflection from a glossy surface.

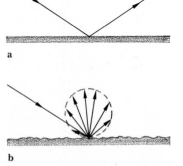

a

b

c

$$A + \alpha + \beta = 180°$$

But $\alpha = 90° - \theta_2$, and $\beta = 90° - \phi_2$, so

$$A + (90° - \theta_2) + (90° - \phi_2) = 180°$$

or

$$\phi_2 = A - \theta_2$$

Thus, in this example, $\phi_2 = 60° - 25.4° = 34.6°$. The angle of refraction ϕ_1 with which the ray emerges back into air is found by using Eq. 13.3 once more:

$$\sin \phi_1 = \frac{n_2}{n_1} \sin \phi_2 = 1.5 \sin 34.6° = 0.853$$

or

$$\phi_1 = 58.6°$$

The *angle of deviation* θ_D can be shown by simple geometry to be

$$\theta_D = (\theta_1 - \theta_2) + (\phi_1 - \phi_2) = \theta_1 + \phi_1 - A$$

which in this example is $\theta_D = 40° + 58.6° - 60° = 38.6°$. □

REMARK This example shows that the angle of deviation depends on the apex angle A, the angle of incidence θ_1, and the index of refraction n_2 of the glass. Since the index of refraction of glass is slightly different for different wavelengths of light, the angle of deviation depends also on wavelength. White light is a mixture of light of different wavelengths, so when a beam of white light is passed through a prism, the different wavelengths of which it is composed are deviated through different angles. Consequently, the light is broken into its component wavelengths. The eye senses these different wavelengths as different colors, and so we see the light emerge with its spectrum of colors. Similarly, a rainbow is produced by sunlight refracting through drops of water in the sky.

TOTAL INTERNAL REFLECTION

A ray of light is refracted toward the normal when it goes into a medium of higher index of refraction, and it is refracted away from the normal when it goes into a medium of lower index of refraction. This latter case is especially interesting. Suppose, for example, that a ray of light in glass ($n_1 = 1.5$) is incident on the glass-air boundary at an angle $\theta_1 = 60°$. From Snell's law the angle of refraction θ_2 is given by

$$\sin \theta_2 = \frac{n_1}{n_2} \sin \theta_1 = \frac{1.5}{1} \sin 60° = 1.3$$

But there is no angle with a sine of 1.3, and in fact no light emerges into the air. All the light incident on the glass-air boundary is reflected back into the glass. This phenomenon, called *total internal reflection*, occurs whenever the angle of incidence θ_1 produces a value of $\sin \theta_2$ that is greater than 1. Thus the condition for total internal reflection is

$$\frac{n_1}{n_2} \sin \theta_1 > 1$$

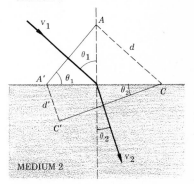

FIGURE 13.11

A plane wavefront refracted at the plane boundary between two media.

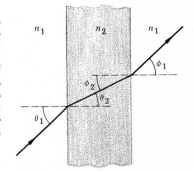

FIGURE 13.12

A light ray passing through a sheet of glass with parallel sides.

FIGURE 13.13

A light ray passing through two sides of a prism.

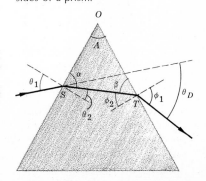

which can be satisfied only if $n_1 > n_2$. The smallest, or *critical*, angle θ_{1c} for internal reflection occurs when

$$\sin \theta_{1c} = \frac{n_2}{n_1} \qquad\qquad 13.4$$

For instance, the sine of the critical angle at the glass-air boundary is

$$\sin \theta_{1c} = \frac{1}{1.5} = 0.667$$

so that the critical angle is

$$\theta_{1c} = 42°$$

Figure 13.14 shows a light ray incident at right angles to one face of a 90–45–45 prism. Since the angle of incidence is 0°, the ray is undeviated as it enters the glass and is incident at 45° on the second face. This is greater than the critical angle for glass, so the ray is totally reflected from the second face. In Fig. 13.14a the ray then emerges from the third face, having been reflected through 90°. Such an arrangement is sometimes used in periscopes to raise the line of sight (Fig. 13.15a). In Fig. 13.14b the ray makes a second total internal reflection from the third face of the prism and emerges from the first face, having been reflected through 180°. Such an arrangement is used in binoculars (Fig. 13.15b) to increase the distance a ray travels from the front to the back lens and to produce an erect (noninverted) image. Without these prisms the binoculars would have to be much longer and objects viewed through it would appear upside down.

A ray going into a medium of higher index of refraction has an angle of refraction less than the angle of incidence. There is thus a maximum angle of refraction, which occurs when the angle of incidence is 90°. This is shown in Fig. 13.16, where light is going from a substance with index of refraction n_1 into a prism with a greater index of refraction n_2. The maximum angle of refraction θ_{2m} of a ray in the prism is given by

$$n_1 \sin 90° = n_2 \sin \theta_{2m}$$

or

$$\sin \theta_{2m} = \frac{n_1}{n_2} \qquad\qquad 13.5$$

This is the same as Eq. 13.4 with n_1 and n_2 interchanged, since now $n_1 < n_2$.

A *refractometer* is an instrument for measuring the index of refraction of a substance. In one form of refractometer the substance is placed on one face of a prism with a higher index of refraction and the light that emerges from the second face is viewed through a telescope mounted on a circular track (Fig. 13.16). From measurement of the angle ϕ of the last emerging ray, and from the known index of refraction n_2 of the prism, the unknown index of refraction n_1 of the substance can be calculated.

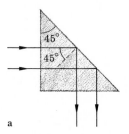

a

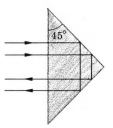

b

FIGURE 13.14

Light being totally reflected inside a right-angle prism.

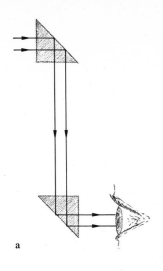

a

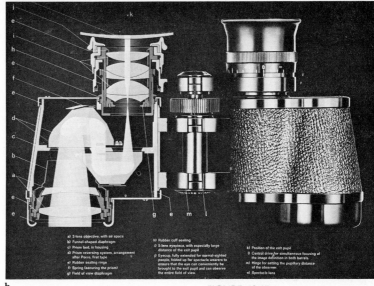

b

FIGURE 13.15

(a) A periscope uses two right-angle prisms to change the line of sight of an observer. (b) Each side of a pair of binoculars uses two porro prisms to increase the effective distance between the front and back lenses. (*Carl Zeiss, Inc.*)

Example 13.5 A substance of unknown index of refraction n_1 is placed on the prism of a refractometer. The index of refraction of the prism is 1.52, and the angle ϕ of the last emerging ray is 38°. What is n_1?

From Snell's law, the angle of incidence ϕ_2 of the last ray on the second face of the prism is

$$\sin \phi_2 = \frac{1.0}{n_2} \sin \phi = \frac{1.0}{1.52} \sin 38° = 0.405$$

or

$$\phi_2 = 23.9°$$

Figure 13.16 shows that $\theta_{2m} = 90° - \phi_2 = 66.1°$. So from Eq. 13.5 the index of refraction n_1 is

$$n_1 = n_2 \sin \theta_{2m} = (1.52) \sin (66.1°) = 1.39 \qquad \square$$

FIBER OPTICS

Total internal reflection finds important application in the field of fiber optics. An *optical fiber* is a long, narrow filament of glass or transparent plastic. Light entering one end of the fiber at an angle θ_1 that is less than a certain value will strike the inside wall of the fiber at an angle ϕ_2 that is greater than the critical angle for total internal reflection (Fig. 13.17). The light will then travel through the fiber, making a succession of reflections as it goes. As long as these reflections are at angles greater than the critical angle, the light can be reflected thousands of times without any loss due to refraction. Because the fiber is very narrow, it can be bent and still remain nearly straight over the short distance between two consecutive reflections. Only when the fiber is bent too sharply, so that light totally reflected from one side strikes the opposite side at less than the critical angle, will light be refracted out of the fiber. But as long as sharp bends are

FIGURE 13.16

A substance of unknown index of refraction n_1 on the prism of a refractometer.

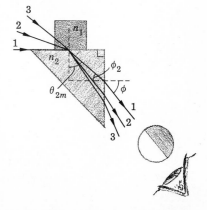

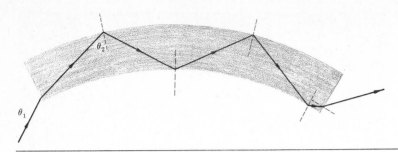

FIGURE 13.17

An optical fiber. Light incident at angle θ_1 at one end of the fiber strikes the side of the fiber at angle θ_2. If θ_2 is greater than the critical angle, the light will be totally reflected.

avoided, light can pass through a long curving optical fiber without appreciable loss.

A flexible light guide is made by bundling together hundreds of thousands of extremely narrow optical fibers (generally about 2 μm in diameter). Figure 13.18a shows how such a bundle can transmit the image of an object around a corner. Each individual fiber in the bundle transmits the light incident on it from a small region of the object, and the thousands of fibers together form an image of the entire object. A *fiberscope* is a flexible optical bundle fitted with a suitable viewing lens (Fig. 13.18b). It is commonly used in industry to inspect otherwise inaccessible locations, such as the interiors of boilers, pumps, and engines. (A light source mounted in the viewing instrument transmits light to the object through some of the fibers. This light is reflected from the object and transmitted back to the viewer through the other fibers.)

Figure 13.19a shows a bundle of 24 hair-thin light guides developed by the Bell System to transmit voice, data, and video signals. By suitably modulating the light in each fiber, a 1-cm-wide optical cable can carry more information than a conventional electric cable (Fig. 13.19b). The success of the experimental installation of light cables in Chicago in 1977 is expected to

FIGURE 13.18

(a) A bundle of optical fibers transmits an image around a corner. (b) A fiberscope being used to inspect an inaccessible location. (*American Optical Corp.*)

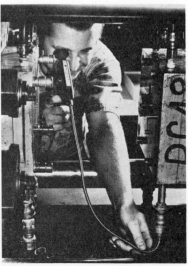

a   b

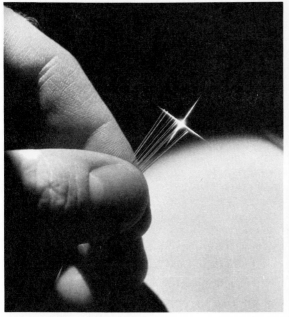

a

b

FIGURE 13.19

(a) A bundle of 24 light guides used for optical communication.
(b) Comparison of an optical cable to an electric cable. (*Bell System.*)

result in the widespread utilization of this technology in the 1980s. When this happens, the production of optical fibers will become a major industry.

13.3   Interference and Diffraction

INTERFERENCE

The principle of superposition (Sec. 11.3) says that when two or more waves coexist at the same point, the displacement of the medium is the sum of the displacements that each wave would produce separately. This has the surprising consequence that two waves which separately would produce equal and opposite displacements at a point superimpose to produce zero displacement at that point. This has already been seen in the case of a standing wave (Sec. 11.5), where two sine waves with the same wavelength and amplitude but traveling in opposite directions always cancel each other at certain points called *nodes*.

To demonstrate the wave nature of light, Thomas Young devised an experiment in which two light beams are caused to cancel each other at some points. His experiment can best be understood by first considering a series of small water waves (ripples) moving toward a screen that has two small holes in it (Fig. 13.20a). The crests, or maxima, of the waves are shown as solid lines, and the troughs, or minima, are shown as dashed lines halfway between the crests. Each line is called a *wavefront* because all the points on a given line have the same displacement. The wavelength λ is the distance between the solid lines.

When the wave hits the screen, a disturbance is produced at each hole, so that each hole acts as a point source of waves. That is, semicircular waves emanate from each hole. Wherever the crest of a wave from one hole intersects the trough of a wave from the other hole, a region of *destructive interference* is formed, in which the elements of the medium do not oscillate, i.e., in which their amplitude is zero. These regions are indicated by the broken lines radiating from the screen in Fig. 13.20*a*. Between these lines a region of *constructive interference* is formed in which the crests (and troughs) of the two waves coincide, so that the elements of the medium oscillate with large amplitudes. These regions are indicated by the solid lines radiating from the screen in Fig. 13.20*a*. Thus the interference of the waves from the two holes produces alternating regions of zero and large-amplitude oscillation. This interference pattern is identical with that produced by circular waves emanating from two points, as shown in Fig. 13.20*b*.

Young demonstrated the wave nature of light by producing a similar interference pattern with light. Today this demonstration is easily performed with a laser. A screen with a pair of narrow slits in it is placed in front of a laser beam, and the light emerging from the slits is allowed to fall on a viewing screen (Fig. 13.21*a*). The pattern on the viewing screen consists of alternating bright and dark bands, called *fringes* (Fig. 13.21*b*). The bright central fringe corresponds to point A in Fig. 13.20*a*. This fringe is equally distant from the two slits, so that a crest from one slit always coincides with a crest from the other, resulting in constructive interference. On either side of the central fringe there are dark fringes, where the lights from the two slits cancel each other. Points in these regions correspond to points B and B' in Fig. 13.20*a*. They are exactly one-half wavelength farther from one slit than from the other, so that a crest from one slit always falls on a trough from the other, resulting in destructive interference. In general, a dark fringe occurs at a point that is a distance $(n + \frac{1}{2})\lambda$ farther from one slit than the other, where n is an integer, and λ is the wavelength of the light.

Away from the central fringe, on either side of the dark fringe, are two more bright fringes. These points, which correspond to points C and C' in Fig. 13.20*a*, are one wavelength farther from one slit than from the other, so that again a crest from one slit always coincides with a crest from the other slit, resulting in constructive interference. In general, a bright fringe occurs at a point that is a distance $n\lambda$ farther from one slit than the other.

REMARK In order to get interference, portions of the same wave must illuminate both slits. *Coherent light,* such as the light produced by a laser, consists of a single long continuous wave. Double-slit interference is thus easily demonstrated with laser light because the two slits are illuminated by the same intense wave. *Incoherent light,* such as the light from the sun or a light bulb, consists of billions of independent short waves. To get interference with incoherent light, a very narrow light source must be used, so that each wave illuminates both slits simultaneously. Because the light source is narrow, the pattern produced is faint and must be viewed in the dark. A broad incoherent light source will not produce a visible interference pattern because independent waves from different points of the source will arrive separately at each slit at the same time.

Figure 13.22 shows the distances from two slits to a point P in a bright fringe. The distance *d* between the slits is usually very

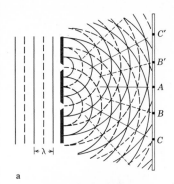

a

b

FIGURE 13.20

(a) Interference of circular waves emanating from two holes. The radial lines indicate the alternate regions of constructive and destructive interference. (b) Interference of circular water waves emanating from two points. (*From PSSC Physics, 4th ed., copyright © 1976 by Education Development Center, Inc., used with permission.*)

small compared with the distance D from the slits to the screen, so that the lines SP and $S'P$ are almost parallel and make approximately the same angle θ with the perpendicular to the screens. The line ST is the perpendicular from slit S to the line $S'P$. The distance from T to P is approximately equal to the distance from S to P, so that the distance a between S' and T is the difference between the distances from S and S' to P. From the right triangle STS' we get the relation

$a = d \sin \theta$

The condition for a bright fringe is $a = n\lambda$. Therefore, the bright fringes occur at angles given by

$$\sin \theta = \frac{a}{d} = \frac{n\lambda}{d} \qquad\qquad 13.6$$

where n is an integer.

For $n = 0$, Eq. 13.6 gives $\theta = 0°$, which is the central fringe. The first bright fringe on either side of the central fringe corresponds to $n = 1$, the second bright fringe corresponds to $n = 2$, and so on. The distance x_n between the central fringe and the nth fringe is approximately

$x_n = D \tan \theta$

Because θ is very small, $\tan \theta$ is approximately equal to $\sin \theta$ (Appendix IV). Then, using Eq. 13.6 for $\sin \theta$, we get

$$x_n = D \sin \theta = \frac{Dn\lambda}{d}$$

or

$$\lambda = \frac{x_n d}{nD} \qquad\qquad 13.7$$

Example 13.6 A neon-helium laser forms a pattern of fringes on a viewing screen that is 3.0 m from a screen with two slits separated by 0.02 cm. The distance between adjacent bright fringes on the viewing screen is 0.95 cm. What is the wavelength of the laser light?

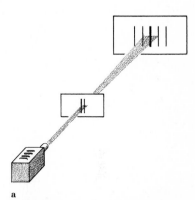

a

b

FIGURE 13.21

(a) A laser beam incident on a pair of slits produces an interference pattern on a screen. (b) Interference pattern produced on a screen.

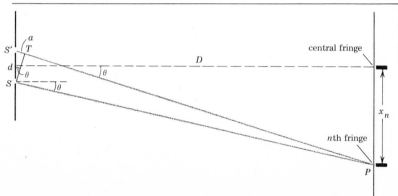

FIGURE 13.22

Distances from two slits S and S' to a point P on a screen. Destructive interference occurs when the difference a between these distances is $(n + \frac{1}{2})\lambda$.

From Eq. 13.7 the wavelength of the light is

$$\lambda = \frac{x_n d}{nD} = \frac{(0.95 \text{ cm})(0.02 \text{ cm})}{(1)(3.0 \times 10^2 \text{ cm})} = 6.3 \times 10^{-5} \text{ cm} = 630 \text{ nm} \quad \square$$

The double-slit experiment provides convincing proof of the wave nature of light, because the pattern of bright and dark fringes it produces can result only from superposition, which is a unique wave property. The experiment also provides a method for measuring the wavelength of a light, because the positions x_n of the bright fringes depend on the wavelength. Light containing more than one wavelength will be broken into its component wavelengths after passing through a double slit because the fringes for the different wavelengths will have different positions on the screen. The frequency composition, or spectrum, of light is measured this way. *Spectroscopy,* the study of the spectra of light emitted by different atoms, is the basis of our present knowledge of atomic structure (Sec. 20.2).

A *spectrometer* is a device for separating and measuring the wavelengths of the components of a light. Instead of a double slit, a spectrometer uses a *diffraction grating,* which is a piece of glass or plastic on which thousands of slits have been carefully etched. The large number of slits increases the intensity of the bright fringes over that obtained from a double slit, though the spacing of the fringes still depends on the distance between adjacent slits in accordance with Eq. 13.7.

Interference is used to reduce the reflection from camera lenses and other optical surfaces. Normally about 4 percent of the light incident near 0° on an air-glass boundary will be reflected. Figure 13.23 shows how some of the light focused on the film is reflected from the film to the lens and from the lens back onto the film. This reflected light produces undesirable background fog on the film.

To reduce reflection below 1 percent, quality lenses are coated with a one-quarter-wavelength-thick layer of a material with a low index of refraction. The index of refraction of the coating is between that of air and glass, so about equal amounts of light are reflected from the glass-coating and the coating-air boundaries. Figure 13.24 shows that the displacements of the

FIGURE 13.23

Some of the light focused on a photographic film can be reflected from the film to the lens and back to the film.

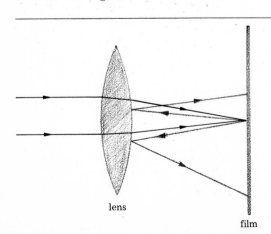

lens

film

glass coating air

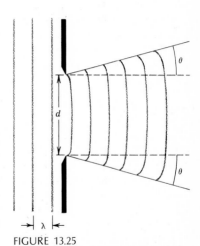

FIGURE 13.24

A one-quarter-wavelength-thick coating on a glass surface reduces the amount of reflection, because light reflected from the outer surface of the coating destructively interferes with the light reflected from the inner surface.

two reflected waves are one-half wavelength apart,* so they interfere destructively, reducing the magnitude of the reflected light. The coating can be made one-quarter wavelength thick for only one particular wavelength of light (usually around 550 nm). At other wavelengths the destructive interference is not as complete, and some light is reflected. With multiple coatings, reflection at all wavelengths can be greatly reduced.

DIFFRACTION

In Fig. 13.20 it is assumed that semicircular waves emanate from the narrow slits when the incident waves reach them. This in itself is an interesting wave phenomenon. Certainly, if the slits are very wide, we would not expect the incident waves to be modified very much as they pass through. However, the wavefronts of a plane wave are always bent slightly when passing through a slit, as shown in Fig. 13.25. Consequently, the wave no longer travels in one direction but instead has an angular divergence θ. This phenomenon is called *diffraction*. It can be shown that θ is related to the wavelength λ and the slit width d by

$$\sin \theta = \frac{\lambda}{d} \qquad 13.8$$

FIGURE 13.25

Diffraction of a plane wave passing through a slit. When the slit is much wider than the wavelength of the wave, the diffraction angle θ is small.

If d is very large compared with λ, the angle θ is very small, and the incident wave is little affected by the slit. On the other hand, if d is comparable to λ, the angle θ is large, and the wave radiates in all directions from the slit. This latter condition must hold for the slits in a double-slit experiment in order that the light from the slits overlaps on the viewing screen.

The wavelength of light is about 5×10^{-7} m, which is small compared with normal-sized objects. Therefore, light is not noticeably diffracted by ordinary objects but appears to travel in straight lines without bending. In contrast, the wavelengths of sound are comparable to normal-sized objects, so sound is highly diffracted. For example, the wavelengths of the human voice are

*This is so because the wave reflected from the glass has traveled one-half wavelength farther than the wave reflected from the coating. The thickness of the coating is one-quarter of the wavelength of the wave in the coating material, which is less than the wavelength of the wave in air. (Both reflected waves are shifted an additional half wavelength relative to the incident wave, but this is a detail of the reflection process that does not concern us here.)

large compared with the size of the mouth, so sound is diffracted in all directions as it leaves the mouth. If this were not the case, you would have to be directly in front of a speaker in order to hear him or her. Ultrasonic waves, which have much shorter wavelengths than sound, do travel in straight lines without much diffraction, just like light.

RESOLUTION

Definition The *resolving power* of an optical instrument is a measure of the degree to which the instrument can separate light coming from different points. Ideally a lens should focus a plane wave to a point. For instance, the portion of a plane wave that passes through the lens of the eye should be focused at a point P on the retina (Fig. 13.26). However, the plane wave is diffracted as it passes through the entrance to the eye (or any other lens), so the focal point is enlarged into a small circle. This limits the amount of detail the eye can see, because light coming from two different directions will not be resolved if the two focal spots overlap too much (Fig. 13.27).

The size of the focal spot depends on the diameter d of the pupil, the distance f from the lens to the retina, and the wavelength λ of the light. The maximum diameter of the pupil is about 7 mm, so with $\lambda = 550$ nm, the diffraction angle given by Eq. 13.8 is

$$\sin \theta = \frac{\lambda}{d} = \frac{550 \times 10^{-9} \text{ m}}{7 \times 10^{-3} \text{ m}} = 7.9 \times 10^{-5}$$

or

$$\theta = 8 \times 10^{-5} \text{ rad}$$

The distance f between the lens and the retina is 1.7 cm, so the radius of the focal spot is

$$r = \theta f = (8 \times 10^{-5})(1.7 \text{ cm}) = 1.4 \times 10^{-4} \text{ cm} = 1.4 \ \mu\text{m}$$

If the centers of two focal spots are less than 1.4 μm apart, the spots cannot be resolved. This has two consequences. First, light coming from two points O and O' that subtend an angle less than 8×10^{-5} rad cannot be resolved (Fig. 13.27). This corresponds to a separation of 0.8 mm at 10 m. Second, the size of the light-sensitive cells in the retina need not be smaller than 1.4 μm. In fact, the smallest cells in the center of the retina are about 1.5 μm in diameter, indicating that these cells are matched

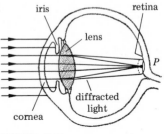

FIGURE 13.26

Diffraction circle formed on the retina of the eye by a plane wave incident on the eye.

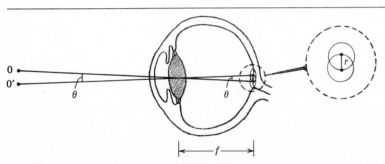

FIGURE 13.27

Two objects O and O' cannot be resolved if the diffraction circles they form on the retina overlap.

to the size of the focal spot. Away from the center, the cells are much larger (5 to 8 μm), indicating that in this region the resolving power is limited by factors other than diffraction. As we shall see in Sec. 14.6, away from the center of the retina, lens aberrations limit the quality of the image.

13.4 Polarization

Sound is a longitudinal mechanical wave (Sec. 12.1); i.e., in a sound wave, the elements of the medium vibrate parallel to the direction of propagation of the wave. There are also transverse waves (Sec. 11.1), such as those of a vibrating cord, in which the elements of the medium vibrate perpendicular to the direction of propagation of the wave. The question naturally arises: Is light a longitudinal or a transverse wave?

It may at first seem impossible to answer this question without probing more deeply into the nature of an electromagnetic wave. Surprisingly, however, there are simple experiments which indicate that light is a transverse wave. The easiest experiment uses two identical Polaroid sheets,* such as are used in polarizing sunglasses. Each sheet is fairly transparent, and when one sheet is placed over the other, the two together are still transparent. However, as one sheet is slowly rotated with respect to the other (Fig. 13.28), the region of overlap darkens until it is completely dark when the one sheet has been rotated 90° with respect to the other; i.e., the two sheets together are now opaque. As the one sheet is rotated further, the region of overlap starts to lighten, and when the sheet has been rotated through another 90°, the region of overlap is again transparent.

Centuries before the invention of Polaroid sheets, a similar phenomenon was observed using certain naturally occurring crystals, such as calcite. In fact, Newton considered this phenomenon to be evidence against the wave theory of light because everyone then assumed that if light was a wave, it was a longitudinal wave, like sound. However, no one could explain how the intensity of a longitudinal wave could possibly be affected by rotating something around an axis parallel to the direction of motion of the wave. In 1817 Young first suggested the natural explanation of the phenomenon—that light is a transverse wave.

A longitudinal wave can vibrate in only one direction, whereas a transverse wave can vibrate in any direction that lies in a plane perpendicular to the direction of motion. In a beam of *polarized* light, all the wavetrains vibrate in the same transverse direction, so that the beam can be represented by a single amplitude **A**, as illustrated in Fig. 13.29. The amplitude of polarized light is treated as a vector quantity because it is characterized by both a magnitude and a direction.

There exist materials, such as Polaroid sheets, that transmit polarized light according to the angle between the amplitude **A** of the light and the *axis* of the sheet. This axis is a fixed direction in the sheet determined by the arrangement of the molecules in the sheet. In Fig. 13.29 a sheet of such material (called the

FIGURE 13.28

As one Polaroid sheet is rotated with respect to another, the amount of transmitted light changes.

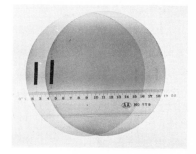

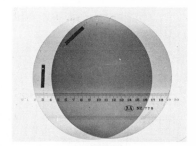

---

*Polaroid sheets are the patented invention of Edwin H. Land (1909–   ).

analyzer) is placed in front of a beam of polarized light. When **A** is parallel to the axis of the sheet, all the light is transmitted, and when **A** is perpendicular to the axis, none of the light is transmitted.

When the amplitude **A** of polarized light makes an angle θ with the axis of an analyzer (Fig. 13.30), the amplitude **A′** of the light transmitted through the analyzer is the component of **A** parallel to the axis; i.e., the magnitude A' of the transmitted amplitude is

$$A' = A \cos \theta \qquad\qquad 13.9$$

and the direction of **A′** is parallel to the axis of the analyzer. As with sound, the intensity of light is proportional to the square of the amplitude, so that the intensity of the light incident on the analyzer is proportional to A^2, and the intensity of the transmitted light is

$$I' = I \cos^2 \theta \qquad\qquad 13.10$$

For example, when $\theta = 0°$, all the light is transmitted ($\cos^2 0° = 1^2 = 1$); when $\theta = 45°$, half the light is transmitted [$\cos^2 45° = (0.707)^2 = 0.5$]; when $\theta = 90°$, none of the light is transmitted ($\cos^2 90° = 0^2 = 0$). Note that the transmitted light is polarized in the direction of the axis of the analyzer and not in the direction of polarization of the incident light. The analyzer thus changes the direction of polarization as well as the intensity of the transmitted light.

Ordinary (unpolarized) light consists of billions of wavetrains, each vibrating in a different direction. This is illustrated in Fig. 13.31, where the amplitudes of several wavetrains are shown. When unpolarized light passes through a Polaroid sheet (called the *polarizer*), only the component of amplitudes parallel to the axis of the sheet are transmitted. The result is a beam of polarized light whose amplitude **A** is parallel to the axis of the polarizer. If the intensity of the incident unpolarized light is I_0, the resulting polarized light has the intensity $I = \frac{1}{2}I_0$, since on the average, only half the wavetrains of the incident light are transmitted.

Example 13.7 Unpolarized light of intensity I_0 is incident on the polarizer-analyzer system shown in Fig. 13.31. If the axis of the analyzer is oriented at 35° to the axis of the polarizer, what is the intensity I' of the light transmitted through the analyzer?

Let I be the intensity of the polarized light incident on the analyzer. From Eq. 13.10 the intensity I' of the transmitted light is

$$I' = I \cos^2 \theta = I(\cos 35°) = (0.67)I$$

But $I = \frac{1}{2}I_0$, where I_0 is the intensity of the unpolarized light incident on the polarizer, so that

$$I' = (0.335)I_0$$

The final light is 0.335 times the intensity of the incident unpolarized light, and its is polarized in the direction of the axis of the analyzer. ☐

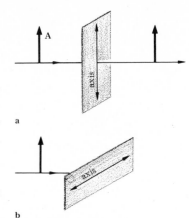

a

b

FIGURE 13.29

Polarized light of amplitude **A** incident on a Polaroid sheet. (a) **A** is parallel to the axis of the sheet, so the light is transmitted. (b) **A** is perpendicular to the axis of the sheet, so the light is not transmitted.

FIGURE 13.30

Amplitude **A** of a beam of polarized light at an angle θ to the axis of a Polaroid sheet.

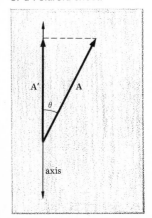

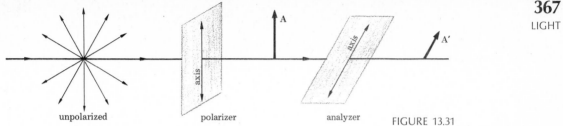

FIGURE 13.31

A beam of unpolarized light incident on a polarizer and an analyzer. The axis of the analyzer is oriented at 35° to the axis of the polarizer.

PARTIALLY POLARIZED LIGHT

Partially polarized light is a mixture of polarized and unpolarized light that can vary from 0 to 100 percent polarization. The wavetrains in partially polarized light vibrate in all directions, but the amplitudes of these vibrations are greatest in one direction, as illustrated in Fig. 13.32. Unpolarized light becomes partially polarized when it is reflected from a nonmetallic surface, the direction of polarization being parallel to the reflecting surface (Fig. 13.32).

The lenses of Polaroid sunglasses consist of Polaroid sheets oriented with their axes vertical. This orientation is perpendicular to the direction of polarization of light reflected from a horizontal surface. Consequently, these glasses eliminate much of the spectrally reflected light that contributes to glare.

Blue skylight consists of sunlight that is reflected (or, more accurately, scattered) by the molecules in the upper atmosphere. As a result, skylight is partially polarized, the magnitude and direction of the polarization at any point in the sky depending on the position of the point relative to the sun. The percentage of polarization is zero near the sun and reaches a maximum value of 70 percent at points 90° from the sun.

ELLIPTICALLY POLARIZED LIGHT

Linearly polarized light consists of a transverse displacement that vibrates in a straight line with amplitude **A**. As the wave moves, the displacement remains in a single plane that is parallel to the direction of propagation (Fig. 13.33a). *Elliptically polarized light* consists of a transverse displacement that rotates about the direction of propagation. As the wave moves, the plane of polarization rotates about the direction of propagation (Fig. 13.33b).

Elliptically polarized light results from the superposition of two linearly polarized waves of the same wavelength but with mutually perpendicular planes of polarization and displacements that peak at different times. Figure 13.34a shows two such waves, which are one-quarter wavelength apart in phase. This means that one wave has its maximum displacement A_1 while the other wave is passing through zero, and vice versa. As a consequence of this phase difference, the vector sum of the displacements of the two waves is never zero; its magnitude oscillates between the magnitudes A_1 and A_2 of the individual waves, and its direction rotates about the direction of propagation with the frequency of the wave. Figure 13.34b shows that as

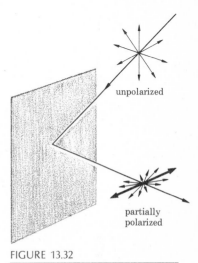

FIGURE 13.32

Unpolarized light reflected from a surface. The reflected light is partially polarized in a direction parallel to the surface.

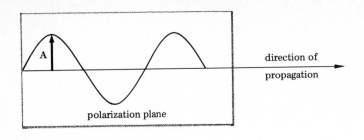

a

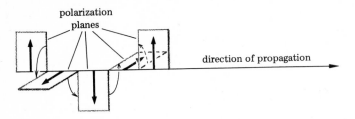

b

FIGURE 13.33

(a) Linearly polarized light. The displacement is confined to a single plane parallel to the direction of propagation. (b) Elliptically polarized light. The displacement rotates about the direction of propagation.

the wave passes through a fixed plane, the vector sum of the displacements traces out an ellipse.

Elliptically polarized light is produced by passing linearly polarized light through a retarder plate made of birefringent material. *Birefringence* is a property common to many crystals, including calcite and quartz, in which the index of refraction of the crystal depends on the direction of polarization of the light

FIGURE 13.34

(a) Two linearly polarized waves with mutually perpendicular planes of polarization. These waves are one-quarter wavelength apart in phase, so the sum of their displacements rotates about the direction of propagation. (b) As the wave in (a) passes through a plane, the vector sum of the displacements traces out an ellipse.

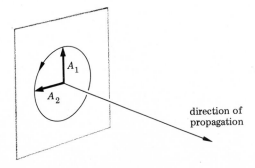

a

b

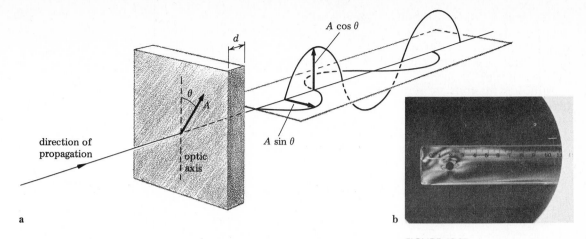

a

b

FIGURE 13.35

(a) A quarter-wave plate converts linearly polarized light into elliptically polarized light. The incident wave is split into two mutually perpendicular linearly polarized waves that are one-quarter wavelength apart in phase. (b) A plastic ruler placed between crossed polarizers. Stresses in the ruler make it birefringent. This converts the linearly polarized light transmitted through it into elliptically polarized light.

passing through it. A *retarder plate* is a plate made of birefringent material, such that the index of refraction has the value $n_\parallel$ for light polarized parallel to a certain direction in the plate (called the *optic axis*) and the value $n_\perp$ for light polarized perpendicular to this direction.

Figure 13.35a shows linearly polarized light incident on a retarder. The amplitude **A** of the light makes an angle θ with the optic axis of the retarder. Thus the component $A \cos \theta$ that is parallel to the optic axis travels through the plate with the speed $c/n_\parallel$, while the component $A \sin \theta$ that is perpendicular to the axis travels with the speed $c/n_\perp$. In a *quarter-wave plate,* the thickness d is chosen so that the faster component emerges from the plate exactly one-quarter wavelength ahead of the slower component. In this way, the two components will combine to produce elliptically polarized light, as in Fig. 13.34a.

Photoelastic stress analysis is a method of studying the stresses produced in structural elements under different loading conditions. It is based on the fact that certain plastics become birefringent when stressed. A plastic model made of the element under study is placed between crossed polarizers (Fig. 13.35b). When the model is stressed, birefringence is produced in it, which converts the incident linearly polarized light into elliptically polarized light. Since elliptically polarized light has some polarization in all directions, some fraction of the light passes through the second polarizer. The amount of light transmitted at each point depends on the magnitude of the birefringence at that point, and hence on the amount of stress. Figure 13.35b shows an ordinary plastic ruler placed between crossed polarizers. The stress pattern in this case shows the stresses that were set into the ruler when it was molded. Note how extra stress was developed around the hole.

13.5   Color Vision

Light coming from a point of the visual field is focused by the eye into a corresponding spot on the retina. The surface of the retina contains light-sensitive cells, the rods and cones. There are 125 million rods distributed over the whole retina, and 5.5 million cones limited mainly to the central region of the retina. The rods are more sensitive to light than the cones, the level of illumination of a moonlit night being sufficient to stimulate the rods but not the cones.

Although the rods are sensitive to low levels of illumination, they do not respond differently to different wavelengths, and so cannot distinguish colors. Colors cannot be seen on a moonlit night, because only the rods are functioning. The cones, on the other hand, contain pigments that preferentially absorb light of different wavelengths. There are three different pigments, and each cone cell contains one of the three. Consequently, the cones can make a crude three-dimensional analysis of a light signal.

The 130.5 million rods and cones are connected to the brain by 900,000 nerve fibers, so on the average, each fiber carries information from a hundred or so cells. This means that the eye can simultaneously process 900,000 independent light signals impinging on different spots on the retina, but that relatively little information is obtained about the frequency composition of each signal. This is in sharp contrast to the ear, which uses 20,000 basilar fibers (Fig. 12.7) to process any one sound signal received at any one time. In other words, the eye is a low-fidelity, multichannel analyzer, whereas the ear is a high-fidelity, single-channel analyzer.

Light of a single wavelength is perceived by an observer with normal color vision to have one of the *spectral* colors. These colors range from bluish purple for light with a wavelength of 420 nm, through green for light with a wavelength of 520 nm, to red for light with a wavelength of 700 nm. However, most light we perceive, such as the light reflected from a blue book, consists of more than one wavelength. Spectral analysis of such a light would show that it contains light of all wavelengths, although there would probably be a greater intensity of light from the shorter-wavelength (blue) end of the spectrum than from the longer (red) end.

Color is a psychological attribute of light, and the relation of this attribute to the physical attributes of light lies in the domain of psychophysics. Unlike the other senses, however, color vision is reducible to exact mathematical laws, which were formulated in 1853 by Hermann Grassmann (1809–1877). These laws are the basis of all industrial color processes and standards.

The First Law of Color *The normal human eye perceives only three attributes of light, often referred to as brightness, saturation, and hue.* From this psychological fact, physiologists have inferred that there must be three, and only three, kinds of pigments in the cones. From color-matching tests with subjects who have different forms of color blindness, it has even been possible to infer the color of these pigments. But only in the last few decades has more direct evidence of these pigments been found.

The three attributes of color can best be understood by a simple demonstration. A projector fitted with a red filter projects a red spot on a screen. *Hue* is the attribute of the spot that distinguishes it from a blue or green spot. The *brightness* of the spot depends on its intensity,* which can be changed by changing the power of the projection bulb. Brightness is thus analogous to loudness: it is the psychological sensation that corresponds to the physical attribute of intensity. If a second projector without a filter projects a white spot over the red one, the result is a pink spot. The pink spot differs from the red one in *saturation*. Saturation describes the strength or purity of the hue. If the amount of the white light is very small compared with the amount of the red light, the spot has high saturation. As the relative amount of white light increases, the saturation of the spot decreases. The reason colored photographs often seem more vivid than real life is that their colors are more saturated than natural colors. The photograph reproduces the hues correctly but raises the saturation.

The word *color* refers collectively to hue, brightness, and saturation. That is, two images have the same color if and only if they have the same hue, brightness, and saturation. This is a psychological definition of color and makes no reference to the physical attributes of a light.

The physical attributes of a light are conveniently displayed by plotting the relative intensity, or irradiance, of each of the wavelengths in it. Figure 13.36 gives an example of such a plot, called the *spectrum of the light*. In this figure, each dot represents the relative irradiance per nanometer of the light in a 10-nm interval. (The dots are connected by a solid line for clarity.) Since each wavelength in the spectrum is a different physical attribute of the light, there is much more physical variability in a light than is detected by the eye. That is, lights with different spectra can be perceived by the eye as having the same color.

Because color has only three attributes, it is possible to match a sample color by mixing, in appropriate amounts, three standard, or *primary*, colors. The laws governing this are conveniently studied using four projectors, each one with a control for varying the power of its bulb. One projector is used to project the sample color S, and the other three are fitted with filters so they project the primary colors. The three primary colors are superimposed on a spot next to a spot of the sample color (Fig. 13.37), and the amounts of the primaries are varied in order to match the color of the primary mixture to the color of the sample. If a match is obtained, the sample color is uniquely specified by the amounts X, Y, Z of the primary colors required to match it. These amounts, called the *tristimulus values*, depend, of course, on the three colors used as primaries.

The *chromaticity* of a sample color is specified by its *chromaticity coordinates* (x, y, z), which are defined in terms of its tristimulus values by

$$x = \frac{X}{X + Y + Z} \qquad y = \frac{Y}{X + Y + Z} \qquad z = \frac{Z}{X + Y + Z}$$

*In illumination engineering the term *irradiance* is preferred to intensity. This is discussed more fully in Sec. 13.6.

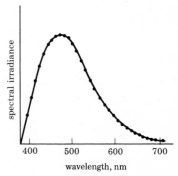

FIGURE 13.36

Spectrum of a light. Each point represents the irradiance (intensity) of the light in a 10-nm-wavelength interval.

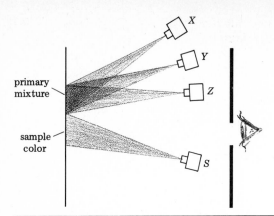

primary
mixture

sample
color

FIGURE 13.37

Color-matching experiment. The
sample color is matched by varying
the amounts X, Y, and Z of three
primary colors projected onto a
single spot.

These coordinates are the fractional amounts of each primary color required to match the sample. Two colors with the tristimulus values X, Y, Z and 2X, 2Y, 2Z have the same chromaticity. Since these colors differ mainly in brightness, chromaticity corresponds mainly to hue and saturation.

The sum $x + y + z$ of the chromaticity coordinates of a color always equals 1, so that only two of these coordinates are needed to specify the chromaticity of a color. Consequently, the chromaticity of every color that can be matched by the three primaries can be represented as a point in a plot of y against x.

For example, suppose the primaries are chosen to be the spectral colors red (700 nm), green (520 nm), and blue (420 nm). Then x, y, and z are the fractions of red, green, and blue required to match the sample. The chromaticity of every color that can be matched by a mixture of these colors is a point in the chromaticity diagram shown in Fig. 13.38. The primaries themselves are the corners of the triangle. For instance, green (G) has the coordinates ($x = 0$, $y = 1$), and so the value of z at this point is zero, because $x + y + z = 1$. The coordinates of blue (B) are ($x = 0$, $y = 0$), because the value of z at this point is 1.

The Second Law of Color *Any color that can be obtained by mixing two specific colors lies on the straight line that connects these colors in a chromaticity diagram.* For instance, all colors that are obtained by mixing the colors represented by the points P and Q in Fig. 13.38 lie on the line PQ. Furthermore, the ratio d_P/d_Q of the distances of a color H on the line PQ from P and Q is equal to the ratio f_Q/f_P of the fractions of Q and P in the mixture. That is, if H is obtained by a mixture of $\frac{1}{4}P$ and $\frac{3}{4}Q$, the point representing H will be located one-fourth of the distance from Q to P. If the color H is mixed with a third color T, the resulting color J will lie on the line HT. Thus, all colors represented by points inside the triangle QPT can be obtained by a mixture of the colors Q, P, and T, just as all the colors in the chromaticity diagram itself can be obtained by a mixture of the primaries R, B, and G.

As shown in Fig. 13.38, the color H can also be obtained by mixing the colors P' and Q'. This H, when mixed with T, yields the same colors as were obtained by mixing T with the H made

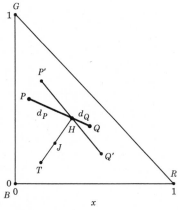

FIGURE 13.38

Chromaticity diagram based on three real primaries: red (700 nm), green (520 nm), and blue (420 nm).

from P and Q. This fact is not self-evident, because the light from the two H's may have different spectral compositions.

The Third Law of Color *The same colors have the same effect in mixtures, even though their spectral compositions are different.* Therefore, color can be treated as a mathematical entity, because it obeys the axioms of addition; e.g., equal colors added to equal colors yield equal colors.

Only the colors that can be obtained from a mixture of the primaries are represented on the chromaticity diagram. Because no set of three "real" primaries will reproduce all colors, it is not possible to represent all colors this way. However, there is a sense in which any color can be matched. When the sample color in Fig. 13.37 cannot be matched by any mixture of the primaries, one of the primaries is mixed with the sample, and the amounts of the three primaries are again varied in an effort to match the two colors. That is, if no values of X, Y, and Z match the sample S, one tries to match (say) $X + Y$ and $S + Z$. This is done by simply projecting the Z primary light onto the sample color. If a match is obtained, the tristimulus values are designated X, Y, and $-Z$. In some cases, two primaries may have to be added to S, so that, for instance, X is matched to $S + Y + Z$. In this case the tristimulus values are X, $-Y$, and $-Z$. The minus sign on a tristimulus value thus indicates that the corresponding primary was added to the sample rather than to the other primaries. With this extension of the notion of color matching it is found that every color can be matched by a suitable combination of three other colors. Furthermore, any three colors can be used as primaries, provided only that they do not lie on the same straight line in a chromaticity diagram, i.e., provided that none of them can be matched by a mixture of the other two.

The introduction of negative mixing enables any color to be represented by a set of tristimulus values. However, negative values prevent displaying chromaticity in a simple two-dimensional diagram. In 1931 the International Commission on Illumination established a system that enables every color to be specified with positive chromaticity coordinates. Known as the CIE system,* it assigns positive chromaticity coordinates to the spectral colors. Since no set of real primaries can match all the spectral colors with positive coordinates, the CIE coordinates are sometimes said to be based on imaginary primaries. It is simpler, perhaps, just to accept these coordinates as an arbitrary representation of the spectral colors, sanctioned by international agreement.

Every color is represented by a point on a plot of the x and y CIE coordinates, as shown as Fig. 13.39. The spectral colors lie on the curve, and all other colors lie inside the closed region bounded by this curve and the straight line connecting its extremities. This follows from the laws of color and the fact that every color is a mixture of spectral colors, so that all such mixtures lie in the closed region. For example, all colors obtained by

*After the French name (Commission Internationale d'Eclairage) of this international organization.

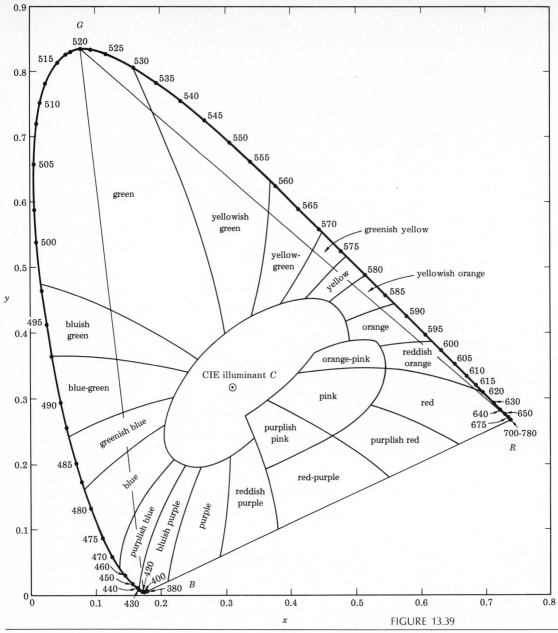

FIGURE 13.39

CIE chromaticity diagram. Colors are presented by points inside the region bounded by the curve and the straight line that connects its extremities. Colors represented by the diagram in Fig. 13.38 lie inside triangle BGR in this diagram. Points (0,0), (0,1), and (1,0), which represent the primaries on which the diagram is based, lie outside the region of real color. (After D. B. Judd and G. Wyszecki, Color in Business Science and Industry, 2d ed., copyright © 1963, John Wiley & Sons, Inc., used with permission.)

a mixture of the 420- and 520-nm spectral colors lie along the straight line BG that connects the two colors. All colors that can be obtained by a mixture of the 420-, 520-, and 700-nm spectral colors lie inside the triangle BGR. These are the same colors represented by the chromaticity diagram in Fig. 13.38.

REMARK In Fig. 13.38 the points with the coordinates (x = 0, y = 0), (x = 0, y = 1), and (x = 1, y = 0) correspond to the primaries on which the diagram is based. In Fig. 13.39 these same points lie outside the region of real color, which is why they are said to represent imaginary primaries.

The point C in Fig. 13.39 is a standard white, resembling daylight on an overcast day. The region around it includes all other whites, including those of incandescent and fluorescent lamps. Two colors are *complementary* if they can be mixed together to give white. This means they must lie on opposite sides of the neutral region. For example, C can be obtained by a mixture of the 470- and 573-nm colors. All the colors that are obtained by mixing white with a spectral color lie on the line connecting C to the spectral color. The spectral color has maximum saturation, and the saturation decreases along the line toward C. That is, saturation increases as one moves radially outward from C toward the periphery of the diagram, whereas hue varies as one moves around the periphery. Since the purples are the only colors that cannot be obtained by mixing C with a spectral color, they are sometimes referred to as *nonspectral colors.*

All color-reproduction systems, such as color television, color photography, and color printing, use three primaries. The primary colors used in a particular system can be represented by three points in a CIE diagram, and the gamut of colors the system yields lies inside the triangle that connects these points. It is evident from the CIE diagram that no set of three real primaries will reproduce all colors.

The CIE diagram is used to predict how light of a given spectral composition will appear to a human observer. For instance, consider a light that contains only the two spectral colors 490 and 605 nm (Fig. 13.40). From Fig. 13.39 the coordinates of these colors are found to be

$$x_{490} = 0.045 \qquad y_{490} = 0.29$$

and

$$x_{605} = 0.65 \qquad y_{605} = 0.35$$

If the mixture contains 56 percent 490-nm light and 44 percent

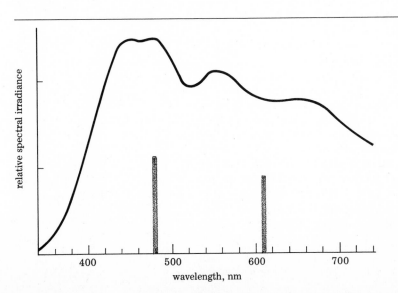

FIGURE 13.40

The continuous spectrum of daylight and the discrete spectrum of a two-wavelength light that has the same chromaticity as daylight.

relative spectral irradiance

wavelength, nm

400   500   600   700

605-nm light, the coordinates of the mixture are

$$x = (0.56)x_{490} + (0.44)x_{605}$$
$$= (0.56)(0.045) + (0.44)(0.65) = 0.31$$

and

$$y = (0.56)y_{490} + (0.44)y_{605}$$
$$= (0.56)(0.29) + (0.44)(0.35) = 0.32$$

These are the same as the coordinates of C, the color of daylight. Thus this two-spectral-component light has the same appearance to a human observer as daylight, even though daylight has the continuous spectrum shown in Fig. 13.40.

Because the eye is a poor judge of the spectral composition of a light, two lights that appear the same to an observer can have very different effects under certain situations. For instance, the responses of the pigments in color film are different from the responses of the pigments in the human eye, so a color photograph taken in daylight will be different from a photograph taken with the two-spectral-component light in Fig. 13.40. Photographers are well aware that all "white" light is not the same, and for best results the spectral composition of the light used must be matched to the spectral responses of the pigments in the film.

13.6 Radiometry and Photometry

Radiometry is the measurement of the electromagnetic energy (radiant energy) emitted by a source or incident on a detector. The radiation can be in the infrared and ultraviolet regions of the electromagnetic spectrum (Fig. 13.1) as well as in the visible region.

Photometry is the measurement of visible light as it appears to a human being with normal vision. In the last section we discussed the photometric measurement of hue and saturation, the chromaticity of a light. In this section we shall discuss the photometric measurement of brightness. These measurements are important for the design of suitable interior and exterior lighting, and generally involve white light.

The response of the human eye to the brightness of an illuminated surface depends on the wavelength of the light coming from the surface. That is, for the same amount of illumination, different spectral colors appear to have different degrees of brightness. In 1924, after psychophysical measurements on many subjects in different countries, the CIE adopted standard values for the relative efficiency V_λ of the human eye for light of different wavelengths. These values, plotted in Fig. 13.41, officially define the *standard observer*. The efficiency curve of the standard observer is used to convert radiometric measurements into photometric measurements, just as the chromaticity diagram is used to convert a measurement of the spectral composition of a light into a measurement of its hue and saturation.

REMARK Figure 13.41 is the luminous counterpart of Fig. 12.8, which shows the sound intensity required at different frequencies to produce the same loudness. Many curves are shown in Fig. 12.8 because the ear's response

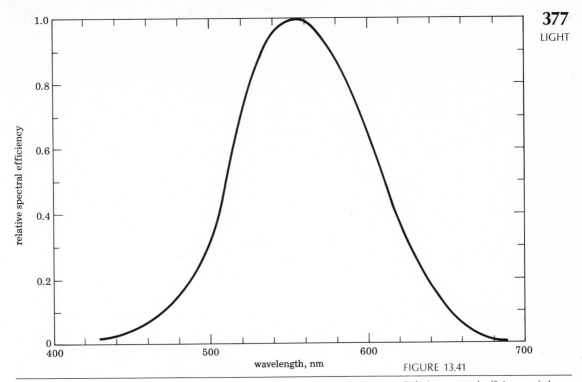

FIGURE 13.41

Relative spectral efficiency of the standard observer. This curve is based on tests made of many individuals under photopic (moderately bright) illumination.

characteristics depend on the loudness. The same is true of the eye. The efficiency curve in Fig. 13.41 is for *photopic vision,* which is vision by the normal eye when it is adapted to moderately bright illumination. A different curve is used for *scotopic vision,* which is vision by the normal eye when it is adapted to low levels of illumination.

RADIOMETRY

Definition *Radiant power,* or *radiant flux,* P_e is the rate at which electromagnetic energy is emitted by a source. Its units are joules per second (J/s), or watts (W). Spectral radiant power $P_{e\lambda}$ is the radiant power per wavelength at a particular wavelength λ. Its units are watts per nanometer (W/nm). Radiant power is analogous to acoustic power P introduced in Sec. 12.3.

Definition *Irradiance* E_e is the electromagnetic energy per second per unit area incident on a surface. Its units are watts per square meter (W/m²). Spectral irradiance $E_{e\lambda}$ is the irradiance per wavelength at a particular wavelength λ. Its units are watts per square meter per nanometer [W/(m² · nm)]. Irradiance is the radiometric term for what we call intensity.

> **REMARK** Except in this section we use the terms *irradiance* E_e and *intensity I* to mean the same thing: energy per second per unit area. In the specialized fields of radiometry and photometry, however, intensity has a different meaning. See the remark at the end of this section.

Radiant power and irradiance are related by the inverse-square law. If a point source is emitting electromagnetic radiation uniformly in all directions at the rate P_e, the irradiance E_e on a spherical surface of radius r centered on the source is

$$E_e = \frac{P_e}{a} = \frac{P_e}{4\pi r^2}$$ 13.11

This is the electromagnetic analog of Eq. 12.18.

Example 13.8 Table 13.2 lists the spectral radiant power of a point source at 50-nm intervals. (a) Find the total radiant power of the source. (b) What is the irradiance at a point 5 m from the source?

(a) Each entry in Table 13.2 gives the radiant power per nanometer in a 50-nm interval, so the power in each interval $\Delta\lambda$ is $P_{e\lambda}\Delta\lambda$. The total power P_e is the sum over all intervals, so

$$P_e = \Sigma P_{e\lambda}\Delta\lambda$$
$$= (7 + 10 + 9 + 9 + 8 + 7)(50)\text{ mW} = 2500\text{ mW} = 2.5\text{ W}$$

(b) From Eq. 13.11 the irradiance at 5 m from the source is

$$E_e = \frac{P_e}{4\pi r^2} = \frac{2500\text{ mW}}{4\pi(5\text{ m})^2} = 8.0\text{ mW/m}^2$$ □

TABLE **13.2**

Spectral radiant power of a source

Wavelength, nm	Spectral radiant power, mW/nm
400 to 450	7
450 to 500	10
500 to 550	9
550 to 600	9
600 to 650	8
650 to 700	7

Irradiance is best measured by a thermal detector, such as a *bolometer,* which absorbs all the radiation incident on a black nonreflecting element and converts it into internal thermal energy, thus increasing the temperature of the element. In a bolometer the element is a blackened strip of platinum, the electrical resistance of which varies with temperature. The change in temperature of the strip is measured by measuring its resistance, and this, together with knowledge of the strip's heat capacity, determines the total energy absorbed. Bolometers and other thermal detectors are accurate devices for measuring irradiance because their response is independent of the wavelength of the incident radiation. Other electrical detectors, such as photoconductors, are more convenient to use, but their responses are highly sensitive to wavelength. We shall discuss photoconductors shortly.

PHOTOMETRY

Definition *Luminous power,* or *luminous flux,* P_v is the rate at which visible light is emitted by a source. Its unit is the lumen (lm), which is a new unit not related to the other fundamental units of physics. *Spectral luminous power* $P_{v\lambda}$ is the luminous power per wavelength at a particular wavelength λ. Its units are lumens per nanometer (lm/nm).

The luminous power of a source depends on its radiant power, i.e., on the amount of electromagnetic radiation emitted, and on the efficiency of this radiation to produce the sensation of brightness in the human eye. Figure 13.41 shows that light at 555 nm is most efficient. At this wavelength the ratio K_{555} of the spectral luminous power to the spectral radiant power is

$$K_{555} = \left(\frac{P_{v\lambda}}{P_{e\lambda}}\right)_{\lambda=555} = 673 \text{ lm/W} \qquad\qquad 13.12$$

The constant K_{555} is used together with Fig. 13.41 to convert from radiant power to luminous power.

Example 13.9 Find the luminous power of the source with the spectral radiant power given in Table 13.2.

The luminous power in each interval $\Delta\lambda$ is $K_{555}V_\lambda P_{e\lambda}\Delta\lambda$, where $P_{e\lambda}$ is the spectral radiant power in the interval, and V_λ is the average efficiency of the interval. For instance, in the 450- to 500-nm interval, the average efficiency estimated from Fig. 13.41 is 0.15, so the luminous power in this interval is

$$K_{555}V_\lambda P_{e\lambda}\Delta\lambda = (673 \text{ lm/W})(0.15)(0.01 \text{ W/nm})(50 \text{ nm})$$
$$= 50.5 \text{ lm}$$

The total luminous power P_v is the sum over all intervals, so

$$\begin{aligned}
P_v &= K_{555}\Delta\lambda\Sigma V_\lambda P_{e\lambda} \\
&= (673 \text{ lm/W})(50 \text{ nm})[(0.002)(0.007 \text{ W/nm}) \\
&\quad + (0.15)(0.01 \text{ W/nm}) + (0.75)(0.009 \text{ W/nm}) \\
&\quad + (0.80)(0.009 \text{ W/nm}) + (0.35)(0.008 \text{ W/nm}) \\
&\quad + (0.004)(0.007 \text{ W/nm})] \\
&= 615 \text{ lm} \qquad\qquad\qquad\qquad\qquad\qquad\qquad\quad \square
\end{aligned}$$

Definition The *luminous efficacy K* of a light source is the ratio of its luminous power P_v to its radiant power P_e:

$$K = \frac{P_v}{P_e} \qquad\qquad 13.13$$

The luminous efficacy of a source is always less than the maximum efficacy $K_{555}(= 673 \text{ lm/W})$.

Example 13.10 (a) What is the luminous efficacy of the source given in Table 13.2? (b) What is the efficacy of an incandescent light bulb rated at 60 W and 855 lm?

(a) In Examples 13.8 and 13.9 we found that the radiant and luminous powers of the source in Table 13.2 are

$$P_e = 2.5 \text{ W} \qquad \text{and} \qquad P_v = 615 \text{ lm}$$

So from Eq. 13.13 the luminous efficacy K of the source is

$$K = \frac{P_v}{P_e} = \frac{615 \text{ lm}}{2.5 \text{ W}} = 246 \text{ lm/W}$$

(b) The radiant and luminous powers of the light bulb are

$$P_e = 60 \text{ W} \qquad \text{and} \qquad P_v = 855 \text{ lm}$$

so its luminous efficacy is

$$K = \frac{855 \text{ lm}}{60 \text{ W}} = 14.2 \text{ lm/W}$$

This is much smaller than the efficacy of the source in part a because an incandescent light bulb emits most of its radiant energy at infrared wavelengths, where the relative efficiency V_λ is zero. In fact, there is no practical light source that emits all its radiant power as visible light. Only 7.5 percent of the radiant power of a 60-W incandescent light bulb is in the visible region of the spectrum. A 40-W fluorescent lamp, on the other hand, emits 20 percent of its radiant power as visible light and has an efficacy of 48 lm/W. □

Definition   *Illuminance* E_v is the illuminous power per unit area incident on a surface. Its SI units are lumens per square meter (lm/m²), which is called a *lux* (lx). Other units of illuminance in common use are the lumen per square foot (lm/ft²), called a *footcandle* (fc), and the lumen per square centimeter (lm/cm²), called a *phot* (ph).

   Illuminance has the same relation to luminous power as irradiance has to radiant power. Thus if P_v is the luminous power of a point source, the illuminance E_v on a spherical surface of radius r centered on the source is

$$E_v = \frac{P_v}{a} = \frac{P_v}{4\pi r^2} \qquad\qquad 13.14$$

This is the same as Eq. 13.11, with the subscript e (for electromagnetic radiation) replaced by v (for visible light).

Example 13.11   What is the illuminance on a working surface 1.5 m below an unshielded light bulb with a luminous power of 1630 lm?

   From Eq. 13.14 the illuminance is

$$E_v = \frac{P_v}{4\pi r^2} = \frac{1630\text{ lm}}{4\pi\,(1.5\text{ m})^2} = 57.6\text{ lm/m}^2 = 57.6\text{ lx}$$

Since 1 m² = (3.28 ft)² = 10.7 ft², we have

$$1\text{ lx} = \frac{1\text{ lm}}{1\text{ m}^2} = \frac{1\text{ lm}}{10.7\text{ ft}^2} = 0.0935\text{ lm/ft}^2$$

$$= 0.0935\text{ fc}$$

So the illuminance, in footcandles, is

$$E_v = 57.6\text{ lx} = (57.6)(0.0935\text{ fc}) = 5.4\text{ fc}$$

   This calculation includes only the light that reaches the surface directly from the bulb. In practice, some additional light will reach the surface after being reflected from the ceiling and walls. A reflector is usually placed over a light bulb to increase the amount of reflected light reaching the working surface. An illuminance of 300 to 700 lx (30 to 70 fc) is recommended for prolonged close work, such as studying or drafting. □

   Accurate electrical measurements of illuminance are difficult because no electric device has the same response to radiation as the human eye. Some photoconductors, however, have response characteristics similar enough to that of the eye to make them valuable for monitoring illumination. A *photoconductor* is

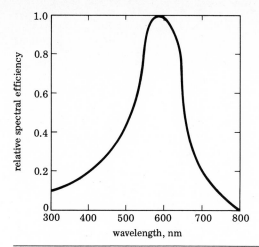

FIGURE 13.42

Relative spectral efficiency of cadmium sulfide. Since this curve is similar to the spectral efficiency of the standard observer (Fig. 13.41), a cadmium sulfide meter gives an approximate measurement of illuminance.

a material whose electric resistance increases when exposed to radiation of certain wavelengths. Cadmium sulfide (CdS) is a widely used photoconductor, often employed in photographic light meters. Its response characteristics, shown in Fig. 13.42, are similar to those of the standard observer (Fig. 13.41), except for the tails in the ultraviolet and infrared regions. With the use of appropriate filters, these regions can be largely eliminated, and the response of the photoconductor can be made to better approximate that of the standard observer. Photoconductors are discussed further in Sec. 19.2.

REMARK The luminous intensity, or candlepower, I_v of a source is the luminous power per unit solid angle.* Its units are lumens per steradian (lm/sr), called a *candela* (cd). (The candela is a fundamental SI unit which replaces an older unit called a *candle*.) For a point source that is emitting light uniformly in all directions, the luminous intensity I_v is related to the luminous power P_v by

$$I_v = \frac{P_v}{4\pi}$$

Thus from Eq. 13.14 we see that the relation between illuminance E_v and luminous intensity I_v is

$$E_v = \frac{I_v}{r^2}$$

The illuminance 1 m from a source with a luminous intensity of 1 cd is 1 lux, and the illuminance 1 ft from the source is 1 fc. Although luminous intensity is convenient for some purposes, it will not be used in this book.

*The unit of solid angle is the steradian (sr), just as the unit of plane angle is the radian. There are 4π sr in a sphere, just as there are 2π rad in a circle.

GUIDE TO MAJOR TOPICS

Topic	Reference	Problems
Electromagnetic waves	Sec. 13.1; Fig. 13.1	1 to 4
Speed of light	Sec. 13.1; Eq. 13.1; Example 13.1; Table 13.1	5 to 8
Reflection	Sec. 13.2; Eq. 13.2; Example 13.2	9 to 12
Refraction	Sec. 13.2; Eq. 13.2; Examples 13.3 and 13.4	13 to 18
Total internal reflection	Sec. 13.2; Eqs. 13.4 and 13.5; Example 13.5	19 to 26
Interference	Sec. 13.3; Eqs. 13.6 and 13.7; Example 13.6	27 to 30
Diffraction	Sec. 13.3; Eq. 13.8	31 to 34
Polarization	Sec. 13.4; Eqs. 13.9 and 13.10; Example 13.7	35 to 38
*Color	Sec. 13.5; Fig. 13.39	39 to 42
*Radiometry	Sec. 13.6; Eq. 13.11; Example 13.8	43 to 44
*Photometry	Sec. 13.6; Eqs. 13.12 to 13.14; Examples 13.9 to 13.11; Fig. 13.41	45 to 48

*Optional topics, which may be omitted.

PROBLEMS

1 Give names of electromagnetic waves with these wavelengths and frequencies: (*a*) 10^{21} Hz; (*b*) 10^{-2} m; (*c*) 2×10^{15} Hz; (*d*) 10^9 Hz; and (*e*) 5×10^{-10} m
Ans. (*a*) gamma rays; (*b*) microwaves; (*c*) ultraviolet; (*d*) UHF; (*e*) x-rays

2 Express in millimeters the smallest and longest wavelengths of infrared radiation.

3 (*a*) What is the wavelength of the radio waves broadcast by an FM station operating at 90 MHz? (*b*) The wavelength of the radio waves broadcast by an AM station is 0.35 km. What is the frequency of these waves?
Ans. (*a*) 3.33 m; (*b*) 857 kHz

4 What is the frequency of the electromagnetic wave that has the same wavelength as a sound wave of frequency 8000 Hz?

5 The speed of light in tetrachloromethane (CCl_4) is 2.06×10^8 m/s. What is the index of refraction of this liquid? *Ans.* 1.46

6 What is the speed of light in a diamond? (See Table 13.1.)

7 When a wave goes from one medium to another, its speed and wavelength change, but its frequency does not. What is the wavelength in the glass ($n = 1.50$) of a green light whose wavelength in air is 520 nm?
Ans. 347 nm

8 Show that if λ is the wavelength of a light wave in air, its wavelength λ' in a medium with index of refraction n is $\lambda' = \lambda/n$ (see Prob. 7).

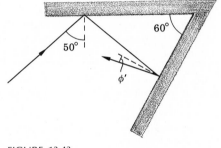

FIGURE 13.43
Problem 9.

9 Two mirrors are arranged at 60° to each other, as shown in Fig. 13.43. A ray of light is incident at 50° on the upper mirror. Find the angle of reflection ϕ' of the ray from the second mirror. *Ans.* 10°

10 Repeat Prob. 9 for two mirrors at a 50° angle, with a ray incident at 60° on the first mirror. Draw a diagram showing ray's path.

11 Two mirrors, each of length $L = 160$ cm, are facing each other as shown in Fig. 13.44.

FIGURE 13.44
Problem 11.

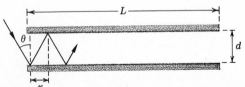

The distance d between the mirrors is 20 cm. A light ray is incident on one end of one mirror at $\theta = 30°$. How many times is the ray reflected before it reaches the other end? (*Hint:* Find the horizontal distance x that the ray travels between reflections.)
Ans. 14

12 Design a periscope using two mirrors. Show how, with different arrangements of the mirrors, the view through the periscope will be either erect or inverted.

13 A beam of light in air is incident on water at an angle of 30°. What is the angle of the beam inside the water?
Ans. 22°

14 A ray of light is incident at an angle of 53° on a block of clear plastic. The angle of refraction of the ray is 34°. What is the index of refraction of the plastic?

15 At what angle must a ray of light be incident on ethanol to be refracted into the liquid at 25°?
Ans. 35.1°

16 Plot the angle of refraction against the angle of incidence of light passing from air into glass ($n = 1.50$).

17 A ray of light is incident at $\theta = 30°$ on a prism with apex angle $A = 55°$ and index of refraction $n = 1.50$, as shown in Fig. 13.13. Calculate the angles the ray makes with the sides of the prism. With a protractor, draw the prism and trace the ray through it. What is the angle of deviation?
Ans. 35.5°

18 Figure 13.45 shows a ray incident at $\theta_1 = 52°$ on a glass plate with index of refraction $n = 1.50$. The ray is reflected and refracted several times, as shown. Find all the labeled angles.

19 What is the critical angle for total internal reflection between water and air?
Ans. 48.5°

20 The laws of reflection and refraction are the same for sound as for light. The index of refraction of a medium (for sound) is defined as the ratio of the speed of sound in air to the speed of sound in the medium. (a) What is the index of refraction (for sound) of water (use Table 12.1)? (b) What is the critical angle θ_c for total reflection of sound from water?

REMARK Sound must be incident on water at an angle of less than θ_c in order for any of it to be refracted into the water. Because θ_c is small, most

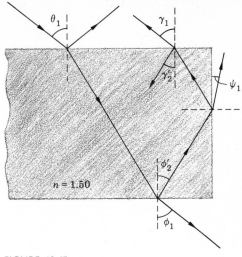

FIGURE 13.45

Problem 18.

sound is totally reflected; this is why it is so quiet under water.

21 Figure 13.46 shows a light ray incident on one face of a 45-45-90 prism with index of refraction $n = 1.52$. Find the maximum angles of incidence θ_1, above and below the normal to the vertical face, for which the ray will be totally reflected from the hypotenuse.
Ans. 90°, 5.9°

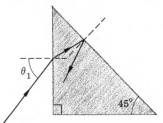

FIGURE 13.46

Problem 21.

22 Figure 13.47 shows two plates of glass ($n_1 = 1.50$) separated by a liquid film (n_2). Show that if the liquid is water ($n_2 = 1.33$), a ray of light incident on the upper glass-liquid surface at an angle $\theta_1 = 64°$ will be totally reflected, but if the liquid is alcohol ($n_2 = 1.36$), some of the light will be refracted through to the lower glass plate.

23 Figure 13.48 shows a ray of light incident at an angle θ_1 on one end of an optical fiber. Its angle of refraction is θ_2, and it strikes the side of the fiber at an angle ϕ_2. If the index of refraction of the fiber is 1.40, what is the largest angle of incidence θ_1 that a ray can have and

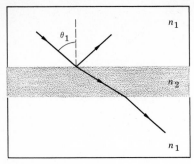

FIGURE 13.47

Problem 22.

still be totally reflected from the side of the fiber?
Ans. 78.5°

24 The optical fiber in Fig. 13.48 is 2 m long and has a diameter of 2×10^{-3} cm. If a ray of light is incident on one end of the fiber at an angle $\theta_1 = 40°$, how many reflections does it make before emerging from the other end? (The index of refraction of the fiber is 1.40.)

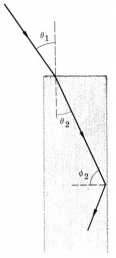

FIGURE 13.48

Problems 23 and 24.

25 In practice, optical fibers have a coating (called a *cladding*) of glass ($n_3 = 1.52$) to protect the optical surface of the fiber. If the fiber itself has an index of refraction $n_2 = 1.62$, what is the critical angle for total reflection of a ray inside the fiber?
Ans. 69.8°

26 Repeat Probs. 23 and 24 for the cladded fiber described in Prob. 25.

27 Light incident on a pair of slits produces

an interference pattern on a screen 2.5 m from the slits. If the slit separation is 0.015 cm and the distance between the bright fringes in the pattern is 0.76 cm, what is the wavelength of the light?
Ans. 4.56×10^{-5} cm

28 Light from a neon-helium laser (630 nm) is incident on a pair of slits. In the interference pattern on a screen 1.5 m from the slits, the bright fringes are separated by 1.35 cm. What is the slit separation?

29 Light of wavelength 589 nm incident on a pair of slits produces an interference pattern in which the separation between the bright fringes is 0.53 cm. A second light produces an interference pattern with a separation of 0.64 cm between fringes. What is the wavelength of this second light?
Ans. 711 nm

30 Use a compass to make an accurate drawing similar to Fig. 13.20a. Place the slits 2 cm apart, and let the wavelength of the incident wave be 1 cm. Find the positions of the nodes and antinodes on a screen 12 cm from the slits. Measure the distance between two antinodes and compare with Eq. 13.7. Explain any discrepancy.

31 Find the radius of the focal spot on the retina when the diameter of the pupil is 2 mm. Use $\lambda = 550$ nm. (The diameter of the pupil varies between 2 and 7 mm, depending on the level of illumination.)
Ans. 4.7 μm

32 Light from a neon-helium laser (630 nm) is incident on a screen with a 0.1-mm-wide slit. (a) What is the diffraction angle of the light after it passes through the slit? (b) What is the width of the spot the light makes on the wall 3 m beyond the slit?

33 A plane wave of ultrasonic sound with frequency $f = 75$ kHz is incident on a blocking screen with a slit 1 cm wide. (a) What is the diffraction angle of the wave after it passes through the slit? (b) What is the width of the wavefront 1.5 m beyond the slit? (*Hint:* Do not use the small angle approximation.)
Ans. (a) 27°; (b) 1.53 m

34 What is the smallest sized object on the moon that can be resolved by the 200-in Mt. Palomar telescope? Use $\lambda = 500$ nm. (The diameter of the telescope's reflecting mirror is 200 in.)

35 The amplitude of a beam of polarized light makes an angle of 65° with the axis of a Polar-

oid sheet. What fraction of the beam is transmitted through the sheet?
Ans. 0.179

36 The axes of a polarizer and an analyzer are oriented at 30° to each other. (a) If unpolarized light of intensity I_0 is incident on them, what is the intensity of the transmitted light? (b) Polarized light of intensity I_0 is incident on this polarizer-analyzer system. If the amplitude of the light makes an angle of 30° with the axis of the polarizer, what is the intensity of the transmitted light?

37 The axes of a polarizer and an analyzer are oriented at right angles to each other. A third Polaroid sheet is placed between them with its axis at 45° to the axes of the polarizer and analyzer. (a) If unpolarized light of intensity I_0 is incident on this system, what is the intensity of the transmitted light? (b) What is the intensity of the transmitted light when the middle Polaroid sheet is removed?
Ans. (a) $0.125I_0$; (b) zero

38 Polarized light of intensity I_0 is incident on a pair of Polaroid sheets. Let θ_1 and θ_2 be the angles between the amplitude of the light and the axes of the first and second sheets, respectively. Show that the intensity of the transmitted light is

$$I = I_0 \cos^2 \theta_1 \cos^2 (\theta_1 - \theta_2)$$

39 (a) What is the color P with CIE chromaticity coordinates ($x = 0.400$, $y = 0.200$)? (b) What are the coordinates of the color which when mixed in equal proportions with P gives standard white (C)?
Ans. (a) red-purple; (b) $x = 0.22$, $y = 0.43$

40 What are the chromaticity coordinates of the color obtained by mixing 2 parts of 480-nm light with 1 part of 555-nm light?

41 What proportions of the spectral colors with wavelengths 485, 520, and 600 nm are required to produce the green color whose chromaticity coordinates are (0.25, 0.40)?
Ans. 0.45, 0.23, and 0.32

42 (a) What is the wavelength of the spectral color that is the complement of 485 nm? (b) Is there a spectral color that is the complement of 520 nm?

43 Table 13.3 lists the spectral radiant power of a light-emitting diode (LED) at 50-nm intervals. (LEDs are the light sources in pocket calculators.) Find the total radiant power of the source.
Ans. 1.06 mW

44 The irradiance of the sun just above the earth's atmosphere is 1400 W/m². (This number is called the *solar constant*.) What is the radiant power of the sun?

45 (a) Find the luminous power of the light-emitting diode (LED) whose spectral radiant power is given in Table 13.3. (b) What is the luminous efficacy of this source (see Prob. 43)?
Ans. (a) 0.042 lm; (b) 40 lm/W

REMARK The relatively high efficacy calculated in Prob. 45b is somewhat misleading, because an LED converts only 1 percent of the electric input power into radiant power. Thus the overall *efficiency* of the LED is actually only 0.4 lm/W.

46 Calculate the luminous efficacy of a source which radiates 168 mW at 490 nm and 132 mW at 605 nm. (This is the two-spectral-component light in Fig. 13.40 that produces white light.)

47 A 1500-W incandescent lamp of the type used to illuminate a baseball field has a luminous power of 33,000 lm. (a) What is the illuminance on a surface 60 m from the lamp, assuming the light is emitted uniformly in all directions? (b) In practice the lamp is mounted inside an aluminum reflector, so that 80 percent of the light is emitted in a cone with an opening angle of 30° (Fig. 13.49). (The other 20 percent is absorbed by the reflecting surface.) What is the illuminance in the middle of the cone at a distance of 60 m from the lamp?
Ans. (a) 0.73 lx; (b) 32.5 lx

TABLE **13.3**

Wavelength, nm	Spectral radiant power, μW/nm	Spectral radiant power of a light-emitting diode (LED)
Less than 600	0	
600 to 650	3.5	
650 to 700	9.0	
700 to 750	6.0	
750 to 800	1.8	
800 to 850	0.7	
850 to 900	0.2	
Greater than 900	0	

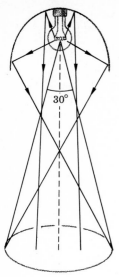

30°

FIGURE 13.49

Problem 47.

REMARK A major league baseball field uses about 1000 such lamps for a night game.

48 Just above the earth's atmosphere the illuminance of the sun is 1.3×10^5 lx. (a) What is the luminous power of the sun? (b) What is the luminous efficacy of the sun (see Prob. 44)? (c) What is the illuminance on Mars?

BIBLIOGRAPHY

JUDD, DEANE B., and GUNTER WYSZECKI: *Color in Business, Science and Industry*, 3rd ed., John Wiley & Sons, Inc., New York, 1975. Comprehensive account of all aspects of color, including measurement, specification, reproduction, and vision.

NUCKOLLS, JAMES L.: *Interior Lighting for Environmental Designers*, John Wiley & Sons, Inc., New York, 1976. This book describes the calculations involved in the design of indoor lighting, as well as the many factors, such as contrast and color, that affect the visual environment. The book also describes the various types of light sources and gives a table of the level of illumination recommended for various tasks.

STIMSON, ALLEN: *Photometry and Radiometry for Engineers*, John Wiley & Sons, Inc., New York, 1974. A clear explanation of the relation between photometry and radiology. The book contains a very useful glossary that gives complete definitions of the many unusual, and often confusing, terms used in photometry and radiometry.

Optics is the control and manipulation of light using lenses, mirrors, and prisms. It involves the design and manufacture of both the lenses themselves and the instruments in which they are used. This chapter does not begin by considering how lenses are made but assumes that ideal lenses with certain characteristics are available and discusses the principles of various optical instruments in terms of these characteristics. Since many optical instruments in common use, such as the microscope, the telescope, the camera, the projector, and the eye, are just different combinations of lenses, they can be understood in terms of the function of a single lens and the general rules for determining the effects of combinations of lenses. In the last section we discuss some of the principles involved in making real lenses.

14

OPTICS

14.1 Lenses

A simple lens is a thin circular piece of transparent material, usually glass or plastic, the thickness of which varies from its center to its edge. The two surfaces of a lens may be either plane, concave, or convex, giving rise to the five possible lens shapes shown in Fig. 14.1. However, in spite of these differences in shape, there are only two basic types of lenses: *converging* (positive) and *diverging* (negative) lenses. Convex and plano-convex lenses (Fig. 14.1*a* and 14.1*b*) are always converging; concave and plano-concave lenses (Fig. 14.1*c* and 14.1*d*) are always diverging; a meniscus lens (Fig. 14.1*e*) can be either converging or diverging depending on the relative curvature of its concave and convex surfaces.

A lens is similar to a prism in that a ray of light is deviated as it passes through it. It differs from a prism, however, in that the angle of deviation of the ray depends on where the ray enters the lens. Figure 14.2 shows several parallel rays incident on a convex lens. The ray *aa'*, which passes through the center of the lens, is undeviated because the front and back surfaces of the lens are parallel to each other at the points where the ray enters and leaves. The ray *bb'* is deviated as though it passed through a prism with an apex angle *B*, where *B* is the angle that the front and back surfaces make with each other at the points where the ray *bb'* enters and leaves. Similarly, the ray *cc'* is deviated as though it passed through a prism with an apex angle *C*. Because of the curvature of the lens, the angle of deviation increases with the distance from the center of the lens.

A *converging*, or *positive*, lens is shaped so that all parallel rays incident on it are deviated through the same point, as shown in Fig. 14.3. The line through the center *C* of the lens, perpendicular to the plane of the lens, is called the *optic axis*. All incident rays that are parallel to the optic axis are deviated so that they pass through the point *F'* on the optic axis, called the *focal point* of the lens.

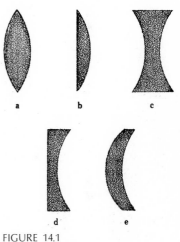

FIGURE 14.1

Five types of lenses: (a) convex, (b) plano-convex, (c) concave, (d) plano-concave, and (e) meniscus.

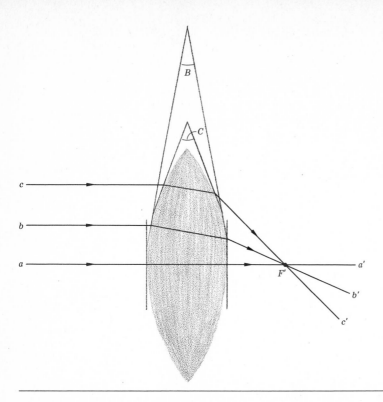

FIGURE 14.2

Parallel rays incident on a positive lens. The farther a ray is from the center of the lens, the more the ray is bent.

Definition The *focal length* f of a lens is the distance from the center C of the lens to the focal point F'. It is the primary characteristic of a lens.

For example, the rays of light coming from the sun are nearly parallel because the sun is so far from the earth. Therefore, all the sun's rays incident on a lens are brought to a focus at the focal point. If a piece of paper is placed one focal length from the lens, the concentrated rays at the focal point will be sufficient to burn the paper.

The deviation of a ray is the same regardless of the side of the lens on which it is incident. Therefore, parallel rays incident from the right on the lens in Fig. 14.3 intersect the optic axis at the point F, which is the same distance f to the left of C as F' is to the right. Consequently, a lens has two focal points, one on each side, which are the same distance f from the center of the lens.

A given ray follows the same path through an optical system regardless of its direction of motion through the system. This principle of *reversibility* is important for the analysis of optical systems. For example, suppose a source of light is placed at the focal point F' of the lens in Fig. 14.3. Since rays diverge in all directions from this source, some of them will pass through the lens to the left. By the principle of reversibility, these rays will emerge parallel to the optic axis, following the paths shown in Fig. 14.3 but in the reverse direction. Although we shall continue to put arrowheads on rays for clarity, is should be remembered that the light can travel in either direction along a ray.

As an immediate application of reversibility, we shall consider

FIGURE 14.3

Rays parallel to the optic axis of a positive lens are brought to a focus at the focal point F'.

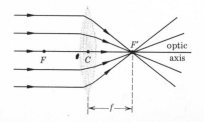

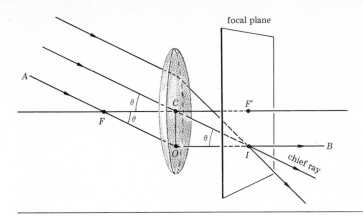

focal plane

FIGURE 14.4

Parallel rays incident on a lens at an angle θ to the optic axis are brought to a focus at a point I in the focal plane.

what happens to a beam of parallel light that is incident on a lens at an angle θ to the optic axis (Fig. 14.4). Of all the incident rays, one ray, called the *chief ray,* passes undeviated through the center of the lens. Another ray AO passes through the front focal point F of the lens. We know that if a ray BO parallel to the optic axis were incident on the lens from the right, it would follow the path OA through F. Therefore, by reversibility, the ray AO follows the path OB to the right and intersects the chief ray at I. If the angle θ is small, all other incident rays parallel to AO also intersect at I.

The triangles FCO and IOC in Fig. 14.4 are both right triangles, with a common side OC and the same angle θ at F and I. Therefore, these triangles are congruent, and so the distance OI is equal to the focal length CF of the lens. This means that I and F' lie in a plane, called the *focal plane,* that is perpendicular to the optic axis at F'. All parallel rays are brought to focus in this plane at a point whose position depends on the angle the rays make with the optic axis.

REAL IMAGES

A positive (converging) lens produces a *real image* of a distant object in the focal plane; i.e., if a screen is placed in the focal plane, an image of the object will be projected on the screen. This is so because all the light that arrives at the lens from a single point of the object is brought to a focus at a single point in the focal plane.

This is shown in detail in Fig. 14.5. Under normal illumination, each point of an object diffusely reflects light in all directions.

FIGURE 14.5

The real image of a distant object produced by a positive lens. The rays that reach the lens from the same point on a distant object are nearly parallel to each other and so are focused on the same point in the focal plane.

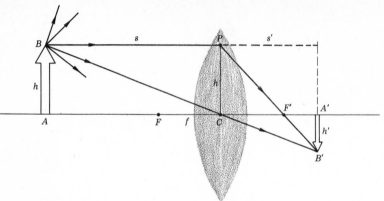

FIGURE 14.6

The real image of an object a distance s from a positive lens.

However, if the object is very far from a lens, all the rays that reach the lens from a single point of the object are nearly parallel to each other and so are brought to a focus at a single point in the focal plane. Since rays from different points of the object make different angles with the optic axis, they are focused at different points in the focal plane. Thus, if a screen is placed in the focal plane, a real inverted image of the object is projected onto it. The image is called *real* to distinguish it from a *virtual* image, which is discussed in the next section.

A lens can form a real image of an object even if the object is not distant from the lens. In fact, if the object is at any distance s from the lens (called the *object distance*), there is a distance s' (called the *image distance*) at which a real image will be formed, provided only that s is greater than f. The image distance s' is found by tracing a pair of rays from a point of the object through the lens. For example, consider point B of the object in Fig. 14.6. Since light rays are reflected in all directions from this point, one of these rays goes toward the lens, parallel to the optic axis. Consequently, it is deviated by the lens so that it passes through the focal point F'. Another ray from B goes through the center C of the lens and is undeviated. The intersection of these two rays determines the point B' where the image of B is located. The image of any other point of the object is found similarly. Thus by this simple graphical procedure, called *ray tracing,* the size h' and position s' of the image can be determined.

THE LENS FORMULA

Ray tracing is useful for understanding the function of a lens system, but it is often not precise enough for calculating the image distance. It is therefore necessary to obtain a formula relating s, s', and f. This formula is derived from Fig. 14.6 with the help of a little geometry.

1 Since the right triangles CAB and CA'B' are similar, we have

$$\frac{h'}{h} = \frac{s'}{s} \qquad\qquad 14.1$$

2 Since the right triangles F'CP and F'A'B' are similar, we have

$$\frac{h'}{h} = \frac{\overline{F'A'}}{\overline{CF'}}$$

But the distance $\overline{CF'}$ is the focal length f, and the distance $\overline{F'A'}$ is $s' - f$, so the last equation is

$$\frac{h'}{h} = \frac{s' - f}{f}$$

3 Using Eq. 14.1, we get

$$\frac{s'}{s} = \frac{s' - f}{f}$$

which, with a little algebraic manipulation, can be rewritten

$$\frac{1}{s} + \frac{1}{s'} = \frac{1}{f} \qquad \text{lens formula} \qquad\qquad 14.2$$

This formula gives the exact relation between object distance s, image distance s', and focal length f.

When the object distance s is very large, the term $1/s$ is nearly zero, and Eq. 14.2 reduces to

$$\frac{1}{s'} = \frac{1}{f} \qquad \text{or} \qquad s' = f$$

This is our original observation: a real image of a distant object is formed at a distance f from a lens. Equation 14.2 shows that as the object is moved toward the lens, its image is formed farther from the lens.

Example 14.1 What is the image distance of an object that is 100 cm in front of a converging lens of focal length 10 cm?

From the lens formula (Eq. 14.2) the image distance s' is given by

$$\frac{1}{100\,\text{cm}} + \frac{1}{s'} = \frac{1}{10\,\text{cm}}$$

Solving this for s', we get

$$\frac{1}{s'} = \frac{1}{10\,\text{cm}} - \frac{1}{100\,\text{cm}} = \tfrac{9}{100}\,\text{cm}^{-1} \qquad \text{or} \qquad s' = \tfrac{100}{9}\,\text{cm} = 11.1\,\text{cm}$$

In this case, the image is formed 1.1 cm farther from the lens than it is when the object is very far away. The image distance increases still more as the object distance decreases. For instance, when $s = 20$ cm, the image distance is $s' = 20$ cm, and when $s = 11.1$ cm, the image distance is $s' = 100$ cm. You should verify these results using the lens formula. □

Definition The *magnification m* of an image is the ratio of the image height h' to the object height h:

$$m = \frac{h'}{h} \qquad\qquad 14.3$$

According to Eq. 14.1, this is equal to

$$m = \frac{s'}{s} \qquad\qquad 14.4$$

The relation between object and image distance can be graphically displayed by introducing the *reduced* distances $\bar{s}$ and $\bar{s}'$, which are the object and image distances divided by the focal length:

$$\bar{s} = \frac{s}{f} \quad \text{and} \quad \bar{s}' = \frac{s'}{f}$$

In terms of $\bar{s}$ and $\bar{s}'$, the object and image distances are

$$s = \bar{s}f \quad \text{and} \quad s' = \bar{s}'f$$

Substitution of these expressions into Eq. 14.2 gives

$$\frac{1}{\bar{s}f} + \frac{1}{\bar{s}'f} = \frac{1}{f}$$

The focal length f is common to all the terms, so it can be canceled. The result is

$$\frac{1}{\bar{s}} + \frac{1}{\bar{s}'} = 1 \qquad\qquad\qquad 14.5$$

which is the relation between the reduced object and image distances. This relation is interesting because it does not depend on the focal length of the lens. It is a universal relation between $\bar{s}$ and $\bar{s}'$ that is valid for all positive lenses. For example, when $\bar{s} = 3$, Eq. 14.5 gives

$$\frac{1}{\bar{s}'} = 1 - \frac{1}{\bar{s}} = 1 - \tfrac{1}{3} = \tfrac{2}{3} \quad \text{or} \quad \bar{s}' = 1.5$$

This means that for any lens, if the object is located 3 focal lengths in front of the lens, the image is formed 1.5 focal lengths in back of the lens. In particular, if $f = 10\,\text{cm}$, then $s = 30\,\text{cm}$ and $s' = 15\,\text{cm}$.

Figure 14.7 is a graph of $\bar{s}'$ versus $\bar{s}$ obtained by plotting, for each value of $\bar{s}$, the value of $\bar{s}'$ given by Eq. 14.5. Thus, for instance, at $\bar{s} = 3$, the curve is at $\bar{s}' = 1.5$.

Figure 14.7 demonstrates a number of important characteristics of a positive lens.

1 The curve is symmetric about the straight line drawn from the origin through the point $\bar{s} = 2$, $\bar{s}' = 2$. This is a consequence of the principle of reversibility, which says that if an object a distance s from a lens is focused at a distance s', then an object a distance s' from the lens will be focused at a distance s.

2 The closer the object is to the lens, the farther away the image.

3 If $\bar{s} > 2$, then $\bar{s}' < 2$, and if $\bar{s} < 2$, then $\bar{s}' > 2$. This means that s and s' cannot simultaneously be less than $2f$, nor can they simultaneously be greater than $2f$.

4 It is not possible to form a real image if the object distance is less than f, and by reversibility, no real image is ever formed at a distance less than f.

In the next section we discuss a number of one-lens optical instruments. These instruments differ from each other in the relative values of $\bar{s}$ and $\bar{s}'$ at which they operate, as indicated in Fig. 14.7.

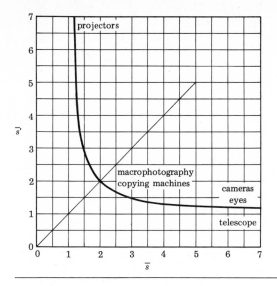

FIGURE 14.7

Graph of the reduced image distance $\bar{s}'$ against the reduced image distance $\bar{s}$. Various optical instruments operate at different values of $\bar{s}'$ and $\bar{s}$.

14.2 One-Lens Instruments

THE PROJECTOR

In a projector (Fig. 14.8), the object, such as a transparent slide, is located close to the lens, and the image is formed a large distance from the lens. The projector thus operates in the region of small $\bar{s}$ and large $\bar{s}'$ in Fig. 14.7.

Example 14.2 A slide projector with a lens of focal length $f = 15$ cm projects an image on a screen a distance $s' = 4$ m from the lens. How far should the slide be from the lens?

With $f = 15$ cm and $s' = 400$ cm, the lens formula gives

$$\frac{1}{s} = \frac{1}{f} - \frac{1}{s'} = \frac{1}{15\text{ cm}} - \frac{1}{400\text{ cm}} = \frac{77}{1200}\text{ cm}^{-1}$$

and so

$$s = \frac{1200}{77}\text{ cm} = 15.6\text{ cm}$$

The slide is placed just a fraction of a centimeter in front of the focal point of the lens. ☐

When a projector is being focused, the position of the lens is adjusted until it is the proper distance s from the slide. A projec-

FIGURE 14.8

A projector. The lens forms an image of the film on a distant screen.

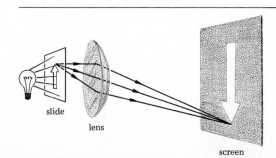

slide

lens

screen

tor with a 15-cm-focal-length lens can focus an image at any distance greater than 244 cm just by changing the distance between the lens and the slide from 16 to 15 cm.

Since the object distance s in a projector is always approximately f, the magnification (Eq. 14.4) is approximately

$$m = \frac{s'}{f} \quad \text{projector} \qquad\qquad 14.6$$

and is always greater than 1.

Example 14.3 What is the magnification of the image formed by the projector in Example 14.2?

From Eq. 14.6 the magnification is

$$m = \frac{s'}{f} = \frac{400\ \text{cm}}{15\ \text{cm}} = 26.7$$

Thus, if the object is a slide of width $h = 35$ mm, the width h' of the image on the screen is

$$h' = mh = (26.7)(3.5\ \text{cm}) = 93.3\ \text{cm}$$

Equation 14.6 shows that the image size can be increased either by increasing the distance between the projector and the screen or by using a lens with a shorter focal length. ☐

THE CAMERA

Eyes and cameras operate in the region of large $\bar{s}$ and small $\bar{s}'$ in Fig. 14.7. All the rays that reach the eye from a single point of a distant object are focused by the lens of the eye onto a single point on the retina (Fig. 14.9a), so that a real inverted image of the object is formed on the retina. The retina is covered with light-sensitive cells that send signals to the brain in response to the light incident on them. Since it is these signals that the brain interprets as a picture, a person "sees" only what is imaged on his or her retina.

Similarly, the lens of a camera focuses an image of a distant object onto light-sensitive film mounted in back of it (Fig. 14.9b). The focal length of a standard camera lens is about 50 mm, so the film must be at least 50 mm in back of the lens. The lens of an inexpensive fixed-focus camera is permanently mounted at the distance $s' = f$ from the film. Only distant objects are focused on the film of such a camera. For example, an object 4 m from the lens is brought to focus at a distance $s' = 50.6$ mm in back of the lens, or 0.6 mm in back of the film. The image on the film will be in fair focus in this case, but the image becomes very blurry if the object distance is much less than 4 m.

The lens of a more expensive camera is movable, so that its distance from the film can be adjusted to give a sharply focused image on the film. For instance, most 35-mm cameras* can focus

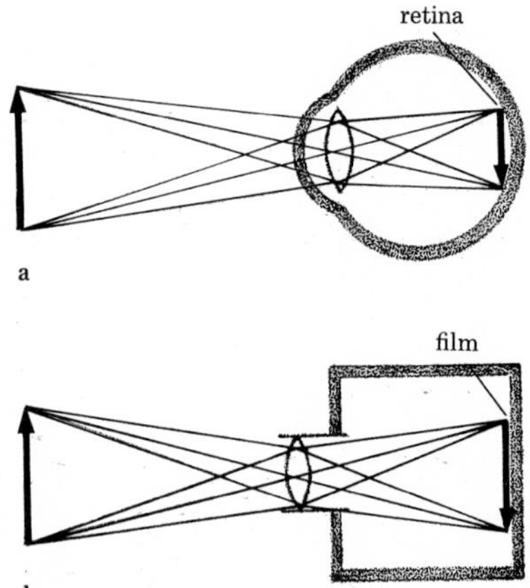

FIGURE 14.9

An eye and a camera. (a) The lens of the eye forms an image of a distant object on the retina. (b) The lens of a camera forms an image of a distant object on the film.

---

*The 35 mm refers to the width of the film used and not to the focal length of the lens. However, the choice of focal length depends on the film size, because a camera with a large film requires a longer-focal-length lens than a camera with smaller film if both cameras are to photograph the same scene.

on objects as close as 0.8 m. With a 50-mm lens, this requires an image distance s' given by

$$\frac{1}{80 \text{ cm}} + \frac{1}{s'} = \frac{1}{5.0 \text{ cm}} \qquad \text{or} \qquad s' = 5.33 \text{ cm}$$

Thus, to be able to focus a camera on objects at any distance from 0.8 m on up, the distance of the lens from the film need vary only between 5.33 and 5.00 cm.

Since the image distance s' in a camera is always approximately f, the magnification (Eq. 14.4) is approximately

$$m = \frac{f}{s} \qquad \text{camera} \qquad \qquad 14.7$$

and is always much less than 1.

Example 14.4 What is the image size of a 1.6-m-tall woman standing 4 m in front of a camera with a 50-mm-focal-length lens?

From Eq. 14.7 the magnification is

$$m = \frac{f}{s} = \frac{5 \text{ cm}}{400 \text{ cm}} = \frac{1}{80} = 0.0125$$

and so the height h' of her image on the film is

$$h' = hm = (1.6 \text{ m})(0.0125) = 0.020 \text{ m} = 2.0 \text{ cm}$$

Equation 14.7 shows that the image size can be increased either by moving the camera closer to the object or by using a lens with a longer focal length. ☐

Definition The *field of view* of a camera is the angle θ subtended by the scene imaged on the film. It is the angle between the chief rays that reach opposite sides of the film, as shown in Fig. 14.10. If w is the width of the film, Fig. 14.10a shows that

$$\tan \tfrac{1}{2}\theta = \frac{\tfrac{1}{2}w}{f}$$

or

$$\theta = 2 \arctan \frac{w}{2f} \qquad \qquad 14.8$$

Example 14.5 The standard image on a 35-mm film is a rectangle 37 by 25 mm. (a) Find the horizontal and vertical fields of view obtained when the film is used with a standard lens ($f = 50$ mm). (b) Repeat the calculations for a wide-angle lens ($f = 35$ mm) and a telephoto lens ($f = 125$ mm).

(a) When a 35-mm camera is held in its normal (horizontal) position, the horizontal dimension of its image is 37 mm. Thus, with $w = 37$ mm and $f = 50$ mm, the horizontal field of view θ_h is

$$\theta_h = 2 \arctan \frac{w}{2f} = 2 \arctan \frac{37 \text{ mm}}{2(50 \text{ mm})} = 40.6°$$

The vertical field of view θ_v is

FIGURE 14.10

The fields of view on a film of width w for lenses of different focal length. (a) Normal field of view. (b) Wide-angle field of view given by a short-focal-length lens. (c) Telephoto field of view given by a long-focal-length lens.

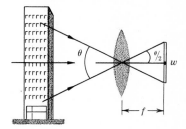

a

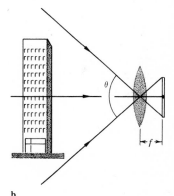

b

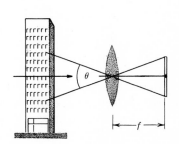

c

$$\theta_v = 2 \arctan \frac{25 \text{ mm}}{2(50 \text{ mm})} = 29.0°$$

(b) Figure 14.10b shows that with a shorter-focal-length lens, a larger field of view is obtained. With $f = 35$ mm, Eq. 14.8 gives

$$\theta_h = 55.7° \quad \text{and} \quad \theta_v = 39.3°$$

The term *wide angle* is used for any lens with a focal length shorter than normal. Although a shorter-focal-length lens gives a larger field of view, Eq. 14.7 shows that it also gives smaller magnification. This is not a contradiction. In order to squeeze a large scene onto the film, the image of each object in the scene must be reduced.

Figure 14.10c shows that with a longer-focal-length lens, a smaller field of view is obtained. With $f = 135$ mm, Eq. 14.8 gives

$$\theta_h = 16.8° \quad , \text{and} \quad \theta_v = 11.4°$$

The term *telephoto* is used for any lens with a focal length longer than normal. Since a longer-focal-length lens focuses a smaller scene onto the film, the image of each object in the scene is larger. Telephoto lenses are invaluable in sports and news photography because they allow the photographer to get a closeup picture of distant events. □

Example 14.6 A photographer using a lens with focal length $f_1 = 250$ mm takes a picture of a baseball pitcher from a distance $s_1 = 16$ m. If the photographer uses a lens with focal length $f_2 = 50$ mm, what distance s_2 from the pitcher would he have to be in order to obtain the same size image of the pitcher?

The images of the pitcher will be the same size if the magnifications are the same in both cases. From Eq. 14.7 the two magnifications are

$$m_1 = \frac{f_1}{s_1} \quad \text{and} \quad m_2 = \frac{f_2}{s_2}$$

So when these are equal, we have

$$\frac{f_1}{s_1} = \frac{f_2}{s_2}$$

or

$$s_2 = \frac{f_2}{f_1} s_1 = \frac{50 \text{ mm}}{250 \text{ mm}} (16 \text{ m}) = 3.2 \text{ m} \qquad □$$

The field of view of a camera depends on the focal length of the lens and the size of the film, but not on the diameter of the lens. This surprising fact can be verified by using a screen with a hole in it to mask part of a lens (Fig. 14.11). With the screen in place the image has the same size as before, but it is less bright. This is because rays from each point of the object pass through all points of the lens. The screen prevents some of these rays from reaching the film, but all the original image points receive some light.

A lens is a light-gathering device as well as a focusing device.

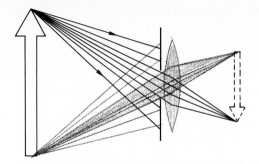

FIGURE 14.11

A screen masking part of a lens. The screen affects the brightness of the image, but not the field of view.

Since all the light incident on a lens from an object is focused onto the image, the intensity of the illumination reaching the image is proportional to the area of the lens and inversely proportional to the area of the image. The area of the lens is proportional to d^2, where d is the diameter of the lens. The area of the image is proportional to m^2, where m is the linear magnification. But from Eq. 14.7, m is proportional to f, so the image area is proportional to f^2. The intensity of illumination reaching the image is thus proportional to d^2/f^2.

Definitions The *relative aperture* of a lens is the ratio d/f of its diameter to its focal length. The intensity of the illumination reaching the image is proportional to the square of the relative aperture. The *f-number* is the reciprocal of the relative aperture:

$$f\text{-number} = \frac{f}{d} \qquad\qquad 14.9$$

The intensity of the illumination reaching the image is inversely proportional to the square of the f-number. In photography it is customary to write an *f-number* of (say) 5.6 as $f/5.6$. Do not confuse *f-number* with focal length f.

The *f-number* of a lens can always be increased, i.e., the effective diameter made smaller, by masking the lens. All quality cameras are equipped with an adjustable aperture, called an *iris,* to control the brightness of the image. The usual sequence of *f-number* settings on the iris is 2.8, 4, 5.6, 8, 11, 16, and 22. Each number is approximately $\sqrt{2}$ times the previous number, so the squares of these numbers are in the ratios $1:2:4:8:\cdots$. A change of the aperture setting by one stop (e.g., from 4 to 5.6 or from 16 to 11) changes the brightness of the image by a factor of 2. An increase in the *f-number* decreases the brightness and a decrease in the *f-number* increases the brightness.

Definition The *shutter speed* of a camera is the length of time the shutter remains opened while taking a picture. The usual sequence of shutter speeds available on a quality camera is 1, 1/2, 1/4, 1/8, 1/16, 1/30, 1/60, 1/125, 1/250, 1/500, and 1/1000 s. Each setting is half the time of the previous one, so each exposes the film to half as much light.

Shutter speed and *f-number* are used together to control the total amount of light reaching the film. A light meter might indicate that the camera should be set for 1/60 s at $f/8$. But the same amount of light will reach the film if the shutter speed is increased to 1/125 s while the lens is opened to $f/5.6$. In this latter case, the film is exposed to twice the intensity for half the time. Other settings equivalent to these are 1/30 s at $f/11$, 1/250 at $f/4$, and 1/500 s at $f/2.8$.

The choice of which combination of shutter speed and f-number to use depends on the situation. A sports photographer needs a fast shutter speed to freeze the action, so he would choose 1/500 s at $f/2.8$. A school photographer taking a class picture needs to focus on people who are standing at different distances from the camera. This photographer will choose a small aperture and a correspondingly longer shutter speed (e.g., 1/30 s at $f/11$) in order to achieve the required depth of field.

Definition The *depth of field* is the range of distances at which an object will be in satisfactory focus for a given camera setting. A large depth of field is desirable when there are many equally important objects at different distances from the camera, as in a class picture. A short depth of field is desirable when all the interest is to be concentrated on a single object, as in a portrait photograph.

Figure 14.12 shows that the rays that emerge from a point P that is more than the correct object distance s from a lens are focused at a point P' in front of the film. The rays that emerge from P' form a circular spot on the film. The size of this spot decreases as the diameter of the iris is decreased, i.e., as the f-number is increased. With a small enough aperture, the circular spot approximates a point, and an acceptable image will be formed of an object located at P. Thus the depth of field increases with increasing f-number. This is illustrated by the two photographs in Fig. 14.13. In both pictures the camera was focused on the woman, but the picture in Fig. 14.13a was taken at 1/1000 s and $f/2.8$, giving a short depth of field, whereas the picture in Fig. 14.13b was taken at 1/30 s and $f/16$, giving a longer depth of field.

In the limit of a tiny pinhole aperture, all the rays go through the center of the lens, where they are not deviated. In this case the lens can be removed, and a satisfactory image still obtained. Figure 14.14 shows a picture taken with a *pinhole camera,* which is a simple homemade device consisting of a light-tight box with a pinhole opening in one side. A sheet of film is mounted inside the box, on the side opposite the pinhole. The pinhole is kept covered until the picture is to be taken. The pinhole camera has an unlimited depth of field, but its f-number is so large ($f/400$ is typical) that a long exposure time is needed (about 1 min). Only

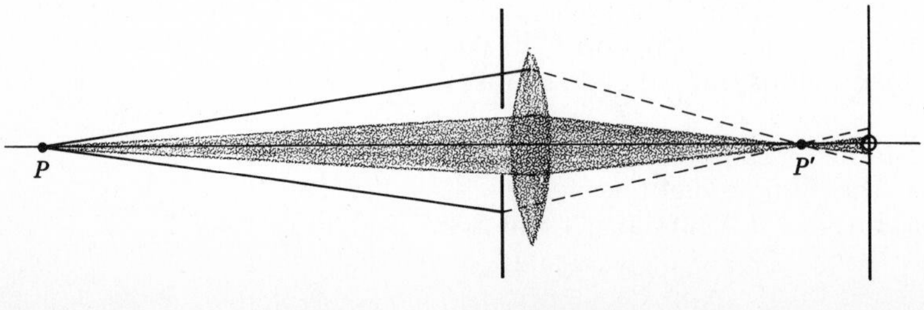

FIGURE 14.12

Rays emerging from a point P that is more than the correct distance from a camera lens are brought to focus in front of the film. As the rays diverge from the image point, they form a circular image spot on the film.

a

b

FIGURE 14.13

Two photographs taken with different *f*-numbers. In both pictures the camera was focused on the woman on the blanket. (*a*) At *f*/2.8 $\frac{1}{1000}$ s the background is out of focus. (*b*) At *f*/16 and $\frac{1}{10}$ s the background and foreground are in focus, showing that the depth of field increases with decreasing aperture. (*Jack Rutledge.*)

an absolutely stationary scene can be photographed with a pinhole camera.

The pinhole camera illustrates the importance of the light-gathering function of a lens. A small-diameter camera lens with a large *f*-number is easy and cheap to manufacture, but it can be used only to photograph brightly lit scenes. A large-diameter camera lens with a small *f*-number is difficult and expensive to manufacture, but it can be used to photograph scenes under poor lighting conditions.

Figure 14.15 shows the construction of a high-quality single-lens reflex camera. The lens system is composed of six separate elements which act together as a single lens. This eliminates the aberrations (Sec. 14.6) present in a single lens. The iris, located between the third and fourth elements, is fully open until the picture is taken. A mirror reflects the incident light onto a ground glass screen, where it forms the same image it would form on the film. A pentaprism takes the light scattered from the screen and bends it out through the viewer. The photographer looking through the viewer sees the image on the ground glass. After focusing the image, and setting the *f*-number* and shutter

*In cameras with automatic exposure control, the *f*-number is set automatically on the basis of the light incident on a built-in light meter.

FIGURE 14.14

Pinhole camera self-portrait. The camera was an empty $8\frac{1}{2}$ X 11 photographic-paper box. One sheet of photographic paper was attached face up to the bottom of the box, and a hole was made in the cover with a No. 10 sewing needle. After uncovering the hole, the photographer ran in front and sat motionless for several minutes. Then she ran back to cover the hole. Because of the long exposure time required for a pinhole picture, her rapid movements do not appear in the picture. Note the large depth of field. (*Maria Mastras, student in Mr. Rutledge's photography class at Wellesley High School, Wellesley, Massachusetts.*)

speed, the photographer presses the shutter-release button. As this is done, the mirror lifts out of the way, the iris closes to its preset value, and the shutter opens and closes, exposing the film. Immediately after the shutter closes, the mirror swings back down and the iris opens completely. The single-lens reflex camera is a marvel of mechanical and optical ingenuity.

THE TELESCOPE

A telescope can be used either visually or photographically. A visual telescope is a two-lens instrument like a microscope and will be discussed in Sec. 14.5. In most important astronomical work, however, the telescope is used photographically, and in this mode it acts as a giant camera.

A *refracting telescope* uses a lens to focus the light from a distant object, whereas a *reflecting telescope* uses a curved

FIGURE 14.15

Single-lens reflex camera, showing how the image formed by the lens is focused on a piece of ground glass. When the shutter release is depressed, the mirror lifts up, causing the same image to fall on the film. (*Minolta Corp.*)

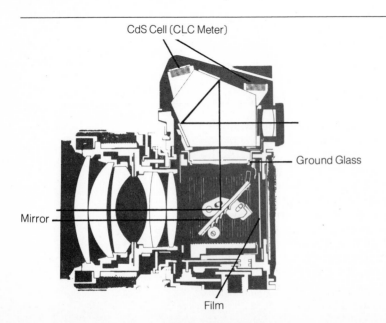

CdS Cell (CLC Meter)

Ground Glass

Mirror

Film

a   b

FIGURE 14.16

Photographs taken with the Hale 200-in telescope. (a) A cluster of galaxies over 4 billion light years (3.8×10^{25} m) away. These are among the most distant objects that can be seen from earth. (b) The planet Jupiter. (*Hale Observatories.*)

mirror (Sec. 14.4). The primary function of an astronomical telescope is to detect and analyze the light coming from stars, and in this capacity it acts as a light-gathering instrument. The larger the lens or mirror of a telescope, the greater is its light-gathering power. The lens of the largest refracting telescope (at the Yerkes Observatory of the University of Chicago) has a diameter of 40 in (102 cm), and the mirrors of the two largest reflecting telescopes (at the Russian Observatory in the Crimea and the Hale Observatory on Mt. Palomar) have diameters of 236 in (600 cm) and 200 in (508 cm), respectively. Large reflecting telescopes are used to study the light coming from the most remote objects in the universe (Fig. 14.16a).

A useful measure of the light-gathering power L of a telescope is the ratio of the area of the telescope lens (or mirror) to the area of the pupil of the human eye. This ratio is

$$L = \frac{d^2}{(0.63 \text{ cm})^2} = 2.5d^2 \qquad\qquad 14.10$$

where d is the diameter of the telescope lens in centimeters, and 0.63 cm is the diameter of the pupil. For the 200-in telescope, the light-gathering power is

$$L = 2.5d^2 = 2.5(508)^2 = 6.5 \times 10^5$$

A secondary function of an astronomical telescope is to study the features of nearby objects, such as the moon and planets (Fig. 14.16b). The magnification m of an object photographed by a telescope is the same as the magnification of an object photographed by any other camera (Eq. 14.7). A useful measure of the magnifying power M of a telescope is the ratio of the magnification of the telescope to the magnification of the human eye. This ratio is

$$M = \frac{f}{(1.7 \text{ cm})} = (0.69)f \qquad\qquad 14.11$$

where f is the focal length of the telescope (in centimeters), and 1.7 cm is the focal length of the human eye.

Example 14.7 The focal length of the 200-in Hale telescope is 17 m. (a) What is the magnifying power of this telescope? (b) What is the diameter of the image of Jupiter that this telescope focuses on a photographic plate? (Use Appendix VI.) (c) The nearest star is 4×10^{16} m from earth. If its diameter is the same as the sun, what is the diameter of the image of the star that the telescope focuses on a photographic plate?

(a) From Eq. 14.11 we have

$$M = (0.69)f = (0.69)(1700) = 1173$$

This means that the diameter of the image focused on a photographic plate by the Hale telescope is 1173 times the diameter of the image of the same object focused on the retina of the eye.

(b) If d is the diameter of an object and d' is the diameter of its image, Eqs. 14.4 and 14.7 give

$$d' = md = \frac{f}{s}d \qquad\qquad 14.12$$

where s is the distance of the object from earth. From Appendix VI we have that the diameter of Jupiter is $d = 144 \times 10^6$ m, and its closest distance from earth is

$$s = 77.8 \times 10^{10} \text{ m} - 14.9 \times 10^{10} \text{ m}$$
$$= 63 \times 10^{10} \text{ m}$$

Therefore, the diameter d' of Saturn's image is

$$d' = \frac{f}{s}d = \frac{17 \text{ m}}{63 \times 10^{10} \text{ m}}(144 \times 10^6 \text{ m})$$
$$= 3.9 \times 10^{-3} \text{ m} = 3.9 \text{ mm}$$

The picture in Fig. 14.16b is an enlargement of a photographic image of this size.

(c) The sun is a typical star. From Appendix VI we find that its diameter is 1.4×10^9 m. The diameter of the image of a star of this diameter located 4×10^{16} m (4 light-years) from earth is

$$d' = \frac{f}{s}d = \frac{17 \text{ m}}{4 \times 10^{16} \text{ m}}(1.4 \times 10^9 \text{ m}) = 6 \times 10^{-7} \text{ m}$$

This is smaller than the image spot of a point object caused by the diffraction of the object's light as it enters the telescope (Sec. 13.3). Consequently, no detail of the star's external structure can be seen in the image. Except for the sun, all stars appear only as points of light to even the largest telescopes. However, by spectroscopically analyzing the light gathered by a telescope, it is possible to determine the chemical composition, temperature, and speed of individual stars. From these data, models of the internal structure of stars are made. □

14.3 Virtual Images

A positive lens does not form a real image of an object when the object distance s is less than the focal length f. This is seen in Fig. 14.17, which traces the chief and the parallel ray from point B of an object that is located less than 1 focal length from the lens. Like all other rays, these rays are diverging as they radiate from B,

FIGURE 14.17

The virtual image formed by an object located inside the front focal point of a positive lens. The rays from point B of the object do not converge after passing through the lens, so a real image is not formed. However, when the refracted rays are projected backwards, they appear to come from point B', which is the virtual image of B.

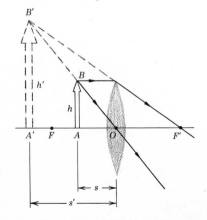

but they are still diverging even after they pass through the lens, so that no real image is formed.

However, if the diverging rays that emerge from the lens are extended backward, they intersect at point B' on the object side of the lens. This means that the rays that emerge from the lens appear to be coming from a point B' located a distance s' in front of the lens. Similarly, after passing through the lens, rays from other points of the object appear to be coming from corresponding points on the line $A'B'$. That is, if one looks at the light coming from the lens, the image formed on the retina of one's eyes is the same as the image formed by a real object of height h' located between A' and B'. In this sense, the rays that emerge from the lens are said to form a *virtual image* of the object on the object side of the lens. It is not a real image because there is no light at the position of the virtual image and no image would be formed on a screen placed there.

The virtual image in Fig. 14.17 is larger than the object itself and it is not inverted, so the eye sees an erect, magnified image of the object. The magnification $m = h'/h$ is still equal to s'/s, as given by Eq. 14.5. (This is seen by noting that the right triangles $OA'B'$ and OAB in Fig. 14.17 are similar.) However, the relation between s, s', and f is no longer given by Eq. 14.2. When the image is virtual, the correct relation is

$$\frac{1}{s} - \frac{1}{s'} = \frac{1}{f} \qquad \text{virtual image} \qquad\qquad 14.13$$

This can be derived from Fig. 14.17 in the same way that Eq. 14.2 was derived from Fig. 14.6.

> **REMARK** The lens formula (Eq. 14.2) can be used for both real and virtual images if the image distance s' is taken to be negative for virtual images. However, it may be simpler to treat all distances as positive quantities and use different formulas for real and virtual images.

THE MAGNIFYING GLASS

A *magnifying glass* is just a simple converging lens used to examine small objects. The object is positioned inside the focal point of the lens, as in Fig. 14.17, so that an erect, magnified virtual image is formed. The magnification $m = s'/s$ depends on the distance s between the lens and the object. For calculations it is usually assumed that s is adjusted so that the virtual image is formed 25 cm in front of the lens, because this is the comfortable reading distance for a person with normal vision. Using $s' = 25\,\text{cm}$ in Eq. 14.13, we get

$$\frac{1}{s} = \frac{1}{25\,\text{cm}} + \frac{1}{f} = \frac{f + 25\,\text{cm}}{(f)(25\,\text{cm})} \qquad\qquad 14.14$$

and so

$$m = \frac{s'}{s} = \frac{25\,\text{cm}}{s} = \frac{f + 25\,\text{cm}}{f} = 1 + \frac{25\,\text{cm}}{f} \qquad\qquad 14.15$$

This gives the magnifying power of a magnifying glass in terms of the focal length of the lens. In Eq. 14.15, the focal length f must be in centimeters, of course.

Example 14.8 (a) What is the magnifying power of a lens of focal length 5 cm? (b) How far from the lens must an object be to achieve this magnification?

(a) From Eq. 14.15 the magnification is

$$m = 1 + \frac{25 \text{ cm}}{5 \text{ cm}} = 6$$

(b) From Eq. 14.14 the distance s from the lens to the object is

$$\frac{1}{s} = \frac{5 \text{ cm} + 25 \text{ cm}}{(5 \text{ cm})(25 \text{ cm})} = 0.24 \text{ cm}^{-1}$$

or

$$s = \frac{1}{0.24 \text{ cm}^{-1}} = 4.16 \text{ cm}$$

This means that the object is held 0.84 cm inside the front focal point of the lens. ☐

> **REMARK** To achieve the theoretical magnification given in Eq. 14.15, the lens must be held directly in front of the eye. Thus with a 5-cm-focal-length lens, the object is held only 4.16 cm from the eye. Without a lens, an object held 4.16 cm from the eye appears $(25 \text{ cm})/(4.16 \text{ cm}) = 6$ times larger than when held 25 cm from the eye. But without a lens, the eye cannot focus on an object only 4.16 cm away. Thus, in a sense, the magnifying glass does not magnify at all, but simply enables the eye to focus on an object held very close to it.

READING GLASSES

The retina of the eye is a fixed distance in back of the eye's lens, and objects at different distances from the eye are focused on the retina by changing the focal length of the lens. This process is called *accommodation*. The normal adult eye cannot accommodate to objects that are less than 25 cm from the eye, whereas children can focus on objects as little as 10 cm from their eyes. Many people over forty can no longer focus on objects held at the normal reading distance (25 cm) and so require corrective lenses. The purpose of these lenses is to produce a virtual image of an object at a distance s' at which the eye can accommodate.

Example 14.9 What should be the focal length of the reading glasses worn by a person who can focus only on objects 100 cm or more away?

The closest distance at which a person can focus is called the *near point*. The reading glasses enable the person to hold the object at the normal reading distance ($s = 25$ cm) while viewing a virtual image at the near point ($s' = 100$ cm). From Eq. 14.13 the focal length required for this is

$$\frac{1}{f} = \frac{1}{s} - \frac{1}{s'} = \frac{1}{25 \text{ cm}} - \frac{1}{100 \text{ cm}} = \frac{3}{100 \text{ cm}} \quad \text{or} \quad f = 33.3 \text{ cm}$$

With eyeglasses of this focal length, the person can read a book held at the normal reading distance because after passing through the glasses, the rays from a book held 25 cm from the eyes have the same divergence as rays coming from a book held 100 cm from the eyes. ☐

REMARK Note that the magnification of the image is

$$m = \frac{s'}{s} = \frac{100\,\text{cm}}{25\,\text{cm}} = 4$$

Thus the virtual image is 4 times the size of the object. But because the image is 4 times farther away than the object, its apparent size is the same.

Definition The *power* of a lens is the reciprocal of its focal length f. The unit of power is the inverse meter (m^{-1}), called a *diopter*. Optometrists use power rather than focal length to specify a particular lens. The power of the lens in Example 14.8 is

$$\frac{1}{f} = \frac{1}{0.333\,\text{m}} = 3\,\text{m}^{-1} = 3\ \text{diopters}$$

NEGATIVE LENSES

A *positive* (converging) *lens* always decreases the divergence of the rays that pass through it. If an object is more than 1 focal length from a positive lens, the divergence of the rays reaching the lens is small enough for the lens to convert them into converging rays, forming a real image. A negative (diverging) lens always increases the divergence of the rays that pass through it. Figure 14.18 shows that parallel rays incident on a negative lens diverge after passing through the lens. When extended backward, these diverging rays intersect at a point F on the optic axis, which is the focal point of the negative lens. The distance from F to the center of the lens is the focal length and is taken to be a negative quantity.

A *negative lens* always forms a virtual image of an object any distances s from the lens (Fig. 14.19). The image distance s' is given by Eq. 14.13 if a negative focal length is used. Negative lenses are used to correct myopia (nearsightedness), which is a visual defect in which the eye cannot accommodate to objects farther than some distance d (the *far point*). A negative lens forms a virtual image in its focal plane of very distant objects. Thus, by wearing eyeglasses with negative lenses of focal length equal to d, a virtual image of a distant object is formed at a distance to which the eye can accommodate.

Example 14.10 What is the power (in diopters) of the lens required to correct myopia in a person with a far point of 250 cm?

The lens is required to form an image at $s' = 250\,\text{cm}$ when s is very large. Since $1/s$ is essentially zero in this case, the focal length given by Eq. 14.13 is

$$\frac{1}{f} = \frac{1}{s} - \frac{1}{250\,\text{cm}} = -\frac{1}{250\,\text{cm}}$$

or

$$f = -250\,\text{cm} = -2.50\,\text{m}$$

The power of the lens is

$$\frac{1}{f} = \frac{1}{-2.5\,\text{m}} = -0.4\,\text{m}^{-1} = -0.4\ \text{diopter}$$

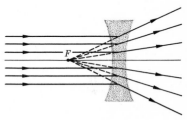

FIGURE 14.18

Rays parallel to the optic axis diverge after passing through a negative lens. When the diverging rays are projected backwards, they appear to come from the front focal point F of the lens.

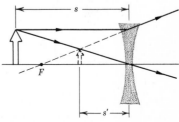

FIGURE 14.19

The virtual image of an object produced by a negative lens. The rays from a point on the object are diverging after they pass through the lens. When these rays are projected backwards, they appear to come from a point on the virtual image.

The minus sign indicates that it is a negative (diverging) lens. □

14.4 Mirrors

A *mirror* is a smooth, highly reflecting surface, usually made by vacuum evaporating a coating of aluminum onto a plate of polished glass. The reflecting surface is made as smooth as possible to eliminate any diffuse reflection. All rays incident on a mirror are therefore reflected in accordance with the law of specular reflection (Eq. 13.2).

> **REMARK** Mirrors found in the home are back plated; i.e., the aluminum coating is applied to the back surface of the glass. This protects the surface from damage, but about 4 percent of the incident light is reflected from the front surface, and the rest is refracted into the glass, reflected from the coated surface, and refracted out again (Fig. 14.20a). A back-coated mirror thus produces two reflected rays, one much fainter than the other. Optical instruments use front-coated mirrors which have only one reflecting surface (Fig. 14.20b). All diagrams in this section are drawn for front-coated mirrors.

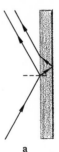

a

PLANAR MIRRORS

A *planar mirror* is a mirror with a plane (uncurved) surface. Figure 14.21 shows that the rays emanating from a point P, after being reflected from a planar mirror, appear to emanate from a point P' in back of the mirror. The point P' is the virtual image of P. For a planar mirror, the image distance $s' = \overline{OP'}$ is equal to the object distance $s = \overline{OP}$:

$$s' = s \qquad \text{planar mirror} \qquad 14.16$$

> **REMARK** To prove Eq. 14.16, consider the ray reflected from an arbitrary point A on the mirror in Fig. 14.21. From the law of reflection ($\theta_1 = \theta_1'$) the angle $\phi = 90° - \theta_1$ is equal to the angle $\phi' = 90° - \theta_1'$. The right triangles POA and $P'OA$ are therefore congruent, because they have a common side and equal angles. This proves that $s = \overline{OP}$ is equal to $s' = \overline{OP'}$.

When an object is viewed through a mirror, an erect virtual image is seen behind the mirror (Fig. 14.22). Since $s = s'$, the magnification of the image is $m = s'/s = 1$. The parts of the object farthest from the mirror appear to be the farthest in back of the mirror, in accordance with Eq. 14.16. This has the peculiar effect of interchanging the handedness of an object; i.e., the mirror image of a right-handed object is left-handed, and vice versa. For instance, the mirror image of a man who parts his hair on the left is that of a man who parts his hair on the right.

b

FIGURE 14.20

(a) Light incident on a back-plated mirror is reflected from both the front and back surfaces of the mirror. (b) Light incident on a front-plated mirror is reflected only from the front surface.

FIGURE 14.21

The virtual image P' of a point P in front of a plane mirror. The rays from P, after being reflected from the mirror, appear to come from P'.

Example 14.11 A photographer standing 75 cm in front of a mirror wishes to photograph herself taking a photograph (Fig. 14.23). (a) For what distance should the camera be set to form a focused image on the film? (b) If she is holding the camera to her right eye when she takes the picture, to which eye is the camera held in the image formed on the film?

(a) If the photographer is 75 cm in front of the mirror, her image will be 75 cm in back of the mirror, or 150 cm from the camera. The light reaching the camera from the mirror has the same divergence as light coming from an object 150 cm away, so the camera should be set for 150 cm (1.5 m).

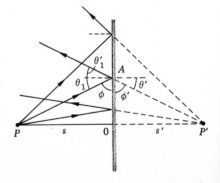

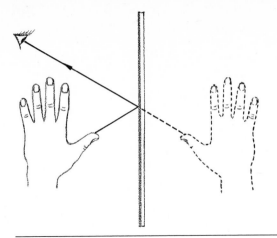

FIGURE 14.22

The mirror image of a left hand is a right hand.

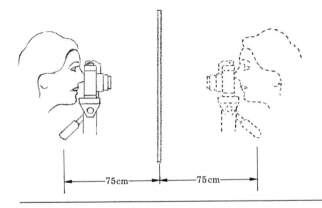

—75cm— —75cm—

FIGURE 14.23

A photographer photographing herself in a mirror.

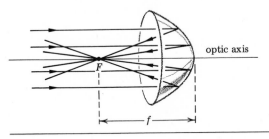

optic axis

F

f

FIGURE 14.24

Rays parallel to the optic axis of a parabolic mirror are reflected through the focal point *F* of the mirror.

(*b*) In the mirror image, the camera is held to the left eye, and this is the way the image will be formed on the negative. (It is possible to reverse the negative in the printing process, and produce a positive print in which the camera is back on the right eye.)

PARABOLIC MIRRORS

A mirror with a properly curved surface can focus light just like a lens. The surface has the shape of a bowl formed by rotating a parabola about an axis through the bottom of the bowl (Fig. 14.24). This axis is the optic axis of the mirror. All incident rays parallel to the optic axis are reflected through the same point,

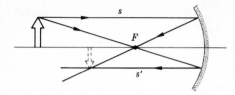

FIGURE 14.25

The real image of an object produced by a parabolic mirror.

called the *focal point F*. The distance from the focal point to the center of the mirror is the *focal length f*.

As with a lens, light from a very distant object is focused onto the plane perpendicular to the optic axis at *F* (the *focal plane*). Light from an object a finite distance *s* from the mirror forms an image a distance *s'* from the mirror (Fig. 14.25), where *s* and *s'* are related by the lens formula (Eq. 14.2). The image is inverted, and its magnification is $m = s'/s$, the same as the image formed by a lens.

All large telescopes use a parabolic mirror instead of a lens to gather the light from distant stars. A mirror is preferred for large telescopes because it has only one surface that must be optically perfect, whereas a lens has two. Furthermore, a large lens will sag under its own weight because it can be supported only at its edge. A mirror, on the other hand, can rest on a firm mechanical support, since light does not pass through it. The largest telescope lens is the 40-in lens at the Yerkes Observatory; the largest telescope mirrors are the 200-in mirror at the Hale Observatory in California and the 236-in mirror at the Russian Observatory in the Crimea.

Stars emit electromagnetic radiation at wavelengths other than that of visible light. Most of the nonvisible radiation is absorbed by the earth's atmosphere and does not penetrate to the earth's surface. However there is a region in the radio part of the electromagnetic spectrum, between 0.1 and 10 m, which reaches the earth's surface along with the visible light. Since 1932 radio telescopes have been built to detect and study this radiation.

All radio telescopes are large parabolic antennas that reflect incident radiation onto a radio receiver located at the focal point (Fig. 14.26). These reflectors must be much larger than optical mirrors because diffraction effects are much larger for this longer-wavelength radiation. The world's largest radio telescope fills a bowl-shaped valley at the Arecibo Observatory in Puerto Rico (Fig. 14.27). The receiver for this telescope is suspended at the focal point by cables strung across the valley.

A parabolic reflector can be used to transmit a signal as well as receive one. The simplest example is a flashlight, which has a small light bulb mounted at the focal point of a parabolic reflector. Light is emitted in all directions by the bulb, but those rays which strike the reflector are reflected into a parallel beam. The reflector in effect channels a certain portion of the light into one direction.

REMARK The beam of a flashlight is not exactly parallel because the bulb's filament is not a point. Light from those parts of the filament not at the focal point is reflected at a small angle to the optic axis. Thus a flashlight, like a

FIGURE 14.26

An 85-ft-diameter radio telescope used to track spacecraft. (*National Aeronautics and Space Administration.*)

FIGURE 14.27

The world's largest radio telescope at the Arecibo Observatory in Puerto Rico. A natural hollow in the mountains was shaped into a spherical surface and lined with steel mesh to reflect radio waves. These waves are focused by the bowl onto the receiver suspended over the bowl. (*Cornell University and Russ Hamilton.*)

spotlight, produces a slightly diverging beam. Only a laser produces a beam that is parallel to within the limits allowed by diffraction.

Much of the interstate communication traffic in the United States, both telephone and television, is carried by microwaves transmitted and received by parabolic reflectors mounted on towers. Some of the signal fed into a transmitter located at the focal point of a parabolic dish is reflected from the dish into a beam aimed at the parabolic dish of a receiver (Fig. 14.28). When the beam strikes the receiving dish it is reflected into the detector. The signal is then amplified and transmitted by a second transmitter. A sequence of towers, spaced 15 to 30 km apart, carry microwave signals across the continent.

REMARK Light rays parallel to the optic axis of a parabolic mirror are reflected through the focal point no matter how far they are from the axis. A spherical mirror approximates a parabolic one for rays close to the optic axis, but does not focus rays far from the axis. Since it is easier to grind a spherical surface than a parabolic one, spherical mirrors are sometimes used where critical focusing is not required.

14.5 Two-Lens Instruments

THE MICROSCOPE

A *microscope* uses two lenses to produce a highly magnified image of a small object. One lens, called the *objective,* is at the lower end of the microscope tube, just above the object on the slide (Fig. 14.29). The objective forms a real image of this object. The other lens, called the eyepiece, or *ocular,* is at the upper end of the microscope tube. It acts as a magnifying lens, forming a magnified virtual image of the image formed by the objective. When the microscope is in focus, the virtual image is located 25 cm below the ocular, so that it can be focused by a normal eye looking through the ocular.

Figure 14.30 traces some rays through such a system of lenses. Notice that the object is placed just outside the front focal point F_1 of the objective in order to produce a real image. On the other hand, this real image is placed inside the front focal point F_2 of the ocular in order to produce the desired virtual image.

REMARK The rays passing through the ocular are focused by the lens of the eye onto the retina. The size of this retinal image is the same as the size of

FIGURE 14.28

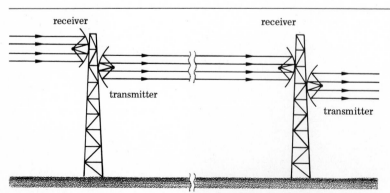

Radar relay towers. Each tower receives radar signals in one parabolic dish and transmits them to the next tower from another parabolic dish.

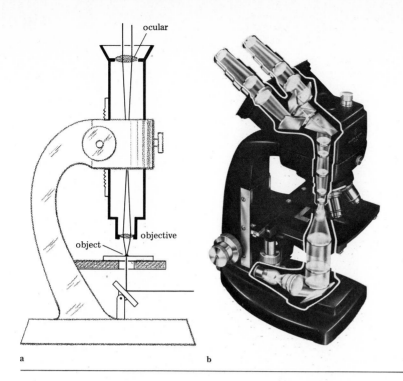

ocular

object

objective

a   b

FIGURE 14.29

The microscope. (a) The essential features of a microscope are two lenses, the objective and the ocular, arranged so that the objective forms a real image of the object just inside the front focal point of the ocular. (b) In a real microscope, both the objective and the ocular are composed of several lenses to improve the quality of the image. In addition, lenses are used to focus light onto the object, and in the binocular microscope shown here, mirrors and prisms are used to divide the light and direct it through two independent oculars. (*Bausch & Lomb.*)

the retinal image that would be formed by a real object that had the size and position of the virtual image.

Example 14.12 The focal lengths of the objective and ocular of a microscope are $f_1 = 0.5$ cm and $f_2 = 3.0$ cm, respectively, and the lenses are a distance $d = 18$ cm apart. Locate the positions of the real and virtual images, and determine the overall magnifying power of the instrument.

From Eq. 14.15 the magnifying power of the ocular is

$$m_2 = 1 + \frac{25 \text{ cm}}{f_2} = 1 + \frac{25 \text{ cm}}{3.0 \text{ cm}} = 9.3$$

(This is the number engraved next to $\times$ on the barrel of the ocular.) The ocular produces a virtual image, at a distance $s_2' = 25$ cm, of the real image that is located a distance s_2 from the ocular. The distance s_2 is found from Eq. 14.13:

$$\frac{1}{s_2} = \frac{1}{s_2'} + \frac{1}{f_2} = \frac{1}{25 \text{ cm}} + \frac{1}{3.0 \text{ cm}} = 0.373 \text{ cm}^{-1}$$

or

$$s_2 = \frac{1}{0.373 \text{ cm}^{-1}} = 2.68 \text{ cm}$$

That is, when the microscope is in focus, the real image produced by the objective must be 2.68 cm below the ocular.

Since the distance d between the ocular and the objective is 18 cm, the image distance s_1' of the real image from the objective is

$$s_1' = d - s_2 = 18.00 \text{ cm} - 2.68 \text{ cm} = 15.32 \text{ cm}$$

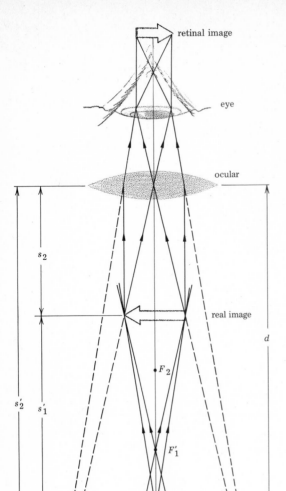

retinal image

eye

ocular

s_2

real image

d

F_2

s'_2

s'_1

F'_1

objective

s_1

F_1

object

virtual image

FIGURE 14.30

The optics of the microscope. The objective forms a real image of the object just inside the front focal point F_2 of the ocular. The ocular forms a virtual image of the real image. The virtual image is located 25 cm below the ocular.

The object distance s_1 can now be obtained from Eq. 14.2:

$$\frac{1}{s_1} = \frac{1}{f_1} - \frac{1}{s'_1} = \frac{1}{0.5 \text{ cm}} - \frac{1}{15.32 \text{ cm}} = 1.935 \text{ cm}^{-1}$$

or

$$s_1 = \frac{1}{1.935 \text{ cm}^{-1}} = 0.517 \text{ cm}$$

The object distance s_1 is just a little larger than f_1, so that the

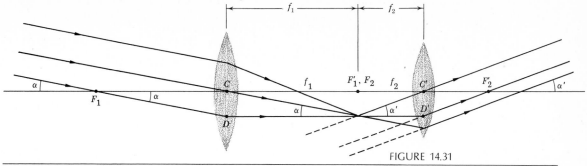

FIGURE 14.31

The optics of the telescope. The objective forms a real image of a distant object at the front focal point of the ocular. The rays that emerge from this point are rendered parallel by the ocular, so they appear to come from a distant virtual image. If $f_2 < f_1$, the emergent angle α' will be greater than the incident angle α, so the virtual image will appear closer than the object.

magnification $m_1 = s_1'/s_1$ of the real image is approximately

$$m_1 = \frac{s_1'}{f_1} = \frac{15.34 \text{ cm}}{0.5 \text{ cm}} = 30.7$$

The real image is larger than the object by a factor m_1, and the virtual image is larger than the real image by a factor m_2. Therefore, the virtual image is M times larger than the object, where

$$M = m_1 m_2 = \frac{s_1'}{f_1}\left(1 + \frac{25 \text{ cm}}{f_2}\right) \qquad 14.17$$

This is the magnifying power of a microscope. In the present example we have

$$M = m_1 m_2 = (30.7)(9.3) = 285 \qquad \square$$

THE TELESCOPE

We have seen in Sec. 14.2 that a one-lens telescope forms a magnified image of a distant object in the focal plane of the lens. This image can be detected by photographic film, but it cannot be viewed directly by the eye. For visual work a small secondary lens, called the *ocular*, is required.

The large primary, or *objective*, lens is the light-gathering lens of the system. It focuses the light from a distant object onto its focal plane. The rays then diverge from this plane and are refracted by the ocular (Fig. 14.31). For comfortable viewing, the focal point of the ocular coincides with the focal point of the objective, so the rays emerging from the ocular are parallel.

Figure 14.31 shows that rays entering the system from above the optic axis emerge from below the axis. Therefore, an observer looking at the parallel rays that emerge from the ocular will see an inverted image of a distant object. If the focal length f_2 of the ocular is less than the focal length f_1 of the objective, the angle α' at which these rays emerge is greater than the angle α at which they were incident. This has the effect of making the image seen through the instrument appear closer than the object seen by the unaided eye.

The ratio α'/α of these angles is approximately equal to the ratio $(\tan \alpha')/(\tan \alpha)$ of their tangents. But from Fig. 14.31 we have

$$\tan \alpha = \frac{\overline{DC}}{f_1} \qquad \text{and} \qquad \tan \alpha' = \frac{\overline{D'C'}}{f_2} = \frac{\overline{DC}}{f_2}$$

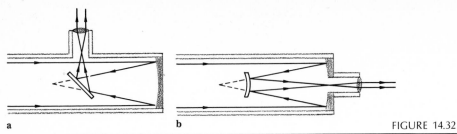

a   b   FIGURE 14.32

Two types of reflecting telescopes.
(a) Newtonian telescope. A plane
mirror diverts the reflected rays
through an ocular lens mounted on
the side of the telescope tube.
(b) Cassegrainian telescope. A
hyperbolodial mirror diverts the reflected rays through a hole in the
parabolic mirror.

Consequently, the ratio of these tangents, called the *angular magnification a* of the telescope, is

$$a = \frac{\tan \alpha'}{\tan \alpha} = \frac{\overline{DC}/f_2}{\overline{DC}/f_1} = \frac{f_1}{f_2} \qquad\qquad 14.18$$

Large magnification requires a very long focal length objective, which is why telescopes are so long.

> **REMARK** It is believed that the Dutch spectaclemaker Hans Lippershey was the first to combine two lenses in the microscope and telescope configurations between 1590 and 1608. However, with the quality of lenses then available, the early microscope was not superior to the magnifying lens. The microscope did not replace the magnifying lens for another century, when it became possible to produce high-quality lenses. The history of the telescope is more dramatic.
>
> Galileo happened to be in Venice about May 1609, where he heard rumors of a perspective instrument. The day after he got back to Padua, Galileo made his first telescope and soon learned how to build superior instruments, with magnifications of up to 30. Galileo was the first person to look at the stars with a telescope, and the world was never the same. He saw the craters on the moon, the spots on the sun, and the moons around Jupiter. These observations contradicted Aristotle's old idea that the heavenly bodies were composed of incorruptible matter, different in substance from the matter on earth. As a consequence, the authority of Aristotle was shaken, and people began to open their minds to new ideas.

In a reflecting telescope, the objective lens is replaced by a parabolic mirror. For visual work, a small secondary mirror, either planar or curved, is used to reflect light out of the telescope barrel. Figure 14.32 shows two common arrangements. The Newtonian telescope uses a planar mirror to reflect light out the side of the barrel. The Cassegrainian telescope uses a hyperboloidal mirror to reflect the light through a hole in the center of the parabolic mirror. The hyperboloidal mirror increases the effective focal length of the parabolic mirror, making for a more compact instrument.

A terrestial telescope is an instrument, such as a surveyor's telescope, used to view objects on earth. It is similar to an astronomical telescope, except that it is designed to give an upright image. The simplest terrestial telescope, called a *Galilean telescope,* uses a negative lens for the ocular (Fig. 14.33). Rays emerging from the objective are focused on the back focal plane of the ocular, so that the ocular renders them parallel. As Fig. 14.33 shows, rays entering the objective from above the optic axis emerge from above the axis as they leave the ocular. There-

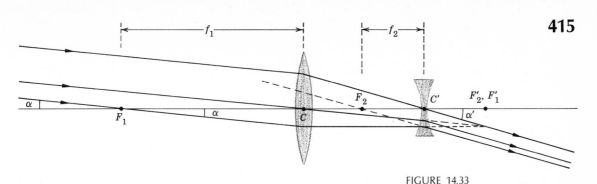

FIGURE 14.33

Galilean telescope. With a negative lens for the ocular, this telescope produces an upright image.

fore, a viewer looking through the ocular will see an upright image.

Each barrel of a pair of binoculars is a compact terrestial telescope which uses a pair of prisms to increase the distance the rays travel from the objective to the ocular (Fig. 13.15b). The reflection of the rays inside the prism also reinverts the image, so objects appear upright when viewed through the instrument.

COMPOUND LENSES

Two lenses placed close together act like a single lens. Most quality optical instruments use several lenses this way to eliminate the distortions (aberrations) produced by a single lens. In a microscope, for instance, the objective and the ocular are each composed of two or more lenses (Fig. 14.29b).

Figure 14.34 shows parallel rays incident on two lenses of focal lengths f_1 and f_2. The distance d between the lenses is less than the focal length f_1 of lens 1, so the rays reach lens 2 before they are focused in the focal plane of lens 1. The dotted lines show how, in the absence of lens 2, the rays passing through the center C_1 and the front focal point F_1 of lens 1 would intersect at point A in the focal plane of lens 1. The effect of lens 2 is to bend the rays further, so that they are focused at a point A' which is closer to lens 1 than point A is.

To locate A', first note that the ray that is parallel to the optic axis between lenses 1 and 2 will pass through the focal point F_2' of lens 2. Second, in the absence of lens 2 there is a ray that goes through point C_2 to A. But this ray is not affected by the presence of lens 2 because it passes through its center. Conse-

FIGURE 14.34

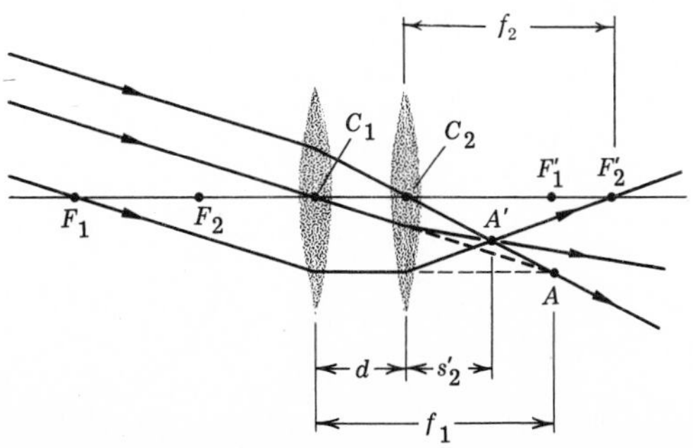

Compound lens. Parallel rays incident on this combination are brought to focus at point A', which is closer than the focal point of either lens. The position of A' is found graphically by first finding the point A at which the rays would be focused by the first lens alone.

quently, the intersection of these two rays at A' determines the position of the focal plane of this two-lens system.

This graphical construction shows that the focal length f of the two lenses together is less than that of either lens alone. Equation 14.2 can be used to find f by noting that the image that would have been formed by lens 1 alone acts as the object for lens 2. Since this image is a distance $s_2 = f_1 - d$ in back of lens 2, it must be treated as a negative distance in Eq. 14.2. With this understanding, the distance s_2' from lens 2 at which the image is formed is given by

$$-\frac{1}{f_1 - d} + \frac{1}{s_2'} = \frac{1}{f_2} \quad \text{or} \quad \frac{1}{s_2'} = \frac{1}{f_2} + \frac{1}{f_1 - d}$$

If the distance between the two lenses is very small (d approximately zero), this becomes

$$\frac{1}{f} = \frac{1}{f_1} + \frac{1}{f_2} \qquad\qquad 14.19$$

where we have set s_2' equal to f because the rays incident on the system are parallel.

Equation 14.19 is used by optometrists when fitting eyeglasses. They adjust the strength of a lens combination by adding or subtracting lenses in the system.

Example 14.13 What is the focal length of a 0.50-m lens in combination with a 0.75-m lens?

From Eq. 14.19 the focal length of the combination is

$$\frac{1}{f} = \frac{1}{0.50 \text{ m}} + \frac{1}{0.75 \text{ m}} = 2.0 \text{ m}^{-1} + 1.33 \text{ m}^{-1} = 3.33 \text{ m}^{-1}$$

or

$$f = 0.30 \text{ m}$$

The optometrist works with the power $1/f$ of a lens, rather than with the focal length f, because the power of a lens combination is given by simply adding the power of the individual lenses. In this example, the powers of the two lenses are 2.0 and 1.33 diopters, and their combination has a power of 3.33 diopters. ☐

14.6 Aberrations

SPHERICAL LENSES

The surfaces of most lenses are spherical, which means that each surface of the lens is a section of a sphere, as shown in Fig. 14.35. The *radius of curvature r* of a spherical surface is the radius of the sphere of which the surface is a section; it is positive if the surface is convex (Fig. 14.35a) and negative if the surface is concave (Fig. 14.35b). For instance, the radius of curvature of the left surface in Fig. 14.35a is r_1, and the radius of curvature of the left surface in Fig. 14.35b is $-r_2$.

The two surfaces of a lens can have different radii of curvature, but we shall consider the special case of a symmetric lens for which both surfaces have the same radius of curvature. The focal length f of a symmetric lens with radius of curvature r is

FIGURE 14.35

The surfaces of a spherical lens are segments of spheres: (a) convex lens and (b) concave lens.

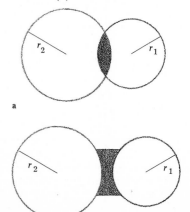

a

b

$$f = \frac{r}{2(n-1)} \qquad\qquad 14.20$$

where n is the index of refraction of the glass. This equation is called the *spherical lens,* or *lens maker's, formula.* It applies to both convex (positive) and concave (negative) symmetric lenses.

Example 14.14 What is the radius of curvature of a symmetric lens of focal length 50 mm made of glass with index of refraction 1.55?

From Eq. 14.20 the radius of curvature is

$$r = 2(n-1)f = (2)(0.55)(5.0\ \text{cm}) = 5.5\ \text{cm} \qquad \square$$

Lenses can be made with surfaces which are spherical to within a fraction of a wavelength of light. In fact it is because nearly perfect spherical lenses can be easily manufactured that they are used in almost all optical instruments. However, even a perfect spherical lens is not an ideal lens; i.e., it does not form an undistorted, sharply focused image.

Any departure from ideal behavior is called an *aberration.* There are seven distinct aberrations of a spherical lens, two chromatic and five monochromatic. We shall discuss these aberrations briefly and give some indication of how they are corrected in practice.

CHROMATIC ABERRATIONS

Chromatic aberrations are failures of a lens to focus light of different color at the same point. They arise from the variation of the index of refraction n with wavelength λ, a phenomenon called *dispersion.* Figure 14.36 shows the variation of n with λ for crown glass. In general n decreases as λ increases; i.e., n is less for red light than for blue.

Since the focal length f of a lens varies inversely with n (Eq. 14.20), f increases as λ increases; i.e., f is greater for red light than for blue. Thus when a lens focuses the white light emanating from a point, the different wavelength components of the light are focused at different points, as shown in Fig. 14.37. As a result, the image of a point on the axis is not a point but a colored circle.

> **REMARK** The variation of focal length with wavelength is called *longitudinal aberration* to distinguish it from *lateral aberration,* which is a similar variation of magnification with wavelength. These two types of chromatic aberration are so closely associated, however, that we shall not bother with the distinction. Usually they are corrected together.

Chromatic aberration is corrected by using a *doublet,* which is a pair of lenses which function as a single lenses. Figure 14.37 shows how the colored rays are oppositely refracted by a positive and a negative lens, so a doublet consisting of opposite lenses can eliminate chromatic aberration altogether (Fig. 14.38).

Alternatively, single *achromatic* lenses which are free from chromatic aberration can be made by cementing a concave and a convex lens together to form a meniscus lens (Fig. 14.39). If the

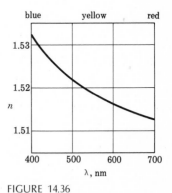

FIGURE 14.36

Variation of the index of refraction n of crown glass with wavelength λ.

FIGURE 14.37

Chromatic aberration. (a) A positive lens focuses red light at a different point from blue light, so the image of a point is a colored circle. (b) A negative lens has the opposite effect on the refracted rays.

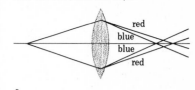

a

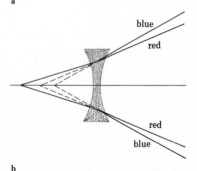

b

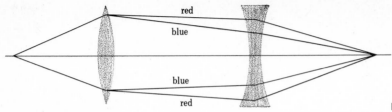

FIGURE 14.38

Chromatic aberration is corrected by using a doublet consisting of a positive lens and a negative lens.

two component lenses had equal but opposite focal lengths, the focal length of the combination would be infinite (zero power), according to Eq. 14.19. To achieve different focal lengths for the components, they are made of different glasses, usually crown glass and flint glass, which have different indexes of refraction.

MONOCHROMATIC ABERRATIONS

There are five monochromatic aberrations which arise, not from the dispersion of glass, but from the failure of a spherical lens to form an ideal image.

Spherical aberration is the failure of a lens to focus the rays emanating from a point on the axis into a point. In general, the more diverging rays are focused closer to the lens than the less diverging rays. Consequently, the image of the point in the image plane is a small circle (Fig. 14.40a).

Spherical aberration can be reduced, though not entirely eliminated, by making the radii of curvature of the two surfaces of a lens unequal. The aberration can be completely eliminated by using a doublet composed of lenses of opposite sign.

Coma is the failure of a lens to focus rays emanating from a point near the axis into a point. In a lens system that has been corrected for spherical aberration but not coma, off-axis rays are brought to a focus at different points in the image plane (Fig. 14.40b).

Astigmatism is the failure of a lens to focus rays emanating from a point far from the axis into a point. Instead the point is focused into two perpendicular line segments separated by a small distance (Fig. 14.40c). Astigmatism can be reduced by properly spacing the lenses in a doublet.

> **REMARK** The term *astigmatism* is also used in optometry to describe a defect of the eye in which the cornea is not spherical. In optics astigmatism refers to an aberration that exists even for a perfectly spherical lens. The two uses of this word should not be confused.

Curvature of field is the failure of the image surface to be a plane. Even when an optical system has been corrected for spherical aberration, coma, and astigmatism, off-axis points are not focused in the same plane as on-axis points (Fig. 14.40d). This aberration is particularly important in photography, where one wants the image surface to coincide with the plane of the film. The retina of the eye, on the other hand, has been curved by evolution to compensate for this aberration.

Distortion is the variation of the magnification of the image with distance from the optic axis. As a result, the image of a

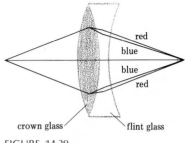

FIGURE 14.39

An achromatic lens consists of a negative and a positive lens of different indexes of refraction cemented together.

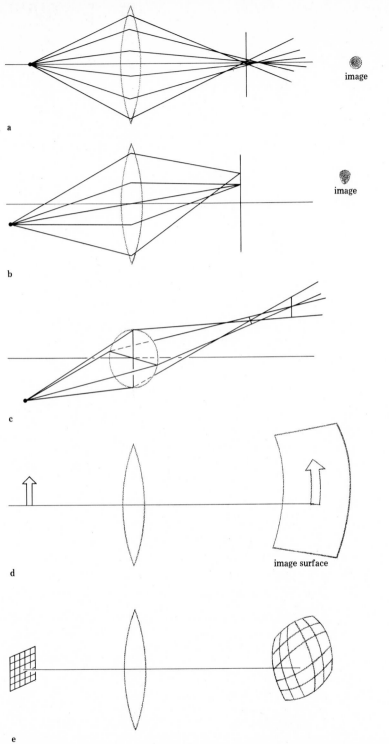

image

image

image surface

FIGURE 14.40

The five monochromatic aberrations: (a) spherical, (b) coma, (c) astigmatism, (d) curvature of field, and (e) distortion.

rectangular grid, though in perfect focus, is distorted in shape (Fig. 14.40e).

No optical system can be fully corrected for all seven aberrations. Every system is specially designed to reduce those aberrations which are most bothersome for its purposes. For example, Fig. 14.41 shows a high-quality-camera-lens system consisting of a plano-convex, plano-concave, and achromatic lens combined to eliminate chromatic and spherical aberration, astigmatism, and curvature of field.

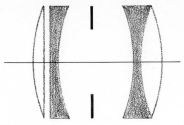

FIGURE 14.41

The lenses in a high-quality camera-lens system.

GUIDE TO MAJOR TOPICS

Topic	References	Problems
Lens formula	Sec. 14.1; Eqs. 14.2 to 14.4; Example 14.1; Fig. 14.7	1 to 4, 30
One-lens instruments	Sec. 14.2	
Projector	Eq. 14.6; Examples 14.2 and 14.3	5 to 8
Camera	Eqs. 14.7 to 14.9; Examples 14.4 to 14.6	9 to 16
Telescope	Eqs. 14.10 to 14.12; Example 14.7	17 and 18
Virtual images	Sec. 14.3; Eq. 14.13	19 and 20
Magnifying glass	Eqs. 14.14 and 14.15; Example 14.8	21 and 22
Reading glasses	Example 14.9	23 to 26
Negative lenses	Example 14.10	27 to 30
Mirrors	Sec. 14.4	
Planar	Eq. 14.16; Example 14.11	31 and 32
Parabolic		
Two-lens instruments	Sec. 14.5	
Microscope	Eq. 14.17; Example 14.12	33 and 34
Telescope	Eq. 14.18	35 to 38
Compound lenses	Eq. 14.19; Example 14.13	39 and 40, 43
Spherical lenses	Sec. 14.6; Eq. 14.20; Example 14.14	41 to 43
Aberrations	Sec. 14.6	43

PROBLEMS

1 An object is 21 cm in front of a lens of focal length 14 cm. (a) Find the image distance s' from the lens formula, from the curve in Fig. 14.7 and by graphical construction. (b) What is the magnification of the image?
Ans. (a) 42 cm? (b) 2

2 Consider a positive lens of focal length 12 cm. Make a table of the values of the image distance s' for the following values of the object distance s: 15, 18, 21, 24, 30, 36, 48, and 60 cm. Convert these values of s and s' to reduced distances $\bar{s}$ and $\bar{s}'$, and plot them on the curve in Fig. 14.7.

3 An object is 12 cm in front of a positive lens. The image of the object is formed 6 cm in back of the lens. What is the focal length of the lens?
Ans. 4 cm

4 An object is placed a distance s in front of a lens with a focal length of 5 cm. If the magnification of the image is 4, what is s?

5 A slide projector with a lens of focal length 10 cm projects an image on a screen that is 2.5 m from the lens. (a) What is the distance s between the slide and the lens? (b) What is the magnification of the image? (c) What is the width of the image of a slide 35 mm wide?
Ans. (a) 10.4 cm; (b) 25; (c) 87.5 cm

6 A classroom slide projector is 12 m from a screen that is 1.5 m wide. What focal length lens is required for the image of a 35-mm slide to just fill the screen?

7 The distance between the lens of a projector and the slide can be varied from 20 to 25 cm. If the focal length of the lens is 21 cm,

what is the smallest distance (between the lens and the screen) at which an image can be focused?
Ans. 131 cm

8   A slide projector with a lens of focal length 20 cm projects the image of a 35-mm slide onto a screen that is 0.8 m wide. How far should the screen be from the lens for the image to just fill the screen?

9   A camera with a lens of focal length 50 mm takes a picture of a child 1.2 m tall standing 3.0 m away. (a) What must be the distance s' between the film and the lens to get a properly focused picture? (b) What is the magnification of the image? (c) What is the height of the image of the child on the film?
Ans. (a) 5.08 cm; (b) 0.0167; (c) 20.0 mm

10   A camera with a 50-mm lens takes a picture of a tree 25 m tall. How far must the camera be from the tree for its image on the film to be 25 mm high?

11   A camera is fitted with a 125-mm telephoto lens. The distance of this lens from the film can be varied from 125 to 130 mm. What is the closest distance that an object can be from this camera and still be focused on the film?
Ans. 3.25 m

12   A lens of focal length 40 mm is used to take extreme close-ups (macrophotography). (a) If the lens can be at most 5.20 cm from the film, what is the closest distance (from the lens) at which an object can be focused? (b) What is the magnification in this case? (c) If the lens cannot be less than 5.0 cm from the film, what is the farthest distance (from the lens) at which an object can be focused?

13   A camera with a telephoto lens of focal length 450 mm takes a picture of an object 60 m away. How far from the object would a camera with a 50-mm lens have to be to form an image of the same size on the film?
Ans. 6.67 m

14   What are the horizontal and vertical fields of view obtained using a 25-mm wide-angle lens on a 35-mm camera?

15   A high-quality camera has an f/1.7 lens with a focal length of 50 mm. What is the diameter of the lens?
Ans. 29.4 mm

16   If a camera will take a properly exposed picture at f/16 and 1/60 s, what f-number is required to get the same exposure at 1/500 s?

17   An amateur astronomer takes a photograph of the full moon using a small telescope with a focal length of 1.2 m. What is the diameter of the moon's image on the film?
Ans. 1.1 cm

18   What is the f-number of the 200-in Hale telescope?

> **REMARK**   The f-number is not a meaningful measure of the intensity of the light that a telescope focuses on the image of a star, because the size of the image is determined solely by diffraction.

19   An object is 4 cm in front of a lens of focal length 6 cm. (a) Locate the position of the virtual image. (b) What is the magnification of the image?
Ans. (a) 12 cm in front of lens; (b) 3

20   Use Eq. 14.13 to plot the reduced virtual image $\bar{s}'$ against the reduced object distance $\bar{s}$ for a positive lens when $\bar{s}$ is less than 1.

21   (a) What is the focal length of a magnifying glass with a magnifying power of 10? (b) How far must the object be from this lens to obtain this magnification?
Ans. (a) 2.78 cm; (b) 2.50 cm

22   What is the magnifying power of a lens with a power of 15 diopters?

23   (a) What is the focal length of the reading glasses required by a person whose near point is 150 cm? (b) What is the power (in diopters) of this lens?
Ans. (a) 30 cm; (b) 3.33 diopters

24   The power of a farsighted person's reading glasses is 2.5 diopters. At what distance must the person hold a book in order to read it without glasses?

25   A man with normal vision puts on a pair of reading glasses with a focal length of 40 cm. (a) At what distance should he hold a book in order to read it comfortably? (b) What is the magnification of the print?
Ans. (a) 15.4 cm; (b) 1.6

26   At age forty a woman requires eyeglasses with lenses of 2 diopters power in order to read a book at 25 cm. At forty-five she finds that while wearing these glasses she must hold a book 40 cm from her eyes. What power lenses does she require at forty-five to read a book at 25 cm?

27   An object is 10 cm in front of a negative lens with a focal length of −5 cm. (a) Find the virtual image of this object by graphical construction. (b) Find the image distance using the lens formula. (c) What is the magnification of the image?
Ans. (b) 3.33 cm; (c) 0.33

TABLE **14.1**

Object distance	Positive lens	Negative lens	Image characteristics for six
$s < f$			possible combinations of object
$f < s < 2f$			distances and lens types (Prob. 30)
$s > 2f$			

28 Plot a curve of $\bar{s}'$ against $\bar{s}$ for a negative lens. Compare with Fig. 14.7.

29 A nearsighted person wears eyeglasses with negative lenses of focal length -2 m. (a) When she looks at an object 6 m away, where is the virtual image produced by the glasses? (b) Locate the image graphically.
Ans. (a) 1.5 m

30 An image can be characterized as either real or virtual, erect or inverted, and magnified or minified (m greater or less than 1). Table 14.1 lists six situations, depending on the object distance s and the type of lens. For each situation, enter the three words that characterize the image.

31 Locate all the images of the point P in the corner mirror shown in Fig. 14.42.
Ans. There are three images.

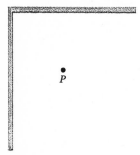

FIGURE 14.42

Problem 31.

32 Figure 14.43 shows a man of height h standing in front of a mirror of length L. Show that the man can see his entire body if $L > \frac{1}{2}h$.

33 A microscope has an objective of focal length 0.3 cm and an ocular of focal length 2.0 cm. (a) Where must the image formed by the objective be for the ocular to produce a virtual image 25 cm in front of the ocular? (b) If the lenses are 20 cm apart, what is the distance of the objective from the object on the slide? (c) What is the total magnification of the microscope?
Ans. (a) 1.85 cm in front of the ocular; (b) 0.305 cm; (c) 804

34 A dissecting microscope is designed to

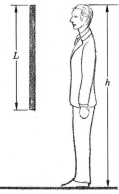

FIGURE 14.43

Problem 32.

have a large distance between the object and the objective. Suppose the focal length of the objective of a dissecting microscope is 5.0 cm, the focal length of the ocular is 4.0 cm, and the distance between these lenses is 17.0 cm. (a) What is the distance between the object and the objective? (b) What is the total magnification? Note that Eq. 14.17 cannot be used in this case.

35 A 3- and a 30-cm-focal-length lens are used to build a small telescope. (a) Which lens should be the objective? (b) What is the magnifying power of the telescope? (c) How far apart are the two lenses in the telescope?
Ans. (b) 10; (c) 33 cm

36 Trace the rays through a telescope with an objective of focal length 10 cm and an ocular of focal length 2 cm.

37 If the telescope in Prob. 35 were used to look at an object 6 m away, how far apart should its two lenses be?
Ans. 34.6 cm

38 Trace the rays through a Galilean telescope with an objective of focal length 10 cm and an ocular of focal length -2 cm.

39 A lens of focal length 500 mm is mounted in front of a 50-mm camera lens. (a) What is the focal length of the combination, assuming zero distance between the lenses? (b) If the lens-to-film distance can be varied from 5.00 to 5.22 cm, what are the closest and farthest ob-

ject distances at which the camera can focus with this attachment?
Ans. (a) 45.6 mm; (b) 36 and 52 cm

40 An object is placed 12 cm in front of a lens of focal length 5 cm. Another lens of focal length 4 cm is placed 2 cm in back of the first lens. Find the image produced by this two-lens system by tracing rays. (*Hint:* First find the image produced by the front lens alone, and use it to find the image formed by the second lens.)

41 What is the focal length of a positive symmetric lens with an index of refraction of 1.62 and a radius of curvature of 20 cm?
Ans. 16.1 cm

42 Show that the minimum possible *f*-number of a symmetric lens is $\frac{1}{4}(n-1)$. (*Hint:* What is the maximum diameter of a symmetric lens of focal length *f*?)

43 An achromatic lens is composed of a symmetric positive lens made of flint glass ($n = 1.65$) and a symmetric negative lens made of crown glass ($n = 1.52$). If both components have a radius of curvature of 10 cm, what is the focal length of the combination?
Ans. 38.3 cm

BIBLIOGRAPHY

HECHT, EUGENE, and ALFRED ZAJEC: *Optics,* Addison-Wesley Publishing Company, Reading, Mass., 1974. This well-written textbook contains much interesting information about optics and optical instruments.

PRICE, WILLIAM H.: "The Photographic Lens," *Scientific American* **235**:72(August 1976). An interesting account of aberrations and how they are corrected in a modern camera-lens system. Colored photographs of the different types of aberrations appear on page 73.

PART IV

ELECTRICITY AND MAGNETISM

Early electric motor (c. 1880).

When certain substances, notably amber and glass, are rubbed with a material such as silk or fur, they acquire the capacity to attract small bits of paper and cork. This phenomenon is a manifestation of *electricity*, one of the fundamental forces of nature. Similarly, the ability of certain iron ores, such as lodestone to attract small bits of iron is a manifestation of *magnetism,* another fundamental force.

Although these simple electric and magnetic phenomena have been known since ancient times, most of the basic quantitative laws of electricity and magnetism were discovered between 1784, when Charles Coulomb investigated the forces between charged objects, and 1831, when Michael Faraday discovered magnetic induction. Prior to this 50-year period of discovery, the only practical electric invention was the lightning rod of Benjamin Franklin (1752). After this period, the practical utilization of electricity increased rapidly, with the development of the telegraph (1844), the telephone (1877), incandescent lighting (1880), and electric motors (1887). Uses of electricity have continued to expand to this day, with the current revolution in microelectronics giving us ever-increasing control over the machines that electricity powers.

Electricity is a fundamental force of nature, analogous to gravity. But whereas the gravitational force between two objects depends on their mass, the electric force between two objects depends on their *charge*. Charge is a basic property of the elementary particles (electrons, protons, and neutrons) that compose ordinary matter. In fact, it is the electric force between the protons and electrons in an atom that holds the atom together.

The practical utilization of electricity is made possible by our ability to generate and control a flow of charged particles. In this chapter we discuss the principles of electricity needed to understand devices, such as x-rays and cathode-ray tubes, that utilize the flow of high-speed electrons inside an evacuated vessel. In subsequent chapters we discuss the instruments and devices which use the flow of electrons in wires and semiconductors.

15
ELECTRICITY

15.1   The Fundamental Forces

The forces we encounter in everyday life, such as normal forces, frictional forces, and spring forces, are the observable effects of the forces between the atoms of the objects involved. These observable forces are not considered to be fundamental because, in principle, they can be analyzed in terms of atomic forces. Even the forces between atoms are not fundamental because they can be analyzed further in terms of the forces between the particles that compose the atoms.

There are three kinds of particles inside an atom: *protons, neutrons,* and *electrons*. The protons and neutrons are tightly packed together to form the dense central core, or *nucleus,* of the atom. The nucleus contains over 99.95 percent of the atom's mass, but it occupies only a minute fraction of the atom's volume. The electrons, which swarm about the nucleus, contain the remaining 0.05 percent of the mass and occupy most of the volume. Figure 15.1 shows a schematic diagram of a carbon atom.* The details of the structure of atoms and nuclei are given in Part V.

Because protons, neutrons, and electrons are not themselves composed of still smaller particles, they are said to be *elementary* particles, and the forces between them are the fundamental forces of nature. All ordinary matter is composed of these elementary particles, and so all forces can be ultimately understood in terms of the fundamental forces between them. At present, four distinct fundamental forces are known:

1   Gravitational force (gravity)

2   Electromagnetic force (electricity and magnetism)

3   Nuclear force

4   Weak force

In a sense, the gravitational force is the weakest fundamental force because the magnitude of the gravitational force between

FIGURE 15.1

Schematic diagram of a carbon atom. Six electrons orbit about a nucleus composed of six protons and six neutrons.

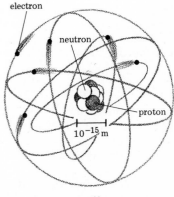

electron

neutron

proton

10^{-15} m

10^{-10} m

*If the atom were drawn to scale, with the nucleus the size shown in Fig. 15.1, the orbits of the outer electrons would be the size of a football stadium.

two elementary particles is far smaller than the magnitude of any other fundamental force. The force of gravity between elementary particles is so weak, in fact, that it has no measurable effect on the behavior of these particles inside an atom. Only an object of astronomical size has sufficient mass to exert a significant gravitational force on an atom. Since this force is attractive, such an object attracts the dust and atoms in the space around it. The accretion of this matter increases the mass of the object, so that the gravitational force it can exert also increases. As a consequence, the object attracts still more matter, which increases its mass even more. This is how immense aggregates of matter, such as stars and galaxies of stars, are formed from the dust and atoms scattered throughout space. Gravity is by far the most important force controlling the structure of astronomical bodies.

The electromagnetic force is the principal force determining the structure of atoms. The electrons are maintained in orbit about the nucleus by electrical attraction, just as the planets are maintained in orbit about the sun by gravitational attraction. The electrons also exert electric forces on each other, and the force between two nearby atoms is just the electric force between their electrons and nuclei. All the forces discussed so far in this book, with the exception of gravity, are the result of this electric force. Thus all the forces of everyday life, with the exception of gravity, are electromagnetic in origin.

The protons and neutrons in the nucleus are held together by the nuclear force. This force is very strong when the particles are close together, but it decreases rapidly with distance. For instance, the nuclear force between two neutrons is essentially zero when they are more than 10^{-14} m apart. This means that the nuclear force does not extend beyond the outer electrons of the atom, which are 10^{-10} m from the nucleus. Thus, although the nuclear force is essential for holding the nucleus together, it plays no direct role in the interaction of atoms with each other.

The weak force is also a short-ranged force confined entirely to the nucleus. It is responsible for some forms of radioactivity, which is the spontaneous transformation of the nucleus of one kind of atom into the nucleus of another (Sec. 22.2).

The electromagnetic force is thus the principal force governing the physics and chemistry of ordinary matter. It is also of great practical importance, since our entire technological civilization is based on electric-power generation and its utilization in motors, lighting, and communications.

15.2 Coulomb's Law

Even though electric and gravitational forces are fundamentally distinct from each other, they have a number of similar properties. In order to understand electricity, therefore, it is helpful to briefly review some of the properties of gravity discussed in Sec. 6.3.

Gravity is an attractive force that exists between two objects with masses m_1 and m_2. This means that the force $\mathbf{F}_1$ exerted by m_2 on m_1 is directed toward m_2 (Fig. 15.2). By Newton's third law

FIGURE 15.2

Gravitational attraction between two masses.

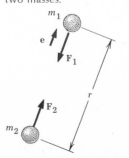

(Property 3, Sec. 2.1), the reaction to F_1 is the force F_2 exerted by m_1 on m_2. Force F_2 has the same magnitude as F_1 but the opposite direction, so it is directed toward m_1, as shown in Fig. 15.2.

Gravity is a force that acts at a distance; i.e., two objects exert their mutual attraction on each other across empty space, without any mechanical connection. The magnitude F_g of the gravitational forces (F_1, F_2) on two objects with masses m_1 and m_2 separated by a distance r is given by Newton's law of gravity,

$$F_g = G\frac{m_1 m_2}{r^2}$$

where G is the universal gravitational constant ($G = 6.67 \times 10^{-11}$ N $\cdot$ m^2/kg^2).

Definition A *unit vector* **e** is a vector whose magnitude is 1 (dimensionless). Any vector **A** can be written

A = A**e**

where A is the magnitude of **A**, and **e** is a unit vector parallel to **A**. The vector $-$**A**, which is antiparallel to **A**, is written

$-$**A** = $-A$**e**

To write a vector expression for the gravitational force F_1 that m_2 exerts on m_1, we introduce the unit vector **e** that points from m_2 to m_1 (Fig. 15.2). Since the magnitude of F_1 is F_g and the direction of F_1 is antiparallel to **e**, we can write

$$F_1 = -F_g\mathbf{e} = -G\frac{m_1 m_2}{r^2}\mathbf{e} \qquad 15.1$$

The minus sign arises because we chose **e** to point away from m_2, whereas the attractive force F_1 points toward m_2.

Electricity is a force acting at a distance between two objects with charges q_1 and q_2. Charge, like mass, is a basic attribute of matter. The SI unit of charge is the *coulomb* (C). This unit is ultimately defined in terms of the force between two current-carrying wires. (See the remark following Prob. 15 in Chap. 17.) But for now, we can take the coulomb to be a fundamental unit, like the kilogram, meter, and second.

Coulomb's law The magnitude F_e of the electric force between two objects with charges q_1 and q_2 separated by a distance r is

$$F_e = K\frac{q_1 q_2}{r^2}$$

where K is a universal electric constant ($K = 9.0 \times 10^9$ N $\cdot$ m^2/C^2).

The electric force F_1 that q_2 exerts on q_1 is

$$F_1 = F_e\mathbf{e} = K\frac{q_1 q_2}{r^2}\mathbf{e} \qquad 15.2$$

where **e** is the unit vector pointing from q_2 to q_1. Notice the similarity between Eqs. 15.1 and 15.2. The electric force depends on the product of the charges of the two objects, just as the gravitational force depends on their masses. Also, the electric and gravitational forces are both inversely proportional to the square of the distance between the objects.

A fundamental difference between gravity and electricity is that gravity is always attractive, whereas the electric force can be either attractive or repulsive because there are two kinds of charge, positive and negative. Objects with the same kind of charge repel each other, and objects with opposite charge attract each other. This is the reason there is no minus sign in Eq. 15.2. When q_1 and q_2 have the same sign, either both positive or both negative, the product $q_1 q_2$ is positive, and so $\mathbf{F}_1$ is parallel to $\mathbf{e}$, indicating a repulsive force (Fig. 15.3a). On the other hand, if q_1 and q_2 have opposite signs, the product $q_1 q_2$ is negative, and so $\mathbf{F}_1$ is antiparallel to $\mathbf{e}$, indicating an attractive force (Fig. 15.3b).

It is standard practice to write the constant K in the form

$$K = \frac{1}{4\pi\epsilon_o} = 9 \times 10^9 \, \text{N} \cdot \text{m}^2/\text{C}^2 \qquad 15.3$$

where $\epsilon_o = 8.85 \times 10^{-12} \, \text{C}^2/(\text{N} \cdot \text{m}^2)$ is called the *permittivity* of space. In terms of ϵ_o, Coulomb's law is written

$$\mathbf{F}_1 = \frac{1}{4\pi\epsilon_o} \frac{q_1 q_2}{r^2} \mathbf{e} \qquad 15.4$$

We shall use this form of Coulomb's law from now on, so that all subsequent equations will be in standard form.

The elementary particles are the ultimate carriers of charge. From Table 15.1, which gives the charges and masses of the elementary particles, we see that a proton has a positive charge $e = 1.602 \times 10^{-19} \, \text{C}$, an electron has a negative charge $-e$ of exactly the same magnitude, and a neutron has zero charge. The charge of an object is the sum of the charges of all its protons and electrons. Thus an object that has an equal number of protons and electrons has zero net charge. Such an object is said to be uncharged or *neutral*.

A neutral object acquires charge by gaining or losing electrons. For instance, when a glass rod is rubbed with silk, electrons are transferred from the glass to the silk. If the glass loses N electrons, it will have N more protons than electrons, so its total charge will be Ne. Similarly, the silk will have N more electrons than protons, so its total charge will be $-Ne$. The total charge of the rod and the silk together is

$$Ne + (-Ne) = 0$$

the same as it was before they were rubbed together. This is an example of the following fundamental law.

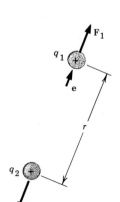

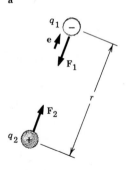

FIGURE 15.3

(a) Electric repulsion between two positive charges. (b) Electric attraction between opposite charges.

TABLE **15.1**

Particle	Mass, kg	Charge, C
Proton	1.673×10^{-27}	$+1.602 \times 10^{-19}$
Neutron	1.675×10^{-27}	0
Electron	9.110×10^{-31}	-1.602×10^{-19}

Mass and charge of the elementary particles

Conservation of Charge *In any physical process, the total charge is not changed.* This law is obviously true for a process that involves merely the transfer of electrons from one object to another. However, the law is much more general than this and is true even for processes, such as the decay of a nucleus, in which protons and electrons are created and destroyed (Sec. 22.2).

Example 15.1 Compare the gravitational and electric forces between an electron and a proton.

From Table 15.1 and Eq. 15.1 we find that the gravitational force between a proton and an electron separated by a distance r is

$$F_g = -G\frac{m_p m_e}{r^2}$$

$$= -\frac{(6.67 \times 10^{-11} \text{ N} \cdot \text{m}^2/\text{kg}^2)(1.67 \times 10^{-27} \text{ kg})(9.11 \times 10^{-31} \text{ kg})}{r^2}$$

$$= -\frac{1.01 \times 10^{-67} \text{ N} \cdot \text{m}^2}{r^2}$$

From Eq. 15.4 we find that the electric force between an electron and a proton is

$$F_e = \frac{1}{4\pi\epsilon_o}\frac{q_p q_e}{r^2}$$

$$= \frac{(9.0 \times 10^9 \text{ N} \cdot \text{m}^2/\text{C}^2)(1.60 \times 10^{-19} \text{ C})(-1.60 \times 10^{-19} \text{ C})}{r^2}$$

$$= -\frac{2.30 \times 10^{-28} \text{ N} \cdot \text{m}^2}{r^2}$$

The ratio of these two forces is

$$\frac{F_e}{F_g} = \frac{2.30 \times 10^{-28}}{1.01 \times 10^{-67}} = 2.28 \times 10^{39}$$

which shows that the electric force is immensely larger than the gravitational force. ☐

Because the gravitational force is so small compared with the electric force, gravity can be completely neglected when calculating the force between elementary particles. However, the force between large aggregates of matter, such as astronomical bodies, is dominated by gravity because these objects tend to be electrically neutral.

INDUCTION

A glass rod that has been positively charged by rubbing with silk will attract a small piece of cork even though the cork is uncharged. The cork is composed of electrons and protons of opposite charge, which are normally distributed uniformly, so that the net charge is zero everywhere in the cork. When a charged object, such as the glass rod, is brought near the cork, the positive charge on the rod attracts the electrons in the cork and repels the protons, causing these particles to shift their positions slightly. The result is that negative charge accumulates on the side of the cork near the rod and positive charge accumulates on the other side. This process is called *induction*. According to the law of conservation of charge, the total charge of the cork must remain zero, since no charge has been added or removed from it. Thus there is as much negative charge induced on one side of the cork as there is positive charge induced on the other.

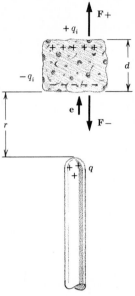

Suppose the glass rod has a charge q and is a distance r from the front side of the cork (Fig. 15.4). If the charge induced on the front side is $-q_i$, the rod attracts it with the force

$$\mathbf{F}_- = \frac{1}{4\pi\epsilon_o}\frac{q(-q_i)}{r^2}\mathbf{e} = -\frac{1}{4\pi\epsilon_o}\frac{qq_i}{r^2}\mathbf{e}$$

At the same time, the rod repels the charge $+q_i$ induced on the back side. This charge is a distance d farther from the rod, where d is the thickness of the piece of cork. Therefore, the repulsive force $\mathbf{F}_+$ on the positive induced charge is

$$\mathbf{F}_+ = \frac{1}{4\pi\epsilon_o}\frac{qq_i}{(r+d)^2}\mathbf{e}$$

which is less than $\mathbf{F}_-$. The total force $\mathbf{F}$ on the cork is the sum of $\mathbf{F}_-$ and $\mathbf{F}_+$, or

$$\mathbf{F} = \mathbf{F}_- + \mathbf{F}_+ = -\frac{1}{4\pi\epsilon_o}qq_i\left[\frac{1}{r^2} - \frac{1}{(r+d)^2}\right]\mathbf{e}$$

$$= -\frac{qq_i}{4\pi\epsilon_o}\frac{(2rd+d^2)}{r^2(r+d)^2}\mathbf{e} \qquad 15.5$$

This shows that the glass rod exerts an attractive force on the cork.

FIGURE 15.4

Induction of charge on a piece of cork. The cork is attracted to the inducing charge q because the attraction between q and the negative induced charge $-q_i$ is greater than the repulsion between q and the positive induced charge $+q_i$.

> **REMARK** If the rod were negatively charged, postive charge would be induced on the side of the cork near the rod and negative charge would be induced on the other side, so the cork would still be attracted by the rod. Thus any charged object, whether positively or negatively charged, exerts an attractive force on an uncharged object.

INSULATORS AND CONDUCTORS
Substances differ in the relative freedom with which electrons can move in them. A substance in which electrons can move very freely is called a *conductor,* and a substance in which electrons can move only slightly is called an *insulator* (or *dielectric*). Metals are all good conductors; glass, rubber, and cork are examples of good insulators.

The difference between insulators and conductors is illustrated by the following experiment. A charge q is brought near an uncharged object, inducing charge on it (Fig. 15.5a). The object is then cut in half, as shown in Fig. 15.5b, and the charge q is removed. Is each half of the object left with the charge that was induced on it? The answer is yes if the object is a conductor and no if the object is an insulator. To understand this, we must compare the atomic structures of insulators and conductors.

In an insulator, every electron is bound to an atom and is not free to move away. The atoms of an insulator are represented in

FIGURE 15.5

(a) Induction of charge on an object by an external charge q.
(b) Object cut to try to isolate the induced charge.

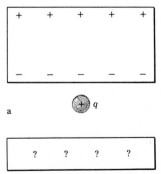

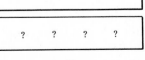

Fig. 15.6a as neutral units composed of positive and negative charges superimposed on each other. A positive charge q brought near an insulator attracts the negatively charged electrons in the atoms and repels the positively charged nuclei. This causes the negative and positive charge in each atom to separate slightly. The atom as a whole remains neutral, but one end becomes positively charged and the other becomes negatively charged. The negative end of each atom faces the inducing charge q, as shown in Fig. 15.6b. As a result, an excess of negative charge forms on the surface of the insulator facing q, and an excess of positive charge forms on the opposite surface. Of course, these are not free charges; they are merely the charged ends of neutral atoms. Figure 15.6c shows that if the insulator is cut in half, induced charge appears on the cut surfaces but the total charge of each half remains zero.

REMARK The excess negative charge on the surface of the insulator near the inducing charge q is attracted to q, and the excess positive charge on the far surface is repelled by q. The result is a net attractive force, as shown by Eq. 15.5. Thus, even though the charges in an insulator are not free, they interact with an external charge as though they were.

In a metal conductor, at least one electron is separated from each atom and is free to move anywhere in the conductor. Atoms from which electrons are missing are called *ions*. In a metal they are positively charged and remain in fixed positions. A metal thus consists of positively charged ions in fixed positions and negatively charged electrons free to move around. Normally the ions and electrons are uniformly distributed, so that the net charge is zero everywhere in the metal (Fig. 15.7a). However, a charged object brought near the conductor changes this distribution. A positive charge q attracts the electrons and repels the ions. Since the electrons are free to move, they accumulate on the side near q, which thus becomes negatively charged (Fig. 15.7b). The opposite side, being deficient in electrons, is positively charged. If the conductor is cut in half, an excess of electrons is trapped on the half facing q, leaving the other half with a shortage of electrons (Fig. 15.7c). Therefore, each half is left charged, even after the inducing charge is removed.

To discharge these conductors it is necessary only to connect a conducting wire from one to the other. The excess electrons on the negatively charged conductor, attracted to the positively charged conductor, flow freely through the wire, until each conductor is again neutral.

INDUCTION MACHINES

The ability to separate charge by induction in a conductor is used in a device called an *induction machine* to produce a continuous separation of charge. Induction machines are primarily used for physics demonstrations today, but in the nineteenth century they were used in electrical research. In the machine shown in Fig. 15.8, metal rods are attached to a wheel, which is rotated by a crank in the direction indicated. The plate P initially has a small positive charge on it, so that charges are

FIGURE 15.6

(a) An insulator. Each atom is a neutral object composed of positive and negative charges. (b) Induction of charge on an insulator. The inducing charge q causes the positive and negative charges in each atom to separate slightly. (c) The charges induced on an insulator are not isolated when the insulator is cut in half.

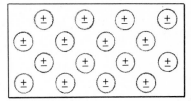

a

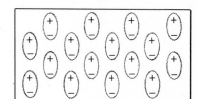

b

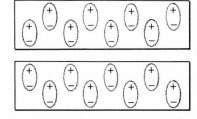

c

induced on the rod in position 1. As this rod is rotated into position 2, it makes momentary contact with the wire W, which simultaneously makes contact with the rod going from position 5 to position 6. During contact, negative charge flows from the lower rod to the upper rod, and when contact is broken, the rod in position 6 is left positively charged, while the rod in position 2 is left negatively charged. These rods deposit their charge to the two conducting spheres by means of contacts at A and B.

Some of the positive charge deposited at A accumulates on plate P. As the charge on P increases, the amount of charge induced on the rod in position 2 increases. This, in turn, increases the charge delivered back to P. Thus this machine uses positive feedback to rapidly build a large charge on the two conducting spheres. When this charge is sufficiently large, the air between the spheres suddenly becomes conducting, and the electrons are able to flow from the negative to the positive sphere. This is seen as a spark jumping between the spheres.

15.3 The Electric Field

A basic electricity problem is to calculate the total force $\mathbf{F}$ that a set of fixed charges $q_1, q_2, \ldots, q_n$ exerts on some other charge q. The fixed charges are called the *source* charges, and the charge on which they act is called the *test* charge. For instance, what is the force $\mathbf{F}$ that the four source charges $q_1, \ldots, q_4$ in Fig. 15.9 exert on the test charge q?

The force $\mathbf{F}_1$ that q_1 alone exerts on q is

$$\mathbf{F}_1 = \frac{1}{4\pi\epsilon_o} \frac{qq_1}{r_1^2} \mathbf{e}_1$$

where r_1 is the distance between q_1 and q, and $\mathbf{e}_1$ is the unit vector pointing from q_1 to q (Fig. 15.9). Similar expressions can be written for the forces $\mathbf{F}_2$, $\mathbf{F}_3$, and $\mathbf{F}_4$ that the charges q_2, q_3, and q_4 individually exert on q. The total force $\mathbf{F}$ on q is the sum of these individual forces, so

$$\mathbf{F} = \mathbf{F}_1 + \mathbf{F}_2 + \mathbf{F}_3 + \mathbf{F}_4$$

$$= \sum_{i=1}^{4} \mathbf{F}_i$$

$$= \sum_{i=1}^{4} \frac{1}{4\pi\epsilon_o} \frac{qq_i}{r_i^2} \mathbf{e}_i$$

Here r_i is the distance from q_i to q, and $\mathbf{e}_i$ is the unit vector pointing from q_i to q. Each term in the sum is a vector, so vector addition must be used to add them.

In general, the force $\mathbf{F}$ that a set of n source charges $(q_1, q_2, \ldots, q_n)$ exerts on a test charge q is

$$\mathbf{F} = \sum_{i=1}^{n} \frac{1}{4\pi\epsilon_o} \frac{qq_i}{r_i^2} \mathbf{e}_i$$

$$= \frac{q}{4\pi\epsilon_o} \sum_{i=1}^{n} \frac{q_i}{r_i^2} \mathbf{e}_i \qquad 15.6$$

FIGURE 15.7

(a) A conductor. The positive charges are the ions, which have fixed positions, and the negative charges are the electrons, which are free to move throughout the material. (b) Induction of charge on a conductor. The inducing charge q causes the free electrons to move toward one side of the conductor. (c) When the conductor is cut in half, an excess of electrons is trapped on one half, and a deficit of electrons is trapped on the other half.

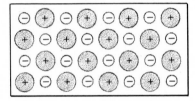

a

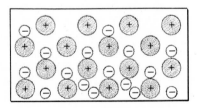

b q

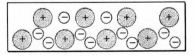

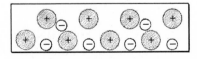

c

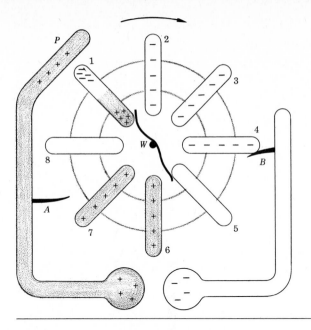

FIGURE 15.8

An induction machine. The rotation of the wheel in a clockwise direction causes positive charge to accumulate on the left and negative charge to accumulate on the right.

This last expression follows because the factor $q/4\pi\epsilon_o$ is common to each term in the sum.

Example 15.2 Find the force **F** on the test charge $q = 3 \times 10^{-5}$ C in Fig. 15.10 resulting from the two source charges $q_1 = q_2 = 5 \times 10^{-4}$ C.

The distance r_1 from q_1 to q is the hypotenuse of a right triangle, so

$$r_1 = \sqrt{(1\ m)^2 + (2\ m)^2} = \sqrt{5}\ m$$

From Eq. 15.4 the magnitude of the force $\mathbf{F}_1$ that q_1 exerts on q is

$$F_1 = \frac{1}{4\pi\epsilon_o} \frac{qq_1}{r_1^2}$$

$$= (9 \times 10^9\ N \cdot m^2/C^2) \frac{(3 \times 10^{-5}\ C)(5 \times 10^{-4}\ C)}{(\sqrt{5}\ m)^2}$$

$$= 27\ N$$

The distance r_2 from q_2 to q is equal to r_1, and q_2 is equal to q_1, so the magnitude of the force $\mathbf{F}_2$ that q_2 exerts on q is equal to F_1:

$$F_2 = F_1 = 27\ N$$

The forces $\mathbf{F}_1$ and $\mathbf{F}_2$ make the same angle θ to the y axis. From Fig. 15.10 we see that

$$\cos\theta = \frac{2\ m}{r_1} = \frac{2}{\sqrt{5}}$$

so the y components of $\mathbf{F}_1$ and $\mathbf{F}_2$ are

$$F_{1y} = F_{2y} = F_1 \cos\theta$$
$$= (27\ N)(2/\sqrt{5}) = 24.1\ N$$

FIGURE 15.9

Four source charges q_1, q_2, q_3, and q_4 and a test charge q.

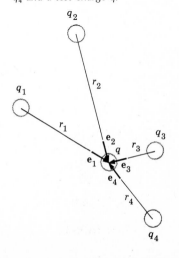

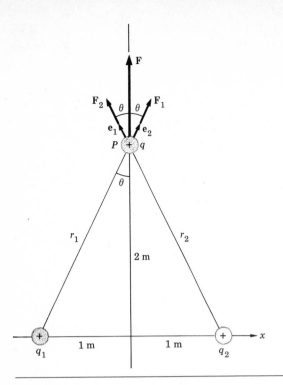

FIGURE 15.10

The forces exerted on a test charge q by two source charges q_1 and q_2.

The y component of the total force **F** is

$$F_y = F_{1y} + F_{2y} = 48.2 \, \text{N}$$

and the x component is zero, because F_{1x} and F_{2x} have opposite signs.

Thus the total force **F** on q has the magnitude $F = 48.2 \, \text{N}$ and is directed along the y axis. ☐

Definition The electric field **E** at a point P resulting from a set of source charges $(q_1, q_2, \ldots, q_n)$ is

$$\mathbf{E} = \frac{1}{4\pi\epsilon_o} \sum_{i=1}^{n} \frac{q_i}{r_i^2} \mathbf{e}_i \qquad\qquad 15.7$$

where r_i is the distance from q_i to P, and $\mathbf{e}_i$ is a unit vector pointing from q_i to P.

From Eqs. 15.6 and 15.7 we see that the force **F** exerted by a set of source charges on a test charge q at P is

$$\mathbf{F} = q\mathbf{E} \qquad\qquad 15.8$$

where **E** is the electric field at P due to the source charges. Equation 15.8 expresses the vector **F** as the product of a number q that depends only on the test charge and a vector **E** that depends only on the source charges. The product of a positive number q and a vector **E** is a vector **F** that has the direction of **E** and the magnitude qE. If q is negative, the direction of **F** is opposite the direction of **E**.

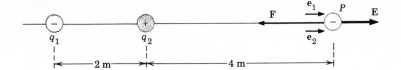

FIGURE 15.11

The electric field **E** at point P resulting from two source charges q_1 and q_2.

Example 15.3 (a) Find the electric field at point P in Fig. 15.11 resulting from the two source charges $q_1 = -6 \times 10^{-5}$ C and $q_2 = 3 \times 10^{-5}$ C. (b) What is the force on a test charge $q = -4 \times 10^{-2}$ C at P?

(a) The unit vectors $\mathbf{e}_1$ and $\mathbf{e}_2$ that point from q_1 and q_2 to P are equal in this case, so Eq. 15.7 gives

$$\mathbf{E} = \frac{1}{4\pi\epsilon_o}\left(\frac{q_1}{r_1^2}\mathbf{e}_1 + \frac{q_2}{r_2^2}\mathbf{e}_2\right)$$

$$= (9 \times 10^9 \text{ N} \cdot \text{m}^2/\text{C}^2)\left[\frac{-6 \times 10^{-5} \text{ C}}{(6 \text{ m})^2} + \frac{3 \times 10^{-5} \text{ C}}{(4 \text{ m})^2}\right]\mathbf{e}_1$$

$$= (1.87 \times 10^3 \text{ N/C})\mathbf{e}_1$$

(b) From Eq. 15.8 the force on the test charge is

$$\mathbf{F} = q\mathbf{E}$$

$$= (-4 \times 10^{-2} \text{ C})(1.87 \times 10^3 \text{ N/C})\mathbf{e}_1$$

$$= -(75 \text{ N})\mathbf{e}_1$$

The minus sign indicates that **F** is antiparallel to $\mathbf{e}_1$, as shown in Fig. 15.11. ☐

Example 15.4 Find the electric field at point P in Fig. 15.10 resulting from the two source charges $q_1 = q_2 = 5 \times 10^{-4}$ C.

In Example 15.2 we found the force **F** on a test charge $q = 3 \times 10^{-5}$ C at P to be $(48.2 \text{ N})\mathbf{e}_y$, where $\mathbf{e}_y$ is a unit vector parallel to the y axis. Then from Eq. 15.8 the electric field at P is

$$\mathbf{E} = \frac{\mathbf{F}}{q} = \frac{(48.2 \text{ N})\mathbf{e}_y}{3 \times 10^{-5} \text{ C}} = (1.61 \times 10^6 \text{ N/C})\mathbf{e}_y$$ ☐

LINES OF FORCE

The electric field is an important concept because it enables us to describe the effect of a given configuration of source charges without having to specify a particular test charge on which they act. Figure 15.12 shows the electric field at various points resulting from a positive source charge. At each point the field is directed away from the positive charge, because the field at a point is equal to the force that would be exerted on a positive 1-C test charge placed at that point. The magnitude of the field decreases with distance from the source charge, because the force it exerts on a 1-C test charge decreases inversely as the square of the distance (Coulomb's law).

Definition A *line of force* is a line (or curve) that is everywhere tangent to the electric field produced by a given configuration of source charges. Figure 15.13 shows the lines of force for a

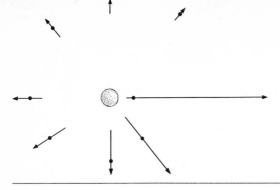

FIGURE 15.12

The electric field at various points resulting from a positive point charge.

positive source charge. Since at every point the electric field points away from the source charge, the lines of force are straight lines radiating outward from the source. Furthermore, the magnitude of the field is stronger where the lines of force are closer together, so that the lines of force graphically display both the direction and relative magnitude of the electric field surrounding the source charge.

The lines of force surrounding a negative source charge are shown in Fig. 15.14. They are identical to the lines of force surrounding a positive source charge, except they are directed inward. This is the direction of the force that the negative source charge would exert on a positive test charge.

Figure 15.15 shows the lines of force produced by two equal and opposite charges separated by a distance d. The lines of force all start on the positive charge and end on the negative charge. The electric field at any point P is tangent to the line of force at P and is equal to the vector sum of the electric fields resulting from each charge separately.

The charge configuration in Fig. 15.15 is called a *dipole*. Even though the total charge of the configuration is zero, it will exert an electric force on another charge. This is just the situation already encountered in discussing the force between a charged object and an uncharged piece of cork (Sec. 15.2). The charged object induced positive and negative charges on opposite sides of the cork. These induced charges form a dipole that exerts a force on the charged object.

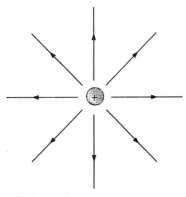

FIGURE 15.13

The lines of force of a positive point charge.

FIGURE 15.14

The lines of force of a negative point charge.

ELECTROSTATIC PRECIPITATOR

An *electrostatic precipitator* is a device that uses an electric field to remove dust and other finely divided particles from a gas. It is widely used to remove particulate pollutants from the smoke and gases released by power plants, steel mills, and cement plants. It is also used to clean the ventilating air in plants where grinding and polishing operations produce a dusty environment.

The precipitator is a chamber containing a number of large plates arranged in parallel rows 25 cm apart (Fig. 15.16). Midway between each pair of plates is a row of wires that runs parallel to the plates. The plates are positively charged, while the wires are negatively charged, so there is an electric field directed from the

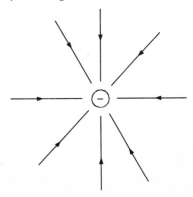

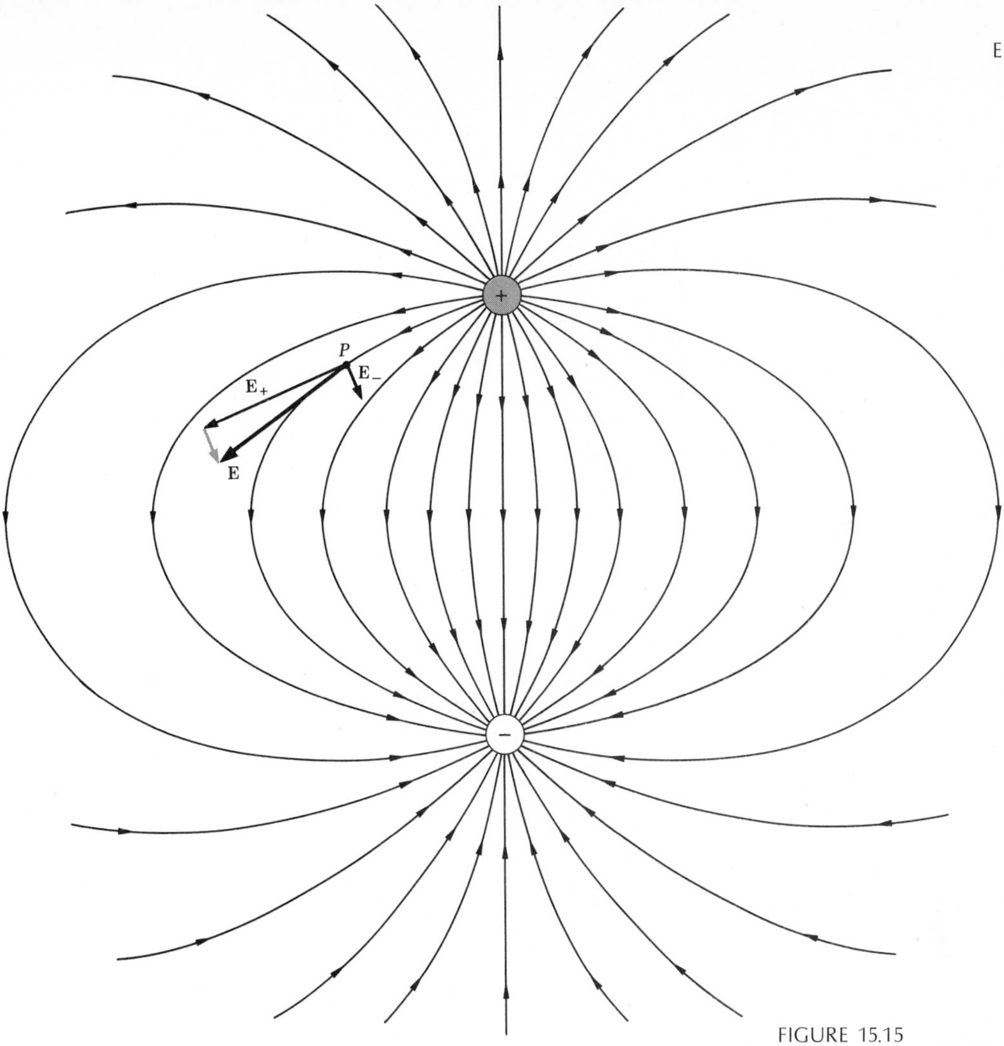

FIGURE 15.15

The lines of force of a dipole. The electric field at any point *P* is tangent to the line of force at that point.

plates to the wire. The top view of the chamber in Fig. 15.17 shows that the lines of force are closest near each wire, indicating that the electric field is strongest there. In fact, near a wire the electric field is 5×10^6 N/C or more, which is large enough to pull electrons out of the atoms of oxygen and nitrogen in the air, thus producing free electrons and positively charged ions. The ions are attracted to the negatively charged wire (from which they recover their lost electrons), and the free electrons are attracted to the plates.

The gas to be cleaned is circulated between the plates. Electrons on their way to the plates attach themselves to the particles in the gas, giving the particles a negative charge. As a consequence, the negatively charged particles are attracted to the plates, where they stick. In this way the particulate matter is removed from the circulating gas. Periodically the plates are rapped to dislodge the collected matter, which then falls into bins from which it is later removed.

FIGURE 15.16

An electrostatic precipitator. Rows of negatively charged wires alternated with positively charged plates. (*Environmental Elements Corp.*)

15.4 Electric Potential

Consider an object of mass m and charge q in the presence of a fixed charge q_1 (Fig. 15.18). If both charges are positive, q_1 exerts a repulsive force on the object. If the object is initially at rest at point A, it will accelerate outward along a line of force. At point B it will have a speed v_B and a kinetic energy $K_B = \frac{1}{2}mv_B^2$. According to the work-energy theorem (Sec. 5.3), the work W_{AB} done on q by the force $\mathbf{F}$ in moving the object from A to B is equal to the change in its kinetic energy,

$$W_{AB} = K_B - K_A \qquad 15.9$$

FIGURE 15.17

Top view of an electrostatic precipitator, showing the lines of force between the negatively charged wires and the positively charged plates.

Since the object starts from rest, K_A is zero in this case.

Like gravity, the electric force is conservative (Sec. 5.2). This means that the work W_{AB} can be written as the difference of the *potential energy* U of the object at A and B,

$$W_{AB} = U_A - U_B \qquad 15.10$$

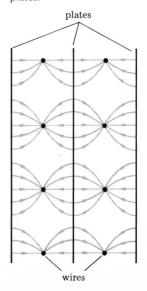

That is, a number U can be assigned to every point such that the work done by the electric force in moving the object between any two points is given by Eq. 15.10. If the same number is added to the value of U at every point, Eq. 15.10 will not be changed. This freedom in the definition of U is used to set U equal to zero at some convenient point in a given situation.

From Eqs. 15.9 and 15.10 we get the result

$$U_A - U_B = K_B - K_A$$

or

$$U_A + K_A = U_B + K_B \qquad 15.11$$

This says that the sum of the kinetic and potential energies of the object at A is equal to the sum of the energies at B. Since A and B are arbitrary points, the sum of these energies is the same at all points; i.e., it is a constant.

The electric potential energy of a point charge q a distance r from a point charge q_1 is

$$U = \frac{1}{4\pi\epsilon_o} \frac{qq_1}{r} \qquad \text{15.12}$$

The potential energy is positive if q and q_1 have the same sign, and it is negative if q and q_1 have the opposite sign. Figure 15.19 gives a plot of U against r for both cases. The magnitude of U decreases as r increases, but U is zero only when the charges are infinitely far apart. Since U is negative in the attractive case (i.e., when q and q_1 have opposite signs), the potential energy of oppositely charged particles actually increases (becomes less negative) as the distance between them increases. This means that work must be done to separate oppositely charged particles, and that once separated, the electric force between them will do positive work in accelerating the particles toward each other.

Example 15.5 (a) What is the potential energy of an electron at a point A, a distance $r_A = 0.53 = 10^{-10}$ m from a proton?* (b) What is the minimum speed v_A required for the electron to escape completely from the proton?

(a) Using Table 15.1 and Eq. 15.12, we find the potential energy of the electron at A to be

$$U_A = \frac{1}{4\pi\epsilon_o} \frac{q_e q_p}{r_A}$$

$$= \frac{(9.0 \times 10^9 \text{ N} \cdot \text{m}^2/\text{C}^2)(-1.6 \times 10^{-19} \text{ C})(1.6 \times 10^{-19} \text{ C})}{5.3 \times 10^{-11} \text{ m}}$$

$$= -4.3 \times 10^{-18} \text{ N} \cdot \text{m} = -4.3 \times 10^{-18} \text{ J}$$

The energy is negative because the electron is attracted to the proton.

(b) Equation 15.11 can be used to calculate the minimum speed v_A that the electron must have at A in order to escape completely from the proton. An electron with the minimum escape speed can move infinitely far from the proton, but its speed at infinity will be zero. Thus when the electron is infinitely far from the proton, both its potential energy U_∞ and its kinetic energy K_∞ are zero. Therefore, from Eq. 15.11 we have

$$K_A + U_A = K_\infty + U_\infty = 0$$

or

$$K_A = \tfrac{1}{2}mv_A^2 = -U_A = 4.3 \times 10^{-18} \text{ J}$$

where m is the mass of the electron. Solving this for v_A, we get

$$v_A^2 = \frac{4.3 \times 10^{-18} \text{ J}}{\tfrac{1}{2}m} = \frac{4.3 \times 10^{-18} \text{ J}}{(0.5)(9.1 \times 10^{-31} \text{ kg})} = 9.4 \times 10^{12} \text{ J/kg}$$

*This is the average distance between an electron and a proton in a hydrogen atom (Sec. 20.2).

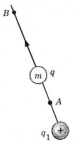

FIGURE 15.18

A test charge q of mass m moving from A to B in the presence of a fixed charge q_1.

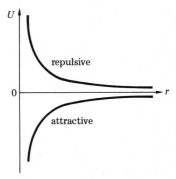

FIGURE 15.19

Plot of the potential energy U of a test charge q against its distance r from a point charge q_1. The upper (repulsive) curve applies when q and q_1 have the same sign, and the lower (attractive) curve applies when they have the opposite sign.

so that

$$v_A = 3.1 \times 10^6 \text{ m/s} \qquad \square$$

The potential energy U of a test charge q resulting from a set of source charges $(q_1, q_2, \ldots, q_n)$ is the sum of the potential energies U_i resulting from each charge individually, so

$$U = \sum_{i=1}^{n} U_i$$

$$= \sum_{i=1}^{n} \frac{1}{4\pi\epsilon_o} \frac{qq_i}{r_i}$$

or

$$U = \frac{q}{4\pi\epsilon_o} \sum_{i=1}^{n} \frac{q_i}{r_i} \qquad 15.13$$

where r_i is the distance from q_i to q.

Example 15.6 Find the potential energy U of the charge $q = 3 \times 10^{-5}$ C in Fig. 15.10 resulting from the two source charges $q_1 = q_2 = 5 \times 10^{-4}$ C.

The distance r_1 and r_2 are both $\sqrt{5}$ m, so from Eq. 15.13 we get

$$U = \frac{q}{4\pi\epsilon_o} \left(\frac{q_1}{r_1} + \frac{q_2}{r_2} \right)$$

$$= \frac{2qq_1}{4\pi\epsilon_o r_1} = 2(9 \times 10^9 \text{ N} \cdot \text{m}^2/\text{C}^2) \frac{(3 \times 10^{-5} \text{ C})(5 \times 10^{-4} \text{ C})}{\sqrt{5} \text{ m}}$$

$$= 121 \text{ J}$$

The problem of finding the potential energy of a test charge is easier than the problem of finding the force on it (Example 15.2) because no vector addition is required. $\qquad \square$

Definition The *electric potential* V at a point P resulting from a set of source charges $(q_1, q_2, \ldots, q_n)$ is

$$V = \frac{1}{4\pi\epsilon_o} \sum_{i=1}^{n} \frac{q_i}{r_i} \qquad 15.14$$

where r_i is the distance from q_i to P.

From Eqs. 15.13 and 15.14 we see that the potential energy U of a test charge q at P resulting from a set of source charges is

$$U = qV \qquad 15.15$$

UNITS Equation 15.15 shows that the unit of electric potential is the joule per coulomb (J/C), which is called a *volt* (V):

$$1 \text{ V} = 1 \text{ J/C}$$

Similarly, Eq. 15.8 shows that the unit of electric field is newtons per coulomb (N/C). But we can write

$$1 \text{ N/C} = 1\frac{\text{N} \cdot \text{m}}{\text{C} \cdot \text{m}} = 1\frac{\text{J}}{\text{C} \cdot \text{m}} = 1\frac{\text{J/C}}{\text{m}} = 1 \text{ V/m}$$

which shows that the units of electric field can equally be taken to be newtons per coulomb or volts per meter.

Example 15.7 (a) Find the electric potential at point P in Fig. 15.11 resulting from the two source charges $q_1 = -6 \times 10^{-5}$ C and $q_2 = 3 \times 10^{-5}$ C. (b) What is the potential energy of a test charge $q = -4 \times 10^{-2}$ C at P?

(a) From Eq. 15.14 we have

$$V = \frac{1}{4\pi\epsilon_o}\left(\frac{q_1}{r_1} + \frac{q_2}{r_2}\right)$$

$$= (9 \times 10^9 \text{ N} \cdot \text{m}^2/\text{C}^2)\left(\frac{-6 \times 10^{-5} \text{ C}}{6 \text{ m}} + \frac{3 \times 10^{-5} \text{ C}}{4 \text{ m}}\right)$$

$$= -2.25 \times 10^4 \text{ V}$$

(b) From Eq. 15.15 the potential energy of the test charge is

$$U = qV$$

$$= (-4 \times 10^{-2} \text{ C})(-2.25 \times 10^4 \text{ V})$$

$$= 900 \text{ J}$$

The positive value of U means that the negative test charge is repelled by the source charges, since in moving away the potential energy of the test charge will decrease and its kinetic energy will increase. ☐

Example 15.8 Find the electric potential at point P in Fig. 15.10 resulting from the source charges $q_1 = q_2 = 5 \times 10^{-4}$ C.

In Example 15.6 we found the potential energy U of a test charge $q = 3 \times 10^{-4}$ C at P to be 121 J, so from Eq. 15.15 the electric potential at P is

$$V = \frac{U}{q} = \frac{121 \text{ J}}{3 \times 10^{-4} \text{ C}} = 4.0 \times 10^5 \text{ V} \qquad ☐$$

EQUIPOTENTIAL SURFACES

Definition An *equipotential surface* is a surface on which the potential is everywhere the same. In a two-dimensional diagram the equipotentials are closed curves formed by the intersections of the plane of the diagram and the surfaces. As an example, the dashed curves in Fig. 15.20 are the equipotential curves of a dipole charge distribution. The equipotential surfaces are the surfaces swept out by these curves when the diagram is rotated about the vertical axis.

The electric field does no work on a test charge that moves along an equipotential, since the potential energy of the charge does not change. This means that the lines of force must be perpendicular to the equipotentials, as shown in Fig. 15.20, for if this were not the case, the electric force would have a component parallel to an equipotential, and some work would be done on a test charge moving along it. The force on a positive test charge is always directed from a region of high potential to a region of low potential, perpendicular to the equipotentials. The force on a negative test charge is the reverse; it is directed from a region of low potential to a region of high potential.

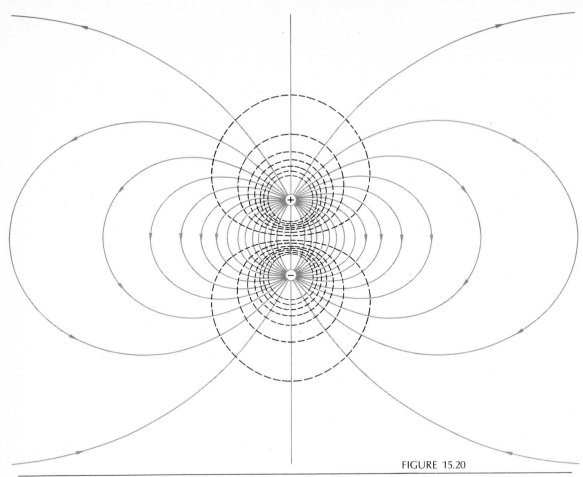

FIGURE 15.20

Equipotentials and lines of force of a dipole. The lines of force are everywhere perpendicular to the equipotentials.

When a metal conductor is placed in a static electric field, different parts of the conductor may be momentarily at different potentials. If this is the case, the negatively charged electrons, which are free to move in the metal, will flow from the regions of low potential to the regions of high potential. The electrons redistribute themselves until, in less than a millionth of a second, the potential is the same everywhere in the metal. Thus, when there is no flow of charge in it, a metal or other conducting object is an equipotential region. Furthermore, since any electric field in a conductor would cause the free electrons to move, the electric field in a conductor must be zero when there is no flow of charge.

In the case of a hollow conducting shell that encloses a charge-free region of space (Fig. 15.21), the electric field is zero both in the shell itself and in the space it encloses. If the shell is charged, all the charge will be on the outside surface; no charge resides on the inside surface of a hollow conducting shell. This is shown in Fig. 15.21, where all the lines of force end on charges on the surface of the shell, and none of the lines of force penetrate the shell. A conducting shell thus shields its interior from all external electric fields.

A *Van de Graaff generator* is a device for generating a high electric potential. It consists of a large conducting shell mounted on an insulated column (Fig. 15.22). A conveyor belt passes inside the shell through a small opening. At the lower end of the belt an auxiliary source raises the potential of a sharp needle to several thousand volts. As a result, the electric field around the point becomes strong enough to ionize the surrounding air, causing a stream of positive ions to be sprayed onto the belt. The belt, driven by a motor, carries this charge into the shell, where it is transferred from the belt to the shell through a conducting wire.

Because the electric field is zero inside the shell, the positive charge on the belt is not repelled by the charge already on the shell. The shell, in effect, shields the new charge from the charge on its surface. When the charge on the belt contacts the wire, it is repelled by the other charge on the belt and flows onto the shell. Once on the shell, the charge immediately spreads over the outside surface. In this manner a large amount of charge can be deposited onto the shell, which raises the shell to a very high potential. A small classroom Van de Graaff generator can reach a potential of 250,000 V, and commercial generators reach 2 to 8 million volts.

The purpose of a Van de Graaff generator is to accelerate charged particles. For instance, if a source of positive ions is located just outside the shell, each ion will have a potential energy $U = qV$, where q is the charge of the ion, and V is the potential of the shell. These ions are repelled by the shell and will move to a region of lower potential. According to Eq. 15.11, the sum of the potential and kinetic energies of a charged particle remains constant. If the kinetic energy K of the ion is zero near the shell, we have

$$U + K = qV$$

so when the ion reaches a region where its potential energy is U', its kinetic energy K' will be

$$K' = qV - U' \qquad\qquad 15.16$$

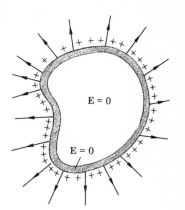

FIGURE 15.21

A hollow conducting shell. If there is no charge in the region enclosed by the shell, all the charge on the shell will be on the exterior surface, and the electric field in the interior of the shell and in the enclosed region will be zero.

Example 15.9 What is the maximum speed that can be obtained by protons accelerated in a 4-million-V Van de Graaff generator?

We assume that the initial kinetic and potential energies of the proton are

$$K = 0$$

and

$$\begin{aligned} U = qV &= eV \\ &= (1.6 \times 10^{-19}\,\text{C})(4 \times 10^6\,\text{V}) \\ &= 6.4 \times 10^{-13}\,\text{J} \end{aligned}$$

When the potential energy is U', the kinetic energy is

$$K' = 6.4 \times 10^{-13}\,\text{J} - U'$$

which is largest when $U' = 0$. Thus the maximum kinetic

a

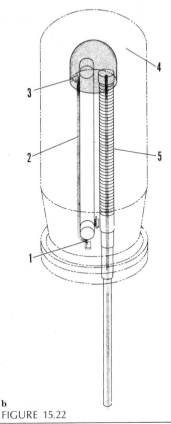

b
FIGURE 15.22

Van de Graaff generator. (a) Four-million-volt positive-ion accelerator, with its outer casing raised to show the high-voltage shell mounted on a stack of guide rings. The rings provide a uniform potential gradient from the shell to ground. (b) Schematic diagram. At (1) positive charge is sprayed onto the conveyor belt (2), which carries the charge inside the shell. At (3) the charge is transferred to the shell. The outer casing (4) encloses the entire machine in a pressurized atmosphere of carbon dioxide and nitrogen that prevents arc-over. Ions inside the shell are accelerated down the evacuated tube (5) surrounded by a second set of metal rings. (*High Voltage Engineering Corporation.*)

energy is

$$K' = \tfrac{1}{2}mv'^2 = 6.4 \times 10^{-13}\,\text{J}$$

and the maximum speed is

$$v' = \sqrt{\frac{2(6.4 \times 10^{-13}\,\text{J})}{1.67 \times 10^{-27}\,\text{kg}}} = 2.77 \times 10^7\,\text{m/s} \qquad \square$$

15.5 Electron Beams

A number of important devices, such as oscilloscopes, television sets, x-ray machines, electron microscopes, and electronic vacuum tubes, use a beam of electrons accelerated by an electric field. In all cases the beam is contained in an evacuated glass tube, as shown in Fig. 15.23. Metal plates, or *electrodes*, are mounted inside the tube, and wires attached to the electrodes pass out of the tube through airtight seals. One electrode, called the *cathode*, is heated by a wire filament in which there is an electric current. When the temperature of the cathode is sufficiently high, some of its free electrons have enough energy to escape from the metal, just like molecules evaporating from a liquid. These evaporated electrons form an electron cloud about the cathode; when it becomes dense enough, the cloud prevents further evaporation.

If the other electrode, called the *anode,* is maintained at a positive potential relative to the cathode, the electrons in the cloud will be attracted to it. Because there is no air in the tube, these electrons move freely toward the anode without colliding with air molecules. As they move away from the electron cloud, they are replaced by more electrons evaporated from the cathode.

The kinetic energy K of an electron as it emerges from the cathode is nearly zero, so Eq. 15.16 gives the kinetic energy K' of the electron when it reaches the anode. The charge of an electron is $q = -e$, so from Eq. 15.16 we have

$$K' = qV - U' = q(V - V')$$
$$= -e(V - V') = e(V' - V)$$

or

$$K' = e\, \Delta V \qquad\qquad 15.17$$

where $\Delta V = V' - V$ is the potential difference between the anode and the cathode.

X-RAY TUBE

In an x-ray tube the electrons collide with the anode, as shown in Fig. 15.23. The sudden deceleration of the electrons generates the x-rays, which are electromagnetic waves of very short wavelength (Sec. 13.1). The wavelength of the x-rays produced becomes shorter as the potential difference ΔV becomes larger (Sec. 20.1). Because shorter-wavelength x-rays are more penetrating than longer-wavelength x-rays, x-ray machines with very large potential differences are now used.

> **UNITS** The *electron volt* (eV) is a unit of energy equal to the kinetic energy gained by an electron in passing through a potential difference of one volt. From Eq. 15.17 we see that when $\Delta V = 1\,V$, the kinetic energy gained is
>
> $$K' = e\, \Delta V$$
> $$= (1.6 \times 10^{-19}\,C)1\,V = 1.6 \times 10^{-19}\,J$$
>
> so
>
> $$1\,eV = 1.6 \times 10^{-19}\,J \qquad\qquad 15.18$$

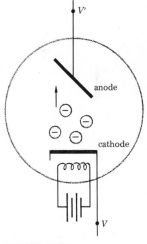

FIGURE 15.23

A vacuum tube. Electrons that evaporate from the heated cathode are accelerated toward the anode, which is maintained at a positive potential with respect to the cathode.

Example 15.10 What potential difference is required to accelerate electrons to a kinetic energy K' of 1.5×10^{-15} J?

We use Eq. 15.18 to convert K' into electronvolts:

$$K' = 1.5 \times 10^{-15}\,J = (1.5 \times 10^{-15})\frac{1}{1.6 \times 10^{-19}}\,eV$$

$$= 9375\,eV$$

But by definition, the increase in the kinetic energy of an electron (in electronvolts) is equal to the potential difference through which the electron passed, so the required potential difference in this case is

$\Delta V = 9375\ V$ □

CATHODE-RAY TUBE

The cathode-ray tube (Fig. 15.24) is used in both oscilloscopes and television sets to provide an electrically controlled picture. Electrons evaporated from the cathode are accelerated to the anode, as in an x-ray tube. However, there is a hole in the anode of a cathode-ray tube through which some of the electrons pass. These electrons then go on to strike the inside surface of the flattened end of the tube. This surface is coated with a phosphorescent material that produces a bright spot where the beam hits it.

The position of the spot is controlled by two pairs of deflecting plates oriented at right angles to each other. As the electrons pass between a pair of plates, they are deflected toward the plate at the higher potential. The amount of deflection is controlled by varying the potential difference between the plates. One pair of plates produces horizontal deflection, and the other pair produces vertical deflection. The two pairs together can move the spot anywhere on the screen.

An oscilloscope (Fig. 15.25) is used to display a time-varying potential, such as those produced in many electronic circuits. By means of wires connected to metal probes, the potential of interest is applied to the vertical plates of an oscilloscope, causing vertical deflections of the spot. To display the time variation of the pulses, the beam is simultaneously swept horizontally from left to right at constant speed. At the end of its sweep, it is rapidly returned to the left and swept across again. This sweep requires that the potential applied to the horizontal plates have the sawtooth waveform shown in Fig. 15.26. This sweep potential is provided by electronic circuits in the oscilloscope, and there are controls for varying the frequency of the sweep.

In operation, the sweep frequency is adjusted to equal the frequency at which the pulses arrive on the vertical plates. Then, with each horizontal sweep, a new pulse is displayed in the same position as the previous pulse, giving a stationary image of a single pulse. Figure 15.27 shows the time-varying potential in an electronic circuit as it appears on an oscilloscope screen.

FIGURE 15.24

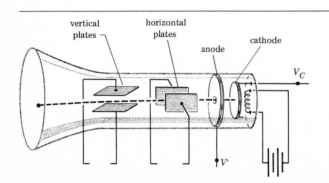

Cathode-ray tube. Some electrons pass through the hole in the anode and go on to strike the phosphorescent screen. The position of the beam on the screen is controlled by the potentials applied to the deflection plates.

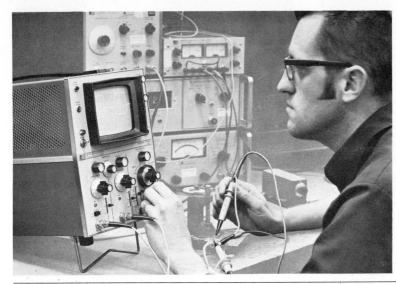

FIGURE 15.25

An oscilloscope. (*Hewlett-Packard.*)

TELEVISION TUBE

A television picture tube is very similar to a cathode-ray tube, except that the beam is deflected magnetically rather than electrically. In the United States and Canada the picture is composed of 525 individual horizontal lines, and it is changed 30 times a second. This means that the beam is swept horizontally across the screen $(525)(30) = 15,750$ times a second. As the beam is swept horizontally, it is also swept vertically, at a rate of 60 times a second. It takes two vertical sweeps to form one picture because only half the lines are displayed on each sweep.* The horizontal and vertical sweeps are produced by

*The eye sees a single image because the eye retains the image of a line for about $\frac{1}{20}$ s after the line vanishes (persistence of vision). Thus the eye still sees the first line of an individual picture when the last line is formed $\frac{1}{30}$ s later.

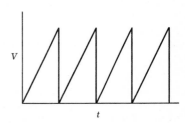

FIGURE 15.26

Time variation of the sweep potential applied to the horizontal deflection plates of an oscilloscope.

FIGURE 15.27

A time-varying potential displayed on an oscilloscope. (*Northeastern University.*)

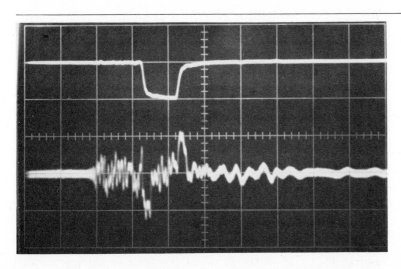

electronic circuits in the set, but they are synchronized to the signal transmitted by the station. This signal also controls the intensity of the electron beam and hence the brightness of the spot produced as the beam is swept across the screen. It is the variation of the beam intensity as the beam sweeps back and forth across the screen that produces a particular picture.

The television camera, another device that utilizes the flow of electrons in a vacuum, is described in Sec. 19.2.

GUIDE TO MAJOR TOPICS

Topic	Reference	Problems
Fundamental forces	Sec. 15.1	
Coulomb's law	Sec. 15.2	
One source charge	Eqs. 15.2 to 15.4; Example 15.1; Table 15.1	1 to 6
More than one		
source charge	Sec. 15.3; Eq. 15.6; Example 15.2	7 to 12
Electric field	Sec. 15.3; Eqs. 15.7 and 15.8; Examples 15.3 and 15.4	13 to 20
Lines of force		21 and 22
Electric potential	Sec. 15.4	
Potential energy	Eqs. 15.9 to 15.13; Examples 15.5 and 15.6	23 to 28
Electric potential	Eqs. 15.14 and 15.15; Examples 15.7 and 15.8	29 to 34
Equipotential surfaces		35 to 38
Van de Graaff generator	Eq. 15.16; Example 15.9	39 and 40
Electron beams	Sec. 15.5; Eq. 15.17; Example 15.10	41 and 42

PROBLEMS

1 (a) What is the magnitude of the force exerted on a $+7$-C charge by a -3-C charge 2 m away? (b) Is the force attractive or repulsive?
Ans. (a) 4.72×10^{10} N; (b) attractive

2 Suppose that the charge on the metal spheres of an induction machine (Fig. 15.8) are $+1.4 \times 10^{-8}$ and -1.4×10^{-8} C, respectively. What is the force that one sphere exerts on the other when the spheres are 5 cm apart?

3 (a) What is the mass of a group of protons with a total charge of 1 C? (b) What is the total charge of 1 kg of protons?
Ans. (a) 1.04×10^{-8} kg; (b) 0.96×10^{8} C

4 (a) Find the electric force between a kilogram of protons and a kilogram of electrons separated by 6×10^{6} m (the radius of the earth). (b) What is the gravitational force between these same objects?

5 A glass rod rubbed with silk acquires a charge of $+3 \times 10^{-10}$ C. How many electrons were transferred from the glass to the silk?
Ans. 1.9×10^{9}

6 Find the force on a charge of 5×10^{-8} C exerted by a charge of 3×10^{-9} C for the following values of r: 0.5, 1.0, 2.0, 2.5, and 3.0 m.

Plot the force against r, and connect the points with a smooth curve.

7 Find the force (magnitude and direction) that the dipole in Fig. 15.28 exerts on a test charge $q = +10^{-10}$ C at point P.
Ans. -0.176 N

8 Find the force (magnitude and direction) that the dipole in Fig. 15.28 exerts on a test charge $q = +10^{-10}$ C at point Q.

9 Find the force (magnitude and direction) that the dipole in Fig. 15.28 exerts on a test charge $q = +10^{-10}$ C at point R. (Vector addition is required.)
Ans. 1.8 N

10 Find the force (magnitude and direction) that the dipole in Fig. 15.28 exerts on a test charge $q = +10^{-10}$ C at point S.

11 Find the force (magnitude and direction) that the four charges in Fig. 15.29 exert on a test charge $q = 10^{-10}$ C at point P.
Ans. 5.09 N (down)

12 Find the force (magnitude and direction) that the four charges in Fig. 15.29 exert on a test charge $q = 10^{-10}$ C at point Q.

13 A charge q_1 exerts a force of 100 N on a

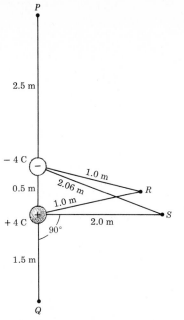

FIGURE 15.28

Problems 7, 8, 9, 10, 29, 30, and 35.

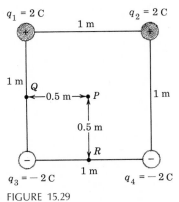

FIGURE 15.29

Problems 11, 12, 17, 18, 21, 27, 31, 32, and 36.

test charge $q_2 = 2 \times 10^{-5}$ C located at a point 0.20 m from q_1. (a) What is the electric field resulting from q_1 at the point? (b) What is the magnitude of q_1?
Ans. (a) 5×10^6 N/C; (b) 2.2×10^{-5}C

14 Find the magnitude of the electric field at 0.2, 0.5, and 0.8 m from a charge of 2×10^{-10} C. Make a scale drawing similar to Fig. 15.12 by drawing arrows to represent the field at these points.

15 (a) What are the magnitude and direction of the total force on charge $q_3 = +5$ C exerted by charges q_1 and q_2 in Fig. 15.30? (b) What is

FIGURE 15.30

Problems 15 and 16.

the electric field at point P resulting from q_1 and q_2?
Ans. (a) 1.06×10^{11} N; (b) 2.12×10^{10} N/C

16 (a) What are the magnitude and direction of the total force on charge $q_2 = 10$ C exerted by charges q_1 and q_3 in Fig. 15.30? (b) What is the electric field at point Q resulting from q_1 and q_3?

17 Find the electric field at point P in Fig. 15.29 (see Prob. 11).
Ans. 5.09×10^{10} N/C (down)

18 Find the electric field at point Q in Fig. 15.29 (see Prob. 12).

19 (a) What are the magnitude and direction of the total force on charge $q_3 = +3$ C exerted by charges q_1 and q_2 in Fig. 15.31? (Vector addition is required.) (b) What is the electric field at point P resulting from q_1 and q_2?
Ans. (a) 1.3×10^{10} N; (b) 4.3×10^9 N/C

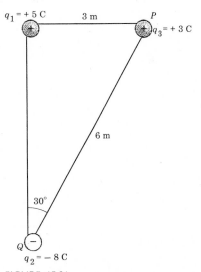

FIGURE 15.31

Problems 19, 20, and 28.

20 (a) What are the magnitude and direction of the total force on charge $q_2 = -8$ C exerted by charges q_1 and q_3 in Fig. 15.31? (b) What is the electric field at point Q resulting from q_1 and q_3?

21 (a) Draw the line of force that passes through point P in Fig. 15.29. (b) Try the same for point Q in Fig. 15.29.

22 Show why two lines of force cannot intersect at a charge-free point. (An exception is a point, such as P in Fig. 15.32, where the field is zero.)

23 An object of mass $m = 2$ kg and charge $q = 5 \times 10^{-5}$ C is moving away from a fixed charge $q_1 = 8 \times 10^{-5}$ C. When q is 2 m from q_1 its speed is 5 m/s. (a) What is the potential energy of q when it is 2 m from q_1? (b) What is the speed of q when it is very far away from q_1? Ans. (a) 18 J; (b) 6.6 m/s

24 Repeat Prob. 23 when the fixed charge is $q_1 = -8 \times 10^{-5}$ C.

25 (a) What is the potential energy of a charge $q = 3$ C that is 3 m from a fixed charge $q_1 = 15 \times 10^{-6}$ C? (b) How much work is done on q by the electric field in moving q to a point 5 m from q_1? Ans. (a) 1.35×10^5 J; (b) 5.4×10^4 J

26 (a) What is the potential energy of an electron that is 20 cm from a fixed charge of 6×10^{-8} C? (b) How much work is required to move the electron very far from the fixed charge?

27 What is the potential energy of charge q_4 in Fig. 15.29 resulting from charges q_1, q_2, and q_3? Ans. -5.1×10^{10} J

28 What is the potential energy of q_2 in Fig. 15.31 resulting from charges q_1 and q_3.

29 (a) Find the potentials at points P, Q, R, and S in Fig. 15.28. (b) How much work is required to move a charge of 7.5×10^{-7} C from point P to point Q? Ans. (a) -2.4×10^9, $+6.0 \times 10^9$, 0, and $+0.52 \times 10^9$ V; (b) 6300 J

30 Find several points in Fig. 15.28 where the potential is -4.0×10^9 V.

31 (a) Find the potentials at points P, Q, and R in Fig. 15.29. (b) What is the work done by the electric field in moving a test charge $q = 2 \times 10^{-9}$ C from Q to R? Ans. (a) 0, 0, -3.8×10^9 V; (b) 7.6 J

32 Calculate the work required to assemble the four charges in Fig. 15.29. Start with q_1. No work is required to put it into position, because no other charge is present. Next, calculate the work required to move q_2 from infinity to its final position, given that q_1 is already in place. Then, with q_1 and q_2 in place, calculate the work required to bring in q_3. Finally, calculate the work required to bring in q_4. The sum

of these works is the total work required to assemble the charge configuration. Does the result depend on the order in which the charges are assembled?

33 A set of source charges $(q_1, q_2, \ldots)$ is located along the x axis at the points $x_i = 2^i$ m (2 m, 4 m, 8 m, 16 m, $\ldots$). Each source charge is 3×10^{-8} C. Find the electric field and potential at the origin $(x = 0)$ resulting from all the source charges. Ans. 90 V/m, 270 V

34 A set of source charges $(q_1, q_2, \ldots)$ is located along the x axis at the points $x_i = i$ m (1 m, 2 m, 3 m, $\ldots$). On the odd integer points the charge is $+3 \times 10^{-8}$ C, and on the even integers the charge is -3×10^{-8} C. Find the electric field and potential at the origin $(x = 0)$ resulting from all the source charges.

35 In Fig. 15.28 sketch the $V = 4.0 \times 10^9$ V equipotential surface (see Prob. 30).

36 In Fig. 15.29 draw the equipotential surface that passes through point P (see Prob. 31).

37 Figure 15.32 shows the equipotential surfaces surrounding two positively charged spheres of different radii. Draw the lines of force on the figure.

38 Figure 15.33 shows the lines of force surrounding two oppositely charged parallel plates. Sketch the equipotential surfaces on the figure.

39 When a charge q is uniformly distributed over the surface of a conducting sphere of radius r, the potential of the sphere is the same as though all the charge were located at the center of the sphere. The spherical shell of a Van de Graaff generator is 60 cm in radius and is charged to a potential of 1.5×10^6 V. What is the charge on the shell? Ans. 1.0×10^{-4} C

40 What must be the potential of the shell of a Van de Graaff generator in order to accelerate a proton to a speed of 1.2×10^7 m/s?

41 The power needed to operate an x-ray tube is equal to the total gain in kinetic energy per second of the electrons accelerated by the tube. What is the power required to move 2×10^{16} electrons per second through a potential difference of 35,000 V? Ans. 112 W

42 The anode of an x-ray tube is maintained at a potential of 12,000 V with respect to the cathode. What is the speed of the electrons when they hit the anode?

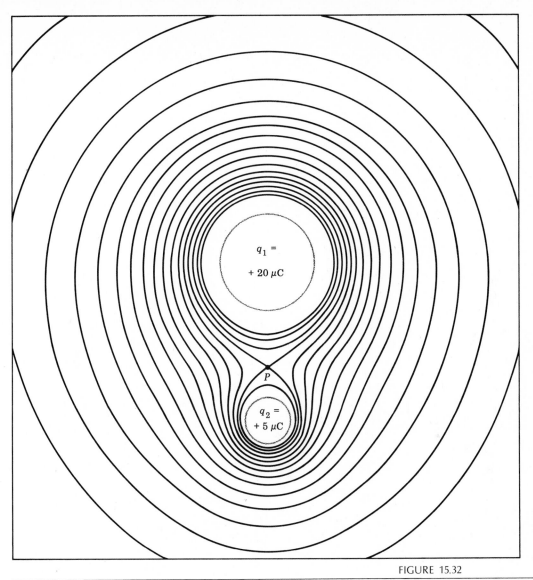

$q_1 =$

$+ 20 \, \mu C$

P

$q_2 =$

$+ 5 \, \mu C$

FIGURE 15.32

Problem 37.

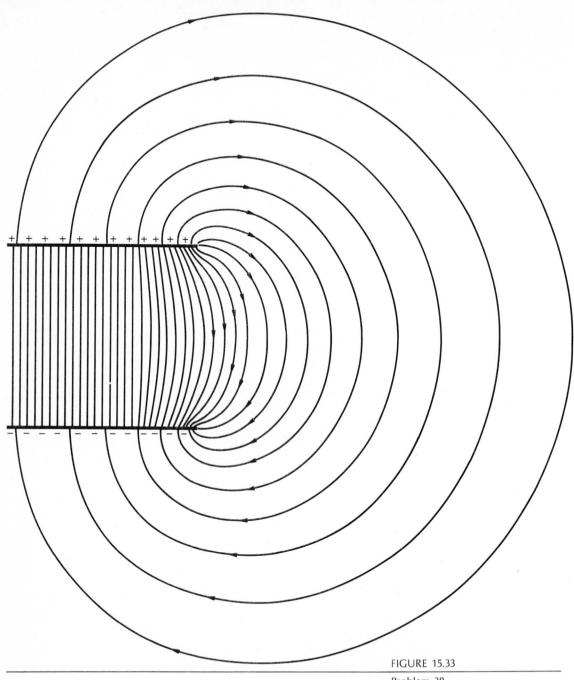

FIGURE 15.33
Problem 38.

A *current* is a flow of charge. This flow can be through vacuum, as when electrons flow from the cathode to the anode of an x-ray or cathode-ray tube (Sec. 15.5), or the flow can be through a conducting medium, as when electrons flow in an electric wire. In either case, as the charge flows, its potential energy is transformed into other forms of energy. In a cathode-ray tube the potential energy of the electrons is first transformed into kinetic energy as the electrons are accelerated by the electric field, and then this kinetic energy is transformed into internal and radiant energy as the electrons strike the anode. Charge flow in a wire is more versatile. In a heating coil the potential energy of the moving charge is transformed into infrared radiation; in a light bulb it is transformed into light and infrared radiation; and in a motor it is transformed into mechanical energy (the kinetic energy of the rotor). By pulsing and modulating a current, it can be used for communications, as in telegraphy and television, and for information control, as in computers. In fact, all electric and electronic devices use currents in one way or another.

16

DIRECT CURRENT

16.1 Current in a Conductor

A metal conductor consists of a lattice of immobile positive ions surrounded by a swarm of mobile electrons (Fig. 16.1). The ions are the atoms of the metal minus one or more of their outer electrons. These detached electrons move freely throughout the metal, forming a reservoir of mobile charge. On the average, the positive ions and the negative electrons neutralize each other, so there is no electric field in the metal.

If a conductor is placed in an electric field $\mathbf{E}_1$, the force $-e\mathbf{E}_1$ is momentarily exerted on each electron. This causes the mobile electrons to shift their average positions until there is an excess of negative charge at one end of the conductor and an excess of positive charge at the other (Fig. 16.2). This charge separation produces its own electric field $\mathbf{E}_2$. The mobile electrons cease moving when the total field $\mathbf{E}_1 + \mathbf{E}_2$ inside the conductor is zero. We already pointed out in Sec. 15.4 that the field is zero inside a conductor when there is no motion of charge.

Suppose the field $\mathbf{E}_1$ is produced by two oppositely charged metal spheres supported on insulated stands, and suppose that

FIGURE 16.1

A metal conductor consists of a lattice of immobile positive ions surrounded by a swarm of mobile electrons.

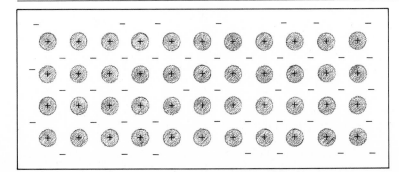

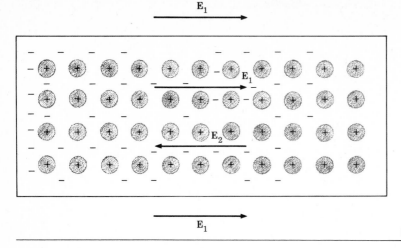

E_1

E_1

E_1

E_2

E_1

FIGURE 16.2

An electric field E_1 applied to a metal conductor. This field causes the electrons to shift position, creating a charge separation in the conductor. This charge separation in turn produces the electric field E_2 that cancels the applied field E_1 inside the conductor.

the conductor is a wire running between the spheres (Fig. 16.3a). Before the wire touches the spheres, the field E_1 produces a charge separation in the wire, and this separated charge produces the field E_2 which cancels E_1 inside the wire. But when the wire touches both spheres (Fig. 16.3b), electrons flow from the negative sphere into the positive end of the wire, canceling the positive charge. At the same time, electrons at the negative end of the wire flow into the positive sphere, leaving the wire uncharged. Thus, once the wire is connected to the spheres, it becomes neutral, and so the field E_2 vanishes. There is then only the field E_1 in the wire, which exerts the force $-eE_1$ on each mobile electron. This force, directed from the negative to the positive sphere, generates a flow of electrons in the same direction.

Definitions The *current* in a wire is the charge per second that flows past a fixed point in the wire. That is, if charge q flows through a wire in time t, the current I in the wire is

$$I = \frac{q}{t} \qquad\qquad 16.1$$

A current is said to be *direct* if the charge always has the same direction of flow; a current is said to be *alternating* if the charge repeatedly alternates its direction of flow. (Alternating current is the subject of Chap. 18.)

UNITS The unit of current is coulombs per second (C/s), which is called an *ampere* (A):

$1\,A = 1\,C/s$

The current between the two spheres in Fig. 16.3 persists until all the excess electrons on the negative sphere reach the positive sphere, a period typically of about 1 microsecond (1 $\mu s = 10^{-6}$ s). Thus, if the initial charge on the spheres is $q = 10^{-8}$ C and the current persists for the time $t = 10^{-6}$ s, the average current I

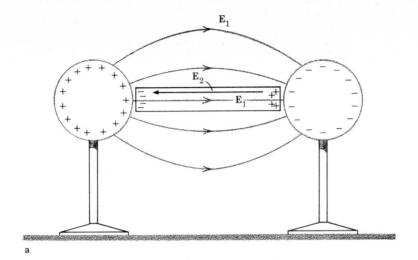

a

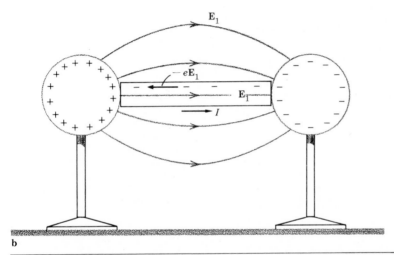

b

FIGURE 16.3

(a) Two oppositely charged spheres applying the field $\mathbf{E}_1$ on a conducting wire. (b) When the wire touches the spheres, the separated charge on the wire passes to the spheres, leaving only the field $\mathbf{E}_1$ inside the wire. This field exerts the force $-e\mathbf{E}_1$ on each mobile electron, causing these electrons to flow through the wire.

during this time is

$$I = \frac{q}{t} = \frac{10^{-8}\,\text{C}}{10^{-6}\,\text{s}} = 10^{-2}\,\text{C/s} = 10^{-2}\,\text{A}$$

The wire itself remains uncharged, even with a current in it, because as many electrons flow into it from the negative sphere as flow out from it to the positive sphere. There is no accumulation of charge in the wire.

By convention, the direction of a current is defined as the direction in which positive charge would flow. Since the moving charges in a metal happen to be negatively charged electrons, the conventional current in a wire is directed opposite to the real flow of charge. This causes no difficulty, however, because a flow of negative charge in one direction is equivalent to a flow of positive charge in the other direction. Thus we can avoid unnecessary minus signs by considering a current to consist of positive charge flowing in one direction, although in fact it consists of negative charge flowing in the opposite direction.

The current in Fig. 16.3b is directed from the positive (high-potential) sphere to the negative (low-potential) sphere and persists only until the potential difference between the spheres vanishes. To maintain a steady current in the wire, an external mechanism is required to transport positive charge back to the positive sphere at the same rate at which it flows to the negative sphere through the wire.

SEAT OF emf

Definition A *seat of emf** is a device that transports positive charge from the negative end of a conductor to the positive end. Since this direction is opposite to the direction of the electric force on the charge, the seat of emf does work on the charge and the charge gains potential energy.

For example, the induction machine in Sec. 15.2 (Fig. 15.8) is a seat of emf that uses mechanical energy to separate negative and positive charge. This mechanical energy is supplied by the work done in rotating the wheel and is transformed into the potential energy of the charges on the metal spheres. When the spheres are connected by a wire, charge flows through the wire, and the potential energy is transformed into internal and other forms of energy. A steady current can be maintained in the wire by steady rotation of the wheel.

A power-generating station similarly converts the mechanical energy of the generator's turbine into electric energy. The process, however, uses the principle of magnetic induction (Sec. 17.4) rather than electric induction.

A battery is a seat of emf that uses chemical energy to maintain a potential difference between its two terminals. When the terminals are not connected, a potential difference $\mathcal{E}$, called the *emf of the battery*, is maintained between them. The emf of a flashlight battery is 1.5 V. That is, the positive terminal is 1.5 V above the negative terminal. No energy is expended when there is no current, but when the terminals are connected by a conducting wire, charge flows through the wire to equalize the potential difference. The battery then expends chemical energy to separate the charges as rapidly as they combine. This can be seen from Fig. 16.4, which shows that inside the battery positive charge is moved toward the positive terminal. Chemical energy is required to move the charge against the repulsive electric force.† (If the current is large, the potential difference V between the terminals may be less than $\mathcal{E}$. However, we shall neglect this possibility here.)

The energy expended by the battery is released in the external circuit. Thus suppose a current I is maintained in the circuit in Fig. 16.4. From Eq. 16.1, the amount of positive charge transferred from the positive to the negative terminal in time t is

$$q = It$$

The work done on this charge in moving it through the wire is

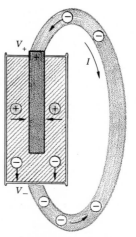

FIGURE 16.4

Inside a battery, chemical energy is used to move positive charge to the positive terminal and negative charge to the negative terminal.

---

*Emf is an abbreviation for *electromotive force,* but the term itself is inaccurate and is seldom used anymore.

†The chemistry of batteries is discussed in Sec. 16.5.

equal to the change in its potential energy. From Eqs. 15.10 and 15.15 we find

$$W = U_+ - U_- = qV_+ - qV_- = q(V_+ - V_-) = qV$$

where $V = V_+ - V_-$ is the potential difference between the terminals. The work done per second, or power P, is

$$P = \frac{W}{t} = \frac{qV}{t} = \frac{ItV}{t}$$

or

$$P = IV \qquad\qquad 16.2$$

Example 16.1 What is the power output of a 1.5-V battery which is delivering a current of 0.2 A?

The potential difference V is just the emf of the battery, and so from Eq. 16.2 we get

$$P = I\mathcal{E} = (0.2\ A)(1.5\ V) = 0.3\ A \cdot V$$

But $1\ A = 1\ C/s$ and $1\ V = 1\ J/C$, so

$$1\ A \cdot V = (1\ C/s)(1\ J/C) = 1\ J/s = 1\ W$$

That is, the product of the potential difference in volts and the current in amperes is equal to the power in watts. Therefore, the power delivered to the external circuit is

$$P = 0.3\ A \cdot V = 0.3\ W$$

This power is supplied at the expense of the internal chemical energy of the battery.

REMARK The SI has been adopted primarily because it gives a simple relation between the common electrical units (volts and amperes) and the mechanical units of energy and power (joules and watts). □

RESISTANCE
A steady current persists in a conductor as long as a constant potential difference V is maintained between the ends of the conductor. This potential difference is called the *voltage* across the conductor. In general, the current in a conductor increases as the voltage increases. This is illustrated in Fig. 16.5, which is a graph of the voltage across the filament of a light bulb against the current in the filament.

Definition The *resistance* R of a conductor is the ratio of the voltage V across the conductor to the current I in it:

$$R = \frac{V}{I} \qquad\qquad 16.3$$

UNITS From Eq. 16.3 we see that the unit of resistance is volts per ampere (V/A), which is called an *ohm* (Ω):*

$$1\ \Omega = 1\ V/A$$

*Ω is the Greek (capital) letter omega.

FIGURE 16.5

Plot of the voltage V across the filament of a light bulb against the current I in the filament.

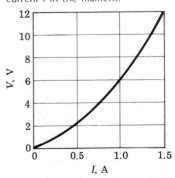

Example 16.2 (a) From the *VI* curve in Fig. 16.5, find the resistance of the light-bulb filament at I = 0.5, 1.0, and 1.5 A. (b) Plot the resistance of the filament against I.

(a) From Fig. 16.5 we see that the voltage across the filament is 2 V when the current is 0.5 A, so the resistance is

$$R = \frac{V}{I} = \frac{2\,V}{0.5\,A} = 4\,V/A = 4\,\Omega$$

At I = 1 A the voltage is 6 V, so R = 6 Ω, and at I = 1.5 A the voltage is 12 V, so R = 8 Ω.

(b) The resistance of the filament is plotted against I in Fig. 16.6. The resistance increases with current because as the filament gets hotter, its conductive properties change. □

Definition An *ohmic conductor* is a conductor whose resistance is independent of the current in it. From Eq. 16.3, the relationship between the voltage and the current in an ohmic conductor is

$$V = RI \qquad\qquad 16.4$$

where R is a constant independent of I. This equation is called *Ohm's law*, although it is really a definition of a class of conductors. Because of the simplicity of Eq. 16.4, and because many conductors are ohmic, Ohm's law is extensively used in circuit analysis.

ELECTRIC CIRCUITS

The simplest electric circuit consists of a seat of emf connected to a resistive element, or *resistor*. For example, Fig. 16.7 shows a battery connected by wires to a light bulb. The resistance of the connecting wires is very small and can be neglected. The entire resistance R of the circuit comes from the extremely thin filament of the bulb.

The circuit in Fig. 16.7 is diagramed in Fig. 16.8. The resistanceless wires are represented by straight lines; the resistance of the light bulb is represented by the symbol —⋀⋀—, and the battery is represented by the symbol —+|ı|ı—. The long vertical line on the battery symbol is the positive (high potential) side. Positive charge flows from the positive side of the battery through the external circuit to the negative side, and then from the negative side through the battery to the positive side.

The basic problem in circuit analysis is to determine the current in the circuit given the resistance and emf. The following rules should be remembered when analyzing a circuit:

1 All points connected by resistanceless stretches of wire are at the same potential. This follows from Eq. 16.4. Thus points a and b in Fig. 16.8 are at the same potential, and points d and e are at the same potential, but points b and c are at different potentials.

2 A battery or other seat of emf is assumed to maintain a constant potential difference $\mathcal{E}$ between its terminals, independent of the current. This is only approximately true, because all real batteries have some internal resistance which causes a drop in potential at large current. In most situations, however, this internal resistance can be neglected.

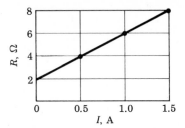

FIGURE 16.6

Plot of the resistance R of a light-bulb filament against current I. For each value of I, the value of R is the ratio V/I taken from Fig. 16.5.

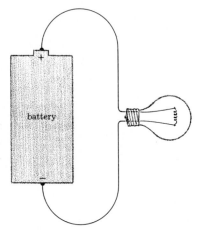

FIGURE 16.7

A battery connected to a light bulb.

FIGURE 16.8

Schematic diagram of a circuit consisting of a battery connected to a resistor (light bulb).

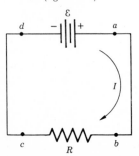

3. The resistive elements of the circuit are assumed to be ohmic, unless stated otherwise. Thus in analyzing a light-bulb circuit, we shall often assume a constant filament resistance.

Let us apply these rules to the circuit in Fig. 16.8. According to rule 2, the potential difference between points d and a is equal to the emf of the battery:

$$V_a - V_d = \mathcal{E}$$

According to Eq. 16.4, the potential difference between points c and b is equal to the resistance R times the current I:

$$V_b - V_c = RI$$

But according to rule 1, $V_a = V_b$ and $V_c = V_d$, so

$$V_a - V_d = V_b - V_c$$

and thus

$$\mathcal{E} = RI \qquad\qquad 16.5$$

In other words, the potential difference across the resistor is equal to the emf of the battery.

Example 16.3 A 4-Ω light bulb is connected to a 12-V battery. What is the current in the bulb?

From Eq. 16.5, the current is

$$I = \frac{\mathcal{E}}{R} = \frac{12\,\text{V}}{4\,\Omega} = 3\,\text{A}$$

This calculation assumes that $4\,\Omega$ is the actual resistance of the bulb's filament when the current in it is 3 A. □

According to Eq. 16.2, a battery increases the potential energy of the charge flowing through it at the rate

$$P = VI \qquad\qquad 16.6$$

where V is the potential difference across the battery (assumed equal to the emf $\mathcal{E}$), and I is the current. Conservation of energy requires that all this energy be dissipated in the external circuit, so Eq. 16.6 also gives the rate at which energy is dissipated in the resistor. Using Eq. 16.4 to replace V by RI, Eq. 16.6 can be written

$$P = RI^2 \qquad\qquad 16.7$$

On the other hand, if I is replaced by V/R, Eq. 16.6 can be written

$$P = \frac{V^2}{R} \qquad\qquad 16.8$$

All three of these expressions for the rate of energy dissipation in a resistor are useful. It is necessary to remember only one expression, however, because the other two are easily derived from it using Ohm's law.

Example 16.4 What is the power dissipated in the light bulb in Example 16.3?

The potential V across the bulb is 12 V and the current I is 3 A, so from Eq. 16.6 the power P is

$$P = VI = (12\,\text{V})(3\,\text{A}) = 36\,\text{W}$$

Since in this case we know that the resistance R is 4 Ω, we can also use Eq. 16.7 to get the same result:

$$P = RI^2 = (4\,\Omega)(3\,\text{A})^2 = 36\,\text{W}$$

This energy is released by chemical reactions in the battery and is delivered to the bulb filament, where it appears in the form of visible and infrared radiation. The electric current does not create this energy; it only transports it from one place to another and transforms it from one form to another. □

Example 16.5 A flashlight bulb dissipates 2 W when connected to a 3-V battery. (a) What is the resistance of the bulb? (b) What is the current in the bulb?

(a) Since we know $P = 2\,\text{W}$ and $V = 3\,\text{V}$, we can use Eq. 16.8 to find R:

$$R = \frac{V^2}{P} = \frac{(3\,\text{V})^2}{2\,\text{W}} = 4.5\,\Omega$$

(b) From Eq. 16.5 the current is

$$I = \frac{\mathcal{E}}{R} = \frac{3\,\text{V}}{4.5\,\Omega} = 0.67\,\text{A} \qquad □$$

16.2 Circuit Analysis

Circuits often consist of a network of interconnected resistors, like the one shown in Fig. 16.9. The basic problem of circuit theory is to find the current in each branch of the network, given the values of the resistors. The analysis of this or any other network uses only two principles, known as *Kirchhoff's laws*.

Kirchhoff's First Law *The total current entering any point of the circuit is equal to the total current leaving the point.* This is a consequence of the fact that no charge accumulates at a point in a circuit, so that as much charge must flow out from a point as flows into it. This means, for instance, that the current I entering point a in Fig. 16.9 is equal to the sum $I_1 + I_2$ of the currents leaving point a.

Kirchhoff's Second Law *The potential difference between any two points in a circuit is the same along any path connecting the points.* This means, for instance, that the potential difference between the points a and b in Fig. 16.9 is the same along the path through R_1 as it is along the path through R_2 and R_3 or the path through R_2 and R_4.

RESISTORS IN SERIES

Definition Two or more resistors are said to be in *series* when they are connected so that the current is the same in each of them. Figure 16.10 shows two resistors R_1 and R_2 in series. From

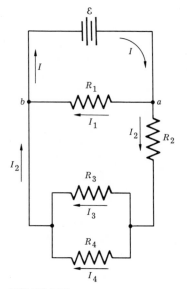

FIGURE 16.9

A complex circuit network.

FIGURE 16.10

Two resistors connected in series to a battery.

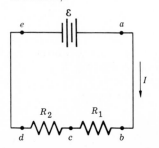

Kirchhoff's first law, the current I that enters point c from R_1 is equal to the current that leaves point c and passes through R_2. In fact the current I is the same everywhere in the circuit because there is only one path through every point.

Let us calculate the potential difference between points a and e along the path abcde through the two resistors. From rule 1 (Sec. 16.1), points a and b are at the same potential, so

$$V_a - V_b = 0$$

From Ohm's law, the potential across R_1 is

$$V_b - V_c = R_1 I$$

and the potential across R_2 is

$$V_c - V_d = R_2 I$$

Points d and e are at the same potential, so

$$V_d - V_e = 0$$

When these four equations are added together, we get

$$(V_a - V_b) + (V_b - V_c) + (V_c - V_d) + (V_a - V_e)$$
$$= 0 + R_1 I + R_2 I + 0$$

or

$$V_a - V_e = (R_1 + R_2)I$$

which is the potential difference between a and e along the path abcde.

From Kirchoff's second law, the potential difference between a and e along the path abcde is equal to the potential difference between a and e along the path through the battery. But this is just the emf $\mathcal{E}$ of the battery, so we have

$$\mathcal{E} = (R_1 + R_2)I$$

Comparing this with Eq. 16.5, we see that two resistors R_1 and R_2 connected in series act like a single resistor of magnitude

$$R = R_1 + R_2$$

This can be generalized to the following rule for finding the total resistance of any number of resistors connected in series.

Resistors in Series *The total resistance R of n resistors R_1, R_2, ..., R_n connected in series is equal to the sum of the individual resistances:*

$$R = \sum_{i=1}^{n} R_i \qquad\qquad 16.9$$

Example 16.6 A battery with $\mathcal{E} = 12\,\text{V}$ is connected in series with two resistors $R_1 = 2\,\Omega$ and $R_2 = 4\,\Omega$ (Fig. 16.10). (a) Find the current in the circuit. (b) Find the voltage across each resistor. (c) Find the power dissipated in each resistor.

(a) From Eq. 16.9 the total resistance R of the circuit is

$$R = R_1 + R_2 = 2\,\Omega + 4\,\Omega = 6\,\Omega$$

and so the current is

$$I = \frac{\mathcal{E}}{R} = \frac{12\,\text{V}}{6\,\Omega} = 2\,\text{A}$$

The current is the same in both resistors.

(b) The voltage across R_1 is

$$V_1 = R_1 I = (2\,\Omega)(2\,\text{A}) = 4\,\text{V}$$

and the voltage across R_2 is

$$V_2 = R_2 I = (4\,\Omega)(2\,\text{A}) = 8\,\text{V}$$

The sum of the voltages across the two resistors in series is equal to the emf of the battery.

(c) From Eq. 16.7 the power P_1 dissipated in R_1 is

$$P_1 = R_1 I^2 = (2\,\Omega)(2\,\text{A})^2 = 8\,\text{W}$$

and the power P_2 dissipated in R_2 is

$$P_2 = R_2 I^2 = (4\,\Omega)(2\,\text{A})^2 = 16\,\text{W}$$

The total power P dissipated in both resistors is

$$P = P_1 + P_2 = 24\,\text{W}$$

and is equal to the power

$$P = \mathcal{E}I = (12\,\text{V})(2\,\text{A}) = 24\,\text{W}$$

delivered by the battery. ☐

RESISTORS IN PARALLEL

Definition Two or more resistors are said to be in *parallel* when they are connected so that the voltage is the same across each of them. Figure 16.11 shows two resistors R_1 and R_2 in parallel. From Kirchoff's second law, the voltage $V_a - V_b$ is the same across each resistor because each resistor is a path between points a and b. Compare Figs. 16.10 and 16.11. In the series circuit, all the current passes through both R_1 and R_2, whereas in the parallel circuit, the current divides, some passing through R_1 and some through R_2.

Let I be the current in the battery, and let I_1 and I_2 be the currents in R_1 and R_2, respectively. The current entering point a is I, and the currents leaving point a are I_1 and I_2, so from Kirchhoff's first law we have

$$I = I_1 + I_2 \qquad\qquad 16.10$$

Furthermore, the voltage across each resistor is equal to the emf $\mathcal{E}$ of the battery, because the battery is another path between a and b. Thus we have

$$\mathcal{E} = R_1 I_1 \qquad \text{and} \qquad \mathcal{E} = R_2 I_2$$

or

$$I_1 = \frac{\mathcal{E}}{R_1} \qquad \text{and} \qquad I_2 = \frac{\mathcal{E}}{R_2}$$

Substituting these equations into Eq. 16.10 gives

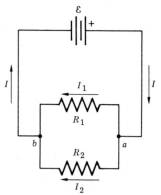

FIGURE 16.11

Two resistors connected in parallel to a battery.

$$I = I_1 + I_2$$

$$= \frac{\mathcal{E}}{R_1} + \frac{\mathcal{E}}{R_2} = \mathcal{E}\left(\frac{1}{R_1} + \frac{1}{R_2}\right)$$

or

$$I = \mathcal{E}/R$$

where R, the equivalent resistance of the two resistors in parallel, is given by

$$\frac{1}{R} = \frac{1}{R} + \frac{1}{R_2}$$

This can be generalized to the following rule for finding the equivalent resistance of any number of resistors connected in parallel.

Resistors in Parallel *The equivalent resistance R of n resistors $R_1, R_2, \ldots, R_n$ connected in parallel is given by*

$$\frac{1}{R} = \sum_{i=1}^{n} \frac{1}{R_i} \qquad\qquad 16.11$$

That is, the reciprocal of R is equal to the sum of the reciprocals of the individual resistances.

Example 16.7 A battery with $\mathcal{E} = 12\,\text{V}$ is connected in parallel with two resistors $R_1 = 2\,\Omega$ and $R_2 = 4\,\Omega$ (Fig. 16.11). (a) Find the current I in the battery. (b) Find the current in each resistor. (c) Find the power dissipated in each resistor.

   (a) From Eq. 16.11 the equivalent resistance R of the circuit is given by

$$\frac{1}{R} = \frac{1}{2\,\Omega} + \frac{1}{4\,\Omega} = \frac{3}{4\,\Omega} \qquad \text{or} \qquad R = \tfrac{4}{3}\,\Omega$$

so the current in the battery is

$$I = \frac{\mathcal{E}}{R} = \frac{12\,\text{V}}{\tfrac{4}{3}\,\Omega} = 9\,\text{A}$$

   The value of $\tfrac{4}{3}\,\Omega$ for the equivalent resistance of two resistors in parallel is to be compared with the value of $6\,\Omega$ found in Example 16.6 for the sum of these same resistors in series. The sum of resistors in series is always greater than the largest individual resistance, whereas the equivalent resistance of resistors in parallel is always less than the smallest individual resistance.

   (b) The currents I_1 and I_2 in resistors R_1 and R_2 are

$$I_1 = \frac{\mathcal{E}}{R_1} = \frac{12\,\text{V}}{2\,\Omega} = 6\,\text{A}$$

and

$$I_2 = \frac{\mathcal{E}}{R_2} = \frac{12\,\text{V}}{4\,\Omega} = 3\,\text{A}$$

The sum of these currents is equal to the current $I = 9\,\text{A}$ in the battery.

(c) The power P_1 dissipated in R_1 is

$$P_1 = R_1 I_1{}^2 = (2\,\Omega)(6\,A)^2 = 72\,W$$

and the power P_2 dissipated in R_2 is

$$P_2 = R_2 I_2{}^2 = (4\,\Omega)(3\,A)^2 = 36\,W$$

As in the case of the series circuit in Example 16.6, the total power P dissipated in both resistors

$$P = P_1 + P_2 = 108\,W$$

is equal to the power

$$P = \mathcal{E}I = (12\,V)(9\,A) = 108\,W$$

delivered by the battery. ☐

COMPLEX CIRCUITS

Kirchhoff's laws can be used to analyze any circuit, no matter how complex. Some complex circuits, like the one shown in Fig. 16.9, are composed entirely of series and parallel connections in various combinations. These circuits can be analyzed by repeated use of Eqs. 16.9 and 16.11 for resistors in series and parallel. Other circuits, like the one in Fig. 16.12, have connections that are neither series nor parallel. Analysis of these circuits requires the use of Kirchhoff's laws to write a set of simultaneous equations for the currents in the different branches of the circuit. We shall not consider such circuits further in this book, rather we shall restrict ourselves to circuits composed only of series and parallel connections.

Example 16.8 Find the currents $I, I_1, I_2, I_3,$ and I_4 in the circuit in Fig. 16.9 when the resistances are $R_1 = 15\,\Omega, R_2 = 8\,\Omega, R_3 = 3\,\Omega,$ and $R_4 = 6\,\Omega$ and the emf of the battery is 3 V.

To analyze a circuit composed of series and parallel connections, one starts with the innermost connection. In this case, the innermost connection is the parallel connection of R_3 and R_4. From Eq. 16.11, the equivalent resistance R' of R_3 and R_4 is

$$\frac{1}{R'} = \frac{1}{3\,\Omega} + \frac{1}{6\,\Omega} \quad \text{or} \quad R' = 2\,\Omega$$

The resistance R' is in series with R_2, so the effective resistance of R' and R_2 together is

$$R'' = R_2 + R' = 8\,\Omega + 2\,\Omega = 10\,\Omega$$

The circuit is redrawn in Fig. 16.13, with $R_2, R_3,$ and R_4 replaced by R''. This shows that R_1 and R'' are in parallel, and so the equivalent resistance R of the entire circuit is

$$\frac{1}{R} = \frac{1}{R_1} + \frac{1}{R''} = \frac{1}{15\,\Omega} + \frac{1}{10\,\Omega} \quad \text{or} \quad R = 6\,\Omega$$

Since the emf of the battery is 3.0 V, the current I is

$$I = \frac{\mathcal{E}}{R} = 0.5\,A$$

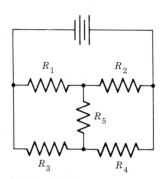

FIGURE 16.12

A complex circuit that is not a combination of series and parallel connections.

FIGURE 16.13

The circuit in Fig. 16.9 redrawn with $R_2, R_3,$ and R_4 replaced by the equivalent resistance R''.

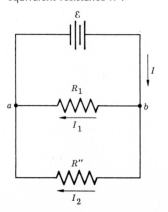

The potential across R_1 and R'' is also 3.0 V, so the currents I_1 and I_2 are

$$I_1 = \frac{3.0\,V}{R_1} = \frac{3.0\,V}{15\,\Omega} = 0.2\,A$$

and

$$I_2 = \frac{3.0\,V}{R''} = \frac{3.0\,V}{10\,\Omega} = 0.3\,A$$

Note that $I_1 + I_2 = I$, in accordance with Kirchhoff's first law.
   The potential difference between points a and b can be written (see Fig. 16.9)

$$V_a - V_b = 3.0\,V = I_2R_2 + I_3R_3 = (0.3\,A)(8\,\Omega) + (I_3)(3\,\Omega)$$

Solving this for I_3, we get

$$(I_3)(3\,\Omega) = 3.0\,V - 2.4\,V = 0.6\,V$$

and so

$$I_3 = \frac{0.6\,V}{3\,\Omega} = 0.2\,A$$

Finally, from the condition that $I_2 = I_3 + I_4$, the current in R_4 is found to be

$$I_4 = I_2 - I_3 = 0.3\,A - 0.2\,A = 0.1\,A$$

This shows how the currents in all the resistors of a circuit can be found by a systematic use of Eqs. 16.9 and 16.11. ☐

16.3   Resistivity

Figure 16.14 shows a cylinder of some material connected to a seat of emf. Experiments show that the resistance R of this cylinder is related to its length L and cross-sectional area A by the equation

$$R = \rho\frac{L}{A} \qquad\qquad 16.12$$

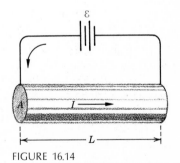

FIGURE 16.14

A conducting cylinder of length L and cross-sectional area A connected to a battery.

where ρ is a characteristic property of the material, called the *resistivity*. Equation 16.12 is very reasonable, since it states that for a given substance the resistance increases with length L and decreases with cross-sectional area A. That is, a short, fat cylinder has less resistance than a long, thin one.

> **REMARK** The *conductance* C of a cylinder is defined as $1/R$, and the conductivity σ is defined as $1/\rho$. In terms of conductance and conductivity, Eq. 16.12 becomes
>
> $$C = \sigma\frac{A}{L}$$
>
> This has the same form as Eq. 9.16 for the thermal conductance of a sheet of material. Thermal conductivity and electric conductivity are closely related. In general, a good electric conductor is also a good thermal conductor.

> **UNITS** In SI units, where length is in meters (m) and area is in square meters (m²), the unit of resistivity is the *ohm-meter* ($\Omega\cdot$m). In English engineering units, length is in feet (ft) and cross-sectional area of a wire is in circular mils

(CM), so the unit of resistivity is the *ohm-circular mil per foot* ($\Omega \cdot$ CM/ft). (A *circular mil* is the area of a circle 0.001 in in diameter.)

Table 16.1 gives the resistivity of various substances. The values range from $1.7 \times 10^{-8}\ \Omega \cdot$ m for copper to over $10^{17}\ \Omega \cdot$ m for fused quartz. Perhaps no other physical quantity has such a wide range of magnitudes. Substances with resistivities less than $10^{-4}\ \Omega \cdot$ m are considered conductors; substances with resistivities greater than $10^{6}\ \Omega \cdot$ m are considered insulators, and substances with resistivities in between (10^{-4} to $10^{6}\ \Omega \cdot$ m) are considered semiconductors. Thus the distinction between conductors and insulators is quantitative. The best conductors have some resistance,* and the best insulators conduct some current. Nevertheless, the large ratio of the resistivity of an insulator to that of a conductor (10^{18} to 10^{25}) makes the distinction between insulators and conductors very important in practice.

Example 16.9 Gauge No. 8 copper wire has a diameter of 0.326 cm. (a) What is the resistance per meter of this wire? (b) If the rate of energy loss in the wire is to be less than 3 W/m, what is the maximum current the wire can carry? (c) Find the diameter of the copper wire that can carry a current of 100 A with a rate of energy loss of 3 W/m.

(a) The cross-sectional area A of No. 8 wire is

$$A = \pi(\tfrac{1}{2}d)^2 = \frac{\pi}{4}(0.326 \times 10^{-2}\ \text{m})^2$$

$$= 8.35 \times 10^{-6}\ \text{m}^2$$

* At temperatures below 10 K ($-263°$C) some metals and alloys become superconducting, which means their resistivity is zero as far as measurements can determine ($\rho < 10^{-25}\ \Omega \cdot$m). A current once established in a superconducting circuit will persist for months or years without a battery or other seat of emf. In spite of the difficulties involved in maintaining the required low temperatures, superconducting wires are being employed in some magnets that use very large currents (Sec. 21.2).

TABLE **16.1**

Resistivity of various substances

Substance	Temperature, °C	Resistivity, $\Omega \cdot$ m
Conductors:		
Copper, annealed	20	1.72×10^{-8}
Aluminum, commercial	20	2.80×10^{-8}
Tungsten	20	5.51×10^{-8}
	1227	4.14×10^{-7}
	2727	9.89×10^{-7}
	3227	1.18×10^{-6}
Graphite	0	1.38×10^{-5}
Semiconductors (pure):		
Germanium	27	4.3×10^{-1}
Silicon	27	2.6×10^{3}
Insulators:		
Slate	20	10^{6}
Marble	20	10^{7} to 10^{9}
Wood	20	5×10^{9} to 5×10^{11}
Glass	20	9×10^{11}
Amber	20	5×10^{14}
Quartz (fused)	20	7×10^{17}

so from Eq. 16.12 and Table 16.1, the resistance R of a 1-m-long piece of this wire is

$$R = \rho\frac{L}{A} = (1.72 \times 10^{-8}\,\Omega \cdot m)\frac{1\,m}{8.35 \times 10^{-6}\,m^2}$$

$$= 2.06 \times 10^{-3}\,\Omega$$

(b) If the power P dissipated in a 1-m-long piece of No. 8 wire is to be kept below 3 W, the current I in the wire must satisfy the inequality

$$P = RI^2 < 3\,W$$

or

$$I < \sqrt{\frac{3\,W}{R}} = \sqrt{\frac{3\,W}{2.06 \times 10^{-3}\,\Omega}} = 38\,A$$

Thus the maximum current the wire can carry is 38 A if it is not to exceed its power rating.

(c) To carry a current I of 100 A at a power of less than 3 W, the resistance of a 1-m-long piece of copper wire must satisfy the inequality

$$RI^2 < 3\,W$$

or

$$R < \frac{3\,W}{I^2} = \frac{3\,W}{(100\,A)^2} = 3 \times 10^{-4}\,\Omega$$

The cross-sectional area A of the wire must then satisfy the inequality

$$R = \rho\frac{L}{A} < 3 \times 10^{-4}\,\Omega$$

so

$$A > \frac{\rho L}{3 \times 10^{-4}\,\Omega} = \frac{(1.72 \times 10^{-8}\,\Omega \cdot m)(1\,m)}{3 \times 10^{-4}\,\Omega}$$

or

$$A > 5.73 \times 10^{-5}\,m^2$$

Finally, the diameter d of the wire must satisfy the inequality

$$A = \tfrac{1}{4}\pi d^2 > 5.73 \times 10^{-5}\,m^2$$

so

$$d > \sqrt{\frac{4(5.73 \times 10^{-5}\,m^2)}{\pi}}$$

or

$$d > 8.54 \times 10^{-3}\,m = 0.854\,cm \qquad \square$$

TEMPERATURE VARIATION

The resistivity of a substance varies with temperature. For most conductors, the resistivity increases with temperature, as is illustrated by the resistivity of tungsten in Table 16.1. For most semiconductors, the resistivity decreases with temperature. The reason for this contrary behavior is that in a conductor there are many free conduction electrons, whereas in a semiconductor there are relatively few. Increasing the temperature of a semiconductor shakes loose some bound electrons, greatly increasing the number of available conduction electrons and thus decreasing the resistivity of the material. Increasing the temperature of a metal does not produce more conduction electrons, but it does increase the thermal motion of the ions. The flow of electrons in a metal is impeded by their collisions with these ions, and the frequency of these collisions increases as the thermal motion of the ions increases. Hence the resistivity of a metal increases with temperature. The difference between conductors and semiconductors is discussed further in Chap. 21.

The temperature variation of resistance provides a convenient method for measuring temperature over a wide range. In a *resistance pyrometer* (Fig. 16.15), the temperature probe consists of a coil of platinum wire. Leads from this coil connect to a resistance meter. If R_0 is the resistance of the coil at 0°C, its resistance R_t at the Celsius temperature t is given by the equation

$$R_t = R_0(1 + 3.97 \times 10^{-3}t - 0.585 \times 10^{-6}t^2) \qquad 16.13$$

Thus, if R_0 is known, a measurement of R_t determines the temperature of the probe. Equation 16.13 is accurate to 1°C from 0 to 1000°C and is accurate to 2°C from −183 to 0°C. Electrical measurements of temperature are valuable because the information is in a form that can be immediately processed by appropriate control circuits.

16.4 Capacitors

Definition A *capacitor* is a circuit element composed of two conducting surfaces separated by an air space or a thin sheet of dielectric (insulating) material (Fig. 16.16). Wires attached to the conducting surfaces enable the capacitor to be connected to an electric circuit. In a commercial capacitor, the sheets of conducting and dielectric material may be rolled into a compact cylinder and embedded in a protective jacket. However, for purposes of discussion, we shall consider the simple parallel-plate arrangement shown in Fig. 16.16.

RESISTOR-CAPACITOR (RC) CIRCUIT

Figure 16.17a is a diagram of resistor-capacitor circuit, which is a circuit in which a capacitor, symbolized by ⊣⊢, is connected in series with a resistor R and a battery $\mathcal{E}$. Because of the insulation between the plates of the capacitor, charge cannot flow through it. However, when the switch S is closed, there will be a transient current through the resistor, as electrons flow from one plate of the capacitor to the other (Fig. 16.17b). Consequently, positive charge q will accumulate on one plate, while an equal amount of negative charge $-q$ accumulates on the other, until the voltage $V = V_b - V_c$ across the capacitor is equal to the emf $\mathcal{E}$ of the battery. At the instant the switch S is closed, the transient current I has its maximum value I_m, but I rapidly decreases to

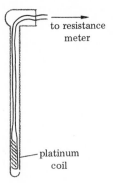

FIGURE 16.15

A resistance pyrometer. The resistance of the plantinum coil in the probe varies in a known way with temperature. Thus the temperature of the probe can be determined electrically by measuring its resistance.

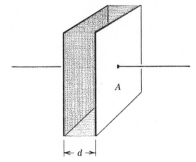

FIGURE 16.16

A capacitor composed of two conducting plates of area A separated by an insulating layer of thickness d.

zero as the capacitor becomes charged. At the same time, the magnitude of the charge q on each plate of the capacitor increases from zero to its maximum value q_m.

Definition The *capacitance* C of a capacitor is the ratio of the magnitude of the charge q on either plate to the voltage V across the plates:

$$C = \frac{q}{V} \qquad \qquad 16.14$$

For a given capacitor, C is a constant independent of V. That is, as a capacitor is being charged, the instantaneous charge q on either plate is CV, where V is the instantaneous voltage.

> **UNITS** The unit of capacitance is coulombs per volt (C/V), which is called a *farad* (F). Because the farad is such a large unit, most commercial capacitors are measured in microfarads (1 μF = 10^{-6} F) and picofarads (1 pF = 10^{-12} F).

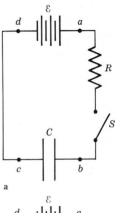

If the capacitor in Fig. 16.17 is initially uncharged, the current I_m in the resistor at the instant ($t = 0$) that the switch S is closed is

$$I_m = \frac{\mathcal{E}}{R} \qquad \qquad 16.15$$

That is, at $t = 0$ charge has not yet accumulated on the capacitor, so the voltage across R is $\mathcal{E}$. At a later time t, the capacitor has a charge q on it, so the voltage across the capacitor is q/C, and the voltage across the resistor is $\mathcal{E} - q/C$. The current in the resistor at this instant, then, is

$$I = \frac{\mathcal{E} - q/C}{R} = \frac{\mathcal{E}}{R} - \frac{q}{RC}$$

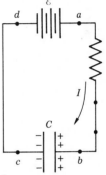

which is less than I_m. As charge accumulates on the capacitor, the current steadily decreases, eventually going to zero. At this point the charge has its maximum value q_m, which we see is given by

$$0 = \frac{\mathcal{E}}{R} - \frac{q_m}{RC}$$

or

$$q_m = C\mathcal{E} \qquad \qquad 16.16$$

FIGURE 16.17

A capacitor connected in series to a battery and a resistor. (a) Initially the switch is opened and the capacitor is uncharged. (b) Immediately after the switch is closed, there will be a transient current in the circuit as charge flows onto the plates of the capacitor. The current ceases when the potential across the capacitor is equal to the emf of the battery.

The current I in the resistor at any time t after the switch is closed is

$$I = I_m e^{-t/RC} \qquad \qquad 16.17$$

where e = 2.718 ... is the base of the natural logarithm. The charge q on the capacitor at time t is

$$q = q_m(1 - e^{-t/RC}) \qquad \qquad 16.18$$

These equations are plotted in Fig. 16.18. Note that since $e^0 = 1$,

$$I = I_m \qquad \text{and} \qquad q = 0 \qquad \text{at} \qquad t = 0$$

and since $e^{-\infty} = 0$,

$I = 0$ and $q = q_m$ at $t = \infty$

Definition The *time constant τ* of an *RC* circuit is

$$\tau = RC \qquad\qquad 16.19$$

From Eqs. 16.17 and 16.18 we see that at $t = \tau$ the current is

$$I = I_m e^{-1} = I_m \frac{1}{2.718} = 0.37\, I_m$$

and the charge is

$$q = q_m(1 - e^{-1}) = q_m(1 - e^{-1}) = 0.63\, q_m$$

The time constant is the typical time it takes for the current or charge to change by a significant amount. A time large compared with τ can be considered infinite for all practical purposes.

> **UNITS** The unit of the time constant is the second (s), which can be shown by writing out the units of R and C:
>
> $[\tau] = [RC] = \Omega \cdot F$
>
> $\qquad = \dfrac{V}{A} \cdot \dfrac{C}{V} = \dfrac{1}{C/s} \cdot \dfrac{C}{1} = s$

Example 16.10 A 4000-Ω resistor and a 50-μF capacitor are connected in series to a 9-V battery. (a) What is the current at the instant $t = 0$ when the circuit is closed? (b) What is the time constant of the circuit? (c) What is the current at time $t = 0.5$ s? (d) What is the maximum charge q_m? (e) What is the charge at time $t = 0.5$ s? (f) At what time does the charge reach 99 percent of its maximum value?

(a) At $t = 0$ the current has its maximum value I_m given by Eq. 16.15:

$$I_m = \frac{\mathcal{E}}{R} = \frac{9\text{ V}}{4000\ \Omega} = 2.25 \times 10^{-3}\,\text{A} = 2.25\text{ mA}$$

(b) From Eq. 16.19 the time constant is

$$\tau = RC = (4000\ \Omega)(50 \times 10^{-6}\text{ F})$$
$$= 0.2\text{ s}$$

Note that the capacitance in microfarads had to be converted to farads in order to get the time constant in seconds.

(c) Substituting the values of I_m and RC in Eq. 16.17 we get

$$I = I_m e^{-t/RC} = (2.25\text{ mA})e^{-0.5/0.2}$$

To evaluate this on a calculator with algebraic logic, press the keys in the following sequence:

$\boxed{.}\ \boxed{5}\ \boxed{\div}\ \boxed{.}\ \boxed{2}\ \boxed{=}$	Display: 2.5
$\boxed{+/-}$	Display: -2.5
$\boxed{e^x}$	Display: .08208 . . .
$\boxed{\times}\ \boxed{2}\ \boxed{.}\ \boxed{2}\ \boxed{5}\ \boxed{=}$	Display: .18469 . . . .

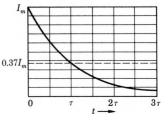

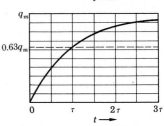

FIGURE 16.18

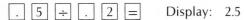

(a) Plot of the transient current I against time t in an *RC* circuit. The unit of time on the horizontal axis is $\tau = RC$, the time constant of the circuit. (b) Plot of the charge q on either plate of a capacitor against time t in an *RC* circuit.

The answer thus is

$I = 0.185\ \text{mA}$

(d) From Eq. 16.16 the maximum charge is

$q_m = C\mathcal{E} = (50\ \mu\text{F})(9\ \text{V})$
$\qquad = 450\ \mu\text{C}$

Note that if the capacitance is left in microfarads, the charge will be in microcoulombs.

(e) Substituting the values of q_m and RC in Eq. 16.18 we get

$q = q_m(1 - e^{-t/RC}) = (450\ \mu\text{C})(1 - e^{-0.5/0.2})$

To evaluate this on a calculator with reverse Polish logic, press the keys in the following sequence:

$\boxed{.}\ \boxed{5}\ \boxed{\text{ENTER}}\ \boxed{.}\ \boxed{2}\ \boxed{\div}$	Display: 2.5
$\boxed{+/-}$	Display: -2.5
$\boxed{e^x}$	Display: .08208 . . .
$\boxed{+/-}\ \boxed{1}\ \boxed{+}$	Display: .9179 . . .
$\boxed{4}\ \boxed{5}\ \boxed{0}\ \boxed{\text{X}}$	Display: 413.06 . . .

The answer thus is

$q = 413\ \mu\text{C}$

(f) The time t at which q is $(0.99)q_m$ is given by the equation

$$(1 - e^{-t/RC}) = \frac{q}{q_m} = 0.99$$

or

$e^{-t/RC} = 1 - 0.99 = 0.01$

The exponent x in e^x is, by definition, the natural logarithm of e^x. Therefore, to find the exponent we have only to take the natural logarithm:

$$\ln e^{-t/RC} = -\frac{t}{RC} = \ln(0.01)$$

or

$t = -(RC)\ln(0.01)$
$\quad = -(0.2\ \text{s})\ln(0.01)$

To evaluate this on a calculator with algebraic logic, press the keys in the following sequence:

$\boxed{.}\ \boxed{0}\ \boxed{1}\ \boxed{\ln x}\ \boxed{\text{X}}\ \boxed{.}\ \boxed{2}\ \boxed{=}\ \boxed{+/-}$
Display: .9210 . . .

Thus the charge reaches 99 percent of its maximum value in 0.92 s, or 4.6 time constants. ☐

PARALLEL-PLATE CAPACITOR

When the plates of a capacitor are charged, an electric field **E** is established between them. In the case of two parallel plates separated by a small distance d (Fig. 16.19), the field is nearly constant. If the space between the plates is empty, the magnitude of this field is

$$E = \frac{q}{A\epsilon_o} \qquad 16.20$$

where q is the charge on each plate, A is the area of each plate, and ϵ_o [8.85×10^{-12} C^2/(N $\cdot$ m^2)] is the permittivity of space (Sec. 15.2).

The electric field exerts the force

$$\mathbf{F} = q'\mathbf{E}$$

on a test charge q' placed between the plates (Fig. 16.19) and does the work

$$W = Fd = q'Ed$$

in moving the charge from the positive to the negative plate. In the process the charge loses the potential energy

$$U = q'V = W$$

where V is the potential difference between the plates. Thus the potential between the plates is

$$V = \frac{W}{q'} = Ed = \frac{qd}{A\epsilon_o}$$

and so the capacitance (Eq. 16.14) of a parallel-plate capacitor is

$$C = \frac{q}{V} = \frac{A\epsilon_o}{d} \qquad 16.21$$

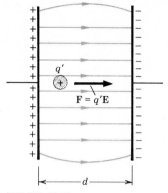

FIGURE 16.19

The lines of force of the electric field between the plates of a parallel-plate capacitor. Except near the edges of the plate, the lines of force are parallel to each other and evenly spaced, indicating that the electric field is constant in direction and magnitude.

Example 16.11 The area of each plate of a parallel-plate capacitor is 25 cm^2, and the distance between the plates is 1 mm. Find the capacitance of this capacitor, assuming the space between the plates is empty (or air filled).

From Eq. 16.21, the capacitance is

$$C = \frac{A\epsilon_o}{d} = \frac{(25 \times 10^{-4} \text{ m}^2)[8.85 \times 10^{-12} \text{ C}^2/(\text{N} \cdot \text{m}^2)]}{10^{-3} \text{ m}}$$

$$= 2.21 \times 10^{-11} \text{ F} = 22.1 \text{ pF}$$

Let us check the units in this case. Since a joule (J) is a newton-meter (N $\cdot$ m), we find that the units in the last expression can be written

$$\frac{(\text{m}^2)[\text{C}^2/(\text{N} \cdot \text{m}^2)]}{\text{m}} = \frac{\text{C}^2}{\text{N} \cdot \text{m}} = \frac{\text{C}^2}{\text{J}} = \frac{\text{C}}{\text{J}/\text{C}} = \frac{\text{C}}{\text{V}} = \text{F}$$

The units of ϵ_o are often given simply as farads per meter (F/m), since

$$1\frac{\text{C}^2}{\text{N} \cdot \text{m}} = 1 \text{ F} \qquad \square$$

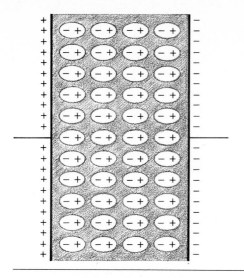

FIGURE 16.20

When an insulating material is placed between the plates of a parallel-plate capacitor, the electric field causes charge separation (polarization) on the molecules of the material. These induced charges partially cancel the true charges on the plates, thus reducing the electric field between the plates.

The capacitance of a capacitor is greatly increased by filling the space between the plates with a dielectric material. The term *dielectric* means the same as *insulator*. It is used to emphasize the fact that in an electric field the molecules of the material develop a positive and negative side (Sec. 15.2, Fig. 15.6b). This process is called *polarization*. When a dielectric material is placed between the plates of a capacitor, polarization causes a negative layer of induced charge to form along the positive plate and a positive layer of induced charge to form along the negative plate (Fig. 16.20). These induced charges partially cancel the effect of the true charge q on the plates, reducing the magnitude of the electric field between the plates.

Definition The *dielectric constant** κ of a dielectric is the ratio of the electric field E_0 in a capacitor when the space between the plates is empty to the electric field E_κ in the capacitor when the space between the plates is filled with the dielectric:

$$\kappa = \frac{E_0}{E_\kappa} \qquad\qquad 16.22$$

Equation 16.20 gives the field E_0 between the plates of a parallel-plate capacitor when the space is empty, so the field E_κ between the plates when the space is filled with a material of dielectric constant κ is

$$E_\kappa = \frac{E_0}{\kappa} = \frac{q}{A\epsilon_o\kappa} \qquad\qquad 16.23$$

The capacitance of the capacitor when it is filled with the dielectric is

$$C = \frac{A\kappa\epsilon_o}{d} \qquad\qquad 16.24$$

*κ is the Greek (lowercase) letter *kappa*.

TABLE **16.2**

Substance	Temperature, °C	Dielectric constant	Dielectric constants of various substances
Air	20	1.00059	
Paraffin	20	2.0 to 2.5	
Epoxy resin	25	3.6 to 3.7	
Quartz		4.3	
Neoprene	24	6.6	
Cupric oxide		18.1	
Water	0	87.9	
	20	80.2	
	50	69.9	
	100	55.6	

The dielectric constant is a unitless number greater than 1. Table 16.2 gives the dielectric constants of various substances.

CAPACITORS IN SERIES AND PARALLEL

Definition Two or more capacitors are said to be in *series* when they are connected so that the charge is the same on each of them.

Figure 16.21 shows n capacitors $C_1, C_2, \ldots, C_n$ in series. When charges q and $-q$ are placed on the outside plates of this array of capacitors, corresponding charges are induced on all the other plates.

The voltage V_i across capacitor C_i in Fig. 16.21 is

$$V_i = \frac{q}{C_i}$$

so the total voltage V across the entire array is

$$V = \sum_{i=1}^{n} V_i = \sum_{i=1}^{n} \frac{q}{C_i} = q \sum_{i=1}^{n} \frac{1}{C_i} = \frac{q}{C}$$

where the equivalent capacitance C of the array is given by

$$\frac{1}{C} = \sum_{i=1}^{n} \frac{1}{C_i} \quad \text{series} \qquad\qquad 16.25$$

Thus capacitors in series combine like resistors in parallel. The equivalent capacitance of a series of capacitors is less than the smallest individual capacitance. This is so because the amount of charge stored is not increased by combining capacitors in series, but the total voltage across the array is.

FIGURE 16.21

Series connection of n capacitors $C_1, C_2, \ldots, C_n$.

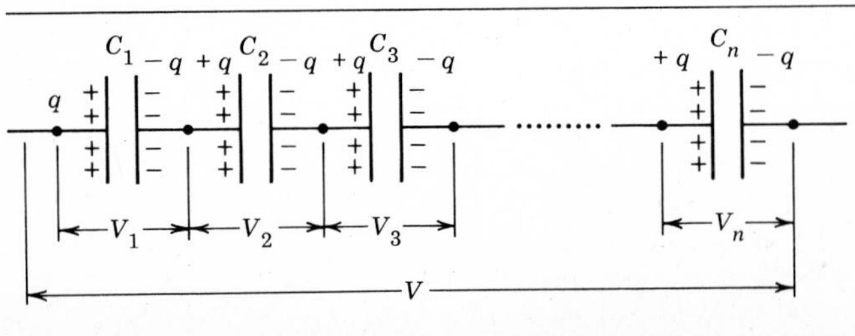

Definition Two or more capacitors are said to be in *parallel* when they are connected so that the voltage is the same across each of them.

Figure 16.22 shows n capacitors $C_1, C_2, \ldots, C_n$ in parallel. All the positive plates connect to point a, and all the negative plates connect to point b, so the voltage $V = V_a - V_b$ is the same across each capacitor.

The charge q_i on capacitor C_i in Fig. 16.22 is

$$q_i = C_i V$$

So the total charge q on the array is

$$q = \sum_{i=1}^{n} C_i V = V \sum_{i=1}^{n} C_i = CV$$

where the total capacitance C of the array is given by

$$C = \sum_{i=1}^{n} C_i \qquad \text{parallel} \qquad \qquad 16.26$$

Thus capacitors in parallel combine like resistors in series. The total capacitance of a parallel array of capacitors is greater than the largest individual capacitance. This is so because the amount of charge stored is increased by combining capacitors in series, but the voltage across the array is not.

Example 16.12 Figure 16.23 shows three capacitors connected to a resistor and a battery. The values of these elements are

$$C_1 = 20\ \mu F \qquad C_2 = 40\ \mu F \qquad C_3 = 100\ \mu F$$
$$R = 2500\ \Omega \qquad \mathcal{E} = 24\ V$$

(a) What is the equivalent capacitance of the three capacitors? (b) What is the time constant of the circuit? (c) What are the charges q_1, q_2, and q_3 on the capacitors 1 s after switch S is closed?

(a) Capacitors C_2 and C_3 are in parallel, so from Eq. 16.26 their total capacitance C' is

$$C' = C_2 + C_3 = 140\ \mu F$$

The capacitor C_1 is in series with C', so from Eq. 16.25 the equivalent capacitance C of the entire array is

$$\frac{1}{C} = \frac{1}{C_1} + \frac{1}{C'} = \frac{1}{20\ \mu F} + \frac{1}{140\ \mu F} = \frac{8}{140\ \mu F}$$

or

$$C = 17.5\ \mu F$$

(b) From Eq. 16.19 the time constant of the circuit is

$$\tau = RC = (2500\ \Omega)(17.5 \times 10^{-6}\ F)$$
$$= 0.0438\ s$$

(c) Since 1 s is more than 20 time constants, the capacitors are fully charged by this time. After 1 s there is no longer a current in R, so the voltage across the equivalent capacitance C

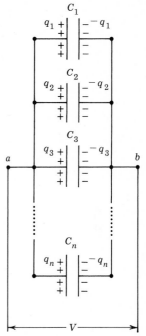

FIGURE 16.22

Parallel connection of n capacitors $C_1, C_2, \ldots, C_n$.

FIGURE 16.23

Three capacitors connected to a resistor and a battery.

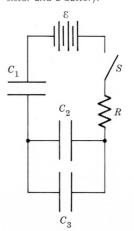

is equal to the emf $\mathcal{E}$ of the battery, and the total charge q on C is given by Eq. 16.16:

$$q = C\mathcal{E} = (17.5\ \mu F)(24\ V) = 420\ \mu C$$

The total charge q on a series array of capacitors is equal to the charge on each capacitor. Since C_1 and C' are in series, $q = 420\ \mu C$ is the charge on both C_1 and C':

$$q = q_1 = q' = 420\ \mu C$$

Thus the voltage across C_1 is

$$V_1 = \frac{q_1}{C_1} = \frac{420\ \mu C}{20\ \mu F} = 21\ V$$

and the voltage across C' is

$$V' = \frac{q'}{C'} = \frac{420\ \mu C}{140\ \mu F} = 3\ V$$

The sum of the voltages across C_1 and C' is equal to the emf of the battery.

Since C_2 and C_3 are in parallel, the voltage is the same across each. Thus the charge q_2 on C_2 is

$$\frac{q_2}{C_2} = 3V \quad \text{or} \quad q_2 = (3\ V)(40\ \mu F) = 120\ \mu C$$

and the charge q_3 on C_3 is

$$\frac{q_3}{C_3} = 3\ V \quad \text{or} \quad q_3 = (3\ V)(100\ \mu F) = 300\ \mu C$$

The sum of the charges on C_2 and C_3 is equal to the charge on the capacitor C_1 that is in series with them. $\square$

16.5 Electrochemistry

ELECTROLYSIS

Pure water consists mostly of neutral H_2O molecules, and as such, it is not electrically conducting. However, if a salt, such as copper sulfate ($CuSO_4$), is dissolved in water, the salt dissociates into positive and negative ions. In the case of copper sulfate, the positive ions are copper atoms that have lost two outer electrons and so carry a net charge of $+2e$; these ions are symbolized Cu^{2+}, where Cu is the chemical symbol for copper and the $2+$ superscript denotes the charge of the ion. The negative ions are sulfate radicals (SO_4^{2-}). A *radical* is a stable group of atoms that can combine with a number of different ions to form different molecules. The sulfate ion, as its symbol indicates, is a combination of one sulfur atom, four oxygen atoms, and two extra electrons.

The copper sulfate solution is conducting because the copper and sulfate ions are free to move in an electric field, just like the electrons in a metal. Any solution with enough ions to be electrically conducting is called an *electrolyte*. The chemical process that accompanies the flow of charge through an electrolyte is called *electrolysis*.

Figure 16.24 shows two electrodes connected by wires to opposite sides of a battery. When these electrodes are dipped into a solution of copper sulfate, a complete electric circuit is formed. In the outer part of the circuit, electrons flow from the positive electrode, called the *anode,* through the battery, over to the negative electrode, called the *cathode.* In the electrolyte, the positive copper ions are attracted to the negative cathode. As electrons flow onto the cathode, copper ions react with them to become copper atoms. This reaction is written

$$Cu^{2+} + 2e^- \longrightarrow Cu$$

where e^- is the symbol for an electron in a chemical reaction. At the same time, negative charge is transferred to the anode, but the reaction involved depends on the nature of the electrode itself. If the electrode is an inert metal like platinum, the reaction is

$$H_2O \longrightarrow O + 2H^+ + 2e^-$$

That is, two electrons are transferred to the anode by dissociating water into an oxygen atom, two hydrogen ions, and two electrons. The oxygen atoms immediately combine into oxygen molecules (O_2) which bubble out of solution. The net effect of the current in the copper sulfate solution is to release oxygen gas at the anode and to deposit copper at the cathode.

Electroplating is a widely used industrial process in which a metal coating is applied by electrolysis. For example, to copper plate a steel part, the part is made the negative electrode (cathode) of an electrolytic cell in which copper sulfate is the electrolyte. The copper ions flowing to the cathode are deposited atom by atom onto the part, resulting in a very thin, even coating of copper. The anode in this case is a plate of copper, which releases electrons by the reaction

$$Cu \longrightarrow Cu^{2+} + 2e^-$$

That is, copper is dissolved at the anode at the same rate at which it is deposited at the cathode, so that the concentration of copper ions in the electrolyte remains constant.

In other electroplating processes, the anode is inert and does not dissolve. Chromium plating, for instance, uses a nonreactive lead anode; the chromium ions are all supplied by the chromic acid in the electrolye, which must be replenished as the process progresses. By using a mixture of electrolytic solutions, alloys (mixtures of metals) can be plated. Printed circuits are made by electroplating with an electrolyte containing a mixture of lead and tin ions.

Electroplating is now used to apply a protective coating to automobile bodies. Figure 16.25 shows a body part as it is being dipped into an electrolyte containing the material to be coated. This process ensures a tightly bonded uniform coating over the entire part.

FARADAY'S LAW

Definition *Faraday's constant F* is the quantity of charge in one mole of protons. Since a mole contains Avogadro's number N_A

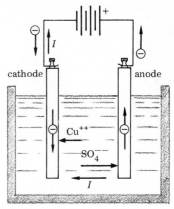

FIGURE 16.24

Two electrodes connected to opposite sides of a battery. The electrodes are immersed in copper sulfate solution, which completes the circuit.

FIGURE 16.25

Electrolytic coating of an automobile body. (*General Motors.*)

of particles, Faraday's constant is

$$F = eN_A = (1.6022 \times 10^{-19}\,\text{C})(6.022 \times 10^{23})$$
$$= 96{,}484\,\text{C} \qquad\qquad 16.27$$

For problems we shall round this off to 96,500 C.

> **UNITS** The *faraday* (Fa) is a unit of charge equal to 96,500 C:
>
> $1\,\text{Fa} = 96{,}500\,\text{C}$
>
> Faraday's constant F is a physical quantity, whereas the faraday (Fa) is a unit. The distinction is probably one that concerns the author more than the reader.

Definition The *valence** ν of an ion is the charge of the ion in units of the proton charge e. It is always a small whole number, either positive or negative. For instance, the valence of a copper ion (Cu^{2+}) is $\nu = 2$, and the valence of a sulfate ion (SO_4^{2-}) is $\nu = -2$.

During electrolysis, each positive ion transfers the charge νe to the cathode, where ν is the valence of the ion. The total charge q transferred in time t is

$$q = It$$

where I is the current in the electrolytic circuit. The number N of atoms deposited on the cathode in time t is thus

$$N = \frac{It}{\nu e}$$

But the number N of atoms is related to the number n of moles by

$$N = nN_A$$

where Avogadro's number N_A is the number of atoms in a mole. Thus the number of moles deposited on the cathode is

*ν is the Greek (lowercase) letter *nu*.

$$n = \frac{N}{N_A} = \frac{It}{\nu e N_A}$$

$$= \frac{It}{\nu F} \qquad\qquad 16.28$$

where F is Faraday's constant (Eq. 16.27).

The mass m of material deposited on the cathode during electrolysis is related to the number n of moles and the atomic mass M of the substance by

$$m = nM \qquad\qquad 16.29$$

since the mass of 1 mol is equal to the atomic mass in grams (Sec. 8.1). When Eqs. 16.28 and 16.29 are combined, we get the following important result:

Faraday's Law *The mass m of material (in grams) deposited during electrolysis is*

$$m = \frac{MIt}{\nu F} \qquad\qquad 16.30$$

where I is the current, t is the duration of the process, M is the atomic mass of the substance (in grams), ν is the valence, and $F = 96,500 \, C$ is Faraday's constant.

Example 16.13 Two platinum electrodes, each weighing 50 g, are connected to opposite poles of a battery and immersed in a silver nitrate ($AgNO_3$) solution. A current of 3.1 A is maintained for 40 min. Afterwards the cathode (negative electrode) is found to weigh 58.3 g. What is the valence of the silver ions?

The cathode is coated with silver during the electrolysis, indicating that the silver ions are positively charged. The mass m of this silver is 8.3 g, the amount by which the mass of the electrode increased. From the table of atomic masses, the atomic mass of silver is $M = 107.9$ g. Thus from Eq. 16.30 the valence ν of a silver ion is

$$\nu = \frac{MIt}{mF} = \frac{(107.9 \text{ g})(3.1 \text{ A})(40 \text{ min})(60 \text{ s/min})}{(8.3 \text{ g})(96,500 \text{ C})}$$

$$= 1.0$$

From this we can conclude that when dissolved in water, silver nitrate dissociates into singly charged silver ions (Ag^+) and singly charged nitrate ions (NO_3^-). The nitrate ion carries the electron lost by the silver ion. ☐

VOLTAIC CELL

Electrochemical seats of emf, or *batteries,* were the sole commercial source of continuous current until central magnetic-induction generators were installed for electric lighting in the 1880s. The first such battery, or cell, was the voltaic cell, invented by Alessandro Volta (1745–1827) in 1800. It consisted of a plate of zinc and a plate of copper immersed in an electrolyte, usually dilute sulfuric acid (H_2SO_4). In such a cell, when the plates are connected by a wire, a current is established in the circuit (Fig. 16.26). With only two plates, the emf of the cell is low (about 0.5 V), but by connecting many cells in series, the emf can be increased accordingly. More important, by connecting the plates of many cells in parallel, a large current can be delivered. With the invention of the voltaic cell, experiments in electrolysis, electromagnetism, and electric discharge became possible for the first time.

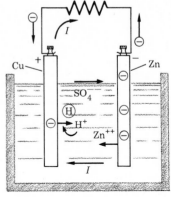

FIGURE 16.26

Voltaic cell. A copper and a zinc electrode immersed in an electrolyte of dilute sulfuric acid.

> **REMARK** Originally the term *battery* meant a grouping of two or more cells, but now it is commonly used to refer to a single cell as well.

When a zinc electrode is immersed in water, some zinc ions (Zn^{2+}) go into solution, leaving two electrons each on the electrode. The reaction is

$$Zn \longrightarrow Zn^{2+} + 2e^-$$

It proceeds until the electrode becomes sufficiently charged to attract back Zn^{2+} ions as rapidly as it loses them. A state of dynamic equilibrium is established in which Zn^{2+} ions go in and out of solution at the same rate.

Similarly when a copper electrode is immersed in water, some copper ions (Cu^{2+}) go into solution. The reaction is

$$Cu \longrightarrow Cu^{2+} + 2e^-$$

It too proceeds until the copper electrode can attract back Cu^{2+} ions as rapidly as it loses them. At equilibrium both the copper and the zinc, or any other metals, are negatively charged.

The important point is that metals differ in the ease with which positive ions go into solution. The more easily this happens, the more negatively charged the electrode becomes. In the case of zinc and copper, the zinc is more negatively charged than the copper. Consequently, if these two metals are immersed in an electrolyte and connected by an external wire, negative electrons will flow to the copper through the wire (Fig. 16.26). The zinc is thus the negative terminal of the battery, and the copper is the positive terminal. However, if zinc and aluminum are used as the terminals of a voltaic cell, the aluminum, being more reactive than the zinc, will be the negative terminal, and zinc will be the positive terminal.*

In the zinc-copper voltaic cell, the energy required to maintain a current is supplied by the zinc reaction. The copper reaction does not take place at all, because the buildup of positive ions in the electrolyte forces positive ions to flow to the copper where they can combine with electrons flowing to the copper through the wire. In the reaction

$$H^+ + e^- \longrightarrow H$$

hydrogen ions from the dissociation of the sulfuric acid form hydrogen atoms at the copper electrode. These atoms immediately combine into hydrogen molecules (H_2), which gather into

*In spite of its light weight, low cost, and high reactivity, aluminum has not found a practical use in commercial batteries because the thin film of aluminum oxide [Al_2O_3] which forms on its surface interferes with the necessary electrolytic reactions.

small bubbles on the electrode. After a time, the electrode is coated with a thin layer of hydrogen bubbles, which prevents further reaction from taking place.

The interference of electrode reactions by the buildup of reaction products on an electrode or the depletion of reactants near an electrode is called *polarization*.* The rapid polarization that takes place in a voltaic cell limits the amount of current that can be drawn at any one time. This limitation led to the eventual replacement of the voltaic cell by more sophisticated devices, but improved versions of the voltaic cell were used as late as 1875 for plating and telegraphy, including use in the operation of the transatlantic cable.

DANIELL CELL

John Daniell developed a two-fluid nonpolarizing battery in 1836. The two fluids are solutions of copper sulfate ($CuSO_4$) and zinc sulfate ($ZnSO_4$), separated by a porous partition. The partition allows some ions to diffuse back and forth between the solutions, but prevents any massive mixing. A copper electrode is immersed in the copper sulfate solution, and a zinc electrode is immersed in the zinc sulfate solution (Fig. 16.27). When the circuit is closed, the zinc electrode loses zinc ions to the zinc sulfate solution, while copper from the copper sulfate solution is deposited on the copper electrode. The reactions are

$$Zn \longrightarrow Zn^{2+} + 2e^- \quad \text{and} \quad Cu^{2+} + 2e^- \longrightarrow Cu$$

This battery is nonpolarizing because the deposition of copper on a copper electrode does not change the electrical or chemical properties of the electrode. A further advantage of this battery is that the reactions at both electrodes contribute to the emf. Consequently, this battery develops a maximum emf of 1.1 V. This simple but effective system was used well into the twentieth century for telegraphy and railway signaling.

ELECTROCHEMICAL POTENTIAL

The Daniell cell can be thought of as two half-cells connected by a porous partition. Each half-cell consists of a single electrode and the electrolyte with which it interacts. Many different half-cells have been constructed. Some, like the copper–copper ion and zinc–zinc ion half-cells, consist of a metal immersed in a solution containing ions of the metal. Others, like the hydrogen-gas half-cell, consists of a chemically inert electrode on which occur reactions that do not involve the electrode itself. In the hydrogen-gas electrode, for instance, gaseous hydrogen is bubbled over a platinum electrode. The reaction

$$H_2 \longrightarrow 2H^+ + 2e^-$$

gives up electrons to the electrode, but otherwise does not involve the electrode in the reaction.

Every chemical reaction can proceed in either direction, depending on the circumstances. When the reaction goes in the

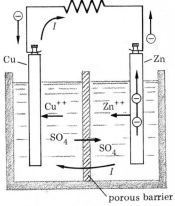

FIGURE 16.27

Daniell cell. The copper electrode is immersed in a copper sulfate solution, and the zinc electrode is immersed in a zinc sulfate solution. The two solutions are separated by a porous barrier.

*This use of the word "polarization" is not to be confused with the polarization of a dielectric discussed in Sec. 16.4.

direction that produces electrons, the process is called *oxidation*. When the reaction goes in the direction that consumes electrons, the process is called *reduction*. Table 16.3 gives examples of some of the oxidation-reduction reactions that occur in half-cells.

A battery is made by coupling two half-cells through a porous partition, as in the Daniell cell. The half-cell with the stronger oxidation reaction is the negative terminal. Oxidation occurs at the negative terminal, while reduction occurs at the positive terminal.

Definition The *electrochemical potential* of a half-cell is the emf of the battery made by coupling the half-cell to a hydrogen-gas half-cell. A negative electrochemical potential means that the oxidation reaction of the half-cell is stronger than the oxidation reaction of the hydrogen-gas half-cell. A positive electrochemical potential means that the oxidation reaction is weaker than the oxidation reaction of the hydrogen-gas half-cell. Table 16.3 lists the electrochemical potentials of some half-cell reactions in order of decreasing oxidation strength.

The emf of the battery made by coupling two half-cells in Table 16.3 is found by subtracting the electrochemical potentials of the half-cells. The half-cell higher in the table will be the negative terminal. For example, the half-cells in the Daniell battery are $Zn|Zn^{2+}$ and $Cu|Cu^{2+}$. Since the $Zn|Zn^{2+}$ half-cell is above the $Cu|Cu^{2+}$ half-cell in Table 16.3, the zinc terminal is negative. Subtracting the electrochemical potentials of these half-cells, we find that the emf of the battery is

$$\mathcal{E} = 0.337\,V - (-0.763\,V) = 1.100\,V$$

The original voltaic cell is something like a $Zn|Zn^{2+}$ half-cell coupled with a hydrogen-gas half-cell. No partition is required between the cells because the hydrogen gas bubbles out of solution. From Table 16.3 the expected emf of the cell is

$$\mathcal{E} = 0 - (-0.763\,V) = 0.763\,V$$

which is more than the 0.5 V actually obtained. The emfs of practical batteries are usually somewhat less than the values calculated from the electrochemical potentials, because of a number of uncontrollable factors that occur in the cell.

Example 16.14 (a) What is the expected emf of a battery made by coupling a $Zn|Zn^{2+}$ half-cell with a $Mg|Mg^{2+}$ half-cell? (b) Which half-cell forms the negative terminal of the battery?

(a) The emf of the battery is found by subtracting the electrochemical potentials of the half-cells:

$$\mathcal{E} = (-0.763\,V) - (-2.363\,V) = 1.600\,V$$

(b) The $Mg|Mg^{2+}$ half-cell, being higher in the table, will be the negative terminal. □

CONTEMPORARY BATTERIES

In the 180 years since the invention of the voltaic cell, many types of batteries have been developed. Often trial and error

TABLE 16.3

Oxidation-reduction reactions and their electrochemical potentials

The left-hand columns describe the electrode and electrolyte of the half-cell in which these reactions occur. The notation $Pt|X_2$ denotes a platinum electrode around which gas X is bubbled.

Description of half-cell		Oxidation reaction	Reduction reaction	Electrochemical potential, V
Electrode	Electrolyte			
Li	Li^+	$Li \longrightarrow Li^+ + e^-$	$Li^+ + e^- \longrightarrow Li$	−3.045
Ca	Ca^{2+}	$Ca \longrightarrow Ca^{2+} + 2e^-$	$Ca^{2+} + 2e^- \longrightarrow Ca$	−2.866
Mg	Mg^{2+}	$Mg \longrightarrow Mg^{2+} + 2e^-$	$Mg^{2+} + 2e^- \longrightarrow Mg$	−2.363
Al	Al^{3+}	$Al \longrightarrow Al^{3+} + 3e^-$	$Al^{3+} + 3e^- \longrightarrow Al$	−1.662
Mn	Mn^{2+}	$Mn \longrightarrow Mn^{2+} + 2e^-$	$Mn^{2+} + 2e^- \longrightarrow Mn$	−1.180
Zn	Zn^{2+}	$Zn \longrightarrow Zn^{2+} + 2e^-$	$Zn^{2+} + 2e^- \longrightarrow Zn$	−0.763
Cd	Cd^{2+}	$Cd \longrightarrow Cd^{2+} + 2e^-$	$Cd^{2+} + 2e^- \longrightarrow Cd$	−0.403
Pb$\|$PbSO$_4$	SO_4^{2-}	$Pb + SO_4^{2-} \longrightarrow PbSO_4 + 2e^-$	$PbSO_4 + 2e^- \longrightarrow Pb + SO_4^{2-}$	−0.359
Pb	Pb^{2+}	$Pb \longrightarrow Pb^{2+} + 2e^-$	$Pb^{2+} + 2e^- \longrightarrow Pb$	−0.126
Pt$\|$H$_2$	H^+	$H_2 \longrightarrow 2H^+ + 2e^-$	$2H^+ + 2e^- \longrightarrow H_2$	0.000
Cu	Cu^{2+}	$Cu \longrightarrow Cu^{2+} + 2e^-$	$Cu^{2+} + 2e^- \longrightarrow Cu$	0.337
Ag	Ag^+	$Ag \longrightarrow Ag^+ + e^-$	$Ag^+ + e^- \longrightarrow Ag$	0.799
Pt$\|$Cl$_2$	Cl^-	$2Cl^- \longrightarrow Cl_2 + 2e^-$	$Cl_2 + 2e^- \longrightarrow 2Cl^-$	1.359
Pt	Mn^{2+}, MnO_4^-	$Mn^{2+} + 4H_2O \longrightarrow MnO_4^- + 8H^+ + 5e^-$	$MnO_4^- + 8H^+ + 5e^- \longrightarrow Mn^{2+} + 4H_2O$	1.510
PbO$_2$	SO_4^{2-}	$PbSO_4 + 2H_2O \longrightarrow PbO_2 + SO_4^{2-} + 4H^+ + 2e^-$	$PbO_2 + SO_4^{2-} + 4H^+ + 2e^- \longrightarrow PbSO_4 + 2H_2O$	1.682
Pt$\|$F$_2$	F^-	$2F^- \longrightarrow F_2 + 2e^-$	$F_2 + 2e^- \longrightarrow 2F^-$	2.870

rather than scientific principle has been used to discover an improved system. Brief descriptions of a few commonly used batteries are given here.

Carbon Battery   The modern carbon dry cell, or common flashlight battery, is based on a wet cell developed by Georges Leclanché in the 1860s. The negative electrode is an amalgamated zinc* cylinder that also serves as the battery casing. The positive electrode is a carbon rod located along the axis of the cylinder. The electrolyte is a conductive paste consisting of zinc chloride ($ZnCl_2$), ammonium chloride (NH_4Cl), and manganese dioxide (MnO_2). The Leclanché carbon battery is similar to a voltaic cell in that the negative terminal is a $Zn \mid Zn^{2+}$ half-cell and hydrogen reduction takes place at the positive terminal. However, the manganese dioxide reacts with the H^+ ions in the reaction

$$2MnO_2 + 2H^+ \longrightarrow Mn_2O_3 + H_2O$$

which largely prevents the positive electrode from being polarized by the evolution of gaseous hydrogen. Furthermore, this reaction has an electrochemical potential of about $+0.8\,V$, so the emf of the battery is between 1.5 and 1.6 V.

> **REMARK**   If a carbon battery is made to deliver a substantial current for a long time, some polarization will occur. This can be observed in the gradual dimming of a flashlight. If the battery is not used for a while, depolarization will occur spontaneously, and the battery will again deliver full power.

Alkaline Battery   The small, flat battery used in watches and other miniaturized systems is based on a zinc–potassium hydroxide (KOH) system invented by Samuel Rubin. Figure 16.28 shows the structure of a flat cell. The negative electrode is a pellet of compressed, powdered, amalgamated zinc, and the electrolyte is a solution of potassium hydroxide and zinc oxide.

*Amalgamated zinc, an alloy of zinc with a small amount of mercury, is used because it resists corrosion much better than pure zinc.

FIGURE 16.28

Alkaline battery. The electrodes are amalgamated zinc and a pressed mixture of mercuric oxide and graphite.

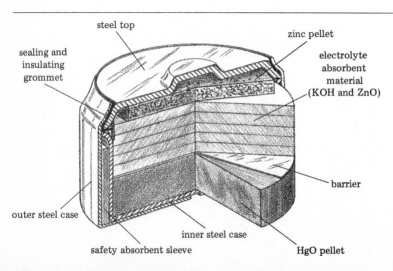

steel top

zinc pellet

sealing and
insulating
grommet

electrolyte
absorbent
material
(KOH and ZnO)

barrier

outer steel case

inner steel case

safety absorbent sleeve

HgO pellet

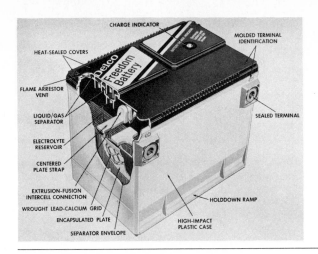

FIGURE 16.29

Lead storage battery. One of the six cells is exposed to view. Each cell contains a number of positive and negative grids connected in parallel (to increase the current capacity of the cell), and the six cells are connected in series (to increase the voltage). (*Delco Remy Division of General Motors.*)

The zinc is oxidized in the reaction

$$Zn + 2H_2O \longrightarrow ZnO_2^{2-} + 4H^+ + 2e^-$$

which has an electrochemical potential of -1.33 V. The positive electrode is a pressed mixture of mercuric oxide (HgO) and graphite. This serves to depolarize the cell by the reaction

$$HgO + 2H^+ + 2e^- \longrightarrow Hg + H_2O$$

This reaction has an electrochemical potential of 0.01 V, so the emf of the cell is about 1.34 V. The two half-cells are separated by a layer of microporous plastic.

Storage Battery The lead storage battery used in automobiles is capable of being "restored" to full power by reversing the current in it. When "charged," the negative electrode is highly porous lead (sponge lead) pressed into a lead grid, the electrolyte is sulfuric acid, and the oxidation reaction is

$$Pb + SO_4^{2-} \longrightarrow PbSO_4 + 2e^-$$

As it proceeds, lead sulfate, which is insoluble, builds up on the grid, and sulfate ions come out of solution. The positive terminal is highly porous lead dioxide (PbO_2) pressed into a lead grid. The electrolyte is also sulfuric acid, and the reduction reaction

$$PbO_2 + SO_4^{2-} + 4H + 2e^- \longrightarrow PbSO_4 + 2H_2O$$

also produces lead sulfate and removes sulfate ions from the electrolyte.

From Table 16.3 the emf of this cell is

$$\mathcal{E} = 1.682\,V - (-0.359\,V) = 2.04\,V$$

A 12-V car battery (Fig. 16.29) contains six such cells connected in series. (Each cell contains a number of positive and negative grids separated by a porous barrier.) As the battery discharges, the density of the sulfuric acid drops, because of the loss of sulfate ions. Thus the state of the battery can be tested with a hydrometer that measures the specific gravity of the electrolyte.

When current is passed in the opposite direction through a

storage battery, the reactions reverse. At the negative terminal lead sulfate is reduced to lead and sulfate ions, and at the positive terminal lead sulfate is oxidized to lead dioxide and sulfate ions. This restores the cell to its charged state.

GUIDE TO MAJOR TOPICS

Topic	References	Problems
Current in a conductor	Sec. 16.1	
Current	Eq. 16.1	1 to 4
Seat of emf	Eq. 16.2; Example 16.1	5 to 8
Resistance	Eq. 16.3; Example 16.2	9 and 10
Circuits	Eqs. 16.4 to 16.8; Examples 16.3 to 16.5	11 to 16
Circuit analysis	Sec. 16.2	
Series circuits	Eq. 16.9; Example 16.6	17 to 22, 28
Parallel circuits	Eq. 16.11; Example 16.7	23 to 28
Complex circuits	Example 16.8	29 to 36
Resistivity	Sec. 16.3; Eq. 16.12; Example 16.9; Table 16.1	37 to 40
Capacitance	Sec. 16.4	
RC circuits	Eqs. 16.14 to 16.19; Example 16.10	41 to 44
Parallel-plate capacitor	Eq. 16.24; Example 16.11; Table 16.2	45 and 46
Capacitors in series and parallel	Eqs. 16.25 and 16.26; Example 16.12	47 to 50
Electrochemistry	Sec. 16.5	
Faraday's law	Eqs. 16.27 to 16.30; Example 16.13	51 to 54
Electrochemical potential	Example 16.14; Table 16.3	55 and 56

PROBLEMS

1 A charge of 75 C flows through a wire in 120 s. (a) What is the current in the wire during this time? (b) How many electrons pass through the wire during this time?
Ans. (a) 0.625 A; (b) 4.69×10^{20}

2 (a) In 30 s, how much charge passes through a wire in which there is a current of 4.5 A? (b) How many electrons pass through the wire in this time?

3 The faraday (Fa) is a unit of charge equal to the charge of 1 mol of protons. (a) What is the value of the faraday in coulombs? (b) How long would it take for 1 Fa to pass through a wire in which the current is 15 A?
Ans. (a) 9.65×10^4 C; (b) 107 min

4 A battery can deliver a total of 0.40 Fa (see Prob. 3). How long can the battery maintain a current of 0.7 A in a circuit?

5 A current of 0.3 A is maintained by a battery with an emf of 1.5 V. (a) What is the power output of the battery? (b) How much energy does the battery expend in 5 min?
Ans. (a) 0.45 W; (b) 135 J

6 A battery with an emf of 3.0 V dissipates energy at the rate of 0.80 W in an external circuit. (a) What is the current in the circuit?

(b) If the total electric energy that the battery can deliver is 3.5×10^4 J, how long can the battery maintain this current?

7 A fully "charged" automotive storage battery with an emf of 12 V can deliver a total of 2.25×10^6 J. (a) How long can this battery maintain a current of 25 A? (b) How much charge, in faradays, is transferred in this time? (See Prob. 3.)
Ans. (a) 125 min; (b) 1.94 Fa

8 A battery supplies energy at the rate of 200 W when the current is 4 A. What is the rate at which energy is supplied when the current is 10 A?

9 What is the resistance of a light bulb when the current is 0.5 A and the voltage is 6 V?
Ans. 12 Ω

10 A conductor has a resistance of 25 Ω when the current in it is 250 mA (1 mA = 10^{-3} A). What is the voltage across the conductor?

11 Suppose the battery in Fig. 16.8 has an emf of 5.0 V and the resistance is 10 Ω. Find (a) the current in the circuit and (b) the power dissipation in the resistor.
Ans. (a) 0.5 A; (b) 2.5 W

12 Suppose the current in Fig. 16.8 is 0.3 A and the resistance is 20 Ω. Find (a) the emf of the battery and (b) the power dissipation in the resistor.

13 Suppose the emf of the battery in Fig. 16.8 is 4.5 V and the current in the circuit is 0.2 A. Find (a) the resistance and (b) the power dissipation in the resistor.
Ans. (a) 22.5 Ω; (b) 0.9 W

14 Suppose the current in Fig. 16.8 is 0.4 A and the power dissipation in the resistor is 1.2 W. Find (a) the emf of the battery and (b) the resistance of the circuit.

15 Suppose the emf of the battery in Fig. 16.8 is 3.0 V and the power dissipation in the resistor is 2.0 W. Find (a) the current in the circuit and (b) the resistance.
Ans. (a) 0.67 A; (b) 4.5 Ω

16 Suppose the resistance in Fig. 16.8 is 27 Ω and the power dissipation in it is 3.0 W. Find (a) the emf of the battery and (b) the current in the circuit.

17 Suppose the emf of the battery in Fig. 16.10 is 7.5 V and the resistors are $R_1 = 8 Ω$ and $R_2 = 12 Ω$. Find (a) the current in the circuit and (b) the power dissipation in each resistor.
Ans. (a) 0.375 A; (b) 1.125 and 1.69 W

18 Suppose the emf of the battery in Fig. 16.10 is 9.0 V and the resistor R_1 is 100 Ω. What must the resistor R_2 be for the current to be 20 mA?

19 A battery usually has a small internal resistance of its own. This is indicated by the resistor r in Fig. 16.30. If the emf of the battery is 3.0 V, $r = 0.5 Ω$, and $R = 5 Ω$, what is the potential difference between the terminals a and b of the battery?
Ans. 2.73 V

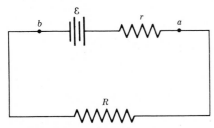

FIGURE 16.30

Problems 19 and 20.

   REMARK Because of the internal resistance of a battery, the potential across its terminals can be less than the emf of the battery.

20 The circuit in Fig. 16.30 has a current of 0.5 A when R is 10 Ω and a current of 0.27 A

when R is 20 Ω. Find (a) the internal resistance r and (b) the emf $ε$ of the battery.

21 Figure 16.31 shows a potential difference of 120 V placed across a circuit that has a lamp with resistance $R_1 = 144 Ω$ connected in series to a variable resistor R_2. The brightness of the lamp is controlled by changing the magnitude of R_2. Find the power dissipations in the lamp (a) when R_2 is zero and (b) when $R_2 = 144 Ω$. (c) What must R_2 be for the power dissipation in the lamp to be 50 W?
Ans. (a) 100 W; (b) 25 W; (c) 59.7 Ω

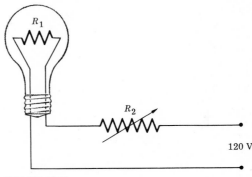

FIGURE 16.31

Problem 21.

22 A three-way light bulb has two filaments which are connected to three wires, as shown in Fig. 16.32. By turning the socket switch, 120 V is put across either ab, bc, or ac. (a) If $R_1 = 144 Ω$ and $R_2 = 216 Ω$, what are the three possible power dissipations of the light bulb? (b) A different three-way light bulb can operate at 300, 100, and 75 W. What are the resistances of its two filaments?

FIGURE 16.32

Problem 22.

23 Suppose the emf of the battery in Fig. 16.11 is 7.5 V and the resistors are $R_1 = 8 Ω$ and

$R_2 = 12\,\Omega$. What are the currents I, I_1, I_2?
Ans. 1.56, 0.94, and 0.62 A

24 Suppose the emf of the battery in Fig. 16.11 is 9.0 V and the resistor R_1 is 100 Ω. What must the resistor R_2 be in order for the current I to be 150 mA?

25 Find the equivalent resistance of three resistors $R_1 = 5\,\Omega$, $R_2 = 10\,\Omega$, and $R_3 = 20\,\Omega$ connected in parallel.
Ans. 2.86

26 Find the equivalent resistance of three resistors $R_1 = 2\,\Omega$, $R_2 = 3\,\Omega$, and $R_3 = 100\,\Omega$ connected in parallel.

27 Six Christmas tree lights are arranged in a parallel circuit, as shown in Fig. 16.33. Each bulb dissipates 10 W when operated at 120 V. (a) What is the resistance R of each bulb? (b) What is the resistance of the entire array of bulbs? (c) What is the total power consumption of the array? (d) What are the currents at points a, b, c, and d?
Ans. (a) 1440 Ω; (b) 240 Ω; (c) 60 W; (d) 0.50, 0.083, 0.167, and 0.083 A

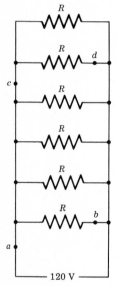

FIGURE 16.33

Problem 27.

28 Six bulbs are arranged in a series circuit, as shown in Fig. 16.34. (Old-fashioned Christmas tree lights were connected this way.) Each bulb has the same resistance R, and the entire array is designed to dissipate 60 W when operated at 120 V. (a) What is the resistance of each bulb? (b) What is the current in the circuit when it is operated at 120 V? (c) What would be the power dissipated in a single bulb operated at

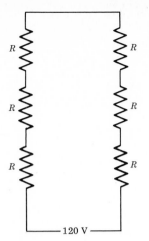

FIGURE 16.34

Problem 28.

120 V? (d) Compare this circuit to the circuit in Prob. 27, and discuss the disadvantages of arranging Christmas tree lights in a series.

29 Figure 16.35 shows the three resistors $R_1 = 5\,\Omega$, $R_2 = 15\,\Omega$, and $R_3 = 25\,\Omega$ in four different circuits. For each circuit find the currents I_1, I_2, and I_3 in each resistor and the current I in the battery.
Ans. (a) $I_1 = I_2 = I_3 = I = 0.067$ A; (b) 0.6, 0.2, 0.12, and 0.92 A; (c) 0.15, 0.15, 0.12, and 0.27 A; (d) 0.209, 0.130, 0.079, and 0.209 A

30 For each circuit in Fig. 16.35, find the power dissipation in each resistor and the total power output of the battery (see Prob. 29).

31 Find all the resistances that can be made with three 10-Ω resistors in various combinations. Not every combination need use all three resistors.
Ans. 3.33, 5, 6.66, 10, 15, 20, and 30 Ω

32 Arrange a 5-, a 10-, and a 20-Ω resistor in a combination that has a total resistance of 14 Ω.

33 Find all the resistances that can be made with a 6-, a 9-, and a 15-Ω resistor in various combinations. Not every combination need use all three resistors.
Ans. 2.90, 3.6, 4.28, 4.8, 5.625, 6, 6.3, 7.5, 9, 11.625, 13.28, 15, 18.6, 21, 24, and 30 Ω

34 Arrange an 8-, a 12-, and a 16-Ω resistor in a combination that has a total resistance of 8.89 Ω.

35 A single-pole, single-throw (SPST) switch is a two-terminal switch that opens and closes one circuit. A single-pole, double-throw (SPDT) switch is a three-terminal switch that connects one terminal to either of two other terminals. Figure 16.36 shows a circuit with an

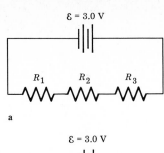

a

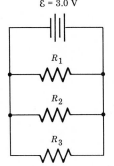

b

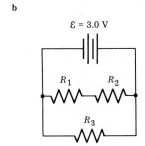

c

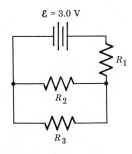

d

FIGURE 16.35

Problems 29 and 30.

SPST and an SPDT switch. Find the current in the battery for all four possible switch combinations.
Ans. 0, 1.5, 1.8, and 3.0 A

36 A stairway light is controlled by two SPDT switches (see Prob. 35), one at the top and one at the bottom of the steps. Each switch can separately control the light. Figure 16.37 shows part of the diagram for wiring these switches to the light bulb. Add the necessary wires and switches to complete the diagram.

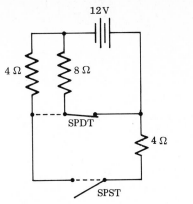

FIGURE 16.36

Problem 35.

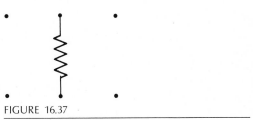

FIGURE 16.37

Problem 36.

37 Find the resistance of a 1-km length of No. 12 copper wire, which has a diameter of 0.205 cm.
Ans. 5.2 Ω

38 An aluminum transmission cable is designed to carry a current of 350 A with a loss of 2 W/m. What is the diameter of the cable?

39 The tungsten filament of a light bulb, when fully uncoiled, is 50 cm long and has a diameter of 50 μm. (a) Find the resistance of this filament at 20 and 2727°C. (b) When the filament is at room temperature (20°C), a voltage of 120 V is placed across it. What is the initial current in the filament? What is the current when the temperature of filament reaches 2727°C? (c) What is the power dissipated in the filament at 2727°C?
Ans. (a) 14, 252 Ω; (b) 8.6 and 0.48 A; (c) 57 W

40 (a) A measure of merit of a transmission cable of length L, resistance R, and mass m is its length per unit resistance (L/R) divided by its mass per unit length (m/L). Show that this

491

measure is equal to

$$\frac{1}{\rho_e \rho_m}$$

where ρ_e is the resistivity, and ρ_m is the mass density of the cable. (b) Calculate this measure for copper and aluminum.

41   A 2-μF capacitor, initially uncharged, is connected in series with a 1000-Ω resistor and a 3-V battery. (a) What is the current in the circuit at the instant the circuit is closed? (b) What is the maximum charge on the capacitor? (c) What is the charge on the capacitor at the instant the current is one-third of its maximum value?
Ans. (a) 3 mA; (b) 6 μC; (c) 4 μC

42   A 400-μF capacitor is connected through a resistor to a battery. Find (a) the resistance R and (b) the emf of the battery $\mathcal{E}$ if the time constant of the circuit is 0.5 s and the maximum charge on the capacitor is 0.024 C.

43   A 50-μF capacitor, initially uncharged, is connected through a 300-Ω resistor to a 12-V battery. (a) What is the magnitude of the final charge q_m on the capacitor? (b) How long after the capacitor is connected to the battery will it be charged to $\frac{1}{2}q_m$? (c) How long will it take for the capacitor to be charged to $0.90q_m$?
Ans. (a) 6×10^{-4} C; (b) 10.5 ms; (c) 37.5 ms

44   A capacitor, initially uncharged, is connected through a resistor to a battery. The time constant of the circuit is 0.2 ms, and the maximum current is 30 mA. (a) Find the current 0.3 ms after the circuit is closed. (b) What is the maximum charge on the capacitor? (c) What is the charge on the capacitor 0.3 ms after the circuit is closed?

45   The tuning capacitor of a radio is an air-filled, parallel-plate capacitor with a plate separation of 0.6 mm. What is the area of each plate if the capacitance is 50 pF?
Ans. 34 cm^2

46   A 250-pF capacitor is made with plates 9 cm^2 in area separated by a 0.15-mm-thick sheet of plastic. What is the dielectric constant of the plastic?

47   (a) What is the equivalent capacitance of the three capacitors in Fig. 16.38? (b) If the potential difference between points a and b is 25 V, what is the charge on each capacitor?
Ans. (a) 20 μF; (b) 100, 150, and 250 μC

48   (a) What is the equivalent capacitance of the three capacitors in Fig. 16.39? (b) If the charge on each capacitor is 50 μC, what is the

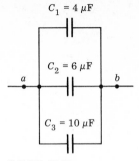

FIGURE 16.38

Problem 47.

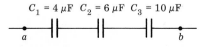

FIGURE 16.39

Problem 48.

voltage across each capacitor? (c) What is the potential difference between points a and b?

49   (a) What is the equivalent capacitance of the three capacitors in Fig. 16.40? (b) If the charge on C_3 is 50 μC, what are the charges on C_1 and C_2?
Ans. (a) 12.4 μF; (b) 12 μC

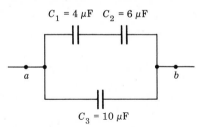

FIGURE 16.40

Problem 49.

50   (a) What is the equivalent capacitance of the three capacitors in Fig. 16.41? (b) If the potential difference between points a and b is 12 V, what is the charge on each capacitor?

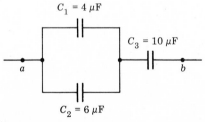

FIGURE 16.41

Problem 50.

51 How long would it take to plate 4 g of silver on a piece of jewelry with a current of 15 A? The electrolyte is silver chloride (AgCl), which contains Ag^+ and Cl^- ions.
Ans. 238 s

52 A current of 7.5 A is maintained for 2 h in a copper sulfate solution. How many grams of copper are deposited on the negative electrode in this time?

53 When an automobile storage battery is "charged," lead is plated on the negative grid. The electrolyte contains Pb^{2+} ions. How much lead is plated in 75 min at a current of 50 A.
Ans. 241 g

54 What current is required to plate 25 g of zinc in 10 min? The electrolyte contains Zn^{2+} ions.

55 (a) What is the emf of a battery made by coupling a $Cu|Cu^{2+}$ half-cell to an $Ag|Ag^+$ half-cell? (b) Which half-cell is the negative terminal?
Ans. (a) 0.462 V; (b) Cu

56 To what half-cell in Table 16.3 should a $Mn|Mn^{2+}$ half-cell be coupled to make a battery with an emf of about 1.7 V?

BIBLIOGRAPHY

BARROW, GORDON M.: *Physical Chemistry*, 3rd ed., McGraw-Hill Book Co., New York, 1973. Chapter 23 describes the chemistry of electrolytic cells.

HEISE, GEORGE W., and CAHOON N. COREY (eds.): *The Primary Battery*, Vol. 1, John Wiley & Sons, Inc., New York, 1971. Chapter 1, by George Heise, gives the history of the development of the battery. Subsequent chapters describe the different cell systems currently in commercial use.

M agnetism is a fundamental force of nature, closely related to electricity. Because of the existence of naturally occurring magnet materials, some simple magnetic effects have been known since antiquity. However, the important electromagnetic phenomena that established the connections between electricity and magnetism were discovered only in the nineteenth century.

The magnetic effect of a current (electromagnetism) was discovered in 1820, and soon resulted in the first widespread application of electricity: telegraphy. The electric effect of a magnet (magnetic induction) was discovered in 1831. It resulted in the invention of the electric generator, which is a more economical source of current than a battery. In 1882 the first central generating plant began distributing electricity for incandesent lighting—the age of electricity had begun.

17
MAGNETISM

17.1 The Magnetic Field

The most familiar example of magnetism is the attraction of small pieces of iron by the ends, or *poles,* of a magnet. This phenomenon is similar in some respects to the attraction of a small piece of cork by an electrified rod (Sec. 15.2). However, iron is one of the few substances attracted by a magnet, whereas any substance is attracted by an electrified rod. Furthermore, a rod remains electrified for only a short time, whereas a magnet retains its magnetism indefinitely.

A compass needle is a long, thin magnet suspended at its center of gravity in such a manner that it is free to rotate in the horizontal plane (Fig. 17.1). If no other magnetic material is nearby, the needle will align itself in a nearly north-south direction. The end of the needle that always points north is called the north (N) pole of the magnet; the other end is called the south (S) pole. *North* and *south* are used to distinguish opposite magnetic poles, just as *positive* and *negative* are used to distinguish opposite electric charge.

Furthermore, the force between two magnetic poles depends on the poles in the same way that the force between two electric charges depends on the charges: opposite poles attract each other, and like poles repel each other. This is easily demon-

FIGURE 17.1

A magnetized needle suspended at its center of gravity.

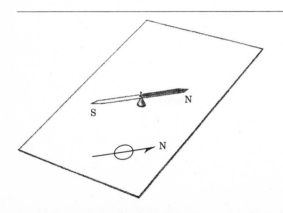

strated with two bar magnets whose poles have been deter-mined from their north-seeking (or south-seeking) tendencies. When the north pole of one magnet is brought near the south pole of the other magnet, the two poles pull together, whereas when the north pole of one magnet is brought near the north pole of the other magnet, the two poles push apart.

Because of the attraction of unlike poles, a compass needle brought near a bar magnet will align itself with its north pole directed toward the south pole of the magnet (Fig. 17.2). The earth itself is a giant magnet whose south pole is located near the earth's north geographic pole. Therefore, in the absence of other magnets, the north pole of a compass needle points north toward the south magnetic pole of the earth. However, the earth's magnetism is so weak that its effect on a compass needle is negligible compared with that of a nearby magnet. Conse-quently, we shall neglect the earth's magnetism when consider-ing the effects of a magnet on a nearby compass.

Unlike electric charge, isolated magnetic poles do not exist. For instance, if we try to separate the north and south poles of a long magnet by breaking it in half (Fig. 17.3), we find that north and south poles immediately appear on the broken ends, so that each fragment remains a complete magnet with both a north and south pole. In fact, no matter how many pieces the magnet is broken into, each piece has a north and south pole of equal strength. Furthermore, even elementary particles, such as elec-trons, protons, and neutrons, act as complete magnets with both north and south poles. Because elementary particles with only one magnetic pole have never been found, magnetic poles lack the fundamental significance that electric charges have.

Nevertheless the *magnetic field* **B** can be defined in the same way as we defined the electric field **E** in Sec. 15.3. Recall that the electric field at any point in space is the force that a system of charges would exert on one unit of positive electric charge placed at that point. Likewise, the magnetic field at any point in space is the force that a magnetic system (a bar magnet, for instance) would exert on one unit of north magnetic pole placed at that point. Since isolated poles do not exist, we have to use little compass needles to measure the field.

For example, consider a small compass needle located at a point P in the magnetic field produced by a bar magnet (Fig. 17.4). The two poles of the magnet exert a net force $\mathbf{F}_1$ on the north pole of the compass needle and a net force $\mathbf{F}_2$ on the

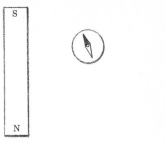

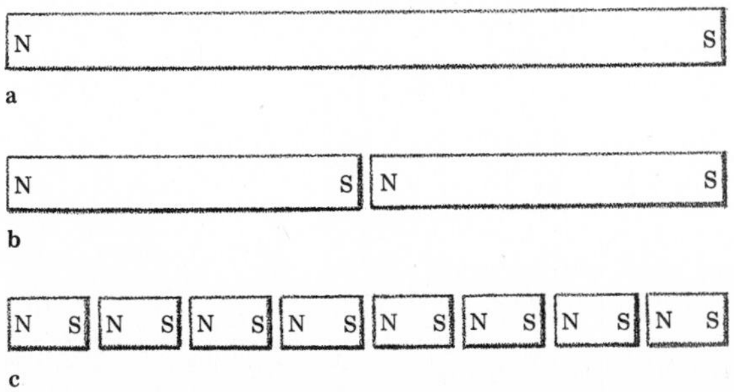

FIGURE 17.2

A compass needle near a bar mag-net aligns itself so that its north pole (dark end) points toward the south pole of the magnet.

FIGURE 17.3

A bar magnet broken in half and then into smaller fragments. Each fragment is a complete magnet with both a north and a south pole.

N | S

a

N | S N | S

b

N S N S N S N S N S N S N S N S

c

south pole. If the compass needle is not already aligned with the magnetic field at P, Fig. 17.4a shows that these forces exert a torque that will rotate the needle about the pivot until it is aligned with the field. Because the torque on a compass needle is zero only when it is aligned with the magnetic field, at equilibrium a compass needle points in the direction of the magnetic field in its vicinity. Consequently, the direction of the magnetic field at any point is given by the direction of a compass needle placed at that point.

Figure 17.5 shows the magnetic field around a bar magnet. As with the electric field, the lines of force are drawn closer together wherever the magnitude of the magnetic field is stronger. A comparison of Figs. 17.5 and 15.15 shows that the magnetic field around a bar magnet is identical to the electric field around an electric dipole. This is not surprising when you realize that a magnet is a magnetic dipole. That is, it is a system composed of two opposite poles separated by a distance d.

> REMARK Although isolated magnetic poles do not exist in nature, magnetic dipoles do. Elementary particles are the ultimate carriers of magnet dipoles, just as they are the ultimate carriers of charge. Normally the magnetic dipoles of the electrons in matter are oriented in all directions, so that they cancel each other (Fig. 17.6a). In a permanent magnet, however, many of these dipoles are aligned in the same direction, so that the substance as a whole acts as a single, large magnetic dipole that is the vector sum of the elementary dipoles (Fig. 17.6b). This is discussed more fully in Sec. 17.2.

Figure 17.7 shows the magnetic field of a C magnet, which is a magnet bent so that its two poles face each other. This configuration has many important applications because it produces a nearly uniform field in the gap between the two poles.

17.2 Electromagnetism

Most of the basic facts about static electricity, currents, and permanent magnets were known by the end of the eighteenth century. Because of the similarities between electric and magnetic phenomena, many scientists suspected a connection between them, but none had yet been discovered. Then in 1820 the Danish physicist Hans Christian Oersted (1777–1851) made an epochal discovery that changed civilization. While performing some demonstrations for a small group of students, Oersted noticed that a compass needle was deflected whenever there was a current in a nearby wire. Experimenting further, he discovered that a current-carrying wire has the same effect on a compass needle as a magnet. That is, magnetism is produced by a current. This phenomenon, called *electromagnetism*, was one of the long-sought connections between electricity and magnetism.

> REMARK Until Oersted's discovery of electromagnetism, electricity had been a laboratory curiosity, with few practical applications. The discovery of electromagnetism was the breakthrough that led to the widespread utilization of electricity. For example, the telegraph, which initiated the modern era of high-speed communication, was developed within 20 years of Oersted's discovery.

FIGURE 17.4

The forces on a compass needle near a bar magnet. (a) When the needle is not aligned with the magnetic field, the forces F_1 and F_2 exerted by the magnet on the needle cause the needle to rotate. (b) Only when the needle is aligned with the field does F_1 and F_2 exert no torque about the pivot.

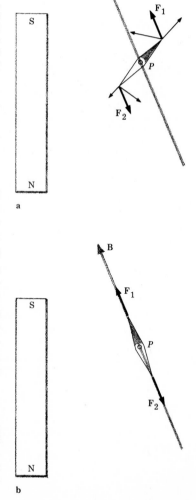

Upon learning of Oersted's discovery, André Marie Ampère (1775–1836) began a mathematical investigation of the relationship between currents and magnetism. He formulated a law (Ampère's law) from which the magnetic field resulting from any distribution of currents could be calculated. However, for our purposes, we shall need only the formulas for the fields resulting from a long, straight wire and a circular coil.

FIELD OF A STRAIGHT WIRE

Ampère's law shows that the magnetic lines of force near a current-carrying wire are in the form of concentric circles about the wire, as shown in Figure 17.8. This can be demonstrated by placing a small compass in a plane perpendicular to the wire. The compass needle, which points in the direction of the field where it is located, will align itself perpendicular to the radius r drawn from the wire to the needle, indicating that the field forms a circle about the wire. The relationship between the direction of the field and the direction of the current is given by the following rule.

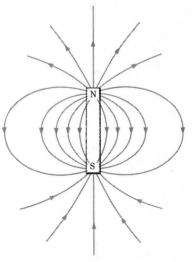

The Right-Hand Rule *When a wire is grasped with the right hand in such a way that the thumb points in the direction of the current, the fingers encircle the wire in the same sense as the magnetic field.*

FIGURE 17.5

Magnetic field around a bar magnet.

The magnitude B of the magnetic field at a point near a very long current-carrying wire is proportional to the current I and inversely proportional to the perpendicular distance r from the point to the wire. In symbols this relationship is

FIGURE 17.6

(a) In ordinary matter, the magnetic dipoles of the electrons are randomly arranged. (b) In magnetized material, the magnetic dipoles of the electrons are aligned.

$$B = k\frac{I}{r} \qquad\qquad 17.1$$

where k is a constant of proportionality.

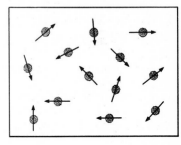

> **UNITS** The SI unit of magnetic field strength is the tesla (T). As we shall show in Sec. 17.3, the tesla is related to the newton, ampere, and meter by
>
> $1\,\text{T} = 1\,\text{N}/(\text{A}\cdot\text{m})$
>
> The gauss (G) is another common unit of magnetic field strength, equal to 10^{-4} T:
>
> $1\,\text{G} = 10^{-4}\,\text{T}$

a

The constant k in Eq. 17.1 is

$$k = \frac{\mu_0}{2\pi} = 2 \times 10^{-7}\,\text{T}\cdot\text{m/A}$$

where

$$\mu_0 = 4\pi \times 10^{-7}\,\text{T}\cdot\text{m/A}$$

is called the *magnetic permeability*. In terms of μ_0, Eq. 17.1 becomes

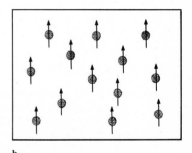

$$B = \frac{\mu_0 I}{2\pi r} \qquad\qquad 17.2$$

b

Example 17.1   What is the magnetic field at a point 5 cm from a wire carrying a current of 3 A?

From Eq. 17.2 the field is

$$B = \frac{\mu_0 I}{2\pi r} = \frac{(4\pi \times 10^{-7} \text{ T} \cdot \text{m/A})(3 \text{ A})}{(2\pi)(0.05 \text{ m})} = 1.20 \times 10^{-5} \text{ T}$$

$$= 0.12 \text{ G}$$

In the middle latitudes, the magnitude of the earth's magnetic field is about 0.5 G, so that the effect of this field on a compass needle is small compared with the effect of the earth's field. ☐

FIELD OF A CIRCULAR COIL

The magnetic field produced by a current-carrying wire is greatly increased if the wire is formed into a circular coil with many turns. This can be understood by considering the coil with two turns shown in Fig. 17.9. Each turn produces a field proportional to the current I in the wire, so that the two turns together produce a field proportional to $2I$. Similarly, a coil with n turns produces a field proportional to nI. Since the center of a coil of radius a is the distance a from the wire, we might expect from Eq. 17.2 that the field at the center is

$$B = \frac{\mu_0 n I}{2\pi a}$$

This is not quite correct, however, because Eq. 17.2 is valid only for a very long, straight wire, whereas here we are dealing with a circular wire. The correct expression for a circular coil is

$$B = \frac{\mu_0 n I}{2a} \qquad 17.3$$

which differs from the previous expression only by a factor of π, because of the difference in geometry between a straight and a circular wire. The field at the center of the coil is perpendicular to the plane of the coil.

Example 17.2   A coil of radius 5 cm has 100 turns of wire. What is the magnetic field at the center of this coil when the current in the wire is 3 A?

From Eq. 17.3, the magnetic field is

$$B = \frac{\mu_0 n I}{2a} = \frac{(4\pi \times 10^{-7} \text{ T} \cdot \text{m/A})(100)(3 \text{ A})}{(2)(0.05 \text{ m})} = 3.77 \times 10^{-3} \text{ T}$$

$$= 37.7 \text{ G}$$

which is 75 times the earth's magnetic field. ☐

Figure 17.10 shows a compass at the center of a coil containing many turns. When the switch S is closed, the compass aligns itself parallel to the magnetic field produced by the current. When the switch is opened, the field vanishes and the needle swings back to a north-south orientation. Thus the orientation of the needle indicates whether there is a current in the wire.

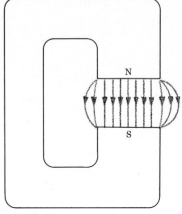

FIGURE 17.7

Magnetic field in the gap of a C magnet.

FIGURE 17.8

Magnetic field around a current-carrying wire. When the thumb of the right hand points in the direction of the current, the fingers of the right hand coil around the wire in the direction of the magnetic field.

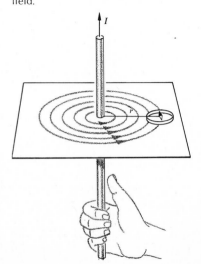

REMARK The first telegraph line, built by William Fothergill Cooke (1806–1879) and Charles Wheatstone (1802–1875) in 1838, used a coil and compass device to receive messages from a distant transmitter. A message was transmitted by an operator opening and closing a switch in accordance with a predetermined code and was received by another operator noting the corresponding deflections of the compass needle (Fig. 17.10).

The line, built in 1838, ran 13 mi from Paddington Station (London) to West Drayton. It was extended farther in the following years, but the general public first became aware of the importance of the new device in 1845. A man suspected of murder was seen to board a train in Slough for Paddington Station. His description was telegraphed to Paddington, where he was apprehended as he got off the train.

The first telegraph line in the United States was built by Samuel Morse (1791–1872) in 1844. It ran 40 mi from Washington to Baltimore. The Morse receiver is described later in this section.

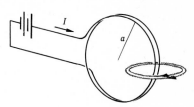

FIGURE 17.9

The magnetic field produced by a coil with two turns is the sum of the fields produced by each turn separately.

THE ELECTROMAGNET

In many practical applications of magnetism the magnetic field of a current loop is enhanced by using a peculiar property of iron, called *ferromagnetism*. Other transition elements (Table 20.3), such as nickel and cobalt, also exhibit this property, but iron is the only abundant ferromagnetic material.

Ferromagnetism is the tendency of the magnetic dipoles of the outer electrons of an atom to align themselves parallel to the magnetic dipoles of the corresponding electrons in a neighboring atom. In a permanent magnet, the dipoles of the outer electrons of all the atoms in a large volume of the material are aligned, so that their magnetic fields add together to produce a large magnetic field outside the magnet. In unmagnetized iron, the dipoles are aligned within a small volume, or *domain,* but the direction of alignment is different in different domains (Fig. 17.11). Consequently, the magnetic fields of these domains cancel each other, and there is no field outside the iron. Each domain, which is typically a few micrometers in width, acts like a tiny permanent magnet, so that unmagnetized iron can be thought of as a random arrangement of billions of these magnets.

When unmagnetized iron is placed in an external magnetic field, domains aligned parallel to the field increase in size at the expense of the other domains. Figure 17.12 shows this for a piece of iron near one pole of a magnet. Because of the increase in the size of the domains of iron parallel to the external field, the iron develops a magnetization in the direction of the field. As seen from Fig. 17.12, the end of the iron nearest the magnet's north pole develops a south pole, so that the iron is attracted to the magnet. The situation is analogous to that of a piece of cork near an electrified rod (Fig. 15.4).

An *electromagnet* consists of a wire coil wrapped around an iron cylinder (Fig. 17.13). The magnetic field produced by a current in the coil increases the size of the domains in the iron that are magnetized in the direction of the field. The iron thus develops its own magnetic field, which adds to the field of the current. The field of the iron can easily be thousands of times larger than the field of the current alone, so that the iron in effect magnifies the magnetic field of the current. Furthermore, if the current is not too large, the domains instantly return to

FIGURE 17.10

A compass needle at the center of a circular coil of wire. (a) When switch S is closed, the needle aligns itself with the magnetic field at the center of the coil. (b) When the switch is opened, the needle swings back to a north-south orientation.

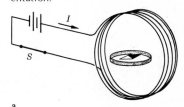

a

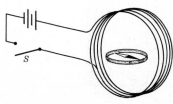

b

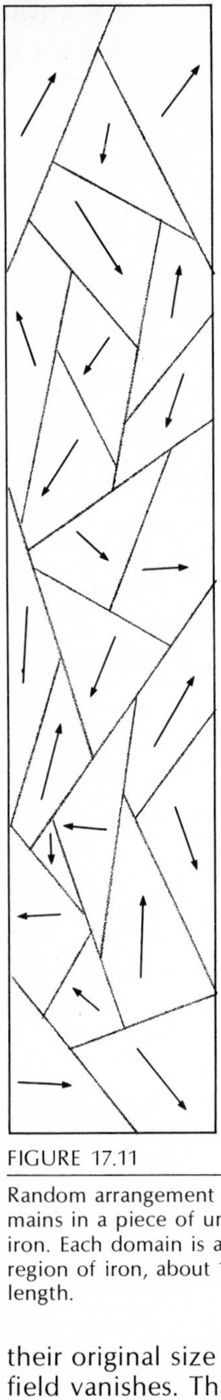

FIGURE 17.11

Random arrangement of the domains in a piece of unmagnetized
iron. Each domain is a magnetized
region of iron, about 1 μm in
length.

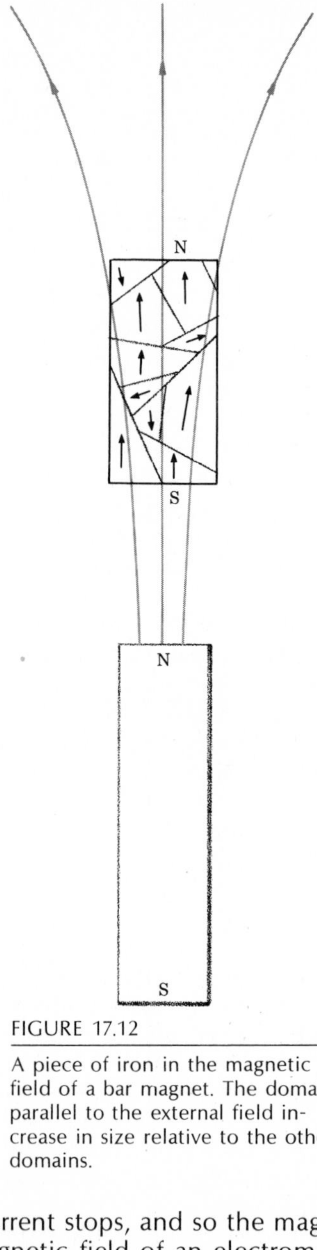

FIGURE 17.12

A piece of iron in the magnetic
field of a bar magnet. The domains
parallel to the external field increase in size relative to the other
domains.

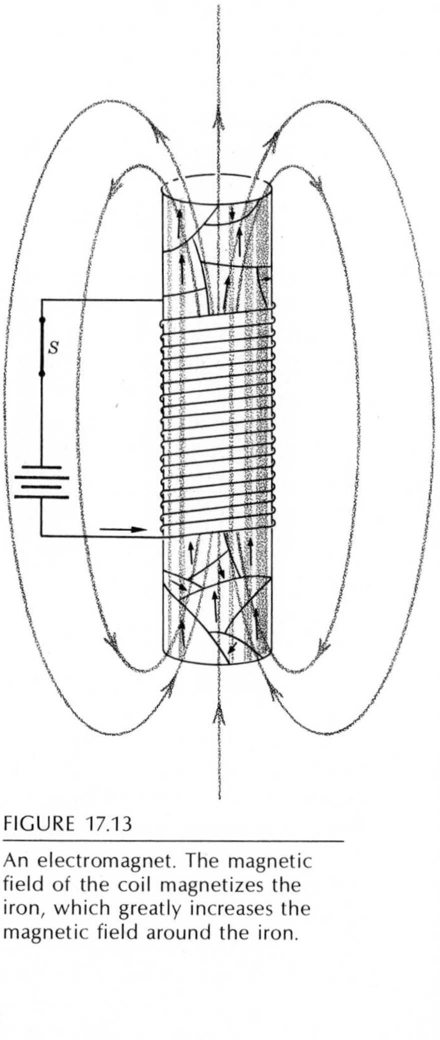

FIGURE 17.13

An electromagnet. The magnetic
field of the coil magnetizes the
iron, which greatly increases the
magnetic field around the iron.

their original size when the current stops, and so the magnetic
field vanishes. That is, the magnetic field of an electromagnet
exists only as long as there is a current in the coil.

> **REMARK** The domains of a piece of iron placed in a very large magnetic
> field become permanently aligned in the direction of the field, so that the
> iron retains its magnetization after the external field is removed. Permanent
> magnets are made simply by placing a bar of iron in the magnetic field of a
> coil in which there is a very large current. In most applications, however,
> electromagnets are operated at values of current too small to magnetize the
> iron permanently.

pivot
spring
clapper
electromagnet
battery
key switch

FIGURE 17.14

Many important applications of electricity are based on the ability to turn a large magnetic field on and off with an electromagnet. We illustrate this by considering the following devices.

Morse telegraph receiver. When the key is depressed, the electromagnet attracts the clapper.

Morse Telegraph The receiver of the Morse telegraph (Fig. 17.14) consists of an electromagnet mounted just below an iron bar, called the *clapper*. One end of the clapper is pivoted, and the other is held above the pole of the electromagnet by a spring. The circuit connecting the electromagnet to a battery is closed whenever the key at the transmitting end of the line is depressed. This causes a current in the coil of the electromagnetic, which then attracts the clapper. When the key is released, the current stops and the spring returns the clapper to its resting position. Thus a coded message tapped on the key at one end of the line causes the clapper to make a corresponding sequence of clicking sounds at the other end of the line.

Electric Relay An *electric relay* is a switch that can be controlled by another circuit. It is similar to a telegraph receiver in that the position of a hinged clapper is controlled by a spring and electromagnet. In one position the clapper makes contact with an electric terminal, closing a circuit; in the other position the clapper breaks contact with the terminal, opening the circuit. The relay shown in Fig. 17.15 is closed when the switch S in

FIGURE 17.15

A relay. The position of the switch in the power circuit is controlled by the position of the switch in the control circuit.

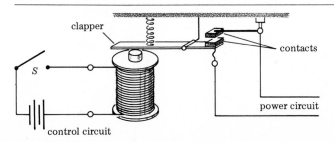

clapper

S

contacts

control circuit

power circuit

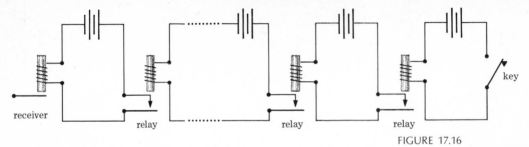

FIGURE 17.16

A series of relay circuits used to transmit a telegraph signal a great distance. Since each section has its own battery, the strength of the signal does not decrease with the distance of transmission.

the controlling circuit is closed, and the relay is open when S is open. Relays are extensively used for the remote control of a circuit.

> **REMARK** The idea of using an electromagnet to control a circuit seems to have originated with Joseph Henry (1797–1878), the foremost American physicist of his time. Independently Morse rediscovered the idea and used it to extend the distance he could transmit a telegraph signal. Figure 17.16 shows a series of relay circuits, each with its own battery. When the transmitting key is closed, it closes the first relay, which in turn closes the second, and so on, until the signal reaches the telegraph receiver.

Electric Bell An electric bell (Fig. 17.17) is similar to a telegraph receiver in that it consists of a clapper and an electromagnet that attracts the clapper when the circuit is closed. Each time the clapper moves toward the magnet, it strikes a metal bell, which makes a ringing sound. Moreover, the clapper is made part of the circuit in such a way that the movement of the clapper toward the magnet opens the circuit. Thus, just as the clapper hits the bell, the circuit opens and the spring returns the clapper to its resting position. Then, just as the clapper reaches its resting position, the circuit is closed and the clapper is again attracted to the magnet. Therefore, as long as the main switch S is closed, the clapper will move back and forth, repeatedly striking the bell. This arrangement is one of the simplest mechanisms for producing continuous mechanical motion from electricity.

Electric Motor A simple *direct-current* (dc) *motor*, like that used in a mechanical toy, consists of an electromagnet on a shaft (the *armature*) that rotates between the poles of a C magnet (Fig. 17.18). The ends of the wire of the electromagnet are connected to two metal contacts (*commutators*) on the shaft. The circuit is completed whenever the commutators touch the external contacts (*brushes*) that are connected to a battery. As the armature rotates, the commutators repeatedly reverse the connection between the battery and the electromagnet, so that the direction of the current in the electromagnet is repeatedly reversed.

When the armature is in the position shown in Fig. 17.18a, the current is directed in such a way that pole A of the electromagnet is south and pole B is north. Consequently, the attraction of the C magnet causes the armature to rotate counterclockwise. When the armature reaches the position shown in Fig. 17.18b, the brushes no longer make contact with the commutators, so the electromagnet is momentarily turned off and thus is able to

FIGURE 17.17

An electric bell. The magnet attracts the clapper, causing it to strike the bell. But just as the clapper strikes, the circuit is opened, and the spring returns the clapper to its original position. This closes the circuit, so the process repeats itself.

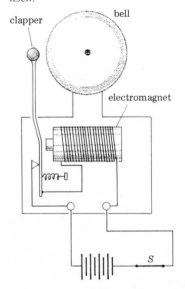

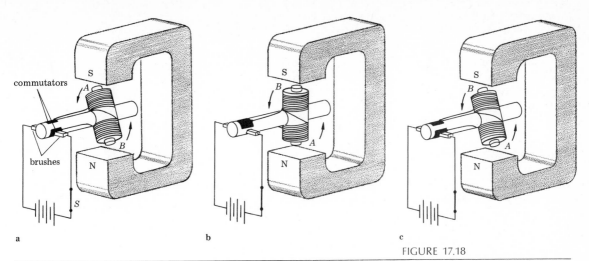

FIGURE 17.18

A dc motor. (a) In this position, the commutators connect the coil to the battery in such a way that side A of the electromagnet is a south pole and side B is a north pole. As a consequence, the rotor is rotated counterclockwise by the forces exerted on it by the C magnet. (b) In this position, the commutators disconnect the battery, so there is no torque on the rotor as it rotates past the poles of the C magnet. (c) In this position, the commutators reverse the connections to the battery, so side A is now a north pole and side B is a south pole. As a consequence, the rotor continues to be rotated counterclockwise.

rotate past the poles of the C magnet. When the armature reaches the position shown in Fig. 17.18c, the brushes again contact the commutators, but with the connection to the battery reversed. Thus pole A of the electromagnet is now north, and pole B is south. Consequently, the attraction of the C magnet continues to rotate the armature in a counterclockwise direction.

> **REMARK** The battery converts chemical energy into electric energy, which the electric motor converts into the mechanical energy of the rotating armature.

17.3   Magnetic Forces

FORCES ON A CURRENT

Oersted's discovery of electromagnetism showed that an electric current exerts a force, through its magnetic field, on a magnet. Therefore, according to Newton's third law (Property 3, Sec. 2.1), a magnet must exert a force, through its magnetic field, on a current. More generally, any external magnetic field exerts a force on a current.

Consider, for instance, a section of wire of length L in which there is a current I (Fig. 17.19). (This section of wire is part of a larger circuit that is not shown.) If the wire makes an angle θ with a uniform magnetic field **B,** the magnitude of the magnetic force $\mathbf{F}_m$ on the wire is

$$F_m = BIL \sin \theta \qquad\qquad 17.4$$

When the field is parallel to the wire, the force is zero because $\sin 0° = 0$. When the field is perpendicular to the wire, the magnitude of the force is

$$F_m = BIL \qquad\qquad 17.5$$

because $\sin 90° = 1$. These are the only cases we shall have to consider.

REMARK From Eq. 17.5 we see that the tesla, the unit of magnetic field strength, is related to the newton, ampere, and meter by

$$1\,T = \frac{1\,N}{(1\,A)(1\,m)} = 1\,N/(A \cdot m) \qquad 17.6$$

The magnetic force on a current is perpendicular to the plane in which the magnetic field and the current lie. The following rule is used to find the direction of the force.

The Left-Hand Rule *When the thumb of the left hand is pointed in the direction of the field and the index finger is pointed in the direction of the current, the middle finger points in the direction of the force.* This is illustrated in Fig. 17.19.

Of course, a current-carrying piece of wire cannot exist by itself; it must be part of a larger circuit. Let us consider, therefore, the forces on the complete circuit shown in Fig. 17.20. Part of the circuit consists of a rectangular loop of wire of length L and width d placed between the poles of a C magnet. The loop is connected to a source of emf through wires that lie outside the magnet. Consequently, there is no magnetic force on these external wires, and only the forces on the loop itself have to be considered.

Furthermore, two of the four straight sections of the loop are parallel to the magnetic field, so the magnetic force on them is zero also. The other two sections are perpendicular to the field, so the magnitude of the magnetic force on each of them is BIL. However, these forces have opposite directions, since the currents are opposite. Thus the total force on the loop is zero.

There is, however, a total torque on the loop, which tends to rotate it around the axis. The magnitude of this torque about the axis is

$$\tau_m = \tfrac{1}{2}dF_m + \tfrac{1}{2}dF_m = dBIL = BIA$$

where $A = Ld$ is the area of the loop. If the loop contains n turns of wire, the torque is

$$\tau_m = nBIA \qquad 17.7$$

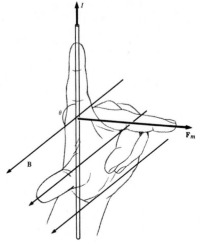

FIGURE 17.19

Magnetic force $\mathbf{F}_m$ on a section of current-carrying wire in a magnetic field $\mathbf{B}$. The directions of the field, current, and force are related to each other as the thumb, index finger, and middle finger of the left hand.

FIGURE 17.20

Forces on the sides of a rectangular current-carrying loop in the field of a C magnet.

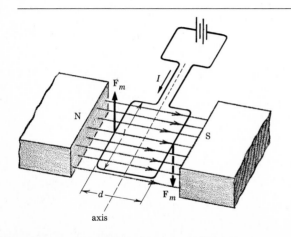

Example 17.3 A loop 6 cm high and 2 cm wide is placed in a magnetic field of 0.02 T. If the loop contains 200 turns and carries a current of 50 mA, what is the torque on it?

From Eq. 17.7 the torque is

$$\tau_m = nBIA = (200)(0.02\ \text{T})(50 \times 10^{-3}\ \text{A})(12 \times 10^{-4}\ \text{m}^2)$$
$$= 2.4 \times 10^{-4}\ \text{T} \cdot \text{A} \cdot \text{m}^2 = 2.4 \times 10^{-4}\ \text{N} \cdot \text{m}$$

where the last equality is obtained by using Eq. 17.6 to convert from electrical to mechanical units. □

METERS

Even though the magnetic torque on a coil is rather small, it is sufficient to rotate a delicately suspended coil. This fact is used in a number of instruments that measure current and potential difference.

Galvanometer A *galvanometer* is an instrument for measuring very small currents. It consists of a many-turn coil suspended between the poles of a C magnet by a fine wire. The suspension is arranged in such a way that when there is no current in the coil, the coil is in the position shown in Fig. 17.20. When there is a small current in the coil, the magnetic torque τ_m on the coil causes it to rotate, thus twisting the suspension wire. As the wire is twisted, it exerts a countertorque τ_w proportional to the angle of rotation θ. This torque can be written

$$\tau_w = k\theta$$

where k is a constant characteristic of the wire. At equilibrium the coil comes to rest at the angle at which τ_w equals τ_m. Therefore, from Eq. 17.7 the condition for equilibrium is

$$k\theta = nBIA \qquad \text{or} \qquad I = \frac{k\theta}{nBA}$$

Thus the current in the galvanometer is determined by measuring the angle of rotation of the coil. With the use of very fine suspension wire, such a device will give measurable deflections with a current of 10^{-6} A. Figure 17.21 shows a table-top galvanometer in which a pointer on the coil indicates its position along a scale.

Ammeter An *ammeter* is a portable current-measuring device similar to a galvanometer except that the coil is mounted on a pivot about which it rotates and the countertorque is applied to the coil by means of a spiral spring, as shown in Fig. 17.22. When there is a current in the coil, the coil rotates until the magnetic torque equals the countertorque exerted by the spring. A needle attached to the coil indicates the current by its position along a calibrated scale.

To measure the current in a branch of a circuit, an ammeter is placed in series with the branch. Figure 17.23a shows the position of an ammeter measuring the current in resistance R_1. The resistance of the ammeter itself must be as small as possible so that the circuit is not altered by its presence. The range of the

FIGURE 17.21

Tabletop galvanometer. The position of the coil is shown by the position of the pointer attached to the suspension wire. (*Sargent-Welch*.)

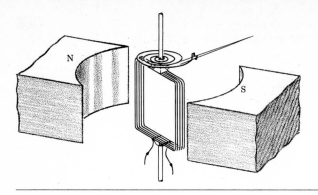

FIGURE 17.22

Suspension mechanism of a portable ammeter. At equilibrium the
spiral spring balances the magnetic
torque on the coil.

ammeter can be varied by putting a resistance, called a *shunt,* in parallel with the ammeter's coil.

Example 17.4 The coil of an ammeter has a resistance of $2\,\Omega$, and the ammeter is fully deflected when there is a current of 1 mA in the coil. (a) What shunt resistance should be used to give the ammeter a full-scale deflection of 10 mA? (b) What is the total resistance of the ammeter in this case?

(a) In Fig. 17.23b the ammeter is represented by a shunt resistance R_s in parallel with the coil resistance R_c. When the total current I_1 in the ammeter is 10 mA, we want the current I_c in the coil to be 1 mA. Therefore, from Kirchhoff's first law (Sec. 16.2) the current I_s in the shunt is

$$I_s = I_1 - I_c = 9\,\text{mA}$$

At full deflection the potential difference across the coil is

$$V = R_c I_c = (2\,\Omega)(1\,\text{mA}) = 2\,\text{mV}$$

which by Kirchhoff's second law is also the potential difference across R_s. Therefore, we have

$$V = R_s I_s \quad \text{or} \quad R_s = \frac{V}{I_s} = \frac{2\,\text{mV}}{9\,\text{mA}} = \tfrac{2}{9}\,\Omega$$

(b) The total resistance R of the ammeter is the equivalent resistance of R_c and R_s in parallel (Eq. 16.11):

FIGURE 17.23

(a) An ammeter A in position to measure the current I_1 in resistor R_1. (b) In an ammeter, the coil resist-ance R_c is in parallel with the shunt resistance R_s.

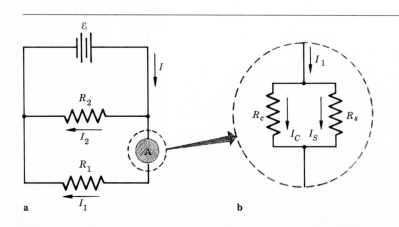

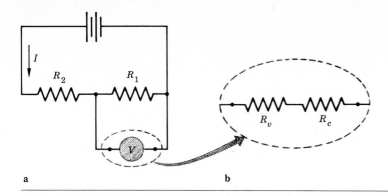

a   b

FIGURE 17.24

(a) A voltmeter V in position to measure the voltage across resistor R_1. (b) In a voltmeter, the coil resistance R_c is in series with a large resistance R_v.

$$\frac{1}{R} = \frac{1}{R_c} + \frac{1}{R_s} = \frac{1}{2\,\Omega} + \frac{1}{\frac{2}{9}\,\Omega} = \frac{10}{2\,\Omega} \qquad \text{or} \qquad R = 0.2\,\Omega \qquad \square$$

Voltmeter A *voltmeter* is an instrument for measuring the potential difference between two points. It consists of an ammeter with resistance R_c in series with a large R_v, as shown in Fig. 17.24. The potential V across the resistor R_1 in Fig. 17.24 is measured by connecting the voltmeter in parallel with R_1. If the total resistance $R_v + R_c$ of the voltmeter is very much larger than R_1, the voltmeter will not significantly change the current in R_1. There will, however, be a small current

$$I_v = \frac{V}{R_v + R_c}$$

through the voltmeter, which causes the meter to deflect. Since I_v is proportional to V, the deflection of the meter is proportional to V. The range of a voltmeter is controlled by the value of the resistance R_v in series with the ammeter.

Example 17.5 What resistance R_v should be placed in series with the ammeter described in Example 17.4 to make a voltmeter with a full-scale deflection of 10 V?

At full deflection there must be 1 mA in the coil and 10 V across the resistances R_c and R_v. Therefore, we must have

$$R_c + R_v = \frac{V}{I} = \frac{10\ \text{V}}{1\ \text{mA}} = 10{,}000\ \Omega$$

or

$$R_v = 10{,}000\ \Omega - R_c = 9998\ \Omega \qquad \square$$

> **REMARK** The magnetic torque on a coil suspended in a magnetic field has been the principal method of measuring current and voltage for over a century. However, these meters are now being replaced by digital ammeters and voltmeters that use modern integrated-circuit technology rather than magnetic effects for these measurements.

FORCES ON A MOVING CHARGE

A magnetic force is exerted on a charged particle moving in a magnetic field. In fact, the magnetic force on a current-carrying wire is just the sum of the magnetic forces on the charges flowing in the wire. The magnetic force on a single charged particle can be found from Eq. 17.5 by considering the average current of a particle of charge q moving with speed v at right angles to a uniform magnetic field $\mathbf{B}$ (Fig. 17.25). The particle travels the distance L from point P to point Q in the time

$$t = \frac{L}{v}$$

so that the average current between P and Q during this time is

$$I = \frac{q}{t} = \frac{q}{L/v} = \frac{qv}{L}$$

But from Eq. 17.5 the magnetic force on this current during this time is

$$F_m = BIL = B\frac{qv}{L}L$$

so the force that a magnetic field exerts on a moving charge is

$$F_m = Bqv \qquad\qquad 17.8$$

Like the magnetic force on a current, this force is perpendicular to both the magnetic field and the direction of motion of the particle.

It was found in Sec. 5.2 that the work done by a force $\mathbf{F}$ on a moving object is $F_d d$, where F_d is the magnitude of the component of $\mathbf{F}$ parallel to the direction of motion of the object, and d is the distance the object moves. Since the magnetic force is always perpendicular to the direction of motion of a charged particle, its component parallel to the direction of motion is zero, and consequently the magnetic force does no work on the particle. Therefore, according to the work-energy theorem (Sec. 5.3), a magnetic field cannot change the particle's speed, although it can change the particle's direction.

Figure 17.26 shows a particle of mass m and charge q moving at right angle to a uniform magnetic field $\mathbf{B}$. Because the magnetic force Bqv is at right angle to the velocity $\mathbf{v}$ of the particle, the particle moves in a circle of radius r. From Newton's second law, the force Bqv on the particle is equal to ma, where $a = v^2/r$ is the particle's centripetal acceleration (Sec. 6.2). Thus a charged particle in a uniform magnetic field moves in a circle of radius r given by the equation

$$Bqv = \frac{mv^2}{r}$$

or

$$mv = Bqr \qquad\qquad 17.9$$

Magnetic fields are used in many devices to change and control the direction of a beam of charged particles. For instance, the electron beam in the picture tube of a television set is moved across the screen by magnetic fields. A television tube differs in this respect from a cathode-ray tube, which uses electric fields to deflect the electrons (Sec. 15.5). Similarly, the elec-

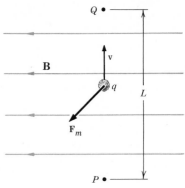

FIGURE 17.25

The magnetic force $\mathbf{F}_m$ on a charged particle moving with velocity $\mathbf{v}$ in a magnetic field $\mathbf{B}$.

FIGURE 17.26

A charged particle moving at right angle to a magnetic field $\mathbf{B}$. Because the magnetic force $\mathbf{F}_m$ on the particle is always perpendicular to the particle's velocity $\mathbf{v}$, the particle moves in a circle of radius r.

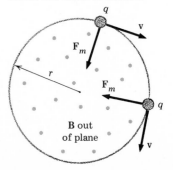

trons scattered by a specimen in an electron microscope are focused by magnetic fields onto a phosphorescent screen. The optics of an electron microscope is similar to that of a light microscope except that magnetic lenses, in the form of current-carrying coils, are used instead of glass lenses.

Another very important instrument based on the circular motion of charged particles in a magnetic field is the mass spectrometer.

MASS SPECTROMETER

A *mass spectrometer* is an instrument that measures the masses of the different atoms and molecules in a gas by bending those particles in a magnetic field. The instrument consists of an ion chamber, an evacuated beam pipe, a magnet, and a detector (Fig. 17.27). A sample of the gas to be analyzed is introduced into the ion chamber, where an electron beam forms positive ions by stripping electrons from some of the atoms and molecules of the gas. This end of the chamber is maintained at a positive voltage V relative to a screen at the other end, so the ions, as they are formed, are accelerated toward the screen.

The ions have essentially zero velocity when they are formed, so their kinetic energy K when they reach the screen is equal to their change of potential energy ΔU:

$$K = \Delta U \qquad \text{or} \qquad \tfrac{1}{2}mv^2 = qV$$

Thus the speed v of an ion of mass m and charge q when it

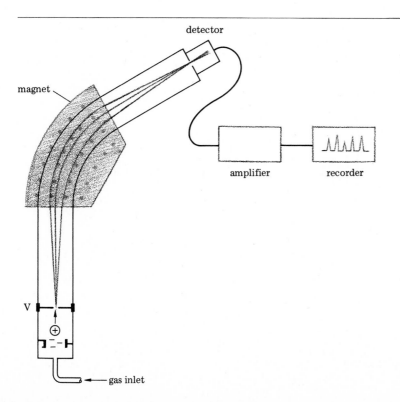

FIGURE 17.27

Schematic diagram of a mass spectrometer. After being accelerated through a voltage V in the ion chamber, ions are bent by the magnetic field of a magnet. Only ions with a certain mass will make it all the way to the detector.

detector

magnet

amplifier   recorder

V

gas inlet

reaches the screen is

$$v = \sqrt{\frac{2qV}{m}}$$ 17.10

Some of the ions reaching the screen pass through a small hole into an evacuated beam pipe. A section of this pipe is bent into an arc of a circle, and a magnet is placed around this section to bend the ions. Only those ions which move in a circle with the same radius as the section will be able to reach the detector. That is, the only ions that can reach the detector are those whose mass and speed satisfy Eqs. 17.9 and 17.10. When the speed given by Eq. 17.10 is substituted into Eq. 17.9, we get the relation

$$m\sqrt{\frac{2qV}{m}} = Bqr$$

Squaring this equation and solving for m gives

$$m = \frac{B^2 r^2 q}{2V}$$ 17.11

A commercial mass spectrometer (Fig. 17.28) measures the ion current reaching the detector for different values of the voltage V. A typical result is shown in Fig. 17.29, where each peak corresponds to an atom or molecule with a mass m related to V by Eq. 17.11. Thus the spectrometer determines the relative abundance of different mass particles in the gas sample. Among the many applications of mass spectrometers are the analysis of

FIGURE 17.28

(a) Mass spectrometer, showing auxiliary equipment such as pumps and vacuum systems. (b) Detailed view of the beam pipe with ion chamber (left), magnet (middle), and detector (right). (*High Voltage Engineering Corp.*)

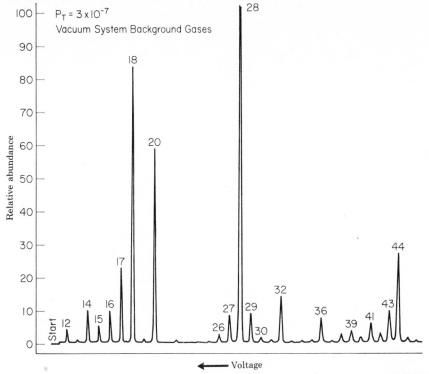

FIGURE 17.29

Typical results of an analysis of a gas sample made with a mass spectrometer. (*High Voltage Engineering Corp.*)

pollutants in the exhaust gas of automobile engines and the monitoring and control of the conditions in vacuum-deposition systems and other high-vacuum technologies.

In mass spectroscopy, masses are usually measured in atomic mass units u (Sec. 8.1), where

$$1\,u = 1.66 \times 10^{-27}\,kg$$

The ions are usually singly charged, so

$$q = e = 1.60 \times 10^{-19}\,C$$

When these values are substituted into Eq. 17.11, the mass in atomic mass units is given by

$$m(1.66 \times 10^{-27}\,kg) = \frac{B^2 r^2 (1.60 \times 10^{-19}\,C)}{2V}$$

or

$$m(\text{in } u) = (4.82 \times 10^7)\frac{B^2 r^2}{V} \qquad\qquad 17.12$$

where B, r, and V are in SI units.

Example 17.6 The radius of the bent section of a mass spectrometer is 5 cm, and the magnetic field in 0.35 T. A peak appears in the ion current at a voltage of 460 V. What is the mass of the molecules in this peak?

From Eq. 17.12 the mass is

$$m = (4.82 \times 10^7)\frac{(0.35)^2(0.05)^2}{460}$$

$$= 32\,u$$

This is the mass of an oxygen molecule (O_2). □

17.4 Magnetic Induction

Oersted's discovery of electromagnetism not only initiated one of the first practical uses of electricity (telegraphy), but also stimulated a great deal of scientific research into electrical and magnetic phenomena. Michael Faraday (1791–1867), one of the nineteenth century's greatest scientists, began his electromagnetic investigations soon after learning of Oersted's work. Faraday reasoned that if an electric current produces a magnetic field, then a magnetic field should somehow be able to produce an electric current. He set about systematically trying one arrangement of wires and magnets after another until, in 1831, he found the effect for which he had been searching.

Faraday discovered *magnetic induction,* the creation of an electric field by a magnetic field. Magnetic induction ranks with electromagnetism as one of the fundamental principles on which our technology is based. For whereas electromagnetism provides the means whereby electric energy can be converted into mechanical work, magnetic induction provides the means whereby mechanical work can be converted into electric energy. Almost all our electric power comes from generators that operate on the principle of magnetic induction.

Magnetic induction can best be understood by a demonstration of one of Faraday's first experiments with induction. A coil of wire is connected to a sensitive ammeter, as shown in Fig. 17.30. As a magnet is moved toward the coil, the ammeter is deflected, which indicates the presence of a current in the coil. The magnitude of the current depends on the speed with which the magnet is moved, and the current stops whenever the magnet is at rest. That is, a magnet at rest near the coil does not generate a current. As the magnet is moved away from the coil, the ammeter is deflected in the opposite direction, which indicates that the direction of the current is reversed.

FIGURE 17.30

Magnetic induction. A current is generated in the coil as the magnet is moved toward it.

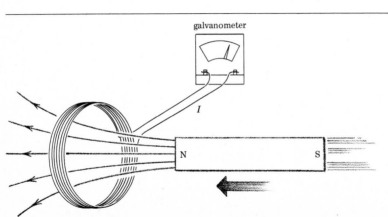

galvanometer

I

N   S

Additional experiments showed that an emf $\mathcal{E}$ is induced in a coil not only by moving a magnet toward (or away from) the coil, but also by creating (or destroying) a current in another nearby coil. The changing magnetic field of the secondary coil has the same effect in inducing an emf as the moving magnet. These and other observations on magnetic induction show that the magnitude of the induced emf is related to the following quantity.

Definition The *magnetic flux** ϕ through a coil is the product of the component B_n of the magnetic field normal (perpendicular) to the coil, the area A of the coil, and the number of turns n:

$$\phi = nB_nA \qquad\qquad 17.13$$

This can also be written

$$\phi = nBA \cos\theta \qquad\qquad 17.14$$

where θ is the angle between the magnetic field **B** and the normal to the coil (Fig. 17.31).

> **UNITS** The unit of magnetic flux is the tesla-square meter $(T \cdot m^2)$, which is called a *weber* (Wb):
>
> $1 \text{ Wb} = 1 \text{ T} \cdot m^2$
>
> From this we see that a tesla is a weber per square meter (Wb/m^2). Until recently, the weber per square meter, rather than the tesla, was the preferred unit of magnetic field strength.

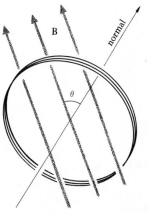

FIGURE 17.31

The magnetic flux ϕ through an n-turn coil of area A is nB_nA, where $B_n = B \cos\theta$ is the component of the magnetic field parallel to the normal to the coil.

The magnitude of the emf $\mathcal{E}$ induced in a coil is given by the following law.

Faraday's Law of Magnetic Induction *The emf $\mathcal{E}$ induced in a coil in a short time interval Δt is*

$$\mathcal{E} = \frac{\Delta\phi}{\Delta t} \qquad\qquad 17.15$$

where $\Delta\phi$ is the change of flux during this interval.

The emf induced in the coil produces an induced current I_i, which in turn gives rise to an induced magnetic field B_i and an induced flux ϕ_i. The direction of the induced current is given by another law.

Lenz's Law *The direction of the induced current in a coil is such that its magnetic field B_i opposes the change of flux $\Delta\phi$ that produced it.*

Example 17.7 The perpendicular component of the external magnetic field through a 10-turn coil of radius 5 cm increases from 0 to 18 T in 0.3 s, as shown in Fig. 17.32. (a) If the resistance of the coil is 2 Ω, what is the magnitude of the induced current? (b) What is the direction of the current?

(a) The initial flux ϕ_1 is zero, and from Eq. 17.13, the final flux ϕ_2 is

$$\phi_2 = nB_nA = (10)(18 \text{ T})(25\pi \times 10^{-4} \text{ m}^2) = 1.41 \text{ Wb}$$

*ϕ is the Greek (lowercase) letter *phi*.

FIGURE 17.32

As the external field **B** through the coil is increased, a current I_i is induced which produces an induced field B_i directed opposite to **B** inside the coil.

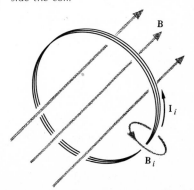

so

$$\Delta\phi = \phi_2 - \phi_1 = 1.41\,\text{Wb}$$

From Eq. 17.15, the induced emf is

$$\mathcal{E} = \frac{\Delta\phi}{\Delta t} = \frac{1.41\,\text{Wb}}{0.3\,\text{s}} = 4.7\,\text{V}$$

and thus the induced current is

$$I_i = \frac{\mathcal{E}}{R} = \frac{4.7\,\text{V}}{2\,\Omega} = 2.35\,\text{A}$$

(b) From Lenz's law, the current is directed in such a way that the magnetic field it produces opposes the change of flux. Using the right-hand rule, we see that a counterclockwise current would produce an induced field $\mathbf{B}_i$ that opposes the increase in the external flux through the coil. Hence the induced current is counterclockwise, as shown in Fig. 17.32. If the external field goes from 18 T to 0, the induced current would be clockwise so as to produce a magnetic field that opposes the decrease in the external flux. ☐

ELECTRIC GENERATOR

An *electric generator* is a seat of emf that uses the motion of a coil in a magnetic field to create a current. Figure 17.33 shows an n-turn coil of area A mounted on a shaft. The coil is located in a uniform magnetic field $\mathbf{B}$, so at any time t the flux ϕ through the coil is

$$\phi = nAB_n = nAB\cos\theta$$

where θ is the angle between $\mathbf{B}$ and the normal to the coil. If the coil is rotated at a uniform angular speed ω (in rad/s), the angle θ (in radians) is

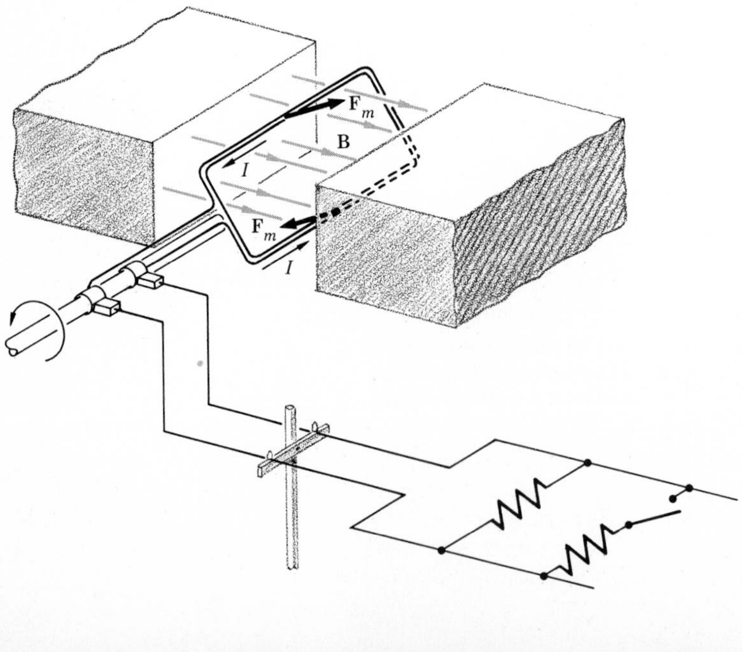

FIGURE 17.33

An electric generator. As the n-turn coil is rotated in the magnetic field, the changing flux generates an alternating emf. This emf is applied to an external circuit by means of brushes that rub against the slip rings mounted on the shaft.

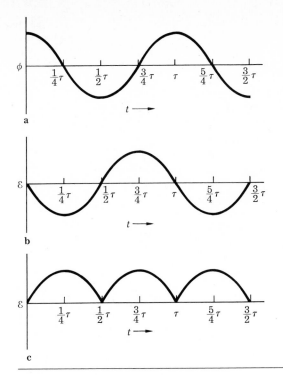

FIGURE 17.34

(a) Plot of the flux ϕ through a generator coil against time. (b) Plot of the emf $\mathcal{E}$ in a generator coil against time. The emf is greatest at times when the flux is changing most rapidly. (c) Plot of the emf $\mathcal{E}$ against time delivered to an external circuit when a commutator is used instead of slip rings.

$$\theta = \omega t$$

and the flux is

$$\phi = nAB \cos \omega t$$

A graph of ϕ against t is shown in Fig. 17.34a. The coil rotates through a complete circle in the time

$$\tau = \frac{2\pi}{\omega}$$

called the *period*. In this time, the flux makes one complete oscillation, as shown in Fig. 17.34a.

According to Faraday's law of induction, the emf $\mathcal{E}$ induced in the coil at time t is equal to the slope $\Delta\phi/\Delta t$ of the ϕ-t curve. That is, at times $0, \frac{1}{2}\tau, \frac{3}{2}\tau, \dots$, when the slope is zero, the induced emf is zero. At times $\frac{1}{4}\tau, \frac{3}{4}\tau, \frac{5}{4}\tau, \dots$, when the slope is greatest, the emf is greatest. Thus $\mathcal{E}$ has the time variation shown in Fig. 17.34b.

The ends of the coil are brought to slip rings on the shaft, where they are connected by brushes to an external circuit (Fig. 17.33). As the coil is rotated, the alternating potential difference between the ends of the coil is applied to the external circuit, creating an alternating current in the circuit.

According to Lenz's law, the direction of the current in the circuit always opposes the motion of the coil. That is, at any instant, the magnetic torque on the current in the coil opposes the angular motion of the coil. Work must be done by an external agent to keep the coil rotating. In a modern generating plant this work is done by a steam turbine, which drives the shaft. The steam, in turn, is produced either from burning fossil fuel (coal, gas, or oil), or from a nuclear reactor.

When you turn on a light, you add a resistor in parallel to the total resistance already in the external circuit (Fig. 17.33). This decreases the equivalent circuit resistance and increases the current in the circuit. The immediate effect of this increased current in the coil is to increase the magnetic torque on the coil. The angular speed of the coil thus starts to decrease. This speed is constantly being monitored at the power station, and as soon as a decrease in speed is detected, more steam is sent to the turbine to overcome the added torque. A power station keeps the coil rotating at constant speed by adjusting the work done by the turbine in response to the changing load. If the demand for electricity exceeds the capacity of the generator, a second generator must be brought on line, or the entire system may fail (a blackout). The power company must anticipate the daily fluctuations in electric demand in order to have the additional generators available when they are needed.

> **REMARK** Before the development of magnetic-induction generators, engineers had worked only with the direct current produced by batteries. Electroplating requires this type of current, and early motors and lighting systems had been designed for it. Many people, led by Thomas Edison (1847–1931), believed that alternating current was either dangerous or useless, so they built generators with commutators like the one on the dc motor in Fig. 17.18. The commutator reverses the connection between the generator and the external circuit each half-cycle, producing an emf that is always positive (Fig. 17.34c). By using several coils mounted at different angles on the same shaft, a nearly constant direct current can be produced. The first network of power plants built by Edison and his followers in the 1880s were of this type. A fierce battle arose when George Westinghouse (1846–1914) and his company introduced alternating current for commercial use. Alternating current, because of its many advantages, ultimately won, but not before execution by electrocution was introduced to try to prejudice people against it.

ELECTROMAGNETIC WAVES

The concept of the electric and magnetic field was developed by Faraday as a pictorial way of viewing electromagnetic phenomena. The field concept was not generally used by other physicists, however, until Maxwell showed how all the basic laws of electricity and magnetism could be described by four equations involving the electric and magnetic fields. Known as *Maxwell's equations*, they are the starting point for all modern discussions of electromagnetism.

As originally formulated, each of Maxwell's equations is a quantitative description of one of the following four laws:

1 Electric charges produce electric fields (*Coulomb's law*, Sec. 15.2).

2 Isolated magnetic poles do not exist (Sec. 17.1).

3 Currents produce magnetic fields (*electromagnetism*, Sec. 17.2).

4 A changing magnetic flux produces an electric field (*magnetic induction*, Sec. 17.4).

Maxwell soon realized, however, that these equations were inconsistent with another fundamental law:

5 The total charge of an isolated system cannot change (*conservation of charge*, Sec. 15.2).

Maxwell could eliminate this inconsistency only by assuming still another law:

6 A changing electric field produces a magnetic field.

This last law, which is the counterpart of magnetic induction, modifies one of Maxwell's equations in such a way that together they are now consistent with the conservation of charge.

Although the effect of a changing electric field is too small to be observed directly, it has profound consequences. For if a changing electric field produces a magnetic field and a changing magnetic field produces an electric field, then the possibility exists that a self-perpetuating disturbance of electric and magnetic fields might be produced. Maxwell showed that such an electromagnetic disturbance would act as a wave that propagates through space with the speed

$$v = \sqrt{\frac{1}{\epsilon_0 \mu_0}}$$

where $\epsilon_0 = 8.85 \times 10^{-12} \, C^2/(N \cdot m^2)$ is the electric permittivity, and $\mu_0 = 4\pi \times 10^{-7} \, N \cdot s^2/C^2$ is the magnetic permeability. With these values of the constants, the calculated speed is

$$v = \frac{1}{\sqrt{\epsilon_0 \mu_0}} = \frac{1}{\sqrt{[8.85 \times 10^{-12} \, C^2/(N \cdot m^2)](4\pi \times 10^{-7} \, N \cdot s^2/C^2)}}$$
$$= \sqrt{9 \times 10^{16} \, m^2/s^2} = 3 \times 10^8 \, m/s$$

which is identical to the measured speed of light.

In an electromagnetic wave, it is the variations in the fields themselves, rather than the displacement of a medium, that propagate through space. Figure 17.35 shows how the electric and magnetic fields vary in a sinusoidal electromagnetic wave. Both fields are perpendicular to the direction of propagation, so the wave is transverse.

FIGURE 17.35

An electromagnetic wave. The electric and magnetic fields are perpendicular to each other and to the direction of propagation of the wave. Both **E** and **B** vary sinusoidally.

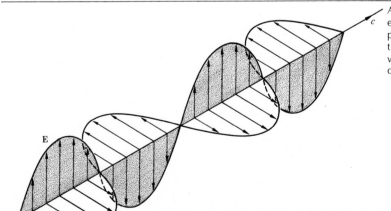

As a result of Maxwell's work, it soon became established that light is just a form of electromagnetic radiation of a certain wavelength. Thus the centuries of research into electric and magnetic phenomena led unexpectedly to the discovery of the true nature of light. This discovery is one of the major triumphs of nineteenth-century physics.

17.5 Inductors

Definition An *inductor* is a circuit element composed of an insulated wire coiled around an air or iron core.

RESISTOR-INDUCTOR (*RL*) CIRCUIT

Figure 17.36 is a diagram of a resistor-inductor circuit, which is a circuit in which an inductor, symbolized by 〰〰, is connected in series with a resistor R and a battery $\mathcal{E}$. Any resistance of the inductor itself is included in the value of R; i.e., the symbol 〰〰 represents a resistanceless element. The maximum current I_m in the circuit is the current

$$I_m = \frac{\mathcal{E}}{R}$$

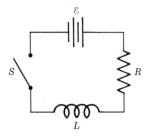

FIGURE 17.36

An inductor L connected in series to a battery $\mathcal{E}$ and a resistor R.

that would exist without the inductor. This is the steady-state current that is established shortly after the switch S is closed. But during the transition period from the closing of S to the establishment of I_m, there is a transient current I that increases with time. At any instant the current I in the inductor produces a magnetic field $\mathbf{B}$ and a flux ϕ through the coil.

Definition The *inductance* L of an inductor is the ratio of the flux ϕ through the inductor coil to the current in it:

$$L = \frac{\phi}{I} \quad \text{or} \quad \phi = LI \qquad\qquad 17.16$$

Inductance, like capacitance, has a characteristic value for each element that is independent of the current.

> **UNIT** The unit of inductance is the weber per ampere (Wb/A), which is called a *henry* (H):
>
> $1\,\text{H} = 1\,\text{Wb/A}$
>
> **REMARK** The flux through an inductor coil is produced by the inductor's own magnetic field. (For this reason L is sometimes called *self-inductance*.) The primary flux through the coil of a generator, on the other hand, is produced by an external magnetic field.

An inductor has no effect on a steady current, but consider what happens immediately after the switch S in Fig. 17.36 is closed. Before the switch is closed, the flux ϕ in the inductor is zero, and some time after the switch is closed, the flux reaches its maximum value

$$\phi_m = LI_m$$

During the transition period, the increasing flux induces an emf $\mathcal{E}$ in the inductor, given by Faraday's law of inductance (Eq.

17.15). This induced emf, in accordance with Lenz's law, produces an induced current I_i that opposes the increase in flux. That is, an inductor resists a change of current by generating an emf to oppose the change. As a consequence, it takes a certain time for a current to be established in a circuit with inductance.

To be specific, if the switch S in Fig. 17.36 is closed at time $t = 0$, the current I at any later time t is given by

$$I = \frac{\mathcal{E}}{R}(1 - e^{-(R/L)t})$$

$$= I_m(1 - e^{-t/\tau}) \qquad\qquad 17.17$$

where

$$I_m = \frac{\mathcal{E}}{R} \qquad\qquad 17.18$$

is the steady-state, or maximum, current in the circuit, and

$$\tau = \frac{L}{R} \qquad\qquad 17.19$$

is the *time constant*. From Eq. 17.17 we see that at $t = 0$, the current is

$$I = I_m(1 - e^0) = 0$$

and at $t = \tau$, the current is

$$I = I_m(1 - e^{-1}) = 0.63\, I_m$$

Thus the time constant is the time required for the current to reach 63 percent of its maximum value. Figure 17.37 is a graph of Eq. 17.17, showing how the current in an RL circuit builds up to its steady-state value.

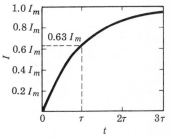

FIGURE 17.37

Plot of the current I in an RL circuit against the time elapsed since the circuit was closed.

Example 17.8 A 15-mH inductor is connected in series to a 240-Ω resistor and a 60-V battery. (a) What is the maximum current in the circuit? (b) What is the time constant of the circuit? (c) What is the current in the circuit 1×10^{-4} s after the switch is closed?

(a) From Eq. 17.18, the maximum current is

$$I_m = \frac{\mathcal{E}}{R} = \frac{60\text{ V}}{240\ \Omega} = 0.25\text{ A}$$

(b) From Eq. 17.19, the time constant is

$$\tau = \frac{L}{R} = \frac{15 \times 10^{-3}\text{ H}}{240\ \Omega} = 6.25 \times 10^{-5}\text{ s}$$

(c) From Eq. 17.17, the current I at time $t = 1 \times 10^{-4}$ s is

$$I = I_m(1 - e^{-t/\tau})$$

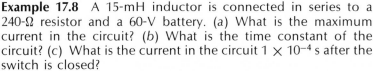

$$= (0.25\text{ A})(1 - e^{-1.6}) = 0.20\text{ A} \qquad \square$$

Just as it requires time to establish a current in a circuit with an inductor, it also requires time to stop a current. When a switch is opened, the existing current cannot stop instantane-ously, because the sudden change in current induces an emf in the inductor that opposes the change. The inductor thus con-tinues to generate a current for a while after the switch is opened. This current is noticeable as the spark that jumps be-tween the contact points of the switch when it is opened.

Inductors of various sizes are manufactured for use in circuits. However, it is important to realize that every circuit element has some inductance, even if the element is made for some other purpose. For instance, a light-bulb filament is a tiny coil of wire, so it has a small inductance. Motors, because of their many windings, have relatively large inductances. A special switch is required to open a circuit with a large current and a large in-ductance because of the damaging spark that is produced.

A *circuit breaker* is a switch that automatically opens when the current in the circuit exceeds a certain value. Small circuit breakers are used to protect household wiring, and large circuit breakers are used to protect major transmission lines. Figure 17.38 shows a circuit breaker used to protect a 135-kV line. The switch is immersed in oil, and special thermal conductors are used to dissipate the energy released by the spark that results when the switch is opened.

FIGURE 17.38

Circuit breaker. (*General Electric.*)

INDUCTORS IN SERIES AND PARALLEL

Definition Two or more inductors are said to be in *series* when they are connected so that the current is the same in each of them. Figure 17.39 shows n inductors $L_1, L_2, \ldots, L_n$ in series.

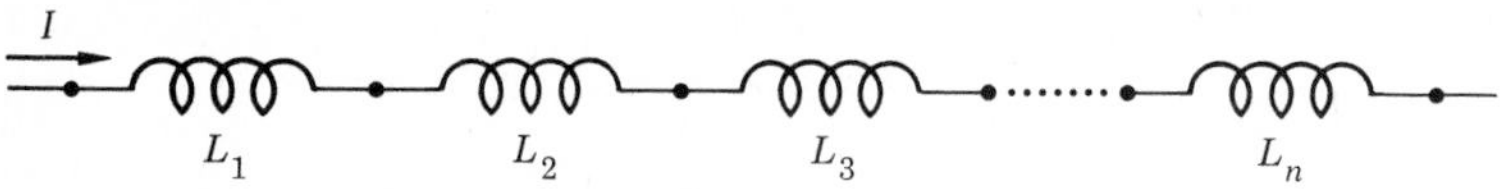

FIGURE 17.39

Series connection of n inductors $L_1, L_2, \ldots, L_n$.

The total inductance L of the inductors in Fig. 17.39 is

$$L = \sum_{i=1}^{n} L_i \qquad \text{series} \qquad \text{17.20}$$

Thus inductors in series combine like resistors in series. The total inductance of a series array of inductors is greater than the largest individual inductance.

Definition Two or more inductors are said to be in *parallel* when they are connected so that the voltage is the same across each of them. Figure 17.40 shows n inductors $L_1, L_2, \ldots, L_n$ in parallel.

The equivalent inductance L of the inductors in Fig. 17.40 is given by

$$\frac{1}{L} = \sum_{i=1}^{n} \frac{1}{L_i} \qquad \text{parallel} \qquad \text{17.21}$$

Thus inductors in parallel combine like resistors in parallel. The total inductance of a parallel array of inductors is less than the smallest individual inductance.

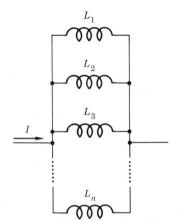

FIGURE 17.40

Parallel connection of n inductors $L_1, L_2, \ldots, L_n$.

GUIDE TO MAJOR TOPICS

PROBLEMS

1 What is the magnitude of the magnetic field at a point 2 cm from a long wire carrying a current of 7 A?
Ans. 0.7×10^{-4} T

2 The magnitude of the magnetic field at a point 8 cm from a long wire is 0.3×10^{-4} T. (a) What is the current in the wire? (b) What is the magnitude of the field at a point 24 cm from the wire?

3 A long wire carrying a current of 15 A produces a magnetic field of 0.35×10^{-4} T at a nearby point. What would be the magnitude of the field at this point if the current were 20 A?
Ans. 0.467×10^{-4} T

4 (a) Show that $1\,T = 1\,kg/(C \cdot s)$. (b) Show that μ_0 has the units $N/A^2 = kg \cdot m/C^2$.

5 Figure 17.41 shows two long parallel wires separated by a distance of 15 cm. There is a current of 5 A in wire 1 and an opposite current of 10 A in wire 2. (a) Find the total magnetic field at point A on the line joining the wires. Point A is 9 cm from wire 1 and 24 cm from wire 2. (b) At what point on the line joining the wires is the magnetic field zero?
Ans. (a) 0.278×10^{-5} T; (b) 15 cm from wire 1

6 Find the total magnetic field at point C produced by the two currents in Fig. 17.41. Point C is 8 cm from wire 1 and 20 cm from wire 2. (*Hint:* The total field at a point is the vector sum of the fields resulting from each wire separately.)

7 Figure 17.42 shows two long parallel wires separated by a distance of 18 cm. There is a

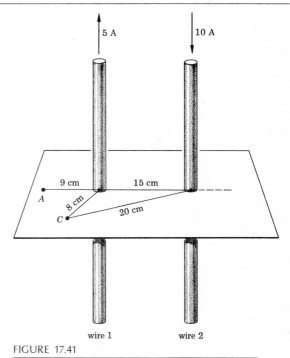

FIGURE 17.41

Problems 5, 6, and 15.

current of 8 A in wire 1 and a current of 12 A in wire 2. (a) Find the total magnetic field at point A, which is on the line joining the wires and 3 cm from wire 1 and 15 cm from wire 2. (b) At what point on the line joining the wires is the magnetic field zero?
Ans. (a) 3.73×10^{-5} T; (b) 7.2 cm from wire 1

521

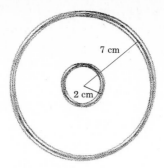

FIGURE 17.43

Problems 11 and 12.

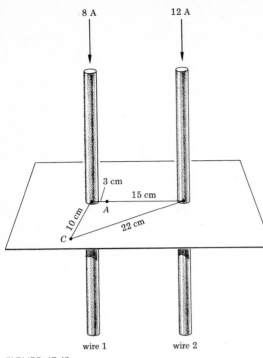

8 A   12 A

3 cm
15 cm
A
10 cm
22 cm
C

wire 1   wire 2

FIGURE 17.42

Problems 7, 8, and 16.

8 Find the total magnetic field at point C produced by the two currents in Fig. 17.42. Point C is 10 cm from wire 1 and 22 cm from wire 2.

9 A circular coil of radius 4 cm consists of 250 turns of wire in which the current is 20 mA. What is the magnetic field in the center of the coil?
Ans. 0.785 G

10 A coil of radius 20 cm is to produce a field of 0.4 G in its center with a current of 0.25 A. How many turns must there be in the coil?

11 Figure 17.43 shows a coil of radius 2 cm concentric with a coil of radius 7 cm. Each coil has 100 turns. With a current of 5 A in the larger coil, find the currents needed in the smaller coil to give the following values for the total magnetic field at the center: (a) 9.0×10^{-3} T, (b) 2.0×10^{-3} T, (c) zero. In each case, determine whether the direction of the current in the smaller coil is the same as the current in the larger coil or opposite.
Ans. (a) 1.44 A, same; (b) 0.793 A, opposite; (c) 1.43 A, opposite

12 Figure 17.43 shows a coil of radius 2 cm concentric with a coil of radius 7 cm. Each coil has 100 turns. With a current of 3 A in the small coil, find the total magnetic field at the center

for each of the following currents in the larger coil: (a) +7 A, (b) −5 A, (c) −12 A. The minus sign indicates that the direction of the current in the larger coil is opposite to the current in the smaller coil.

13 A wire loop is suspended between the poles of a magnet, as shown in Fig. 17.44. The horizontal section of the loop is 8 cm long, and the magnetic field is 3000 G. (a) What is the force on the loop when there is a current of 5 A in it? (b) What current is required to produce a force of 2.5 N on the loop?
Ans. (a) 0.12 N; (b) 104 A

14 A wire loop is suspended between the poles of a magnet, as shown in Fig. 17.44. When the current in the loop is 3 A, there is a magnetic force of 0.25 N on the 8-cm-long horizontal section of the loop. (a) What is the magnitude of the magnetic field? (b) What is the force on the loop when the current is 35 A?

15 (a) Calculate the force that wire 1 in Fig. 17.41 exerts on a 1-m-long section of wire 2. That is, calculate the force per unit length on wire 2. (b) Is this force attractive or repulsive? (*Hint:* Use Eq. 17.2 to find the magnetic field at wire 2 resulting from wire 1. Then use Eq. 17.5 to find the force on wire 2.)
Ans. (a) 6.67×10^{-5} N; (b) repulsive

REMARK The ampere, not the coulomb, is the fundamental electrical unit in the SI. The *ampere* is defined as the current which, when carried by two very long, parallel wires separated by a distance of one meter, results in a magnetic force of 2×10^{-7} newtons per meter on each wire. The *coulomb* is then defined as the amount of charge transported by a current of one ampere in one second.

16 Repeat Prob. 15 for the wires in Fig. 17.42.

17 The coil resistance of an ammeter is 5 Ω, and the ammeter is fully deflected when there is a current of 2 mA in the coil. (a) What shunt resistance should be used to give the ammeter

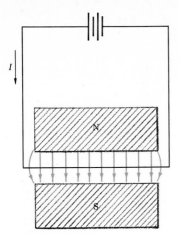

FIGURE 17.44

Problems 13 and 14.

a full-scale deflection of 50 mA? (b) What is the total resistance of the ammeter in this case?

Ans. (a) $0.208\,\Omega$; (b) $0.200\,\Omega$

18   To protect an ammeter from damage from excessive current in the coil, a large protective resistance R_p is sometimes placed in series with the coil resistance R_c, as shown in Fig. 17.45.

FIGURE 17.45

Problem 18.

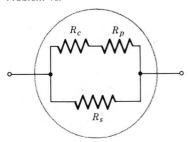

The shunt resistance R_s is made of a material which melts when the current in it exceeds some critical value I_s. Suppose that $I_s = 2\,A$, $R_c = 2\,\Omega$, and the ammeter is fully deflected when there is a current of 5 mA in the coil. (a) Find the values of R_p and R_s that will cause the ammeter to have full-scale deflection with a current of 100 mA and will cause the shunt resistance to melt when a potential difference of 10 V or more is placed across the ammeter. (b) If the ammeter is accidently connected to a 100 V potential difference, what will be the current in the coil?

REMARK   Without the protective resistance the coil would burn out if a large voltage were placed across the ammeter terminals.

19   The coil resistance of an ammeter is $3\,\Omega$, and the ammeter is fully deflected when there is a current of 6 mA in the coil. What resistance R_v should be placed in series with the coil to make a voltmeter with a full-scale deflection of 50 V?

Ans. $8330\,\Omega$

20   A 10-V battery is connected to resistances R_1 and R_2, as shown in Fig. 17.46. (a) If $R_1 = R_2 = 500\,\Omega$, what is the potential difference across R_2? (b) A voltmeter with a total resistance of $5000\,\Omega$ is used to measure the potential across R_2. What does the voltmeter read?

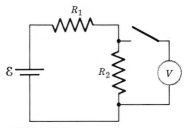

FIGURE 17.46

Problem 20.

REMARK   Because the resistance of a voltmeter is not infinite, it modifies the circuit it is measuring to some extent.

21   An electron is moving with a speed of 5×10^6 m/s at right angles to a magnetic field of 4000 G. (a) What is the force on the electron? (b) What is the centripetal acceleration of the electron? (c) What is the radius of the circle in which the electron moves?

Ans.   (a) 3.2×10^{-13} N;   (b) 3.5×10^{11} m/s^2; (c) 8.4×10^{-3} m

22   Figure 17.47 shows an electric field **E** and a magnetic field **B** at right angles to each other. A

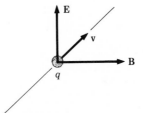

FIGURE 17.47

Problem 22.

particle of charge q moves into the plane of the paper, at right angles to both fields. Show that the particle will not be deflected by these fields if its speed v is equal to E/B.

REMARK   An arrangement of crossed electric and

magnetic fields like the one in Prob. 22 is used to select particles of a given speed.

23 A *cyclotron* is a machine that accelerates protons and other charged particles to high speed. It consists of a large magnet that produces a region of nearly uniform magnetic field strength. The protons, injected into the center of this field, move in a circle perpendicular to the lines of force (Fig. 17.48). At every half-turn the protons are accelerated through a potential difference of several thousand volts, which moves them into a circle of larger radius. After several thousand orbits, the protons emerge at the circumference of the field with an energy of several million electronvolts. (a) The field of a cyclotron has a strength of 1.2 T and a radius of 0.8 m. What is the energy of a proton that emerges from the cyclotron at this radius? (b) If the proton is injected into the center of the cyclotron with an initial energy of 10 keV (1 keV = 10^3 eV = 1.6×10^{-16} J), what is the radius of its initial orbit?
Ans. (a) 44 MeV; (b) 1.2 cm

24 Show that the time τ required for a particle of mass m and charge q to move around one complete circle in a uniform magnetic field **B** is

$$\tau = \frac{2\pi m}{Bq}$$

Note that τ is independent of the radius of the

FIGURE 17.48

Problem 23.

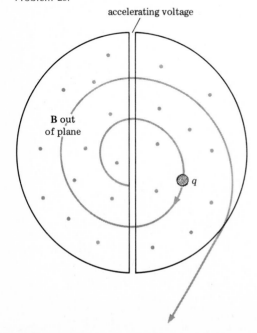

accelerating voltage

B out of plane

q

orbit and the speed of the particle.

REMARK The *cyclotron frequency* f_c of a uniform magnetic field is the number of complete orbits that a particle of charge q and mass m makes in one second. From Prob. 24 we see that

$$f_c = \frac{1}{\tau} = \frac{Bq}{2\pi m}$$

25 A mass spectrometer with a radius of 8 cm is designed to detect oxygen molecules when the voltage across the ionization chamber is 300 V. (a) What is the magnitude of the magnetic field? (b) What voltage is required to detect carbon atoms?
Ans. (a) 0.176 T; (b) 800 V

26 A mass spectrometer with a magnetic field of 0.21 T detects nitrogen molecules when the voltage across the ionization chamber is 600 V. What is the radius of the spectrometer?

27 A mass spectrometer of unknown magnetic-field strength and unknown radius is found to detect nitrogen molecules when the voltage across the ionization chamber is 365 V. What is the voltage at which oxygen molecules are detected?
Ans. 319 V

28 In a mass spectrometer, nitrogen molecules are detected at a voltage 40 V greater than the voltage at which oxygen molecules are detected. (a) What is the voltage at which oxygen molecules are detected? (b) A peak is observed at a voltage 56 V less than the voltage of the oxygen peak. What is the mass of this other peak?

29 A magnetic field of magnitude $B = 0.05$ T is perpendicular to a 200-turn coil of radius 1.5 cm (Fig. 17.49). (a) What is the flux through the coil? (b) If the field decreases to $B = 0.02$ T in 0.1 s, what is the magnitude of the emf induced in the coil? Indicate on Fig. 17.49 the

FIGURE 17.49

Problem 29.

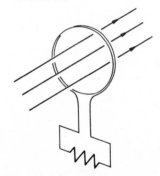

524

direction of the induced current. (c) Repeat (b) for an increase in B from 0.05 to 0.07 T in 0.2 s.

Ans. (a) 7.07 mWb; (b) 42.4 mV; (c) 14.1 mV

30   Figure 17.50 shows a long, narrow rectangular loop parallel to a long, straight wire carrying a current of 25 A. (a) When the center of the loop is 10 cm from the straight wire, what is the flux through the loop? (Use Eq. 17.2 to calculate the field at the center of the loop resulting from the current, and assume that this is the average field across the loop.) (b) The loop is moved 5 cm farther from the wire in 0.02 s. Find the emf induced in the loop during this time, and indicate the direction of the induced current.

31   Figure 17.51 shows a rectangular loop being moved out of a uniform magnetic field **B**. (a) Show that if the loop moves at constant velocity **v** in the direction shown, the emf induced in the loop is BLv. (b) Indicate the direction of the induced current. (c) Indicate the direction of the magnetic force that the magnetic field exerts on the induced current.

REMARK   Whenever a conductor is moved through a magnetic field, currents like the one in Prob. 31 are induced in the conductor. These are called *eddy currents*. The magnetic force on the eddy currents always opposes the motion of the conductor, in accordance with Lenz's law.

32   Figure 17.52 shows a U-shaped wire at right angles to a uniform magnetic field **B**. A straight wire contacts the arms of the U, completing a circuit. The U-shaped wire is resistanceless, and the straight wire has a resistance R. (a) Show that if the straight wire is moved

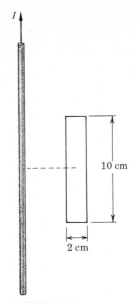

FIGURE 17.50

Problem 30.

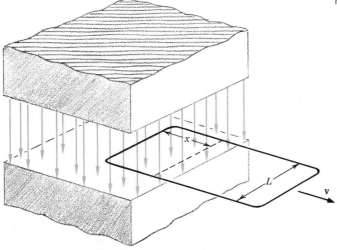

FIGURE 17.51

Problem 31.

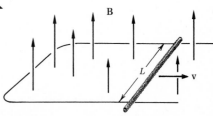

FIGURE 17.52

Problem 32.

to the right at constant velocity **v**, the current I induced in the loop is BLv.

$$I = \frac{BLv}{R}$$

where L is the distance between the arms of the U. (b) Show that the magnitude of the magnetic force on the moving wire is

$$F_m = \frac{B^2 L^2 v}{R}$$

and is directed opposite to **v**. (c) A force $\mathbf{F}_a$ equal to $-\mathbf{F}_m$ must be applied to the straight wire to keep it moving at constant velocity. Show that the power $F_a v$ needed to sustain this force is just equal to the power RI^2 dissipated in the resistance of the wire.

> **REMARK** The last problem illustrates how the mechanical work needed to maintain the motion of an electric generator is just equal to the electric energy delivered by the generator.

33 A 50-mH inductor is connected in series with a 200-Ω resistor and a 25-V battery. (a) What is the steady-state current in the circuit? (b) What is the time constant of the circuit? (c) How long after the circuit is closed does it take for the current to reach 95 percent of its steady-state value?
Ans. (a) 125 mA; (b) 0.25 ms; (c) 0.749 ms

34 A 75-mH inductor is connected in series with a resistor R and a battery $\mathcal{E}$. (a) What values of R and $\mathcal{E}$ are needed for a time constant of 2 ms and a steady-state current of 100 mA? (b) What is the current in the circuit 3 ms after the switch is closed?

35 A 30-mH inductor is connected in series with a 60-Ω resistor and a 120-V battery. (a) Find the current in the circuit at time $t = 0.1\,\tau$, where τ is the time constant of the circuit. (b) What is the change in the flux through the inductor between $t = 0$ and $t = 0.1\,\tau$? (c) What is the emf induced across the inductor during this time interval?
Ans. (a) 0.19 A; (b) 5.7 mWb; (c) 114 V

> **REMARK** In the last problem we found that between $t = 0$ and $t = 0.1\,\tau$ there was an average voltage of 114 V across the inductor. This voltage opposes the emf of the battery, so on the average, only 6 V was available to maintain a current through the 60-Ω resistor. Thus the average current during this period is 0.1 A, which is consistent with the fact that the current increased from 0 to 0.19 A during this period.

36 (a) Repeat Prob. 35 for the time interval from $t = \tau$ to $t = 1.1\,\tau$. That is, find the average voltage induced across the inductor during this period. (b) Average the currents at $t = \tau$ and $t = 1.1\,\tau$ given by Eq. 17.17, and compare the value to the current calculated by dividing the difference between the voltages of the battery and the inductor by the resistance (see the preceding remark).

37 Find the equivalent inductance of the three inductors in Fig. 17.53.
Ans. 13.3 mH

38 Find the equivalent inductance of the three inductors in Fig. 17.54.

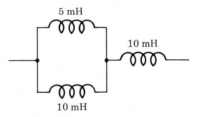

FIGURE 17.53

Problem 37.

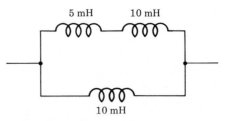

FIGURE 17.54

Problem 38.

BIBLIOGRAPHY

MACLAREN, MALCOLM: *The Rise of the Electrical Industry during the Nineteenth Century*, Princeton University Press, Princeton, N.J., 1943. Informal history of the discoveries, inventions, and enterprises that contributed to the commercial uses of electricity.

SHAMOS, MORRIS H.: *Great Experiments in Physics*, Holt, Rinehart and Winston, Inc., New York, 1959. Excerpts from original articles by Oersted, Faraday, and Maxwell discuss their important contributions to electromagnetism.

Electric utilities generate and distribute alternating current, as opposed to direct current, because the voltage of an alternating current is easily increased or decreased by a transformer. At the generating station, transformers "step up" the voltage for economical transmission over large distances; at substations along the transmission network, transformers "step down" the voltage for local distribution; just before the current reaches the customer, a local transformer further decreases the voltage. Thus alternating current is generated at one voltage, transmitted at much higher voltage, and utilized at relatively low voltage.

The first generating systems of the Edison Electric Company used direct current, because the need for long-distance transmission was not foreseen. The Westinghouse Companies installed the first alternating-current system in Buffalo, New York in 1886. A great struggle followed, in which the direct-current interests tried to discredit the rival system by claiming it was a danger to the public safety. The introduction of execution by electrocution in New York, using Westinghouse equipment, is believed to have been a scheme of the Edison people to promote fear of alternating current. In the end, however, the safety record of the alternating-current industry, and the great technical advantages of the system, led to its universal adoption.

18
ALTERNATING CURRENT

18.1 Generation of Alternating Current

An electric *generator* is a device that converts mechanical energy into electric energy. It operates on the basis of Faraday's law of induction (Eq. 17.15), which states that when the magnetic flux through a coil changes, an emf is generated. The magnetic field can be produced by a permanent magnet, although in all but the smallest generators, the field is produced by an electromagnet supplied with its own direct current. The change of flux results either from the rotation of the coil past the magnetic field or the rotation of the magnetic field past the coil.

The coils in which the emf is generated are mounted in a structure called the *armature*. The coils that produce the magnetic field are mounted in the *field structure*. The structure which rotates is called the *rotor,* and the stationary structure is called the *stator*. In most alternating-current (ac) generators, the armature is the stator and the field structure is the rotor. That is, in an ac generator the change in flux is produced by rotating a magnetic field past a stationary armature. In most direct-current (dc) generators, on the other hand, the reverse is the case; the change in flux is produced by rotating the armature past a stationary magnetic field. A generator of this latter type is described in Sec. 17.4.

Figure 18.1 gives a cross-sectional view of an idealized one-phase, two-pole ac generator. The rotor consists of a single electromagnet (two poles) that is excited by a direct current brought to it through slip rings. The armature is a stationary iron yoke with two slots in which the single armature coil is seated. The armature current passes in one direction in slot a and in the

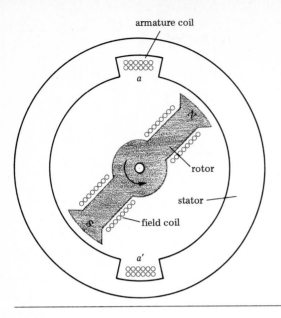

armature coil

a

rotor

stator

field coil

a'

FIGURE 18.1

Cross-sectional view of a one-phase, two-pole ac generator.

opposite direction in slot a'. The wires in the slots are connected in front and back of the armature to form a single coil, the free ends of which carry the generated current to the load. The iron yoke ensures that most of the magnetic field of the rotor passes through the armature coil.

The rotation of the rotor magnet produces an alternating flux through the armature coil, generating an alternating emf in the coil. This emf is given by the equation

$$\mathcal{E} = V_p \cos 2\pi ft \qquad\qquad 18.1$$

and is plotted in Fig. 18.2. Here V_p is the peak voltage, and f is the frequency in hertz. In the generator in Fig. 18.1, the emf goes through one complete cycle for each complete rotation of the

FIGURE 18.2

Plot of the emf against time generated by the ac generator.

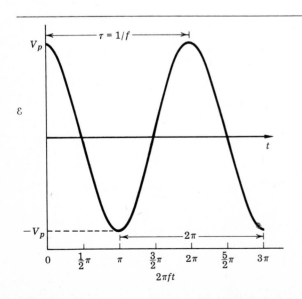

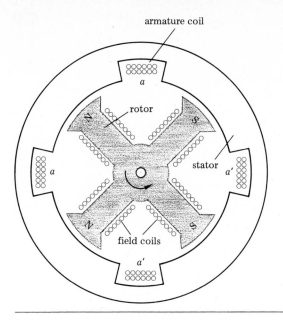

FIGURE 18.3

Cross-sectional view of a one-phase, four-pole ac generator.

rotor, so *f* is equal to the rotational frequency of the rotor. In the United States and Canada the frequency of commercial power is 60 Hz, or 60 c/s. The generator in Fig. 18.1 would have to rotate at 60 r/s (3600 r/min) to produce 60-Hz ac power.

Figure 18.3 shows a one-phase, four-pole ac generator. By mounting two field magnets on the rotor and two sets of armature coils on the stator, the induced emf goes through two complete cycles for each complete revolution of the rotor. Thus this generator has to rotate only at 1800 r/min to generate 60-Hz electricity.

Figure 18.4 shows a three-phase, two-pole ac generator. This

FIGURE 18.4

Cross-sectional view of a three-phase, four-pole ac generator.

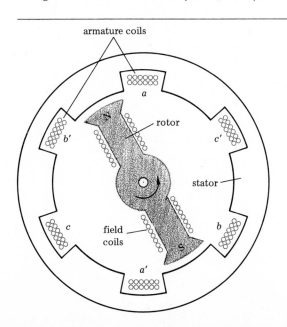

is more realistic than the previous figures. It has three independent armature coils (a, b, and c) mounted 120° apart. The coils produce separate emfs that peak at different times. In fact, because of the 120° separation, the peaks are one-third cycle apart. The separation of these peaks, measured in degrees or radians, is called the *phase separation*. Since 2π radians corresponds to one cycle, the phase separation is $2\pi/3$ rad, or 120°, as shown in Fig. 18.5. All commercial generators produce three-phase electricity this way, although for economical operation the armature of a real generator (Fig. 18.6) has many more coils around its circumference. These coils are connected into three separate groups to produce the three phases.

Thus a commercial generator can be thought of as three separate generators, the output of which is carried by six conductors (Fig. 18.7a). However, because of the fixed phase relation between these generators, the conductors can be joined in either the Y connection (Fig. 18.7b) or the Δ connection (Fig. 18.7c).

In the Y connection, one lead from each phase is brought out, and the other leads are joined to a common connection, which may or may not be brought out as a neutral lead. That is, a Y connection may have three or four output leads. In the Δ connection, the three phases are connected in series, and three leads are brought out, as shown in Fig. 18.7c.

The primary advantage of three-phase electricity is that it can be transmitted using less conducting cable (by weight) than single-phase electricity. A secondary advantage is that it delivers continuous, as opposed to pulsating, power for the operation of heavy machinery. These advantages will be discussed further in the next section.

FIGURE 18.5

Plot of the emfs against time generated by each phase of a three-phase ac generator.

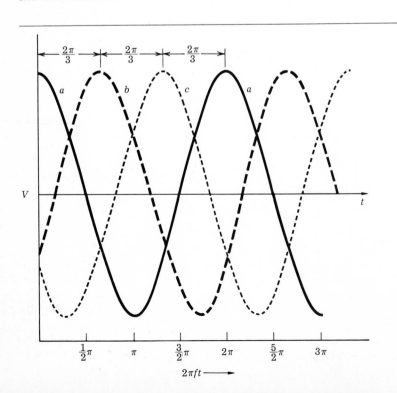

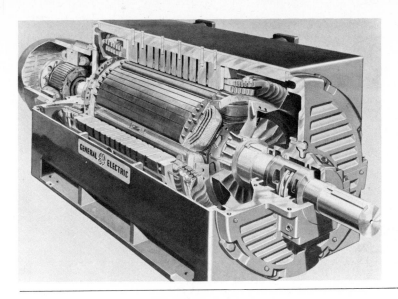

FIGURE 18.6

Cutaway view of a three-phase, eight-pole ac generator. The small generator mounted on the rear supplies the direct current for the rotor coils. (*General Electric.*)

For home and light industrial use, the three-phase current delivered by the utility company is broken into separate single-phase circuits. For instance, leads *a* and *o* in Fig. 18.7*b* deliver the ordinary 120-V current used for lighting and small appliances, while leads *a* and *b* deliver 208-V current used for heavy-duty appliances, such as clothes dryers and air conditioners. Although we shall be mostly concerned with the theory of single-phase alternating current in the remainder of this chapter, mention will be made of the practical use of three-phase current when appropriate.

18.2 Alternating Current in Pure Resistive Circuits

The simplest ac circuit consists of a single-phase generator connected to a resistor. This is shown in Fig. 18.8, where the symbol

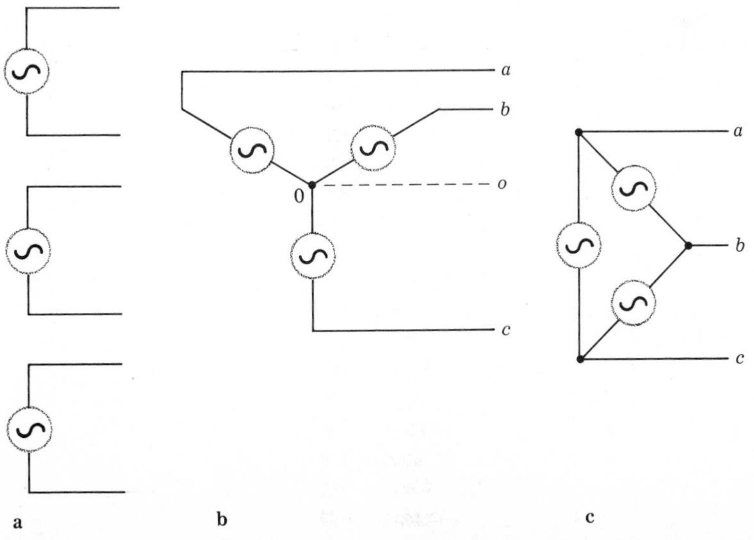

FIGURE 18.7

(*a*) The three phases of a three-phase generator act like three separate one-phase generators. (*b*) Y connection. One lead from each phase is connected to a common lead *o* and the other three leads (*a, b, c*) are brought out. (*c*) Δ connection. The three phases are connected in series and three leads are brought out.

a   b   c

—⌇— represents a single-phase generator, or one phase of a three-phase generator. The generator produces an oscillating voltage across its terminals, which is given by Eq. 18.1 and shown in Fig. 18.2. The instantaneous current I_x in the resistor is given by Ohm's law

$$I_x = \frac{V_x}{R} = \frac{\mathcal{E}}{R} = I_p \cos 2\pi ft \qquad 18.2$$

where

$$I_p = \frac{V_p}{R} \qquad 18.3$$

is the peak current. We shall use the subscript x to denote the instantaneous value of a time-dependent quantity; the subscript p denotes a time-independent quantity.

Example 18.1 Plot the current in the resistor in Fig. 18.8 if the peak voltage of the generator is 170 V, the frequency is 60 Hz, and the resistance is 50 Ω.
From Eq. 18.3 the peak current in the resistor is

$$I_p = \frac{V_p}{R} = \frac{170\ \text{V}}{50\ \Omega} = 3.4\ \text{A}$$

The period of the alternating current is

$$\tau = \frac{1}{f} = \frac{1}{60\ \text{Hz}} = 0.0167\ \text{s}$$

Thus the current oscillates between $+3.4$ A and -3.4 A, with a period of 0.0167 s. The equation for this current is

$$I_x = I_p \cos 2\pi ft = (3.4\ \text{A}) \cos 120\pi t$$

and Fig. 18.9 gives a graph of I_x against t.
When the current is positive, charge flows in one direction, and when the current is negative, charge flows in the other direction. Thus there is no net flow of charge around the circuit, but only a surging back and forth of charge in the resistor. ☐

POWER
At any instant the power P_x dissipated in an ac circuit is

$$P_x = V_x I_x$$

In a purely resistive circuit, where $V_x = RI_x$, the power can be written in three ways:

$$P_x = V_x I_x = RI_x^2 = \frac{V_x^2}{R} \qquad 18.4$$

Substituting Eq. 18.2 for the current in Eq. 18.4, we find that the power is

$$P_x = RI_x^2 = R(I_p \cos 2\pi ft)^2$$
$$= RI_p^2 \cos^2 2\pi ft$$

This is plotted in Fig. 18.10 for the current in Example 18.1.

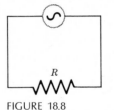

FIGURE 18.8

An ac generator connected to a re-sistor R.

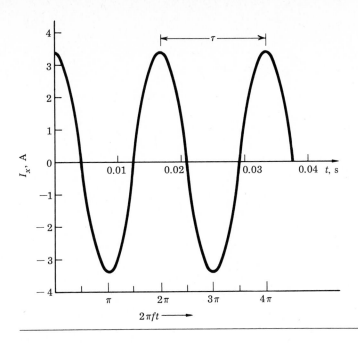

FIGURE 18.9

Plot of the instantaneous current I_x against time in a 50-Ω resistor connected to an ac generator with a frequency of 60 Hz and a peak voltage of 170 V.

Comparing Figs. 18.9 and 18.10, we see that although the current oscillates in sign, the power, which is proportional to the square of the current, is always positive. The peak power P_p in Example 18.1 is

$$P_p = RI_p^2 = (50\,\Omega)(3.4\,\text{A})^2 = 578\,\text{W}$$

and the instantaneous power varies between 0 and P_p twice each cycle.

It can be shown that the average power $\bar{P}$ is equal to half the peak power:

$$\bar{P} = \tfrac{1}{2}P_p \qquad \text{or} \qquad R\bar{I^2} = \tfrac{1}{2}RI_p^2 \qquad\qquad 18.5$$

FIGURE 18.10

Plot of the instantaneous power P_x against time in a 50-Ω resistor connected to an ac generator with a frequency of 60 Hz and a peak voltage of 170 V.

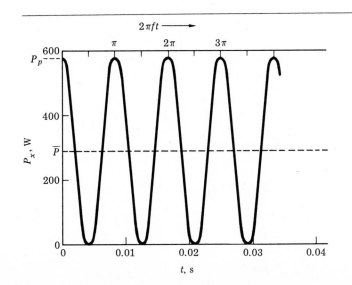

Here $\overline{I^2}$ is the average of the square of the current.

Definition The *root-mean-square*, or *rms*, *current* I_{rms} is the square root of the average of the square of the current:

$$I_{rms} = \sqrt{\overline{I^2}}$$

From Eq. 18.5 we see that

$$I_{rms} = \sqrt{\tfrac{1}{2}I_p^2}$$

or

$$I_{rms} = \frac{I_p}{\sqrt{2}} \qquad\qquad 18.6a$$

Definition The *root-mean-square*, or *rms*, *potential* V_{rms} is

$$V_{rms} = \frac{V_p}{\sqrt{2}} \qquad\qquad 18.6b$$

If we multiply both sides of Eq. 18.3 by $1/\sqrt{2}$, we get the relation

$$I_{rms} = \frac{V_{rms}}{R} \qquad\qquad 18.7$$

That is, the rms current and potential satisfy Ohm's law, just as the peak current and potential do.

In the United States and Canada, the rms potential supplied to a normal electric outlet is 120 V. Therefore, from Eq. 18.6*b*, the peak potential across the outlet is

$$V_p = \sqrt{2}\,V_{rms} = \sqrt{2}\,(120\text{ V}) = 170\text{ V}$$

In Europe the rms potential is 240 V, which corresponds to a peak potential of 340 V.

The advantage of working with rms currents and potentials is that they give the average power when used in Eq. 18.4. For instance, we have

$$RI_{rms}^2 = R\left(\frac{I_p}{\sqrt{2}}\right)^2 = \tfrac{1}{2}RI_p^2 = \tfrac{1}{2}P_p = \overline{P}$$

which shows that RI_{rms}^2 is equal to the average power $\overline{P}$. The other expressions in Eq. 18.4 give the same result, so we have

$$\overline{P} = V_{rms}I_{rms} = RI_{rms}^2 = \frac{V_{rms}^2}{R} \qquad\qquad 18.8$$

Example 18.2 (a) What is the resistance of an American toaster rated at 1200 W? (b) What would the power dissipated in this toaster be if it were plugged into an outlet in England?

(a) The power rating gives the average power at 120 V. Therefore, using Eq. 18.8, we find

$$R = \frac{V_{rms}^2}{\overline{P}} = \frac{(120\text{ V})^2}{1200\text{ W}} = 12\ \Omega$$

(b) If this toaster were plugged into an outlet in England, the average power would be

$$\bar{P} = \frac{V_{rms}^2}{R} = \frac{(240 \text{ V})^2}{12\,\Omega} = 4800 \text{ W}$$

At this power, the heating element in the toaster would melt. In general, an appliance designed to operate at one potential will not operate properly at a significantly different potential. ☐

Single-phase alternating current delivers pulsating power, as seen in Fig. 18.10. This causes lights to flicker 120 times a second, which is too rapid for the eye to detect. In the operation of heavy machinery, however, pulsating power can cause harmful vibrations. To avoid this, heavy electric motors run on three-phase current. As shown in Fig. 18.11, the power peaks of each phase are separated by one-sixth of a cycle. The total power at any instant is the sum of the instantaneous power of the three phases at that instant. It is a peculiarity of three-phase current that this sum is independent of time: it is always equal to 3 times the average power of a single phase. Thus three-phase alternating current delivers steady power like direct current, while retaining the voltage-transforming advantages of single-phase current.

PHASORS
The time variation inherent in alternating current requires special methods for study. The simplest method uses the fact that the x component of a vector **A** is $A \cos\theta$, where θ is the angle between **A** and the x axis.

Definition A *phasor* is a two-dimensional vector used to represent an alternating quantity such as current or voltage. The

FIGURE 18.11

Plot of the instantaneous powers against time delivered by the three phases of a three-phase generator. The sum of these three powers is a constant, independent of time.

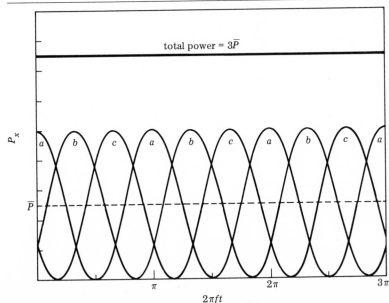

magnitude of the phasor is equal to the peak value of the quantity, and the angle θ between the phasor and the x axis is equal to the time factor $2\pi ft$.

Figure 18.12 shows how the voltage

$$V_x = V_p \cos 2\pi ft$$

is represented by the phasor **V** with magnitude V_p and angle $\theta = 2\pi ft$. The x component of this phasor,

$$V_x = V_p \cos\theta = V_p \cos 2\pi ft$$

is equal to the voltage at time t. If the phasor is imagined to rotate counterclockwise around the origin with constant angular speed

$$\omega = 2\pi f$$

the x component will equal the voltage at each instant.

In a pure resistive circuit, the current and voltage are proportional to each other,

$$I_p = \frac{V_p}{R} \cos 2\pi ft$$

so their phasors are aligned, as shown in Fig. 18.12. We shall see in the next section that this is no longer the case when inductors or capacitors are added to the circuit.

Phasors are very useful in studying three-phase current. In the Y connection (Fig. 18.13a), the line-to-neutral voltage between leads a and o is

$$V_{ao} \cos 2\pi ft$$

and is represented by the phasor **V**$_{ao}$ in Fig. 18.13b. Likewise, the line-to-neutral voltage between leads b and o is

$$V_{bo} \cos(2\pi ft + 2\pi/3)$$

and is represented by the phasor **V**$_{bo}$. The phasor **V**$_{bo}$ has the same magnitude as **V**$_{ao}$ but is always $2\pi/3$, or 120°, ahead of **V**$_{ao}$. The line-to-neutral voltage between leads c and o is

$$V_{co} \cos(2\pi ft + 4\pi/3)$$

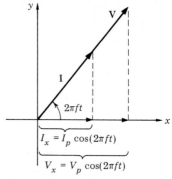

FIGURE 18.12

Phasors **V** and **I** representing the voltage and current in a pure resistive circuit. These phasors rotate in the plane with the angular speed $\omega = 2\pi f$, and the instantaneous voltage V_x and current I_x are the x components of the phasors at each instant.

FIGURE 18.13

(a) Y connection with neutral lead brought out. (b) Phasors representing the line-to-neutral voltages of each phase.

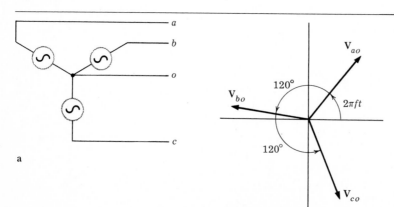

and is represented by the phasor $\mathbf{V}_{co}$, which is $4\pi/3$ radians, or 240°, ahead of $\mathbf{V}_{ao}$. The three phasors rotate like the spokes of a wheel around the origin, their x components at any instant being equal to the line-to-neutral voltages at that instant.

The line-to-line voltage between leads a and b is represented by the phasor $\mathbf{V}_{ab}$, which is the vector difference of $\mathbf{V}_{ao}$ and $\mathbf{V}_{bo}$. The relationship between the magnitudes of these phasors is shown in Fig. 18.14. From the right triangle OXA we see that

$$\cos 30° = \frac{\frac{1}{2}V_{ab}}{V_{ao}}$$

or

$$V_{ab} = (2\cos 30°)V_{ao} = \sqrt{3}\,V_{ao} \qquad\qquad 18.9$$

In other words, the peak line-to-line voltage in a Y connection is $\sqrt{3}$ times the peak line-to-neutral voltage. Likewise, the rms line-to-line voltage is $\sqrt{3}$ times the rms line-to-neutral voltage. For instance, if the rms line-to-neutral voltage is 120 V, the rms line-to-line voltage is $\sqrt{3}\,(120\,\text{V}) = 208\,\text{V}$. Thus a three-phase Y connection can supply either 120- or 208-V power.

Figure 18.15a shows the three phases of a three-phase generator connected to three separate circuits. A Y connection is obtained by having the circuits share the common return line oo' (Fig. 18.15b). The system is balanced when the resistances in the three circuits are equal. For a balanced system the current I in the line oo' is zero, so this line can be omitted altogether.

To understand this strange property of three-phase current, we note that the currents in the resistors have equal magnitudes (because the resistors are equal) and the same phase relations as the voltages. Thus these currents are represented by three phasors $\mathbf{I}_a$, $\mathbf{I}_b$, and $\mathbf{I}_c$ of equal magnitude and 120° apart (Fig. 18.16a). By Kirchhoff's first law, the current in the line oo' is the sum of the currents in the resistors, so the phasor $\mathbf{I}$ that represents current in oo' is the vector sum of $\mathbf{I}_a$, $\mathbf{I}_b$, and $\mathbf{I}_c$:

$$\mathbf{I} = \mathbf{I}_a + \mathbf{I}_b + \mathbf{I}_c$$

Figure 18.16b shows that this sum is zero. What is happening is that charge is surging back and forth between a and o', b and o', and c and o' in such a way that at every instance the total charge arriving at o' is zero. In effect, the outgoing lines ao', bo', and co' act as return lines for each other during different parts of the ac cycle.

Because a three-phase system can power three circuits with only three lines, it can transport power more economically than an equivalent one-phase system. A small neutral line (oo') is sometimes included, however, to carry the small current that results whenever the system is unbalanced.

18.3   Inductance and Capacitance in an ac Circuit

An inductor offers zero resistance to a steady direct current, but because an alternating current induces an emf that opposes the current in an inductor, an inductor does impede an alternating current. Likewise, a capacitor offers zero conductivity (infinite

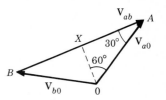

FIGURE 18.14

The phasor $\mathbf{V}_{ab}$ representing the line-to-line voltage between two lines of a three-phase system is the difference between the phasors $\mathbf{V}_{bo}$ and $\mathbf{V}_{ao}$ representing the line-to-neutral voltages.

FIGURE 18.15

(a) The three phases of a three-phase system connected by six leads to three different resistive loads. (b) A Y connection formed by combining three of the leads in (b). If the three resistances are equal, there will be no current in the common return line oo'.

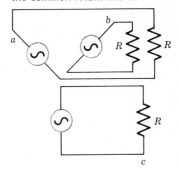

a

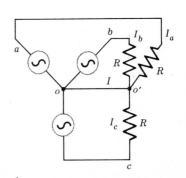

b

resistance) to a steady direct current, but because an alternating current alternately charges one side and then the other of a capacitor (Fig. 18.17), a capacitor does conduct an alternating current. Thus in an ac circuit, resistors, inductors, and capacitors all affect the current produced by a given voltage source.

We shall study the effect of these elements using phasors to represent the current and voltage in a circuit. The current represented by the phasor **I** has a peak value I equal to the magnitude of **I** and an instantaneous value I_x equal to the x component of **I**. The voltage represented by the phasor **V** has a peak value V equal to the magnitude of **V** and an instantaneous value V_x equal to the x component **V**. Both **I** and **V** rotate about the origin with the frequency of the current.

We shall first discuss the effect of resistors, inductors, and capacitors separately, and then show how their combined effect can be calculated.

RESISTIVE CIRCUIT

Figure 18.18a shows a purely resistive ac circuit with frequency f. At every instant the current in a purely resistive circuit with resistance R is related to the voltage by

$$\mathbf{I} = \frac{\mathbf{V}}{R} \qquad \text{resistance} \qquad 18.10$$

This equation indicates that the phasor **I** is aligned with the phasor **V** and has the magnitude $I = V/R$ (Fig. 18.18b). Because the phasors are aligned, the corresponding instantaneous current and voltage (I_x and V_x) peak at the same instant, as shown in Fig. 18.18c. In a purely resistive ac circuit, the current and voltage are said to be *in phase*.

The instantaneous power P_x delivered to the resistor is

$$P_x = V_x I_x$$

and the average power $\bar{P}$ is

$$\bar{P} = \tfrac{1}{2}VI = \tfrac{1}{2}RI^2 = \tfrac{1}{2}\frac{V^2}{R} \qquad 18.11$$

These are the same results we found in Sec. 18.2 (Eq. 18.4); they are repeated here for completeness and because we have dropped the subscript p on peak values.

INDUCTIVE CIRCUIT

Figure 18.19a shows a purely inductive ac circuit with frequency f.

Definition The *inductive reactance* X_L of an inductor with inductance L in an ac circuit with frequency f is

$$X_L = 2\pi fL \qquad 18.12$$

UNIT The unit of inductive reactance can be found from the units of frequency and inductance:

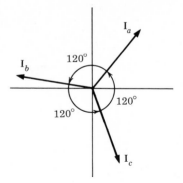

a

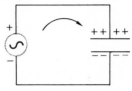

b

FIGURE 18.16

(a) Phasors representing the currents in the three lines in Fig. 18.15b. (b) The sum of these currents is zero.

FIGURE 18.17

A capacitor connected to an ac generator. As the voltage alternates, charge flows back and forth between the plates of the capacitor.

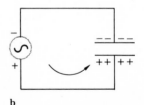

a

b

$$[X_L] = [f][L] = Hz \cdot H = \frac{1}{s} \cdot \frac{Wb}{A}$$

$$= \frac{1}{s} \cdot \frac{T \cdot m^2}{A} = \frac{[N/(A \cdot m)] \cdot m^2}{A \cdot s} = \frac{N \cdot m}{C \cdot A} = \frac{J/C}{A} = \frac{V}{A} = \Omega$$

Thus inductive reactance has the same unit as resistance. In spite of this, however, resistance and reactance cannot be added directly, which is why they are given different names.

At every instant the current in a purely inductive circuit with inductive reactance X_L is related to the voltage by

$$\mathbf{I} = \lceil \frac{V}{X_L} \qquad \text{inductance} \qquad\qquad 18.13$$

The symbol $\lceil$ is introduced to show that the phasor **I** has the magnitude $I = V/X_L$, but is rotated 90° *behind* **V** (Fig. 18.19b). Because the phasors are displaced 90° to one another, the corresponding instantaneous current and voltage peak at different times, as shown in Fig. 18.19c. In a purely inductive circuit, the current is said to lag the voltage by 90° or $\pi/2$ radians.

REMARK When the frequency f is zero, the current is constant, i.e., direct. In this case, Eq. 18.12 shows that the inductive reactance X_L is zero. That is, the voltage drop across an inductor is zero when there is a direct current in it. On the other hand, a large voltage is required to maintain a high-frequency alternating current in an inductor.

Example 18.3 A 75-mH inductor is connected to a voltage source with a peak voltage of 30 V and a frequency of 500 Hz. (a) What is the inductive reactance of the circuit? (b) What is the peak current in the circuit? (c) Write equations for the instantaneous current and voltage. (d) What is the instantaneous current when the instantaneous voltage is 27 V?

(a) From Eq. 18.12 the inductive reactance X_L is

$$X_L = 2\pi f L = 2\pi (500 \text{ Hz})(75 \times 10^{-3} \text{ H})$$
$$= 236 \; \Omega$$

(b) From Eq. 18.13 the peak current I is

$$I = \frac{V}{X_L} = \frac{30 \; \Omega}{236 \; \Omega} = 0.127 \text{ A}$$

(c) Figure 18.19b shows that the instantaneous voltage V_x is

$$V_x = V \cos 2\pi f t$$
$$= (30 \text{ V}) \cos 1000\pi t$$

and the instantaneous current I_x is

$$I_x = I \cos (2\pi f t - \tfrac{1}{2}\pi)$$
$$= (0.127 \text{ A}) \cos (1000\pi t - \tfrac{1}{2}\pi)$$

Using the trigonometric relation

$$\cos (\theta - \tfrac{1}{2}\pi) = \sin \theta$$

the last equation can be written

$$I_x = (0.127 \text{ A}) \sin 1000\pi t$$

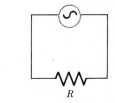

a

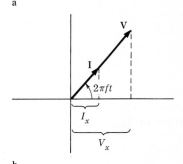

b

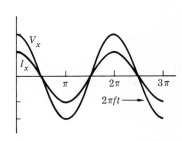

c

FIGURE 18.18

(a) A resistive circuit. (b) Phasors **V** and **I** representing the voltage and current in the circuit. The phasors are aligned. (c) Plot of the instantaneous voltage V_x and current I_x in the circuit. The current and voltage are in phase.

(d) When $V_x = 27$ V we have

$$\cos 1000\pi t = \frac{V_x}{V} = \frac{27 \text{ V}}{30 \text{ V}} = 0.9$$

so

$$1000\pi t = \arccos 0.9 = 0.45 \text{ rad}$$

Using this value of $1000\pi t$ in the equation for I_x, we get

$$I_x = (0.127 \text{ A}) \sin 0.45 = 0.055 \text{ A} \qquad \square$$

The instantaneous power P_x delivered to an inductor is

$$P_x = V_x I_x$$

During one half of the ac cycle, the current and voltage have the same sign, so P_x is positive, but during the other half of the cycle, the current and voltage have opposite sign, so P_x is negative. When P_x is positive, energy from the circuit is being stored in the inductor; when P_x is negative, this energy is returned to the circuit. Over a complete cycle, the average power $\bar{P}$ is zero:

$$\bar{P} = 0$$

Electromagnetic energy flows in and out an inductor; it is not transformed into nonelectric forms of energy, as in a resistor.

CAPACITIVE CIRCUIT

Figure 18.20a shows a purely capacitive ac circuit with frequency f.

Definition   The *capacitive reactance* X_C of a capacitor with capacitance C in an ac circuit with frequency f is

$$X_C = \frac{1}{2\pi fC} \qquad 18.14$$

> **UNIT**   The unit of capacitive reactance can be found from the units of frequency and capacitance to be volts per ampere, or ohms. Thus resistance, inductive reactance, and capacitive reactance have the same units.

At every instant the current in a purely capacitive circuit with capacitive reactance X_C is related to the voltage by

$$I = \overset{\curvearrowleft}{|} \frac{V}{X_C} \qquad \text{capacitance} \qquad 18.15$$

The symbol $\overset{\curvearrowleft}{|}$ is introduced to show that the phasor **I** has the magnitude $I = V/X_C$, but is rotated 90° *ahead* of **V** (Fig. 18.20b). Consequently, the instantaneous current and voltage peak at different times, as shown in Fig. 18.20c. In a purely capacitive circuit, the current is said to lead the voltage by 90°, or $\pi/2$ rad.

> **REMARK**   Direct current corresponds to zero frequency. In this limit Eq. 18.14 shows that the capacitive reactance is infinite. That is, a capacitor does not conduct a direct current. On the other hand, as the frequency of the current increases, the capacitive reactance decreases, so that a high-frequency alternating current can be maintained in a capacitor with a small voltage across it.

Example 18.4   A 125-µF capacitor is connected to a voltage

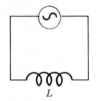

a

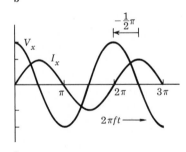

b

c

FIGURE 18.19

(a) An inductive circuit. (b) Phasors **V** and **I** representing the voltage and current in the circuit. The current phasor **I** is 90° behind the voltage phasor **V**. (c) Plot of the instantaneous voltage V_x and current I_x in the circuit. The current lags the voltage by $\frac{1}{2}\pi$.

source with a peak voltage of 170 V and a frequency of 60 Hz. (a) What is the capacitive reactance of the circuit? (b) What is the peak current in the circuit? (c) Write equations for the instantaneous current and voltage. (d) Show that the average power delivered to the capacitor in one cycle is zero.

(a) From Eq. 18.14 the capacitive reactance is

$$X_C = \frac{1}{2\pi f C} = \frac{1}{2\pi(60\text{ Hz})(125 \times 10^{-6}\text{ F})} = 21.2\ \Omega$$

(b) From Eq. 18.15 the peak current I is

$$I = \frac{V}{X_C} = \frac{170\text{ V}}{21.2\ \Omega} = 8.02\text{ A}$$

a

(c) Figure 18.20b shows that the instantaneous voltage V_x is

$$V_x = V \cos 2\pi f t$$
$$= (170\text{ V}) \cos 120\pi t$$

and the instantaneous current I_x is

$$I_x = I \cos (2\pi f t + \tfrac{1}{2}\pi)$$
$$= (8.02\text{ A}) \cos (120\pi t + \tfrac{1}{2}\pi)$$

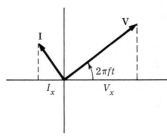

b

Using the trigonometric relation

$$\cos (\theta + \tfrac{1}{2}\pi) = -\sin \theta$$

the last equation can be written

$$I_x = -(8.02\text{ A}) \sin 120\pi t$$

(d) The instantaneous power P_x is

$$P_x = V_x I_x = [(170\text{ V}) \cos 120\pi t][-(8.02\text{ A}) \sin 120\pi t]$$
$$= -(1363\text{ W}) \cos 120\pi t \sin 120\pi t$$

We let $\theta = 120\pi t$ and use the trigonometric relation

$$\cos \theta \sin \theta = \tfrac{1}{2} \sin 2\theta$$

to write

$$P_x = -(682\text{ W}) \sin 2\theta$$

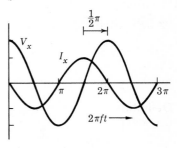

c

FIGURE 18.20

(a) A capacitive circuit. (b) Phasors V and I representing the voltage and current in the circuit. The current phasor I is 90° ahead of the voltage phasor V. (c) Plot of the instantaneous voltage V_x and current I_x in the circuit. The current leads the voltage by $\tfrac{1}{2}\pi$.

In one cycle, θ changes by 2π and 2θ changes by 4π. Thus in one cycle, $\sin 2\theta$ goes through two complete oscillations. Since $\sin 2\theta$ is positive as often as it is negative, its average value in one oscillation is zero. Thus the average power delivered in one cycle is zero. □

FIGURE 18.21

An RLC circuit. A resistor R, inductor L, and capacitor C in series with an ac generator.

A capacitor, like an inductor, stores energy during part of a cycle and delivers energy during another part. Electromagnetic energy flows in and out of a capacitor without being transformed into nonelectric forms of energy.

RLC CIRCUIT

Figure 18.21 shows an RLC circuit. It consists of a resistor R, an inductor L, and capacitor C connected in series to an ac generator with peak voltage V and frequency f.

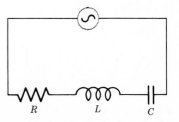

Definitions The *impedance vector* **Z** of an *RLC* circuit is a vector whose x component is equal to the resistance R of the circuit and whose y component is equal to the difference $X_C - X_L$ between the capacitive and inductive reactances of the circuit (Fig. 18.22). The *impedance* Z of an *RLC* circuit is the magnitude of the impedance vector:

$$Z = \sqrt{R^2 + (X_C - X_L)^2} \qquad\qquad 18.16$$

The *phase angle* ϕ is the angle between **Z** and the x axis:

$$\phi = \arctan\frac{X_C - X_L}{R} \qquad\qquad 18.17$$

UNIT Impedance, like resistance and reactance, is measured in ohms (Ω).

At every instant the current in an *RLC* circuit with impedance Z and phase angle ϕ is related to the voltage by

$$I = \phi\rangle\,\frac{V}{Z} \qquad RLC \text{ circuit} \qquad\qquad 18.18$$

The symbol $\phi\rangle$ is introduced to show that the phasor **I** has the magnitude $I = V/Z$, but is rotated ahead of **V** through the angle ϕ (Fig. 18.23a). As **I** and **V** rotate about the origin, their x components equal the instantaneous current and voltage, as shown in Fig. 18.23b.

In a purely resistive circuit ($X_C - X_L = 0$), Z is equal to R, and ϕ is zero. The current and voltage are in phase, as shown in Fig. 18.18b. In a purely inductive circuit ($R = 0$ and $X_C - X_L = -X_L$), Z is equal to X_L, and ϕ is equal to $-\frac{1}{2}\pi$. The current lags the voltage by 90°, as shown in Figure 18.19b. In a purely capacitive circuit ($R = 0$ and $X_C - X_L = X_C$), Z is equal to X_C, and ϕ is equal to $+\frac{1}{2}\pi$. The current leads the voltage by 90°, as shown in Fig. 18.20b. Thus purely resistive, inductive, and capacitive circuits are special cases of the general *RLC* circuit described by Eqs. 18.16, 18.17, and 18.18.

From Fig. 18.23b we see that the instantaneous voltage and current in an *RLC* circuit are

$$V_x = V \cos 2\pi ft \qquad\qquad 18.19a$$

$$I_x = I \cos (2\pi ft + \phi) \qquad\qquad 18.19b$$

The instantaneous power P_x delivered to this circuit is

$$P_x = V_x I_x = VI \cos 2\pi ft \cos (2\pi ft + \phi)$$

and the average power $\overline{P}$ is

$$\overline{P} = \tfrac{1}{2}VI \cos \phi = V_{rms}I_{rms} \cos \phi \qquad\qquad 18.20$$

In a purely resistive circuit, ϕ is zero and $\cos \phi = 1$; in this case, Eq. 18.20 reduces to Eq. 18.11. In a purely inductive or capacitive circuit, $\phi = \pm\frac{1}{2}\pi$ and $\cos \phi = 0$; in this case, $\overline{P}$ is zero. In the general case, $\cos \phi$ is between 0 and 1. The term $\cos \phi$ is called the *power factor* of an ac circuit.

Example 18.5 In an *RLC* circuit, $R = 50\,\Omega$, $L = 200\,mH$, $C = 50\,\mu F$, $V = 300\,V$, and $f = 60\,Hz$. (a) Find the impedance

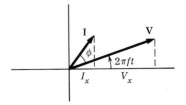

FIGURE 18.22

Impedance vector **Z** of an *RLC* circuit. The x component of **Z** is the resistance R of the circuit, and the y component of **Z** is the difference $X_C - X_L$ between the capacitive and inductive reactances of the circuit. The phase angle ϕ is the angle between **Z** and the x axis.

FIGURE 18.23

Phasors **V** and **I** representing the voltage and current in an *RLC* circuit. The current phasor leads the voltage phasor by the phase angle ϕ given in Fig. 18.22. (b) Plot of the instantaneous voltage V_x and current I_x in the circuit. The current leads the voltage by ϕ.

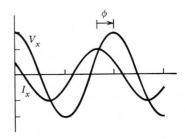

and phase angle of this circuit. (b) Find the rms current and voltage. (c) Draw the phasors **V** and **I**. (d) What is the average power dissipated in the circuit?

(a) From Eqs. 18.12 and 18.14 the inductive and capacitive reactances are

$$X_L = 2\pi fL = 2\pi(60\text{ Hz})(0.20\text{ H}) = 75.4\,\Omega$$

and

$$X_C = \frac{1}{2\pi fC} = \frac{1}{2\pi(60\text{ Hz})(50 \times 10^{-6}\text{ F})} = 53.0\,\Omega$$

Thus from Eq. 18.16 the impedance Z is

$$Z = \sqrt{R^2 + (X_C - X_L)^2}$$
$$= \sqrt{(50\,\Omega)^2 + (53.0\,\Omega - 75.4\,\Omega)^2}$$
$$= \sqrt{3000\,\Omega^2} = 54.8\,\Omega$$

and from Eq. 18.17 the phase angle ϕ is

$$\phi = \arctan\frac{X_C - X_L}{R} = \arctan\frac{53.0\,\Omega - 75.4\,\Omega}{50\,\Omega}$$

$$= \arctan(-0.448) = -24.1°$$

(b) From Eq. 18.18 the peak current I is

$$I = \frac{V}{Z} = \frac{300\text{ V}}{54.8\,\Omega} = 5.47\text{ A}$$

From Eqs. 18.6 and 18.7 the rms voltage and current are

$$V_{\text{rms}} = \frac{V}{\sqrt{2}} = \frac{300\text{ V}}{\sqrt{2}} = 212\text{ V}$$

and

$$I_{\text{rms}} = \frac{I}{\sqrt{2}} = \frac{5.47\text{ A}}{\sqrt{2}} = 3.87\text{ A}$$

(c) The phasor **V** has the magnitude 300 V and is drawn at an arbitrary angle $2\pi ft$ to the x axis (Fig. 18.24). The phasor **I** has the magnitude 5.47 A and is drawn 24.1° behind **V**. The current lags the voltage in this case because ϕ is negative; this happens whenever the inductive reactance is greater than the capacitive reactance.

(d) From Eq. 18.20 the average power is

$$\bar{P} = V_{\text{rms}}I_{\text{rms}}\cos\phi$$
$$= (212\text{ V})(3.87\text{ A})\cos(-24.1°) = 749\text{ W}\qquad \square$$

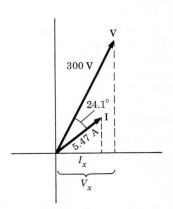

FIGURE 18.24

The phasors found in Example 18.5.

RESONANCE

Definition *Resonance* in an *RLC* circuit occurs when the inductive reactance X_L is equal to the capacitive reactance X_C:

$$X_L = X_C$$

From Eqs. 18.12 and 18.14 this condition is

$$2\pi fL = \frac{1}{2\pi fC} \quad\text{ or }\quad f = \frac{1}{2\pi}\sqrt{\frac{1}{LC}} \qquad\qquad 18.21$$

From Eq. 18.16 we see that at resonance, the impedance Z of the circuit is equal to the resistance.

The tuning circuit of a radio or television receiver is an RLC circuit with a very small resistance and a variable capacitor (Fig. 18.25). The electromagnetic waves radiated by station transmitters induce ac voltages in the antenna of the receiver, which in turn induce voltages of the same frequencies in the inductor of the tuning circuit. The frequencies of the voltages are the frequencies of the waves. To select a particular station's frequency, the variable capacitor is adjusted until Eq. 18.21 is satisfied for that frequency. At this point the impedance of the circuit, for this frequency only, is just the small resistance of the circuit. Consequently, a relatively large current of this frequency is obtained, which can be amplified and processed to produce the audio and video output. The impedances for voltages of nonresonant frequencies are much larger, so the corresponding currents are relatively small. In this way the circuit can select one frequency from the many that impinge on the receiver.

Example 18.6 The inductance of the tuning circuit of an AM radio is 5 mH. What is the value of the capacitance of the circuit when the circuit is tuned to a station broadcasting at 1250 kHz?

Solving Eq. 18.21 for C, we get

$$C = \frac{1}{(2\pi f)^2 L} = \frac{1}{(2\pi)^2(1.25 \times 10^6 \text{ Hz})^2(5 \times 10^{-3} \text{ H})}$$

$$= 3.24 \text{ pF} \qquad \square$$

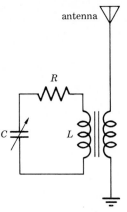

FIGURE 18.25

An antenna coupled to an RLC circuit. The capacitor is varied until the resonant frequency of the circuit is equal to the frequency to be detected.

18.4   The Transformer

A transformer consists of two coils wrapped around a common iron yoke (Fig. 18.26). The first coil, called the *primary*, has N_1 turns and is connected to an ac seat of emf

$$\mathcal{E} = V_1 \cos 2\pi ft$$

The second coil, called the *secondary*, has N_2 turns and is connected through a switch S to a resistance R.

When switch S is open, the secondary circuit is disconnected, and the primary coil acts likes an inductance L. In this case, the primary circuit is a purely inductive circuit, and so the current in it, called the *magnetizing current*, has the peak value

FIGURE 18.26

A transformer. The voltage V_1 applied to the primary coil induces a voltage V_2 across the secondary coil.

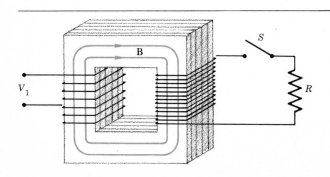

$$I_{mag} = \frac{V_1}{2\pi f L}$$

and is 90° out of phase with the voltage. Consequently, no power is delivered to the primary circuit.

The ac magnetizing current establishes an ac magnetic field $\mathbf{B}$ in the iron. Through coil 1 the flux of this field is

$$\phi_1 = N_1(BA)_1$$

where $(BA)_1$ is the product of the field and the cross-sectional area of coil 1. The emf $\mathcal{E}$ across coil 1 is related to this flux by Faraday's law:

$$\mathcal{E} = \frac{\Delta\phi_1}{\Delta t} \qquad\qquad 18.22$$

or

$$V_1 \cos 2\pi ft = N_1 \frac{\Delta(BA)_1}{\Delta t}$$

Similarly, the emf across the secondary coil is

$$V_2 \cos 2\pi ft = \frac{\Delta\phi_2}{\Delta t} = N_2 \frac{\Delta(BA)_2}{\Delta t}$$

where $\phi_2 = N_2(BA)_2$ is the flux through coil 2. The iron yoke traps the magnetic lines of force so that BA is essentially the same for both coils:

$$(BA)_2 = (BA)_1$$

Consequently, the emf across the secondary coil can be written

$$V_2 \cos 2\pi ft = N_2 \frac{\Delta(BA)_1}{\Delta t} = \frac{N_2}{N_1} V_1 \cos 2\pi ft$$

so

$$V_2 = \frac{N_2}{N_1} V_1 \qquad\qquad 18.23$$

The ratio V_2/V_1 of the voltages across the secondary and primary coils of a transformer is equal to the ratio N_2/N_1 of the number of turns on these coils.

When switch S is closed, a current $I_2 \cos 2\pi ft$ is established in the secondary circuit, where

$$I_2 = \frac{V_2}{R}$$

This current produces its own flux in the iron yoke that opposes the flux resulting from the magnetizing current I_{mag} in coil 1. In order to maintain the net flux required by Eq. 18.22, the generator must send an additional current $I_1 \cos 2\pi ft$ through coil 1. This current is in phase with the voltage, so the average power delivered by the generator is

$$\overline{P}_1 = \tfrac{1}{2} V_1 I_1$$

In an ideal transformer, all this energy is transferred without loss to the load in the secondary circuit. That is,

$$\bar{P}_1 = \tfrac{1}{2}V_1 I_1 \quad \text{and} \quad \bar{P}_2 = \tfrac{1}{2}V_2 I_2$$

are equal, so

$$I_2 = \frac{V_1}{V_2} I_1$$

or, from Eq. 18.23,

$$I_2 = \frac{N_1}{N_2} I_1 \qquad\qquad 18.24$$

REMARK Although Eqs. 18.23 and 18.24 are written for peak voltages and currents, they hold as well for rms voltages and currents.

Example 18.7 A transformer on a utility pole is designed to reduce the power-line voltage from 2400 to 120 V (rms). (a) If the primary coil of the transformer has 400 turns, how many turns does the secondary have? (b) If the transformer delivers 5000 W at 120 V, what are the rms currents in the primary and secondary coils? (Neglect the magnetizing current.)

(a) Using rms values for the voltages in Eq. 18.23, we have

$$\frac{N_1}{N_2} = \frac{V_1}{V_2} = \frac{2400\ \text{V}}{120\ \text{V}} = 20 \quad \text{or} \quad N_2 = \frac{N_1}{20} = \frac{400}{20} = 20$$

(b) The average power $\bar{P}$ is

$$\bar{P} = V_{\text{rms}} I_{\text{rms}}$$

In the primary coil, the current is

$$I_{1,\text{rms}} = \frac{\bar{P}}{V_{1,\text{rms}}} = \frac{5000\ \text{W}}{2400\ \text{V}} = 2.08\ \text{A}$$

In the secondary coil, the current is

$$I_{2,\text{rms}} = \frac{\bar{P}}{V_{2,\text{rms}}} = \frac{5000\ \text{W}}{120\ \text{V}} = 41.6\ \text{A}$$

The ratio of the currents is in agreement with Eq. 18.24:

$$\frac{I_{2,\text{rms}}}{I_{1,\text{rms}}} = \frac{41.6\ \text{A}}{2.08\ \text{A}} = 20 = \frac{N_1}{N_2}$$

Thus the transformer increases the current by a factor of 20 while decreasing the voltage by the same factor, so the output power equals the input power. ☐

The purpose of transmitting power at high voltage is to reduce the power loss resulting from the resistance R of the line. A line carrying an rms current I_{rms} at an rms voltage V_{rms} transmits the power

$$\bar{P} = V_{\text{rms}} I_{\text{rms}}$$

and loses the power

$$\bar{P}_{\text{loss}} = R I_{\text{rms}}^2$$

where R is the total resistance of the line. Substituting $I_{rms} = \bar{P}/V_{rms}$ into the last equation, we get

$$\bar{P}_{loss} = \frac{R\bar{P}^2}{V_{rms}{}^2}$$

which shows that for a given value of $\bar{P}$, the power loss is less the greater the voltage. In the United States and Canada the standard transmission line-to-line voltages are 115, 138, 230, 345, 500, and 735 kV.

Example 18.8 A 345-kV transmission line carries 420 MW a distance of 250 km on three cables of a three-phase system. The resistance of each cable is 0.01 Ω/km. (a) What is the rms current in each cable? (b) What is the total power loss? (c) What would be the power loss if this power were carried by cables of the same resistance in a one-phase system?

(a) In a three-phase system each line carries one-third the power, or 140 MW. Since the line-to-line voltage is 345 kV, from Eq. 18.9 the line-to-neutral voltage is

$$V_{ao} = \frac{V_{ab}}{\sqrt{3}} = \frac{345 \text{ kV}}{\sqrt{3}} = 199 \text{ kV}$$

The rms current in each cable is

$$I_{rms} = \frac{\bar{P}}{V_{rms}} = \frac{140 \times 10^6 \text{ W}}{199 \times 10^3 \text{ V}} = 704 \text{ A}$$

(b) The resistance of each cable is

$$R = (0.01 \ \Omega/\text{km})(250 \text{ km}) = 2.5 \ \Omega$$

so the loss per cable is

$$\bar{P}_{loss} = RI_{rms}{}^2 = (2.5 \ \Omega)(704 \text{ A})^2 = 1.24 \text{ MW}$$

and the total loss on all three cables is 3.72 MW. Even if a fourth neutral cable is present, there is no loss in it, because it carries no current in a balanced system.

(c) If the same power is carried by two cables of a one-phase system, the current is

$$I_{rms} = \frac{\bar{P}}{V_{rms}} = \frac{420 \times 10^6 \text{ W}}{345 \times 10^3 \text{ V}} = 1217 \text{ A}$$

and the loss per cable is

$$\bar{P}_{loss} = RI_{rms}{}^2 = (2.5 \ \Omega)(1217 \text{ A})^2 = 3.72 \text{ MW}$$

The total loss on the two cables is 7.44 MW, or twice the loss in the three-phase system. ☐

REMARK Of course, the three-phase system is using 50 percent more cable than the one-phase system (not counting the neutral line). If the wire in the three cables of the three-phase system were used to make the two cables of the one-phase system, the resistance of each cable would be only 1.67 Ω, and the power loss in the one-phase system would be 3.35 MW, or four-thirds the loss of the three-phase system. That is, per pound of conducting material, the three-phase system suffers only 75 percent of the line loss of a one-phase system.

EDDY CURRENTS

The changing flux inside the iron yoke of a transformer induces a current not only in the wire of the secondary coil, but in the iron itself. Such currents induced in the body of a conductor are called *eddy currents*. They are undesirable in a transformer because the accompanying RI^2 loss heats the transformer and reduces the efficiency with which power is transferred from the primary to the secondary.

As shown in Fig. 18.27, the eddy currents circulate in loops perpendicular to the magnetic fields of force. To reduce these currents, transformer cores are made of thin sheets of steel (laminations) bonded together with insulation between them. The insulation reduces the size of the eddy currents and reduces the power loss inside a transformer. Modern high-power transformers have efficiencies close to 99 percent; that is, all but 1 percent of the electric energy that flows into the primary is transferred to the secondary.

18.5 Motors

A *generator* is a device for converting mechanical energy into electric energy, and a *motor* is a device for converting electric energy into mechanical energy. Although most devices are built specifically for one purpose or the other, the devices are, in principle, reversible. That is, if one drives the shaft of a motor with an outside force, the motor will act like a generator. The generator-motor of a pumped hydroelectric facility (Sec. 5.5) is reversible. During periods of peak demand, water in the upper reservoir flows through a turbine, driving the generator-motor so that it generates electricity. During periods of low demand, excess electric power from base-load plants is used to run the generator-motor as a motor, turning the turbine in reverse, so that it pumps water back to the upper reservoir.

In Sec. 17.2 we showed how a motor uses magnetic force to produce rotational motion. In this section we describe in more detail the major types of motors in use today.

Like a generator, a motor consists of a stationary set of coils, called the *stator,* and a rotating set of coils, called the *rotor* (Fig. 18.28). The *armature* is the set of coils, either rotor or stator, that carries the main current. The other set of coils is called the *field*.

Current from a voltage source is delivered to the rotor coils by means of stationary graphite contacts, called *brushes,* that rub against rotating metal contacts on the rotor shaft. The rotor contacts can be either *commutators* or *slip rings* (Fig. 18.28). Commutators reverse the direction of the current to the rotor as the rotor turns, whereas slip rings do not.

Definition The *efficiency* e of a motor is the ratio of the mechanical output power P_{out} to the electric input power P_{in}:

$$e = \frac{P_{out}}{P_{in}}$$

The efficiency is always less than 100 percent because of RI^2 loss in the coils, frictional loss in the brushes and bearings, and

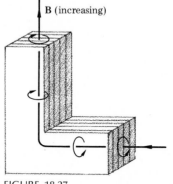

B (increasing)

FIGURE 18.27

Eddy currents in the yoke of a transformer. By making the yoke with alternating layers of iron and insulation, these currents can be greatly reduced.

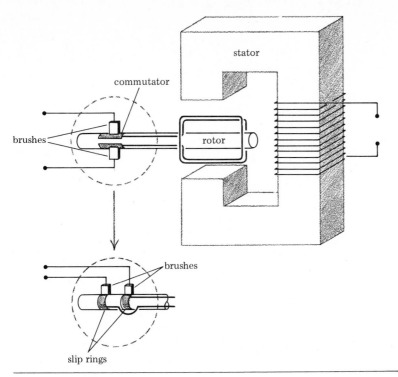

FIGURE 18.28

Basic components of a motor are
the rotor and stator. Current is supplied to the rotor through brushes
that make contact with either a
commutator or slip rings.

eddy-current losses. However, by proper design these losses can
be made very small. Efficiencies vary from about 74 percent for
1-hp motors to 97 percent for 5000-hp motors.

COMMUTATOR MOTORS

In a *commutator motor*, the stator produces a field in a fixed
direction, and the current in the rotor is continuously changed
to keep the torque on the rotor in the direction of motion. This
is illustrated in Fig. 17.18 for the case of a two-pole rotor in the
field of a permanent magnet. Figure 18.29 shows the stator, rotor,

FIGURE 18.29

Industrial commutator motor.
(*General Electric.*)

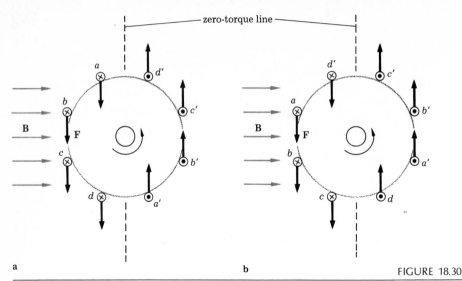

a   b

and commutator of a large, commercial commutator motor. The stator has a set of coil-wound poles mounted in a circle, and the rotor has nearly 100 coils, each connected to its own commutator bar. The coils of the rotor are interconnected in series, so they carry the same current at the same time. As the brushes move from one commutator bar to the next, only the current in the coil that at that instant is passing the zero-torque point in the stator field is reversed.

This is illustrated in Fig. 18.30, which shows a cross-sectional view of a rotor with four coils. A dot represents current out of the plane of the paper and a cross represents current into the plane. From the left-hand rule of Sec. 17.3, the force on each current resulting from the constant field **B** has the direction shown. As a coil passes the dotted line, the commutator reverses its current while keeping the current in the other coils unchanged. (Note that in Fig. 18.30b, only the current in coil *dd'* reverses.) In this way, the constant field of the stator produces continuous rotation of the rotor.

Direct-current commutator motors are used in heavy industry because they can be operated at various speeds and torques. In particular, these motors can be wound to develop a large torque at zero speed, so they can be used in motor vehicles. They are especially important in the diesel-electric locomotive, where they deliver the power to the wheels. The diesel engine itself is connected directly to a dc generator, which powers the motor. This system eliminates the need for a mechanical transmission between the diesel engine and the wheels.

If the stator and rotor of a commutator motor are connected in series to the same voltage source, the motor can operate on alternating current. This is so because the stator field and rotor current alternate together, so the force on the rotor remains fixed. Small commutator motors of this type are called *universal motors,* because they can operate on either alternating or direct current. They are found in many home appliances, such as sewing machines, vacuum cleaners, and power tools.

FIGURE 18.30

Current in the rotor coils of a commutator motor. Each coil is represented by a letter and its prime (*aa'*, *bb'*, etc.), and the direction of the current is indicated by a dot (out of the plane) or a cross (into the plane). The arrows show the direction of the magnetic force on each current segment. (a) All the forces exert torques in the direction of rotation. (b) As coil *dd'* passes the zero-torque line, the commutator reverses the direction of its current. This reverses the force on this coil, so that the torques are still in the direction of rotation.

INDUCTION MOTORS

The armature of an *induction motor* is the stator, which is supplied with alternating current. The rotor has no current supply of its own, but consists of a number of closed conducting loops. The *squirrel-cage rotor*, for instance, is made of conducting bars connected at either end to conducting rings (Fig. 18.31). Wire-wound and solid rotors are also used.

The alternating magnetic field of the stator induces current in the rotor, just as the alternating field of the primary coil of a transformer induces current in the secondary coil. The stator field then interacts with the induced rotor current to produce a torque on the rotor. If the stator field were produced by single-phase altenating current, the torque would oscillate in direction, and the rotor would not turn.

In all large induction motors, three-phase alternating current is used to power a set of coils evenly spaced around the stator. In the simplest case, there are three coils, each connected to one of the three phases (Fig. 18.32). At the center of the stator, each coil produces a field perpendicular to the plane of the coil, and the total field is the vector sum of the individual fields. The magnitude of each individual field oscillates with the frequency of the current, as shown in Fig. 18.33a. The vector sum **B** of these fields, which are 60° apart in space and 120° apart in phase, are shown in Fig. 18.33b; the field **B** is constant in magnitude and rotates with the frequency of the current. If several sets of three coils are used in the stator, the field will rotate with a frequency that is a fraction of the frequency of the current. In any case, the angular speed of the stator field is called the *synchronous speed;* it is a constant for a given stator powered by current of fixed frequency.

The rotating stator field interacts with the currents induced in the rotor, exerting a torque on the rotor. Figure 18.34 gives a typical torque-speed curve for an induction motor. The stator

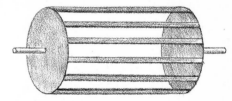

FIGURE 18.31

Squirrel-cage rotor. The magnetic field of the stator induces currents in the bars of the rotor.

FIGURE 18.32

Stator of an induction motor. Each of the three coils of this stator receives a different phase of three-phase current. Consequently, the magnetic fields $\mathbf{B}_a$, $\mathbf{B}_b$, and $\mathbf{B}_c$ produced by these coils have different magnitudes at different times.

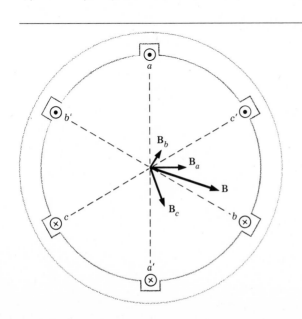

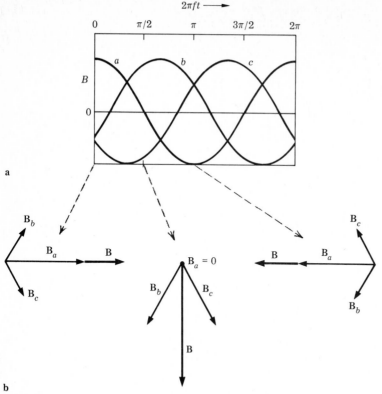

a

b

FIGURE 18.33

(a) Plot of the magnitudes of the magnetic fields produced by each of the coils in the stator in Fig. 18.32. Since each coil receives current of a different phase, the fields are 120° apart in phase.
(b) Sum of the magnetic fields produced by the three coils at different times. This sum is constant in magnitude and rotates with the frequency of the current.

FIGURE 18.34

Plot of the percent of maximum torque against the percent of synchronous speed for a synchronous motor. At synchronous speed, the motor delivers no torque.

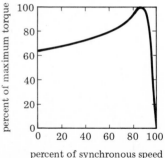

exerts considerable torque on the rotor even when the rotor is at rest; this means that the rotor is self-starting. The torque increases with rotor speed until it reaches a maximum at about 80 percent of the synchronous speed. After that, the torque decreases rapidly, and is zero at synchronous speed.

Industry uses large three-phase induction motors with power outputs from 100 to more than 5000 hp. Smaller single-phase induction motors are used in domestic refrigerators and air conditioners. These motors use a capacitor to shift the phase of the current in some of the stator coils 90° relative to the phase in other coils. This phase shift produces the rotating stator field necessary to run the motor. With a large starting torque and no brushes to wear out, these motors can operate compressors and pumps for decades without servicing.

SYNCHRONOUS MOTORS

The stator of a *synchronous motor* is the same as that of an induction motor; in large motors a set of coils powered by three-phase current produces a rotating magnetic field, whereas in small motors single-phase current is split and phase shifted to produce the rotating field. The rotor of a synchronous motor consists of electromagnets powered by direct current supplied through slip rings (Fig. 18.35). When the rotor has the rotational speed of the stator field (synchronous speed), the rotor magnets lock into the stator field. Thereafter the rotor continues to rotate

FIGURE 18.35

Rotor of a synchronous motor used in industry. (*General Electric.*)

at the synchronous speed regardless of changes in the load. That is, the synchronous motor can deliver a variable torque at constant speed. Synchronous motors are used wherever constant speed under changing load conditions is required.

REMARK The synchronous motor described here is the same as the ac generator described in Fig. 18.4. The same device will generate three-phase current when its rotor is turned by an outside source, and the rotor will turn when the stator is supplied with three-phase current.

Synchronous motors are not self-starting, so some mechanism must be provided to get them up to synchronous speed. Often the rotor has a squirrel cage built into it, so that the motor acts like an induction motor until it nears synchronous speed. At this point, current is supplied to the rotor, and the rotor locks into synchronous speed.

In large industrial motors, the rotor current is often supplied by a dc generator built onto the rotor shaft. In small domestic synchronous motors, such as those found in clocks and phonograph turntables, the rotor has no current source. Instead it is made of thin discs of hardened steel. When the motor is started, these discs act like the rotor of an induction motor; that is, the stator field interacts with the currents it induces in the discs. But as the rotor nears synchronous speed, the discs, which are now stationary with respect to the rotating stator field, become magnetized by that field. The magnetized rotor then locks into the stator field and rotates at the synchronous speed. The efficiency of such motors is only a few percent, but they are inexpensive to build and give reliable service for small-load operations.

GUIDE TO MAJOR TOPICS

PROBLEMS

1 Find the peak current, frequency, and period of the alternating current plotted in Fig. 18.36.
Ans. 2.5 A, 25 Hz, 0.04 s

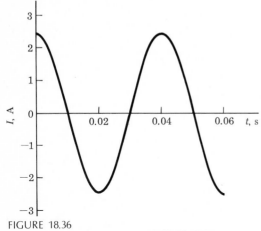

FIGURE 18.36

Problem 1.

2 An ac generator with a peak voltage of 80 V and a frequency of 50 Hz is connected across a 40-Ω resistor. (a) What is the peak current in the resistor and (b) the period of the alternating current? (c) Plot the current against time.

3 The current I_x in a resistor connected to an ac generator is

$$I_x = (3.2 \text{ A}) \cos 80\pi t$$

where t is the time in seconds. (a) What are the peak current and frequency? (b) What is the current at time $t = 0.01$ s?
Ans. (a) 3.2 A, 40 Hz; (b) −2.59 A

4 Write the equation for the current I_x in the resistor in Prob. 2.

5 An ac generator with a peak voltage of 75 V is placed across a 15-Ω resistor. Find (a) The rms current in the resistor and (b) the average power dissipated in the resistor.
Ans. (a) 3.53 A; (b) 187.5 W

6 An alternating current with an rms current of 2.4 A is carried by a 25-Ω resistor. Find (a) the peak voltage across the resistor and (b) the average power dissipated in the resistor.

7 A light bulb dissipates 60 W when placed across an ac voltage source with an rms voltage of 120 V. Find (a) the resistance of the light-bulb filament and (b) the rms and peak current in the filament.
Ans. (a) 240 Ω; (b) 0.5 and 0.707 A

8 An alternating current with a peak current of 5 A delivers 80 W to a resistor. Find (a) the rms voltage across the resistor and (b) the resistance of the resistor.

9 An alternating current delivers 240 W to a 6-Ω resistor. Find (a) the rms current in and (b) the peak voltage across the resistor.
Ans. (a) 6.32 A; (b) 53.7 V

10 When an ac voltage source with a peak voltage of 340 V is placed across a resistor, the peak current is 9 A. Find (a) the resistance of the resistor and (b) the power dissipated in the resistor.

11 An alternating current with a frequency of 25 Hz and a peak current of 6 A is represented at time t_0 by the phasor **I** in Fig. 18.37. (a) What is the current at time t_0? (b) Draw the phasor at time $t_0 + 0.01$ s. (c) What is the current at time $t_0 + 0.01$ s?
Ans. (a) 5.2 A; (c) −3.0 A

12 The alternating voltage across a resistor is

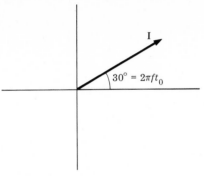

FIGURE 18.37

Problem 11.

$V_x = (12 \text{ V}) \cos 8\pi t$

where t is the time in seconds. Draw the phasor representing this voltage at times 0, 0.1, 0.2, and 0.3 s.

13   An ac generator with a peak voltage of 24 V and a frequency of 60 Hz is placed across a 250-mH inductor. (a) What is the inductive reactance of the circuit? (b) What is the peak current in the inductor? (c) Draw the phasors representing the current and voltage at the instant when the voltage is 12 V.
Ans. (a) 94.2 Ω; (b) 0.255 A

14   When an inductor is placed across a 120-V (rms), 60-Hz alternating voltage source, the current in it has a peak value of 2 A. What is the inductance of the inductor?

15   An ac generator with a peak voltage of 24 V and a frequency of 60 Hz is placed across a 250-μF capacitor. (a) What is the capacitive reactance of the circuit? (b) What is the peak current in the capacitor? (c) Draw the phasors representing the current and voltage at the instant when the voltage is 12 V.
Ans. (a) 10.6 Ω; (b) 2.26 A

16   When a capacitor is placed across a 120-V (rms), 60-Hz alternating voltage source, the current in it has a peak value of 2 A. What is the capacitance of the capacitor?

17   A 2-μF capacitor, a 100-mH inductor, and a 100-Ω resistor are placed in series with a 12-V (rms), 500-Hz voltage source. (a) Find the impedance, rms current, and phase angle of this circuit. (b) What is the power dissipated in the circuit?
Ans. (a) 185 Ω, 0.065 A, $-57°$; (b) 0.42 W

18   (a) Show that the impedance Z of an RLC circuit is related to the resistance R and the phase angle ϕ by

$$Z = \frac{R}{\cos \phi}$$

(Hint: Use Fig. 18.22.) (b) Show that the average power $\bar{P}$ dissipated in an RLC circuit can be written

$$\bar{P} = RI_{rms}^2$$

19   In an RLC circuit the peak current is 5 A, the peak voltage is 20 V, and the power dissipated is 35 W. Find the phase angle, impedance, and resistance of the circuit (see Prob. 18).
Ans. 45.6°, 4 Ω, 2.8 Ω

20   A 50-Ω resistor and a 5-mH inductor are connected in series across a 100-V (peak), variable frequency voltage source. On the same graph draw the phasor representing the voltage when it is at its peak value and the phasors representing the current at frequencies of 0, 500, 1000, 2000, and 10,000 Hz.

21   Find values of the resistance R and the capacitance C of an RC circuit which give an impedance of 500 Ω and a phase angle of 30° at a frequency of 60 Hz.
Ans. 433 Ω and 10.6 μF

22   A voltage source with peak potential V and frequency f is placed across the resistor R and capacitor C shown in Fig. 18.38. The peak voltage across each element is the product of the peak current and the impedance (resistance or capacitive reactance) of the element. (a) Show that the peak voltages V_R and V_C across the resistor and capacitor are

$$V_R = \frac{2\pi fRC}{\sqrt{1 + (2\pi fRC)^2}} V$$

and

$$V_C = \frac{1}{\sqrt{1 + (2\pi fRC)^2}} V$$

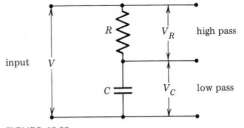

FIGURE 18.38

Problem 22.

(These peak voltages do not add to V because the instantaneous voltages across the resistor and capacitor peak at different times.) (b) Show that V_R goes from 0 to V as f goes from 0 to infinity. This means that a dc or low-frequency input voltage yields a zero or small voltage across R, whereas a high-frequency input voltage yields nearly the full voltage across R. When V_R is taken as the output voltage, this is called a *high-pass filter*, because dc and low-frequency input voltages are filtered out. (c) Show that V_C goes from V to 0 as f goes from 0 to infinity. This means that a dc or low-frequency input voltage yields the full or nearly the full voltage across C, whereas a high-frequency input voltage yields a nearly zero voltage across C. When V_C is taken as the output voltage, this is called a *low-pass filter*, because high-frequency input voltages are filtered out.

23 (a) What is the resonance frequency of the circuit in Prob. 17? (b) What is the impedance at resonance?
Ans. (a) 356 Hz; (b) 100 Ω

24 A radio tuning circuit has an inductor in series with a variable capacitor. The capacitance of the capacitor can vary from C_{min} to C_{max}. Find the ratio C_{max}/C_{min} required if the circuit is to be tunable to frequencies from 530 to 1600 kHz.

25 A transformer is designed to increase the 20-kV (rms) output voltage of a generator to a transmission voltage of 345 kV (rms). If the primary winding has 50 turns, how many turns does the secondary have? (b) If the generator delivers 250 MW, what are the input and output rms currents?
Ans. (a) 862; (b) 1.25×10^4 and 725 A

26 The primary winding of a transformer has 40 turns and the secondary has 5. If the input voltage is 120 V, what is the output voltage?

27 Complete Fig. 18.33 by drawing the vector sum of the magnetic fields when $2\pi ft$ equals $3\pi/2$ and 2π.

28 The stator of a one-phase induction motor has two coils mounted 90° apart (Fig. 18.39a). Coil bb' is powered by current that has been shifted by a capacitor 90° in phase relative to the current in coil aa' (Fig. 18.39b). Show that

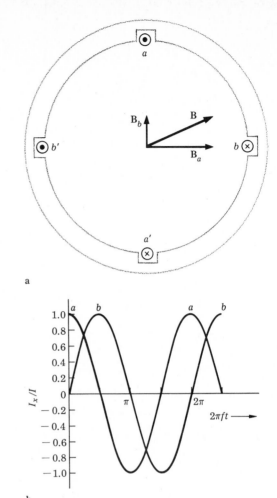

a

b

FIGURE 18.39

Problem 28.

the resulting magnetic field at the center of the stator is constant in magnitude and rotates uniformly with the frequency of the current.

BIBLIOGRAPHY

FITZGERALD, A. E., DAVID E. HIGGINBOTHAM, and ARVIN GRABEL: *Basic Electrical Engineering*, 4th ed., McGraw-Hill Book Company, New York, 1975. The basic principles of electromechanical energy-converting devices—motors and generators—are discussed in Chaps. 14, 15, and 16 of this electrical engineering textbook.

MACLAREN, MALCOLM: *The Rise of the Electrical Industry during the Nineteenth Century*, Princeton University Press, Princeton, N.J., 1943. Brief account of the early development of electric power, including the great battle between Edison and Westinghouse.

E *lectronics* is the study and design of electric circuits used to process information. It thus includes circuits used for communications (radio, television), control (automation), and computation (computers). The term was originally meant to distinguish circuits that contained vacuum tubes from conventional circuits that contained only resistors, capacitors, and inductors. But as semiconductor devices have come to replace vacuum tubes in most electronic circuits, the distinction between conventional and electronic circuits has become blurred. Today almost every circuit, from a simple dimmer switch to a computer-controlled power grid, uses some electronic components.

All the major discoveries needed for the generation and utilization of electric power—generators, transformers, motors—were made before the beginning of this century. Electronics, on the other hand, was developed almost entirely in the twentieth century, with some of the most important discoveries having been made in the last decade or so. Indeed, in spite of the many current uses of electronics—television, calculators, high-fidelity phonographs—it appears that many major uses lie in the future. Industry is only beginning to realize the revolution in control made possible by integrated circuits and microprocessors. Some automobiles, for instance, are now equipped with a microprocessor that monitors and controls the air-gas mixture reaching the engine. This ensures that at any speed and in all operating conditions the mixture gives the best performance with the least pollution.

19
ELECTRONICS

19.1 Communications, Information, and Control

COMMUNICATIONS

Communications is the science and technology of transmitting information from one place to another. The term includes the traditional forms of communication—speech, writing, sign language—as well as electric forms. Although we are primarily concerned with electric communications in this chapter, certain general statements about communications apply to all forms.

Every communications system consists of three elements (Fig. 19.1): (1) a *transmitting transducer,* or *transmitter,* which transforms the source information into a signal capable of being transmitted; (2) a *channel* which carries the signal to the receiver; and (3) a *receiving transducer,* or *receiver,* which transforms the signal into destination information.

The telegraph was the first electric communications system (Fig. 19.2). The transmitter is the telegraph key, which opens and closes an electric circuit. The operator encodes the message as a sequence of taps on the key. The channel is the circuit itself, and the receiver is an electromagnet that moves a clapper in re-

FIGURE 19.1

The elements of a general communication system.

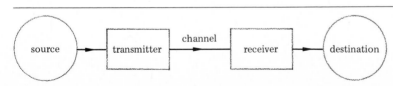

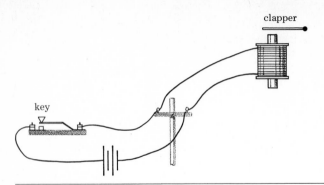

FIGURE 19.2

A telegraphic communication system.

sponse to the movements of the key. An operator at the receiving end decodes the sounds of the clapper back into the original message. Since trained operators are required at both ends to encode and decode the signals, they may be considered to be an essential part of the transmitter and receiver. Although the transmitter has only two active states—open and closed—it is capable of transmitting a complex message because each letter of the alphabet is represented by a different sequence of openings and closings (the Morse code).

The transmitter of a telephone is a resistance microphone consisting of a diaphragm attached to a block of compacted carbon particles (Fig. 19.3). Motion of the diaphragm, caused by the pressure variations in an incident sound wave, changes the compression of the carbon particles. When the particles are more compressed, the area of contact between them is increased, which reduces the electric resistance of the block. Thus the pressure variations in the sound wave produce similar variations in the resistance of the block.

The receiver of a telephone is an earphone consisting of an electromagnet attached to a diaphragm in such a way that the displacement of the diaphragm is proportional to the current in the magnet coil. The channel is the circuit connecting the carbon block and the electromagnet in series with a battery (Fig. 19.3). Sound waves incident on the transmitter produce resistance variations in the block. This causes a corresponding variation in the current in the circuit and hence in the displacement of the receiver diaphragm. Motion of the receiver diaphragm

FIGURE 19.3

A telephone communication system.

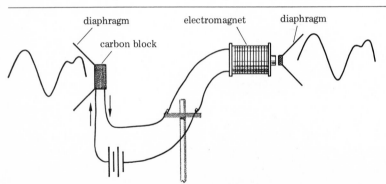

generates a sound wave similar to the one incident on the transmitter.

INFORMATION THEORY

The telephone appears to transmit more information than the telegraph, because besides hearing the message, the listener can identify the caller by voice and determine something of his or her mood. Often, however, voices are mistaken on the telephone that would not be mistaken in person. This means that the telephone does not transmit all the information in the original sound.

A mathematical theory of information was developed by Claude E. Shannon in 1948. This theory defines a measure of the information content of a message that is useful for comparing the information capacities of different communications channels. Information theory has also been applied to such diverse fields as genetics, linguistics, and psychology, but we shall limit our discussion to electronically transmitted information.

In information theory, information does not refer to the meaningfulness of a message in the ordinary sense, but rather to the selection of a specific message from among a set of N possible messages. For instance, the number N of possible two-letter combinations, such as *of*, *at*, *ta*, *xx*, is

$$N = 26 \times 26 = 26^2 = 676$$

A specific two-letter combination has been successfully transmitted when enough information is given to enable the receiver to distinguish it from the 675 other possible combinations. The fact that few of the combinations are actual words does not distract from the information content; a particular combination might be an abbreviation or a part of a longer word.

Definition The basic unit of information is the *bit*, and it is the information contained in a message selected from a set of two possible messages.

In the binary number system, only the symbols 0 and 1 are used. Thus a one-digit binary message, which can be either 0 or 1, carries one bit of information. A two-digit binary message, which can be either 00, 01, 10, or 11, carries two bits of information, one for each digit. A three-digit binary message, which can be either 000, 001, 010, 011, 100, 101, 110, or 111, carries three bits of information. In general, an n-digit binary message is one of $N = 2^n$ possible messages and carries n bits of information.

We can turn this around and say that any message that is selected from a set of 2^n possible messages can be coded by an n-digit binary number* so the message carries n bits of information. That is, to find the information content n of a message selected from a set of N possible messages, we write

$$2^n = N \hspace{4cm} 19.1$$

*The code consists of matching each of the 2^n messages to a different one of the n-digit binary numbers. Since there are 2^n such binary numbers, each message will have its own number.

or

$$n = \log_2 N \qquad\qquad 19.2$$

where $\log_2$ is the logarithm to the base 2. Equation 19.2 is Shannon's measure of information in the case where all the N messages can occur with equal probability.

Example 19.1 What is the information content of a two-letter message?

To solve Eq. 19.1 for n with a calculator that does not have a $\log_2$ key, we must take the logarithm of both sides to the base 10:

$$\log_{10} 2^n = \log_{10} N$$

But from the properties of logarithms we have

$$\log_{10} 2^n = n \log_{10} 2 = (0.301)n$$

so

$$n = 3.32 \log_{10} N \qquad\qquad 19.3$$

Since there are $N = 676$ different two-letter combinations, the information content of a two-letter message is

$$n = (3.32) \log_{10} N = (3.32) \log_{10} (676)$$
$$= (3.32)(2.83) = 9.4 \text{ bits}$$

This means that it takes a 10-digit binary number to code all two-letter combinations. There are $2^{10} = 1024$ different 10-digit numbers, but only $2^9 = 512$ different 9-digit numbers. ☐

Example 9.2 Find the information content of a telegraphic message that is 1000 characters long. Assume that each character can be one of 40 different symbols (letters, digits, space, and punctuation marks).

The number N of distinct 1000-character messages is found by realizing that in each of the 1000 positions we are free to put any one of the 40 possible characters. Changing any character anywhere in the message changes the entire message; i.e., two messages are considered distinct if they differ at any point. Thus the number of distinct messages is

$$N = \underbrace{40 \times 40 \times \cdots \times 40}_{1000 \text{ times}}$$
$$= 40^{1000}$$

This is an immense number, greatly exceeding the capacity (10^{100}) of your calculator.

From Eq. 19.3 the information content n of a 1000-character message is

$$n = 3.32 \log_{10} 40^{1000}$$
$$= 3.32(1000) \log_{10} 40$$
$$= 5319 \text{ bits}$$

Another way to calculate this is to first calculate the information content n_1 of a one-character message. Since there are 40 possible one-character messages, we have

$$n_1 = 3.32 \log_{10} (40) = 5.319 \text{ bits}$$

The information content n of a 1000-character message is therefore just 1000 times the information content of the one-character message:

$$n = (1000)n_1 = 5319 \text{ bits}$$

It is the property of the logarithmic definition of information content that the information content of a message is the sum of the information contents of each element in the message. □

Definition   The *capacity* C of an information channel is the maximum rate (in bits per second) at which information can be transmitted with arbitrarily small error. Shannon proved that every channel has a capacity C that depends on the amount of interference and distortion (*noise*) present in the channel. If information is transmitted at a rate greater than C, some of the message will invariably be lost.

Different media require different capacities. For the sake of comparison, suppose the telegraphic message in Example 19.2 were transmitted in 60 s. (This is about the time it would take to read a 1000-character text.) The rate of transmission is

$$\frac{5319 \text{ bits}}{60 \text{ s}} = 89 \text{ bits/s}$$

and a channel with a capacity of at least this much would be required. A telephone channel, by comparison, has a capacity of 2400 bits/s. A voice communication over a channel with less capacity (such as CB radio) is noticeably more noisy than normal telephonic communication.

Television requires a channel capacity much greater than that needed for voice communication. One might not think of a television picture as a message in the same sense as a telegraph message, but from the point of view of information theory it is. The picture is made by scanning an electron beam across the face of the picture tube. The inside of the tube is coated with a phosphorescent material that glows when hit by the electrons. In the American* television system the picture is made up of 525 horizontal lines which are each divided into 435 spots, so the message in one picture has

$$525 \times 435 = 228{,}000 \text{ elements}$$

That is, as the beam scans across the face of the tube, the incoming signal carries the information about how bright each of these 228,000 elements is to be. The information content of the picture is thus 228,000 times the information content of each element.

The brightest of each element varies continuously from zero to some maximum value, so it would seem to have an infinite number of possible values and hence an infinite information

*North America, South America, and Japan.

content. However, because of the inevitable noise and distortion present in the transmission process, the system cannot distinguish between brightness values that are very close together. In fact, commercial television is capable of distinguishing approximately $32 = 2^5$ different brightness values, so the information content of each element is 5 bits.

Thus the information content of a single television picture is

$$(228,000 \text{ elements})(5 \text{ bits/elements}) = 1.14 \times 10^6 \text{ bits}$$

Since the picture changes 30 times a second, the channel capacity required for the picture is

$$C = (30 \text{ pictures/s})(1.14 \times 10^6 \text{ bits/picture})$$
$$= 3.4 \times 10^7 \text{ bits/s}$$

This is to be compared with the channel capacity of 2.4×10^3 bits/s required for voice communication. To obtain this channel capacity, each television channel occupies a broad band in the electromagnetic spectrum. That is, the television signal is not carried by a single frequency, but by a band of frequencies 6 MHz wide.

> **REMARK** This calculation assumes that each transmitted picture is independent of every other one. Since in fact most of the 228,000 elements of a picture do not change from one picture to the next, it is possible to decrease the capacity requirements of the channel by using a different coding scheme. For example, less information would be required if the signal gave the difference between the brightness values of the current picture and the preceding one. In advanced communication engineering, such schemes are used to optimize the amount of information a given channel can carry.

CONTROL THEORY

Definition A *control* is a device that regulates the output of a mechanical or electrical system. In its simplest form a control is just a switch that can turn a machine on or off or a variable resistor that can regulate the speed of a machine.

Definition *Feedback* is the return of a small fraction of the output of a system to its input. If the output is added to the input, it is called *positive feedback*. If the output is subtracted from the input, it is called *negative feedback*.

When negative feedback is suitably coupled with a control, a stable self-regulating system results. The most familiar example of this is the heating system of a house. The thermostat is the control switch. It contains a bimetal strip made by binding together two metals with different coefficients of thermal expansion (Fig. 19.4). The metals expand by different amounts when the temperature changes, causing the strip to bend. The thermostat can be adjusted so that at a preset temperature (say 18°C), the strip makes contact with an electric terminal, closing the control circuit. The circuit is connected to a relay in the furnace, and when the control circuit is closed, the relay turns on the furnace.

Heat from the furnace (the output) warms the air of the house, and a small portion of this air warms the bimetal strip in the thermostat. This is the feedback. When the temperature reaches another preset value (say 20°C), the strip breaks contact,

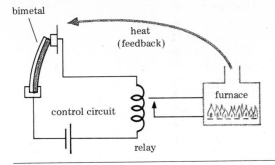

bimetal

heat
(feedback)

control circuit

furnace

relay

FIGURE 19.4

Thermostatic control system of a house. Heat from the furnace causes the bimetal strip to bend, opening the control circuit and turning off the furnace.

opening the control circuit and turning off the furnace. The feedback is negative because the output of the furnace ultimately turns the furnace off. As the room cools, the strip bends until it again closes the control circuit and turns on the furnace. In this way the operation of the furnace is controlled so that it maintains the temperature of the house between 18 and 20°C.

Negative feedback is essential for any stable self-regulating system. Most body functions, including the temperature regulation of the body, rely on negative feedback to maintain stable operation. Sophisticated negative feedback mechanisms are being used in some automobiles to increase engine performance. Sensors in the automobile monitor various parameters of the engine's output: engine temperature, engine speed, exhaust pollutants, etc. This information is fed to a microprocessor—a small computer—that calculates the air-gas mixture required under the current conditions for the best performance with the least pollution. This system uses negative feedback to continuously regulate the carburetor for optimal operation.

Positive feedback, on the other hand, is always unstabilizing. It is occasionally useful, as in the electrostatic induction machine described in Sec. 15.2. There some of the charge induced on the rods is added to the plate which induced the charge (Fig. 15.8). This increases the charge on the rods further, building up a very large voltage. In most instances, however, positive feedback is undesirable, and engineers try to eliminate it when it occurs naturally in a system.

A familiar example of positive feedback is the unpleasant ringing sound sometimes produced by a public address system. When a speaker speaks into the microphone, amplified sound comes out of the loudspeakers. If some of this amplified sound (the output) reaches the microphone (the input), it will be reamplified. Some of this reamplified sound in turn reaches the microphone and is amplified again. A positive feedback is thus created that results in a runaway situation. The sound grows in intensity until it reaches the power limitation of the system. To avoid this, the loudspeakers should be located so that their output does not reach the microphone.

19.2 Transducers

Definition A *transducer* is a device that converts an input signal of one type into an output signal of another. In most transducers of interest in electronics, one of the signals is electric and the other is acoustic, thermal, photo, or mechanical. A microphone, for instance, is an acoustic transducer: it converts an acoustic signal into an electric one. A mercury thermometer, on the other hand, is a nonelectric transducer: it converts a thermal signal (the temperature of a substance) into a mechanical signal (the length of a column of mercury).

Transducers are widely used in industry for the remote sensing of output parameters that describe the state of a process. For instance, in the steel industry, thermal transducers are inserted at many points around the wall of a blast furnace. These transducers produce an electric signal that is proportional to their temperature. The signals from all the transducers are carried by cables to a central computer, which uses them to calculate the temperature in the core of the furnace. If the core temperature is above or below its optimal value, the computer directs appropriate changes in the inputs (air, coke) to the furnace.

In this section we will describe a few of the many transducers in use today. These will serve as examples for all the others.

PHOTOTRANSDUCERS

A *phototransducer* is a transducer that converts a light signal into an electric signal. One type of phototransducer, the *photoresistive cell,* uses a photoconductive material, such as cadmium sulfide, which has the property that its resistance decreases with increasing irradiance.

In the absence of any light, a photoconductive material is an insulator, which means that all its electrons are bound to atoms and so are unable to form a current. Light incident on the material dissociates some of the electrons, which are then available to form a current. When the light is removed, the electrons soon rejoin their atoms, and the material again becomes nonconducting.

A photoresistive cell consists of a deposit of photoconductive material on a ceramic base, with ribbon-shaped electrodes deposited on top of the photoconductive material (Fig. 19.5). In total darkness the resistance of the cell can be as high as $10^{13}\ \Omega$, but in normal room light the resistance is about $10^3\ \Omega$.

To measure irradiance, the photoresistive cell is connected to a battery and an ammeter, as shown in Fig. 19.6. As the irradiance increases, the resistance of the cell decreases, and the current in the circuit increases. Thus the current increases with increasing irradiance.

Photoresistive cells are used in photographic light meters that measure irradiance reflected from a subject and in instruments that measure the concentration of a solution by the amount of light transmitted through it.

THERMAL TRANSDUCERS

A *thermal transducer* is a transducer that generates an electric signal that depends on the temperature of the device. One type of thermal transducer, the *thermoelectric couple,* or *thermocouple,* is based on the *Seebeck effect.* When a circuit is made of two different conductors (Fig. 19.7), a current will be generated if the two junctions of these conductors are at different temper-

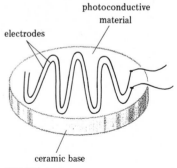

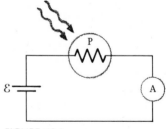

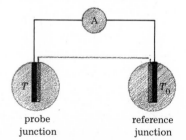

FIGURE 19.5

Photoresistive cell. The resistance between the electrodes depends on the intensity of the light incident on the photoconductive surface.

FIGURE 19.6

A photoresistive cell in a circuit. Light incident on the cell P controls the current in the ammeter A.

FIGURE 19.7

A thermocouple. Two junctions of unlike metals are maintained at different temperatures. The current in the circuit is proportional to the difference of these temperatures.

atures. The magnitude of the current is proportional to the temperature difference between the junctions, so if one junction is held at a known temperature, the current will be a measure of the temperature of the other junction. Thermocouples are used to measure the high temperatures in furnaces and other inaccessible locations. However, because the core temperature of a blast furnace would melt a thermocouple, this temperature must be measured indirectly by mounting thermocouples around the walls of the furnace and using the data to calculate the core temperature.

Other thermal transducers are based on the temperature variation of the resistivity of conductors and semiconductors (Sec. 16.3). The *resistance pyrometer*, described in Sec. 16.3, uses a coil of platinum wire, whereas a *thermistor* uses a metal oxide. The circuitry is similar to that for a photoconductor (Fig. 19.6); the variation in the temperature of the probe causes a variation in the resistance of the circuit and hence in the current.

ACOUSTIC TRANSDUCERS

A *microphone* is an *acoustic transducer* that converts sound into electric signals, and a *loudspeaker* is an *acoustic transducer* that converts electric signals into sound. A variety of microphones are available, each using a different physical principle. The *resistance microphone,* described in Sec. 19.1, uses the changes in the resistance of a block of compacted carbon particles to modulate the current in a dc circuit (Fig. 19.3). Magnetic and crystal microphones, on the other hand, generate their own emf. The *magnetic microphone* consists of a metal ribbon suspended between the poles of a magnet (Fig. 19.8). As the ribbon vibrates in response to sound waves incident on it, a voltage is magnetically induced between the ends of the ribbon. This voltage is proportional to the instantaneous velocity of the ribbon, and hence to the motion of air in the sound wave.

The *crystal* microphone is based on a phenomenon known as the *piezoelectric effect.* When certain asymmetric crystals are compressed, charges of opposite sign appear on the opposing crystal faces (Fig. 19.9a). The amount of charge, and hence the voltage across the crystal, is proportional to the compressive stress on the crystal. If the compressive stress is changed to a tensile stress, the charge and voltage reverse sign (Fig. 19.9b). In a crystal microphone, the pressure of a sound wave passing through a piezoelectric crystal generates a proportional voltage across the crystal. This voltage is the output signal, and it is picked up by leads attached to the crystal faces.

Most loudspeakers consist of a stiff diaphragm that is moved by an electromagnet. Variations in the current in the magnet coil produce variations in the position of the diaphragm, and as the diaphragm moves, it causes corresponding variations in the pressure of the surrounding air. Ultrasonic transducers use a piezoelectric crystal instead of a magnet. When a voltage is placed across a piezoelectric crystal, a reverse piezoelectric effect occurs: the voltage produces stress in the crystal, causing the crystal to expand or contract slightly. In an ultrasonic transducer, a high-frequency voltage is applied to a piezoelectric crystal,

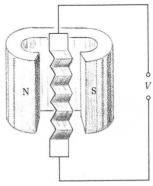

FIGURE 19.8

Magnetic microphone. The vibration of the metal ribbon in a magnetic field produces a voltage between its ends.

FIGURE 19.9

Piezoelectric effect. (a) Compressing a piezoelectric crystal causes opposite charge to form on opposing crystal faces. (b) When the compression is changed to a tension, the charges reverse sign.

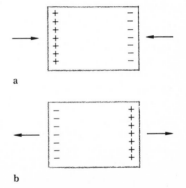

a

b

causing rapid variations in its size. These movements are transmitted to the air by means of a diaphragm, resulting in ultrasonic waves of the same frequency as the driving voltage.

> **REMARK** Electromagnets are impractical at high frequency because their large inductance offers a large impedance to high-frequency current (Eq. 18.14).

MECHANICAL TRANSDUCERS

A *mechanical transducer* is a transducer that converts a mechanical signal (usually a stress or a strain) into an electric signal. Some phonograph cartridges, for instance, use a piezoelectric crystal to convert the stresses exerted on the stylus into electric signals. Another type of transducer, called a *strain gauge,* uses the fact that the resistance of a wire changes when the wire is strained. It is made by printing a conducting path onto a small plastic wafer. When the wafer is strained, either by being stretched or compressed, the resistance of the conducting path changes. By measuring this resistance change, strains as small as 10^{-6} can be measured.

Strain gauges are often attached to structural units, such as a bridge girder, to measure the strain they undergo under different loading conditions. Strain gauges are also used to make electronic manometers and scales. The electronic scale now found in markets and butcher shops uses a strain gauge to convert weight into an electric signal. This signal, after passing through a microprocessor, gives a digital display of the weight and the total price of the product.

VIDEO TRANSDUCERS

A *television picture tube* is a *video transducer* that converts an electric signal into a visual image. A *television camera* is a *video transducer* that converts a visual image into an electric signal. Both work by scanning the image in a sequence of horizontal sweeps, in effect reducing the two-dimensional image to a set of 525 lines (625 in Europe). Each line is broken into 435 spots, so the image is reduced to a mosaic of $525 \times 435 = 228,375$ elements. The camera sweeps these elements in precise order, transmitting a signal that gives the brightness of each element. This signal, reaching a television receiver, controls the brightness of the corresponding elements as they are reproduced on the screen.

The television picture tube is an evacuated glass tube with a narrow neck at one end and a large flat face (the screen) at the other (Fig. 19.10). An electron beam is formed in the neck and focused to a small spot on the screen. The inside of the screen is coated with a phosphorescent material that glows when it is struck by the electron beam. The horizontal and vertical position of the spot on the screen is controlled by magnetic coils surrounding the neck. Circuits in the receiver cause these coils to sweep the beam horizontally 15,750 times a second and vertically 60 times a second (Fig. 19.11). A complete picture of 525 lines is swept out 30 times a second, there being two vertical sweeps per picture (see remark below). As the beam sweeps

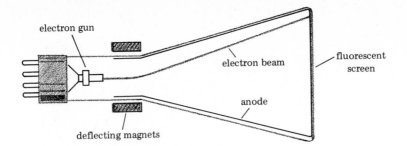

electron gun

electron beam

fluorescent
screen

anode

deflecting magnets

FIGURE 19.10

Television picture tube. An electron
beam, whose position is controlled
by a deflecting magnet, causes a
spot to form on the phosphorescent screen.

across the screen, the incoming signal controls the intensity of
the electron beam, so that the brightness of each spot on the
screen is proportional to the brightness of the corresponding
spot of the image being transmitted. In this way a time-ordered
sequence of electric impulses is converted into a two-
dimensional picture.

> **REMARK** Certain properties of the human eye are essential to all this. For
> instance, the eye retains the image of a line for about $\frac{1}{20}$s after the line
> vanishes (persistence of vision). Thus the eye still sees the first line of a
> picture when the last line is formed $\frac{1}{30}$s later.
>
> The actual scanning pattern used in commercial television is shown in Fig.
> 19.12. For each picture the screen is swept twice: the odd horizontal lines are
> scanned on the first pass, and the even horizontal lines are scanned on the
> second, so that a complete picture is formed from two interlacing half-
> pictures. This peculiar pattern is used because it is found that while 30
> pictures per second is sufficient to display motion smoothly, a perceptible
> flicker would result if the screen went dark 30 times a second. However, to
> transmit 60 complete pictures a second would require twice as much channel

FIGURE 19.11

Scanning pattern of a television
picture. The beam is swept horizontally 15,750 times a second,
while it is swept vertically 60 times
a second.

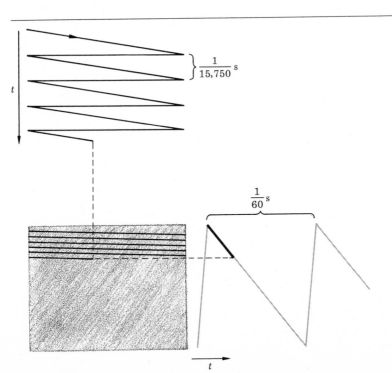

$\dfrac{1}{15,750}$ s

$\dfrac{1}{60}$ s

capacity as is needed. By transmitting 60 half-pictures per second, flicker is eliminated without increasing the rate of information transmission.

A television camera tube is an evacuated glass cylinder with an electron source at one end and a flat face at the other (Fig. 19.13). A lens system in front of the tube focuses an image of the scene being transmitted onto a plate just inside the face of the tube. Magnetic coils in the tube focus the electron beam onto the plate and move it across the plate in the same scanning pattern used in a picture tube.

In the *orthicon camera tube* the plate consists of a thin transparent insulator with a transparent metal coating (the *signal plate*) on one side and a precise array of 228,375 separate squares (the *mosaic*) on the other. Each square is made of a photosensitive material that emits electrons when struck by light (the *photoelectric effect,* Sec. 20.1).

The image formed by the lens passes through the signal plate to the mosaic, where it ejects electrons from each square in proportion to the amount of light reaching the square. Since the squares are insulated from each other, each square builds up a positive charge that is proportional to the light falling on it. Thus at each instant, the total charge on the mosaic is the sum of the charges on each of the 228,375 squares. The mosaic and the signal plate form a parallel-plate capacitor, the voltage of which is proportional to the charge on the mosaic.

As the scanning electron beam hits a particular square, the electrons neutralize the positive charge on the square and then build up a negative charge. When the negative charge reaches a certain value, additional electrons are repelled by the square and return to the other end of the tube where they are collected by a positive electrode. The change in the charge of the square depends on the amount of positive charge originally on the square, and hence on the amount of light that had fallen on the square since the last time the electron beam struck it. This change of charge changes the voltage on the capacitor, and this voltage change, picked up by a wire attached to the signal plate, constitutes the video signal.

An important feature of this system is that the charge on a particular square builds up continuously during the $\frac{1}{30}$-s interval between successive hits by the electron beam. Thus, while the beam is in contact with the square for only about 0.1 μs, the signal strength depends on the total amount of light that reached the square during the previous $\frac{1}{30}$s. This gives this system vastly greater sensitivity than systems that respond only to the light arriving at a point at the instant of interrogation.

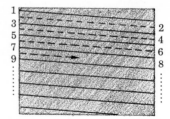

FIGURE 19.12

Two vertical scans are required for each complete television picture. The odd lines are swept out on the first scan and the even lines on the second.

19.3 The Diode

Definition A *diode* is a circuit element that conducts current in one direction only. Originally all diodes were vacuum-tube devices, but now semiconductor diodes have replaced vacuum-tube diodes in most applications. We shall discuss the internal structure of a semiconductor diode in Sec. 21.4, but for now we are interested only in the circuit characteristics of a diode.

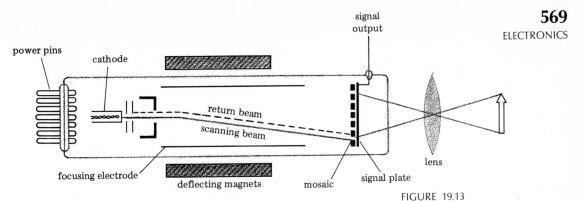

FIGURE 19.13

Orthicon camera tube. An image is formed on a mosaic of photosensitive squares. Each square is charged proportionally to the light incident on it. When the scanning beam hits a square, its charge is neutralized. This changes the voltage on the signal plate. The pattern of voltage changes constitutes the video signal.

An electronic device is characterized by its *IV* curve, which is a plot of the current *I* in the device for different values of the voltage *V* across it. Figure 19.14 gives the *IV* curve for a typical semiconductor diode, together with a similar curve for an ohmic resistor. Note the difference. The *IV* curve of a resistor is a straight line, indicating that the current is proportional to the voltage. Furthermore, when the voltage across a resistor is reversed, the direction of the current is also reversed, but the magnitude of the current is not changed. In contrast, the *IV* curve of a diode rises steeply when the voltage is positive (*forward bias*), indicating that the current is essentially independent of the voltage; i.e., with forward bias the resistance of the diode is nearly zero. When the voltage is reversed (*reverse bias*), there is an extremely small reverse current; i.e., with reverse bias the resistance of the diode is nearly infinite.

An *ideal diode* is a diode that has zero resistance in the forward direction and infinite resistance in the reverse direction. Figure 19.15 shows the *IV* curve of an ideal diode. Since real diodes approximate the ideal diode to a large extent, it is often useful to treat diodes as ideal when studying a circuit.

Figure 19.16a shows the symbol for a diode. The arrow points in the direction of the current when the diode has a forward bias, i.e., when it is conducting (Fig. 19.16b). When the voltage across the diode is reversed (Fig. 19.16c), the diode is nonconducting, and there is little or no current in the reverse direction.

RECTIFICATION

Definition A *rectifier* is a device that converts alternating current into unidirectional current. Rectifiers are used in electronic devices (calculators, TV games, etc.) to convert household alternating current into the direct current needed for the electronic circuitry. They are also used in the demodulating circuits of radios and televisions (Sec. 19.5). A diode is the essential element of any rectifying circuit, since it conducts only during the positive half of an ac cycle.

Figure 19.17 shows a *half-wave rectifying circuit*. Since the diode conducts only during the positive half of each ac cycle, the voltage across the resistor has the time variation shown. This circuit simply eliminates the negative half of the ac cycle.

FIGURE 19.14

IV curves of a diode and an ohmic resistor.

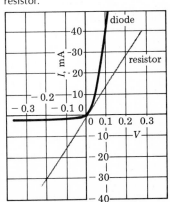

Figure 19.18 shows a *full-wave rectifying circuit*. During one half of the ac cycle, the current follows the path *abdc*; during the other half, it follows the path *cbda*. Since during each half the current is from *b* to *d* in the resistor, the voltage across the resistor is positive during each half of the cycle and has the time variation shown. A rectifying circuit like this one is used in the converters that come with electronic calculators and other electronic products. The power supply of a toy train also uses a full-wave rectifier to convert from alternating to direct current.

REMARK When a rectifier is coupled with a low-pass filter (Chapter 18, Prob. 22), the time variation of the unidirectional voltage can be smoothed out, giving a more constant voltage.

LOGIC CIRCUITS

In the binary number system, every number is thought of as a sum of powers of 2. For instance, the number 13 can be written

$$13 = 8 + 4 + 0 + 1$$
$$= (1 \times 2^3) + (1 \times 2^2) + (0 \times 2^1) + (1 \times 2^0)$$

and so the binary expression for 13 is 1101. That is, each place in the binary expression represents a power of 2, just as each place in a decimal expression represents a power of 10. If the digit in the *n*th position (measured from the right) is 1, then 2^n is in the sum; if the digit is zero, 2^n is not in the sum. In this way, every number can be represented by a unique sequence of zeros and ones.

In a computer, a number is represented by an array of voltages that correspond to the binary expression of the number. For instance, the digit 1 might be represented by a voltage of 3 V, and the digit 0 represented by the absence of a voltage (0 V). In this case, the number 13 (1101) would be represented by the array of voltages shown in Fig. 19.19. Since each register in the array can have only one of two possible values, each register carries one bit of information.

A logic circuit is a circuit designed to perform a specific mathematical or logical operation on one or more arrays of voltages. Although these circuits can be exceedingly complex, they are all built from three basic logic circuits, called *AND, OR,* and *NOT* gates. We shall show how AND and OR gates can be made using diodes,* and how a multiplication circuit is made by connecting AND, OR, and NOT gates in proper sequence.

*Current technology uses transistors rather than diodes in gates. This is discussed further in Sec. 21.4.

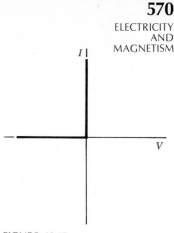

FIGURE 19.15

IV curve of an ideal diode. The resistance of an ideal diode is zero when the bias is forward and infinite when the bias is reversed.

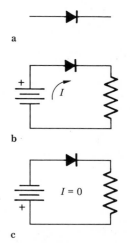

FIGURE 19.16

(a) Symbol of a diode. (b) The arrow on the symbol points in the direction in which the diode conducts. (c) The diode does not conduct when the bias on the diode is reversed.

FIGURE 19.17

Half-wave rectifying circuit.

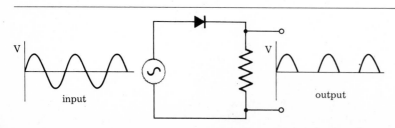

input   output

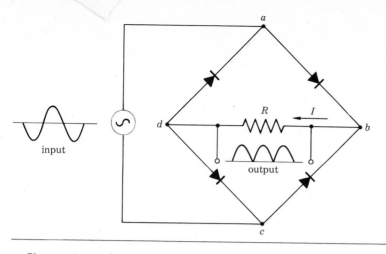

FIGURE 19.18

Full-wave rectifying circuit. During one half of the ac cycle, the current follows the patch *abdc*; during the other half, it follows the path *cbda*.

Figure 19.20*a* shows an AND gate with two inputs IN_1 and IN_2 and one output *O*. The voltage at point *C* is always zero; i.e., all other voltages are measured relative to it. Thus when we say that the input voltage V_1 at IN_1 is zero, we mean that the voltage between IN_1 and *C* is zero. Likewise when $V_1 = 3$ V, the voltage between IN_1 and *C* is 3 V. Now when both V_1 and V_2 are zero, the bias of both diodes is forward, so each diode is conducting. When a diode conducts, the voltage across it is essentially zero. Consequently, the output voltage V_0 at *O* is the same as the input voltages, or zero. If V_1 is raised to 3 V while V_2 remains zero, the bias on diode 1 becomes reversed, but diode 2 remains conducting. The voltage at *O* remains zero, because the voltage across the conducting diode is still zero. Likewise, if $V_1 = 0$ and $V_2 = 3$ V, then $V_0 = 0$. However, if both V_1 and V_2 are 3 V, then both diodes are nonconducting, and so the voltage at *O* is equal to the emf of the battery, or 3 V. The table in Fig. 19.20*b* gives the input-output properties of this gate in terms of the binary digits represented by the voltages. With this table we can forget about the details of the gate and represent it by the symbol in Fig. 19.20*c*. The × indicates that in terms of binary digits, the output is the product of the inputs.

An OR gate with two inputs IN_1 and IN_2 and one output *O* is shown in Fig. 19.21*a*. By the same analysis as we used for the AND gate, you can show that the OR gate has the properties shown in Fig. 19.21*b*. The symbol for the OR gate is shown in Fig. 19.21*c*.

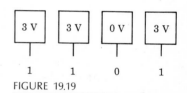

FIGURE 19.19

The number 13 represented by an array of voltages. The number is expressed in binary notation (1101), with 3 V standing for 1 and 0 V standing for 0.

FIGURE 19.20

(a) An AND gate made with two diodes connected in parallel to a battery and a resistor. (b) Input-output table of an AND gate. (c) Symbol of an AND gate.

	IN_2 0	1
IN_1		
0	0	0
1	0	1

a   **b**   **c**

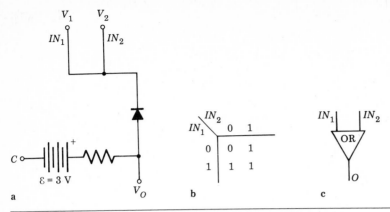

FIGURE 19.21

(a) An OR gate made with two inputs connected in parallel to a
diode, a battery, and a resistor.
(b) Input-output table of an OR
gate. (c) Symbol of an OR gate.

A NOT gate, or *inverter*, has one input and one output, as shown in Fig. 19.22a. The table in Fig. 19.22b shows the two possible input-output combinations in terms of the binary digits they represent. An input of 0 produces an output of 1, and an input of 1 produces an output of 0. Thus the output of a NOT gate is the inverse, or negative, of its input.

REMARK Unlike AND and OR gates, a NOT gate cannot be made with a diode. Figure 21.34 in Sec. 21.4 describes a NOT-gate circuit made with a field-effect transistor.

The next step in building a multiplication circuit is to make an ADD gate by combining the three basic gates into the gate shown in Fig. 19.23a. In this gate, the two inputs IN_1 and IN_2 go simultaneously to an AND and an OR gate. The output of the AND gate, after passing through a NOT gate, joins the output of the OR gate at the input of a second AND gate. It is easy to determine the output of this gate for each combination of inputs by tracing the signals through the individual component gates. This is done in Fig. 19.23a for the case in which IN_1 and IN_2 are both 1; the output in this case is 0. Doing the same analysis for the other three input combinations, we obtain the input-output table shown in Fig. 19.23b. With this table we can again forget about the details of the gate and represent the entire gate by the symbol in Fig. 19.23c. The + indicates that in terms of binary digits, the output is the sum of the inputs. (The output of the ADD gate is 0 when both inputs are 1, because in binary $1 + 1 = 10$. Additional circuitry is required to handle the 1 that is carried over in the sum.)

By combining AND and ADD gates in suitable combinations, it is possible to design circuits that will perform complex arithmetic operations. Figure 19.24 shows a circuit designed to multiply two two-digit binary numbers. The inputs are IN_1, IN_2 and IN'_1, IN'_2; the outputs are O_1,–O_4. Although you may be confused at first by the complexity of this circuit, by studying it for a while you will see how it does essentially what you would do if you multiplied the numbers by hand.

The top two AND gates in Fig. 19.24 multiply IN'_1 with the multiplicand ($IN_1 IN_2$); the second two AND gates multiply IN'_2 with the multiplicand. This second product is shifted one place to the left and added to the first product, just as in a hand

IN

IN	0	1
O	1	0

b

FIGURE 19.22

calculation. The only tricky parts to this circuit are the three shaded circuits, which test for the carryover digits.

REMARK A complex circuit like the one in Fig. 19.24 is a nested hierarchy of simpler circuits and components. The lowest level of the hierarchy is composed of the diodes and resistors. The second level is composed of the basic AND, OR, and NOT gates made from these components. The third level is the ADD gate made from the basic gates. The fourth level is the multiplication circuit made from AND and ADD gates. By combining complex multiplication and addition circuits, fifth-level circuits are made that calculate sines, logarithms, and other special functions. Although each level in the hierarchy is merely an assemblage of the elements from the next lower level, it performs functions that are qualitatively different from the functions of its elements.

Up until 1970 computers were made this way. That is, one manufacturer made diodes and resistors, another used these components to make AND, OR, and NOT gates, while a third wired AND, OR, and NOT gates into computer circuits. Since 1970, however, integrated-circuit technology has made it possible to combine many levels of the hierarchy into a single circuit element, called an *integrated circuit* (IC), or *chip*. A single chip, less than 1 cm across, contains the equivalent of 10,000 diodes and resistors (Fig. 19.25). Integrated circuits may be the most revolutionary electronic development of the past 50 years. They have made possible the manufacture of small and inexpensive calculators and microprocessors, and have opened up the possibility of reaching still higher levels in the circuit hierarchy.

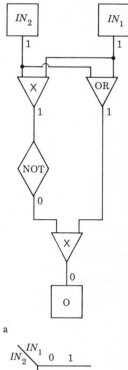

ZENER DIODE

Definition A *Zener diode* is a diode that suddenly becomes conducting when the reverse bias reaches a certain value, called the *breakdown voltage* V_b. Figure 19.26 shows the *IV* curve for a Zener diode with a breakdown voltage of 4 V; the symbol for a Zener diode is shown in Fig. 19.27. Although any diode will conduct in the reverse direction when the reverse voltage is large enough, such a situation destroys an ordinary diode. In a Zener diode, however, the breakdown is not permanent. A Zener diode becomes nonconducting again when the reverse bias is lowered below the breakdown voltage.

Zener diodes are used to provide a constant voltage. In the circuit in Fig. 19.27, for instance, the output voltage V_O across the Zener diode is a constant equal to the breakdown voltage V_b of the diode for all values of the input greater than V_b.

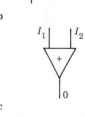

FIGURE 19.23

(a) An ADD gate made by connecting together AND, OR, and NOT gates. (b) Input-output table of an ADD gate. (c) Symbol of an ADD gate.

Example 19.3 In the circuit in Fig. 19.27, the Zener diode has a breakdown voltage of 4 V and $R = 50\,\Omega$. Find the current I in the circuit and the voltage V_O across the diode when V_i is (a) 2 V, (b) 5 V, and (c) 6 V.

(a) When $V_i = 2$ V, the reverse bias across the diode is less than the breakdown voltage, so the diode is nonconducting. In this case $I = 0$ and $V_O = V_i = 2$ V.

(b) When $V_i = 5$ V, the reverse bias across the diode reaches the breakdown voltage, so the diode becomes conducting, and the voltage across it is

$$V_O = V_b = 4\text{ V}$$

The voltage RI across the resistor is $V_i - V_O$, so the current is

$$I = \frac{V_i - V_O}{R} = \frac{1\text{ V}}{50\,\Omega} = 20\text{ mA}$$

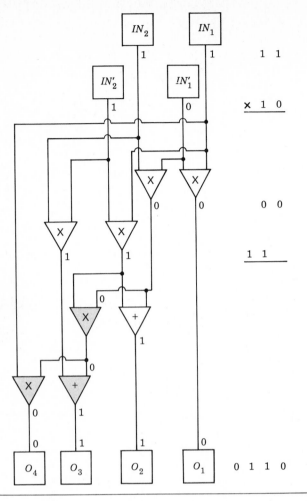

FIGURE 19.24

Circuit to multiply two two-digit binary numbers. The two numbers are represented by the inputs IN_2IN_1 and $IN'_2IN'_1$. The first two rows of AND gates multiply IN_1 and IN_2 by IN'_1 and IN'_2. The second product is shifted and added to the first product by the ADD gate in the middle. The three shaded gates check for carryover. The output of each gate is shown for the case of $3 \times 2 = 6$ ($11 \times 10 = 0110$).

(c) When $V_i = 6\,\text{V}$, the diode is still conducting, and the voltage across it is still 4 V. The voltage RI across the resistor is now 2 V, so the current is

$$I = \frac{2\,\text{V}}{50\,\Omega} = 40\,\text{mA}$$

The circuit in Fig. 19.27 is called a *voltage regulator* because it delivers a constant output voltage in spite of variations in the input voltage. ☐

19.4 The Transistor

Definition A *transistor* is a three-terminal semiconductor device used to control and amplify current. It is made from an especially processed piece of germanium or silicon crystal to which three wire terminals are attached. The symbol for a transistor is shown in Fig. 19.28. The arrow shows the direction of the current between the two end terminals (*emitter* and *collector*). In normal operation, a voltage difference V_{CE} is maintained between the emitter and collector by means of a battery $\mathcal{E}_C$, as shown in Fig. 19.29. When the middle terminal (*base*) is discon-

a   b

FIGURE 19.25

(a) Integrated circuit used in a digital watch (*Hewlett-Packard*).
(b) Photomicrograph of an integrated circuit (*Intel*).

nected, the transistor is nonconducting; i.e., there is no current between the emitter and the conductor. However, if a small current I_B is drawn from the base by means of a second battery $\mathcal{E}_B$, the transistor is turned on, and there is a large current I_C between the emitter and the collector. Thus, the current between the emitter and the collector is controlled by the base current I_B.

The behavior of a transistor is displayed in Fig. 19.30. This graph plots, for different values of the base current I_B, the current I_C in the collector against the voltage difference $V_{CE} = V_E - V_C$ between the emitter and the collector. With $I_B = 0$, the collector current I_C is essentially zero for all values of V_{CE}; that is, the transistor is nonconducting. For nonzero values of I_B, there is a large collector current that increases as V_{CE} increases. For a fixed value of V_{CE}, the collector current increases as the base current increases. For example, with $V_{CE} = 7.5\,\text{V}$ and $I_B = 0.15\,\text{mA}$, the collector current is $I_C = 14.5\,\text{mA}$. If V_{CE} is kept at 7.5 V while the base current is increased to 0.20 mA, the collector current increases to 21 mA. Thus a change in base current of only 0.05 mA produces a 6.5-mA change in the collector current. It is this property that enables a transistor to operate as an amplifying and controlling element.

BASIC AMPLIFIER

Definition An *amplifier* is an electronic device that increases the power of a time-varying signal without changing the shape of the signal. At each instant, the voltage of the output signal is a constant factor g_V times the voltage of the input signal, and the current of the output signal is a constant factor g_I times the

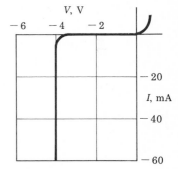

FIGURE 19.26

IV curves of a Zener diode.

FIGURE 19.27

Voltage-regulating circuit made with a Zener diode.

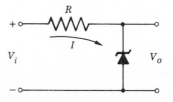

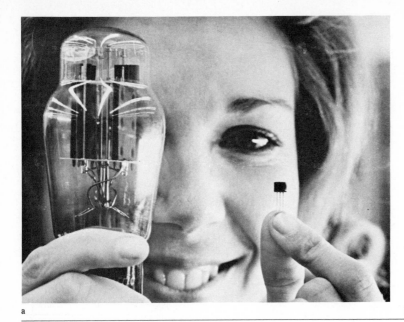

a

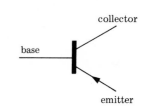

FIGURE 19.28

(a) A transistor (right) being compared with the vacuum tube (left) that it replaces (*Bell Laboratories*). (b) Schematic diagram of a transistor showing its three terminals.

current of the input signal. The factors g_V and g_I are called the *voltage gain* and *current gain* of the amplifier. Since power is the product of voltage and current, the power gain is equal to $g_V g_I$. In an amplifier both g_V and g_I are greater than 1, so the power gain is greater than 1. This is in contrast to a transformer, where the voltage increase of the output is always accompanied by a proportional current decrease, so the power gain is 1.

Figure 19.31 shows a basic amplifier circuit that uses a transistor to amplify an input voltage V_i placed across the input terminals. To understand how this circuit works, we first consider the circuit in Fig. 19.29, which is similar to Fig. 19.31 but without the input voltage. The resistor R_B is chosen in conjunction with the battery $\mathcal{E}_B$ to give a reasonable base current, say, $I_B = 0.25$ mA. The current I_C and the potential V_{CE} are then related by the $I_B = 0.25$ mA curve in Fig. 19.30. For instance, if $V_{CE} = 10$ V, the current is 29 mA.

The voltage V_{CE} across the transistor depends on the battery $\mathcal{E}_C$ and the voltage drop across the resistor R_C. From Kirchhoff's second law we have

$$V_{CE} = \mathcal{E}_C - R_C I_C \qquad 19.4$$

This is a second relation between V_{CE} and I_C. The values of I_C and V_{CE} must simultaneously satisfy this equation and lie on the $I_B = 0.25$ mA curve in Fig. 19.30.

For example, suppose $\mathcal{E}_C = 16$ V and $R_C = 400\ \Omega$. Then Eq. 19.4 becomes

$$V_{CE} = 16\ \text{V} - (400\ \Omega)I_C \qquad 19.5$$

which is the equation of a straight line. To determine I_C and V_{CE}, this equation is plotted on the graph of the $I_C V_{CE}$ curves of the transistor (Fig. 19.30). The line is easily found by observing that

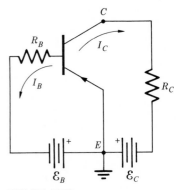

FIGURE 19.29

Basic amplifier circuit. A small current in the base causes a large current in the collector.

when I_C is zero, V_{CE} equals 16 V, and when V_{CE} is zero, I_C equals 16 V/400 Ω = 40 mA. Thus the straight line representing Eq. 19.5, which is called the *load line*, connects the 16-V point on the horizontal axis and the 40-mA point on the vertical axis. All points on this line satisfy Eq. 19.5, and all points on the $I_B = 0.25$ mA curve are possible values of I_C and V_{CE} when the base current is 0.25 mA. The intersection of the line and the curve at Q gives the values of I_C and V_{CE} that simultaneously satisfy both these conditions. From Fig. 19.30 the values of I_C and V_{CE} are found to be 25.5 mA and 5.5 V, respectively. These are the *quiescent* operating values of the circuit.

Now suppose that an alternating input voltage of the form

$$V_i = V_{ip} \sin 2\pi ft$$

is placed across the input terminals of the circuit in Fig. 19.31. This input voltage and the battery $\mathcal{E}_B$ together produce the base current

$$i_B = I_B + I_i$$

which is the sum of the battery current I_B and the input current

$$I_i = \frac{V_i}{R_B} = I_{ip} \sin 2\pi ft$$

where the peak input current is

$$I_{ip} = \frac{V_{ip}}{R_B}$$

For instance, with $V_{ip} = 30$ mV and $R_B = 600$ Ω, the peak input current is

$$I_{ip} = \frac{V_{ip}}{R_B} = \frac{30\ \text{mV}}{600\ \Omega} = 0.05\ \text{mA}$$

If $I_B = 0.25$ mA, the total base current, then, is

$$i_B = 0.25\ \text{mA} + (0.05\ \text{mA}) \sin 2\pi ft$$

and thus oscillates between 0.20 and 0.30 mA. Figure 19.30 shows that as a consequence of this oscillation the operating point oscillates between Q_1 and Q_2 along the load line, and the collector current I_C oscillates between 21 and 30 mA.

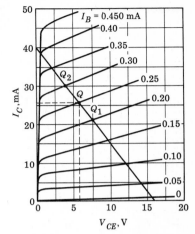

FIGURE 19.30

Characteristics of a transistor. The variation of the collector current I_C with collector voltage V_{CE} is shown for different values of the base current I_B. The load line is drawn for the circuit in Fig. 19.31, with $R_c = 400$ Ω and $\mathcal{E}_c = 16$ V.

FIGURE 19.31

Basic amplifier. When an alternating voltage is placed across the input terminals, a much larger alternating voltage appears across the output terminals.

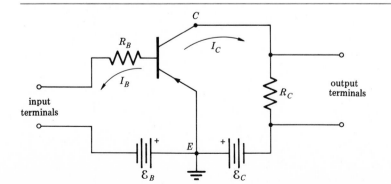

The collector current is a mixture of a direct current of 25.5 mA and an alternating current with a peak of 4.5 mA;

$$i_c = 25.5 \text{ mA} + (4.5 \text{ mA}) \sin 2\pi ft$$

Thus, since a peak input current of 0.05 mA produces a peak output current of 4.5 mA, the current gain g_I of the amplifier is 4.5 mA/0.05 mA = 90. Likewise, the peak input voltage is 30 mV, and the peak output voltage across R_C is

$$(400 \,\Omega)(4.5 \text{ mA}) = 1800 \text{ mV}$$

so the voltage gain is g_I is 1800 mV/30 mV = 60. Consequently, the power gain of the amplifier is

$$g_I g_V = (90)(60) = 5400$$

DIFFERENTIAL AMPLIFIER

The basic amplifier (Fig. 19.31) has a constant voltage $V_c = I_c R_c$ across its output terminals even when the voltage across the input terminals is zero. Furthermore if the voltage of each input terminal is raised the same amount above ground voltage, the base current will change and consequently the output voltage will change as well. That is, the output voltage responds to a common change in the voltages of the input terminals, not just to the difference in their voltages.

For many purposes it is desirable to have what is called a *differential amplifier,* which responds only to the difference in the voltages of its input terminals and is insensitive to their common voltage. Such an amplifier is made from two identical transistors connected as shown in Fig. 19.32. The variable resistor R is adjusted so that the voltage across the output terminals (3 and 4) is zero when the voltage across the input terminals (1 and 2) is zero.

If a voltage difference is applied across the input terminals, half the difference is applied to the base of each transistor, but with opposite sign. Each transistor amplifies its own base voltage, producing amplified voltages of opposite sign. But since the transistors are connected with opposite polarity to resistors R_{C1} and R_{C2}, the voltage across the output terminals is the sum of the magnitudes of the amplified voltages from each transistor.

FIGURE 19.32

Differential-amplifier circuit. The output voltage is proportional to the voltage across the input terminals.

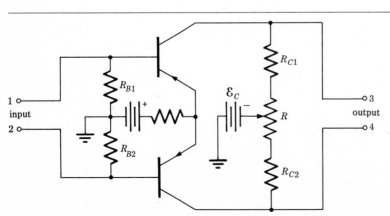

On the other hand, if the same voltage is applied to input terminals 1 and 2, the base voltages have the same sign, and the two transistors produce opposite voltages across R_{C1} and R_{C2}. If the transistors were identical, the voltage across the output terminals would thus be zero. Since the transistors are very similar but not identical, the output voltage will be small but not zero. The ratio of the common input voltage to the actual output voltage is called the *common-mode rejection ratio,* and should be as large as possible.

OPERATIONAL AMPLIFIER

If the output of one amplifier is connected to the input of a second amplifier, the overall gain of the system will be the product of the individual gains of each amplifier. Thus by connecting in series four amplifiers, each with a gain of 10, a system with a gain of 10^4 can be obtained. With integrated-circuit technology, all the components and interconnections of such a high-gain, multistage amplifier can be placed on a single silicon chip only 1 cm² in area.

Definition An *operational amplifier,* or *op-amp,* is a high-gain, integrated-circuit differential amplifier with a very large input impedance. The symbol for an op-amp is a triangle resting on its side, as shown in Fig. 19.33. There are two input terminals on the device: the *inverting input* $(-)$ and the *noninverting input* $(+)$. Because of the high input impedance, there is essentially no input current. At the output of an op-amp, the voltage V_{out} is

$$V_{out} = g_V(V_+ - V_-) \qquad 19.6$$

where g_V is the *open-loop gain* of the op-amp, V_+ is the input voltage applied to the noninverting $(+)$ terminal, and V_- is the input voltage applied to the inverting $(-)$ terminal. All voltages are measured relative to a fourth terminal called *common C.* An op-amp has several other terminals, usually not shown, for connecting the necessary dc voltage supply.

The open-circuit gain g_V is typically 10^4 or more, but its value fluctuates erratically. To obtain stable amplification, a fraction of the output voltage is connected to the inverting-input terminal (Fig. 19.34), where it is subtracted from the input signal $V_+ - V_-$ present at that time. This is another example of negative feedback (Sec. 19.1), and it stabilizes the gain of the amplifier in the same way that a thermostat stabilizes the temperature of a house. Any increase in the open-loop gain g_V produces an increase in the output voltage. As a consequence, the fraction of the output applied to the input is increased. Since this fraction is subtracted from the input voltage, the output voltage is reduced. Conversely, any decrease in g_V produces a decrease in the output voltage and in the fraction of the output voltage applied to the input. In this case, less is subtracted from the input voltage, so the output is increased. These compensating changes occur almost instantaneously, so the result is a stable *closed-circuit gain G.*

Let us calculate the gain G of the op-amp in Fig. 19.34. A

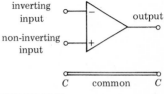

FIGURE 19.33

Symbol of an operational amplifier. The voltage at the output is proportional to the difference between the voltages at the noninverting and inverting inputs. All voltages are measured relative to common (C).

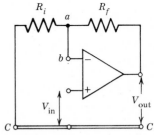

FIGURE 19.34

Negative feedback in an operational amplifier. The fraction $R_f/(R_i + R_f)$ of the output voltage is applied to the inverting input.

voltage $V_+ = V_{in}$ is applied to the noninverting terminal, and the feedback voltage V_- is applied to the inverting terminal. The full output voltage V_{out} is applied across the resistors R_i and R_f. There is no current between a and b because of the high input impedance of the amplifier. Consequently, the feedback voltage is

$$V_- = \frac{R_f}{R_i + R_f} V_{out}$$

When these values of V_+ and V_- are substituted into Eq. 19.6, we get

$$V_{out} = g_V(V_+ - V_-)$$
$$= g_V(V_{in} - \frac{R_f}{R_i + R_f} V_{out})$$

The negative-feedback character of the circuit is shown by the presence of the negative term involving V_{out} on the right-hand side of the equation for V_{out}. Rearranging terms in this equation, we obtain

$$g_V V_{in} = \left(1 + g_V \frac{R_f}{R_i + R_f}\right) V_{out} \quad \text{or} \quad V_{in} = \left(\frac{1}{g_V} + \frac{R_f}{R_i + R_f}\right) V_{out}$$

Since g_V is very large (10^4 or more), the term $1/g_V$ is small, regardless of fluctuations in the exact value of g_V. To the extent that we can neglect $1/g_V$ altogether, the closed-circuit gain G of the circuit is independent of g_V and is given by

$$G = \frac{V_{out}}{V_{in}} = \frac{R_i + R_f}{R_f} \qquad\qquad 19.7$$

Thus the gain is equal to the reciprocal of the feedback fraction. For example, suppose $R_i = 100\,\Omega$ and $R_f = 10\,\Omega$. Then the fraction

$$\frac{R_f}{R_i + R_f} = \frac{10\,\Omega}{100\,\Omega + 10\,\Omega} = \frac{1}{11}$$

of the output voltage is fed back to the input, and the gain is

$$G = \frac{R_i + R_f}{R_f} = 11$$

19.5 Modulation and Demodulation in AM Broadcasting

Radio and television signals are transmitted from the broadcasting station to the individual receivers by electromagnetic waves. In Secs. 19.1 and 19.2 we showed how the original signal (audio or video) is converted into an electric signal by the transmitting transducer (microphone or video camera), and how the electric signal is converted back into a reception signal by the receiving transducer (loudspeaker or picture tube). In this section we show, for the case of AM (*amplitude modulation*) radio, how the electric signal gets from the output of the studio microphone to the input of the radio loudspeaker.

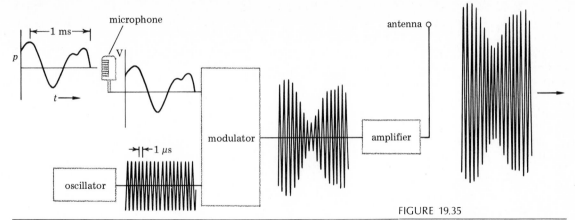

FIGURE 19.35

Amplitude modulation. A microphone converts sound waves into electric signals of the same frequency. These signals are mixed with a high-frequency electric signal, producing a high-frequency signal with an amplitude that varies like the sound wave. This signal is amplified and broadcast as an electromagnetic wave.

Sound is a pressure wave in air consisting of a mixture of frequencies between 20 and 20,000 Hz. For purposes of discussion we shall consider the typical sound frequency to be 1000 Hz, which means that major variations in the pressure occur on a time scale of about 1 ms. A microphone converts the pressure wave into a time-varying voltage of the same shape (Fig. 19.35). This voltage is not converted directly into an electromagnetic wave, however, because a wave of such low frequency (1 kHz) would be rapidly absorbed by the atmosphere. Furthermore, if all stations broadcast at the same frequency, there would be no way to tune in on just one. Each station, therefore, broadcasts a characteristic high-frequency electromagnetic wave at a frequency (typically 1 MHz) assigned to it by the licensing agency. This is called the *carrier wave*. The information about the audio signal is coded into the carrier wave by a process called *modulation*.

Figure 19.35 shows the steps involved in the production of an AM radio wave. An *oscillator* generates a pure sine wave at the station's assigned frequency. This wave, together with the audio wave coming from the microphone, goes to the *modulator*, which mixes the two input signals in such a way that the output signal is a wave with the frequency of the oscillator and an amplitude that varies with the audio signal. The modulated signal is then amplified and sent to the station's antenna, which converts the time-varying voltage into an electromagnetic wave of the same shape.

> **REMARK** In amplitude modulation, the amplitude of the carrier wave varies with the amplitude of the audio signal. In frequency modulation (FM) the frequency of the carrier wave varies with the amplitude of the audio signal.

Figure 19.36 shows the steps involved in the reception and demodulation of a radio wave. The wave is picked up by an antenna coupled to an *LC* circuit. All the broadcast waves coming from different stations induce voltages in the antenna that vary with the frequencies of the carriers. By adjusting the capacitance of the *LC* circuit, its resonant frequency can be tuned to the frequency of a particular station. The voltage variation induced by the wave of this frequency then produces a much

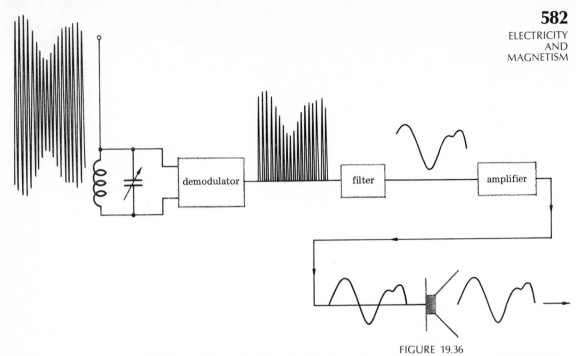

FIGURE 19.36

Detection and demodulation. The broadcast wave is picked up by an LC circuit tuned to the carrier frequency. The electric signal is passed through a half-wave rectifier (demodulator) and then through a high-frequency filter. This process recovers the low-frequency component of the original electric signal. This component is amplified and fed to a loudspeaker, which generates a sound wave.

larger current in the LC circuit than any other wave. This high-frequency alternating current goes to the *demodulator*, which is a half-wave rectifier that transmits only the positive half of each high-frequency cycle. The rectified wave goes through a low-pass filter that removes the high-frequency oscillations, leaving a wave with the shape of the original audio signal. This wave is amplified and sent to the loudspeaker, which transforms it into a pressure wave of the same shape.

BANDWIDTH

The wave generated by the oscillator in Fig. 19.35 is a pure sine wave with the carrier frequency f_c. The modulated wave is not a pure sine wave because its amplitude is not constant. However, according to Fourier's theorem (Sec. 11.4), every wave can be expressed as the sum of pure sine waves. When this is done for the modulated wave, it is found to contain sine waves with all frequencies between $f_c - \frac{1}{2}\Delta f$ and $f_c + \frac{1}{2}\Delta f$. That is, the station does not broadcast a single electromagnetic wave with frequency f_c, but a spectrum of electromagnetic waves with a frequency spread Δf.

Definition The spread in frequencies of a communication signal is called the *bandwidth* Δf of the signal. The more information per second carried by the signal, the greater must be its bandwidth. In radio broadcasting the bandwidth is comparable to the range of frequencies in the audio signal, or about 10 to 20 kHz. In television broadcasting, because of the much greater rate of information transmission (Sec. 19.1), the bandwidth is

6 MHz. That is, each VHF television channel occupies a band 6-MHz wide in the VHF (50 to 500 MHz) region of the electro-magnetic spectrum (Fig. 13.1).

GUIDE TO MAJOR TOPICS

Topic	References	Problems
Communications, information, and control	Sec. 19.1	
Communications		
Information theory	Eqs. 19.1 to 19.3; Examples 19.1 and 19.2	1 to 4
Control theory		5 and 6
Transducers	Sec. 19.2	
Diodes	Sec. 19.3	
Rectification		7 and 8
Logic circuits		9 to 12
Zener diode	Example 19.3	13 and 14
Transistors	Sec. 19.4	
Basic amplifier	Eqs. 19.4 and 19.5	15 to 18
Differential amplifier		
Operational amplifier	Eqs. 19.6 and 19.7	
Modulation and demodulation	Sec. 19.5	

PROBLEMS

1 (a) What is the information content of a 50-character-long message, if each character can be any of the 26 letters of the alphabet? (b) In another message, each character can be any of the 10 digits (0 through 9). How many characters would this second message need in order to carry as much information as the 50-character-long message in (a)?
Ans. (a) 235; (b) 71

REMARK The answer to Prob. 1(b) implies that 50 letters can be coded using only 71 digits. The most obvious code (01 = A, 02 = B, ..., 26 = Z) requires 100 digits, two for each letter. But this code makes no use of the combinations 27 through 99. A better code matches the numbers from 001 through 676 with the 676 two-letter combinations (001 = AA, 002 = AB, ..., 676 = ZZ). With this code, 50 letters could be represented by only 75 digits, three for every two letters.

2 A coding system uses only the four characters A, B, C, and D. (a) Rewrite the message AACBDDCABBC in binary, using only the digits 0 and 1. (b) Show that an N-character-long message in the four-letter system contains the same information as a 2N-character-long message in binary.

3 A printed photograph, called a *halftone*, is actually a dense array of dots. There are 70 points per centimeter in a halftone, and each point is either a dot (black) or a blank (white).

What is the information content of a 6 × 8 cm halftone?
Ans. 2.35 × 10^5

4 Figure 19.37 shows a square divided into 100 boxes. Every picture that can be drawn in the square by completely filling in some of the boxes can be represented by a 100-character-long binary number. The digits in the number, read from left to right, correspond to the boxes in the square, starting with the upper left-hand box and reading across and down in the usual

FIGURE 19.37

Problem 4.

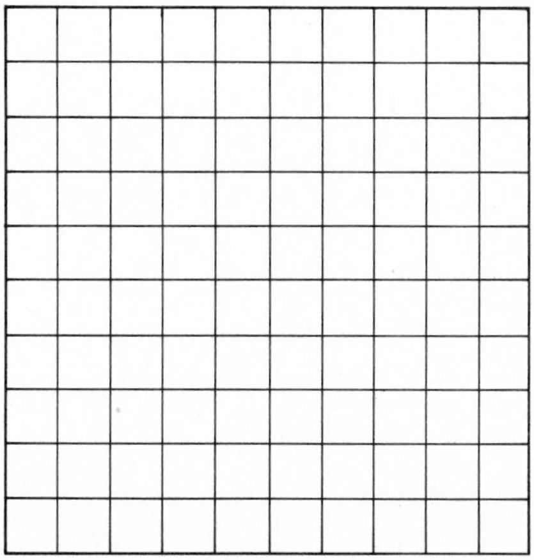

scanning pattern. When a digit is 0, the corresponding box is blank; when the digit is 1, the box is filled in. (a) Draw in Fig. 19.37 the picture represented by the following number: 00000 00000 00000 00000 00110 01100 00110 01100 00000 00000 00000 00000 01000 00010 00100 00100 00011 11000 00000 00000. (b) What is the information content of a picture drawn on the square in this way?

5 On a cold day a window is left open near the thermostat that controls the heating system for an entire house. Explain why this will cause the rest of the house to be too warm.

6 A dual-control electric blanket has separate thermostatic controls for each side of the bed. Suppose these controls are somehow reversed, so that the husband is unknowingly controlling the wife's side of the blanket while the wife is unknowingly controlling the husband's side of the blanket. Describe what happens as they each try to find a comfortable temperature setting.

7 The time-varying voltage shown in Fig. 19.38 is sent through a half-wave rectifier. Draw the output voltage.

FIGURE 19.38

Problems 7 and 8.

8 Repeat Prob. 7 for a full-wave rectifier.

9 Figure 19.39 shows a circuit with three input voltages, V_1, V_2, and V_3, each of which can be either 0 or 3 V. Determine the output for each of the eight possible combinations of inputs. *Ans.* Output is zero if V_1 is zero or both V_2 and V_3 are zero.

10 Determine the input-output characteristics of the circuit shown in Fig. 19.40.

11 Using AND and ADD gates, design a circuit with two pairs of inputs, such that the output is 1 if and only if there is a 1 and a 0 in each input pair.

12 Using AND and OR gates, design a circuit with four inputs, such that the output is 1 if and only if three or four of the inputs are 1.

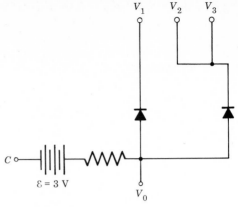

FIGURE 19.39

Problem 9.

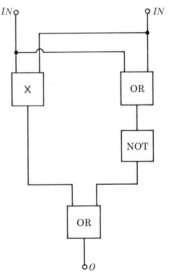

FIGURE 19.40

Problem 10.

13 In the circuit in Fig. 19.27 the Zener diode has a breakdown voltage of 6 V. (a) Find the value of R that gives a current of 60 mA when the emf of the battery is 7.5 V. (b) What is the current when the emf is 8 V? *Ans.* (a) 25 Ω; (b) 80 mA

14 In the circuit in Fig. 19.27 the Zener diode has a breakdown voltage of 10 V and $R = 200\ \Omega$. What is the emf of the battery when the current is 35 mA?.

REMARK The next four problems refer to the transistor with the characteristics given in Fig. 19.30.

15 What is the collector current in a transistor

584

when the collector potential V_{CE} is 7.5 V and the base current is 0.40 mA?
Ans. 45 mA

16 What must the base current be in a transistor to give a collector current of 25 mA when the collector potential V_{CE} is 10 V?

17 Find (a) the maximum and (b) the minimum potential difference across the 400-Ω resistor R_c in the circuit in Fig. 19.31 when the circuit is operating between the points Q_1 and Q_2 in Fig. 19.30.
Ans. (a) 8.4 V; (b) 12 V

18 On Fig. 19.30 draw the load line for the amplifying circuit in Fig. 19.31 with $\mathcal{E}_c = 18$ V and $R_c = 400\ \Omega$. Find the quiescent point when $I_B = 0.30$ mA. What is the gain of an alternating input current whose peak is 0.05 mA?

BIBLIOGRAPHY

FITZGERALD, A. E., DAVID E. HIGGIN-BOTHAM, and ARVIN GRABEL: *Basic Electrical Engineering,* 4th ed., McGraw-Hill Book Company, New York, 1975. A fuller discussion of transistor characteristics and basic amplifying circuits can be found in Chaps. 7, 9, and 10 of this book.

PART V

MODERN PHYSICS

By the year 1875, the development of physics seemed to be complete. The laws of mechanics (Part I), thermodynamics (Part II), and electromagnetism (Part IV) formed a coherent body of fundamental knowledge that appeared to include all the basic laws of the universe. This knowledge, now called *classical physics,* was so confidently felt to explain all physical phenomena that when Max Planck entered the University of Munich in 1875, he was discouraged from pursuing a career in physics because, as he was told, "all the important discoveries in physics have been made."

Contrary to this appraisal, however, the next 50 years proved to be the most exciting and fruitful period in

Model of a uranium atom.
(*The Granger Collection.*)

the history of physics. Extraordinary discoveries were made in the fields of atomic and nuclear physics that could not be understood in terms of classical physics. It was found that fundamentally new principles and concepts were needed to reconcile the behavior of electrons and atoms with the behavior of ordinary-sized objects. The theories that emerged during this period—relativity and quantum mechanics—constitute *modern physics,* a coherent body of fundamental knowledge that replaces classical physics in discussions of atomic and nuclear phenomena, but which is equivalent to classical physics in discussions of macroscopic phenomena.

I n 1875 almost nothing was known about the structure of atoms. Indeed, some physicists of the period still questioned the existence of atoms, while others considered it impossible to ever investigate the inner structure of an atom. But so rapidly did physics develop in the next 50 years that by 1925 atomic structure was known in complete and exhaustive detail.

20
ATOMIC PHYSICS

An important breakthrough came in 1913, when Niels Bohr developed a model of the atom that explained the characteristic spectrum of the radiation emitted by a gas when an electron beam passes through it. Bohr succeeded where others had failed because he was bold enough to introduce several new physical principles, even though they contradicted principles of classical physics. In the period from 1913 to 1925 these new principles were developed into *quantum mechanics*, a comprehensive theory of the motion of atomic particles that provides the necessary framework for the complete understanding of atomic phenomena.

This chapter introduces the basic principles of quantum mechanics and uses them to study atomic structure, the chemical properties of the elements, and the nature of the chemical bond.

20.1 Wave-Particle Duality

Classical physics makes a sharp distinction between waves and particles. A *particle* is localized in space and is characterized by a definite mass m. It can move with any speed v, and its kinetic energy is related to its mass and speed by

$$K = \tfrac{1}{2}mv^2$$

An *electromagnetic wave*, on the other hand, is extended in space and has a constant speed c. It can have any amplitude A and frequency f, and its wavelength λ is related to its frequency by

$$\lambda = \frac{c}{f}$$

According to classical physics, an electron is a particle and light is an electromagnetic wave.

Quantum mechanics blurs the distinction between waves and particles by stating that all entities, such as electrons and light, have both a wavelike and a particlelike nature. This paradoxical concept has no counterpart in classical physics. It is impossible to form a mental picture of an entity that is simultaneously a wave and a particle because our mental pictures are based on our experiences in the macroscopic world of classical physics. Such nonclassical entities must be handled formally with the aid of mathematics.

The first hint that light had particle properties was given by Max Planck (1858–1947) in 1900. On the basis of his study of the radiation emitted by heated objects (Sec. 9.5), he showed that electromagnetic radiation is absorbed or emitted by an object only in discrete bundles of energy. For radiation of a given frequency f this bundle, or *quantum*, of energy is

$$E = hf \qquad\qquad 20.1$$

where $h = 6.63 \times 10^{-34}\,\text{J}\cdot\text{s}$ is a universal constant, called *Planck's constant*.

Example 20.1 What is the energy of a quantum of yellow light with a frequency of $6.0 \times 10^{14}\,\text{Hz}$?

From Eq. 20.1 the energy is

$$E = hf = (6.63 \times 10^{-34}\,\text{J}\cdot\text{s})(6.0 \times 10^{14}\,\text{s}^{-1}) = 4.0 \times 10^{-19}\,\text{J} \qquad \square$$

A *quantum* of yellow light is the smallest amount of (yellow) light which can be absorbed (or emitted) in any process. For instance, it is the smallest amount of light that can be absorbed by a rod cell in the retina of the eye. (A rod cell is so sensitive that it can send an impulse to the brain upon absorbing a single quantum of light. That is, a rod cell is sensitive enough to detect the smallest possible light signal.)

In 1905 Einstein carried the idea of light quanta a step further by proposing that light itself is composed of corpuscular units, called *photons*. A photon travels with the speed of light, and its energy is related to the frequency of the radiation by Eq. 20.1. Thus Newton's suggestion that light is corpuscular in nature was reintroduced into physics less than 100 years after the wave nature of light had been established by Young and Fresnel (Sec. 13.1). Only now the corpuscular nature of light somehow coexists with its wave nature. This wave-particle duality is built into Eq. 20.1, since the frequency f is a wave property, whereas the quantum of energy E is a corpuscular property.

Such a paradoxical concept would be considered completely absurd if it did not rest on incontrovertible experimental evidence. The wave nature of light is firmly established by the interference experiment of Young (Sec. 13.2). The particle nature of light was first clearly established by experiments on the photoelectric effect by Philipp Lenard (1862–1947) in 1902.

PHOTOELECTRIC EFFECT

Definition The *photoelectric effect* is the ejection of electrons from the surface of a metal when the surface is irradiated by electromagnetic radiation of certain frequencies. Lenard made several important discoveries about this phenomenon: (1) For each metal there is a *critical frequency* f_0 for the ejection of electrons: radiation with a frequency less than f_0 never ejects electrons, regardless of its intensity; (2) for radiation with a frequency greater than f_0, the number of electrons per second ejected from a metal is proportional to the intensity of the radiation and is independent of its frequency; and (3) the maximum kinetic energy K_m of an ejected electron is independent of the intensity of the radiation and is related to its frequency f by

$$K_m = h(f - f_0) \qquad\qquad 20.2$$

The wave theory of light is unable to explain why an increase in intensity increases the number of ejected electrons but not their energy, or why the energy of an electron increases with the frequency. However, in 1905 Einstein showed that these facts are easily explained if light is assumed to consist of photons, because then an individual electron can receive only the energy $E = hf$ of a single photon. If we assume that there is a minimum energy E_0 needed to remove an electron from a metal, then the difference $E - E_0$ is the maximum kinetic energy an ejected electron can have. Increasing the intensity of the radiation increases the number of photons hitting the metals, and hence the number of electrons ejected, but it does not increase the energy received by an individual electron. On the other hand, increasing the frequency of the radiation does increase the energy received by an individual electron. Because there is a minimum energy E_0 required to eject an electron from a metal, radiation with a frequency less than $f_0 = E_0/h$ cannot eject electrons.

The quantity E_0 is called the *photoelectric work function.* Table 20.1 gives the values of E_0 for various metals.

> **UNITS** In atomic physics it is convenient to measure energies in electronvolts. An *electronvolt* (eV) is the energy gained by an electron in passing through a potential difference of one volt (Sec. 15.5), so the conversion between electronvolts and joules is
>
> $1 \text{ eV} = (e)(1 \text{ V}) = 1.60 \times 10^{-19} \text{ J}$ and $1 \text{ J} = 6.25 \times 10^{18} \text{ eV}$
>
> Related units are the *kilo-electronvolt* (keV) and the *mega-electronvolt* (MeV), equal to 10^3 eV and 10^6 eV, respectively. In electronvolts, Planck's constant is
>
> $h = 4.14 \times 10^{-15} \text{ eV} \cdot \text{s}$

Example 20.2 (a) What is the critical frequency of cadmium? (b) Ultraviolet radiation with a frequency of 1.5×10^{15} Hz is incident on cadmium. What is the maximum kinetic energy of the ejected electrons?

(a) From Table 20.1 the work function E_0 of cadmium is 4.07 eV, so the critical frequency f_0 is

$$f_0 = \frac{E_0}{h} = \frac{4.07 \text{ eV}}{4.14 \times 10^{-15} \text{ eV} \cdot \text{s}} = 9.8 \times 10^{14} \text{ Hz}$$

(b) From Eq. 20.2 the maximum kinetic energy K_m is

$$K_m = h(f - f_0)$$
$$= (4.14 \times 10^{-15} \text{ eV} \cdot \text{s})(1.5 \times 10^{15} \text{ Hz} - 9.8 \times 10^{14} \text{ Hz})$$
$$= 2.2 \text{ eV}$$

Planck's constant in electronvolts was used in these calculations because the energies were in electronvolts. ☐

In the photoelectric effect, a photon is destroyed and some of its energy is converted into the kinetic energy of the electron. (The rest of the energy is used to remove the electron from the metal.) In the reverse process, called *bremsstrahlung,* the kinetic energy of an electron is converted into a photon. This process is the basis for the production of x-rays, which are high-frequency (short-wavelength) photons. In an x-ray machine, electrons that have been accelerated through an electric potential V suddenly collide with a metal electrode (Sec. 15.5). Just before the collision, the kinetic energy of an electron is

$$K = eV$$

TABLE 20.1

Photoelectric work function of various metals

Metal	Work function, eV
Rubidium	2.09
Sodium	2.28
Calcium	2.71
Cerium	2.84
Cadmium	4.07
Aluminum	4.08
Silver	4.73
Palladium	4.97
Nickel	5.01

where e is the electron's charge. During the collision, this energy may be converted into many low-energy photons or a few high-energy photons. The highest-frequency (shortest-wavelength) radiation is produced when all the energy of an individual electron is converted into a single photon of the same energy. From Eq. 20.1 the frequency of this photon is

$$f = \frac{K}{h} = \frac{eV}{h}$$

and its wavelength is

$$\lambda = \frac{c}{f} = \frac{hc}{eV} \qquad 20.3$$

Example 20.3 What is the shortest-wavelength x-ray produced by a 10,000-V x-ray machine?
From Eq. 20.3 the wavelength is

$$\lambda = \frac{hc}{eV} = \frac{(6.63 \times 10^{-34}\,\text{J}\cdot\text{s})(3.0 \times 10^{8}\,\text{m/s})}{(1.6 \times 10^{-19}\,\text{C})(10^{4}\,\text{V})}$$
$$= 1.2 \times 10^{-10}\,\text{m}$$

Here we have used Planck's constant in joules, because eV is the kinetic energy of the electron in joules. □

de BROGLIE WAVES

The wave-particle duality of light was accepted, although not fully understood, by the scientific community in the decade following Einstein's work on the photoelectric effect. This led Louis de Broglie (1892–   ) in 1923 to speculate that matter, and, in particular, electrons may have wavelike properties and thus have a dual nature similar to that of light. In analogy with Planck's formula (Eq. 21.1), he proposed that the wavelength λ of an electron is related to its mass m and speed v by the equation

$$mv = \frac{h}{\lambda} \qquad 20.4$$

This equation is an expression of the dual nature of the electron, because the mass on the left is a particle property, whereas the wavelength on the right is a wave property. This equation was verified in 1925, when George Thomson (1892–   ) in England and Clinton Davisson (1881–1958) in the United States independently obtained interference patterns using a beam of electrons instead of light. The experiments were similar in principle to Young's first interference experiment using a double slit except that Thomson and Davisson used the spacing between the atoms of a crystal instead of slits.

Example 20.4 What is the wavelength of an electron with a speed of 5×10^{7} m/s?
From Eq. 20.4 the wavlength is

$$\lambda = \frac{h}{mv} = \frac{6.63 \times 10^{-34}\,\text{J}\cdot\text{s}}{(9.1 \times 10^{-31}\,\text{kg})(5 \times 10^{7}\,\text{m/s})}$$
$$= 1.46 \times 10^{-11}\,\text{m} \qquad \qquad □$$

Modern quantum mechanics is based on the principle that all fundamental physical entities, such as electrons, protons, and photons, have a dual (wave-particle) nature. This radical break with classical physics has been the key to unlocking the secrets of the atom. Unfortunately, duality is an abstract concept which, like four-dimensional space, cannot be visualized. However, the implications of duality can be deduced mathematically. This can be done even when it is not completely clear what the symbols in the equations represent. For instance, Eq. 20.4 gives the wavelength of an electron without saying anything about the nature of the wave itself. Nevertheless, Eq. 20.4 is sufficient, as we shall see in the next section, to determine the basic structure of the atom. In Sec. 20.3 we shall say more about the nature of the electron wave.

20.2 Bohr's Model of the Atom

RUTHERFORD'S SCATTERING EXPERIMENT

By the beginning of the twentieth century, it was known that an atom has a diameter of about 10^{-10} m and contains a number of negatively charged electrons. Since the atom as a whole is neutral, it was inferred that it must contain an equal quantity of positive charge. Furthermore, it was known that the mass of an atom is thousands of times greater than the mass of an electron, so that the positive charge must constitute almost all the mass of an atom. One early model of the atom assumed that the positive charge formed a continuous fluid filling the volume of the atom in which the electrons floated, like "plums in a pudding" (Fig. 20.1).

In 1911 Ernest Rutherford (1871–1937) tested this plum-pudding model by bombarding a thin sheet of gold foil with alpha particles, the high-speed positively charged particles emitted by radioactive elements such as radium (Sec. 22.2). The alpha particles, most of which passed through the foil without appreciable deflection, were detected by the light produced when they struck a fluorescent screen (Fig. 20.2). Occasionally, however, a particle was scattered through a large angle, and about 1 particle in 20,000 was scattered by more than 90°. This was very unexpected, because alpha particles are too heavy and move too fast to be appreciably deflected by an atom composed of electrons and a diffuse positively charged fluid.

Rutherford could account for the observed number of large-angle deflections only by assuming that almost all the mass of the atom is concentrated in a central core of radius 10^{-15} m. This core, or *nucleus,* is positively charged, and so exerts a repulsive electric force on an alpha particle that passes near it. Since the radius of the nucleus is only 1/100,000 the radius of the atom, almost all the alpha particles pass through the atom without getting close enough to the nucleus to experience an appreciable force. However, the few that do get close to the nucleus experience a very strong force and consequently are deflected through a large angle (Fig. 20.3).

In the model of the atom that emerged from Rutherford's work, the electrons were thought to revolve in orbits about the nucleus, just as the planets revolve about the sun. Most of the

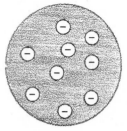

FIGURE 20.1

Pre-1911 plum-pudding model of the atom. The electrons were thought to float in a positively charged fluid that filled the volume of the atom.

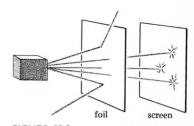

FIGURE 20.2

Rutherford's scattering experiment. Alpha particles from a radioactive source are aimed at a thin gold foil. The scattered particles are detected by the light emitted when they hit a fluorescent screen.

volumn of an atom is filled with orbiting electrons, which are too light to deflect an incident alpha particle. The electrons are maintained in orbits about the nucleus by the attractive electric force exerted on them by the nucleus, just as the planets are maintained in orbits about the sun by the attractive gravitational force exerted on them by the sun.

The simplest atom, hydrogen, has only one electron in orbit about the nucleus. It is tempting to treat this case just as we treated the case of a satellite in orbit about the earth (Sec. 6.3, Example 6.6). Thus suppose that the electron is revolving about the nucleus in a circular orbit of radius r (Fig. 20.4). The force on the electron is the electric force (Eq. 15.4)

$$F_e = \frac{1}{4\pi\epsilon_o} \frac{q_e q_n}{r^2}$$

where q_e and q_n are the charges of the electron and nucleus, respectively. The electron's charge is denoted $-e$, and the nucleus has the opposite charge $+e$, so

$$F_e = -\frac{1}{4\pi\epsilon_o} \frac{e^2}{r^2}$$

The minus sign indicates that this is an attractive force, directed toward the nucleus, as shown in Fig. 20.4.

An electron moving at constant speed v in a circular orbit has the centripetal acceleration (Eq. 6.5)

$$a = \frac{v^2}{r}$$

directed toward the nucleus. From Newton's second law (Eq. 4.6) the force on the electron is equal to its mass times its acceleration, so

$$\frac{1}{4\pi\epsilon_o} \frac{e^2}{r^2} = m\frac{v^2}{r}$$

or

$$mv^2 = \frac{1}{4\pi\epsilon_o} \frac{e^2}{r} \qquad\qquad 20.5$$

This equation gives the speed v of the electron for any orbital radius r.

From Eq. 20.5 we see that the kinetic energy K of the electron is

$$K = \tfrac{1}{2}mv^2 = \frac{1}{8\pi\epsilon_o} \frac{e^2}{r} \qquad\qquad 20.6$$

and thus is inversely proportional to the radius of the orbit. From Eqs. 15.13 and 20.6 the potential energy U of the electron is

$$U = \frac{1}{4\pi\epsilon_o} \frac{q_e q_n}{r}$$

$$= \frac{1}{4\pi\epsilon_o}\left(\frac{-e^2}{r}\right) = -2K \qquad\qquad 20.7$$

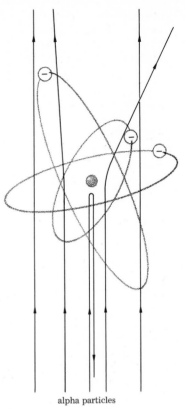

alpha particles

FIGURE 20.3

The tracks of alpha particles traveling through an atom. Only particles that get close to the small massive nucleus are deflected through large angles.

FIGURE 20.4

An electron moving in a circular orbit about a hydrogen nucleus.

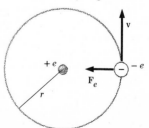

The minus sign indicates that the potential energy of an electron a distance r from the nucleus is less than the potential energy at infinity; i.e., work has to be done on the electron to move it from r to infinity. Combining Eqs. 20.6 and 20.7, the total energy E of the electron is

$$E = K + U = K - 2K$$
$$= -K = -\frac{1}{8\pi\epsilon_o}\frac{e^2}{r} \qquad\qquad 20.8$$

Again the minus sign indicates that the energy of an electron in orbit about the nucleus is less than the energy it would have if it were at rest and infinitely far from the nucleus. Thus $-E$ is the energy required to separate the electron from the nucleus completely, i.e., to *ionize* the atom.

BOHR'S QUANTUM CONDITION

This was the situation in 1912, when Niels Bohr (1885–1962) was a student of Rutherford's at the University of Manchester. Bohr understood that this orbital model, although very appealing, could not be entirely correct, because it did not explain, for instance, why all atoms of hydrogen have identical chemical properties. According to classical physics, the electron can be in an orbit of any radius, and so there is a continuous range of possible energies for the electron. Yet hydrogen behaves as though all its atoms have the same energy. Furthermore, even if the electron of each atom was originally in a specially preferred orbit, in time this orbit would change because of collisions between atoms.

An even more serious objection to this model comes from electromagnetic theory, which says that a charged particle emits electromagnetic radiation whenever it is accelerated. In particular, an orbiting electron is expected to radiate electromagnetic radiation at the expense of its own energy. Consequently, the orbital motion of an electron would be unstable, because the electron would spiral closer and closer to the nucleus as it radiated away its energy.

Bohr worked on these problems after he returned to Copenhagen in 1913, trying to introduce the new quantum ideas of Planck and Einstein into the model. What was needed was some restriction on Eq. 20.5 that would limit the possible orbits. He succeeded in finding this restriction and so obtained, for the first time, a quantitative description of the atom.

The Bohr condition is best understood today in terms of the wave nature of the electron, although this was discovered 10 years after Bohr's original paper. Recall (Sec. 11.5) that a standing wave on a string fixed at both ends can have only certain wavelengths, determined by the condition that there must be an integral number of half-wavelengths on the string (Fig. 20.5). Suppose, similarly, that only standing electron waves exist in an atom. Figure 20.6 shows that only an integral number of whole wavelengths can exist on a circular string because a wave with an odd number of half-wavelengths does not join smoothly together (Fig. 20.6b). Thus the condition for a standing wave on

FIGURE 20.5

Possible standing waves on a string fixed at both ends.

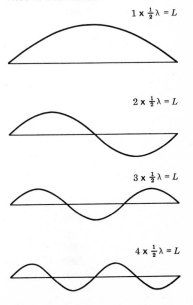

$$1 \times \tfrac{1}{2}\lambda = L$$

$$2 \times \tfrac{1}{2}\lambda = L$$

$$3 \times \tfrac{1}{2}\lambda = L$$

$$4 \times \tfrac{1}{2}\lambda = L$$

a circle of radius r is

$$n\lambda = 2\pi r$$

where n is an integer ($n = 1, 2, 3, \ldots$), and $2\pi r$ is the circumference of the circle. This condition, in conjunction with Eq. 20.4 for the wavelength of an electron wave, gives

$$mv = \frac{h}{\lambda} = \frac{h}{2\pi r/n} = \frac{nh}{2\pi r}$$

which is a nonclassical equation relating the speed of the electron and the radius of the orbit. Squaring both sides of this equation, we get

$$m^2v^2 = \frac{n^2h^2}{4\pi^2r^2}$$

or

$$mv^2 = \frac{n^2h^2}{4\pi^2mr^2} \qquad \text{Bohr's quantum condition} \qquad 20.9$$

This is the additional restriction needed to limit the possible orbits given by Eq. 20.5. Combining Eqs. 20.5 and 20.9, we get

$$\frac{1}{4\pi\epsilon_o}\frac{e^2}{r} = \frac{n^2h^2}{4\pi^2mr^2}$$

or

$$r = \frac{\epsilon_o n^2h^2}{\pi me^2} \qquad\qquad 20.10$$

According to Bohr, the only allowed orbits of an electron are those obtained from Eq. 20.10 with integer values of n.

Example 20.5 What is the radius of the smallest allowed orbit in hydrogen?

The smallest orbit occurs with $n = 1$. Substituting the known values of h, m, e, and ϵ_o into Eq. 20.10, we get the radius

$$r = \frac{\left(8.85 \times 10^{-12}\dfrac{C^2}{N \cdot m^2}\right)(1^2)(6.63 \times 10^{-34}\,J \cdot s)^2}{\pi(9.11 \times 10^{-31}\,kg)(1.60 \times 10^{-19}\,C)^2}$$

$$= 0.53 \times 10^{-10}\frac{J^2 \cdot s^2}{(N \cdot m^2) \cdot kg} = 0.53 \times 10^{-10}\,m$$

This quantity, known as the *Bohr radius,* is consistent with what was known in 1913 about the size of a hydrogen atom. The next orbit occurs when $n = 2$, and its radius is $2^2 = 4$ times the Bohr radius; the radius of the $n = 3$ orbit is $3^2 = 9$ times the Bohr radius, and so on. ☐

The allowed energies E_n of an electron in a hydrogen atom are obtained by substituting the allowed radii from Eq. 20.10 into Eq. 20.8:

$$E_n = -\frac{1}{8\pi\epsilon_o}\frac{e^2}{r} = -\frac{me^4}{8\epsilon_o^2n^2h^2} \qquad\qquad 20.11$$

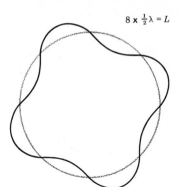

$8 \times \tfrac{1}{2}\lambda = L$

a

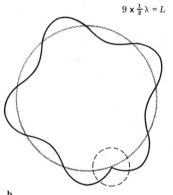

$9 \times \tfrac{1}{2}\lambda = L$

b

FIGURE 20.6

(a) A possible standing wave on a circular string. (b) A wave with an odd number of half-wavelengths is not a possible standing wave, because the wave does not fit smoothly on a circle.

The lowest energy E_1 is given by Eq. 20.11 with $n = 1$:

$$E_1 = \frac{(9.11 \times 10^{-31}\ \text{kg})(1.60 \times 10^{-19}\ \text{C})^4}{8\left(8.85 \times 10^{-12}\dfrac{\text{C}^2}{\text{N} \cdot \text{m}^2}\right)^2 (1^2)(6.63 \times 10^{-34}\ \text{J} \cdot \text{s})^2}$$

$$= -2.17 \times 10^{-18}\ \text{m}^2 \cdot \text{kg/s}^2 = -2.17 \times 10^{-18}\ \text{J} = -13.6\ \text{eV}$$

From Eq. 20.11 we see that the nth energy E_n is

$$E_n = \frac{E_1}{n^2} = -\frac{13.6\ \text{eV}}{n^2} \qquad\qquad 20.12$$

For instance, with $n = 2$, the energy is

$$E_2 = -\frac{13.6\ \text{eV}}{2^2} = -3.4\ \text{eV}$$

which is greater than E_1 since it is less negative.

The allowed energies of the electron in a hydrogen atom form an increasing sequence, starting with the lowest energy ($E_1 = -13.6$ eV) and ending with the highest energy ($E_\infty = 0$). This is shown in Fig. 20.7, where each allowed energy is represented by a horizontal line. An electron can only be in an orbit, or state, corresponding to an allowed energy. Normally the electron is in the lowest, or *ground*, state, from which it cannot radiate away energy since it already has its lowest possible energy.

Furthermore, since there is a 10.2-eV difference between the ground state of hydrogen and the next higher state, a hydrogen atom in its ground state must absorb at least 10.2 eV in order to change its state. The kinetic energy of a molecule at room temperature is only about 0.02 eV, so a hydrogen atom is unaffected by collisions with other gas molecules. Thus an atom in its ground state is completely stable against radiating away energy and is relatively stable against absorbing energy. All hydrogen atoms normally have identical properties because they are all in their ground state.

ATOMIC SPECTRA

Hydrogen atoms can be excited into higher-energy states by bombarding them with a beam of energetic electrons. This is most conveniently done in a gas-discharge tube (Fig. 20.8), which is similar to the vacuum tubes discussed in Sec. 15.5 except that it contains hydrogen (or some other gas) at very low pressure. Metal plates, or electrodes, are mounted inside the sealed tube, and wires attached to the electrodes pass through the wall of the tube. When a potential difference is placed across the electrodes, electrons are pulled out of the negative electrode (the cathode) and are accelerated toward the positive electrode (the anode). The electrons gain kinetic energy as they move toward the anode, and occasionally they collide with a hydrogen atom. In the collision process some of the electron's kinetic energy may be transferred to the atom by promoting the atom's electron to a higher energy state. At least 10.2 eV must be absorbed to excite an atom to the $n = 2$ state, and an atom that absorbs more than 13.6 eV is ionized, its electron being perma-

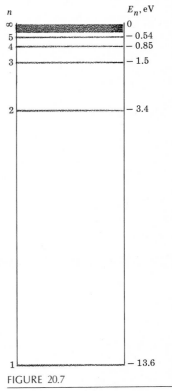

FIGURE 20.7

Energy levels of a hydrogen atom.

FIGURE 20.8

A gas-discharge tube.

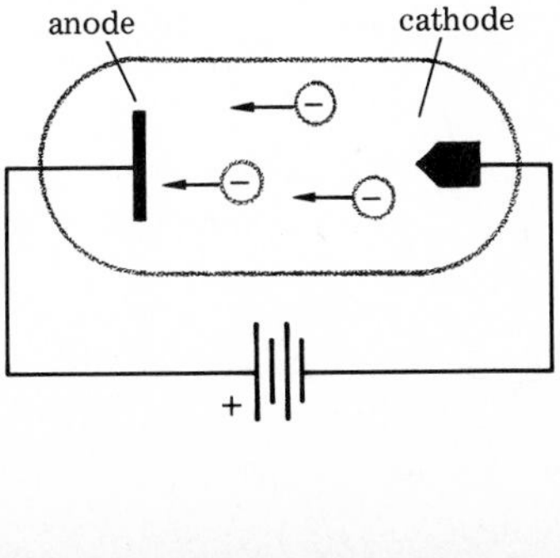

nently separated from its nucleus. As a result of collisions between electrons and atoms, the atoms of the gas in a discharge
tube are in all possible states.

An atom in an excited state soon radiates away some of its
excess energy and falls into a lower energy state. Bohr showed
that this radiation consists of a single photon whose frequency f
is related to the energy change by Eq. 20.1. That is, if the
atom makes a transition directly from a state of energy E_n to a
state of energy $E_{n'}$, a single photon of frequency

$$f = \frac{E_n - E_{n'}}{h} \qquad\qquad 20.13$$

is emitted. For instance, if the atom makes a transition from
the $n = 3$ state to the ground state ($n = 1$), the energy change is

$$E_n - E_{n'} = -\frac{13.6\ \text{eV}}{n^2} - \left(-\frac{13.6\ \text{eV}}{n'^2} \right)$$

$$= -\frac{13.6\ \text{eV}}{3^2} - \left(-\frac{13.6\ \text{eV}}{1^2} \right)$$

$$= -1.5\ \text{eV} - (-13.6\ \text{eV}) = 12.1\ \text{eV}$$

and the frequency of the emitted photon is

$$f = \frac{E_n - E_{n'}}{h} = \frac{12.1\ \text{eV}}{4.14 \times 10^{-15}\ \text{eV} \cdot \text{s}} = 2.92 \times 10^{15}\ \text{Hz}$$

which is ultraviolet radiation (see Fig. 13.1). This transition is
represented in Fig. 20.9 by the arrow drawn from the $n = 3$ state
to the $n = 1$ state.

On the other hand, an atom in the $n = 3$ state may first make
a transition to the $n = 2$ state, emitting a photon of frequency

$$f = \frac{-1.5\ \text{eV} - (-3.4\ \text{eV})}{h}$$

$$= \frac{1.9\ \text{eV}}{4.14 \times 10^{-15}\ \text{eV} \cdot \text{s}} = 0.46 \times 10^{15}\ \text{Hz}$$

and then make a transition from the $n = 2$ state to the $n = 1$
state, emitting a photon of frequency

$$f = \frac{-3.40\ \text{eV} - (-13.6\ \text{eV})}{4.14 \times 10^{-15}\ \text{eV} \cdot \text{s}} = 2.46 \times 10^{15}\ \text{Hz}$$

In this two-step process, two photons are emitted, one in the
visible region and one in the ultraviolet region. These transitions
are represented in Fig. 20.9 by the arrows from the $n = 3$ to the
$n = 2$ state and from the $n = 2$ to the $n = 1$ state.

The spectrum of the radiation emanating from a gas-discharge
tube contains all the frequencies that can be obtained from
transitions between any two energy states. Hundreds of the
frequencies in the spectrum of hydrogen have been measured,
and they all correspond to frequencies calculated from Eqs.
20.12 and 20.13. Thus the spectrum of radiation emitted by a gas
in a discharge tube gives direct information about the energy
levels of an atom. Hydrogen is the simplest atom to analyze, and
its spectrum was used to verify Bohr's model of atomic structure.

FIGURE 20.9

A hydrogen atom in the $n = 3$
state can get to the ground state
either in a single transition or in a
two-step transition.

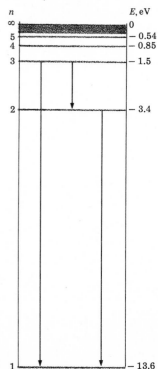

The energy levels of complex atoms cannot be calculated as easily as those of hydrogen, so they are determined experimentally from their spectra. Figure 20.10 shows a few of the energy levels of sodium as deduced from its spectra.

20.3 Quantum Mechanics

Although it successfully describes the atom and correctly predicts the spectrum of radiation emitted by hydrogen, Bohr's model is based on intrinsically contradictory concepts. Specifically, newtonian mechanics, used to derive Eq. 20.5, allows a particle to be in an orbit of any radius, whereas Eq. 20.9 restricts the allowed radii with a condition foreign to newtonian mechanics. This contradiction was resolved only with the development of *quantum mechanics,* which is a comprehensive theory of the motion of particles of atomic size. For particles of ordinary size, quantum mechanics is equivalent to newtonian mechanics; i.e., quantum mechanics is more general than newtonian mechanics and reduces to it when the mass of the particle is large.

One form of quantum mechanics was developed in 1925 by Erwin Schrödinger (1887–1961) in terms of an equation for the electron wave de Broglie had proposed the year before (Sec. 20.1). The *Schrödinger equation* is similar to the classical equations for waves on a string and for electromagnetic waves, but it differs in some essential details. Like these classical equations, it supposes that the particle wave obeys the superposition principle (Sec. 11.3). Unlike these other equations, however, the wave itself represents neither the displacement of a medium nor the magnitude of a physical quantity. Instead, the wave has to be interpreted as a *probability amplitude.*

To understand this, consider a particle of mass m moving in a box with perfectly rigid walls, as shown in Fig. 20.11. According to newtonian mechanics, the particle moves in a straight line at constant speed v until it hits a wall of the box. It then bounces off the wall and moves in the opposite direction with the same speed v until it hits the other wall. At any instant of time the particle has a definite position in the box, and in the course of time it moves back and forth in the box. The particle can have any speed v and any kinetic energy $K = \frac{1}{2}mv^2$.

The quantum mechanics of Schrödinger represents the particle by a wave ψ.* Since the wave is confined in a finite space, only standing waves with certain wavelengths are possible. Like a wave on a string fixed at both ends, the particle wave must be zero at each wall of the box, and so a standing wave must contain an integral number of half-wavelengths (Fig. 20.12). If the box is of length L, the wavelength satisfies the condition

$$n\frac{\lambda}{2} = L \quad \text{or} \quad \lambda = \frac{2L}{n} \qquad\qquad 20.14$$

where n is an integer. This equation is identical to Eq. 11.9. However, the wave ψ has a very different interpretation.

At any point in the box the quantity ψ^2, the square of the

*ψ is the Greek (lower-case) letter psi.

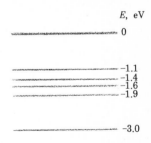

E, eV

0

−1.1
−1.4
−1.6
−1.9

−3.0

ground
state −5.1

FIGURE 20.10

The first few energy levels in a sodium atom.

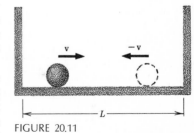

FIGURE 20.11

Classical description of a particle moving back and forth in a box.

FIGURE 20.12

Quantum-mechanical description of a particle moving back and forth in a box.

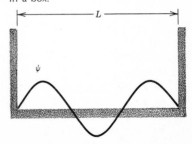

wave, gives the probability of finding the particle at that point. Figure 20.13 shows the square of the wave in Fig. 20.12. From Fig. 20.13 we see that the particle is most likely to be found at points A, B, and C and will never be found at points W, X, Y, and Z. The particle therefore cannot be thought of as having a definite position at each instant of time. All that can be said about the particle is that, at each instant of time, there is a definite probability of its being found at any point in the box.

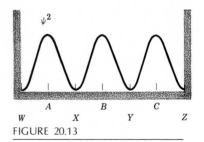

FIGURE 20.13

The square of the quantum-mechanical wave gives the probability of finding the particle at various points in the box.

REMARK This is perhaps the strangest assertion of quantum mechanics, and if you find it hard to believe, you are in good company because Einstein never accepted it. In a famous series of discussions he argued with Bohr that quantum mechanics is not a complete description of nature because it does not give the position of an electron at all times. Bohr argued that quantum mechanics is a complete description because it describes everything that is physically measurable.

To measure the position of an electron, it is necessary to reflect some radiation from it. Since radiation is quantized, at least one photon must strike the electron and bounce into the measuring instrument. In the process, however, the electron's energy is changed and the electron is no longer in its original state. Thus the position of an electron in a particular state can be measured only once because as a consequence of the measurement itself the electron is knocked into another state.

In order for a particle to be said to have a definite position at every instant of time, it must be possible to monitor the particle's position continuously without otherwise disturbing it. For example, a marble rolling on a table has a definite position at every instant because the light used to observe it exerts a negligible force on it. But the photon nature of light limits the ultimate gentleness with which a particle can be observed. The motion of a very light particle, such as an electron, cannot be observed because a single photon bouncing off it is sufficient to alter the particle's motion radically.

It might be thought that the electron can possess a definite position at every instant even though it is not possible to observe it. Bohr, however, argued vigorously against this idea, showing that it is inconsistent with experiments on the interference of electron beams. Besides, a quantity that in principle cannot be measured cannot be physically meaningful. Einstein had maintained much the same point of view regarding relativity, but he never accepted Bohr's viewpoint of quantum mechanics. Today almost all physicists accept Bohr's contention that quantum mechanics provides a complete description of atomic processes. This viewpoint is sometimes referred to as the *Copenhagen interpretation* because it was developed by Bohr and his collaborators at the University of Copenhagen.

A particle moving in a box of length L exists only in the states with wavelengths given by Eq. 20.14. From the de Broglie relation (Eq. 20.4), a particle with wavelength λ and mass m has a speed

$$v = \frac{h}{m\lambda}$$

so that the possible speeds of a particle in a box are

$$v = \frac{h}{2Lm/n} = \frac{nh}{2Lm} \qquad 20.15$$

and the possible kinetic energies are

$$K = \tfrac{1}{2}mv^2 = \frac{n^2h^2}{8L^2m} \qquad 20.16$$

Thus the energies of a particle in a box are quantized: only those energies are allowed which satisfy Eq. 20.16 with an integer value of n.

The lowest allowed energy occurs with $n = 1$. That is, a particle confined to a box of length L must have a kinetic energy of at least $h^2/8L^2m$. This is the so-called *zero-point energy*. For example, an atom has a diameter of about 10^{-10} m, which means the electrons are effectively confined to a region of length $L = 10^{-10}$ m. Inserting this into Eq. 20.16, we obtain (with $n = 1$)

$$K = \frac{n^2h^2}{8L^2m} = \frac{(1^2)(6.63 \times 10^{-34} \text{ J} \cdot \text{s})^2}{(8)(10^{-10} \text{ m})^2(9.11 \times 10^{-31} \text{ kg})}$$

$$= 6.03 \times 10^{-18} \text{ J} = 37.7 \text{ eV}$$

which is the correct order of magnitude of the kinetic energy of an electron in the ground state of hydrogen.* Equation 20.16 is not expected to give a more exact result, since it applies to a particle in a box and not to a particle attracted by a point charge. Nevertheless, Eq. 20.16 correctly shows that the smaller the region of confinement, the larger the kinetic energy.†

For example, a nucleus consists of protons and neutrons confined to a region with a diameter of 10^{-15} m, or 10^{-5} times the diameter of an atom. The mass of a proton is 2×10^3 times the mass of an electron, so from Eq. 20.16 the ratio of the kinetic energy of a proton to the kinetic energy of an electron is

$$\frac{K_p}{K_e} = \left(\frac{L_e}{L_p}\right)^2\left(\frac{m_e}{m_p}\right) = \left(\frac{10^{-10} \text{ m}}{10^{-15} \text{ m}}\right)^2 \times \left(\frac{1}{2 \times 10^3}\right) = 5 \times 10^6$$

The kinetic energy of a proton in a nucleus is more than a million times the energy of an electron in an atom because the nucleus is much smaller than the atom. This is why nuclear reactions involve so much more energy than chemical reactions.

The existence of a zero-point energy seems to contradict the fact that an ordinary-sized particle can have zero energy. A marble, for instance, can sit inside a bowl with no apparent motion, whereas, according to Eq. 20.15, the marble has a zero-point speed. For a marble of mass 10^{-2} kg inside a bowl of diameter 0.2 m, this speed is

$$v = \frac{h}{2Lm} = \frac{6.63 \times 10^{-34} \text{ J} \cdot \text{s}}{(2)(0.2 \text{ m})(10^{-2} \text{ kg})} = 1.7 \times 10^{-31} \text{ m/s}$$

At this speed it would take the marble a thousand billion years to move a distance of one atomic diameter. A speed this small is indistinguishable from zero. It is the small size of h that ensures that special quantum-mechanical effects, such as the zero-point energy, are significant only for particles of atomic size. For ordi-

---

*Equation 20.8 shows that the kinetic energy of an electron in an atom is equal to $-E$. Therefore, the kinetic energy of an electron in the ground state of hydrogen is 13.6 eV.

†This is a special case of a more general quantum principle, known as the *uncertainty principle*, which says that the greater the certainty of a particle's position, i.e., the smaller the region of confinement, the greater the uncertainty in the particle's velocity, i.e., the larger the speed.

nary-sized objects, newtonian mechanics still applies to very good approximation.

20.4   Complex Atoms and the Periodic Table

QUANTUM STATES OF AN ATOM

When the Schrödinger wave equation is solved for an electron bound to a positively charged nucleus, it is found (as in the case of a particle in a box) that only certain states are possible. The energies of these states are identical to the energies that Bohr found earlier using his simpler model. However, quantum mechanics abandons the picture of the electrons revolving in orbits about the nucleus and replaces it by distributions of the probability of finding the electrons at every point around the nucleus.

For example, Fig. 20.14 shows the probability distribution of an electron in the lowest $n = 1$ state, with the old Bohr orbit superimposed for comparison. This is a spherical distribution, and the probability of finding the electron is greatest at the nucleus and decreases rapidly with distance from it. Although this distribution bears no resemblance to the Bohr model of an orbiting electron, the figure shows that the Bohr orbit gives the correct size of the atom.

FIGURE 20.14

Probability distribution of an electron in the ground ($n = 1$) state of hydrogen. The electron is more likely to be found at points where the shading is darkest. The circle gives the position of the Bohr orbit for comparison.

> **REMARK**   An electron in an atom is sometimes said to form a cloud about the nucleus. The word is meant to describe the fuzzy probability distribution shown in Fig. 20.14 and not to imply that the electron itself is smeared out. Any measurement of an electron in an atom always reveals one whole electron with the full charge $-e$, and never a fraction of an electron or a continuous distribution of charge.

The Schrödinger equation gives four different waves for $n = 2$ and nine different waves for $n = 3$. This means there are four states with the energy E_2 and nine states with the energy E_3. The states of the hydrogen atom are designated by the three *quantum numbers n, l, m*. The *principal quantum number n* corresponds to the Bohr orbits and determines the energy of the state; the *orbital quantum number l* determines the general symmetry of the wave; and the *magnetic quantum number m* determines the spatial orientation of the wave. For instance, $l = 0$ is a spherically symmetric wave which has a probability distribution like the one shown in Fig. 20.14. Thus l is zero in the single $n = 1$ state. Furthermore, since orientation is meaningless for a spherical distribution, m is zero also. Therefore the quantum numbers of the $n = 1$ state are:

n	l	m
1	0	0

One of the $n = 2$ states also has $l = 0$, $m = 0$, and the other three have $l = 1$. The $l = 1$ waves are doughnut-shaped rings in two orientations ($m = \pm 1$) and form two isolated blobs in the third orientation ($m = 0$). These distributions are shown in Fig. 20.15, with the $n = 1$ Bohr orbit superimposed to show that the $n = 2$ distributions extend much farther than the $n = 1$ distribution. The distributions with $m = +1$ and $m = -1$ are identical, but the waves corresponding to them differ in sign. Essentially $m = +1$ describes an electron rotating clockwise around the nucleus, and $m = -1$ describes an electron rotating counterclockwise around the nucleus. An $l = 1$ distribution is mostly outside the Bohr orbit, whereas an $l = 0$ distribution has a significant portion inside the orbit. These facts are important for understanding atomic structure. The quantum numbers of the four $n = 2$ states are:

n	l	m
2	0	0
	1	$-1, 0, 1$

In general, for a given value of the principal quantum number n, states exist with orbital quantum numbers l varying in integer steps from 0 to $n - 1$, and for each value of l, states exist with magnetic quantum numbers m varying in integer steps from $-l$ to $+l$. Thus the quantum numbers of the nine $n = 3$ states are:

n	l	m
3	0	0
	1	$-1, 0, 1$
	2	$-2, -1, 0, 1, 2$

A group of states with the same value of n is a *shell,* and a group of states with the same values of n and l is a *subshell.* Table 20.2 lists all the shells and subshells with principal quantum numbers 1 to 5, together with the number of states in each subshell. (Note that there are $2l + 1$ states in a subshell with orbital quantum number l.) A subshell is denoted by its principal quantum number n and a code letter for its orbital quantum number l. The code letters for $l = 0, 1, 2, 3, 4, 5$ are s, p, d, f, g, respectively, as indicated in Table 20.2. Thus the $n = 3$, $l = 2$ subshell is denoted $3d$.

The properties of complex (many-electron) atoms and the structure of the periodic table are determined by the quantum numbers n, l, and m, in conjunction with two other discoveries made in the critical year of 1925.

Spin From analysis of atomic spectra, it was found that atoms have twice the number of quantum states given in Table 20.2. Samuel A. Goudsmit (1902–1978) and George E. Uhlenbeck (1900–) showed that this was because the electron itself has two intrinsic states. The electron can be thought of as spinning on an axis,* and the two states correspond to a spin axis up or down with respect to an arbitrary reference line. These states are designated by the *spin quantum number m_s,* which has only two possible values: $m_s = +\frac{1}{2}$ for spin up and $m_s = -\frac{1}{2}$ for spin

*The spin axis of an electron coincides with the axis of its magnetic dipole (Sec. 17.1).

FIGURE 20.15

Probability distributions of an electron in the first excited ($n = 2$) states of hydrogen. The position of the $n = 1$ Bohr orbit is shown for comparison. (a) The $n = 2$, $l = 0$ state is spherically symmetric. It has two regions with appreciable probability, one of which is inside the Bohr orbit. (b) The $n = 2$, $l = 1$, $m = +1$ states are doughnut-shaped regions that correspond closely to Bohr's original model of orbiting electrons. (c) The $n = 2$, $l = 1$, $m = 0$ state consists of two isolated blobs of high probability on either side of the nucleus.

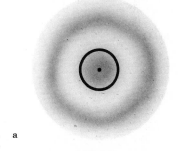

a

$n=2$, $l=0$

b

TOP VIEW

$n=2$, $l=1$, $m=\pm1$

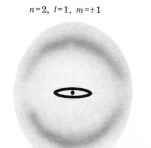

c

SIDE VIEW

$n=2$, $l=1$, $m=0$

TABLE **20.2**

n	l	Subshell symbol	m	Number of states
1	0	$1s$	0	1
2	0	$2s$	0	1
	1	$2p$	$-1,0,1$	3
3	0	$3s$	0	1
	1	$3p$	$-1,0,1$	3
	2	$3d$	$-2,-1,0,1,2$	5
4	0	$4s$	0	1
	1	$4p$	$-1,0,1$	3
	2	$4d$	$-2,-1,0,1,2$	5
	3	$4f$	$-3,-2,-1,0,1,2,3$	7
5	0	$5s$	0	1
	1	$5p$	$-1,0,1$	3
	2	$5d$	$-2,-1,0,1,2$	5
	3	$5f$	$-3,-2,-1,0,1,2,3$	7
	4	$5g$	$-4,-3,-2,1,0,1,2,3,4$	9

States of an electron bound to a positively charged nucleus

The notation used to denote each subshell is given, together with the number of states in the subshell; this number must be doubled to include the two possible spin orientations of the electron.

down. Thus two states exist for every set of quantum numbers n, l, and m, one with $m_s = +\frac{1}{2}$, and one with $m_s = -\frac{1}{2}$.

Exclusion Principle This principle, formulated by Wolfgang Pauli (1900–1958), states that *two or more electrons in an atom cannot occupy the same state at the same time.* Consequently, all the electrons in a many-electron atom are in different states. Two states differ if any of their four quantum numbers (n, l, m, m_s) is different.

PERIODIC TABLE

An atom consists of Z electrons bound to a nucleus of charge $+Ze$, where Z is an integer, called the *atomic number.* The atoms of each chemical element have a specific value of Z. For instance, hydrogen atoms have $Z = 1$, carbon atoms have $Z = 6$, and uranium atoms have $Z = 92$. The periodic table of the elements (Table 20.3) is obtained by arranging the elements in order of increasing Z, with elements with similar chemical properties placed under each other. Dmitri Mendeleev (1834–1907) discovered empirically in 1869 that in such an arrangement elements with similar properties recur at regular (periodic) intervals. Note that in Table 20.3 the atomic number Z increases across each row (period), and elements in the same column (group) have similar properties. For instance, group I contains the *alkali halides,* elements (such as sodium and potassium) that easily form singly charged positive ions, while group VII contains the *halogens,* elements (such as chlorine and fluorine) that easily form singly charged negative ions.

We shall now derive the periodic table from the principles of quantum mechanics. In this construction we assume that the electrons in the ground state of an atom occupy the lowest energy levels consistent with the exclusion principle.

TABLE 20.3

Periodic table of the elements

As shown in the key, each entry gives the atomic number Z, the chemical symbol, the atomic mass, and the electron configuration; a number in parentheses is the mass number of the most stable isotope of that element; the electron configurations are the number of electrons in the last three or four subshells; see Table 20.4 for the order in which the subshells are filled.

Key

88.91	—— Atomic mass
Y	
39	—— Atomic number
6,2,1	

Symbol —— Y
Electron configuration —— 6,2,1

Each cell below is given as: **Symbol**, Z, atomic mass, electron configuration.

Period	Group I	Group II	Group III	Group IV	Group V	Group VI	Group VII	Group 0
1	**H** 1, 1.008, 1							**He** 2, 4.00, 2
2	**Li** 3, 6.94, 2,1	**Be** 4, 9.01, 2,2	**B** 5, 10.81, 2,2,1	**C** 6, 12.01, 2,2,2	**N** 7, 14.01, 2,2,3	**O** 8, 16.00, 2,2,4	**F** 9, 19.00, 2,2,5	**Ne** 10, 20.18, 2,2,6
3	**Na** 11, 22.99, 2,6,1	**Mg** 12, 24.31, 2,6,2	**Al** 13, 26.98, 6,2,1	**Si** 14, 28.09, 6,2,2	**P** 15, 30.97, 6,2,3	**S** 16, 32.06, 6,2,4	**Cl** 17, 35.45, 6,2,5	**Ar** 18, 39.95, 6,2,6
4	**K** 19, 39.10, 2,6,1	**Ca** 20, 40.08, 2,6,2	**Ga** 31, 69.72, 2,10,1	**Ge** 32, 72.59, 2,10,2	**As** 33, 74.92, 2,10,3	**Se** 34, 78.96, 2,10,4	**Br** 35, 79.90, 2,10,5	**Kr** 36, 83.80, 2,10,6
5	**Rb** 37, 85.47, 10,6,1	**Sr** 38, 87.62, 10,6,2	**In** 49, 114.82, 2,10,1	**Sn** 50, 118.69, 2,10,2	**Sb** 51, 121.75, 2,10,3	**Te** 52, 127.60, 2,10,4	**I** 53, 126.90, 2,10,5	**Xe** 54, 131.30, 2,10,6
6	**Cs** 55, 132.91, 10,6,1	**Ba** 56, 137.34, 10,6,2	**Tl** 81, 204.37, 14,10,1	**Pb** 82, 207.19, 14,10,2	**Bi** 83, 208.98, 14,10,3	**Po** 84, (209), 14,10,4	**At** 85, (210), 14,10,5	**Rn** 86, 222, 14,10,6
7	**Fr** 87, (223), 10,6,1	**Ra** 88, (226), 10,6,2						

Transition elements

Period	Group III	Group IV	Group V	Group VI	Group VII	(VIII)	(VIII)	(VIII)	Group I	Group II
4	**Sc** 21, 44.96, 6,2,1	**Ti** 22, 47.90, 6,2,2	**V** 23, 50.94, 6,2,3	**Cr** 24, 52.00, 6,1,5	**Mn** 25, 54.94, 6,2,5	**Fe** 26, 55.85, 6,2,6	**Co** 27, 58.93, 6,2,7	**Ni** 28, 58.71, 6,2,8	**Cu** 29, 63.54, 6,1,10	**Zn** 30, 65.37, 6,2,10
5	**Y** 39, 88.91, 6,2,1	**Zr** 40, 91.22, 6,2,2	**Nb** 41, 92.91, 6,1,4	**Mo** 42, 95.94, 6,1,5	**Tc** 43, (97), 6,2,5	**Ru** 44, 101.1, 6,1,7	**Rh** 45, 102.91, 6,1,8	**Pd** 46, 106.4, 6,0,10	**Ag** 47, 107.87, 6,1,10	**Cd** 48, 112.40, 6,2,10
6	57–71 †	**Hf** 72, 178.49, 2,14,2	**Ta** 73, 180.95, 2,14,3	**W** 74, 183.85, 2,14,4	**Re** 75, 186.2, 2,14,5	**Os** 76, 190.2, 2,14,6	**Ir** 77, 192.2, 2,14,7	**Pt** 78, 195.09, 1,14,9	**Au** 79, 196.97, 1,14,10	**Hg** 80, 200.59, 2,14,10
7	89–103 ‡									

† Lanthanides

La	**Ce**	**Pr**	**Nd**	**Pm**	**Sm**	**Eu**	**Gd**	**Tb**	**Dy**	**Ho**	**Er**	**Tm**	**Yb**	**Lu**
138.91	140.12	140.91	144.24	(145)	150.35	151.96	157.25	158.92	162.50	164.93	167.26	168.93	173.04	174.97
57	58	59	60	61	62	63	64	65	66	67	68	69	70	71
6,2,0,1	6,2,0,2	6,2,3	6,2,4	6,2,5	6,2,6	6,2,7	6,2,7,1	6,2,9	6,2,10	6,2,11	6,2,12	6,2,13	6,2,14	2,14,1

‡ Actinides

Ac	**Th**	**Pa**	**U**	**Np**	**Pu**	**Am**	**Cm**	**Bk**	**Cf**	**Es**	**Fm**	**Md**	**No**	**Lw**
(227)	(232)	(231)	(238)	(237)	(244)	(243)	(247)	(247)	(251)	(254)	(257)	(256)	(254)	(–)
89	90	91	92	93	94	95	96	97	98	99	100	101	102	103
6,2,0,1	6,2,0,2	6,2,2,1	6,2,3,1	6,2,4,1	6,2,5,1	6,2,6,1	6,2,7,1	6,2,8,1	6,2,10	6,2,11	6,2,12	6,2,13	6,2,14	2,14,1

Period 1 The periodic table starts with hydrogen $(Z = 1)$, whose one electron occupies the lowest 1s subshell. Hydrogen is placed in group I because it easily forms an ion of charge $+e$. The exclusion principle allows the two electrons of helium $(Z = 2)$ to occupy the 1s subshell if their spins have opposite orientations. This arrangement has the lowest energy, and so it is the ground-state configuration of helium. Since the helium nucleus has twice the charge of the hydrogen nucleus, each helium electron is more tightly bound than the electron in hydrogen. As a consequence, helium atoms have no tendency to gain or lose electrons and do not react with other atoms to form molecules. Helium is an inert gas, belonging to group 0 of the periodic table. An inert gas is formed whenever the electrons of an atom complete a shell.

Period 2 The exclusion principle allows only two of the three lithium $(Z = 3)$ electrons to occupy the 1s state. The third electron must go into the next higher $(n = 2)$ shell. In a one-electron atom, all the $n = 2$ states have the same energy, but in an atom in which the 1s subshell is filled, the 2s states have slightly lower energy than the 2p states because the electrons in the 1s subshell partially neutralize the charge of the nucleus for electrons outside the 1s cloud. A 2p electron, which is entirely outside the 1s distribution (Fig. 20.15), is attracted by an effective charge of only $+e$, whereas a 2s electron, which is partially inside the 1s distribution, is sometimes attracted by the full $+3e$ charge of the nucleus. Consequently, the 2s state has lower energy (is more tightly bound) than the 2p state, and so the third lithium electron is in the 2s state. Lithium is placed in group I because it can easily lose its 2s electron to form an ion of charge $+e$.

The four electrons of beryllium $(Z = 4)$ fill the 1s and 2s subshells. Beryllium is in group II because it can easily lose the two 2s electrons.

The next six elements $(Z = 5$ to $Z = 10)$ are formed by adding electrons to the 2p subshell. Table 20.4 shows how the 2p subshell can hold up to six electrons with properly oriented spins. Fluorine $(Z = 9)$, which is one electron short of filling the 2p shell, has a strong tendency to fill the last position by picking up an additional electron to form an ion of charge $-e$. Therefore, fluorine is in group VII. The inert gas neon $(Z = 10)$ fills the 2p shell, completing period 2.

Period 3 The next eight elements $(Z = 11$ to $Z = 18)$ are formed similar to the preceding eight by adding electrons in turn to the 3s and 3p subshells (Table 20.4). Thus sodium $(Z = 11)$, which has one electron in the 3s state, has properties similar to lithium, whereas chlorine $(Z = 17)$, which is one electron short of filling the 3p subshell, has properties similar to fluorine. The inert gas argon $(Z = 18)$ fills the 3p subshell, completing period 3.

The order in which succeeding subshells are filled is governed by two competing factors. On the one hand, a subshell with a smaller principal quantum number n tends to have lower energy than a subshell with larger n. On the other hand, a subshell with a smaller orbital quantum number l tends to have lower energy

TABLE 20.4

Order of subshell filling in atoms

Each arrow represents a quantum state, with the point representing the electron's spin orientation; the number under each arrow is the atomic number of the element that fills all the preceding states; an asterisk beside a number indicates an irregularity in subshell filling; in these cases the actual order can be found from Table 20.3; e.g., Table 20.3 shows that niobium ($Z = 41$) has four electrons in the 4d subshell and one electron in the 5s subshell, in contrast to the expected arrangement (three in the 4d and two in the 5s) shown in this table

Period	Sequence	Subshell	Quantum states													
7	Transition	6d	↑ 103	↑	↑	↑	↑	↓	↓	↓	↓	↓				
	Actinides	5f	↑ 89*	↑ 90*	↑ 91*	↑ 92*	↑ 93*	↑ 94*	↓ 95*	↓ 96*	↓ 97*	↓ 98*	↓ 99	↓ 100	↓ 101	↓ 102
		7s	↑ 87	↓ 88												
6		6p	↑ 81	↑ 82	↑ 83	↓ 84	↓ 85	↓ 86								
	Transition	5d	↑ 71	↑ 72	↑ 73	↑ 74	↑ 75	↓ 76	↓ 77	↓ 78*	↓ 79*	↓ 80				
	Lanthanides	4f	↑ 57*	↑ 58	↑ 59	↑ 60	↑ 61	↑ 62	↓ 63	↓ 64	↓ 65	↓ 66	↓ 67	↓ 68	↓ 69	↓ 70
		6s	↑ 55	↓ 56												
5		5p	↑ 49	↑ 50	↑ 51	↓ 52	↓ 53	↓ 54								
	Transition	4d	↑ 39	↑ 40	↑ 41*	↑ 42*	↑ 43	↓ 44*	↓ 45*	↓ 46*	↓ 47*	↓ 48				
		5s	↑ 37	↓ 38												
4		4p	↑ 31	↑ 32	↑ 33	↓ 34	↓ 35	↓ 36								
	Transition	3d	↑ 21	↑ 22	↑ 23	↑ 24*	↑ 25	↓ 26	↓ 27	↓ 28	↓ 29*	↓ 30				
		4s	↑ 19	↓ 20												
3		3p	↑ 13	↑ 14	↑ 15	↓ 16	↓ 17	↓ 18								
		3s	↑ 11	↓ 12												
2		2p	↑ 5	↑ 6	↑ 7	↓ 8	↓ 9	↓ 10								
		2s	↑ 3	↓ 4												
1		1s	↑ 1	↓ 2												

than a subshell with larger *l*. Thus the 2s subshell is lower than the 3s subshell, and the 3s subshell is lower than the 3p subshell. However, there exist competing cases, such as the 2p and the 3s subshells, in which *n* is smaller for one subshell while *l* is smaller for the other. In such an event, the subshell with the smaller *n* has the lower energy unless the difference in the *l* values is greater than the difference in the *n* values. Thus the 2p subshell is lower than the 3s subshell, but the 4s subshell is lower than the 3d subshell. The 4s subshell has lower energy, in spite of its large value of *n*, because its value of *l* is 2 units lower than the *l* value of the 3d subshell. This peculiar inversion of the ordering of the subshells is shown in Table 20.4. It results in the existence of a group of transition elements which have similar properties but do not belong to any of the standard groups in the periodic table. These elements occur for the first time in period 4.

Period 4 The first two elements ($Z = 19$ and $Z = 20$) of period 4 are formed by adding electrons to the $4s$ subshell and so belong to groups I and II. Thereafter, electrons are added to the $3d$ subshell, forming a series of 10 transition elements ($Z = 21$ to $Z = 30$) between groups II and III. These elements have similar properties because they all have two electrons in the $4s$ subshell.* After filling the $3d$ subshell, electrons go into the $4p$ subshell, and the elements formed resemble the $2p$ and $3p$ elements in the preceding periods. The inert gas krypton ($Z = 36$) fills the $4p$ subshell, completing period 4.

Period 5 This period is identical in form to period 4. Electrons fill the $5s$, $4d$, and $5p$ subshells in period 5, just as they fill the $4s$, $3d$, and $4p$ subshells in period 4.

Period 6 The first two elements ($Z = 55$ and $Z = 56$) of period 6 are formed by filling the $6s$ subshell. Thereafter the $4f$ subshell is filled because it has lower energy than the $5d$ subshell. Because the $4f$ electrons are inside the $6s$ electrons, they have little effect on the chemical properties of these elements. Consequently, the next 14 elements ($Z = 57$ to $Z = 70$) form a sequence with nearly identical chemical properties. This sequence, called the *lanthanides* (or rare earths), does not correspond to any previous sequence of elements, so it is given a special place in the periodic table. The $4f$ shell is filled at $Z = 70$. After this, electrons go into the $5d$ subshell, forming a sequence of 10 transition elements analogous to the ones in periods 4 and 5. The first member of the transition sequence ($Z = 71$) is also considered to be the last member of the lanthanide sequence. The inert gas radon fills the $6p$ subshell, completing period 6.

Period 7 This period begins the same as the preceding period, with the filling of the $7s$ and the $5f$ subshells. The filling of the $5f$ subshell forms a new sequence of elements ($Z = 89$ through $Z = 102$), called the *actinides*. The last naturally occurring element is uranium ($Z = 92$), but elements with Z as high as 103 have been produced artificially in nuclear reactions. This last element has one electron in the $6d$ subshell, and so it is the first member of the period 7 transition sequence and the last member of the actinides.

> **REMARK** Elements with Z greater than 103 have been reported but not yet confirmed. The existence of high-Z elements is limited by the ability of their nuclei to hold together. If higher-Z elements are produced, they are expected to fill period 7 in regular sequence.

The quantum-mechanical explanation of the periodic table is one of the great triumphs of twentieth-century physics. It is astonishing that the properties of the chemical elements follow so simply from just a few fundamental physical principles.

*There are some irregularities in the filling of the $3d$ subshell. Some transition elements have only one $4s$ electron, the other electron being in the $3d$ subshell. These and other minor irregularities can be seen by studying the electron configurations shown in Table 20.3

20.5 Chemical Bonds

After the success of quantum mechanics in explaining the structure of single atoms, the same principles were applied to *molecules,* which are aggregates of two or more atoms bonded together by interatomic forces. It was found that quantum mechanics gave a natural explanation of the two basic bonding mechanisms—ionic and covalent bonding—and that it could even account for bonding of an intermediate character. These investigations involve solving the Schrödinger equation for electrons orbiting two or more atoms in a molecule. We shall describe the results of these difficult calculations in qualitative terms, but it must be remembered that any verbal description is only an approximation of the full mathematical result.

IONIC BOND

Ionic bonding occurs in simple molecules, such as NaCl and KF, that are composed of one atom from group I and one atom from group VII. Group I atoms have only one electron in their outer s subshell and $Z - 1$ inner electrons. These inner electrons tend to neutralize most of the charge of the nucleus, so that the outer electron is bound to the atom by an effective charge much less than the full charge $(+Ze)$ of the nucleus. Consequently, the outer electron is easily removed from a group I atom, forming a positively charged ion. Group VII atoms, on the other hand, are one electron short of filling their outer p subshell, so they tend to attract an extra electron, forming a negatively charged ion.

An *ionic bond* is formed when a neutral Na atom is brought close enough to a neutral Cl atom, so that the outer electron of the Na atom attaches itself to the Cl atom. The result is two ions of opposite charge, which attract each other and move closer together. When they get too close, however, their electron clouds start to repel each other, and at a certain separation distance the attractive force just balances the repulsive force. This is the equilibrium separation distance for the ions in an ionic bond. At this distance the ions are prevented from moving closer together by the repulsion of their electron clouds and from moving farther apart by the attraction of their opposite charges. Thus a stable molecule is formed, with a fixed distance between the ions.

COVALENT BOND

A *covalent bond* is formed by two atoms sharing electrons rather than exchanging them. Figure 20.16 shows how the probability distributions of the electrons in two separated hydrogen atoms rearrange themselves to form a covalently bonded hydrogen molecule, H_2, in which the two electrons are shared equally with each nucleus. The probability distribution is denser between the nuclei than outside, which means that an electron is more likely to be between the nuclei than outside. It is this concentration of negative charge between the two nuclei that holds the molecule together. It can be proved from the Schrödinger equation that this distribution forms a stable molecule.

The covalent-bonding properties of carbon are especially

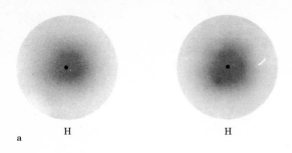

a   H   H

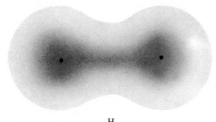

b H_2

FIGURE 20.16

Probability distributions of the electrons in two hydrogen atoms (a) before and (b) after the atoms are bound together to form a covalently bonded hydrogen molecule, H_2.

important because they enable carbon to form an infinite variety of compounds. These compounds are called *organic,* because it was once thought that they could be made only inside a living organism. Today organic chemists can synthesize many of the organic compounds found in nature and many others that are entirely new.

The four electrons in the $n = 2$ shell of carbon are all available for covalent bonding to other atoms, including other carbon atoms. In methane (CH_4), these four electrons are shared with four hydrogen atoms, as shown schematically in Fig. 20.17. This figure is drawn flat for clarity, although in fact the hydrogen atoms are located at the corners of a tetrahedron (pyramid), with the carbon atom at the center.

FIGURE 20.17

Methane molecule (CH_4), showing how the $n = 1$ electrons of the four hydrogen atoms are shared with the four $n = 2$ electrons of carbon. The molecule is drawn flat for clarity, but in reality the hydrogen atoms are located at the corners of a tetrahedron.

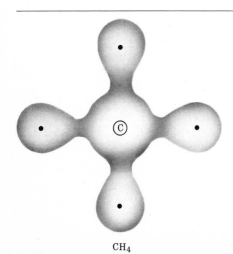

CH_4

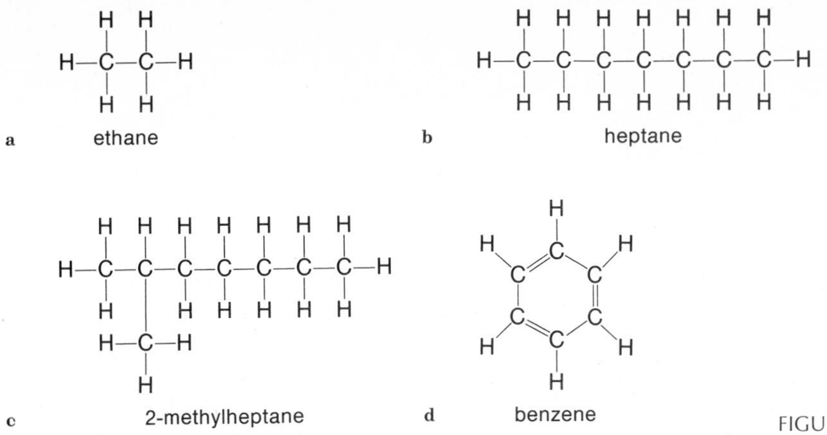

H H
| |
H—C—C—H
| |
H H

a   ethane

H H H H H H H
| | | | | | |
H—C—C—C—C—C—C—C—H
| | | | | | |
H H H H H H H

b   heptane

H H H H H H H
| | | | | | |
H—C—C—C—C—C—C—C—H
| | | | | |
H H H H H H
|
H—C—H
|
H

c   2-methylheptane

H
|
C
H—C   C—H
‖   ‖
C   C
H—C   C—H
‖
C
|
H

d   benzene

FIGURE 20.18

Symbolic representation of different hydrocarbons: (a) ethane (C_2H_6), (b) heptane (C_7H_{16}), (c) 2-methylheptane ($C_7H_{15}CH_3$), and (d) benzene (C_6H_6). 2-Methylheptane is a heptane molecule in which one of the hydrogen atoms bonded to the second carbon atom is replaced by the methyl radical, CH_3. Benzene is one of many molecules in which the carbon atoms form a closed ring.

In ethane (C_2H_6), three electrons from each carbon atom bind three hydrogen atoms, and the two carbon atoms are bound by sharing their fourth electrons. This is shown symbolically in Fig. 20.18a, where each line represents an electron of carbon being shared with a neighboring atom. The line between two carbon atoms represents the mutual sharing of two electrons. This pattern can be extended almost indefinitely, forming a whole family of carbon-hydrogen molecules called *hydrocarbons*. Figure 20.18b, c, and d shows some other members of this family.

The straight, unbranched hydrocarbons, such as methane, ethane, and heptane, form a special class of hydrocarbons called *alkanes*. The smaller alkanes (methane, ethane, and propane) are gases at normal temperature and pressure, and are the major ingredients of natural gas. The heavier alkanes are liquid and, together with more complex hydrocarbons, are the major ingredients of crude oil.

REMARK Gasoline is a complex mixture of hundreds of different hydrocarbons, most of which contain between 4 and 14 carbon atoms. The *octane rating* of a particular gasoline is a measure of the antiknocking characteristics of the gasoline compared with a mixture of heptane and 2-methylheptane (formerly called isooctane). An octane rating of 90, for instance, means the gasoline has the same antiknocking characteristics as a mixture of 10 percent heptane and 90 percent 2-methylheptane.

FIGURE 20.19

Possible bonding mechanisms for hydrogen chloride, HCl. (a) The bond is covalent if the electron from hydrogen has a 50 percent probability of being around either the hydrogen or the chlorine nucleus. (b) The bond is ionic if the electron from hydrogen has 100 percent probability of being around the chlorine nucleus. (c) The bond is in fact intermediate because the electron from hydrogen has a probability between 50 and 100 percent of being around the chlorine nucleus.

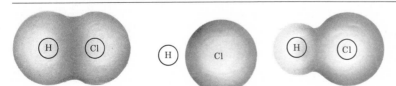

a   **b**   **c**

INTERMEDIATE BOND

Ionic and covalent bonds do not represent qualitatively different bonding mechanisms, but are rather the two extremes of a single mechanism. This can be better understood by considering the *intermediate bonding* that occurs in a molecule like hydrogen chloride (HCl). If the electron from hydrogen had equal probability of being found near the hydrogen and the chlorine atom (Fig. 20.19a), it would form a pure covalent bond. If the electron had a 100 percent probability of being found near the chlorine atom (Fig. 20.19b), it would form a pure ionic bond. In fact, the electron does have a greater probability of being found near the chlorine than near the hydrogen, but its probability of being near the hydrogen atom is not zero (Fig. 20.19c). The wave nature of the electron shows that ionic and covalent bonds are just the extreme ways in which the wave function of an electron can be distributed between two atoms. Intermediate distributions can and do occur.

GUIDE TO MAJOR TOPICS

Topic	References	Problems
Wave-particle duality	Sec. 20.1	
Photons	Eq. 2.1; Example 20.1	1 and 2
Photoelectric effect	Eqs. 20.2 to 20.3; Examples 20.2 and 20.3; Table 20.1	3 to 8
deBroglie waves	Eq. 20.4; Example 20.4	9 to 12
Bohr atom	Sec. 20.2; Eqs. 20.5 to 20.12; Example 20.5	13 to 16
Atomic spectra	Sec. 20.2; Eq. 20.13	17 to 22
Quantum mechanics	Sec. 20.3; Eqs. 20.14 to 20.16	23 and 24
Quantum states of an atom	Sec. 20.4; Tables 20.2 and 20.3	25
Periodic table	Sec. 20.4; Table 20.4	26
Chemical bond	Sec. 20.5	

PROBLEMS

REMARK Express all energies in electronvolts.

1 The wavelengths of visible light are between 400 and 700 mm. What are the corresponding photon energies?
Ans. 3.1 and 1.8 eV

2 Find the wavelength and frequency of a 13.6-eV photon.

3 (a) What is the critical frequency of silver? (b) If radiation of frequency 2.0×10^{15} Hz is incident on silver, what is the maximum kinetic energy of the ejected electrons?
Ans. (a) 1.14×10^{15} Hz; (b) 3.55 eV

4 When radiation with a wavelength of 420 nm is incident on potassium, electrons are ejected with a maximum energy of 0.7 eV. What is the work function of potassium?

5 When radiation with a frequency of 4.0×10^{15} Hz is incident on germanium, electrons with a maximum kinetic energy of 12 eV are ejected. What is the maximum kinetic energy of the electrons ejected when radiation with a frequency of 4.5×10^{15} is incident on germanium?
Ans. 14.07 eV

6 The minimum energy required to remove an electron from magnesium is 3.7 eV. What is the longest-wavelength photon that will eject an electron from magnesium?

7 What is the shortest-wavelength x-ray produced by a 200-kV x-ray machine?
Ans. 6.21×10^{-12} m

8 What potential difference is required to

produce x-rays with a minimum wavelength of 4.5×10^{-11} m?

9 (a) What is the wavelength of an electron with a speed of 5×10^7 m/s? (b) What is the wavelength of a proton with the same speed?
Ans. (a) 1.46×10^{-11} m; (b) 7.94×10^{-15} m

10 What is the kinetic energy of an electron with a wavelength of 5×10^{-10} m?

11 The kinetic energy of an electron is 100 eV. (a) What is its wavelength? (b) What is the wavelength of a 100-eV proton?
Ans. (a) 1.23×10^{-10} m; (b) 2.87×10^{-12} m

12 What are the wavelengths of a 1-eV photon and a 1-eV electron, respectively?

13 What is the speed of an electron in the lowest Bohr orbit of hydrogen?
Ans. 2.18×10^6 m/s

14 What is the kinetic energy of an electron in the $n = 2$ Bohr orbit of hydrogen?

15 What is the radius of the $n = 20$ Bohr orbit of hydrogen?
Ans. 2.12×10^{-8} m

16 How much energy is required to ionize a hydrogen atom in the $n = 2$ state?

17 Find the wavelength of the radiation emitted when a hydrogen atom makes a transition from the $n = 6$ state to the $n = 3$ state.
Ans. 1.096×10^{-6} m

18 A hydrogen atom in its ground state absorbs a photon with a frequency of 2.9×10^{15} Hz. To which level is the atom excited?

19 The spectrum of hydrogen contains four wavelengths in the visible region. Find the transitions that emit these wavelengths, and determine the wavelengths.
Ans. 411, 435, 487, and 657 nm

20 All transitions that go directly to the ground state of hydrogen produce a sequence of wavelengths called the *Lyman series*. (a) What are the longest and shortest wavelengths in the Lyman series? (b) To what region of the electromagnetic spectrum does this radiation belong?

21 All transitions that go directly to the $n = 2$ state of hydrogen produce a sequence of wavelengths called the *Balmer series*. Find (a) the longest and (b) the shortest wavelength in the Balmer series. (c) To what regions of the electromagnetic spectrum do these radiations belong?
Ans. (a) 657 nm; (b) 365 nm; (c) visible and ultraviolet

22 Figure 20.10 shows the first six energy levels of the outer electron in sodium. (a) What is the ionization energy of sodium? (b) What is the wavelength of the radiation emitted in the transition from the third excited state (-1.6 eV) to the first excited state (-3.0 eV)? (c) What is the transition that gives rise to the characteristic yellow light of sodium at 589 nm?

23 What is the minimum kinetic energy of an electron confined to a region the size of a nucleus (10^{-15} m)?
Ans. 3.8×10^{11} eV

24 A 10-g marble moves back and forth in a box 10 cm long at a speed of 10 cm/s. (a) What is the number n of the quantum state of the marble? (b) Why is it apparently possible to vary the energy of the marble continuously even though its energy can change only by discrete amounts?

25 List all the subshells in the $n = 7$ shell. How many states are there in this shell?

26 List the subshells with $n \leq 6$ that are not occupied in any atom.

BIBLIOGRAPHY

BARROW, GORDON M.: *Physical Chemistry*, 3rd ed., McGraw-Hill Book Company, New York, 1973. Schematic pictures are used instead of extensive mathematics to describe electron orbits in atoms and molecules.

BEISER, ARTHUR: *Concepts of Modern Physics*, McGraw-Hill Book Company, New York, 1963. For students with some calculus background, this book provides an excellent introduction to the mathematical treatment of quantum mechanics.

CLINE, BARBARA LOVETT: *Men Who Made a New Physics*, The New American Library, New York, 1969. A simply written account of the development of modern physics, with special emphasis on the ideas and personalities of the men involved.

GAMOW, GEORGE: *Mr. Tomkins in Wonderland: Stories of c, G, and h*, The Macmillan Company, New York, 1940. Mr. Tomkins visits a world in which *h* is so large that quantum effects are everyday occurrences.

21

SOLID-STATE PHYSICS

S olid-state physics is the experimental and theoretical study of condensed matter, i.e., of liquids and solids containing 10^{19} or more atoms in close contact with one another. As with an electron in an individual atom or molecule, the motion of an electron in bulk matter must be treated quantum-mechanically. But because the electron can interact with 10^{19} atoms, an exact solution of the Schrödinger equation is impossible. Consequently, approximation methods must be developed. This has been most successful in the case of crystals, where the orderly arrangement of the atoms makes possible a simple description of electron motion. For this reason, most of the work in solid-state physics has been limited to simple crystalline substances.

In spite of this limitation, solid-state physics has been of immense practical value. The entire semiconductor industry, which has revolutionized electronics, is the result of solid-state research. Solid-state physicists also study the properties of metals and alloys, helping to develop new materials for science and industry. Even superconductivity, which has long been a subject of deep interest to solid-state physicists, is beginning to find practical applications. Furthermore, theoretical methods are being developed to handle the more difficult problem of disordered systems, i.e., liquids and amorphous (noncrystalline) solids. As these systems become better understood, many new products can be expected. The liquid-crystal display on watches and calculators is an early example of this.

21.1 Crystal Structure
LATTICE STRUCTURE

The molecules of a liquid, although in close contact, are not bound to fixed positions. They are free to move about, much like the molecules of a gas. But as the temperature of a liquid is lowered, the thermal motion of the molecules decreases, and at a certain temperature, called the *melting point** (Table 8.3), the motion is slow enough that the attractive forces between the molecules are able to lock the molecules into fixed positions. These positions are usually the ones that give the molecules the tightest possible binding.

A *crystal* is a solid in which the atoms or molecules are arranged in an orderly three-dimensional array, like the one in Fig. 21.1. In even the smallest crystal, there are so many atoms in any one row (10^5 or more) that only a relatively few are near an outer edge. We neglect these edge effects, and consider all crystal arrays to be infinite.

The number of distinct infinite arrays is limited by the way in which space can be filled. This is most easily seen in two dimensions, so we shall at first restrict our discussion to two-dimensional arrays, like the ones in Fig. 21.2.

Definition A *lattice* is an infinite array of points. The arrays in Fig. 21.2 are lattices.

*The melting point is the same as the freezing point. Whether a substance is melting or freezing depends on whether it is being heated or cooled.

FIGURE 21.1

Model of a sodium chloride (NaCl) crystal. (*Ealing Scientific Corp.*)

Definition A lattice is *translationally invariant* if, when the lattice is moved a certain distance in a certain direction, the displaced lattice exactly coincides with the original lattice.

To understand translational invariance, imagine that a piece of tracing paper is laid over Fig. 21.2a and all the points are copied on the paper. If the paper is moved (translated) in the direction **A** a distance a, 2a, 3a, etc., the traced dots will exactly overlay the printed dots. The same thing happens if the paper is moved in the direction **B** a distance b, 2b, 3b, etc. Remember, the lattice is infinite, so we never run out of either printed or traced dots.

There are two important consequences of translational invariance. First, in a lattice with translational invariance, all points are identical. That is, the arrangement of points around one point is the same as the arrangement of points around any other point. Each point has the same environment. Second, all simple translationally invariant lattices can be constructed from identical parallelograms, since this is the general polygon that, by repeated translation, will cover the plane.*

Figure 21.2a is a *general lattice* in which the parallelogram has unequal sides a and b and an arbitrary angle θ. Figure 21.2b is a *rectangular lattice* in which the parallelograms are rectangles with $\theta = 90°$ (and a and b unequal). Figure 21.2c is a *hexagonal lattice* in which $a = b$ and $\theta = 120°$. In this special case the lattice can also be thought of as constructed from identical hexagons.

> **REMARK** The equilateral traingle, the square, and the regular hexagon are the only regular polygons that cover the plane by repeated translation; i.e., they are the only polygons with equal sides and angles that can be used for tiles, if there is to be no space between them. In our discussion of Fig. 21.2 we considered the general parallelogram to include triangles and hexagons as special cases.

Figure 21.3 shows an orderly pentagonal lattice. This is not a crystal lattice, however, because it is not translationally invariant. The point in the center, for instance, is unique, so it cannot be moved to another point like itself. Likewise, the environment of a point in a corner is different from the environment of a point on an edge. Such a structure is not impossible, but it is rare in nature, being limited to certain microscopic arrangements of virus particles.

In three dimensions, as in two, only certain lattice structures have translational invariance. These space lattices are all based on the general parallelepiped (three-dimensional parallelogram) and its special cases, such as the cube. Figure 21.4 shows a few of the 14 distinct space lattices. If a crystal is allowed to grow undisturbed, its shape will reflect its lattice structure. From studies of the shapes of natural crystals, crystallographers of the last

*Complex translationally invariant lattices can be formed by the superposition of two or more simple lattices based on different parallelograms, but we shall not consider such cases here. See Martin Gardner's article in the July 1975 *Scientific American* for a fascinating discussion of tiling, or the problem of covering the plane with identical figures.

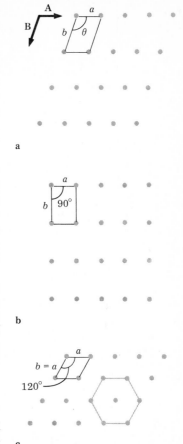

a

b

c

FIGURE 21.2

Two-dimensional translationally invariant lattices. (a) General lattice. Neighboring points form a parallelogram with unequal sides a and b and arbitrary angle θ. (b) Rectangular lattice, in which $a \neq b$, but $\theta = 90°$. (c) Hexagonal lattice, in which $a = b$ and $\theta = 120°$.

FIGURE 21.3

Pentagonal lattice. This is not a possible crystal lattice, because it is not translationally invariant.

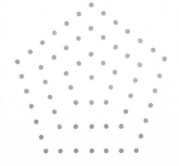

century found that every crystal did have a structure conforming to one of the 14 translationally invariant space lattices.

In a crystal, each point of the lattice is usually the position, not of a single atom, but of a group of atoms, called the *basis* of the crystal. The group has the same composition and orientation at every lattice point, so the lattice is just the abstract pattern of how the basis is repeated throughout the crystal structure. We shall illustrate this concept with two examples.

Cesium Chloride (simple cubic) The space lattice of cesium chloride (CsCl) is the simple cubic lattice shown in Fig. 21.5a. (Two of the front cubes have been removed for clarity.) The basis is a Cs^+ ion and a Cl^- ion. Figure 21.5a shows one unit of this basis placed so that the Cl^- ion is on a lattice point; the Cs^+ ion then is located in the middle of the cube. Figure 21.5b shows the lattice with two more units placed on two other lattice points. The complete crystal structure emerges when all the lattice points are occupied by basis units (Fig. 21.5c). Since Fig. 21.5c is only a small section of an infinite structure, we do not see the Cs^+ ions associated with many of the Cl^- ions. However, as the structure is extended, these ions appear. In an infinite structure, there are equal numbers of Cs^+ and Cl^- ions.

Sodium Chloride (face-centered cubic) The space lattice of sodium chloride (NaCl) is the face-centered cubic lattice shown in Fig. 21.6a. The basis is an Na^+ ion and a Cl^- ion. Figure 21.6a shows one unit of the basis placed so that the Cl^- ion is on a lattice point; the Na^+ ion is then in the middle of the cube, as in the previous case. However, the face-centered cubic lattice has lattice points in the center of each face, as well as at the corners.

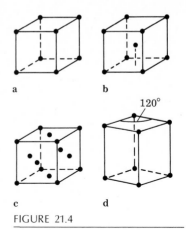

FIGURE 21.4

Three-dimensional crystal lattices: (a) simple cubic, (b) body-centered cubic, (c) face-centered cubic, and (d) hexagonal.

FIGURE 21.5

Cesium chloride (CsCl) crystal. (a) Simple cubic lattice showing the position of the Cs^+–Cl^- basis. (b) The same lattice with two more basis units in position. (c) Complete CsCl crystal built on a simple cubic lattice.

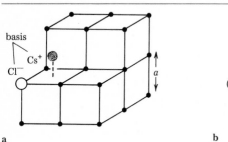

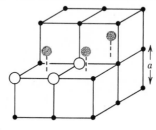

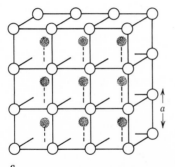

An Na^+–Cl^- basis is located at each of these points too. Figure 21.6b shows the lattice with three more basis units in place. The complete crystal structure emerges when all the lattice points are occupied by basis units (Fig. 21.6c).

A comparison of Figs. 21.5c and 21.6c shows how two different lattices form two different crystal structures. The simple cubic lattice (CsCl) gives a structure in which a square array of Cl^- ions in one plane alternates with a square array of Cs^+ ions in another. The face-centered cubic lattice (NaCl) gives a structure in which each plane of the crystal is a checkerboard pattern of Na^+ and Cl^- ions. Crystallographers have found that every crystal is a repetition of some basis on one of the 14 distinct space lattices.

BONDING IN CRYSTALS

The atoms and molecules of a crystal are bonded together by the electrical attraction between unlike charges. As in the case of molecular bonding (Sec. 20.5), it is useful to classify the different bonding mechanisms according to how tightly the electrons are bonded to individual atoms.

Ionic Bonding In an ionic crystal like CsCl or NaCl, the crystal is held together by the coulomb force between ions of opposite charge. The electrons of each ion are tightly bonded to it and are not free to move about in the crystal. The absence of any mobile charge in ionic crystals make them poor conductors of electricity; i.e., they are insulators.

Covalent Bonding In a covalent crystal like silicon or germanium, the four outer electrons of each atom are shared with four neighboring atoms. In a perfect covalent crystal at absolute zero, all the electrons would be bonded to atoms and not free to move, so the crystal would be an insulator. But in a real covalent crystal at normal temperature (300 K), the thermal motion of the atoms frees some of the bonded electrons. These free electrons, along with other electrons from impurities in the crystal, provide the crystal with a moderate degree of conductivity; i.e., the crystal is a semiconductor.

Metal Bonding In a metal the outer electrons of each atom are completely free to move anywhere inside the metal. These free electrons form a uniform sea of negative charge which bonds the positive ions together to form a crystal lattice. The high conductivity of metals is the result of this plentiful supply of free electrons.

van der Waals Bonding Inert elements like neon and argon and diatomic molecules like hydrogen, nitrogen, and oxygen form crystals at low temperatures. In each case, the crystal is composed of neutral atoms or molecules with tightly bonded electrons, so the bonding is neither ionic nor covalent. Since the electric field outside a neutral atom like neon, which has a spherical distribution of electrons surrounding its nucleus, is

FIGURE 21.6

Sodium chloride (NaCl) crystal. (a) Face-centered cubic lattice showing the position of the Na^+–Cl^- basis. (b) The same lattice with three more basis units in position. (c) Complete NaCl crystal built on a face-centered cubic lattice (see the crystal model in Fig. 21.1).

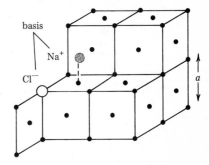

a

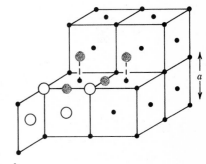

b

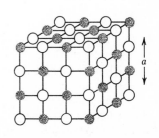

c

expected to be zero, it is not obvious how any attractive force can exist between two such atoms.

The explanation of the force between neutral atoms and molecules is given by quantum mechanics, which interprets the distribution of charge around an atom as a probability distribution. In a spherical atom, an electron has equal probability of being found on any side of the nucleus (Fig. 21.7a), so the average electric field outside the atom is zero. But at any instant, the electron has a probability of being on one side of the nucleus (Fig. 21.7b). In this case, the atom has a dipole charge distribution, and the field outside it is the field of a dipole (Sec. 15.3). This field will induce a dipole charge distribution on a nearby atom (Fig. 21.7c), and the two dipoles will attract each other. When the resulting force is averaged over all possible positions of the electrons in the atom, the average force is still attractive: it is called the *van der Waals force*. This force is much weaker than ionic or covalent forces, so crystals with van der Waals bonding exist only at very low temperatures. At even moderate temperatures, the thermal motion of the atoms is sufficient to overcome the weak attractive force between the atoms.

In the rest of this chapter we shall concentrate on the electrical properties of metals and semiconductors because of their great importance in electronic technology.

21.2 Metals

FREE-ELECTRON MODEL

Metals are characterized by their shiny luster, their ability to be bent (*malleability*) and drawn into wires (*ductility*), and their high electric and thermal conductivity. Most metals are elements, or mixtures of elements, that have one, two, or three electrons in their outer shell. This includes all the elements in groups I, II, and III of the periodic table (with the exception of boron), as well as the transition elements, the lanthanides, and the actinides. Five elements in groups IV, V, and VI (tin, antimony, lead, bismuth, and polonium) are also metals. In all, about three-quarters of the elements are metals.

The outer, or *valence,* electrons of the atoms in a metal are not bonded to any single atom or group of atoms, but are free to move anywhere in the metal. Stripped of their outer electrons, the atoms form a closely packed lattice of positive ions that is held together by the surrounding sea of valence electrons (Fig. 21.8). Each ion can vibrate about its lattice site, but it cannot move away from it. Only the valence electrons are free to move throughout the metal.

The motion of the valence electrons is influenced by the presence of the ions, but not to the extent one might imagine. Because of the periodic arrangement of the ions, standing electron waves can be established in the metal that closely resemble the standing waves of a particle in a box (Sec. 20.3). That is, because of the translational invariance of the crystal structure and the wave nature of the electron, the valence electrons move through the metal very much as though the ions were not there.

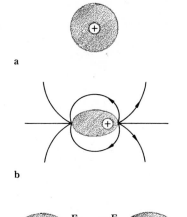

FIGURE 21.7

Origin of the van der Waals force between neutral atoms. (a) In a spherical atom, an electron has equal probability of being found on any side of the nucleus. (b) At any instant, however, the electron may be more on one side than the other. This gives rise to an instantaneous dipole charge distribution. (c) The electric field of the dipole charge distribution of one atom induces a dipole charge distribution in a second nearby atom. The two dipoles then attract one another.

FIGURE 21.8

In a metal, the atoms, stripped of their valence electrons, form a closely packed lattice of positive ions.

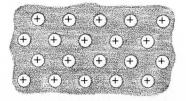

In the *free-electron model* of metals, the lattice ions are ignored altogether, and the electrons are treated like particles in a box.

For simplicity, consider a one-dimensional crystal containing N evenly spaced ions (Fig. 21.9). The distance between adjacent ions is a, so the length L of the crystal is Na. Since N is typically 10^8 or more, L is very much greater than a. In fact, L is a macroscopic length (say, 10^{-2} m), whereas a is a microscopic length (say, 10^{-10} m). As we shall see, this difference in scale between L and a plays an important role in the theory of metals.

Each of the N atoms in a metal contributes ν valence electrons, where ν is usually 1, 2, or 3. These νN electrons travel at constant speed in a straight line until they reach the edge of the crystal, at which point they are reflected back in the opposite direction. The allowed quantum states of the electrons are those corresponding to standing waves. For the case of the one-dimensional crystal, the situation is the same as that of a particle confined to a box of length L. In Sec. 20.3 we showed that the de Broglie wave of such a particle must have a node at each end of the box, so the allowed waves have wavelengths λ given by

$$\lambda = \frac{2L}{n} \qquad\qquad 21.1$$

where n is an integer ($n = 1, 2, 3, \ldots$).

The speed v of an electron is related to its wavelength by the de Broglie relation (Eq. 20.4):

$$mv = \frac{h}{\lambda}$$

Here m is the mass of the electron, and h is Planck's constant. Thus the possible kinetic energies of an electron in a one-dimensional crystal are given by

$$K = \tfrac{1}{2}mv^2 = \frac{1}{2m}\frac{h^2}{\lambda^2} = \frac{n^2h^2}{8mL^2} \qquad\qquad 21.2$$

This is the same as Eq. 20.16 in Sec. 20.3. Each integer value of n corresponds to a possible energy level of an electron, and Eq. 21.2 gives the energy of that level.

A similar result holds for a three-dimensional crystal: each electron must be in one of the energy levels of the crystal. The lowest energy of the crystal is obtained when the electrons occupy the lowest-energy energy levels. As with a single atom, the filling of the energy levels must comply with the Pauli exclusion principle: at most two electrons (with opposite spin) can occupy each energy level. This means that the lowest energy of a crystal with νN valence electrons is obtained by filling the first $\tfrac{1}{2}\nu N$ energy levels (Fig. 21.10a).

An important difference between the energy levels of a single atom and those of a metal is the magnitude of the spacing. In an atom, the levels are typically several electronvolts apart, whereas in a metal the levels are less than 10^{-23} eV apart. This incredible difference is due to the fact that the electrons in an atom are confined to a region of width 10^{-10} m, whereas the electrons in a metal are confined to a region of macroscopic size (say, 10^{-2} m).

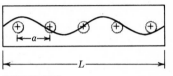

FIGURE 21.9

A one-dimensional crystal lattice containing N evenly spaced ions. The de Broglie wavelength of a valence electron is $2L/n$, where L is the length of the lattice, and n is an integer.

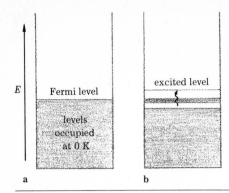

E — Fermi level

levels
occupied
at 0 K

excited level

a   b

FIGURE 21.10

(a) At absolute zero (0 K) the νN valence electrons in a metal occupy the lowest available energy levels. (b) At 300 K, some of the electrons just below the Fermi level have enough energy to move into levels above the Fermi level.

Equation 21.2 shows that the energy spacing is proportional to $1/L^2$, so the smaller the region of confinement, the greater the spacing. In a metal the levels are so close together that the spacing between them can be ignored. As we shall see in Sec. 21.3, the special properties of semiconductors that make them so important in electronics come from the existence of a gap of about 1 eV in their energy levels.

At absolute zero (0 K), the electrons in a metal are in the lowest-energy levels allowed by the exclusion principle. The last occupied level is called the *Fermi level,* and its energy, called the *Fermi energy,* is about 5 eV. As the temperature of a metal is raised, the thermal energy excites some of the electrons to levels above the Fermi level. At 300 K, the available thermal energy is about 0.02 eV per particle. An electron within 0.02 eV of the Fermi level can absorb this energy and move to an unoccupied level above the Fermi level (Fig. 21.10b), but most of the electrons, which are more than 0.02 eV below the Fermi level, cannot. An electron cannot absorb energy that would move it into an occupied state, because this would violate the exclusion principle. Thus the exclusion principle freezes most of the electrons in their lowest-energy states and allows only a relatively few electrons to participate in the random exchange of thermal energy.

CONDUCTION

Two things are required in order for a substance to be a normal conductor: (1) a supply of mobile charge that is free to move when acted upon by an electric field, and (2) factors that limit this motion to some extent. In a metal the valence electrons are the mobile charge, and crystal impurities and lattice vibrations are the factors that limit electron motion.

Even at 0 K, the valence electrons are moving back and forth in a metal at high speed. For instance, the speed of an electron near the Fermi level is around 10^6 m/s. However, as many electrons are moving in one direction as another, so this random motion does not result in a net flow of charge. It is like a swarm of bees that is hovering above the ground. Within the swarm, the bees dart back and forth at high speed, but the center of the swarm is stationary because the average speed in one direction is the same as the average speed in any other direction. For the center of the swarm to start to move with speed v_d, it is necessary only for bees moving in one direction to increase their speed by v_d while bees moving in the opposite direction decrease their speed by v_d. Likewise, if the valence electrons in a metal increase their average speed by v_d in one direction while decreasing it by v_d in the opposite direction, the entire sea of electrons will move in one direction with speed v_d. This is called the *drift speed* of the electrons, and it is very much smaller than their random back-and-forth speed. In studying electric conduction, we can ignore the back-and-forth motion of the electrons and consider just the drift speed. That is, we can view an electric current as the flow of charge resulting from all the valence electrons moving with the same speed v_d.

Figure 21.11 shows a wire of cross-sectional area A and length L in which the electrons are moving to the right with a drift speed v_d. This results in a current I in the opposite direction, because the direction of the conventional current is defined as the direction of positive charge flow. The magnitude of I is

$$I = \frac{q}{t}$$

where Q is the net charge that passes a point in time t.

Definition The *electron density** η of a substance is the number of mobile electrons per unit volume. In a metal, it is the number of valence electrons per unit volume.

The volume of the wire in Fig. 21.11 is AL, so the number of valence electrons in it is ηAL, and their total charge q is

$$q = e\eta AL$$

where η is the electron density of the wire, and apart from sign, e is the charge of an electron. All this charge will pass point O in the time

$$t = \frac{L}{v_d}$$

required for an electron at point P to reach O. Thus the current in the wire is

$$I = \frac{q}{t} = \frac{e\eta AL}{\dfrac{L}{v_d}} = e\eta A v_d \qquad \text{21.3}$$

This is a general equation, applicable to semiconductors as well as metals. In metals the electron density η is high, and the current is limited by the drift velocity. In semiconductors the low value of η limits the current.

Example 21.1 A copper wire with a diameter of 2.5 mm carries a current of 8 A. What is the drift velocity? (A copper atom has one valence electron; see Table 20.2).

Before we can use Eq. 21.3, we must find the electron density of copper. The mass m of a volume V of copper is

$$m = \rho V$$

*η is the Greek (lower-case) letter eta.

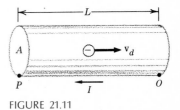

FIGURE 21.11

A wire of cross-sectional area A and length L. Electrons are moving to the right with a drift velocity v_d.

where ρ is the mass density. The mass of 1 mol of copper is equal to the atomic mass M of copper (in grams), so the number n of moles in the volume V is

$$n = \frac{m}{M} = \frac{\rho V}{M}$$

and the number N of atoms in V is

$$N = nN_A = \frac{\rho V N_A}{M}$$

where N_A is Avogardro's number, the number of atoms in a mole. Thus the electron density is

$$\eta = \frac{\nu N}{V} = \frac{\nu \rho N_A}{M} \qquad 21.4$$

From Table 1.6 we find that the density of copper is $8.9 \times 10^3 \text{ kg/m}^2$, or 8.9 g/cm^3; from the table of the elements we find that the atomic mass of copper is 63.5. Using these values in Eq. 21.4 we get

$$\eta = \frac{(1)(8.9 \text{ g/cm}^3)(6 \times 10^{23})}{63.5 \text{ g}}$$
$$= 0.84 \times 10^{23} \text{ electrons/cm}^3$$

The cross-sectional area A of the wire is

$$A = \pi r^2 = \pi(0.125 \text{ cm})^2 = 4.9 \times 10^{-2} \text{ cm}^2$$

With these values of η and A, the drift velocity v_d given by Eq. 21.3 is

$$v_d = \frac{I}{e\eta A}$$
$$= \frac{8 \text{ C/s}}{(1.6 \times 10^{-19} \text{ C})(1.6 \times 10^{23} \text{ cm}^{-3})(4.9 \times 10^{-2} \text{ cm}^2)}$$
$$= 6.4 \times 10^{-3} \text{ cm/s}$$

This demonstrates that the drift speed of an electron is very much smaller than its back-and-forth speed. The electric current in a metal resembles a slow-moving glacier more than it does a rapidly flowing stream. $\square$

RESISTIVITY

In Sec. 16.3 we showed that the resistance R of a conductor of length L and cross-sectional are A can be written

$$R = \rho\frac{L}{A} \qquad 21.5$$

where ρ is a characteristic property of the conductor called the *resistivity*. (In Example 21.1 ρ was used for the mass density; be careful not to confuse these two different uses of the same symbol.) We now want to discuss the physical factors that determine the resistivity of a metal.

From Ohm's law, the voltage V across a conductor is

$$V = RI$$

where I is the current and R is the resistance. If the length of the conductor is L, the electric field in it is

$$E = \frac{V}{L}$$

and this field exerts the force $\mathbf{F}_e$ of magnitude

$$F_e = eE = \frac{eV}{L} \qquad\qquad 21.7$$

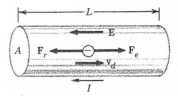

FIGURE 21.12

The electric force $\mathbf{F}_e$ on an electron moving in a metal is balanced by a resistive force $\mathbf{F}_r$.

on each electron, causing it to accelerate (Fig. 21.12). ($\mathbf{E}$ and $\mathbf{F}_e$ have opposite directions in Fig. 21.12 because the electron is negatively charged.) It is the force $\mathbf{F}_e$ that drives the current.

Since each electron moves with constant speed v_d, the total force on it must be zero. As soon as the force $\mathbf{F}_e$ starts to accelerate an electron, a resistive force $\mathbf{F}_r$ develops to oppose $\mathbf{F}_e$. This resistive force is due to collisions between the electrons and the ions of the metal, and it can be written

$$F_r = cv_d \qquad\qquad 21.8$$

where c is a parameter that depends on the frequency of these collisions. When a steady current is established in a conductor, the resistive force $\mathbf{F}_r$ equals the driving force $\mathbf{F}_e$, so from Eqs. 21.7 and 21.8 we have

$$\frac{eV}{L} = cv_d$$

or

$$V = \frac{cLv_d}{e}$$

Solving Eq. 21.3 for v_d and substituting it into this last equation we get

$$V = \frac{cL}{e}\left(\frac{I}{e\eta A}\right)$$

$$= \left(\frac{c}{e^2\eta}\frac{L}{A}\right)I$$

Comparing this with Eq. 21.6 we see that the resistance of the conductor can be written

$$R = \frac{c}{e^2\eta}\frac{L}{A}$$

and so from Eq. 21.5 the resistivity is

$$\rho = \frac{c}{e^2\eta} \qquad\qquad 21.9$$

This is the equation we are after. It shows that the resistivity increases with the frequency of electron-ion collisions and decreases with the density of electrons η.

A metal has a large unvarying electron density, but the collision frequency is highly variable. In fact, in a perfect crystal at absolute zero the collision frequency is zero, because electrons can move freely through a periodic array of ions. It is the departure of the array from perfect periodicity that produces electron-ion collisions.

There are two major causes of crystal imperfection: impurities and lattice vibrations. *Impurities* are atoms of a different element that either occupy lattice sites normally occupied by majority atoms or occupy positions between lattice sites (Fig. 21.13). In either case these foreign atoms disrupt the periodicity of the crystal and cause electron collision. *Lattice vibrations* are the random vibrations of the ions about their lattice sites. This random motion, which varies from one ion to the next, also disrupts the periodicity of the crystal and causes electron collision. We can write

$$c = c_i + c_v$$

where c_i is the contribution to the collision frequency resulting from impurities, and c_v is the contribution from lattice vibrations.

For a given metal sample, the impurity contribution c_i will be the same at all temperatures, since there are a fixed number of impurities in the sample. But lattice vibrations increase with temperature, so the lattice-vibration contribution c_v increases with temperature. At absolute zero, c_v is zero, and all the electron-ion collisions come from the impurities. Figure 21.14 shows a graph of the resistivity of a metal against its absolute temperature T. At $T = 0$ the resistivity is very small, being due entirely to impurities. The resistivity increases with temperature because of the increase in lattice vibration, and at $T = 300$ K the resistivity is 10^3 times its value at $T = 0$ K. A crystal with no impurities would have zero resistivity at absolute zero; i.e., it would be a perfect conductor.

We have discussed the electrical properties of metals in some

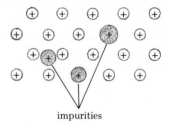

impurities

FIGURE 21.13

Impurity atoms in a crystal occupy sites normally occupied by majority atoms or positions between lattice sites.

FIGURE 21.14

The ratio ρ/ρ_{300} of the resistivity ρ of a metal at temperature T to its resistivity ρ_{300} at 300 K, plotted against T.

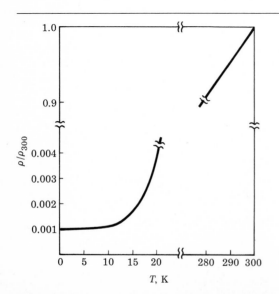

detail, because a proper understanding of conduction in metals is required in order to understand the very different behavior of semiconductors.

SUPERCONDUCTIVITY

In 1911 Heike Kamerlingh Onnes (1853–1926) discovered that the resistivity of certain metals falls abruptly to zero below a certain temperature. Figure 21.15 shows the ρ-T curve for such a metal. Above the *transition temperature* T_c the metal behaves normally, its resistivity decreasing steadily as the temperature decreases. Below the transition temperature, the resistivity is zero;* the metal is then said to be *superconducting*. Table 21.1 gives the transition temperature of some superconducting metals. Note that compounds can have a higher transition temperature than the elements they contain. Although all known transition temperatures are very low by ordinary standards ($-253°C$ or less), they are still substantially above absolute zero.

In a superconductor, the interaction of the electrons with the ions creates a gap of about 10^{-3} eV between the highest filled level and the next higher level. Below the transition temperature the normal scattering mechanisms are not strong enough to excite electrons to states above the gap. Since all the states below the gap are filled, the electrons cannot be scattered. In effect, the scattering mechanisms that cause resistivity in a normal metal are frozen out in a superconductor.

Superconductivity has long held the promise of lossless power transmission, but the technical problems are formidable. Wires have to be made from such exotic metals as niobium and titanium, and they must be maintained at a temperature of 10 K or less. The first large-scale application of superconductivity will probably be the development of superconducting coils for the magnets used in the high-energy accelerators of the Fermi National Accelerator Laboratory and the Brookhaven National Laboratory (Fig. 21.16). Each installation will use 1000 magnets, each carrying 4000 A. The power required to maintain this much current in normal conductors is 20 MW. By using superconducting coils, the power can be reduced drastically. There is a severe price for this, however, in the very complex network of cryogenic equipment needed to keep the coils at superconducting temperatures.† This equipment alone will use 10 MW of power to liquefy 5000 L of helium an hour. The magnets are expected to be in operation by the mid-1980s.

21.3 Semiconductors

BAND STRUCTURE

A *semiconductor* is a substance with a resistivity (at 300 K) of between 10^{-3} and $10^3\ \Omega \cdot m$, intermediate between that of a good conductor ($10^{-8}\ \Omega \cdot m$) and a good insulator ($10^{12}\ \Omega \cdot m$).

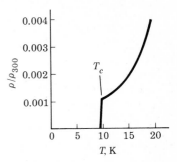

FIGURE 21.15

The temperature variation of the resistivity of a superconductor near the transition temperature T_c.

TABLE **21.1**

Transition temperatures of some superconductors

Element or compound	Transition temperature, K
Titanium	0.39
Molybdenum	0.92
Aluminum	1.18
Tin	3.72
Vanadium	5.38
Niobium	9.20
Nb_3Ti_2	9.8
Nb_3Al	17.5
Nb_3Sn	18.05

---

*The resistivity may not be exactly zero, but it is below any detectable value. A current once established in a superconductor will persist for at least 100,000 years without any power source.

†The coils are made with a niobium-titanium wire (Nb_3Ti_2) that has a transition temperature of 9.8 K.

FIGURE 21.16

Superconducting magnet with some of the cryogenic equipment needed to keep its coils at superconducting temperatures. (*Brookhaven National Laboratory.*)

Commercially important semiconductors are the elements silicon and germanium and the compounds cuprous oxide (Cu_2O), lead sulphide (PbS), silicon carbide (SiC), and gallium arsenide (GaAs). Because silicon is used extensively in the manufacture of microprocessors, we shall use it as the example of a typical semiconductor.

A silicon atom has four outer electrons, and in a silicon crystal these electrons form covalent bonds with four neighboring atoms. These covalent bonds are similar to the metallic bonds in a metal, except that the valence electrons are more closely bonded to the ions in a semiconductor than in a metal. The electron waves still extend over the entire crystal, giving rise to a large number of quantum states. As in a metal, the Pauli exclusion principle requires that each of the valence electrons be in a different state, so that even at absolute zero the electrons occupy a band of energies (Fig. 21.17). The critical difference between a semiconductor and a metal is that because of the tighter bonding of the valence electrons to the ions, there exists a gap E_g between the energy E_0 of the highest energy level occupied at 0 K and the next higher level. An electron in a semiconductor cannot exist with an energy between the energy E_0 of the last occupied level and the energy $E_0 + E_g$ of the next available state, just as an electron in a hydrogen atom cannot exist with an energy between that of the ground-state energy and the energy of the first excited state.

At 0 K all the states with energy below E_0 are occupied by valence electrons; this group of states is called the *valence band*. Above the gap region there is another group of closely spaced states, called the *conduction band*. At 0 K the conduction band is unoccupied.

When a current is established in a normal metal at 0 K, the energy of each electron is increased slightly because of the added drift speed. Electrons at the Fermi level move into unoccupied states just above the Fermi level, and electrons just below the Fermi level move into the vacated states at the Fermi

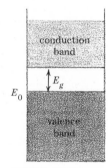

FIGURE 21.17

The valence band of energy levels occupied by the valence electrons of a semiconductor at absolute zero (0 K). An energy gap exists between the last occupied level and the next available level.

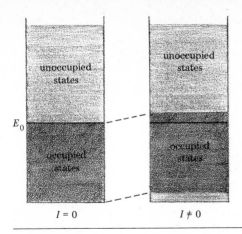

E_0

$I = 0$ $I \neq 0$

FIGURE 21.18

In a metal, there is no gap between the last occupied level and the next available level, so even at 0 K, electrons in the highest occupied levels can move into adjacent higher levels when their energy is increased by an electric field.

level. Figure 21.18 shows how the entire band of electrons is moved up slightly as a result of this process. In a semiconductor, on the other hand, such a slight increase in the energies of all the electrons is not possible, because of the gap. The increase in energy due to conduction is about 10^{-12} eV, much too small to jump the gap, which is typically 1 eV. Thus at 0 K a pure semiconductor cannot conduct at all; its resistivity is infinite. A pure metal, in contrast, has zero resistivity at absolute zero.

As the temperature is increased, metals and semiconductors behave in opposite fashion. In metals, which have an abundance of mobile electrons, the increased lattice vibrations increase the collision frequency of the electrons, increasing the resistivity. In a semiconductor, which has no mobile electrons at 0 K, the increased lattice vibrations excite some of the electrons from the valence band into the conduction band.* Once in the conduction band, the electrons can move freely from one closely spaced energy level to another, and so can acquire a drift speed. The resistivity of a semiconductor decreases with increasing temperature because the number of mobile electrons increases. Figure 21.19 shows the ρ-T curve for silicon. Note the resistivity decrease by a factor of 10 between 300 and 340 K.

*Lattice vibrations also increase electron-ion collisions in semiconductors, just as they do in metals. But in semiconductors the increase in conduction electrons (which decreases resistivity) is more important than the increase in electron-ion collisions (which increases resistivity).

FIGURE 21.19

Resistivity of a semiconductor plotted against absolute temperature T.

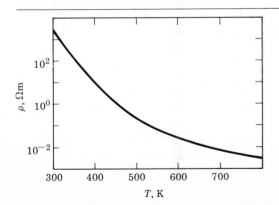

From now on, unless we indicate otherwise, the word "electron" will refer only to electrons in the conduction band of a semiconductor. The density of electrons η_n is the number of conduction electrons per cubic centimeter. (The subscript n is to remind us that electrons are negatively charged.) In a pure semiconductor at 300 K, η_n is about 10^{10} electrons/cm³, whereas in a metal the density is 10^{23} electrons/cm². But in a semiconductor η_n increases rapidly with temperature, whereas in a metal it is independent of temperature.

When an electron is excited to the conduction band of a semiconductor, a vacancy, or *hole,* is created in the valence band (Fig. 21.20a). A valence electron can now acquire a drift speed because there is an empty state for it to occupy. As the valence electron occupies the hole, a new hole opens in the state formerly occupied by the electron (Fig. 21.20b). It is convenient to think of the hole as a particle that moves through the valence band just as an electron moves through the conduction band. Since the motion of the hole is opposite that of a valence electron, it acts like a positively charged particle. Both holes and (conduction) electrons contribute to conduction in a semiconductor.

The density of holes is denoted η_p, where the subscript p reminds us that the holes act like positively charged particles. In a pure semiconductor, each electron in the conduction band has been excited from the valence band, so the number of electrons and holes is equal:

$$\eta_n = \eta_p \qquad \text{pure semiconductor}$$

For silicon at 300 K, these densities are 7×10^9 particles/cm³.

n- AND p-TYPE MATERIAL

To prepare silicon for use in electronic devices, the density of electrons and holes is modified by the addition of impurities. The process is called *doping.*

In *n-type material* a small number of atoms from group V of the periodic table (phosphorous, arsenic, antimony) is added to

FIGURE 21.20

(a) When an electron in the valence band of a semiconductor is excited into the conduction band, the resulting vacancy in the valence band is called a *hole.* (b) A hole can move in the valence band as an electron from one level is excited into the hole at another level.

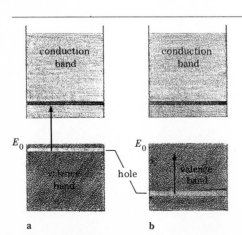

a   b

silicon. These impurity atoms have five outer electrons surrounding an ion with a charge of $+5e$, whereas the silicon
atoms have four outer electrons surrounding an ion with a
charge of $+4e$. These ions are shown in Fig. 21.21a. The four
outer electrons on each atom form covalent bonds with neighboring atoms. These electrons are represented by short, dark
lines attached to the ions, and their bonding to neighboring ions
is represented by long, gray lines. These bonded electrons are in
the valence band and are not free to move. However, the fifth
electron of the impurity atom is unbonded and goes into the
conduction band. This greatly increases the density of electrons, so

$$\eta_n \gg \eta_p \qquad n\text{-type material}$$

(The symbol $\gg$ means "very much greater than.")

In n-type silicon, the electron density is typically around
5×10^{17} electrons/cm^3, which is much larger than the electron

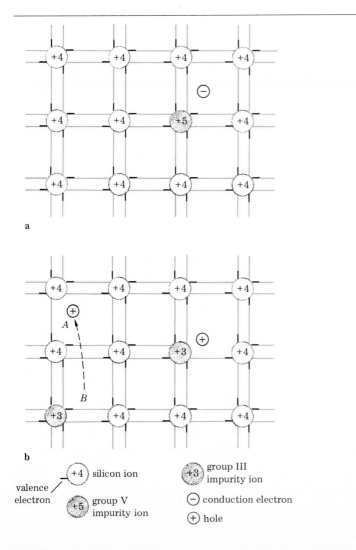

FIGURE 21.21

Doped semiconductor material. (a)
n-type material. Impurity atoms
with five valence electrons replace
some silicon atoms. These impurity
atoms have one unbonded electron, which is free to move. (b) p-
type material. Impurity atoms with
three valence electrons replace
some silicon atoms. The absence of
a fourth bonding electron creates a
hole in the valence band. The hole
moves like a positive charge, because as a valence electron at A
moves into a hole at B, the hole
appears to move from B to A.

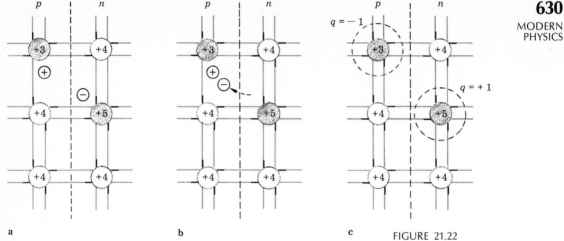

FIGURE 21.22

Electron diffusion across a pn junction. (a) The p region has an excess of holes, and the n region has an excess of electrons. However, initially both regions are electrically neutral. (b) An electron from the n region diffuses into the p region. (c) The electron fills a vacant bonding site in the p region, destroying a hole. As a result, the p region becomes negatively charged, and the n region becomes positively charged.

density in pure silicon, but much smaller than the electron density in a conductor (Example 21.1). Current is carried mainly by the negatively charged electrons in n-type material, with the positively charged holes playing a minor role.

In *p-type material,* a small number of atoms from group III of the periodic table (boron, aluminum, gallium) is added to silicon. These impurity atoms have three outer electrons surrounding an ion with a charge of $+3e$, as shown in Fig. 21.21b. The impurity atoms can form covalent bonds with only three neighboring silicon atoms. The absence of a fourth bonding electron creates a vacancy in the valence band, which is called a *hole.* The hole can move through the crystal because valence electrons can jump from one atom to another. Figure 21.21b shows a hole moving to A, as a result of the valence electron that was at A filling in the vacancy at point B. The addition of group III atoms to silicon greatly increases the density of holes, so

$$\eta_p \gg \eta_n \qquad \text{p-type material}$$

Current is carried mainly by the positively charged holes in p-type material, with the negatively charged electrons playing a minor role.

pn JUNCTION

A semiconductor device, such as a diode or a transistor, consists of adjoining regions of n- and p-type material. The boundary between these regions is called a *pn junction.* All semiconductor devices depend on the very special properties of a *pn* junction.

Figure 21.22 shows some atoms on either side of a *pn* junction. Because there are more electrons in the n region than in the p region, there is a tendency for electrons to diffuse from the n region to the p region (Fig. 21.22b). Once in the p region, the electron will soon find a vacant bonding site and fill it (Fig. 21.22c). This is called *recombination:* both the electron and the hole are destroyed, inasmuch as neither is any longer free to move.

The diffusion of the electrons across the junction leaves the n region positively charged and the *p* region negatively charged. Furthermore, when the electron recombines with a hole in the *p* region, this negative charge becomes fixed; i.e., it is no longer free to move. The result is that a double layer of charge builds up on either side of the junction, as shown in Fig. 21.23. This double charge layer acts as a *barrier* against further electron diffusion from *n* to *p*, since an electron that gets in between the two layers will be attracted back to the *n* region.

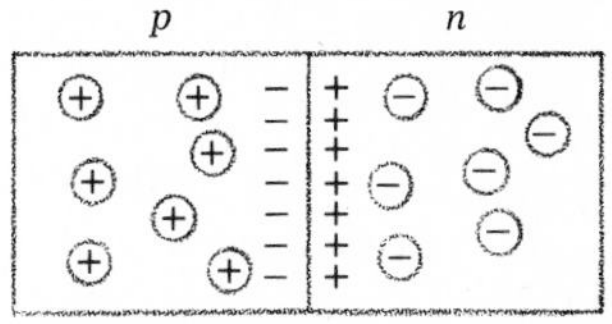

The holes in the *p* region tend to diffuse into the *n* region in the same manner. Figure 21.24*a* shows the diffusion of a hole across the junction by the moving of valence electrons in the opposite direction. The diffusion of the hole leaves the *p* region negatively charged and the *n* region positively charged, just as the diffusion of electrons does. Figure 21.24*b* shows that when a hole recombines with an electron in the *n* region, the positive charge becomes fixed. Thus the diffusion of holes adds to the double charge layer produced by electron diffusion. This layer also acts as a barrier to further hole diffusion from the *p* region to the *n* region, since a hole that gets in between the two layers will be attracted back to the *p* region.

In summary, at a *pn* junction there is a double layer of charge formed which acts as a barrier for the further diffusion of holes from the *p* region to the *n* region and of electrons from the *n* region to the *p* region.

In the next section we shall discuss some important devices that make use of these special properties of a *pn* junction.

21.4 Discrete Semiconductor Devices

Discrete semiconductor devices are individual semiconductor elements, like diodes and transistors, that must be wired together to form a complete circuit. The operating characteristics of diodes and transistors, as well as circuits made from these devices, are described in Secs. 19.3 and 19.4. In this section we shall explain how these devices work.

FIGURE 21.23

Double charge layer formed along the boundary of a *pn* junction. This charge prevents further diffusion of electrons into the *p* region and holes into the *n* region.

FIGURE 21.24

Hole diffusion across a *pn* junction. (a) A hole diffuses from the *p* region to the *n* region. (b) In the *n* region, the hole is filled by an electron. Thus, as with electron diffusion, the *p* region becomes negatively charged and the *n* region becomes positively charged.

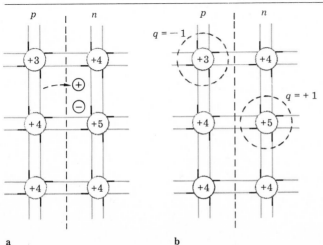

a   b

DIODE

A *diode* is a device that conducts a current in one direction but not the other. A semiconductor diode consists of an *n*-type region adjoining a *p*-type region on the same silicon crystal. Figure 21.25*a* shows the double charge layer that forms along the *pn* junction and prevents the diffusion of electrons and holes between the two regions.

When the *p* side of the diode is connected to the negative terminal of a battery and the *n* side is connected to the positive terminal, the diode is said to have *reverse bias* (Fig. 21.25*b*). In this configuration, the potential of the battery increases the amount of charge along the *pn* junction, which inhibits electron-hole diffusion even more. Since charge cannot flow across the junction, the diode does not conduct a current.

When the *p* side of the diode is connected to the positive terminal of a battery and the *n* side is connected to the negative terminal, the diode is said to have *forward bias* (Fig. 21.25*c*). In this configuration, the potential of the battery reduces the amount of charge along the *pn* junction, which reduces the barrier to electron-hole diffusion. Holes in the *p* region, repelled by the positive potential of the *p* region, flow toward the *n* region, while electrons in the *n* region, repelled by the negative potential of the *n* region, flow toward the *p* region. These two opposite flows of oppositely charged particles constitutes the current in the diode. The electrons and holes recombine as soon as they reach opposite regions, which tends to reestablish the charge barrier. However, with forward biasing, the battery removes this charge as fast as it is formed, so a steady current is maintained.

Figure 21.26 shows the circuit symbol for a diode. The arrow points in the direction in which the diode conducts a current. The magnitude of this current increases very rapidly as the forward voltage is increased, because the rate at which electrons and holes can diffuse across the *pn* junction increases rapidly as the barrier is reduced. This is shown in Fig. 21.27, which is a plot of the current I in a diode against the forward voltage V. (In Sec. 19.3 we sometimes approximated the sharp rise in the diode current by a vertical line, but a more realistic picture of the VI curve of a diode is needed to understand the transistor.)

BIPOLAR TRANSISTOR

The *bipolar transistor* is the basic transistor invented in 1948 by John Bardeen (1908–   ), Walter Brattain (1902–   ), and William Shockley (1910–   ). It is a semiconductor crystal with a narrow *n*-type region sandwiched between two *p*-type regions (a *pnp* transistor), or a narrow *p*-type region sandwiched between two *n*-type regions (an *npn* transistor). Both types of transistors work in essentially the same way, so we shall describe only the *pnp* transistor.

REMARK Bardeen, Brattain, and Shockley were awarded the Nobel Prize in 1956 for the invention of the transistor. Bardeen won the Nobel Prize again in 1972 (together with Leon Copper and John Schrieffer) for the development of the theory of superconductivity.

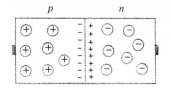

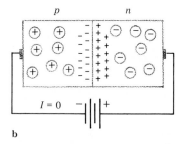

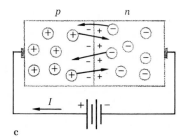

FIGURE 21.25

Semiconductor diode. (a) A *pn* junction with a double charge layer across the boundary. (b) Reverse bias. When the voltage of the *n* region is made positive relative to the *p* region, the charge layer increases, which further inhibits electron-hole diffusion across the junction. (c) Forward bias. When the *n* region is made negative relative to the *p* region, the charge layer is reduced, which allows electron-hole diffusion across the junction.

FIGURE 21.26

Circuit symbol for a diode. The arrow points in the direction in which the diode conducts.

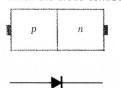

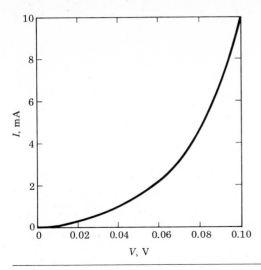

FIGURE 21.27

Plot of the current in a diode against the forward-bias voltage across the diode. The rapidly increasing slope of this curve indicates that the resistance of the diode is decreasing with increasing voltage.

Figure 21.28 shows the structure of a *pnp* transistor together with its circuit symbol. The three regions of the transistor are termed *emitter, base,* and *collector,* as shown in the figure. The base is lightly doped with impurity atoms, the collector is moderately doped, and the emitter is heavily doped. Wire leads are attached to each of the three regions so they can be connected to an external circuit.

It is useful to think of a transistor as two diodes connected back to back. Figure 21.29 shows the collector of a *pnp* transistor connected to the negative terminal of a battery, while the emitter is connected to the positive terminal. In this configuration, the emitter-base junction is forward biased, while the base-collector junction is reverse biased. Holes from the emitter readily diffuse into the base, while holes in the collector do not. The base is made narrow enough so that most of the holes from the emitter can diffuse across to the base-collector junction

FIGURE 21.28

pnp transistor: (a) structure and (b) circuit symbol.

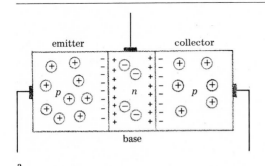

a

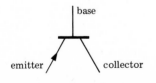

b

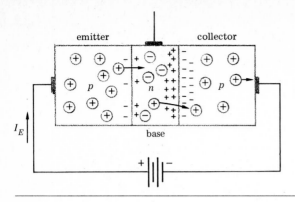

FIGURE 21.29

A *pnp* transistor with the voltage
on its collector made negative relative to the voltage on the emitter.
In this configuration, the emitter-
base junction is forward biased,
while the base-collector junction is
reverse biased.

without recombining. Once at the base-collector junction, these
holes are swept across it into the collector. This is because the
reverse bias on the base-collector junction, which prevents
holes from moving from the collector to the base, accelerates
holes moving in the opposite direction. This flow of holes from
emitter to collector constitutes the emitter current I_E.

Although most of the holes that diffuse from the emitter to
the base reach the collector, a few recombine first. This causes
positive charge to accumulate in the base, which inhibits the
flow of holes from emitter to base. The small percentage of
holes that recombine has a large influence on the flow of the
many holes that do not, because the recombining holes increase
the barrier to hole diffusion from the emitter to the base. With
only the emitter and collector of a transistor connected to a
battery (Fig. 21.29), the current is very small because of the
positive charge in the base.

The current can be greatly increased by drawing off some of
the charge on the base. This is done by connecting the base to a
second battery, as shown in Fig. 21.30. The base current I_B then
in effect controls the collector current I_C, because small changes
in I_B produce large changes in I_C. Figure 21.31 is a plot of I_C
against I_B for a fixed value of voltage between the emitter and
the collector.

Transistors are used to amplify an input signal. Thus suppose

FIGURE 21.30

A *pnp* transistor with its base connected to the negative terminal of
a battery. The collector current I_C is
strongly influenced by the base
current I_B.

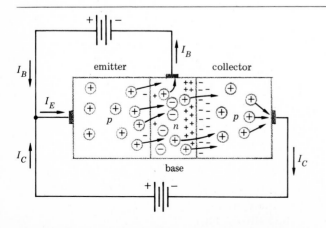

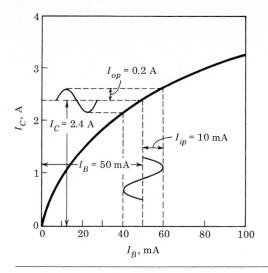

FIGURE 21.31

Plot of the collector current I_C against the base current I_B for a fixed value of the voltage between the emitter and collector.

that an alternating input current

$$I_i = I_{ip} \sin 2\pi ft$$

is added to the direct base current I_B, so that the total base current is

$$i_B = I_B + I_{ip} \sin 2\pi ft$$

The collector current i_C will then also consist of a direct component I_C and an alternating component $I_{op} \sin 2\pi ft$:

$$i_C = I_C + I_{op} \sin 2\pi ft$$

For example, if $I_B = 50$ mA and $I_{ip} = 10$ mA, Fig. 21.31 shows that $I_C = 2.4$ A and $I_{op} = 0.2$ A. This means that an input signal with a peak current of 10 mA produces an output signal with a peak current of 0.2 A. The *current gain* g_I is the ratio of the peak output current to the peak input current:

$$g_I = \frac{I_{op}}{I_{ip}}$$

In the present example, the current gain is

$$g_I = \frac{200 \text{ mA}}{10 \text{ mA}} = 20$$

The design of amplifying circuits using transistors is discussed in more detail in Sec. 19.4.

FIELD-EFFECT TRANSISTOR
The field-effect transistor (FET) controls the current in the main circuit by means of a gating voltage, in contrast with the bipolar transistor, which controls the main current by means of the base current. Field-effect transistors are important components of microcircuits.

An *enhancement-mode* FET, like a transistor, consists of two regions of one type of material separated by a region of the other type. Figure 21.32a shows a *pnp*-FET in which two *p* re-

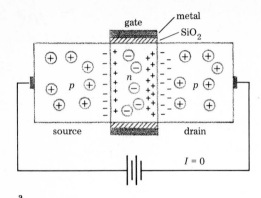

a

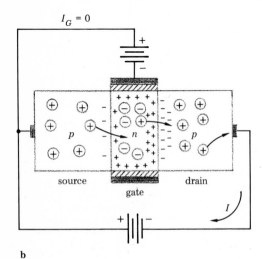

b

FIGURE 21.32

Enhancement-mode *pnp*-MOSFET.
(a) With no voltage on the gate,
the current *I* between the source
and the drain is zero. (b) When
the voltage on the gate is made
negative relative to the voltage on
the source, the source-drain current *I* is not zero.

gions are separated by an *n* region. The three regions are called
source, gate, and *drain,* as shown in the figure. In normal operation, the source is connected to the positive terminal of a battery
and the drain is connected to the negative terminal. As in a *pnp*
transistor, some of the holes that diffuse from the source to the
gate recombine before reaching the drain. This causes positive
charge to accummulate in the gate, which increases the barrier
at the source-gate junction and prevents further diffusion of
holes. An enhancement-mode FET does not conduct when there
is no voltage on the gate.

A voltage is applied to the gate through a layer of metal that
surrounds the gate. The metal is separated from the gate by a
thin insulating layer of silicon dioxide (SiO_2), so there is no
current between the gate and the metal layer. Because of the
material used in its construction, this is called a *metal-oxide-
semiconductor field-effect transistor,* or MOSFET.

When a negative potential is applied to the gate, some of the
positive charge along the source-gate junction is attracted to the
underside of the insulating layer (Fig. 21.32*b*). This decreases the
charge layer along the junction, so that holes can diffuse into
the gate and flow over to the drain. The more negative the

voltage on the gate, the larger the current. Thus an MOSFET acts like a resistor whose resistance can be controlled by a voltage.

An enhancement-mode FET made using p-type material for the gate and n-type material for the source and drain is called an *npn*-MOSFET. An *npn*-MOSFET conducts when a positive voltage is applied to the gate.

A *depletion-mode* FET is made entirely of one type of material. Figure 21.33a shows a depletion-mode *p*-MOSFET in which the three regions are made of *p*-type material. The gate is made narrower than the source and drain and is surrounded by an insulating layer of SiO_2 and a conducting layer of metal. In normal operation, the source is connected to the positive terminal of a battery and the drain is connected to the negative terminal. When no voltage is applied to the gate, the FET conducts, because holes can pass freely from the source to the drain through the gate. When a positive voltage is applied to the gate, holes are driven out of the gate into the source and drain (Fig. 21.33b). As a result, a hole-free region, called the *depletion region,* forms on the underside of the insulating layer. Since there is no mobile charge in the depletion layer, it cannot conduct a current, and the current is therefore restricted to the channel in the middle of the gate.

As the voltage on the gate increases, the depletion grows, and

FIGURE 21.33

Depletion-mode *p*-MOSFET.
(a) With no voltage on the gate, charge can flow from the source to the drain, so the current *I* is not zero. (b) When a positive voltage is applied to the gate, holes are driven out of the narrow gate channel, so it becomes nonconducting. The current *I* decreases to zero as the gate voltage increases.

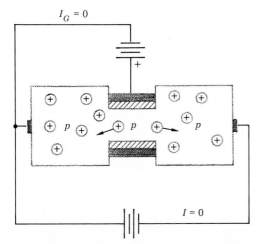

a

b

the conducting channel through the gate narrows. This increases the resistance between the source and the drain, and thus decreases the current. At a certain value of the gate voltage the depletion layer closes off the gate completely, and the current is reduced to zero. As with an enhancement-mode FET, a depletion-mode FET acts like a resistor whose resistance can be controlled by the gate voltage. In a depletion-mode FET the resistance increases with gate voltage, whereas in an enhancement-mode FET it decreases with gate voltage.

A depletion-mode FET made using n-type material is called an n-MOSFET. The resistance of an n-MOSFET increases when a negative voltage is applied to the gate.

Field-effect transistors are the basic components of logic gates in microelectronic circuits. The inputs to a logic gate are voltages that represent the binary digits 0 and 1. Zero voltage might represent the digit 0, and a positive voltage might represent the digit 1. The output is also either a zero or a positive voltage, representing 0 or 1. Section 19.3 showed how circuits that add and multiply can be made from three basic gates, called NOT, AND, and OR. In Sec. 19.3 AND and OR gates were constructed with diodes, but transistors are actually used in practice. Here we shall show how NOT, AND, and OR gates can be made from npn-MOSFETs, which conduct when a positive voltage is applied to their gates.

Figure 21.34 shows a NOT gate. All voltages are relative to the voltage at C, which can be taken to be zero. Terminals marked C are connected by resistanceless wires, but these are omitted for clarity. When a zero voltage is applied to the input terminal IN, the FET is nonconducting, so the voltage at the output terminal O is just the positive voltage of the battery. When a positive voltage is applied to IN, the FET conducts. In this case the voltage at O is zero, because the voltage of the battery is dropped across the resistor. The characteristic of a NOT gate is that it converts a zero voltage to a positive voltage and a positive voltage to a zero voltage. In terms of binary numbers, it converts a 0 to a 1 and a 1 to a 0.

An OR gate with two input voltages V_1 and V_2 is shown in Fig. 21.35a. When both V_1 and V_2 are zero, neither FET is conducting, so the voltage at A is the positive voltage of the battery. This positive voltage makes the third transistor conducting, so that the output voltage at O is zero. If either V_1 or V_2 is positive, at least one of the input transistors will be conducting, and there will be a voltage drop across R_1. In this case, the voltage at A will be zero, the third transistor will not conduct, and the output will be positive. An OR circuit gives a positive output if either or both of its inputs are positive; otherwise the output is zero. These characteristics are given in binary terms in Fig. 21.35b.

An AND gate with two inputs V_1 and V_2 is shown in Fig. 21.36a. By the same analysis we gave for the OR gate, you can show that the AND gate has the characteristics shown in Fig. 21.36b.

By coupling together many NOT, AND, and OR gates, circuits can be built to perform any arithmetic or logical operation.

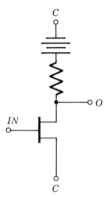

FIGURE 21.34

A NOT gate made with an npn-MOSFET that conducts only when a positive voltage is applied to the gate.

FIGURE 21.35

(a) An OR gate made with three npn-MOSFETs. (b) Input-output characteristics of an OR gate.

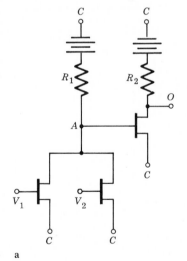

a

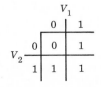

		V_1	
		0	1
	0	0	1
V_2			
	1	1	1

b

Examples of this are given in Sec. 19.3. In the next section we discuss the techniques that are used to fabricate these circuits.

21.5 Microelectronics

HISTORY OF COMPUTERS

The principle of the general-purpose automatic computer was invented by Charles Babbage (1797–1871) in 1835. Although never built, he developed plans for a machine that would use punched cards to input data and control the sequence of operations. Data cards would enter numbers into a memory unit consisting of thousands of counter wheels, and control cards would transfer these numbers to a mechanical calculator. His machine had many of the features of a modern computer, but its functions could not be reliably executed by mechanical means.

In the late 1930s and early 1940s, automatic calculating machines were successfully built using electromechanical relays as logic gates, but these were soon replaced by electronic computers. The first all-electronic computer was the ENIAC, built in 1946. This cumbersome machine weighed 30 tons, occupied 6000 cubic feet, and required 100 kW to operate its 18,000 vacuum tubes.

Each of the thousands of logic gates in the ENIAC used one or more vacuum tubes. Each vacuum tube is a sealed glass bulb several inches tall containing a heating filament to drive electrons from the cathode (Sec. 15.5). The heating filament uses several watts of power and is subject to burn out, which makes the vacuum tube a hot, unreliable device.

For these reasons, the invention of the transistor in 1948 was a boon to the computer industry. A transistor 0.5-cm-high is more reliable than a 6-cm-high vacuum tube, and it uses only one-thousandth as much power. As they became commercially available in the 1950s, transistors replaced vacuum tubes in computers and many other electronic instruments.

By 1960 a still newer method of constructing electronic circuits was developed, called *integrated-circuit technology*. Instead of manufacturing individual resistors, capacitors, and transistors and then wiring them together, this new technology fabricates a complete circuit, containing resistors, capacitors, transistors, and their interconnections, on a 0.5-cm chip of silicon. At first only simple circuits, like individual logic gates, were included on a chip, but by 1970 this low-scale integration was replaced by large-scale integration, in which the thousands of logic gates needed for a computer are placed on a single chip.

These remarkable developments have produced a revolution in electronics that promises to be as significant as the invention of printing. The rapid decrease in the cost and size of electronic circuitry since 1950 now makes it possible to use computer logic to control almost any process, from automobile engines to cooking. More important, perhaps, it will make it possible for everyone to own a personal computer, which may radically change the way human beings acquire and process information.

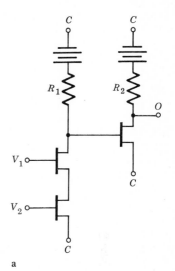

a

b

FIGURE 21.36

(a) An AND gate made with three *npn*-MOSFETs. (b) Input-output characteristics of an AND gate.

In this section we shall describe some of the techniques used in making an integrated-circuit chip and how a chip functions in a small calculator.

THE SILICON SUBSTRATE

The connecting leads, resistors, capacitors, and transistors in an integrated circuit are made by depositing thin films of various materials onto a silicon substrate. Silicon is universally used as the substrate on which an integrated circuit is built, because when it is heated to over 1000°C in an atmosphere of pure oxygen, a thin layer of SiO_2 forms on its surface. Since SiO_2 is an excellent insulator, this technique provides a simple and inexpensive method of insulating parts of the circuit.

Silicon is obtained from the reduction of silicon dioxide, which is the main ingredient of sand. The silicon is purified by a sequence of chemical processes until there is only one impurity atom for every 10^9 silicon atoms; i.e., "pure" silicon is 99.9999999 percent silicon. N-type or p-type silicon is made by adding the

FIGURE 21.37

Some steps in the manufacture of an integrated circuit chip.
(a) Growing a silicon crystal (*Siltec Corp.*). (b) Etching a wafer containing several hundred chips (*Electronics Research Laboratory at the University of California, Berkeley, and W. G. Oldham*). (c) Single chip mounted to its connecting pins (*Texas Instruments*).

appropriate impurity atoms (usually boron and phosphorous) to molten silicon. Doped silicon contains about one impurity atom for every 10^5 silicon atoms; i.e., doped silicon is still 99.999 percent silicon.

A large, single crystal of doped silicon is formed by inserting a small "seed" crystal into molten silicon and slowly turning and withdrawing it (Fig. 21.37a). As the seed is withdrawn, the molten silicon freezes around the seed, forming a single crystal 12 cm in diameter and up to 1 m long. The crystal is ground into a cylinder 10 cm in diameter and sliced with a diamond saw into 0.5-mm-thick wafers. The wafers are ground, polished, and chemically cleaned, until their surfaces are perfectly smooth and free from dust, scratches, and chemical contaminants.

The doped silicon wafer is the substrate on which the microelectronic chips are made. About 250 chips, each 5×5 mm square, can be formed on a single wafer (Fig. 21.37b), and many wafers can be processed at the same time. When the circuit-forming processes are completed, each chip is electronically tested, and the defective ones are marked. The individual chips are then cut out of the wafer; the defective ones are discarded, and the good ones are mounted on a metal fitting containing the connecting pins which connect the chip to its power supply, input and output terminals, and other auxiliary units (Fig. 21.37c). The final unit is sealed inside a plastic container.

THIN-FILM TECHNOLOGY

The various circuit components, such as resistors, capacitors, and transistors, are formed on the silicon substrate by the deposition of insulating, conducting, and semiconducting films. If, for example, the substrate were made of p-type silicon, a resistor would consist of a narrow strip of n-type silicon deposited on the substrate (Fig. 21.38). A chip is always operated with reverse bias on its substrate, so charge cannot flow between the substrate and the opposite type film above it. This electrically isolates the elements on the chip.

Example 21.2 What is the resistance R of a semiconducting strip 10 μm wide, 3 μm thick, and 25 μm long? The resistivity ρ of the semiconducting material is 0.1 $\Omega \cdot$ m.

Because the substrate is reverse biased, the current in the strip does not penetrate into the substrate. The cross-sectional area of

FIGURE 21.38

A resistor in an integrated circuit consists of a narrow strip of n-type material deposited on a substrate of p-type material.

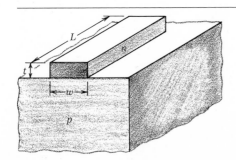

the strip is

$$A = wt = (10 \times 10^{-6}\,\text{m})(3 \times 10^{-6}\,\text{m})$$
$$= 30 \times 10^{-12}\,\text{m}^2$$

From Eq. 21.5, the resistance of the strip is

$$R = \frac{\rho L}{A} = \frac{(0.1\,\Omega \cdot \text{m})(25 \times 10^{-6}\,\text{m})}{(30 \times 10^{-12}\,\text{m}^2)}$$
$$= 83 \times 10^3\,\Omega \qquad \square$$

REMARK The cost of each circuit element on a chip depends primarily on the area it occupies. Resistors usually require more area than transistors, so they are more expensive. This is the reverse of discrete-component technology, where transistors are more expensive than resistors. As a consequence, microelectronic circuits are designed using more transistors and fewer resistors than their discrete-component counterparts.

Figure 21.39a shows an *npn* transistor built on a *p*-type substrate, and Fig. 21.39b shows an *n*-MOSFET on the same substrate. Note that the substrate is 50 times thicker than the films that are deposited on it. In thin-film technology, all the components are built on the upper surface of the substrate—nothing passes through it—so all connections must be made on top.

The fabrication of a microelectronic chip begins with the circuit design. Then, with the help of a computer (Fig. 21.40a), the two-dimensional pattern of each layer of material is drawn 500 times actual size (Fig. 21.40b). This pattern is photographi-

FIGURE 21.39

(a) An *npn* transistor deposited on a *p*-type substrate. (b) An *n*-MOSFET deposited on a *p*-type substrate.

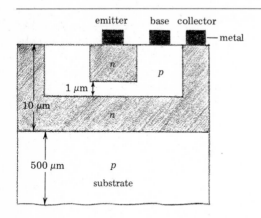

a

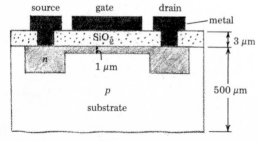

b

FIGURE 21.40

(a) Computer-assisted design of a microelectronic circuit (*Applicon, Inc.*). (b) Inspection of a complete circuit diagram (*Intel*).

cally reduced in several stages, until a lifesize image is obtained. This image is duplicated 250 times on a photographic plate the size of a wafer. About six such plates are required for a complete circuit, one plate for each layer of material that is deposited on the wafer.

The techniques for forming a patterned film of material on a wafer involve photolithography, etching, vacuum evaporation, and chemical-vapor deposition. For example, to form a layer of n-type material, the wafer is heated in an atmosphere of dilute silicane (SiH_4), which causes a uniform layer of polycrystalline silicone to form over the wafer. The wafer is next heated at a temperature of 1100°C in an atmosphere containing phosphorus, which causes some phosphorus atoms to diffuse into the silicon. Under these conditions a 1-μm-thick layer of p-type silicon forms in an hour (Fig. 21.41a). An insulating layer of SiO_2 is next placed over the doped layer by heating the wafer in an atmosphere of SiH_4 and H_2O.

The wafer is prepared for photolithography by coating it with a polymerizing photoresist chemical. This chemical has the property that it polymerizes; i.e., its molecules form covalent bonds with one another, when it is exposed to ultraviolet light. The photographic plate with the pattern of the desired p-type layer is laid over the wafer, and the wafer is exposed to ultraviolet light. The plate exposes the wafer where the p-type material is to remain and masks the wafer where it is to be removed (Fig. 21.41b). After exposure, the wafer is placed in a solvent which dissolves the unexposed photoresist.

The wafer is then etched in a bath of hydrofluoric acid (HF), which dissolves the SiO_2 but not the polymerized photoresist or the underlying silicon layer (Fig. 21.41c). This exposes the silicon layer, so next a nitric acid bath is used that dissolves the p-type silicon and the photoresist, but not the protective layer of SiO_2. Finally, the SiO_2 is removed with another hydrofluoric acid bath (Fig. 21.41d). The SiO_2 layer is necessary because the nitric acid required to etch the silicon also attacks the photoresist. However, since many wafers can be treated at once and each wafer contains hundres of chips, the process is very economical. A metal layer, which can be etched with a weaker acid solution, does not require the protective SiO_2 coating.

A succession of layers is built up on the wafer in this fashion. Since structures are less than 1-μm across, each layer must be positioned with extreme accuracy relative to the one below it. Figure 21.42 is a photomicrograph of a MOSFET gate. The gate is 0.7 μm wide and rests on a 0.07-μm-thick layer of SiO_2. Also, a single dust particle can ruin a chip, so the air in a processing facility must be electrostatically cleaned (Fig. 21.43). Although this reduces the dust to less than 1 percent of the dust found in a normally clean environment, more than half the chips on a wafer will ultimately be found to be defective.

MEMORY UNITS

In addition to the circuitry required to perform arithmetic calculations, a computer needs units for storing data and instructions. Both data and instructions are stored as binary numbers, so every memory unit consists of cells which can be in one of two states, corresponding to the binary digits 0 and 1. Microelectronic memories have capacitor-FET cells. The capacitor can be either charged or uncharged, corresponding to 1 or 0. A 5 × 5 mm chip containing 16,384 capacitors and FETs is capable of storing 16,384 bits of information.

The capacitors and FETs are arranged in a square array (Fig. 21.44). One side of each capacitor is connected to a common ground line (C), and the other side is connected to the source of an FET. The gates of all the FETs in a row are connected to the same *selection line,* and the drains of all the FETs in a column are connected to the same *data line.* To read the unit in a given row and column (say, row 2, column 3), a positive voltage is applied to the selection line of the row (2 in this case), which turns on all the FETs in that row. That is, all the capacitors in row 2 are connected to their data line. The voltage on data line 3 then is the voltage of the capacitor in row 2 column 3. If the capacitor is charged, the voltage will be positive; if the capacitor is uncharged, the voltage will be zero. To add data, a positive voltage is applied to the selection line, and the data line is either grounded or charged. Such a device is called a read/write random access memory (RAM), because each unit can be located with equal speed.

The charge on the capacitors in this type of memory tends to leak off rapidly, so additional circuitry is required to keep them charged. Pocket calculators use this type of memory for their key-controlled memory locations, but they also have a permanent read-only memory (ROM) that contains all their calculation instructions. This is similar to the read/write memory, except each capacitor is replaced by either an open circuit or a connection to ground, depending on whether the cell is to represent 1 or 0 (Fig. 21.45). The data lines are all connected through a resistor to a positive voltage source, so when a particular cell is read, the voltage at the bottom of the data line will be positive if the cell has an open circuit and zero if the cell has a connection to ground. Since these connections are part of the

FIGURE 21.41

Thin-film technology. (a) A 1-μm-thick layer of p-type silicon is formed by heating the water in an atmosphere containing phosphorus. (b) A protective layer of SiO_2 is deposited over the p layer. This is covered with a layer of photoresist, which is exposed to ultraviolet light through a mask of the desired pattern of the p layer. (c) The unexposed photoresist and its underlying layer of SiO_2 are disolved in solvents. (d) The exposed p-type silicon and the photoresist are dissolved in nitric acid, and the remaining SiO_2 is dissolved in hydrofluoric acid.

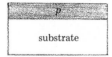

a

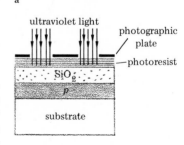

b

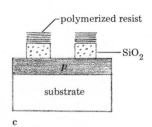

c

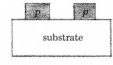

d

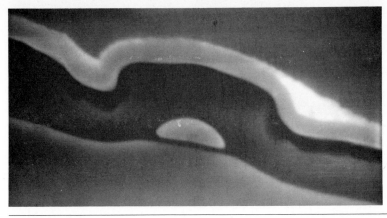

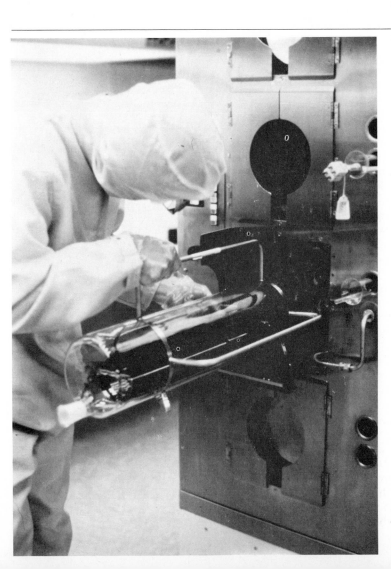

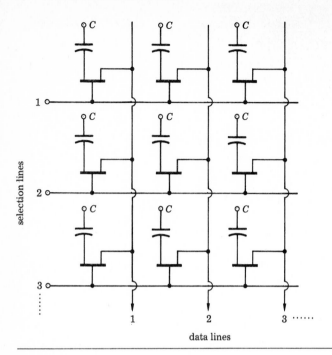

selection lines

1

2

3

1   2   3 ......

data lines

FIGURE 21.44

Random-access memory array. Each capacitor-FET unit in the array can store one bit.

FIGURE 21.45

Read-only memory. The capacitor is replaced by an open-circuit or short-circuit to ground.

circuit design of the unit, all the calculator's instructions are built into the unit during the fabrication process.

ELECTRONIC CALCULATOR

The small, pocket-sized electronic calculator was one of the first microelectronic products developed for the consumer market. In the simplest models, all the logic circuits and memory units are contained on a single chip (Fig. 21.46). In more complex models, two chips may be used. In addition to the chips, a calculator has a keyboard for inputing data and instructions, a display, an oscillator, and a power supply.

The *keyboard* is just a set of switches which are closed whenever a key is depressed. The simplest board has 20 or more switches.

The *display* consists of 11 segmented arrays, like the one in Fig. 21.47. Each array has 8 segments, 7 for forming the digits and 1 for the decimal point. The segments are either light-emitting diodes (LEDs) or liquid-crystal displays (LCD). A *light-emitting diode* is a special diode which emits a colored light (often red) when a voltage is placed across its two input leads. A *liquid-crystal display* consists of a layer of liquid crystal sandwiched between a pair of transparent electrodes. The liquid crystal has the property that it becomes opaque whenever a voltage is placed across it. Thus, when a voltage is placed across the electrodes of a segment, the liquid crystal under the segment turns dark. Since an LCD uses reflected light rather than generating its own light, it consumes much less power than an LED.

Each segment of each digit of the display is separately controlled by the chip. In an 11-digit display, this requires $11 \times 8 = 88$ output leads. Each key on the board also requires a

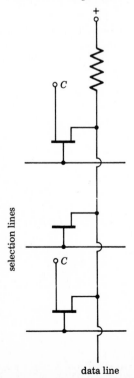

selection lines

data line

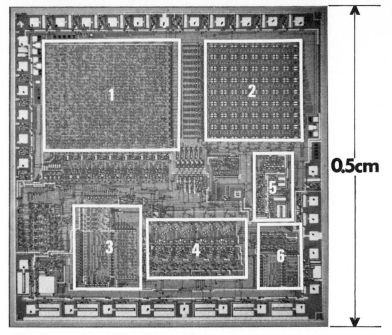

FIGURE 21.46

Enlarged view of a complete MOS microprocessor showing (1) read-only memory, (2) random-access memory, (3) control decode, (4) arithmetic logic unit, (5) clock, (6) input-output decode. (*Texas Instruments.*)

lead to the chip, so a 20-key, 11-digit calculator needs 108 leads in all. But a typical chip has only 28 pins by which it can be connected to the outside world.

The problem of connecting 108 leads to 28 pins is solved by time-sequencing the connections. An oscillator in the calculator generates a clock signal which synchronizes the various operations of the calculator. Eight pins of the chip are used for the display. Each pin connects to the same segment on all 11 digits. The clock scans each display digit in turn. For 132 μs the eight pins control the segments of the first digit; the display turns off for 24 μs while the scan switches to the second digit; then for the next 132 μs the eight pins control the second digit. In a period of 11 × (132 + 24) = 1716 μs, each digit has been displayed once, for a period of 132 μs. Thus, although the display appears constant to the eye, each digit is flickering on and off 583 times a second.

Four pins connect the 20 or more keys to the chip. One pin is connected to all 10 digits (0 through 9); the other pins are connected to groups of instruction keys. The clock scans each digit and instruction key at the same rate as it scans the display. This means that a certain key, say, the 3, is connected to the chip for only 132 μs out of every 1716 μs. When the 3 is depressed, it must wait until the scan cycle reaches it before it can send a signal to the chip. Since the cycle is only 1.7 ms, the response appears to be instantaneous.

A small calculator uses a mixed numerical system called *binary coded decimal* (BCD). Each decimal digit is stored in the chip as a 4-digit binary number; thus 3 is 0011 and 9 is 1001. A decimal number like 39 is stored as 0011 1001, rather than in its pure binary form (100111). When operations are performed on

FIGURE 21.47

Eight-segment display. The numerals 0 through 9 are displayed by lighting different segments in the display.

two decimal numbers, binary arithemtic is used for the individual digits, and then special circuitry is used for carryover. This system simplifies the conversion between decimal and binary at the expense of slower arithmetic routines. This is efficient for a calculator, because much of the time is spent on input/output operations. In a full-size computer, however, where most of the time is spent on internal computations, the more efficient binary arithmetic is used.

The incredible complexity and sophistication of the microelectronic chip reflects the ingenuity and talent of thousands of technicians, engineers, and scientists. Perhaps never have so many concepts, ideas, and inventions been focused on such a small piece of material.

GUIDE TO MAJOR TOPICS

Topic	References	Problems
Lattice structure	Sec. 21.1	
Bonding in crystals	Sec. 21.1	
Metals	Sec. 21.2	
Free-electron model	Eqs. 21.1 and 21.2	1 and 2
Conduction	Eqs. 21.3 and 21.4; Example 21.1	3 and 4
Resistivity	Eqs. 21.5 to 21.9	5 and 6
Superconductivity	Table 21.1	
Semiconductors	Sec. 21.3	
Band structure		
Electrons and holes		
n- and p-type material		
pn junction		
Discrete semiconductor devices	Sec. 21.4	
Diode		
Bipolar transistor		
Field-effect transistor		7 to 10
Microelectronics	Sec. 21.5	
History		
Silicon substrate		
Thin-film technology	Example 21.2	11 and 12
Memory units		13 and 14
Electronic calculator		

PROBLEMS

1 The Fermi energy in copper is 7.0 eV. (a) What is the speed of an electron that occupies the Fermi level? (*Hint:* Uses $K = \frac{1}{2}mv^2$ after converting from electronvolts to joules.) (b) What is the wavelength of this electron? *Ans.* (a) 1.57×10^6 m/s; (b) 4.6×10^{-10} m

2 A one-dimensional crystal of length $L = 10$ cm contains 3×10^8 valence electrons. What is the Fermi energy of this crystal? *Ans.* 3.4 eV

3 Calculate the electron density of aluminum. Aluminum has 3 valence electrons.

Ans. 1.8×10^{23} electrons/cm^3 or 1.8×10^{29} electrons/m^3

4 An aluminum wire with a diameter of 2 cm carries a current of 750 A. What is the drift velocity? (Use the result of Prob. 3.)

5 (a) Use Eq. 21.9, Table 16.1, and the result of Prob. 3 to calculate the parameter c for aluminum at 20°C. (b) It can be shown that $c = mf$, where m is the mass of the electron, and f is the collision frequency, i.e., the number of times per second an electron collides with an ion. What is f for aluminum at 20°C?

Ans. (a) 1.29×10^{-16} kg/s; (b) 1.4×10^{14} collisions/s

6 The speed of an electron in the Fermi level of aluminum is 2×10^6 m/s. How far does this electron travel between collisions? (Use the result of Prob. 5.)

> **REMARK** In aluminum the distance between ions is 4×10^{-10} m, which is $\frac{1}{35}$ of the distance between collisions found in Prob. 6. Thus an electron passes 35 ions before colliding with one.

7 Figure 21.48 shows a gate with two input voltages V_1 and V_2 and four *npn*-MOSFETs. All terminals marked $+$ are at the same positive voltage, and all terminals marked C are at zero voltage. Find the input-output characteristics of this gate.

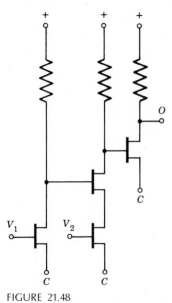

FIGURE 21.48

Problem 7.

8 Figure 21.49 shows a gate with three input voltages V_1, V_2, and V_3 and four *npn*-MOSFETs. Find the input-output characteristics of this gate.

9 Show that the gate in Fig. 21.50 has the input-output characteristics of the ADD gate in Fig. 19.23.

10 Using four *npn*-MOSFETs, design a gate with two input voltages V_1 and V_2 that gives an output of 1 only when $V_1 = 0$ and $V_2 = 1$.

11 A microelectronic capacitor consists of two conducting layers, each with an area of 200 $(\mu m)^2$, separated by a 0.2-μm-thick layer of SiO_2. Use Eq. 16.24 and Table 16.2 to find the capacity C of this capacitor.

Ans. 38×10^{-15} F, or 38 fF

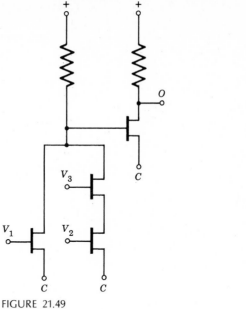

FIGURE 21.49

Problem 8.

FIGURE 21.50

Problem 9.

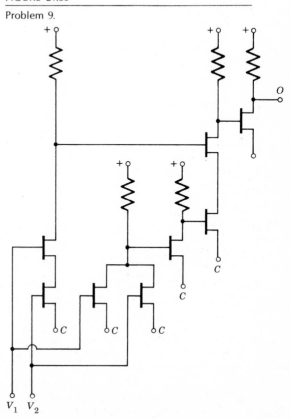

12 In a microelectronic chip, a conducting layer of aluminum is 0.5 μm thick. How many layers of aluminum ions are in the conducting layer? The distance between aluminum ions is 4×10^{-10} m.

13 The capacity C of the capacitor in a memory cell is 50 fF. (a) What is the charge on the capacity when the voltage across it is 8 V? (b) How many excess electrons are on the negative plate of the capacitor?
Ans. (a) 400 fC; (b) 2.5×10^6

14 The time constant τ of a circuit is a measure of the time required to move charge on or off the plates of a capacitor. It determines the speed with which a memory cell can be charged. Find the time constant of a circuit consisting of the capacitor in Prob. 13 connected in series with a 30-kΩ resistor.

BIBLIOGRAPHY

FITZGERALD, A. E., DAVID E. HIGGIN-BOTHAM, and ARVIN GRABEL: *Basic Electrical Engineering,* 4th ed., McGraw-Hill Book Company, New York, 1975. The basic principles of semiconductor devices and integrated circuits are discussed in Chap. 7.

GARDNER, MARTIN: "Mathematical Games: On Tessellating the Plane with Convex Polygon Tiles," *Scientific American* **233,** 112 (July 1975). Fascinating account of the kinds of convex polygons that can cover the plane. The complex patterns shown in this article are the superposition of many simple lattices.

KITTEL, CHARLES, *Introduction to Solid State Physics,* 4th ed., John Wiley & Sons, Inc., New York, 1971. This advanced textbook has a very readable account of crystal lattices (Chap. 1). Also, the discussions of conductivity in metals, semiconductors, and superconductors give deep insight into the nature of these phenomena.

McWHORTER, EUGENE W.: "The Small Electronic Calculator," *Scientific American* **234,** 88 (March 1976). Description of the time-sequencing of operations in a one-chip calculator.

Scientific American, September 1977. Entire issue devoted to articles on microelectronics. Of special interest are the articles by James D. Meindl ("Microelectronic Circuit Elements"), William C. Holton ("The Large-Scale Integration of Microelectronic Circuits"), and William G. Oldham ("The Fabrication of Microelectronic Circuits").

T he *nucleus* is the tiny, massive, postively charged core of an atom. It has a complex internal structure of its own, but this structure has almost no effect on the electron orbits because the nucleus is so much smaller than the orbits. For this reason, all the properties of the atom, including the chemical properties of the elements, are largely independent of the internal structure of the nucleus. This is fortunate, since otherwise nuclear physics would have had to be developed before atomic physics.

Much was known empirically about the nucleus before 1930, but it was only after the discovery of the neutron by James Chadwick (1891–1974) in 1932 that a systematic understanding of the internal structure of the nucleus could be obtained. While atomic physics has provided a deeper understanding of the known properties of atoms and molecules, nuclear physics has discovered new and unexpected properties of matter. These discoveries have had the most profound consequences, resulting in major advances in medical diagnosis, radiation therapy, and power generation, as well as in military weapons of great destructive capabilities.

22
NUCLEAR PHYSICS

22.1   The Structure of the Nucleus

A nucleus is composed of two kinds of particles, *protons* and *neutrons,* bound together by nuclear forces. The proton is an elementary particle (Sec. 15.1) with charge $+e$ and a mass 1836 times the mass of an electron. The neutron is an elementary particle with no charge and a mass 1839 times the mass of an electron.

Definition   The *atomic number Z* of an element is equal to the number of protons in a nucleus of that element. For instance, a carbon nucleus ($Z = 6$) contains six protons, and a uranium nucleus ($Z = 92$) contains 92 protons.

Definition   The *mass number A* of a nucleus is the sum of the proton number Z and the neutron number N:

$$A = Z + N$$

The mass number is the total number of *nucleons* (neutrons and protons) in the nucleus.

All nuclei with the same value of Z are denoted by the chemical symbol of the corresponding element. All nuclei with the same values of Z and A constitute a particular nuclear species, or *nuclide.* A particular nuclide is denoted by affixing the mass number A as a superscript to the chemical symbol. For example, the nuclide composed of six protons and six neutrons is written ^{12}C, the nuclide composed of six protons and eight neutrons is written ^{14}C, and the nuclide composed of seven protons and seven neutrons is written ^{14}N. This notation completely specifies the nuclide if the atomic numbers and symbols of the elements are known. To avoid confusion, however, the atomic number Z

is sometimes affixed as a subscript to the symbol, for example, $^{12}_{6}C$, $^{14}_{6}C$, $^{14}_{7}N$.

Nucleons are bound together in a nucleus by a fundamental force (Sec. 15.1) that differs from either the electric or gravitational force. At small distances the nuclear force is much stronger than the electric force, but it decreases rapidly as the distance between two nucleons increases. The nuclear force is best illustrated by plotting the nuclear potential energy U_n between two nucleons against the distance r between them. Figure 22.1 shows the neutron-proton (n-p) and the neutron-neutron (n-n) potentials. Both potentials rapidly approach zero as r increases, and they have minima near $r = 1.0 \times 10^{-15}$ m. The negative potential energy indicates an attractive force between the nucleons because positive work must be done to separate them. The n-p force is more attractive than the n-n because its potential is more negative. The sharp rise in the potential energy at distances less than 1.0×10^{-15} m indicates a strong repulsive force at small distances. Nucleons behave as though they had a hard core preventing them from getting closer than 0.5×10^{-15} m.

The nuclear force between two protons is the same as the n-n force, but in addition there is the electrical repulsion between the protons. Figure 22.2 shows the nuclear and electric potential energies between two protons and the combined potential energy.

Two factors govern the ratio of neutrons to protons in a nucleus: (1) the n-p force, being the most attractive, favors an equal number of neutrons and protons, and (2) the repulsive electric force between the protons, which increases as the num-

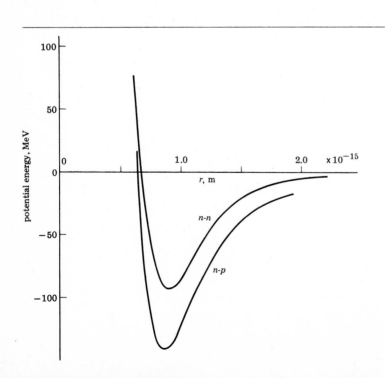

FIGURE 22.1

Nuclear potential energy U_n between two neutrons (n-n) and between a neutron and a proton (n-p), plotted against the distance r between them.

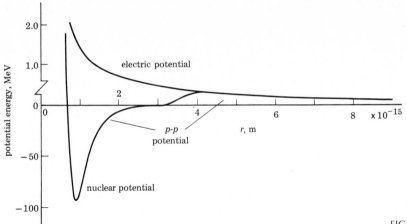

FIGURE 22.2

Potential energy between two protons plotted against the distance r between them. The potential energy is the sum of the negative nuclear potential energy and the positive electric potential energy. Note that the positive vertical scale is different from the negative vertical scale.

ber of protons increases, favors an excess of neutrons over protons. In light nuclides ($A \leq 40$), the first factor dominates, and stable nuclides have approximately equal numbers of protons and neutrons. In heavier nuclides, however, the second factor dominates, and N exceeds Z. This is shown in Fig. 22.3, which is a plot of N against Z for the stable nuclides. The $Z = N$ line is shown for comparison. Since no stable nuclide exists with $Z > 84$, the points with $Z \geq 85$ represent the longer-lived radioactive nuclides (Sec. 22.2).

Definition Nuclides with the same Z but different A are called *isotopes*. Isotopes form atoms with identical chemical properties but very different nuclear properties. There are many isotopes for every Z, only a few of which are stable and naturally occurring. The others are radioactive and must usually be produced artificially by a nuclear reaction. Table 22.1 is a selected list of some important isotopes. A complete table of the properties of all the known isotopes can be found in the *Handbook of Chemistry and Physics*.

The mass of an atom is measured in atomic mass units (u). An *atomic mass unit* is defined as one-twelfth the mass of a ^{12}C atom:

$$1\,u = \tfrac{1}{12}m_{12_C} \quad \text{or} \quad m_{12_C} = 12.00000\,u$$

The masses of some common nuclides are given in Table 22.1. These are the masses of the entire atom, including the electrons. For instance, the mass of ^1H is the combined mass of a proton and an electron. Table 22.1 also gives the proton, neutron, and electron masses in atomic mass units.

The mass of an atom is approximately equal to the mass of Z hydrogen atoms and N neutrons. Since the masses of ^1H and n are both approximately 1.0 u, all atomic masses are approximately equal to the mass number A. For example, the mass of eight ^1H atoms and eight neutrons is

$$(8)(1.007825\,u) + (8)(1.008665\,u) = 16.1319\,u$$

which is 0.137 u more than the mass of ^{16}O given in Table 22.1.

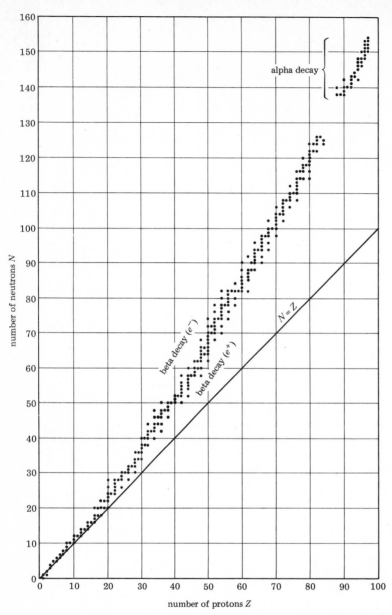

number of neutrons N

number of protons Z

alpha decay

beta decay (e^-)

beta decay (e^+)

$N = Z$

FIGURE 22.3

Plot of the number of neutrons N against the number of protons Z in the stable nuclides. The points with $Z \geq 85$ represent long-lived radioactive nuclides. The line $N = Z$ is shown for comparison.

In fact, the mass of every atom is slightly less than the sum of the masses of its constituent parts.

Definition The *mass defect* Δm of an atom is the difference between the masses of its parts and the mass of the atom. For instance, the mass defect of ^{16}O is

$$\Delta^{16}O = 16.1319\,u - 15.9949\,u = 0.1370\,u$$

Example 22.1 What is the mass defect of ^{32}S?
From the table of the elements inside the back cover we find

TABLE **22.1**

Z	A	Symbol	Atomic mass	Percent abundance	Half-life	Decay mode
		e	0.0005486			
0	1	n	1.008665		10.8 min	e^-
1	1	p	1.007277			
1	1	^1_1H	1.007825	99.985		
1	2	^2_1H	2.01402	0.015		
1	3	^3_1H	3.01605	†	12.26 y	e^-
2	4	^4_2He	4.00260	100		
6	12	$^{12}_6C$	12.00000	98.89		
6	13	$^{13}_6C$	13.00335	1.11		
6	14	$^{14}_6C$	14.00324	†	5730 y	e^-
7	13	$^{13}_7N$	13.00574		10 min	e^+
7	14	$^{14}_7N$	14.00307	99.63		
7	15	$^{15}_7N$	15.00011	0.37		
8	16	$^{16}_8O$	15.99491	99.759		
8	17	$^{17}_8O$	16.99913	0.037		
8	18	$^{18}_8O$	17.99916	0.204		
15	31	$^{31}_{15}P$	30.97376	100		
15	32	$^{32}_{15}P$	31.97391		14.3 d	e^-
16	32	$^{32}_{16}S$	31.97207	95.0		
19	39	$^{39}_{19}K$	38.96371	93.70		
19	40	$^{40}_{19}K$	39.97400	0.00118	1.28×10^9 y	e^-, e^+
19	41	$^{41}_{19}K$	40.96183	6.88		
27	59	$^{59}_{27}Co$	58.93319	100		
27	60	$^{60}_{27}Co$	59.93381		5.26 y	e^-
38	88	$^{88}_{38}Sr$	87.90564	82.56		
38	90	$^{90}_{38}Sr$	89.90775		28.1 y	e^-
43	99	$^{99}_{43}Tc^m$	98.90625		6.0 h	γ
43	99	$^{99}_{43}Tc$	98.90625		2.12×10^5 y	e^-
83	209	$^{209}_{83}Bi$	208.98039	100		
83	214	$^{214}_{83}Bi$	213.99869		19.7 min	e^-, α
92	235	$^{235}_{92}U$	235.04392	0.7	7.1×10^8 y	α
92	238	$^{238}_{92}U$	238.05077	99.3	4.51×10^9 y	α
92	239	$^{239}_{92}U$	239.05430		23.5 min	e^-
93	239	$^{239}_{93}Np$	239.05292		2.35 d	e^-
94	239	$^{239}_{94}Pu$	239.05215		2.44×10^4 y	α

Properties of selected nuclides

Besides the atomic number, mass number, and atomic mass of selected nuclides, the table gives the percentage of abundance of the naturally occurring nuclides and the half-life and decay mode of the radioactive nuclides.

†Occurs naturally in trace amounts.

that the atomic number of sulfur (S) is $Z = 16$. Consequently, the neutron number of ^{32}S is

$$N = A - Z = 32 - 16 = 16$$

The mass of the separated nucleons is

$$Zm_{1_H} + Nm_n = (16)(1.007825 \text{ u}) + (16)(1.008665 \text{ u})$$
$$= 32.26384 \text{ u}$$

From Table 22.1 the mass of ^{32}S is 31.97207 u, so the mass defect is

$$\Delta^{32}S = 32.26384 \text{ u} - 31.97207 \text{ u} = 0.292 \text{ u} \qquad \square$$

Definition The *binding energy* ΔE of a bound system is the energy of the separated parts of the system minus the energy of the bound system. It is equal to the work required to separate the system into its parts. For instance, the energy of a hydrogen atom is 13.6 eV less than the energy of a separated proton and electron pair, which means it requires 13.6 eV to ionize hydrogen. Similarly, an oxygen nucleus has 127.6 MeV (1 MeV = 10^6 eV) less energy than its separated parts, which means it requires 127.6 MeV to break an ^{16}O nucleus into 16 separated nucleons. The scale of nuclear binding is a million times greater than that of atomic binding, but the principle is the same.

According to the theory of relativity, mass is a form of energy. Consequently, a bound system, which has less energy than its separated parts, has less mass as well. Thus the mass defect is just a manifestation of the lower energy of the bound system. Einstein showed that the mass defect Δm is related to the binding energy ΔE by

$$\Delta m = \frac{\Delta E}{c^2} \qquad\qquad 22.1$$

where c is the speed of light. To use this equation, m, E, and c must be in SI units. In nuclear physics, however, mass is usually given in atomic mass units and energy is given in million electron volts (MeV). The conversions between these units and the SI units are

$$1\,u = 1.66053 \times 10^{-27}\,kg$$
$$1\,MeV = 1.60219 \times 10^{-13}\,J$$

So if Δm and ΔE are expressed in atomic mass units and MeV, respectively, Eq. 22.1 becomes

$$\Delta m\,(1.66053 \times 10^{-27}\,kg/u) = \frac{\Delta E\,(1.60219 \times 10^{-13}\,J/MeV)}{(2.99792 \times 10^8\,m/s)^2}$$

or

$$\Delta m = (1.07356 \times 10^{-3}\,u/MeV)\,\Delta E$$

$$\Delta E = (931.48\,MeV/u)\,\Delta m \qquad\qquad 22.2$$

Example 22.2 (a) What is the binding energy of ^{32}S? (b) What is the average binding energy per nucleon in ^{32}S?

(a) From Eq. 22.2 and the mass defect found in Example 22.1, the binding energy is

$$\Delta E = (931.48\,MeV/u)\,\Delta m = (931.48\,MeV/u)(0.292\,u)$$
$$= 272.0\,MeV$$

(b) The average binding energy per nucleon in ^{32}S is

$$\frac{\Delta E}{A} = \frac{272.0\,MeV}{32} = 8.5\,MeV$$

which is typical of most nuclei. This is the average energy required to remove one nucleon from the nucleus. ☐

Very accurate measurements of the masses of the nuclides have been made with mass spectrometers (Sec. 17.3). From these data the mass defect Δm and the average binding energy per nucleon $\Delta E/A$ can be determined. The results of many such measurements are summarized in Fig. 22.4, which is a plot of $\Delta E/A$ against A. This shows that although $\Delta E/A$ is approximately 8 MeV for almost all nuclei, it increases gradually with A to a maximum of 8.8 MeV around $A = 58$ and then decreases gradually to 7.6 MeV at $A = 238$. This behavior is of critical importance for the production of nuclear energy (Sec. 22.4).

REMARK   Because the energy of a hydrogen atom is 13.6 eV less than the energy of a separated proton and electron, the mass of a hydrogen atom is less than the mass of a proton and an electron. According to Eq. 22.2, this mass difference is

$$\Delta m = (1.073 \times 10^{-3} \text{ u/MeV})(13.6 \times 10^{-6} \text{ MeV}) = 1.46 \times 10^{-9} \text{ u}$$

which is too small to be measured. In general, the energy changes in chemical reactions, which are of the order of electronvolts, are too small to result in measurable mass changes.

22.2   Radioactivity

Only a few hundred of the thousands of known nuclides are stable. An unstable, or *radioactive*, nuclide is one that spontaneously transforms itself (decays) into another nuclide. If the daughter nuclide is also radioactive, it decays further until a stable (nonradioactive) nuclide is formed. Although the decay process itself is instantaneous, a radioactive nucleus can survive hours, days or years before suddenly decaying. Radioactive nuclides with sufficiently long lifetimes have found important applications in technology, medicine, and biological research.

MODES OF DECAY

The two principal modes of decay are called *alpha* and *beta decay*. Each radioactive nuclide has a characteristic decay mode,

FIGURE 22.4

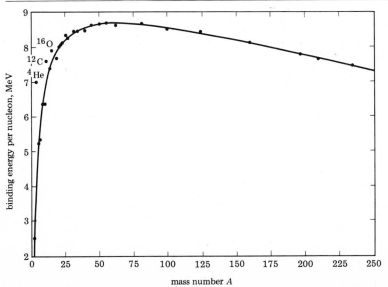

The binding energy per nucleon $\Delta E/A$ plotted against the mass number A of the stable nuclides.

as listed in Table 22.1, although a few nuclides, such as ^{214}Bi, decay through either mode. Alpha decay is confined primarily to isotopes of the heavier elements ($Z \geq 78$), whereas beta decay occurs in isotopes of all elements. As a rule, alpha decay occurs for nuclides near the line marked alpha decay in Fig. 22.3, and beta decay occurs for nuclides above and below the line of stable nuclides.

Beta Decay In beta decay, a neutron inside a nucleus spontaneously converts into a proton by emitting an electron e$^-$ and a chargeless, massless particle called a *neutrino ν*. For instance, the decay of ^{14}C is

$$^{14}_{6}\text{C} \longrightarrow {}^{14}_{7}\text{N} + e^- + \nu$$

Electrons produced this way are called *beta particles*. Notice that the total charge ($+6e$) and the total number of nucleons (14) are unchanged in this reaction. It is a general result, true for any nuclear reaction, that *the total charge and the total nucleon number of a system never change*. This is a statement of the laws of *conservation of charge* (Sec. 15.2) and *conservation of nucleon number*. These laws have been found to hold without exception in all physical processes.

From Table 22.1 it is seen that the mass of ^{14}C is greater than the mass of ^{14}N by 0.00017 u. In the decay process this excess mass is converted into 0.156 MeV of energy, which appears as the kinetic energy of the electron and neutrino.* In any particular decay, the electron has a kinetic energy between 0 and 0.156 MeV, and the neutrino has the remainder. The neutrino seldom interacts with other matter once it is produced, so that it has no effect on people or property. It is important in astronomy, however, because it is one of the ways a star loses energy.

Beta decay is the process by which a nuclide that lies above the stability line, i.e., a nuclide with an excess of neutrons, is converted into a stable nuclide. A nuclide that lies below the stability line, i.e., a nuclide with an excess of protons, undergoes an analogous decay process in which a proton converts into a neutron by emitting a positron e$^+$ and a neutrino. A *positron* is a particle identical to an electron in all respects except that it is positively charged. For example, the decay of ^{13}N is

$$^{13}_{7}\text{N} \longrightarrow {}^{13}_{6}\text{C} + e^+ + \nu$$

Note again that charge and nucleon number are conserved.

Alpha Decay In alpha decay, a ^{4}He nucleus is spontaneously emitted by a heavy nucleus. For example, the decay of ^{238}U is

$$^{238}_{92}\text{U} \longrightarrow {}^{234}_{90}\text{Th} + {}^{4}_{2}\text{He}$$

Helium nuclei produced in this way are called *alpha particles*. The alpha particles used by Rutherford in his famous scattering

---

*The mass of the beta particle is already included in the mass of $^{14}_{7}$N because atomic masses include the mass of the Z orbital electrons in addition to the mass of the nucleus. That is, the mass of $^{14}_{7}$N includes the mass of one more electron than the mass of $^{14}_{6}$C does.

experiment (Sec. 20.2) were obtained from the decay of ^{214}Bi. The mass difference between the parent nuclide and the sum of the masses of ^{4}He and the daughter nuclide appears as the kinetic energy of the alpha particle. As the alpha particle moves through matter, it quickly transfers this energy to the surrounding atoms, increasing the internal energy of the material.

Gamma rays are short-wavelength photons that sometimes accompany an alpha or beta decay. The motion of the nucleons inside a nucleus is governed by the same quantum-mechanical laws that govern the motion of electrons in an atom, so that a nucleus, like an atom, can exist only in certain discrete quantum states. When a radioactive nucleus decays, the daughter nucleus is not necessarily produced in the state of lowest energy (*ground state*). But like an atom, a nucleus in an excited state rapidly decays to the ground state by the emission of photons. The photons produced this way are called *gamma rays*. Because the energy difference between quantum states of a nucleus is often millions of electronvolts, gamma rays have very high energy and very short wavelengths. Gamma rays are physically the same as short-wavelength x-rays, and they have the same use in radiation therapy.

DECAY RATE

Definition The *decay rate,* or *activity,* R is the number of nuclei in a radioactive sample that radioactively decay in unit time. In a sample containing N identical radioactive nuclei, the decay rate is proportional to N:

$$R = \lambda N \qquad\qquad 22.3$$

The parameter λ is called the *decay constant,* and it is a characteristic property of each nuclide.

> **UNITS** Various time units are used in radioactive studies, such as seconds (s), hours (h), days (d), years (y), and even billions of years (Gy). The corresponding units of the decay rate and the decay constant are disintegrations per second (s^{-1}), disintegrations per hour (h^{-1}), and so on. It is important to keep the same time unit throughout a particular calculation. Another unit of activity is the curie (Ci):
>
> $1 \text{ Ci} = 3.70 \times 10^{10}$ disintegrations/s

The quantum-mechanical nature of radioactive decay makes it completely uncertain when a particular nuclei will decay. All that can be predicted is the probability that a particular nuclei will decay in a given time. The decay constant is, in effect, the probability that a single nucleus will decay in unit time. When this probability is multiplied by the number N of nuclei in a sample, we get the number of nuclei that will decay in unit time. This is the meaning of Eq. 22.3.

Example 22.3 The probability that a ^{32}P nuclei will decay in any 1-h interval is 0.00291. (*a*) What is the decay constant of ^{32}P? (*b*) What is the decay rate of a sample containing 2×10^{-6} mol of ^{32}P?

(*a*) Since the probability of decay in 1 h is 0.00291, the decay constant is

$\lambda = 0.00291 \text{ h}^{-1}$

(*b*) The number N of nuclei in a sample is

$N = nN_A$

where n is the number of moles, and N_A is Avogadro's number. Thus the number of ^{32}P nuclei is

$N = nN_A = (2 \times 10^{-6})(6 \times 10^{23}) = 12 \times 10^{17}$

and the decay rate is

$R = \lambda N = (0.00291 \text{ h}^{-1})(12 \times 10^{17})$

$\quad = 3.49 \times 10^{15} \text{ h}^{-1}$

$\quad = \dfrac{3.49 \times 10^{15}}{3600 \text{ s}} = 9.69 \times 10^{11} \text{ s}^{-1}$

$\quad = \dfrac{9.69 \times 10^{11} \text{ s}^{-1}}{3.70 \times 10^{10} \text{ s}^{-1}/\text{Ci}} = 26.2 \text{ Ci}$ $\qquad \square$

If a sample contains N_0 radioactive nuclei at time $t = 0$, the number N of undecayed nuclei at time t is given by the equation

$N = N_0 e^{-\lambda t}$ $\hfill$ 22.4

Substituting this into Eq. 22.3, we see that the decay rate R at any time is given by

$R = \lambda N = \lambda N_0 e^{-\lambda t}$

or

$R = R_0 e^{-\lambda t}$ $\hfill$ 22.5

where

$R_0 = \lambda N_0$ $\hfill$ 22.6

is the decay rate at $t = 0$.

Figure 22.5 is a graph of R against t, showing that the decay rate of a sample decreases steadily with time. This is just because as nuclei decay, fewer undecayed nuclei are left in the sample. Equation 22.5 is used to determine the decay constant of a nuclide by measuring the decay rate at two different times.

FIGURE 22.5

Plot of the decay rate R against time, for a radioactive nuclide with a decay constant of 1 h^{-1}.

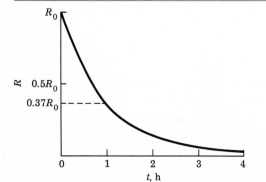

Example 22.4 The decay rate of a radioactive sample is $1250 \ s^{-1}$ at $t = 0$ and $825 \ s^{-1}$ at $t = 30$ min. (a) What is the decay constant of the nuclide? (b) What was the initial number N_0 of nuclei in the sample? (c) What will be the decay rate at $t = 2$ h?

(a) From Eq. 22.5 we have

$$e^{-\lambda t} = \frac{R}{R_0}$$

To solve this equation for λ, we take the natural logarithm of both sides:

$$\ln e^{-\lambda t} = \ln \frac{R}{R_0}$$

Therefore, we have

$$-\lambda t = \ln \frac{R}{R_0}$$

or

$$\lambda = - \frac{\ln (R/R_0)}{t} \qquad\qquad 22.7$$

In the present example, this gives

$$\lambda = - \frac{\ln (825/1250)}{30 \ \text{min}} = 0.0139 \ \text{min}^{-1}$$

(b) From Eq. 22.6, the initial number N_0 of nuclei is

$$N_0 = \frac{R_0}{\lambda}$$

However, we must be careful to express both R_0 and λ in the same units. Thus if we write

$$R_0 = 1250 \ s^{-1} = (1250)(60) \ \text{min}^{-1}$$
$$= 7.5 \times 10^4 \ \text{min}^{-1}$$

we get

$$N_0 = \frac{7.5 \times 10^4 \ \text{min}^{-1}}{0.0139 \ \text{min}^{-1}} = 5.4 \times 10^6$$

(c) From Eq. 22.5 the decay rate at time $t = 2 \ h = 120$ min is

$$R = R_0 e^{-\lambda t}$$
$$= (1250 \ s^{-1}) \exp \left[- (0.0139 \ \text{min}^{-1})(120 \ \text{min}) \right]$$
$$= (1250 \ s^{-1}) \ e^{-1.668} = 236 \ s^{-1}$$

REMARK The initial number N_0 of nuclei in this sample is much smaller than the number of atoms that can be detected by chemical means. Radioactivity provides an important method of detecting minute quantities of material. ☐

HALF-LIFE

Definition The *half-life* τ of a radioactive nuclide is the time in which half the nuclei in a given sample will decay. That is, if N_0 is the number of nuclei at $t = 0$, $N = \frac{1}{2}N_0$ will be the number present at $t = \tau$. Putting these values into Eq. 22.4, we get

$$\tfrac{1}{2}N_0 = N_0 e^{-\lambda\tau}$$

or

$$e^{-\lambda\tau} = \tfrac{1}{2}$$

Taking the natural logarithm of both sides of this last equation, we find

$$-\lambda\tau = \ln \tfrac{1}{2} = -\ln 2$$

or

$$\tau = \frac{\ln 2}{\lambda} = \frac{0.693}{\lambda} \qquad\qquad 22.8$$

Thus the half-life of a nuclide is the reciprocal of its decay constant times 0.693. Table 22.1 gives the half-lives of some radioactive nuclides.

Example 22.5 What is the activity of 1 kg of ^{239}Pu?

From Table 22.1 the half-life of ^{239}Pu is found to be 2.44×10^4 y. So from Eq. 22.8 the decay constant λ of ^{239}Pu is

$$\lambda = \frac{0.693}{\tau} = \frac{0.693}{2.44 \times 10^4 \text{ y}} = 2.84 \times 10^{-5} \text{ y}^{-1}$$

The number n of moles of ^{239}Pu in 1 kg is

$$n = \frac{1000 \text{ g}}{239 \text{ g}} = 4.18$$

since the atomic mass of ^{239}Pu is 239 u. Thus the number N of nuclei in 1 kg is

$$N = nN_A = (4.18)(6 \times 10^{23}) = 2.5 \times 10^{24}$$

and so the activity R is

$$R = \lambda N = (2.84 \times 10^{-5} \text{ y}^{-1})(2.5 \times 10^{24})$$
$$= 7.1 \times 10^{19} \text{ y}^{-1}$$
$$= \frac{7.1 \times 10^{19} \text{ y}^{-1}}{3.15 \times 10^7 \text{ s/y}} = 2.25 \times 10^{12} \text{ s}^{-1}$$
$$= \frac{2.25 \times 10^{12} \text{ s}^{-1}}{3.70 \times 10^{10} \text{ s}^{-1}/\text{Ci}} = 61 \text{ Ci} \qquad\qquad \square$$

The half-lives of the nuclides vary from a fraction of a second to billions of years. For instance, ^{238}U has a half-life of 4.5 billion years, which is about the age of the solar system. This means that of all the ^{238}U present when the solar system was formed, about half still exists. The isotope ^{235}U, on the other hand, has a half-life of only 0.7 billion years, so that over 6 half-lives have elapsed since the formation of the solar system. Consequently, only the fraction

$$\frac{1}{2^6} = \frac{1}{64} = 0.0156$$

of the original ^{235}U still exists. At present ^{235}U constitutes only

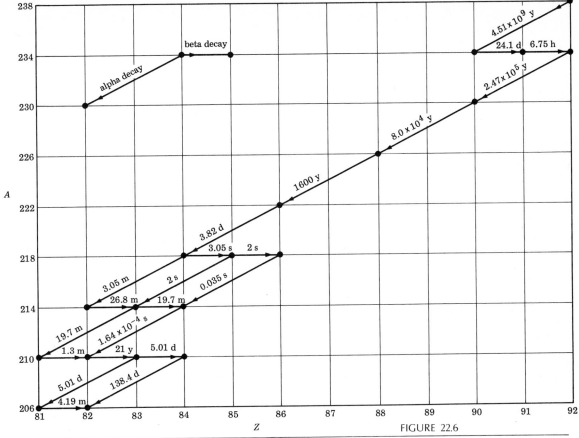

FIGURE 22.6

0.7 percent of naturally occurring uranium, while ^{238}U constitutes the other 99.3 percent.

Nuclides with half-lives less than 0.1 billion years will not have survived their original formation in any detectable amounts. Short-lived nuclides that do occur naturally have been recently formed by one of two processes.

Decay of a Long-Lived Parent For example, ^{238}U is the longest-lived member of a long chain of decays, which ends with the stable isotope ^{206}Pb. The nuclides in this chain, along with their half-lives and decay modes, are displayed on a plot of A versus Z in Fig. 22.6. In a sample of naturally occurring uranium, all the decay products are present in proportion to their half-lives. Note that only nuclides with $A = 238 - 4n$ can be formed from ^{238}U by alpha and beta decay. Other decay chains exist starting with ^{235}U, ^{232}Th, and ^{237}Np.

Production by Cosmic Rays Cosmic rays are high-energy particles, mostly electrons and protons, that bombard the earth from outer space. The collisions of cosmic-ray protons with nuclei in the upper atmosphere, if violent enough, produce a number of secondary protons, neutrons, and alphas by breaking the nuclei apart. The secondary neutrons in turn create radioactive nuclides through collisions with other nuclei. One of the most important reactions of this type is

$$^{14}_{7}N + ^{1}_{0}n \longrightarrow ^{14}_{6}C + ^{1}_{1}H$$

in which a neutron converts ^{14}N to ^{14}C by changing places with a proton. This process occurs continuously in the atmosphere, and a dynamic equilibrium exists between the rate of ^{14}C production and the rate of ^{14}C decay.

ACTIVATION ANALYSIS

The beta decay of a nuclide is usually accompanied by gamma rays of definite energy. By measuring the energies of the gamma rays emitted from a mixture of radioactive nuclides, the identity of the nuclides in the mixture can be determined. Even nuclides present in extremely small amounts can be readily identified in this way.

The presence of trace amounts of nonradioactive nuclides is detected by *activation analysis*. In this technique, a sample of material is bombarded with neutrons, or other elementary particles, to transform some of the nonradioactive nuclides into radioactive nuclides. The spectrum of gamma rays that emerge from the sample is then measured to identify the radioactive nuclides created by the bombardment. From this, the nonradioactive nuclides in the sample can be inferred.

For example, the small amount of oxygen in steel can be measured by bombarding the steel with energetic neutrons, which converts some of the ^{16}O nuclei to ^{16}N nuclei through the reaction

$$^{16}_{8}O + ^{1}_{0}n \longrightarrow ^{16}_{7}N + ^{1}_{1}H$$

The ^{16}N beta decays with a half-life of 7.2 s, emitting gamma rays with energies of 6.129 and 7.117 MeV. By monitoring the number of gamma rays with these energies emitted by the irradiated steel, the quantity of oxygen in the steel is determined.

Activation analysis is commonly used in industry because it is a simple, nondestructive technique for identifying trace quantities of material. For instance, it is used in the semiconductor industry to measure the purity of the silicon wafers on which microelectronic circuits are fabricated (Sec. 21.5). It is also used in criminal investigations, where the origin of a piece of mud or a spot of paint can sometimes be determined from the characteristic grouping of trace elements found in it.

22.3   Nuclear Radiation

Nuclear radiation refers collectively to the high-energy particles (alpha particles, beta particles, protons, neutrons) and the electromagnetic radiation (gamma rays, x-rays) that accompany nuclear decay and nuclear reactions. All these radiations penetrate through matter, ionizing atoms and breaking apart molecules in their path. As a consequence they have deleterious effects on living cells and may be a health hazard to people who work with radioactive materials or near nuclear reactors. At the same time, the ability of nuclear radiation to destroy cell function makes it very effective in the treatment of cancer and related diseases.

Although all nuclear radiations have the same general effect on living matter, each radiation has its own characteristic properties that affect its potency and usefulness in different situations. Thus it is necessary to consider the individual properties of each type of radiation.

Alpha Particles Because they are so much heavier than electrons, alpha particles travel through matter in a straight line, knocking aside the electrons in their path.* In each alpha-electron collision, the alpha loses about 33 eV by knocking an electron out of an atom. Thus an alpha particle with an initial kinetic energy of 5 MeV makes about

$$\frac{5 \text{ MeV}}{33 \times 10^{-6} \text{ MeV}} = 151,000$$

collisions before it comes to rest. Since each collision ionizes an atom or breaks a molecule apart, an alpha particle does considerable damage before it stops.

Definition The *range* of a particle is the distance it travels before stopping. In a particular material, all alpha particles with a given energy have the same range. This range increases with the energy of the alphas and decreases with the density of the material in which they travel. This is seen from Table 22.2, which gives the range of alphas of various energies in air, body tissue, and aluminum. This table also shows that the range of alphas is very small. A 5-MeV alpha, for instance, travels only 0.21 mm through tissue and is completely stopped by a thin sheet of aluminum foil. Consequently, it is very easy to shield against alphas, and even without shielding they cannot penetrate through the skin. However, a radioactive alpha emitter, if swallowed, may accumulate in certain parts of the body, where it can deliver very harmful doses of radiation.

*The Rutherford scattering experiment (Sec. 20.2) studied the rare instances in which an alpha gets close enough to a nucleus to suffer a large-angle deflection.

TABLE **22.2**

Range of alpha and beta particles of various energies in air, body tissue, and aluminum

Energy, MeV	Range, cm		
	Air	Body tissue	Aluminum
Alpha particles:			
1.0	0.55	0.33×10^{-2}	0.32×10^{-3}
2.0	1.04	0.63×10^{-2}	0.61×10^{-3}
3.0	1.67	1.00×10^{-2}	0.98×10^{-3}
4.0	2.58	1.55×10^{-2}	1.50×10^{-3}
5.0	3.50	2.10×10^{-2}	2.06×10^{-3}
Beta particles:			
0.01	0.23	0.00027	
0.10	12.0	0.0151	0.0043
0.50	150	0.18	0.059
1.0	420	0.50	0.15
2.0	840	1.00	0.34
3.0	1260	1.50	0.56

Beta Particles Being simply high-speed electrons, beta particles are more easily deflected by collisions with atomic electrons than alphas are. Nevertheless, like alphas, betas continuously lose energy by ionizing the atoms they pass and come to rest after they have traveled a definite distance through a particular material. Table 22.2 gives the range of beta particles of various energies. Note that the range increases rapidly with energy and that a 3-MeV beta particle travels 150 times farther through tissue than a 3-MeV alpha particle.

In addition to losing energy through collisions with electrons, beta particles produce electromagnetic radiation (x-rays) whenever they suffer a sudden deceleration. This braking radiation, or *bremsstrahlung,* is the source of the x-rays produced when an electron beam collides with the anode of an x-ray tube (Sec. 20.1). X-rays penetrate much farther into matter than beta particles, so that even after all the beta particles have stopped in a substance, the x-rays they have generated continue to travel through it.

Figure 22.7 shows the tracks of beta particles incident on a lead plate thick enough to stop them. However, the x-rays generated by the beta particles in the lead pass through to the other side, where they may still present a radiation hazard. Thus, to provide adequate protective shielding from beta particles, the barrier must be much thicker than the range of the beta particles.

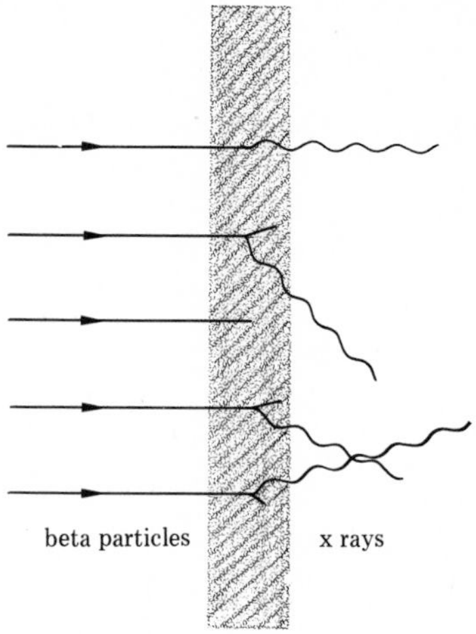

Gamma Rays Unlike charged particles, gamma rays can lose all, or at least a substantial part, of their energy in a single interaction, rather than continuously in a series of collisions. For instance, when a gamma-ray photon interacts with an atomic electron, all the photon's energy can be transferred to the electron, which consequently is ejected from the atom with a large kinetic energy. This process, called the *photoelectric effect* (Sec. 20.1), results in the destruction of the photon. Alternatively, the gamma ray can scatter from the electron, losing a substantial fraction of its initial energy, without being destroyed. This process is called *Compton scattering.* A third process, called *pair production,* is important for gamma rays with energy greater than 1.02 MeV. Such gamma rays can interact with a nucleus to produce an electron-positron pair (Fig. 22.8):

$$\gamma \longrightarrow e^- + e^+$$

All these processes produce high-energy electrons which behave like beta particles, ionizing other atoms as they travel through matter. (These electrons can in turn generate lower-energy x-rays, as shown in Fig. 22.7.)

The distance a photon travels before it interacts with matter cannot be predicted. All that can be predicted is the distance in which a photon has a 50 percent chance of interacting. This is called the *half-value layer* and is analogous to the half-life of a radioactive nuclide. Table 22.3 gives the half-value layer for gamma rays of various energies in tissue and lead.

For example, the half-value layer for a 0.1-MeV gamma ray in tissue is 4.05 cm. This means that after it has traveled through

beta particles   x rays

FIGURE 22.7

Beta particles incident on a lead plate that is thick enough to stop them. As the particles stop, they generate x-rays that pass through the plate.

TABLE **22.3**

	Half-value layer, cm	
Energy, MeV	Body tissue	Lead
0.01	0.131	0.00076
0.05	3.12	0.012
0.10	4.05	0.012
0.50	7.20	0.42
1.0	9.80	0.89
5.0	23.0	1.52

Half-value layer for gamma rays of various energies in body tissue and lead

4.05 cm of tissue, the intensity of a gamma-ray beam is half its original value. After it has traveled through another 4.05 cm, the intensity of the remaining beam is reduced by one-half again. Thus, after the beam has traveled through 8.10 cm of tissue, its intensity is only a quarter of its original value. To reduce a beam to less than 1 percent of its original value, it has to travel through at least seven half-value layers, or 28.35 cm in this case. Gamma rays and x-rays are thus far more penetrating than charged particles.

Neutrons Like gamma rays, neutrons lose a substantial part of their energy in a single collision, although neutrons collide with nuclei rather than with electrons. Neutrons are very penetrating, and it requires massive amounts of lead and concrete shielding to protect workers from the neutrons produced by a reactor.

DETECTORS

Nuclear radiation is detected by the ionization it produces as it passes through matter. The ionization trail of a charged particle is made visible in a *bubble chamber* by the formation of bubbles around the individual ions (Fig. 22.8). Similarly, the ionization produced by a charged particle impinging on the photosensitive emulsion of photographic film is revealed when the film is developed. Radiation workers wear film badges to monitor their exposure to radiation. The badge contains a piece of photographic film wrapped in opaque paper. Once a week the film is developed and the level of exposure to radiation determined by the degree of exposure of the film.

For the instantaneous detection of individual particles, electronic detectors are used. Some detectors can also measure the energy of individual particles.

Geiger Tube The *Geiger-Müller counter,* or *Geiger tube,* is the oldest and simplest electronic detector. It consists of a metallic gas-discharge tube with a wire running through its center (Fig. 22.9). The wire is maintained at a large positive potential (about +500 V) relative to the wall of the tube, and the tube contains a gas at low pressure (about 0.1 atm). When a single charged particle passes through the gas in the tube, it ionizes one or more atoms in the gas, releasing several free electrons. These electrons are then accelerated toward the wire, but before they reach it, they usually ionize other atoms in their path, thereby

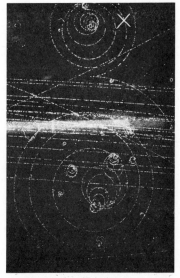

FIGURE 22.8

Bubble-chamber tracks of protons and an electron-positron pair. A magnetic field perpendicular to the plane of the picture causes the electrons and positrons to move in opposite spirals. The protons have much greater momentum, and so are little affected by the magnetic field. (*Brookhave National Laboratory.*)

releasing more electrons. These additional electrons in turn ionize still more atoms as they accelerate toward the wire. A chain reaction (or *avalanche*) develops, which rapidly ionizes most of the gas in the tube. As a consequence, the gas becomes electrically conducting, and there is a momentary surge of current through the tube. This current pulse goes to an amplifier, and the output of the amplifier operates a counter which records the discharge produced each time a charged particle passes through the tube.

Geiger tubes do not detect photons and neutrons directly. However, if the tube is surrounded by the proper material, some of the incident gamma rays and neutrons will knock electrons and protons out of the atoms of the material, and these charged particles will be detected. Of course, only a fraction of the incident photons and neutrons are detected this way, since only a fraction of the incident particles interact with the surrounding material.

Geiger tubes are relatively inexpensive and easy to operate. Their main disadvantages are their inability to measure the energy of the incident particle and their rather long *dead time,* or time between the passage of one charged particle through a detector and the moment when the detector is able to detect a second particle. In a Geiger tube this is the time required for all the ions and electrons produced by the first particle to recombine into neutral atoms, so that the tube is back in its resting state. The dead time of a Geiger tube is of the order of milliseconds, which means that the tube is limited to a few hundred counts per second. If particles are incident on a Geiger tube at a faster rate, it will fail to count them all since many will arrive during a dead period.

Scintillation Counter This detector is an electronic version of the method used by Rutherford to detect alpha particles in 1911 (Sec. 20.2). It is based on the fact that certain materials convert some of the energy of an ionizing particle into light and ultraviolet radiation. In Rutherford's time this light was detected by eye, but in a modern scintillation counter it is detected by a *photomultiplier tube.*

The scintillation material most often used today is a transparent crystal of sodium iodide, NaI. When a charged particle stops in the crystal, a quantity of light proportional to the energy of the particle is generated. Some of this light falls on the first electrode of a photomultiplier tube (Fig. 22.10), from which it knocks out electrons by the photoelectric effect (Sec. 20.1). These electrons are accelerated to the second electrode, which is maintained at a potential of several hundred volts above the first electrode. Upon hitting this electrode, each electron knocks out several electrons, which in turn accelerate toward a third electrode maintained at a potential several hundred volts above the second electrode. At each stage the number of electrons released is 2 to 10 times the number incident, so the total number of electrons rapidly increases. For instance, in a phototube with 10 electrodes, 256 electrons will reach the last electrode for each electron leaving the first electrode if the multiplication

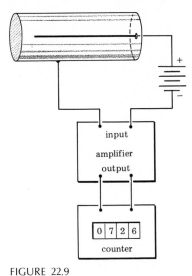

FIGURE 22.9

Geiger-Müller counter.

factor is 2. Thus only a few photons of light have to arrive at the first electrode to produce a measurable signal at the last electrode.

Since the amount of light generated when a charged particle stops in the NaI crystal is proportional to the energy of the particle, and since the current at the last electrode is proportional to the number of photons incident on the first electrode, the output of the photomultiplier tube is proportional to the energy of the particle. Thus a scintillation counter not only detects charged particles, but once calibrated, it can measure their energy as well. Furthermore, the dead time of a scintillation counter is the order of microseconds, so it is capable of much higher counting rates than a Geiger tube.

Semiconductor Detector A *pn* junction (Sec. 21.3) maintained in a reverse-bias condition makes an excellent radiation detector. A charged particle passing through the junction creates a large number of electrons and holes in the region between the two layers of immobile charge (Fig. 22.11). The holes are immediately swept to the *p* side of the junction, while the electrons are swept to the *n* side, giving rise to a sudden current, which can be amplified and measured. The number of electron-hole pairs produced, and hence the magnitude of the current, is proportional to the energy of the incident particle, so this type of detector measures particle energy. In addition, it has a very short dead time, because the electrons and holes recombine in less than 1 μs. On the other hand, the crystal itself is subject to deterioration resulting from damage caused by the radiation.

DOSIMETRY

Radiation medicine involves the protection of radiation workers and the general population from the harmful effects of radiation, as well as the use of radiation for the diagnosis and treatment of disease. All these applications require a quantitative measure of the unit of absorbed radiation dose.

UNIT The unit of absorbed radiation dose is the *rad* (rd), defined as the absorption of 10^{-2} J of ionizing radiation per kilogram of absorbing material. Only the energy actually absorbed is counted; a gamma ray that passes through the body without interaction does not contribute to the dose. Since dense material, such as bone, has a smaller half-value layer for gamma rays than soft material, such as tissue, the same intensity of radiation results in a greater dose in bone than in tissue.

The rad is a physical unit of dose. However, equal doses (in rads) of different types of radiation have different biological effects. The *relative biological effectiveness* (RBE) of a particular type of radiation is the ratio of the dose (in rads) of x-rays or gamma rays to the dose (in rads) of the radiation that produces the same biological effect. From Table 22.4, which gives the RBE of various radiations, it is seen that beta particles have the same effectiveness as gamma rays but that alpha particles are 10 to 20 times more effective.

UNIT The biological unit of dose is the *rad equivalent man* (rem), which is

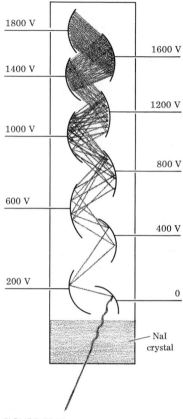

FIGURE 22.10

Scintillation counter, consisting of a photomultiplier on top of an NaI crystal.

FIGURE 22.11

A semiconductor detector consisting of a *pn* junction in reverse bias.

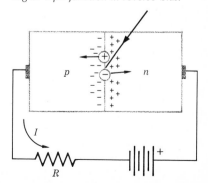

defined in terms of the RBE by

$$rem = RBE \times rad$$

For instance, a dose of 50 rd of alpha particles is equal to 500 to 1000 rem. Thus a 50-rd whole-body dose of alpha particles would be as biologically effective as a 500-rd (= 500-rem) dose of gamma rays. Of course, because of the short range of alpha particles, whole-body exposure is possible only by ingesting an alpha emitter that diffuses throughout the entire body.

The general population is exposed to radiation from a variety of natural and artificial sources. Table 22.5 lists the average annual dose received by an individual in the United States from all sources. Note that exposure for medical purposes (chest x-rays, cancer therapy, etc.) is the largest source of artificial radiation, far exceeding the negligible dose from nuclear power reactors.

The Environmental Protection Agency (EPA) sets 0.17 rem/y as the maximum whole-body dose (exclusive of natural and medical radiation) permissible for a member of the general population to receive.* Table 22.5 shows that the only sources of radiation covered by this limit are fallout (from atmospheric testing of atomic bombs) and nuclear power reactors, which together contribute only 0.004 rem/y. Although some critics have argued that the EPA limit is too high, it is clear that as long as nuclear power reactors are built according to current regulations, the actual dose from nonmedical artificial sources will remain far below 0.17 rem/y.†

The main concern with chronic exposure to low levels of radiation is that it may cause an increase in cancer in the exposed population and an increase in genetic defects in the next generation. All evidence about this, however, comes from studies of a few populations, such as Japanese survivors of the atom bomb, that received acute high-level doses of radiation and from studies of animals experimentally exposed to high-level doses. By extrapolating the results of these studies, a maximum risk has been assigned to low-level doses. Although this is the properly conservative approach to take for the protection of the general population, it should be remembered that there is no

*The maximum permissible whole-body dose for a radiation worker is 5 rem/y.

†The operation failure of the Three Mile Island nuclear plant in March 1979 resulted in the unscheduled release of short-lived radioactive nuclides into the atmosphere. This gave a total dose of about 0.1 rem to people living near the plant. Thus in spite of the emotionalism that surrounds nuclear energy, the actual harm it has done the public has been negligible.

TABLE **22.4**

Relative biological effectiveness (RBE) of different types of radiation

Radiation	RBE
Gamma rays, x-rays, beta particles	1.0
Fast neutrons and protons	10
Slow neutrons	4 to 5
Alpha particles	10 to 20

TABLE **22.5**

Source	Dose, rem
Cosmic rays	0.044
Radioactive material naturally present in rocks	0.058
Medical	0.073
Fallout	0.004
Nuclear power reactors	3×10^{-6}
Other	0.003
Total	0.182

Sources of radiation

The average annual dose received by a member of the general population from various sources.

TABLE **22.6**

Dose, rd	Effect
0 to 25	No observable effect
25 to 100	Slight blood change
100 to 200	Moderate blood change; vomiting in 5 to 50 percent of cases within 3 h; complete recovery within a few weeks (except for blood-forming system)
200 to 600	Severe blood changes; vomiting in 50 to 100 percent of cases within 3 h; loss of hair within 2 weeks; hemorrhaging and infection; death in 0 to 80 percent of cases within 2 months
600 to 1000	Severe blood changes; vomiting within 1 h; loss of hair; hemorrhaging and infection; death in 80 to 100 percent of cases within 2 months

Short-term effects of acute doses of whole-body gamma radiation

direct evidence that there are any harmful effects from exposure to radiation at the level of less than 1 rem/y.

The problem of assessing the effects of low levels of radiation is complicated by the known ability of the body to recover from radiation exposure. Thus, while a single acute dose of 1000 rem is inevitably fatal, cancer patients are routinely given daily doses of 100 rem or more for 2 weeks without serious short-term effect. Table 22.6 shows the short-term effects of a single exposure to various doses of whole-body gamma radiation. Conclusive evidence does exist that exposure to radiation at the 100-rem level does increase the long-term risk of cancer. Thus such a level of exposure is justified only in the treatment of a serious illness.

MEDICAL USES OF RADIATION
Nuclear radiation has found important applications in medical research and in the treatment and diagnosis of disease.

Treatment The efficacy of radiation for the treatment of cancer depends on the fact that ionizing radiation is even more lethal to rapidly growing cancer cells than it is to healthy cells. Treatment involves subjecting a patient to a dose of radiation high enough to kill the cancer cells without causing extreme harm to normal cells.

One source of radiation for cancer therapy is ^{60}Co. This nuclide beta-decays to an excited state of ^{60}Ni, which then immediately decays to its ground state by emitting in succession a 1.17-MeV gamma ray and a 1.33-MeV gamma ray. The energies of these gamma rays are much higher than the energies of the x-rays produced by all but the largest x-ray machines, so that ^{60}Co is a relatively inexpensive and convenient source of very penetrating radiation. The required amount of ^{60}Co (about 10 g) is kept inside a heavy lead box, mounted above a table. A patient lying on the table is exposed to radiation by opening a small hole in the box. The room itself is heavily shielded from the rest of the hospital. Since no personnel can be in the room during the treatment, the procedure is monitored on closed-circuit television.

Example 22.6 What is the dose received by a 60-kg patient exposed for 30 s to an 8-g ^{60}Co source? Assume that 1 percent of the gamma rays emitted by the source reach the patient.

From Table 22.1 the half-life of ^{60}Co is

$$\tau = 5.26\,y = 1.65 \times 10^8\,s$$

so from Eq. 22.8 the decay constant is

$$\lambda = \frac{\ln 2}{\tau} = \frac{0.693}{1.65 \times 10^8\,s} = 4.2 \times 10^{-9}\,s^{-1}$$

Then from Eq. 22.3 the decay rate R of 8 g ($\frac{8}{60}$ mol) of ^{60}Co is

$$\begin{aligned}R = \lambda N &= \lambda n N_A \\ &= (4.2 \times 10^{-9}\,s^{-1})(\tfrac{8}{60})(6 \times 10^{23}) \\ &= 3.4 \times 10^{14}\,s^{-1}\end{aligned}$$

Since each ^{60}Co decay produces two gamma rays, there are 6.8×10^{14} gamma rays produced per second. These gamma rays are emitted uniformly in all directions, and only about 1 percent, or 6.8×10^{12}, fall on the patient per second. From Table 22.3 we see that the half-layer value of a 1-MeV gamma ray in tissue is about 10 cm, which is about the front-to-back thickness of the body. This means that approximately half the gamma rays incident on the patient pass through without interaction and the other half are absorbed by the body. Thus about 3.4×10^{12} gamma rays per second are absorbed by the body.

The average energy of a ^{60}Co gamma ray is 1.25 MeV, so the energy absorbed per kilogram per second is

$$\frac{(3.4 \times 10^{12}\,s^{-1})(1.25\,MeV)}{60\,kg} = 7.08 \times 10^{12}\,MeV/(kg \cdot s)$$

$$= 1.1 \times 10^{-2}\,J/(kg \cdot s) = 1.1\,rad/s$$

since the rad is defined to be 10^{-2} J/kg of absorbed radiation. Thus a 30-s exposure to the ^{60}Co results in a whole-body dose of 33 rd. ☐

Radiopharmaceuticals are also used for radiation therapy because certain elements, e.g., iodine, are preferentially absorbed by certain organs in the body, such as the thyroid. Thus some diseases of the thyroid are treated by administering pharmaceuticals containing radioactive ^{131}I. Since the iodine concentrates in the thyroid, the beta rays that accompany the decay of ^{131}I are delivered directly to the diseased organ without too much radiation being absorbed by healthy tissue.

Diagnosis Conventional x-ray pictures of the body are only able to distinguish organs of differing density. In particular, dense bone is clearly seen in an x-ray picture in contrast with the soft surrounding tissue because less radiation passes through the bone than through the tissue. However, x-ray pictures are unable to detect an abnormal mass of soft tissue, such as a tumor, embedded in normal tissue. For this reason, radioactive pharmaceuticals have been developed that are preferentially absorbed by the tumor. Special devices detect the radiation coming from a patient who has been injected with such a radiopharmaceutical to determine whether a tumor is present.

The radioactive nuclide usually used in these studies is $^{99}_{43}Tc^m$ (the superscript m stands for *metastable* state). This particular technetium isotope is obtained from ^{99}Mo, which beta-decays to an excited state of ^{99}Tc rather than to the ground state. This is common in many decays. However, unlike most other excited states, the excited state of ^{99}Tc is semistable (metastable). The half-life for the decay of $^{99}Tc^m$ to the ground state is 6.0 h, rather than 10^{-8} s, which is the usual half-life for such a decay. Thus $^{99}Tc^m$ lives long enough to be used in a medical procedure. Furthermore, since it decays to its ground state by emitting a gamma ray, no charged particles are involved. (The ground state beta-decays to ^{99}Ru with a half-life of 212,000 y, so the technetium is eliminated from the body long before any significant number of beta particles is produced.) The absence of charged particles minimizes the radiation hazard involved in using $^{99}Tc^m$, because many of the gamma rays pass out of the body before interacting.

REMARK Technetium ($Z = 43$) and promethium ($Z = 61$) are the only elements with $Z < 84$ that have no stable isotope. Neither element occurs naturally on earth; like plutonium, both must be produced artificially from nuclear reactions.

The $^{99}Tc^m$ is obtained in the form of the pertechnetate ($^{99}Tc^mO_4^-$) by washing a salt of $^{99}MoO_4^-$ with saline solution. The $^{99}MoO_4^-$ continuously decays into $^{99}Tc^mO_4^-$, which remains mixed with the permolybdenate until it is dissolved in the saline. The molybdenum salt is referred to as the generator since it can be stored for several days and washed with saline solution whenever $^{99}Tc^mO_4^-$ is needed. Because the half-life of $^{99}Tc^m$ is only 6 h, the pertechnetate must be injected into the patient immediately and the patient examined within a few hours after injection.

The blood-brain barrier, which normally prevents foreign chemicals from diffusing into the brain, does not exist in a brain tumor, so that the pertechnetate diffuses into a brain tumor but not into the surrounding tissue. Consequently, more gamma rays are given off by a brain tumor than by the surrounding tissue, making possible the detection of the tumor. However, because gamma rays cannot be focused, special techniques are required to get a meaningful image.

A *rectilinear scanning device* consists of a single scintillation counter with a lead collimator covering the front of the crystal (Fig. 22.12). The collimator is perforated by a number of holes angled in such a way that only gamma rays coming from a small volume of space can pass through them all. Thus the counter "sees" only this small volume.* A complete picture of the head is obtained point by point by passing the counter back and forth across the head. At each point the number of counts recorded

*Each hole detects gamma rays coming from anywhere along the line of sight of the hole. However, since only gamma rays coming from the point of intersection of all the lines of sight can pass through all the holes, more gamma rays will be detected from this point than from any other.

FIGURE 22.12

Radiation detector in a rectilinear scanning device.

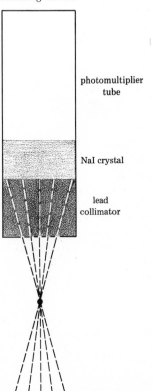

photomultiplier tube

NaI crystal

lead collimator

by the counter measures the amount of $^{99}Tc^m$ at that point.

A picture of the head, or *brain scan*, is obtained indirectly by having a small light move back and forth across a photographic plate in synchronization with the counter. The intensity of the light at each point is adjusted to be proportional to the number of counts recorded at the corresponding point on the patient's head, so that a visual record of the radioactivity detected by the counter is obtained. A tumor appears as a light spot on the plate. By taking scans from different angles, the tumor can be located in three dimensions.

Because the rectilinear scanning device constructs an image point by point, it takes about 15 min to get a single picture. Faster results are possible with the Anger gamma-scintillation camera (Fig. 22.13). This device consists of a single, large scintillation crystal with 19 phototubes mounted in back of it (Fig. 22.14). The front of the crystal is covered by a large lead collimator containing thousands of parallel collimating holes. The de-

FIGURE 22.13

An Anger camera examining a patient's chest. (*Ohio-Nuclear, Inc.*)

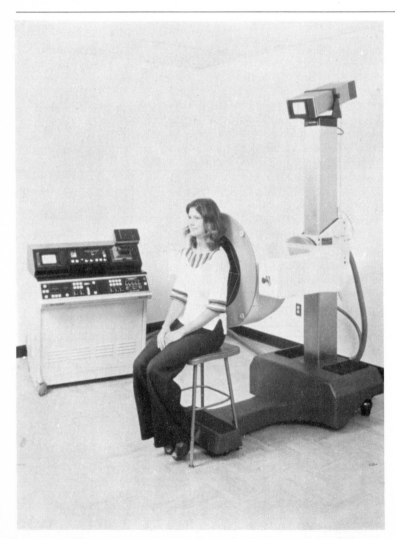

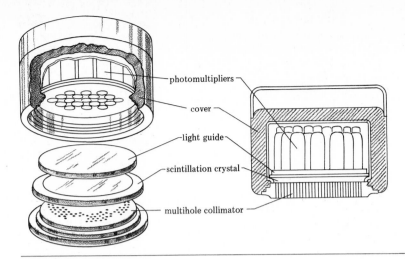

photomultipliers

cover

light guide

scintillation crystal

multihole collimator

FIGURE 22.14

The arrangement of phototubes in an Anger camera.

vice is called a camera because it receives gamma rays simultaneously from all parts of the field of view. Each collimating hole views a different point in the field because a gamma ray coming from a particular point can pass through the collimator only directly above it.

The light given off when a gamma ray interacts in the crystal is detected simultaneously by each of the 19 phototubes, but with an intensity that depends on the distance of the gamma ray from the phototube. The position of the gamma ray in two dimensions is obtained from an analysis of the relative intensities detected by the phototubes. This analysis is performed electronically in less than a microsecond, and a spot corresponding to the position of each gamma ray is displayed on an oscilloscope. A camera mounted in front of the oscilloscope screen records each spot as it is displayed until a complete image is formed on the film. Figure 22.15 is a set of pictures of the oscilloscope display of a brain scan taken with an Anger camera. A satisfactory image can be taken in only a few minutes with an Anger camera.

22.4 Nuclear Energy

In a chemical reaction, such as the oxidation of carbon

$$C + O_2 \longrightarrow CO_2 \qquad\qquad 22.9$$

the energy released is the difference between the binding energies of the final and initial reactants. A higher binding energy corresponds to a more tightly bound molecule, because binding energy is equal to the work required to break a molecule apart. The work required to break CO_2 into C and O_2 is equal to the energy released when C and O_2 react to form CO_2.

Similar concepts govern the release of energy in a nuclear reaction. For instance, the nuclear reaction

$$^2_1H + ^2_1H \longrightarrow ^3_2He + ^1_0n \qquad\qquad 22.10$$

releases energy because the nucleons in ^3He are more tightly bound than they are in ^{2}H. In fact, the nuclear reaction in Eq. 22.10 releases 3.2 MeV for each ^{3}He nucleus formed, whereas the chemical reaction in Eq. 22.9 releases 4.1 eV for each CO_2 molecule formed.

It has been known since the early 1900s that nuclear reactions release a million times more energy than chemical reactions. However, most nuclear reactions require more energy to initiate than they release, so they are not practical sources of energy. At present the only reaction capable of providing a controlled source of nuclear energy is the fission of ^{235}U and ^{239}Pu. Intense research efforts in many countries are devoted to developing nuclear energy from the controlled fusion of ^{2}H and ^{3}H, but the most optimistic estimates do not see the commercial utilization of fusion before the year 2030. Thus, in spite of the extreme emotionalism that has accompanied the development of nuclear power, the fission of ^{235}U is destined to be a vital part of our energy resources far into the twenty-first century.

FISSION

No nuclear reaction capable of providing a self-sustaining release of energy was known until German scientists discovered in 1938 that when the uranium isotope ^{235}U absorbs a neutron, the newly formed ^{236}U nucleus, instead of undergoing alpha or beta decay, immediately breaks apart into two nearly equal fragments. This type of nuclear decay is called *fission*.

The fission fragments vary in mass number between 75 and 160, but a typical fission reaction is

$$^{235}_{92}U + {}^{1}_{0}n \longrightarrow {}^{236}_{92}U \longrightarrow {}^{95}_{38}Sr + {}^{139}_{54}Xe + 2\,{}^{1}_{0}n$$

There are a number of important features to note about this reaction.

1 The ^{236}U nucleus exists for less than 10^{-12} s, so that the fission process can be considered to be instantaneous.

2 Figure 22.4 shows that the average binding energy per nucleon of a stable nuclide with A between 75 and 160 is about 8.5 MeV, whereas the binding energy per nucleon of ^{235}U is 7.6 MeV. This means that the 234 nucleons are more tightly bound in the fission fragments than in ^{235}U. The binding energy per nucleon is the energy required to remove a nucleon from a nucleus, or alternatively, it is the energy released when a nucleon is bound to a nucleus. Therefore, increasing the binding energy from 7.6 to 8.5 MeV releases 0.9 MeV per nucleon, or

$$(234)(0.9 \text{ MeV}) = 211 \text{ MeV}$$

per fission event. About 86 percent of this energy is released immediately in the form of kinetic energy of the fragments and the neutrons.

3 Because the neutron-to-proton ratio of ^{235}U is greater than that of stable nuclides in the $A = 125$ mass region, the fission fragments have an excess of neutrons and lie above the stability curve in Fig. 22.3. These fragments, therefore, beta-decay in a series of steps until stable nuclides with the same mass numbers are formed. The ^{236}U nucleus breaks apart in many different ways, and each fragment forms a series of three or four radioactive nuclides, so that over 200 different radioactive nuclides are produced by fission. Most of these do not occur naturally on earth.

FIGURE 22.15

Anterior, right lateral, and left lateral brain scans, taken with an Anger camera. No tumor is present in these pictures. (*Ohio-Nuclear, Inc.*)

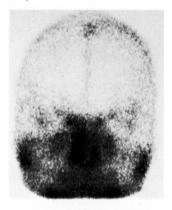

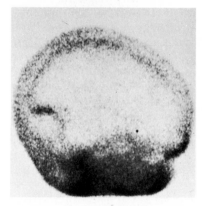

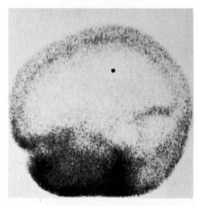

REMARK These radioactive nuclides are a dangerous and undesirable by-product of fission. During the 1950s, when both the United States and the Soviet Union exploded many atomic bombs, a large amount of radioactive material was released in the atmosphere from the fission process. The most notorious of these was ^{90}Sr, which is chemically similar to calcium (Table 20.2) and has a relatively long half-life (Table 22.1). As a consequence, it lived long enough to enter the food chain, where it was deposited along with calcium in the bones of growing children. The worldwide radiation hazard that this presented led eventually to the ban on atomic tests in the atmosphere.

4 The fission process, which requires a neutron to initiate, releases (on an average) 2.5 neutrons per fission. It is this fact that immediately attracted the attention of scientists, because it meant that a chain of self-sustaining nuclear reactions was possible if the extra neutrons released in one fission event could be made to initiate additional fission events.

News of the discovery of fission reached the United States in January 1939. In July Einstein wrote to President Roosevelt about fission, urging that research into the military potential of this discovery be started at once. In September 1939 Germany invaded Poland, plunging Europe into war. The United States, which was then in the process of general rearmament, embarked on an ambitious program to build a fission bomb.

The first chain of self-sustaining fission reactions (a *chain reaction*) was achieved in a reactor built by Enrico Fermi (1901–1954) and his colleagues at the University of Chicago on December 2, 1942. This date is usually taken to mark the beginning of the Atomic Age,* since it is only through a chain reaction that usable amounts of energy can be obtained from a nuclear reaction.

The average number of neutrons released in a fission event that go on to initiate other fission events is called the *multiplication factor f*. The maintenance of a chain reaction requires that f be greater than or equal to 1. Since, on average, 2.5 neutrons are released in each fission event, it is not necessary that every neutron initiate another fission. This is fortunate, because there are competing processes that remove neutrons from the reaction region. Of these, the most important are:

1 Escape of neutrons from the reactor

2 Capture of neutrons by ^{238}U in the reaction

$$^{238}U + {}^1n \longrightarrow {}^{239}U \qquad\qquad 22.11$$

3 Capture of neutrons by other nuclides

Process 1 is minimized by making the reactor so large that a neutron is very likely to be captured before it escapes. Process 3 is minimized by the proper choice of the materials placed in the reactor.

Process 2 is minimized by encasing the uranium in metal rods which are immersed in a moderating material, often ordinary

*In this book *atomic* usually refers to the electronic properties of an atom, and *nuclear* refers to the nuclear properties. In common usage, however, atomic is synonymous with nuclear.

water. Because of the geometry of this arrangement, a high-energy neutron from a fission reaction in a fuel rod is most likely to escape into the moderator before it is captured by a ^{238}U nucleus. In the moderator its energy is degraded by successive collisions with moderator atoms, until it eventually reenters a fuel rod. By this time its energy is too small to be captured by a ^{238}U nucleus, but it can still be captured by a ^{235}U nucleus to induce another fission reaction.

Several types of nuclear reactors are currently in use in electric-power plants throughout the world. Figure 22.16 is a schematic diagram of a *pressurized-water reactor* (PWR) common in the United States. A quite different design is used in the Canadian CANDU (Canada-deuterium-uranium) reactor, but the same four components are found in all reactors.

1 *Fuel* Pellets of the uranium oxide U_3O_8 encased in long, thin rods made of a zirconium-steel alloy. Enriched uranium containing 2 to 4 percent of the fissionable* ^{235}U must be used in a PWR, whereas a CANDU reactor can use natural uranium containing only 0.7 percent ^{235}U.

2 *Moderator* A material that completely surrounds the fuel rods, and which can slow neutrons without absorbing them. In a PWR, the moderator is ordinary water at high pressure (100 atm or more), whereas in a CANDU reactor, it is heavy water at ordinary pressure. *Heavy water* is water in which all the hydrogen is the isotope ^{2}H (deuterium) rather than the common isotope ^{1}H. Heavy water, which is a much better moderator than ordinary water, allows the CANDU reactor to operate with a much lower concentration of ^{235}U.

3 *Control rods* Rods made of a material like boron which readily

*A nuclide is fissionable only if it can sustain a chain reaction. There are nuclides, such as ^{238}U, which can be made to fission under certain circumstances but which cannot sustain a chain reaction. ^{235}U is the only naturally occurring fissionable nuclide. Two other fissionable nuclides, ^{233}U and ^{239}Pu, are produced artificially.

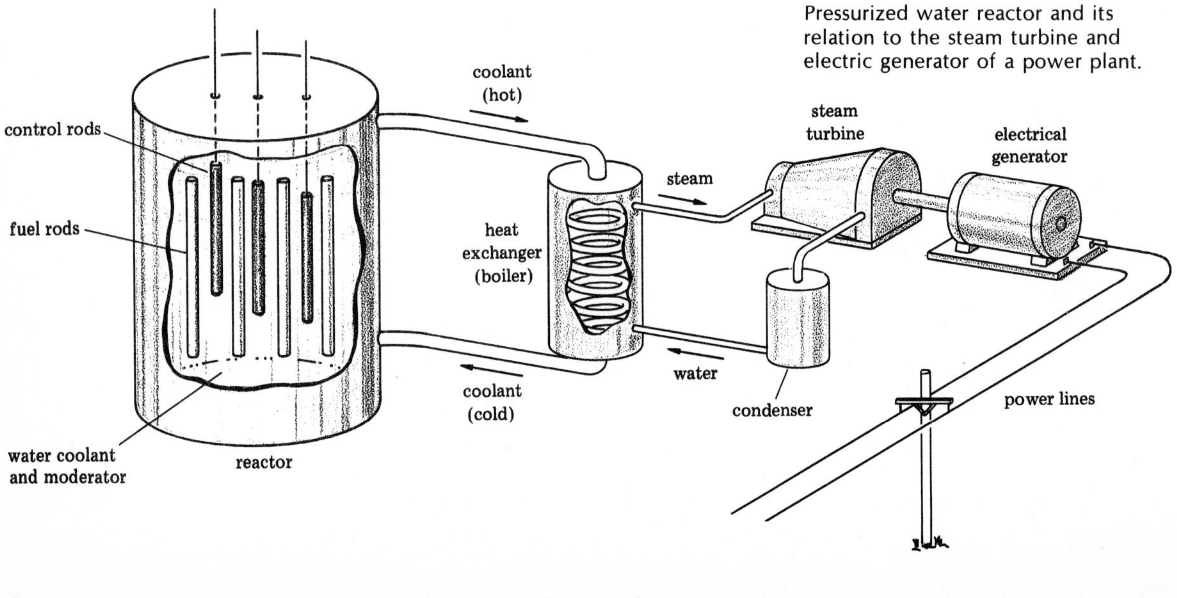

FIGURE 22.16

Pressurized water reactor and its relation to the steam turbine and electric generator of a power plant.

absorbs neutrons. The multiplication factor of a reactor is increased or decreased by removing or inserting these rods.

4 *Coolant* A fluid which removes the heat generated by the reactor and transfers it to a boiler to make steam for the generator turbines. In a PWR, the coolant is the same pressurized water that is used as the moderator. The water is pressurized so that it can be raised to high temperature without boiling. The high temperature is essential in order to obtain a thermodynamically efficient conversion of heat into work (Sec. 10.2). In a CANDU reactor, the coolant is a separate system of pressurized heavy water that is circulated through the fuel rods. Thus in a CANDU reactor, the high pressure is contained by hundreds of relatively small structures, whereas in a PWR, the pressure is contained by the walls of the reactor itself.

To start a reactor, the control rods are gradually withdrawn. This increases the multiplication factor by decreasing the opportunity for neutrons to be captured by nonfissionable boron. When the multiplication factor exceeds 1, the reactor is *critical*. At this stage a single fission event, initiated by a stray cosmic-ray neutron, can rapidly multiply until billions of fissions per second are occurring. For instance, starting with a single fission event and a multiplication factor $f = 1.007$, there will be $f^n = 1.007^n$ fissions after n generations, which amounts to 10^3 fissions by the thousandth generation. Since the time interval between generations is only 10^{-3} s, the fission rate builds up rapidly once the reactor becomes critical.

When the reactor reaches the desired level of activity, the control rods are adjusted to reduce f to 1.00000. Thereafter the reactor maintains a constant rate of activity. Through collisions with other nuclei, the kinetic energy of the fission fragments and neutrons released by the fission process is converted into internal energy, which raises the temperature of the interior of the reactor. This energy is removed by the coolant that circulates through the reactor and is transferred to a boiler, where it is used to make the steam to turn the turbine to run the generator to produce electricity. In this way the energy of the nucleus is converted into usable electric energy. The only difference between a nuclear power plant and a conventional power plant is the fuel used to produce the steam. Once the steam is produced, both types of plants are identical in their operation.

The only naturally occurring fissionable material, ^{235}U, constitutes only 0.7 percent of natural uranium. Thus the world supply of fissionable material is very limited and is probably not sufficient to sustain the fully developed nuclear power industry expected in the next century. Fortunately two other fissionable nuclides, ^{233}U and ^{239}Pu, can be produced in a nuclear reactor. Plutonium 239 is a natural decay product of ^{239}U, which is produced from nonfissionable ^{238}U by the capture of a neutron (Eq. 22.11). The beta decay of ^{239}U produces ^{239}Np, a new element with atomic number 93. The reaction is

$$^{239}_{92}U \longrightarrow\ ^{239}_{93}Np + e^- + \nu$$

and the half-life of ^{239}U is 23.5 min. Neptunium 239 in turn beta-decays with a half-life of 2.35 d into ^{239}Pu, a new element with atomic number 94. The reaction is

$$^{239}_{93}\text{Np} \longrightarrow {}^{239}_{94}\text{Pu} + e^- + \nu$$

Thus the ^{239}U produced in a reactor decays within a few days to ^{239}Pu. Since ^{239}Pu has a half-life of 24,400 y, it is stable enough for industrial purposes.

In conventional nuclear reactors, the amount of ^{239}Pu produced is less than the amount of ^{235}U consumed. However, *breeder reactors* now being developed in several countries will produce more fissionable material than they consume. In these reactors there is no moderator to slow the neutrons, which are captured either by ^{235}U (causing fission) or by ^{238}U (producing ^{239}Pu). Of the 2.5 neutrons released in each fission event, 1 is required to initiate another fission and 1.5 are available to produce ^{239}Pu. Since ^{238}U (from which ^{239}Pu is made) is 140 times more plentiful than ^{235}U, breeders open up the possibility of a virtually unlimited fuel supply.

With only minor modification, the existing CANDU reactors can be made to breed fissionable ^{233}U from ^{232}Th. The reaction is

$$^{232}_{90}\text{Th} + {}^{1}_{0}n \longrightarrow {}^{233}_{90}\text{Th}$$

The $^{233}_{90}\text{Th}$ beta-decays to $^{233}_{91}\text{Pa}$, with a half-life of 22.2 min, and the $^{233}_{91}\text{Pa}$ beta-decays to $^{233}_{92}\text{U}$, with a half-life of 27 d. The half-life of ^{233}U is 1.6×10^5 y, so it accumulates in the reactor. Since thorium is more plentiful than uranium, this process is another possible method of extending our limited fissionable-fuel resources.

The isotopes of uranium are chemically identical, so the enriched uranium needed for PWRs must be obtained from a process that depends on the mass difference of ^{235}U and ^{238}U. In the gas-diffusion process, gaseous uranium, in the form of uranium hexafluoride (UF_6), is allowed to diffuse through a porous membrane. At the same temperature, a UF_6 molecule composed of ^{235}U diffuses more rapidly. Thus the diffusion of UF_6 through a porous membrane results in a minute enrichment of ^{235}U. By repeatedly passing the UF_6 through porous membranes, the percentage of ^{235}U can be increased to any desired value.

After enrichment, the uranium is converted into U_3O_8 and formed into pellets. A single fuel rod consists of many pellets encased in a stainless-steel or zirconium-steel cladding. After 3 years in a reactor, the percentage of ^{235}U in a fuel rod has decreased to the point where the rod must be replaced. A staggered refueling cycle is generally followed, in which one-third of the reactor's fuel rods are replaced each year. Because of the many radioactive fission fragments they contain, the spent fuel rods are very radioactive. Upon removal from the reactor, they are stored under water at the power plant for several months, by which time their activity has decreased to a level where they can be safely transported to a reprocessing plant.

At the reprocessing plant, the pellets are dissolved in acid and chemically processed to recover the uranium and plutonium. The remaining liquid waste is highly radioactive, and its safe disposal is still a matter of controversy. At present this material is stored in steel tanks, but eventually it is hoped the waste can be converted into a solid and permanently deposited in deep mines. The recovered uranium is enriched and formed into fuel pellets for use in new fuel rods. The recovered plutonium is already suitable for fuel, without further processing. It is also pure enough for use in a nuclear weapon, so safeguards are required to ensure that it is not diverted to military purposes.

An atomic bomb, unlike a reactor, requires fissionable material in pure form. Since the gas-diffusion process requires an immense industrial facility, the conversion of enriched uranium to pure ^{235}U cannot be done by a small country, much less a terrorist group. However, the plutonium produced in a reactor can be separated by relatively simple chemical methods that are within the capabilities of many countries. A terrorist group, on the other hand, would probably find it simpler, and certainly more dramatic, to steal one of the thousands of existing weapons sitting in arsenals around the world than to fabricate its own weapon from stolen reactor plutonium.

In a block of pure ^{235}U or ^{239}Pu, a single fission event can be started by a stray cosmic-ray neutron. The neutrons released by the event will then either escape from the block or initiate another fission event. Thus the multiplication factor depends primarily on the size, and hence on the mass, of the block. If the mass is too small, too many neutrons escape before initiating fission, so f is less than 1. Such a mass is said to be *subcritical*. If the mass is large enough, however, enough neutrons initiate fission events before they reach the surface, so f is greater than 1. Such a mass is said to be *critical*. An uncontrolled chain reaction will spontaneously erupt in a block of critical mass.

In an atomic bomb, two or more subcritical masses of ^{235}U, or ^{239}Pu, are suddenly forced together by conventional explosives, so that a critical mass is formed. Before this mass can disintegrate, enough fission takes place to generate an enormous amount of energy in a very small volume, which results in an immense explosion.

The energy released in the fission of 1 mol (235 g) of ^{235}U is about

$$(6 \times 10^{23})(200\,\text{MeV}) = 1.2 \times 10^{26}\,\text{MeV} = 2 \times 10^{13}\,\text{J}$$

In the bombs dropped on Japan, approximately 1 kg (4.25 mol) of ^{235}U (or ^{239}Pu) underwent fission. The energy released, therefore, was

$$(4.25)(2 \times 10^{31}\,\text{J}) = 8.5 \times 10^{13}\,\text{J}$$

which is equivalent to the energy released in the explosion of 20,000 tons of TNT.

The energy of the sun, and hence ultimately all nonnuclear energy on earth, comes from the fusion of four hydrogen nuclei (protons) to form helium. The overall reaction is

$$4\,{}^{1}_{1}\text{H} \longrightarrow {}^{4}_{2}\text{He} + 2e^+ + 2\nu \qquad\qquad 22.12$$

From the mass change

$$\Delta m = 4\,m_{{}^{1}\text{H}} - m_{{}^{4}\text{He}} = (4)(1.007825\text{ u}) - 4.0026\text{ u} = 0.0287\text{ u}$$

the energy released in each fusion event is found to be 26.7 MeV. The masses of the positrons are not included in the calculation of Δm because these masses are soon converted into energy in the reaction

$$e^+ + e^- \longrightarrow 2\gamma \qquad\qquad 22.13$$

where each γ is a high-energy gamma ray. The energy released by the fusion of 1 g of hydrogen is equal to the energy released by burning 16 tons of oil.

Equation 22.12 is just the overall reaction, but in reality four protons can never all come together at once. Instead the helium is built up in a series of reactions, each one of which involves the fusion of only two nuclei. In 1938 Hans Bethe (1906–   ) suggested two reaction sequences, the *carbon cycle* and the *proton-proton chain*, which are believed to be the primary processes taking place in the sun and similar stars. These sequences are shown in Table 22.7.

In the carbon cycle, hydrogen nuclei, instead of combining directly with each other, combine with a ^{12}C nucleus. After four hydrogen nuclei have attached themselves to the ^{12}C nucleus, it breaks apart into ^{4}He and ^{12}C. Thus ^{12}C is just a catalyst for this

TABLE **22.7**

Reaction	MeV	Carbon cycle and proton-proton chain
Carbon cycle:		
${}^{12}_{6}\text{C} + {}^{1}_{1}\text{H} \longrightarrow {}^{13}_{7}\text{N}$	+ 2.0	
${}^{13}_{7}\text{N} \longrightarrow {}^{13}_{6}\text{C} + e^+ + \nu$	+ 2.2†	
$\tau \doteq 10$ min		
${}^{13}_{6}\text{C} + {}^{1}_{1}\text{H} \longrightarrow {}^{14}_{7}\text{N}$	+ 7.5	
${}^{14}_{7}\text{N} + {}^{1}_{1}\text{H} \longrightarrow {}^{15}_{8}\text{O}$	+ 7.3	
${}^{15}_{8}\text{O} \longrightarrow {}^{15}_{7}\text{N} + e^+ + \nu$	+ 2.7†	
$\tau = 2$ min		
${}^{15}_{7}\text{N} + {}^{1}_{1}\text{H} \longrightarrow {}^{12}_{6}\text{C} + {}^{4}_{2}\text{He}$	+ 5.0	
	26.7	
Proton-proton chain:		
${}^{1}_{1}\text{H} + {}^{1}_{1}\text{H} \longrightarrow {}^{2}_{1}\text{H} + e^+ + \nu\ +\ 1.5\dagger$ (twice)	= 3.0	
${}^{2}_{1}\text{H} + {}^{1}_{1}\text{H} \longrightarrow {}^{3}_{2}\text{He}\qquad\ \ + 5.4$ (twice)	= 10.8	
${}^{3}_{2}\text{He} + {}^{3}_{2}\text{He} \longrightarrow {}^{4}_{2}\text{He} + 2\,{}^{1}_{1}\text{H} + 12.9$ (once)	= 12.9	
	26.7	

†Includes the energy obtained from the ultimate annihilation of the positron (Eq. 22.13).

reaction, since it is not ultimately consumed. The overall reaction is given by Eq. 22.12.

In the proton-proton chain, two hydrogen nuclei first combine to form deuterium, ^{2}H. This then reacts with normal hydrogen to form ^{3}He, and finally two ^{3}He nuclei react with each other to produce a ^{4}He nucleus and two protons, ^{1}H. The overall reaction is again given by Eq. 22.12.

These fusion reactions do not occur at normal temperature because the electrical repulsion between the nuclei keep them from getting close enough to react. Figure 22.2 shows how the repulsive electric force between two protons dominates the potential energy at large distances, while the attractive nuclear force dominates at small distances. Only at temperatures of 20 million degrees or more, such as exist in the interior of stars, do nuclei have sufficient kinetic energy to overcome the electrical repulsion. The situation is similar to any combustion process, which occurs only above a certain temperature.

The fusion of the hydrogen isotopes *deuterium* (^{2}H) and *tritium* (^{3}H) in the reaction

$$^3_1H + {}^2_1H \longrightarrow {}^4_2He + {}^1_0n + 17.6\,\text{MeV} \qquad 22.14$$

has been achieved on earth. Deuterium is a naturally occurring nonradioactive nuclide which is easily separated from the more abundant ^{1}H. Tritium is a radioactive nuclide that is produced by placing ^{6}Li inside a reactor, where it absorbs a neutron in the reaction

$$^6_3Li + {}^1_0n \longrightarrow {}^4_2He + {}^3_1H \qquad 22.15$$

The reaction in Eq. 22.14 is called a *thermonuclear reaction* because it occurs only at a temperature of millions of degrees. In a hydrogen bomb, an uncontrolled fusion reaction is produced by placing a deuterium-tritium mixture around an ordinary fission bomb. When the fission bomb explodes, temperatures of millions of degrees initiate the ^{3}H–^{2}H fusion reaction, resulting in an explosion of astronomical size.

Because of the enormous energy potential of fusion, extensive research has been directed toward achieving a controlled thermonuclear reaction. The problem is that no material can contain a gas at the temperatures required for controlled fusion. The most successful approach so far has been the *tokamak reactor* invented in Russia. In a tokamak, a plasma of ionized particles is trapped by a magnetic field in a toroidal (donut-shaped) region. Since the ions never touch the walls of the region, their temperature can be raised to millions of degrees.

So far a fusion reactor has not achieved a temperature high enough, or a confinement time long enough, to yield more energy than it consumes. But enough is known about the possibilities of this approach to begin tentative design studies for a commercial reactor. Figure 22.17 shows the cross section of a commercial tokamak reactor envisioned by scientists at the Oak Ridge National Laboratory. Superconducting coils (Sec. 21.2) would supply the magnetic field needed to contain the plasma at a temperature of 10^8 K. Pellets of deuterium-tritium would be injected into the plasma, where they would undergo fusion.

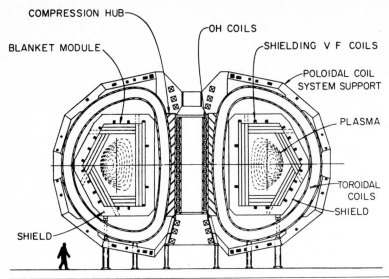

FIGURE 22.17

Tokamak reactor of the twenty-first century, as envisioned by scientists at the Oak Ridge National Laboratory. (*Oak Ridge National Laboratory.*)

Liquid sodium nitrate, or some other salt, would remove the heat generated by the fusion reaction and transfer it to a boiler to make steam for the generator turbines. A blanket of liquid lithium surrounding the reactor would absorb excess neutrons and produce tritium through the reaction in Eq. 22.15. Controlled fusion offers the promise of unlimited power in the future, but the design in Fig. 22.17 is still decades away from being built.

GUIDE TO MAJOR TOPICS

Topic	References	Problems
Structure of the nucleus	Sec. 22.1	
Nuclides		1 to 4
Binding energy	Eqs. 22.1 and 22.2; Examples 22.1 and 22.2; Table 22.1	5 to 8
Radioactivity	Sec. 22.2	
Modes of decay		9 to 12
Decay rate	Eqs. 22.3 to 22.7; Examples 22.3 and 22.4	13 to 16
Half-life	Eq. 22.8; Example 22.5	17 to 20
Nuclear radiation	Sec. 22.3	
Properties	Tables 22.2 and 22.3	21 to 24
Detectors		25
Dosimetry	Table 22.4	26 to 28
Medical uses	Example 22.6	
Nuclear energy	Sec. 22.4	
Fission		29 to 32
Fusion	Eqs. 22.12 to 22.15; Table 22.7	33 to 34

PROBLEMS

REMARK Refer to Table 22.1 and the Table of the Elements for data concerning nuclides mentioned in these problems.

1 How many neutrons are there in ^{159}Tb? *Ans.* 94

2 What is the symbol for the nuclide with 51 protons and 72 neutrons?

3 Mirror nuclides are two nuclides with the same atomic mass A and with atomic numbers Z_1 and Z_2 related by

$Z_1 + Z_2 = A$

(a) What nuclide is the mirror of ^{13}N?
(b) Show that the neutron number of a nuclide is equal to the proton number of its mirror.
Ans. (a) ^{13}C

4 Which of the following unidentified nuclides are isotopes of each other (a) $^{175}_{71}X$, (b) $^{71}_{32}X$, (c) $^{175}_{74}X$, (d) $^{167}_{71}X$, (e) $^{71}_{30}X$, (f) $^{180}_{74}X$?

5 Find the mass defect and the binding energy of (a) ^{2}H (deuterium), and (b) ^{3}H (tritium).
Ans. (a) 2.5×10^{-3} u, 2.3 MeV; (b) 9.1×10^{-3} u, 8.5 MeV

6 The mass defect of ^{3}He is 0.0083 u. Find the binding energy and the atomic mass of this nuclide.

7 What is the binding energy per particle of ^{88}Sr?
Ans. 8.73 MeV

8 Show that the decay ^{16}O $\longrightarrow$ ^{12}C + ^{4}He is not possible. (Hint: Consider the masses of the nuclei involved.)

9 (a) What nuclide alpha-decays to ^{238}U?
(b) To what nuclide does ^{135}Xe beta-decay?
Ans. (a) ^{242}Pu; (b) ^{135}Cs

10 Which of the following nuclides are e^- emitters and which are e^+ emitters: (a) $^{50}_{20}$Ca, (b) $^{88}_{40}$Zr, (c) $^{151}_{60}$Nd, (d) $^{189}_{80}$Hg?

11 Show that a free neutron can beta-decay into a proton and an electron but that a free proton cannot beta-decay into a neutron and a positron.

REMARK A free neutron beta-decays with a half-life of 10.8 min, whereas a free proton is stable. However, a proton inside a proton-rich nucleus can beta-decay.

12 *Electron capture* is a nuclear transformation in which a proton in a nucleus combines with an orbital electron to form a neutron. For example, a typical electron-capture reaction is $^{168}_{69}$Tm + $e^- \rightarrow$ $^{168}_{68}$Er. Do nuclei that undergo electron capture lie above or below the band of stable nuclei in Fig. 22.3?

13 A sample of ^{65}Zn has an activity of 0.2 Ci. The decay constant of ^{65}Zn is 3.29×10^{-8} s^{-1}. (a) How many ^{65}Zn nuclei are in the sample? (b) What is the mass of the sample?
Ans. (a) 2.25×10^{17}; (b) 24.4 μg

14 The decay rate of 1 g of ^{36}Cl is 1.18×10^9 s^{-1}. (a) What is the decay constant of ^{36}Cl? (b) What is the probability that a ^{36}Cl nucleus will decay in 1 y? (Hint: If p is the

probability that an event will occur in time t, then np is the probability the event will occur in time nt, provided $np \ll 1$.)

15 A sample of ^{86}Rb has an activity of 3.2×10^6 s^{-1} at $t = 0$ and an activity of 2.2×10^6 s^{-1} at $t = 10$ d. (a) What is the decay constant of ^{86}Rb? (b) How long will it take for the activity to reach 2×10^5 s^{-1}?
Ans. (a) 0.0375 d^{-1}; (b) 73.9 d

16 The decay rate of a sample of ^{177}Yb is measured for 1 min every hour. The results are given in Table 22.8. (a) Plot these data on semilogarithmic paper, and draw the straight line that passes as close as possible to all the points. (b) Show that

$$\log R = \log R_0 - (\lambda \log e)t$$

where log is the logarithm to the base 10, and $e = 2.718$ is the natural logarithm base. Thus the slope of the straight line in part a is $-\lambda \log e$. (c) Determine the decay constant λ from the slope of the line.

TABLE **22.8**

Decay rate of a sample of ^{177}Yb at 1 h intervals	
Time, h	Decay rate, min^{-1}
0	721
1	510
2	323
3	230
4	182
5	120
6	71

17 (a) What is the decay constant of ^{214}Bi? (Use data from Table 22.1.) (b) What is the activity of 1 μg of ^{214}Bi?
Ans. (a) 5.86×10^{-4} s^{-1}; (b) 44.4 Ci

18 In the atmosphere, and in living organisms which exchange CO_2 with the atmosphere, there is one ^{14}C nucleus for every 7.7×10^{11} ^{12}C nuclei. (a) What is the decay rate in 1 g of carbon taken from a living organism? (b) When an organism dies, it no longer exchanges CO_2 with the atmosphere, so its supply of ^{14}C gradually decays away. This fact is used to date archeological specimens. What is the decay rate in 1 g of carbon taken from a 2000-year-old-bone? (c) One gram of carbon from a bone of unknown age is found to have a decay rate of 5.6 min^{-1}. What is the age of the bone?
Ans. (a) 15 min^{-1}; (b) 11.8 min^{-1}; (c) 8140 y

19 Heavy elements like uranium are believed to be formed in the interior of massive stars.

These stars ultimately explode, releasing these elements into space. This interstellar matter recondenses into new stars and planets. Thus the uranium on earth was once inside a star. Assuming that at the time the star exploded there were equal amounts of ^{235}U and ^{238}U, how long ago did the explosion occur?
(*Hint:* Use the half-lives of ^{235}U and ^{238}U to compute the time required for these nuclides to have the relative abundance now found on earth.)
Ans. 6.0×10^9 y

20 From the Table of the Isotopes in the *Handbook of Chemistry and Physics,* draw a chart similar to Fig. 22.6 starting with ^{232}Th.

21 An alpha particle has a range in air of 1.67 cm. What is the range in aluminum of a beta particle of the same energy?
Ans. 0.56 cm

22 How thick must a lead wall be to stop 75 percent of the 1-MeV gamma rays incident on it?

23 After traveling a distance x in matter, the intensity I of a gamma ray beam is

$$I = I_0 e^{-\mu x}$$

where I_0 is the initial intensity, and μ is the *absorption coefficient* characteristic of the material. (a) Show that μ is related to the half-value layer λ by

$$\mu = \frac{\ln 2}{\lambda}$$

(b) What fraction I/I_0 of a 1-MeV gamma ray beam remains after passing through 5 cm of body tissue?
Ans. (b) 0.70

24 The intensity of a gamma ray beam is $0.27\, I_0$ after passing through 0.8 cm of lead, where I_0 is the incident intensity. (a) What is the absorption coefficient of the gamma rays? (See Prob. 23.) (b) What is the half-value layer? (c) What is the energy of the gamma rays?
Ans. (a) 1.64 cm^{-1}; (b) 0.42 cm; (c) 0.5 MeV

25 In a photomultiplier tube with 10 electrodes, how many electrons reach the last electrode for every electron incident on the second electrode if the multiplication factor is 8?

26 A 50-keV (0.050-MeV) x-ray beam is used to take a chest x-ray. What is the ratio of the dose received by the chest skin to the dose received by the back skin? (Assume the chest is 10 cm thick.)

27 A patient having a brain scan is injected with 20 mCi of $^{99}Tc^m$. The energy of the gamma ray emitted when this nuclide decays to its ground state is 0.143 MeV. Assuming that half of the gamma rays escape from the body before interacting, what is the radiation dose received by a 60-kg patient? (*Hint:* Because of its short half-life, all the $^{99}Tc^m$ decays while still in the body.)
Ans. 0.44 rd

28 If the intensity of a 50-keV x-ray beam is 300 W/m^2, estimate the dose received by a patient during a 0.25-s exposure. (*Hint:* Take the body to be a 10-cm thick slab with the density of water.)

29 Determine the atomic number and mass number of the unknown nuclide in the following reactions:
(a) $^{1}_{0}n + ^{16}_{8}O \rightarrow X$
(b) $^{4}_{2}He + ^{118}_{50}Sn \rightarrow X + ^{1}_{0}n$
(c) $^{1}_{1}H + ^{127}_{53}I \rightarrow ^{50}_{21}Sc + X$
(d) $^{235}_{92}U + ^{1}_{0}n \rightarrow ^{107}_{43}Tc + X + 5\,^{1}_{0}n$
Ans. (a) ^{17}O; (b) ^{121}Te; (c) ^{78}As; (d) ^{124}In

30 Nuclei, such as ^{95}Sr and ^{139}Xe, which are the direct fragments of a fission event, undergo a sequence of beta decays until stable nuclei are formed. (a) From the Table of the Isotopes in the *Handbook of Chemistry and Physics,* trace the sequence of decays that ^{95}Sr and ^{139}Xe each follows. (b) What is the stable nuclide formed from each fragment? (c) What is the approximate half-life of each sequence?

31 The average person in the United States consumes about 2.5×10^{10} J of energy per year. How much ^{235}U is required to generate this energy?
Ans. 0.29 g

32 Approximately what percent of the mass of ^{235}U is converted into energy by fission?

33 Calculate the energy released in the "oxygen burning" reaction

$$^{16}_{8}O + ^{16}_{8}O \longrightarrow ^{28}_{14}Si + ^{4}_{2}He$$

The mass of ^{28}Si is 27.97693 u.
Ans. 9.58 MeV

REMARK Although the reaction in Prob. 33 is exothermic, it can occur only at temperatures of many billions of degrees because of the strong electrical repulsion between two oxygen nuclei. If temperatures this high were produced in an atomic explosion, all the oxygen in the atmosphere would convert to silicon. However, the temperature in an atomic explosion is only a few million degrees.

34 Calculate the energy released in the fusion

reaction

$$^2_1H + ^2_1H \longrightarrow ^3_1H + ^1_1H$$

REMARK This fusion reaction is favored by some researchers over the deuterium-tritium reaction (Eq. 22.14) because it does not produce damaging neutrons.

BIBLIOGRAPHY

GLASSTONE, SAMUEL: *Sourcebook on Atomic Energy,* 3d ed., D. Van Nostrand Company, Inc., Princeton, N.J., 1968. The clearest, simplest, and most complete account of the basic principles of nuclear physics and the applications of nuclear physics to science, technology, and medicine.

JUNGK, ROBERT: *Brighter than a Thousand Suns,* Harcourt, Brace, and Company, Inc., New York, 1968. Popular history of the development of the atomic bomb, including the history of Germany's atomic-bomb effort.

OLANDER, DONALD R.: "The Gas Centrifuge," *Scientific American* **239:**37 (August 1978). The gas centrifuge is an alternative to gaseous diffusion for the separation of ^{235}U.

PARKINS, W. E.: "Engineering Limitations of Fusion Power Plants," *Science* **199:**1403 (1978). A sober account of the effect of radiation damage on the materials used in a fusion reactor.

ROBERTSON, J. A. L.: "The CANDU Reactor System: An Appropriate Technology," *Science* **199:**657 (1978). Description of the Canadian heavy-water reactor system.

STEINER, DON, and JOHN F. CLARKE: "The Tokamak: Model T Fusion Reactor," *Science* **199:** 1395 (1978). Report of a design study of a commercial-size fusion reactor that could be built by the year 2000.

WEAST, ROBERT C. (ed.): *Handbook of Chemistry and Physics,* 59th ed., CRC Press, Inc., West Palm Beach, Florida, 1978. This well-known reference book contains a 300-page table listing the properties of all the known nuclides. A new edition is published every year or so.

APPENDIXES

I

THE INTERNA- TIONAL SYSTEM OF UNITS

The International System of Units, abbreviated SI in all languages, uses a single basic unit for each physical quantity, and multiples and fractions of this unit are formed by adding a prefix. For example, the SI unit of length is the meter (m); the millimeter (mm), centimeter (cm), and kilometer (km) are formed by adding the prefixes *milli-* (m), *centi-* (c), and *kilo-* (k) to it. The same prefixes are used for all physical quantities. These prefixes and their symbols are listed in Table I.1. Table I.2 lists some common SI units and their symbols.

The following rules should be observed when using symbols of units:

1 Do not put a period after a symbol. Thus the notation 7 km. is not correct, but 7 km is.

2 Do not put an s at the end of a symbol. Thus, the notation 7 kms is not correct, but 7 km is.

3 The symbol for a unit is a lowercase letter except when the unit is derived from a proper name. Thus the symbol for the meter is m, whereas the symbol for the newton is N. (An exception to this is the symbol for liter, which is now taken to be L.)

4 A symbol with a prefix is treated as a new unit which can be raised to a power without using brackets. Thus the notation cm^3 means $(10^{-2} \text{ m})^3 = 10^{-6} \text{ m}^3$, and not 10^{-2} m^3.

TABLE I.1

SI prefixes

Prefix	Symbol	Value
exa-	E	10^{18}
peta-	P	10^{15}
tera-	T	10^{12}
giga-	G	10^{9}
mega-	M	10^{6}
kilo-	k	10^{3}
hecto-	h	10^{2}
deca-	da	10^{1}
deci-	d	10^{-1}
centi-	c	10^{-2}
milli-	m	10^{-3}
micro-	μ	10^{-6}
nano-	n	10^{-9}
pico-	p	10^{-12}
femto-	f	10^{-15}
atto-	a	10^{-18}

TABLE **1.2**

Some common SI units

Quantity	Name of unit	Symbol of unit	Unit expressed in terms of fundamental units
Fundamental units:			
Length	Meter	m	
Mass	Kilogram	kg	
Time	Second	s	
Temperature	Kelvin	K	
Electric current	Ampere	A	
Derived units:			
Frequency	Hertz	Hz	s^{-1}
Force	Newton	N	$kg \cdot m/s^2$
Pressure	Pascal	Pa	$N/m^2 = kg/(m \cdot s^2)$
Viscosity	Poiseuille	Pl	$N \cdot s/m^2 = kg/(m \cdot s)$
Energy	Joule	J	$kg \cdot m^2/s^2$
Power	Watt	W	$J/s = kg \cdot m^2/s^3$
Electric charge	Coulomb	C	$A \cdot s$
Electric potential	Volt	V	$J/C = kg \cdot m^2/(A \cdot s^3)$
Electric resistance	Ohm	Ω	$V/A = kg \cdot m^2/(A^2 \cdot s^3)$
Electric capacitance	Farad	F	$C/V = (A^2 \cdot s^4)/(kg \cdot m^2)$
Magnetic field	Tesla	T	$kg/(A \cdot s^2)$

CONVER-SIONS BETWEEN SI, ENGLISH, AND CUSTOMARY UNITS

The following tables give the conversions between various units of length, volume, time, mass, speed, force, energy, power, pressure, and specific heat. Each entry gives the value of the unit in the left-hand column in terms of the unit in the top row. For example, from the length table, we find that 1 mi = 5280 ft.

Length

	Meter	Kilometer	Foot	Mile	Inch	Mil
Meter	1	0.001	3.281	6.214×10^{-4}	39.37	3.937×10^4
Kilometer	10^3	1	3.281×10^3	0.6214	3.937×10^4	3.937×10^7
Foot	0.3048	3.048×10^{-4}	1	1.894×10^{-4}	12	1.2×10^4
Mile	1.609×10^3	1.609	5.280×10^3	1	6.336×10^4	6.336×10^7
Inch	0.0254	2.54×10^{-5}	0.08333	1.578×10^{-5}	1	10^3
Mil	2.54×10^{-5}	2.54×10^{-8}	8.333×10^{-5}	1.578×10^{-8}	0.001	1

Volume

	Cubic meter	Cubic foot	Liter	Gallon (U.S.)
Cubic meter	1	35.31	10^3	264.2
Cubic foot	0.02832	1	28.32	7.480
Liter	10^{-3}	0.03531	1	0.2642
Gallon (U.S.)	3.785×10^{-3}	0.1337	3.785	1

Time

	Second	Minute	Hour	Day	Year
Second	1	0.01667	2.778×10^{-4}	1.157×10^{-5}	3.169×10^{-8}
Minute	60	1	0.01667	6.944×10^{-4}	1.901×10^{-6}
Hour	3600	60	1	4.167×10^{-2}	1.141×10^{-4}
Day	8.64×10^4	1440	24	1	2.738×10^{-3}
Year	3.156×10^7	5.260×10^5	8766	365.26	1

Mass

	Kilogram	Slug	Pound
Kilogram	1	0.06852	2.205
Slug	14.59	1	32.17
Pound (mass)	0.4536	3.108×10^{-2}	1

Force

	Newton	Pound	Kilogram
Newton	1	0.2248	0.1020
Pound	4.448	1	0.4536
Kilogram (force)	9.807	2.205	1

Speed

	Meter per second	Foot per second	Kilometer per hour	Mile per hour
Meter per second	1	3.281	3.600	2.237
Foot per second	0.3048	1	1.087	0.682
Kilometer per hour	0.278	0.911	1	1.609
Mile per hour	0.447	1.467	0.6214	1

Energy

	Joule	Foot-pound	Kilocalorie	Btu	Kilowatt hour
Joule	1	0.7376	2.390×10^{-4}	9.478×10^{-4}	2.778×10^{-7}
Foot-pound	1.356	1	3.240×10^{-4}	1.285×10^{-3}	3.766×10^{-7}
Kilocalorie	4.184×10^{3}	3.086×10^{3}	1	3.966	1.162×10^{-3}
Btu	1.055×10^{3}	778.2	0.2521	1	2.931×10^{-4}
Kilowatt hour	3.600×10^{6}	2.655×10^{6}	860.4	3.412×10^{3}	1

Power

	Watt	Kilowatt	Foot-pound per second	Horsepower	Btu per hour
Watt	1	0.001	0.7376	1.34×10^{-3}	3.41
Kilowatt	1000	1	737.6	1.34	3.41×10^{3}
Foot-pound per second	1.356	1.356×10^{-3}	1	$1.818 \times ^{-3}$	4.63
Horsepower	746	0.746	550	1	2.54×10^{3}
Btu per hour	0.293	2.93×10^{-2}	0.216	3.93×10^{-4}	1

Pressure

	Pascal	Pound per square foot	Torr	Atmosphere	Bar
Pascal	1	2.088×10^{-2}	7.502×10^{-3}	9.87×10^{-6}	10^{-5}
Pound per square foot	47.88	1	0.3591	4.725×10^{-4}	4.788×10^{-4}
Torr (mmHg)	133.3	2.784	1	1.316×10^{-3}	1.333×10^{-3}
Atmosphere	1.013×10^{5}	2.116×10^{3}	760	1	1.013
Bar	10^{5}	2.088×10^{3}	750.1	0.9869	1

Specific Heat

	Joule per kilogram per degree Celsius	Kilocalorie per kilogram per degree Celsius	Btu per pound per degree Fahrenheit
Joule per kilogram per degree Celsius	1	2.390×10^{-4}	2.390×10^{-4}
Kilocalorie per kilogram per degree Celsius	4184	1	1
Btu per pound per degree Fahrenheit	4184	1	1

Any positive number x can be written as a power:

$x = a^y$

The constant a is called the *base*, and the exponent y is called the *logarithm of x to the base a*, written

$y = \log_a x$

When $a = 10$, y is the *common logarithm*, written

$y = \log x$

LOGARITHMS

If $x_1 = a^{y_1}$ and $x_2 = a^{y_2}$, then from the law of exponents (Sec. 1.3) the product $x_1 x_2$ is

$x_1 x_2 = a^{y_1} a^{y_2} = a^{y_1 + y_2}$

and the logarithm of $x_1 x_2$ is

$$\log x_1 x_2 = y_1 + y_2 = \log x_1 + \log x_2 \qquad \text{III.1}$$

In other words, the logarithm of a product is equal to the sum of the logarithms of its factors. Similarly, the logarithm of a quotient is equal to the difference of the logarithms of the numerator and denominator:

$$\log \frac{x_1}{x_2} = \log x_1 - \log x_2 \qquad \text{III.2}$$

Table III.1 gives the common logarithms of numbers between 1.0 and 9.9. To find the logarithm of a number outside this range, the number is written in scientific notation and Eq. III.1, or a calculator, is used.

Example III.1 What is the logarithm of 510?
We write

$510 = 5.1 \times 10^2$

and so from Eq. III.1

$\log 510 = \log (5.1 \times 10^2) = \log 5.1 + \log 10^2$

From Table III.1, $\log 5.1 = 0.708$, and by definition, $\log 10^2 = 2$, so

$\log 510 = 0.708 + 2 = 2.708$ □

Example III.2 What is the logarithm of 1.8×10^{-5}?
Using Table III.1 we find

$\log (1.8 \times 10^{-5}) = \log 1.8 + \log 10^{-5}$
$\qquad\qquad = 0.255 + (-5) = -4.745$

Alternatively, on a calculator we press the keys

and the display is

$-4.7447 \ldots$ □

The relation

$$\log a^y = y \log a \qquad\qquad \text{III.3}$$

is useful for calculating the logarithm of a power.

Example III.3 What is the cube root of 2200?

From Table III.1 and Eq. III.3 the logarithm of $\sqrt[3]{2200} = 2200^{1/3}$ is

$$\log 2200^{1/3} = \tfrac{1}{3}\log 2200 = \tfrac{1}{3}(\log 2.2 + 3) = 1.114$$

and so

$$\sqrt[3]{2200} = 10^{1.114} = 10^{1} \times 10^{0.114}$$

From Table III.1 the number whose logarithm is 0.114 is found to be 1.3, so

$$\sqrt[3]{2200} = 10^{1} \times 1.3 = 13$$

Alternatively, on a calculator we press the keys

and the display is

13.0059 . . .   □

TABLE **III.1**

x	.0	.1	.2	.3	.4	.5	.6	.7	.8	.9	**Common logarithms**
1.	.000	.041	.079	.114	.146	.176	.204	.230	.255	.279	This table gives the logarithms of numbers from 1.0 to 9.9; to find the logarithm of a number x outside this range, write $x = x' \times 10^n$, where x' is between 1.0 and 9.9; then $\log x = \log x' + n$.
2.	.301	.322	.342	.362	.380	.398	.415	.431	.447	.462	
3.	.477	.491	.505	.519	.531	.544	.556	.568	.580	.591	
4.	.602	.613	.623	.633	.643	.653	.663	.672	.681	.690	
5.	.699	.708	.716	.724	.732	.740	.748	.756	.763	.771	
6.	.778	.785	.792	.799	.806	.813	.820	.826	.833	.839	
7.	.845	.851	.857	.863	.869	.875	.881	.886	.892	.898	
8.	.903	.908	.914	.919	.924	.929	.934	.940	.944	.949	
9.	.954	.959	.964	.968	.973	.978	.982	.987	.991	.996	

PROBLEMS

1   Find the logarithms of the following numbers:
(a) 120   (b) 0.027
(c) 6.0×10^{23} (d) 1.6×10^{-24}
Ans.   (a) 2.079;   (b) -1.569;   (c) 23.778;
(d) -23.796

2   Prove that

$$\log \frac{1}{x} = -\log x$$

3   Evaluate the following quantities using logarithms:

(a) $\sqrt{870}$ (b) $\left(\dfrac{3}{2}\right)^{2/3}$

(c) 2^{π} (d) $(3.5 \times 10^{-5})^{-1/2}$

Ans.   (a) 29.5;   (b) 1.310;   (c) 8.825;
(d) 1.69×10^{2}

4   Repeat Prob. 3 using a calculator.

Geometry

It is useful to think of the apex of an angle as coinciding with the center of a circle, as shown in Fig. IV.1. Then, if the circumference of the circle is divided into equal divisions, the angle can be measured by the number of divisions included inside it. In Fig. IV.1 the circle is divided into 16 equal parts and the angle includes 2.5 of them.

Of course in practice a circle is divided into 360 equal parts, called *degrees* (°). The number 360 is convenient because it has so many divisors, but other than this it has no special significance. The division of the circle into 360 parts seems to be of very ancient origin, probably dating back to the Babylonians, who used a number system based on 60.

A more meaningful measure of angle is the radian (rad). Figure IV.2 shows an angle θ drawn at the center of a circle of radius r. The angle subtends the arc length s of the circle. The angle θ in radians is defined as the ratio of s to r.

$$\theta = \frac{s}{r} \quad \theta \text{ in radians} \qquad\qquad \text{IV.1}$$

Since s and r both have the dimension of length, θ is dimensionless.

The arc length of a complete circle is its circumference $2\pi r$, so the angle of a complete circle is

$$\frac{2\pi r}{r} = 2\pi \text{ rad} = 360°$$

This gives the conversion from radians to degrees:

$$1 \text{ rad} = \frac{360°}{2\pi} = 57.3° \qquad\qquad \text{IV.2}$$

Table IV.1 gives some angles in both degrees and radians.

Angles are easily measured and drawn with a semicircular scale, called a *protractor*, whose circumference is divided into 180°. To measure an angle with it, you first place the center mark of the protractor over the apex of the angle (Fig. IV.3). (The center mark is a point on the diameter of the protractor that marks the center of the circle.) Then the protractor is rotated until one side of the angle falls on the 0° division of the protractor. The division that coincides with the other side of the angle is the number of degrees in the angle.

REMARK Be sure you know where the center mark is on your protractor. A common mistake in using a protractor is to place the wrong point of the protractor over the apex of an angle. Another common mistake is to read the wrong scale. Some protractors have two scales, one starting with 0° from the left side of the protractor and one starting with 0° from the right side. The angle in Fig. IV.3 is 50°, not 130°, because the scale that starts with 0° along one side of the angle must be used.

To draw a given angle, say, 35°, you first draw a straight line and mark a point on it. Next, place the diameter of the protractor along the line with its center mark over the point. Then locate 35° on the circumference (being sure to use the correct

IV

GEOMETRY AND TRIGO-NOMETRY

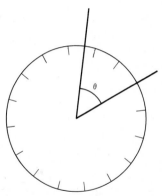

FIGURE IV.1

An angle coinciding with the center of a circle divided into 16 equal parts.

FIGURE IV.2

An angle θ subtending the arc length s on a circle of radius R.

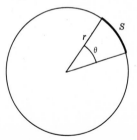

TABLE **IV.1**

Angle		Sine	Cosine	Tangent
deg	rad			
0	0	0	1.000	0
30	$\frac{1}{6}\pi$	0.500	0.866	0.577
45	$\frac{1}{4}\pi$	0.707	0.707	1.000
60	$\frac{1}{3}\pi$	0.866	0.500	1.732
90	$\frac{1}{2}\pi$	1.000	0	∞
180	π	0	−1.000	0
270	$\frac{3}{2}\pi$	−1.000	0	−∞
360	2π	0	1.000	0

Trigonometric functions of some angles in degrees and radians

scale), and mark a point there. Finally, draw a straight line between the two points. With a little practice you can learn to draw angles quickly and accurately.

You should be familiar with the elementary properties of angles and triangles. A few important theorems from geometry are given here for review.

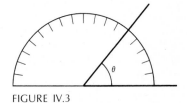

FIGURE IV.3

A protractor measuring an angle.

Theorem 1 *If two straight lines intersect, the opposite angles are equal.* In Fig. IV.4 the two lines AB and QP intersect at O. The angles α and β are opposite and therefore equal. Likewise, the angles θ and ϕ are equal.

Theorem 2 *If a line intersects two parallel lines, the opposite interior angles are equal.* In Fig. IV.5 the line AB intersects the two parallel lines at O and P. Here α and β are opposite interior angles and therefore equal. Likewise, the angles θ and ϕ are equal.

Theorem 3 *If the corresponding sides of two angles are perpendicular to each other, the angles are equal.* In Fig. IV.6 the side AO of angle α is perpendicular to the side QP of angle β,

FIGURE IV.4

Two straight lines intersecting at O.

FIGURE IV.5

A straight line intersecting two parallel lines.

FIGURE IV.6

Two angles with corresponding sides perpendicular to each other.

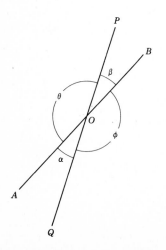

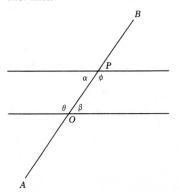

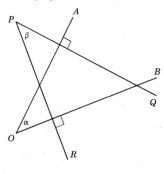

while the side *BO* of α is perpendicular to the side *RP* of β. Therefore, the angles α and β are equal. This theorem is used several times in this book.

Theorem 4 *The sum of the interior angles of a triangle equals 180°.* In Fig. IV.7 the interior angles of the triangle *ABC* are α, β, and γ. The sum of these angles is

$$\alpha + \beta + \gamma = 180°$$

Theorem 5 *If the interior angles of two triangles are equal, the triangles are similar.* In Fig. IV.8 the triangles *ABC* and *PQR* are similar because α = θ, β = φ, and γ = ψ.

Trigonometry

A *right triangle* is a triangle in which one of the angles is 90°. The sum of the other two angles must, by Theorem 4, equal 180° − 90° = 90°. This means that if one of these two angles is given, the other angle is determined. For example, if one angle is 35°, the other must be 55°.

Consider the angle θ in the right triangle in Fig. IV.9. The side across from θ is called the *opposite* side, whereas the side across from the right (90°) angle is called the *hypotenuse*. By Theorem 5 all right triangles with the same angle θ are similar, since the other angles are also equal. Consequently, the ratio of corresponding sides is the same for all these triangles. For example, in all right triangles in which θ = 30°, the ratio of the opposite side to the hypotenuse, opposite/hypotenuse, is 0.500. This ratio is called the *sine* of θ, written sin θ. The sine can be calculated for any angle.

Ratios of other sides of a right triangle define other *trigonometric functions*. The three most important ones are the sine, cosine, and tangent. These three functions are defined as follows:

$$\text{Sine} \quad = \frac{\text{opposite}}{\text{hypotenuse}} \qquad \text{written sin } \theta$$

$$\text{Cosine} = \frac{\text{adjacent}}{\text{hypotenuse}} \qquad \text{written cos } \theta$$

$$\text{Tangent} = \frac{\text{opposite}}{\text{adjacent}} = \frac{\text{sine}}{\text{cosine}} \qquad \text{written tan } \theta$$

Table IV.1 gives the values of the sine, cosine, and tangent for some angles, but in general you should use a calculator to find the trigonometric function of an arbitrary angle.

When θ is very small, the sine and tangent are given by the small-angle approximation

$$\sin \theta = \tan \theta = \theta \qquad \theta \text{ in radians} \qquad \text{IV.3}$$

where θ is the angle in radians. If θ is in degrees, Eq. IV.2 can be used to rewrite Eq. IV.3 in the form

$$\sin \theta = \tan \theta = \frac{\theta}{57.3} = 0.0174\,\theta \qquad \theta \text{ in degrees} \qquad \text{IV.4}$$

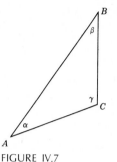

FIGURE IV.7

The interior angles of a triangle.

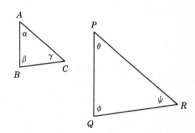

FIGURE IV.8

Two similar triangles.

FIGURE IV.9

A right triangle.

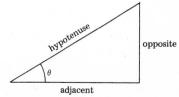

Most calculators have an R-D slide key. The key should be set to D when you want to calculate the trigonometric function of an angle that is given in degrees. Also, when in this position, the D/R key converts a number from radians to degrees.

Example IV.1 (a) Find the sine of 26.3°. (b) What is the angle whose tangent is 0.476? (c) Convert 1.67 rad to degrees. (Set the R-D slide key of your calculator to D.)

(a) On an algebraic calculator, press the keys

| 2 | 6 | · | 3 | sin |

and the display is

0.44307 . . .

(b) Press the keys

| · | 4 | 7 | 6 | arc | tan |

and the display is

25.454 . . .

which is the angle in degrees.

> **REMARK** On some calculators the inversion key is marked INV instead of ARC.

(c) Press the keys

| 1 | · | 6 | 7 | D/R |

and the display is

95.683 . . .

which is the angle in degrees. ☐

The R-D slide key should be set to R when you want to calculate the trigonometric function of an angle that is given in radians. Also, when the slide key is in this position, the D/R key converts a number from degrees to radians.

Example IV.2 (a) Calculate cos 0.2π. (b) Find the angle whose sine is −0.32. (c) Convert 27° to radians. (Set the R-D slide key of your calculator to R.)

(a) On a calculator with reverse Polish logic, press the keys

| · | 2 | ENTER | π | × | cos |

and the display is

.80901 . . .

(b) Press the keys

·	3	2	+/−	ENTER	inv	sin

and the display is

−.32572 . . .

which is the angle in radians. Note that the angle in radians is close to the value of the sine when both numbers are small.
   (c) Press the keys

2	7	ENTER	R/D

and the display is

0.47123 . . .

which is the angle in radians.   □

    A protractor and a calculator provide alternative methods for solving problems involving triangles. These are the graphical and trigonometric methods discussed in Sec. 2.4. If you feel uncomfortable with trigonometry, you should first use the graphical method. It has the advantages of being concrete and of helping you grasp the geometrical relations involved. The trigonometric method should be used only when you are confident that you can save time with it.

PROBLEMS

1 Express the following angles in radians:
(a) 10°   (b) 35°
(c) 150°   (d) 290°
Ans. (a) 0.174; (b) 0.611; (c) 2.618; (d) 5.061

2 Express the following angles in degrees:
(a) 0.5 rad   (b) $\frac{3}{4}\pi$ rad
(c) 2 rad   (d) 10 rad

3 What is the arc length subtended by a 32° angle centered on a circle of radius 25 cm?
Ans. 13.96 cm

4 What is the sum of the interior angles of a quadrilateral (four-sided) figure? (Hint: Divide the figure into triangles.)

5 In Fig. IV.10 the angle θ is 23°. What is ϕ?
Ans. 57°

6 Using a protractor, draw a right triangle with $\theta = 35°$. Determine sin 35° and cos 35° by measuring the sides of the triangle with a ruler. Compare your results with the values given by your calculator.

7 Prove that sin θ is equal to cos $(90° − \theta)$. Check this with your calculator.

8 Plot graphs of sin θ and cos θ against θ. Plot at least 16 points between 0 and 360° for each graph and connect them with a smooth curve.

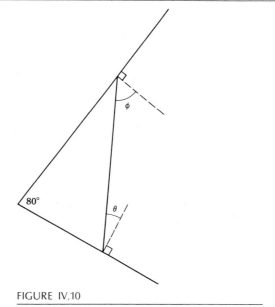

FIGURE IV.10

Problem 5.

9 Find the length of the sides of a right triangle, given that the hypotenuse is 15 cm and $\theta = 25°$.
Ans. 6.35 and 13.6 cm

10　In a 37° right triangle, the opposite side is 2.4 cm. What are the lengths of the hypotenuse and the adjacent side?

11　The sides of a right triangle are 30 and 50 cm. (a) What is the angle opposite the 30-cm side? (b) What is the hypotenuse?
Ans. (a) 31°; (b) 58.2 cm

12　A man starting from home walks 1.2 km south and then 1.6 km east to reach his office. (a) What is the straight-line distance between his home and office? (b) What angle does this line make with a north-south line?

13　Use the small-angle formula to calculate sin 3°. Compare the result with the value given by your calculator.

14　What are the angles whose sines are (a) 0.485, (b) 0.338, and (c) 0.025?

Equation 3.12 gives the distance h that an I beam sags at its middle when it is loaded at its middle with a weight W. To derive this equation, one must know the shape of the deformed beam, which cannot be calculated by elementary means. However, if we assume (incorrectly) that the shape of the deformed beam is an arc of a circle, we can derive a formula very similar to Eq. 3.12. Although not a legitimate derivation of Eq. 3.12, this calculation does demonstrate how the change of length ΔL of the upper and lower flanges is related to the sag.

When a beam of length L is deformed (Fig. V.1), the lower flange expands to length $L + \Delta L$, while the upper flange contracts to length $L - \Delta L$. The length of the midline of the beam remains L. To accommodate these changes, the ends of the beam rotate away from the vertical through a small angle θ.

Let us assume (incorrectly) that the midline of the beam is bent into the arc of a circle of radius R, and that the upper and lower flanges are bent into arcs of concentric circles of radii $R - \tfrac{1}{2}a$ and $R + \tfrac{1}{2}a$, respectively. Since each arc subtends the same angle 2θ, we have

$$2\theta = \frac{L}{R} = \frac{L + \Delta L}{R + \tfrac{1}{2}a} \qquad\qquad \text{V.1}$$

or

$$L(R + \tfrac{1}{2}a) = R(L + \Delta L)$$

Solving this for R gives

$$R = \frac{La}{2\,\Delta L}$$

so from Eq. V.1 the angle θ is

$$\theta = \frac{L}{2R} = \frac{\Delta L}{a}$$

Figure V.1 shows that the distance h of the sag at the midpoint of the beam is

$$h = R - R\cos\theta$$
$$= \frac{La}{2\,\Delta L}(1 - \cos\theta)$$

For small angles (in radians), the cosine is given by

$$\cos\theta = 1 - \tfrac{1}{2}\theta^2$$

so the sag is

$$h = \frac{La}{2\,\Delta L}(\tfrac{1}{2}\theta^2) = \frac{La}{4\,\Delta L}\left(\frac{\Delta L}{a}\right)^2$$
$$= \frac{L\,\Delta L}{4a} \qquad\qquad \text{V.2}$$

This is similar to the last expression in Eq. 3.12, except that Eq. V.2 has a 4 in the denominator instead of a 3. The correct factor is 3 not 4, because the deformed beam is not exactly the arc of a circle.

JUSTIFI-CATION OF EQUATION 3.12

FIGURE V.1

When a beam sags, its ends rotate away from the vertical. Here the midline of the beam is assumed (incorrectly) to be the arc of a circle of radius R.

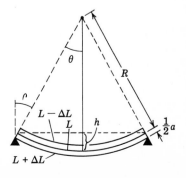

When the expression for ΔL given in Eq. 3.11 is substituted into Eq. V.2, we get

$$W = \frac{WL^3}{32EAa^2}$$

which again differs from the correct equation (Eq. 3.12) by the numerical factor in the denominator. Thus by elementary means we can justify Eq. 3.12, although we cannot derive the correct numerical factor.

VI

SOLAR-SYSTEM DATA

Body	Mass, 10^{24} kg	Radius, 10^6 m	Distance from sun, 10^{10} m	Acceleration of gravity at the surface, m/s²
Sun	1,970,000	696		274
Mercury	0.328	2.57	5.8	3.92
Venus	4.83	6.31	10.8	8.82
Earth	5.983	6.378	14.9	9.80
Mars	0.634	3.43	22.8	3.71
Jupiter	1,880	71.8	77.8	26.5
Saturn	563	60.3	142.6	11.8
Uranus	86.1	26.7	286.9	9.80
Neptune	99.6	24.6	448.5	9.80
Pluto	6	6.3	590	—
Moon	0.0735	1.738	0.038†	1.67

†Distance from earth.

It can be proved that a transverse wave of any shape propagates with the speed

$$v = \sqrt{\frac{T}{\mu}} \qquad \qquad \text{VII.1}$$

along a stretched cord (Sec. 11.2). Here T is the tension in the cord, and μ is the mass per unit length of the cord. Although it is not possible to give the general proof here, we shall estimate the speed of the simple pulse shown in Fig. VII.1. This calculation is instructive because it shows how Newton's second law applies to a mechanical wave.

The pulse in Fig. VII.1 is assumed to be moving to the right with constant speed v along a very long cord. The entire pulse is of length $2l$, and it has a maximum displacement h. In the time

$$t = \frac{l}{v} \qquad \qquad \text{VII.2}$$

the pulse moves a distance l along the cord, reaching the position shown by the dashed pulse. This requires that the segment of the cord between points A and B move up into the position between A' and B in the same time t. This segment starts from rest and is accelerated upward by the forces exerted on it by the remainder of the cord. If its average upward acceleration is a, the distance x it travels in time t, according to Eq. 4.4, is

$$x = \tfrac{1}{2}at^2$$

Of course, different parts of the segment travel different distances, but the average distance is $\tfrac{1}{2}h$. Putting this into the last equation, we get

$$\tfrac{1}{2}h = \tfrac{1}{2}at^2 \qquad \text{or} \qquad h = at^2$$

Using Eq. VII.2 for t, this becomes

$$h = a\left(\frac{l}{v}\right)^2$$

so that

$$v^2 = \frac{al^2}{h} \qquad \qquad \text{VII.3}$$

This is the condition that determines v. It comes from the requirement that the segment of the cord between A and B move into the proper position in the proper time.

The acceleration of the segment is calculated from the total force acting on it. Figure VII.2 shows the forces $\mathbf{F}_1$ and $\mathbf{F}_2$ on the segment at the instant the pulse reaches A. The magnitudes of $\mathbf{F}_1$ and $\mathbf{F}_2$ are both equal to the tension T of the cord. The sum $\mathbf{S} = \mathbf{F}_1 + \mathbf{F}_2$ is shown in Fig. VII.3. Recall that in Sec. 11.2 we said that the theory of mechanical waves is restricted to small displacements. In this case, this means that h must be much smaller than l or that θ is a small angle. Under such a circumstance, $\mathbf{S}$ is directed nearly vertically, and so the triangle in Fig. VII.3 is almost a right triangle. Thus to good approximation we have

VII

THE SPEED OF A TRANSVERSE WAVE ON A CORD

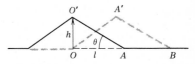

FIGURE VII.1

A pulse moving to the right with constant speed v along a cord.

FIGURE VII.2

The forces $\mathbf{F}_1$ and $\mathbf{F}_2$ on the segment AB of the cord at the instant the pulse reaches it.

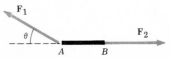

$$\frac{S}{F_2} = \frac{S}{T} = \tan\theta \qquad \text{or} \qquad S = T\tan\theta$$

But from the triangle AOO' in Fig. VII.1 the tangent of θ is seen to be h/l, so that

$$S = T\frac{h}{l} \qquad\qquad\qquad \text{VII.4}$$

The acceleration of the segment is given by Newton's second law,

$$a = \frac{S}{m}$$

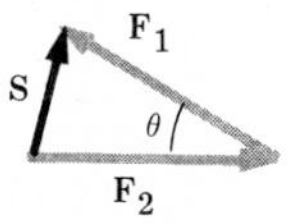

FIGURE VII.3

The sum **S** of the forces F_1 and F_2 in Fig. VII.2. The angle θ is very small, so **S** is nearly vertical.

where S is the total force on the segment, and m is the mass of the segment. The mass is related to μ, the mass per unit length, by $m = \mu l$. Therefore, the acceleration is

$$a = \frac{S}{m} = \frac{Th/l}{\mu l} = \frac{Th}{\mu l^2} \qquad\qquad \text{VII.5}$$

This is the acceleration of the segment at the instant the pulse reaches it. Once the segment moves upward, the forces on it change and its acceleration changes. However, for the purposes of this calculation, we shall assume that the acceleration is given by Eq. VII.5 throughout the motion of the segment.

When the value of the acceleration given by Eq. VII.5 is substituted into Eq. VII.3, we get

$$v^2 = \frac{(Th/\mu l^2)l^2}{h} = \frac{Th/\mu}{h} = \frac{T}{\mu}$$

and so

$$v = \sqrt{\frac{T}{\mu}}$$

This is the correct expression for the speed as given by Eq. VII.1.

This calculation is only an estimate of v. The final result agrees exactly with Eq. VII.1 because the estimates of x and a were made with a bit of hindsight. Had other reasonable estimates been used, a result like $\sqrt{2T/\mu}$ or $\sqrt{T/2\mu}$, which differs from the exact value by a simple numerical constant, would have been obtained. The point is that it is possible to show that v is proportional to $\sqrt{T/\mu}$ from the requirement that the segment AB must accelerate into position $A'B$ in the time it takes the wave to move the distance l. It is more difficult to prove rigorously that v is equal to $\sqrt{T/\mu}$.

REMARK The length l and height h of the pulse canceled out in the final expression for v. This is important because it demonstrates that the speed of a pulse does not depend on its size or shape.

Listed below are the most important symbols used in this book, together with the physical quantities they symbolize and the sections in which the quantities are first defined. Symbols are in alphabetical order, with the Greek alphabet following the Latin alphabet.

VIII
GLOSSARY OF TERMS USED IN THE BOOK

Symbol	Physical quantity	Section
a	Acceleration	4.2
	Angular magnification	14.5
A	Amplitude	6.4
	Area	1.5
	Mass number	22.1
B	Bulk modulus	12.1
	Magnetic field	17.1
c	Specific heat	9.2
	Speed of light	13.1
C	Capacitance	16.4
	Compression	2.2
	Electric conductance	16.3
	Heat capacity	9.2
	Information capacity	19.1
	Thermal conductance	9.5
d	Distance	1.2
	Weight density	1.5
D	Disorder	10.4
e	Coefficient of performance	9.4
	Efficiency	5.1
	Elementary charge	15.2
	Strain	3.4
E	Electric field	15.3
	Energy	5.4
	Young's modulus	3.4
E_e	Irradiance	13.6
E_v	Illuminance	13.6
	Emf	16.1
f	Focal length	14.1
	Frequency	11.4
F	Faraday's constant	16.5
	Force	2.1
g	Acceleration of gravity	4.2
	Gain	19.4
G	Close-loop gain	19.4
	Gravitational constant	6.3
	Shear modulus	3.4
h	Object height	14.1
	Planck's constant	20.1
h'	Image height	14.1
H_f	Heat of fusion	9.4
H_s	Heat of sublimation	9.4
H_v	Heat of vaporization	9.4
I	Electric current	16.1
	Intensity	12.2
	Internal energy	5.4
	Moment of inertia	6.5
I_v	Luminous intensity	13.6
k	Boltzmann constant	8.4
	Spring constant	2.2
K	Electric constant	15.2
	Kinetic energy	5.3
	Luminous efficacy	13.6
	Thermal conductivity	9.5
l	Orbital quantum number	20.4
L	Inductance	17.5

κ (kappa)	Dielectric constant	16.4
λ (lambda)	Decay constant	22.2
	Wavelength	11.4
μ (mu)	Linear mass density	11.2
	Poisson ratio	12.1
μ_k	Coefficient of kinetic friction	2.2
μ_0	Magnetic permeability	17.2
μ_s	Coefficient of static friction	2.2
ν (nu)	Valence	16.5
ρ (rho)	Mass density	1.5
	Resistivity	16.3
σ (sigma)	Electric conductivity	16.3
	Stefan-Boltzmann constant	9.5
τ (tau)	Half-life	22.2
	Period	6.2
	Time constant	16.4
	Torque	3.1
ϕ (phi)	Magnetic flux	17.4
	Phase angle	18.3
ψ (psi)	Quantum-mechanical wave	20.3
ω (omega)	Angular speed	6.5

INDEX

INDEX

TABLE OF THE ELEMENTS

The masses listed are based on $^{12}_{6}C = 12$ u.

A value in parentheses is the mass number of the most stable (long-lived) of the known isotopes.

ELEMENT	SYMBOL	ATOMIC NUMBER Z	AVERAGE ATOMIC MASS	ELEMENT	SYMBOL	ATOMIC NUMBER Z	AVERAGE ATOMIC MASS
Actinium	Ac	89	(227)	Mercury	Hg	80	200.59
Aluminum	Al	13	26.9815	Molybdenum	Mo	42	95.94
Americium	Am	95	(243)	Neodymium	Nd	60	144.24
Antimony	Sb	51	121.75	Neon	Ne	10	20.183
Argon	Ar	18	39.948	Neptunium	Np	93	(237)
Arsenic	As	33	74.9216	Nickel	Ni	28	58.71
Astatine	At	85	(210)	Niobium	Nb	41	92.906
Barium	Ba	56	137.34	Nitrogen	N	7	14.0067
Berkelium	Bk	97	(247)	Nobelium	No	102	(254)
Beryllium	Be	4	9.0122	Osmium	Os	76	190.2
Bismuth	Bi	83	208.980	Oxygen	O	8	15.9994
Boron	B	5	10.811	Palladium	Pd	46	106.4
Bromine	Br	35	79.904	Phosphorus	P	15	30.9738
Cadmium	Cd	48	112.40	Platinum	Pt	78	195.09
Calcium	Ca	20	40.08	Plutonium	Pu	94	(244)
Californium	Cf	98	(251)	Polonium	Po	84	(209)
Carbon	C	6	12.01115	Potassium	K	19	39.102
Cerium	Ce	58	140.12	Praseodymium	Pr	59	140.907
Cesium	Cs	55	132.905	Promethium	Pm	61	(145)
Chlorine	Cl	17	35.453	Protactinium	Pa	91	(231)
Chromium	Cr	24	51.996	Radium	Ra	88	(226)
Cobalt	Co	27	58.9332	Radon	Rn	86	222
Copper	Cu	29	63.546	Rhenium	Re	75	186.2
Curium	Cm	96	(247)	Rhodium	Rh	45	102.905
Dysprosium	Dy	66	162.50	Rubidium	Rb	37	85.47
Einsteinium	Es	99	(254)	Ruthenium	Ru	44	101.07
Erbium	Er	68	167.26	Samarium	Sm	62	150.35
Europium	Eu	63	151.96	Scandium	Sc	21	44.956
Fermium	Fm	100	(257)	Selenium	Se	34	78.96
Fluorine	F	9	18.9984	Silicon	Si	14	28.086
Francium	Fr	87	(223)	Silver	Ag	47	107.868
Gadolinium	Gd	64	157.25	Sodium	Na	11	22.9898
Gallium	Ga	31	69.72	Strontium	Sr	38	87.62
Germanium	Ge	32	72.59	Sulfur	S	16	32.064
Gold	Au	79	196.967	Tantalum	Ta	73	180.948
Hafnium	Hf	72	178.49	Technetium	Tc	43	(97)
Helium	He	2	4.0026	Tellurium	Te	52	127.60
Holmium	Ho	67	164.930	Terbium	Tb	65	158.924
Hydrogen	H	1	1.00797	Thallium	Tl	81	204.37
Indium	In	49	114.82	Thorium	Th	90	232.0381
Iodine	I	53	126.9044	Thulium	Tm	69	168.934
Iridium	Ir	77	192.2	Tin	Sn	50	118.69
Iron	Fe	26	55.847	Titanium	Ti	22	47.90
Krypton	Kr	36	83.80	Tungsten	W	74	183.85
Lanthanum	La	57	138.91	Uranium	U	92	238.03
Lawrencium	Lr	103	(257)	Vanadium	V	23	50.942
Lead	Pb	82	207.19	Xenon	Xe	54	131.30
Lithium	Li	3	6.939	Ytterbium	Yb	70	173.04
Lutetium	Lu	71	174.97	Yttrium	Y	39	88.905
Magnesium	Mg	12	24.312	Zinc	Zn	30	65.37
Manganese	Mn	25	54.9380	Zirconium	Zr	40	91.22
Mendelevium	Md	101	(256)				